MECHANICS
OF MATERIALS
With Applications in Excel®

MECHANICS OF MATERIALS

With Applications in Excel®

Bichara B. Muvdi

Souhail Elhouar

CRC Press
Taylor & Francis Group
Boca Raton London New York

CRC Press is an imprint of the
Taylor & Francis Group, an **informa** business

CRC Press
Taylor & Francis Group
6000 Broken Sound Parkway NW, Suite 300
Boca Raton, FL 33487-2742

© 2016 by Taylor & Francis Group, LLC
CRC Press is an imprint of Taylor & Francis Group, an Informa business

No claim to original U.S. Government works

Printed on acid-free paper
Version Date: 20150603

International Standard Book Number-13: 978-1-4665-7071-9 (Hardback)

Library of Congress Cataloging-in-Publication Data

Muvdi, B. B.
  Mechanics of materials : with applications in Excel / authors, Bichara B. Muvdi and Souhail Elhouar.
    pages cm
  Includes bibliographical references and index.
  ISBN 978-1-4665-7071-9
  1. Materials--Mechanical properties. 2. Materials--Data processing. 3. Microsoft Excel (Computer file) I. Elhouar, Souhail. II. Title.

TA404.8.M88 2016
620.1'10285554--dc23   2015016936

Visit the Taylor & Francis Web site at
http://www.taylorandfrancis.com

and the CRC Press Web site at
http://www.crcpress.com

To our families for their patience and encouragement.

Contents

Preface..xiii
Authors...xv

Chapter 1   Axial Loads...1

    1.1   Introduction 1
    1.2   Internal Axial Force 1
           1.2.1   Concentrated Force 1
           1.2.2   Sign Convention 3
           1.2.3   Distributed Forces 4
    1.3   Normal and Shearing Stresses..8
           1.3.1   Normal Stress 8
           1.3.2   Stress Element 10
           1.3.3   Shearing Stress 10
           1.3.4   Stresses on Inclined Planes 11
           1.3.5   Units 13
           1.3.6   Sign Convention 13
    1.4   Normal Strain and Stress–Strain Diagrams...................................20
           1.4.1   Normal Strain..20
           1.4.2   Units 21
           1.4.3   Sign Convention 21
           1.4.4   Mechanical Properties...21
           1.4.5   Design Considerations...29
    1.5   Load-Deformation Relations 34
    1.6   Statically Indeterminate Members 43
           1.6.1   Temperature Effects 47
   *1.7   Stress Concentration..56
   *1.8   Impact Loading 60
    Review Problems..65

Chapter 2   Torsional Loads 69

    2.1   Introduction 69
    2.2   Internal Torque 69
           2.2.1   Concentrated Torque 69
           2.2.2   Sign Convention 70
           2.2.3   Distributed Torques 71
    2.3   Stresses and Deformations in Circular Shafts................................78
           2.3.1   Shearing Strain..78
           2.3.2   Sign Convention 79
           2.3.3   Shearing Stress and Shearing Deformation 79
           2.3.4   Material Properties in Shear 82
           2.3.5   Stress Element 83
           2.3.6   Stresses on Inclined Planes 83

---

* A star ahead of a chapter or a section number indicates that the associated material is not generally covered in a first mechanics of materials course.

2.4   Statically Indeterminate Shafts 96
2.5   Design of Power-Transmission Shafts 103
2.6   Stresses under Combined Loads 104
    2.6.1   Hooke's Law in Two Dimensions 106
*2.7   Stress Concentration 111
*2.8   Impact Loading 113
*2.9   Shafts of Noncircular Cross Sections 120
    2.9.1   Analytical Solutions 120
    2.9.2   Experimental Solutions 123
    2.9.3   Application to Torsion of a Long, Thin Rectangle 127
    2.9.4   Special Cases 129
        2.9.4.1   Narrow Circular Section with Thin Slit 129
        2.9.4.2   Sections Composed of Narrow Rectangles 129
    2.9.5   Application to Torsion of Thin-Walled Tubes 131
*2.10 Elastoplastic Behavior 139
Review Problems 145

Chapter 3   Bending Loads: Stresses 149

3.1   Introduction 149
3.2   Internal Shear and Moment 150
    3.2.1   Shear and Moment at Specified Positions 150
    3.2.2   Sign Convention 151
    3.2.3   Shear and Moment Functions: Shear and Moment Diagrams 153
3.3   Load, Shear, and Moment Relationships 160
3.4   Bending Stresses under Symmetric Loading 166
3.5   Shearing Stresses under Symmetric Loading 178
    3.5.1   Longitudinal Surfaces Normal to Loads 178
    3.5.2   Shear Flow 180
3.6   Stresses under Combined Loads 192
3.7   Allowable-Stress Design 199
*3.8   Stress Concentration 202
Review Problems 207

*Chapter 4   Bending Loads: Additional Stress Topics 211

4.1   Introduction 211
4.2   Beams of Two or Three Materials Loaded Symmetrically 211
    4.2.1   General Principles 211
    4.2.2   Application to Reinforced Concrete 215
4.3   Bending Stresses under Unsymmetric Loading 223
    4.3.1   Arbitrary Centroidal Axes 223
    4.3.2   Principal Centroidal Axes 225
4.4   Thin-Walled Open Sections: Shear Center 234
    4.4.1   Symmetric Bending 234
    4.4.2   Unsymmetric Bending 240
4.5   Curved Beams 247

---

* A star ahead of a chapter or a section number indicates that the associated material is not generally covered in a first mechanics of materials course.

	4.6	Elastoplastic Behavior: Plastic Hinge	254
	4.6.1	Shape Factor	254
	4.6.2	Plastic Hinge	256
	4.7	Fatigue	265
		Review Problems	275

Chapter 5   Bending Loads: Deflections under Symmetric Loading 279

	5.1	Introduction	279
	5.2	Moment–Curvature Relationship	280
	5.3	Deflection: Two Successive Integrations	282
	5.4	Derivatives of the Deflection Function	295
	5.5	Deflection: Superposition	299
	5.6	Deflection: Area–Moment	303
	5.6.1	Moment Diagrams by Cantilever Parts	307
	5.7	Statically Indeterminate Beams: Two Successive Integrations	319
	5.8	Statically Indeterminate Beams: Superposition	327
	5.9	Statically Indeterminate Beams: Area–Moment	333
		Review Problems	338

*Chapter 6   Bending Loads: Additional Deflection Topics 343

	6.1	Introduction	343
	6.2	Deflection: Singularity Functions	343
	6.3	Deflection: Castigliano's Second Theorem	353
	6.4	Deflection: Unsymmetric Bending Loads	361
	6.5	Statically Indeterminate Beams: Singularity Functions	366
	6.6	Statically Indeterminate Beams: Castigliano's Second Theorem	373
	6.7	Impact Loading	380
		Review Problems	386

Chapter 7   Analysis of Stress 389

	7.1	Introduction	389
	7.2	Stress at a Point	389
	7.3	Components of Stress	390
	7.4	Plane-Stress Transformation Equations	392
	7.4.1	Stresses on Inclined Planes	392
	7.4.2	Principal Stresses	396
	7.4.3	Maximum In-Plane Shearing Stress	398
	7.5	Mohr's Circle for Plane Stress	410
	7.5.1	Construction of Mohr's Circle	410
	7.5.2	Principal Stresses and Maximum In-Plane Shearing Stress	412
	7.5.3	Stresses on Inclined Planes	419
	*7.6	Three-Dimensional Stress Systems	427
	7.6.1	Mohr's Circle for Triaxial Stress Systems: Absolute Maximum Shearing Stress	432
	7.7	Thin-Walled Pressure Vessels	441
	7.7.1	Cylindrical Vessels	441
	7.7.2	Spherical Vessels	443

*7.8   Thick-Walled Cylindrical Pressure Vessels...447
       7.8.1   Stresses..447
       7.8.2   Deformations...450
       7.8.3   Special Cases..450
                7.8.3.1   Internal Pressure Only.......................................451
                7.8.3.2   External Pressure Only......................................451
                7.8.3.3   Shrink-Fitting Operations..................................453
                7.8.3.4   Internal Pressure on Thin-Walled Cylinders...................454
*7.9   Theories of Failure 457
       7.9.1   Brittle Material 457
                7.9.1.1   Maximum Principal Stress Theory 457
                7.9.1.2   Mohr's Theory 458
       7.9.2   Ductile Materials...460
                7.9.2.1   Maximum Shearing Stress Theory....................460
                7.9.2.2   The Energy of Distortion Theory 460
Review Problems...471

Chapter 8   Analysis of Strain.. 475

8.1   Introduction 475
8.2   Strain at a Point: Components of Strain 475
       8.2.1   Units and Sign Conventions 477
8.3   Plane-Strain Transformation Equations 477
       8.3.1   Principal Strains.. 481
       8.3.2   Maximum In-Plane Shearing Strain 482
8.4   Mohr's Circle for Plane Strain... 488
       8.4.1   Principal Strains and Maximum In-Plane Shearing Strain 489
       8.4.2   Inclined Axes 490
       8.4.3   Development of the Relation $G = E/2(1 + \mu)$...................................... 495
*8.5   Three-Dimensional Hooke's Law.. 499
       8.5.1   Summary of Hooke's Laws in One and Two Dimensions.............. 499
       8.5.2   Hooke's Law in Three Dimensions 499
       8.5.3   Volume Change: Bulk Modulus of Elasticity........................ 502
       8.5.4   Strain Energy: Energy of Distortion 503
*8.6   Mohr's Circle for Three-Dimensional Strain Systems 506
*8.7   Strain Measurements: Strain Rosettes.. 510
Review Problems... 517

Chapter 9   Columns 521

9.1   Introduction 521
9.2   Stability of Equilibrium.. 521
       9.2.1   Theoretical Background... 521
       9.2.2   Column Models.. 523
9.3   Euler's Ideal-Column Theory.. 529
9.4   Effect of End Conditions 541
       9.4.1   Critical Load for Column Fixed at One End
                and Free at the Other.. 541

---

* A star ahead of a chapter or a section number indicates that the associated material is not generally covered in a first mechanics of materials course.

| 9.4.2 | Effective Length | 543 |

9.5	Secant Formula	548
	9.5.1	Eccentrically Loaded Pin-Ended Column	548
	9.5.2	Maximum Deflection	549
	9.5.3	Secant Formula	550
	9.5.4	Initially Bent Pin-Ended Columns	552
9.6	Design of Centrically Loaded Columns	561
	9.6.1	Inelastic Column Buckling	561
	9.6.2	Empirical Equations	562
	9.6.3	Structural Steel	563
		9.6.3.1 Short and Intermediate Columns	563
		9.6.3.2 Long Columns	563
9.7	Design of Eccentrically Loaded Columns	578
	9.7.1	Allowable-Stress Method	578
	9.7.2	Interaction Method	579
Review Problems	589

Chapter 10 Excel Spreadsheet Applications 593

10.1	Introduction	593
10.2	Spreadsheet Applications Concepts and Techniques	593
	10.2.1	Using Styles	594
	10.2.2	Protecting Formula Cells	594
	10.2.3	Data Validation	594
	10.2.4	Using Excel Functions	597
	10.2.5	Automatic Real-Time Sorting	602
	10.2.6	Defining New Functions	604
	10.2.7	Conditional Formatting	605
	10.2.8	Using Controls	605
10.3	Example 1: Drawing Shear and Moment Diagrams	607
10.4	Example 2: Drawing Mohr's Circle	614
	10.4.1	Plotting Mohr's Circle	616
	10.4.2	State of Stress on the x' Plane	618
	10.4.3	Representation of the State of Stress on the x Plane, the y Plane, and the x' Plane on Mohr's Circle	619
10.5	Example 3: Principal Stresses in Three-Dimensional Stress Elements	621
10.6	Example 4: Computation of Combined Stresses	626
10.7	Excel Spreadsheet Application Projects	631
	10.7.1	Project 10.1: Cross-Sectional Properties of a General Shape	631
	10.7.2	Project 10.2: Stresses in Statically Indeterminate Systems	632
	10.7.3	Project 10.3: Inelastic Analysis	632

Answers to Even-Numbered Problems 633

Appendix A: SI Units 647

Appendix B: Selected References 649

Appendix C: Properties of Plane Areas 651

Appendix D: Typical Physical and Mechanical Properties of Selected Materials (U.S. Units and SI Units) 677

Appendix E 679

Appendix F: Design Properties for Selected Structural Wood Sections (U.S. Units and SI Units) 693

Appendix G: Beam Slopes and Deflections for Selected Cases 695

Appendix H: Two-Dimensional Supports and Connections 697

Index 701

Preface

This book was written in order to provide the engineering student with a clear and understandable treatment of the concepts underlying the subject known as *mechanics of materials* or *strength of materials*. This subject is concerned with the behavior of deformable bodies when subjected to axial, torsional, and bending loads as well as combinations thereof. In contrast to idealized rigid bodies considered in *statics* and *dynamics*, in *mechanics of materials* the bodies are considered to be deformable, and these deformations are studied when known forces are applied to them. The molecular structure of matter, although well established by experiments, will not be of primary concern in this book because *mechanics of materials* is based upon the concepts of a "continuous media"—a convenient idealization of the true nature of matter that has proved satisfactory for engineering analysis and design.

In view of the fact that the international system of units, referred to as SI (Système International), is gaining acceptance in the United States, the authors decided to use it in this book. However, it is realized that a complete transition from the U.S. Customary to SI units will be a slow and costly process that may take many years. Thus, both systems of units are used in this textbook. Approximately one-half of the examples and one-half of the problems are stated in terms of the U.S. Customary system, whereas the remaining examples and problems are given in terms of the emerging SI system of units.

The objectives of the subject of *mechanics of materials* in this textbook are achieved by presenting the subject matter in nine chapters. The material within each chapter is organized such that the student is led through the theoretical development of the fundamental concepts and the applicable mathematical relations. This is followed by one or more examples that illustrate the application of the mathematical relations to physical situations. Then, a set of Homework Problems is provided that is arranged from the simplest to the most challenging. Finally, a more challenging number of Review Problems are stated at the end of each chapter.

Chapter 10 is comprised of subject matter that makes it stand out with respect to other available textbooks. This chapter presents techniques and skills for developing Microsoft® Excel™ applications for solving intermediate to advanced mechanics of material problems using numerical techniques.

In this textbook, the authors present a *unique approach* for the teaching of mechanics of materials, which they refer to as an *integrated approach* because it consists of *two major components*: the first deals with traditional *printed matter* and the second with an *electronic enhancement* of the first.

The first major component of the book, the *printed matter* component is contained in the 10 chapters of the *textbook* and the companion *Solutions Manual*. It is worth noting that the authors have made use of previous publications by the primary author in the area of mechanics of materials. However, they have made some major changes and improvements which, among other things, include the following:

1. The organization has been completely changed in an effort to respond to suggestions made by users of the original work.
2. While still using the concepts and strong features of the original publications, in most cases the material has been re-written and improved and new topics added as the need arose.
3. Most of the original Examples and Homework Problems have been changed in some fashion and new ones added.
4. A set of Review Problems has been added to each of the nine chapters designed to challenge the students and provide them with a means to review the material.

Each chapter in the textbook contains numerous examples with detailed solutions enabling the student to understand the basic concepts. Also, at the end of each section in a given chapter, a large number of homework problems are included to allow instructors to assign different problems for several semesters. The textbook is written for a first engineering course in mechanics of materials or strength of materials. This course is normally taught during the first or second semester at the sophomore level or first semester at the junior level. The prerequisites include a working knowledge of differential and integral calculus together with rigid-body mechanics, primarily statics.

The second major component of this book, the *electronic enhancement* component, is designed to help, not only the students with their understanding of the subject matter, but also the instructors with their teaching of the material. Its five subcomponents are described briefly below.

1. A complete chapter dealing with Excel spreadsheet applications relating to the concept of mechanics of materials.
2. A set of multimedia-enhanced electronic versions of some of the main concepts introduced in the textbook to be made available to instructors who adopt it.
3. Excel Spreadsheet Collection: All the Examples in the textbook are made available to the instructor in Excel spreadsheet format. This approach enables the instructor to repeat the same Example with different sets of input, thus enhancing the student's understanding of the effect of different parameters on the problem at hand.
4. Textbook web site: This web site has a collection of resources for students and instructors complementing information provided in the text.
5. Model PowerPoint Presentations: This subcomponent consists of a collection of 39 simple PowerPoint presentations (about one for each class period) that would contain headings of the articles to be covered and brief descriptions of key concepts.
6. A Graphics Collection: This subcomponent consists of an electronic collection of all the artwork in the textbook.

The authors wish to acknowledge with deep appreciation the help and many valuable comments and suggestions made over the years by colleagues from Bradley University and by the following engineering educators from other institutions: Amara Loulizi, Ecole Nationale d'Ingénieurs de Tunis, Tunisia, and Mahmoud Farag, the American University in Cairo, Egypt.

Bichara B. Muvdi
Souhail Elhouar
Peoria, Illinois

Authors

Bichara B. Muvdi is a professor emeritus of the Department of Civil Engineering and Construction at Bradley University, Peoria, Illinois, USA. He received his B.M.E. and M.M.E. from Syracuse University, New York, USA, and his Ph.D. from the University of Illinois, Urbana-Champaign, Illinois, USA.

Souhail Elhouar is a professor of Civil Engineering and Construction at Bradley University, Peoria, Illinois, USA. He received his B.Sc. from the National Engineering School of Tunis, University of Tunis - El Manar, Tunis, Tunisia, and his M.Sc. and Ph.D. from the University of Oklahoma, Norman, Oklahoma, USA.

1 Axial Loads

1.1 INTRODUCTION

The basic course in *statics* deals with members that are assumed to be *rigid*, meaning that they do not deform (change shape or size) when subjected to forces. In a course dealing with *mechanics of materials*, this assumption is no longer valid. In other words, the kinds of bodies dealt with in this latter course are those that exhibit changes, in dimension and shape, under the influence of forces that may act on them. However, the assumption of *rigid bodies* is still used in the case of deformable bodies, when applying the conditions of equilibrium.

The concept of the *free-body diagram* is as important in the study of *mechanics of materials* as it was in the case of rigid-body mechanics. It is, therefore, strongly recommended that the student review this concept before embarking on the study of *mechanics of materials*. It is also strongly recommended that the student review the basic concepts of equilibrium which, for convenience, are summarized here by the two vector equations, $\Sigma F = 0$ and $\Sigma M = 0$, keeping in mind that each of these two vector equations represents three scalar equations. Thus, using an x–y–z right-handed coordinate system, these equations may be expressed as follows:

$$\sum F_x = 0, \quad \sum F_y = 0, \quad \sum F_z = 0$$

$$\sum M_x = 0, \quad \sum M_y = 0, \quad \sum M_z = 0$$

Note that in the two-dimensional case, only three of the above six scalar equation are needed to define the equilibrium of a given system. Using an x–y right-handed coordinate system, these three scalar equations become

$$\sum F_x = 0, \quad \sum F_y = 0, \quad \sum M_z = 0$$

In the remainder of this chapter, we will discuss topics related to members subjected to axial loads. Thus, in Section 1.2, we discuss the concept of the internal axial force; in Section 1.3, we deal with the subject of normal and shearing stresses; in Section 1.4, we discuss the topics of normal strain and the behavior of materials and their mechanical properties when subjected to axial loading; in Section 1.5, we develop and discuss the relations that exist between axial loads and the resulting deformations; in Section 1.6, we introduce the concept of statically indeterminate members; in Section 1.7, we deal with the topic of stress concentration and its importance in the design of members subjected to nonstatic (dynamic) axial loads, such as impact; and, finally, in Section 1.8, we deal with the effects of impact-loading.

1.2 INTERNAL AXIAL FORCE

1.2.1 CONCENTRATED FORCE

A *concentrated axial force* may be defined as one applied over a very small area (assumed to be a point) and whose line of action coincides with the *centroidal* axis of a straight two-force member. For example, the straight, prismatic member *AB* in Figure 1.1a is under the action of a concentrated

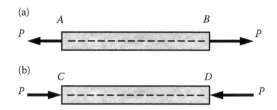

FIGURE 1.1 Members under concentrated axial loads, *AB* in tension and *CD* in compression.

axial force *P* because the line of action of this force passes through the *centroid* of each and every cross-sectional area along the entire straight member.

The member shown in Figure 1.1a is said to be under the action of a concentrated *tensile* axial force because its material fibers are being pulled apart or stretched. On the other hand, the straight prismatic member *CD* in Figure 1.1b is said to be under the action of a concentrated *compressive* axial force because its material fibers are being pushed together or compressed.

There are many applications in engineering practice in which straight prismatic members subjected to axial forces (tension or compression) are encountered. These include members in trusses, columns, struts, cable systems, as well as other structures and machines that contain two-force members. We will study the behavior of these members in the following pages.

The concept of *internal forces* in axially loaded members was first encountered in *statics*, when we learned how to determine the forces in members of a truss. For example, in dealing with the method of *sections*, we cut the required number of members and exposed the forces in them before constructing a suitable free-body diagram. These exposed forces are internal to the cut members. In determining these internal forces, we make use of two basic concepts. The first states that *if an entire member is in equilibrium, any part of this member is also in equilibrium.* The second deals with the adjectives *internal* and *external*. When modifying the noun force, these adjectives are regarded as relative, that is, an internal force for an entire member becomes an external force for a part of this member.

A straight two-force member *AB* is shown in Figure 1.2, subjected to the concentrated external axial forces *P* placing the member in tension. If we imagine the member to be cut at any position *C* into two parts by a cutting plane, each of the two parts, *AC* and *CB*, must be in equilibrium. Thus, if we consider part *AC*, equilibrium dictates that the force F_C, acting at the exposed plane at *C*, must be equal in magnitude to the force *P* applied at *A*. Similarly, the force F_C at *C*, acting on part *CB*, must be equal in magnitude to the force *P* acting at *B*. Thus, we conclude that the force $F_C = P$, acting on part *AC*, represents the force exerted on it by part *CB*. Similarly, the force $F_C = P$, acting on part *CB*, represents the force exerted on it by part *AB*. Therefore, since member *AB* is not really cut, the force $F_C = P$ at *C* acts *internally* at position *C* in the member. It is important to note that the *internal axial force* F_C does not act at one single point in the exposed area at *C* but is distributed in some manner over this area. This internal force distribution will be discussed in more detail in later paragraphs. However, it should be pointed out that the concentrated force F_C represents the resultant of this distribution.

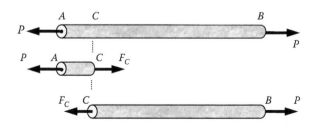

FIGURE 1.2 Member *AB* under axial load *P* showing an imaginary cut at position *C*.

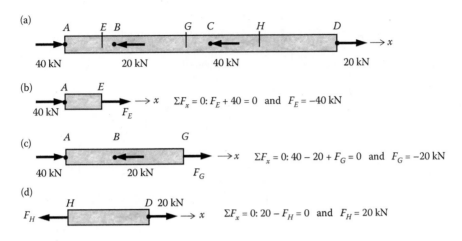

FIGURE 1.3 Member AD under four axial loads, showing internal forces at the imaginary cuts at E, G, and H.

We now consider the case of a member subjected to concentrated axial forces not only at its ends but also at other intermediate locations along its length. The straight member shown in Figure 1.3a is subjected to concentrated axial forces at A, B, C, and D as shown. If we were interested in determining the internal axial force anywhere within segment AB, we would cut the member at any location such as E within this segment, and construct the free-body diagram of either the left part (i.e., the left free body [LFB]) or the right part (i.e., the right free body [RFB]). The LFB diagram is shown in Figure 1.3b. Equilibrium of this part of the member dictates that the internal force F_E at the cut section be a compressive force with a magnitude of 40 kN as indicated by the computations. Similarly, using an LFB diagram as shown in Figure 1.3c, we conclude that the internal axial force at any position such as G within segment BC is compressive with a magnitude of 20 kN. Finally, by using an RFB diagram as shown in Figure 1.3d, we conclude that the internal axial force at any location such as H in segment CD is tensile with a magnitude of 20 kN.

1.2.2 SIGN CONVENTION

In determining the magnitude and character of the internal axial force at any location in a member, we make use of two distinct sign conventions. The first deals with the character of the internal axial force (i.e., tension [+] or compression [−]), and may be termed a *physical sign convention*. The second deals with the application of the equations of equilibrium referred to as an *equations sign convention*. Consider, for example, the free-body diagram of segment ABC of member $ABCD$ shown in Figure 1.4, where an x axis, measured positive to the right, has been established with origin at A. In constructing this free-body diagram, the assumption was made that the unknown internal force

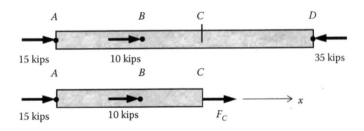

FIGURE 1.4 Free-body diagram of section ABC, showing a coordinate system with origin at A and the x axis positive to the right.

F_C at C is tension, or positive according to the *physical sign convention*. Assuming the unknown internal force to be positive is good practice and is strongly recommended not only in this chapter but also in future chapters. Once the free-body diagram is completed showing a convenient coordinate system, then the *equations sign convention* is applied.

Thus,

$$\sum F_x = 0 \Rightarrow F_C + 10 + 15 = 0$$

Therefore, $F_C = -25$ kips.

The negative sign of the answer simply indicates that the internal force F_C is not tensile as assumed but compressive. Therefore, the answer for the internal force at C may be stated as follows: $F_C = 25$ kips (C), where the symbol (C) is used to signify compression in the member.

In summary, to find the internal axial force in a member subjected to concentrated axial forces, we use the *physical sign convention* to construct an appropriate free-body diagram along with a convenient coordinate system assuming the unknown internal force to be positive (tension). We then use the *equations sign convention* to apply the appropriate equation of equilibrium to obtain the magnitude and character of the unknown internal force. Note that *the equations sign convention may be chosen arbitrarily*.

1.2.3 DISTRIBUTED FORCES

Occasionally, we encounter straight members subjected to distributed rather than concentrated forces. Such is the case with member AB shown in Figure 1.5. The intensity of the force distribution p is expressed as force per unit of length. Note that the intensity of the distributed force acting on member AB of Figure 1.5 is constant with a magnitude of 10 kN/m. This, of course, means that the total force acting on the member is $5 \times 10 = 50$ kN, since the length of the member is 5 m as shown.

The intensity of the distributed load p acting on a straight member is not always constant but may vary along the length of the member in some manner. Consider, for example, the straight member shown in Figure 1.6, which is subjected to a distributed force that varies according to the function $p = f(x)$ as shown. We examine the free-body diagram of a differential length dx at a distance x from point A, the origin of the coordinate system. Here again, the resultant internal axial forces F and $F + dF$ are assumed positive (tension) according to the *physical sign convention*.

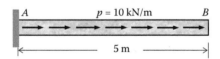

FIGURE 1.5   Member AB under a constant-intensity, uniformly distributed load p.

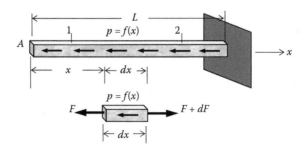

FIGURE 1.6   Member subjected to a variable-intensity load $p = f(x)$.

In order to determine a relationship between the applied distributed force p and the internal axial force F at any point in the member, we apply the equation of equilibrium using the *equations sign convention*. Thus,

$$\sum F_x = 0 \Rightarrow F + dF - pdx - F = 0$$

Thus,

$$dF = pdx$$

and

$$\int_1^2 dF = \int_1^2 pdx$$

Integrating and applying the limits, we obtain

$$\Delta F_{1-2} = F_2 - F_1 = \int_1^2 pdx \tag{1.1}$$

Equation 1.1 enables us to find the change in the internal axial force between any two arbitrary positions such as 1 and 2 in Figure 1.6. If the distributed force p is known as a function of x, then we may perform the integration indicated on the right-hand side of Equation 1.1.

In certain cases of axially loaded members, it is useful to construct what may be called an *internal axial force diagram*. Such a diagram shows at a glance the way that the internal force varies along the member. To develop the diagram, we determine the internal force F at a number of positions along the member, including relations between this force and the variable x in segments where distributed forces exist. Once this is accomplished, a graph is constructed showing the force F on a vertical axis and the variable x, measuring positions along the member, on a horizontal axis. The resulting points are then connected by straight lines and/or curves to complete the construction. The construction of an internal axial force diagram is illustrated in Example 1.2.

EXAMPLE 1.1

Consider the member of length L shown in the illustration, subjected to the distributed force $p = 2x^3$. Determine an expression for the internal axial force F in terms of the variable x.

SOLUTION

Using a cutting plane at a distance x from point A, the origin of the coordinate system, we construct an LFB diagram as shown. Note that the internal axial force F is assumed positive (tension) according to the *physical sign convention*. Using the *equations sign convention*, we apply the equation of equilibrium. Thus,

$$\sum F_x = 0 \Rightarrow F + \int_0^x 2x^3 \, dx = 0$$

Integrating and solving for F, we obtain

$$F = -\left(\frac{x^4}{2}\right)\cdots 0 \le x \ge L \qquad \textbf{ANS.}$$

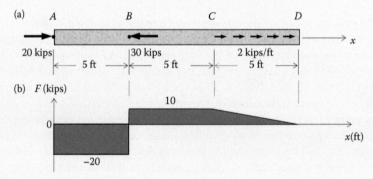

The negative sign of the answer indicates that the internal force F is not tension as assumed but compression.

EXAMPLE 1.2

Construct the internal axial force diagram for the straight member loaded as shown in figure (a) below.

SOLUTION

A convenient placement for the internal axial force diagram is either directly under the loaded member as shown in figure (b) above, if the member is horizontal as is the case here, or adjacent to it (left or right) if the member is vertical. The development of this diagram was accomplished by passing three cutting planes—one within segment AB, a second within segment BC, a third within segment CD—and constructing corresponding free-body diagrams as shown, respectively, in figures (c) through (e) below. Considering figure (c) below, we conclude on the basis of equilibrium that the force F is compressive with a magnitude of 20 kips and constant over the entire segment AB. The free-body diagram in figure (d) below dictates that the force F is tensile and equal to 10 kips in magnitude and constant over the entire segment BC. Finally, the free-body diagram of figure (d) below tells us that the internal axial force may be expressed by the relation $F = 30 - 2x$. This relation represents a straight line that varies from $F = 10$ kips at $x = 10$ ft to $F = 0$ at $x = 15$ ft. All of these computations are shown below adjacent to the free-body diagrams. Note the abrupt change in the magnitude of the internal axial force at point B (from −20 to +10 kips), a change of 30 kips. This sudden change is referred to as a *discontinuity* in the internal axial force and is due to the assumption that a concentrated force can be applied at a single point such as *B*. The magnitude of the internal force at *B* is undefined. However, we can determine this force just to the left ($F = -20$ kips) and just to the right ($F = +10$ kips) of this point.

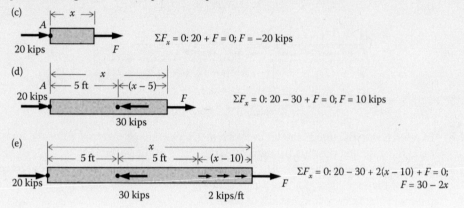

(c)

A

20 kips

x

F

$\Sigma F_x = 0: 20 + F = 0; F = -20$ kips

(d)

A　5 ft　$(x - 5)$

20 kips

30 kips

F

$\Sigma F_x = 0: 20 - 30 + F = 0; F = 10$ kips

(e)

5 ft　　5 ft　　$(x - 10)$

x

20 kips

30 kips

2 kips/ft

F

$\Sigma F_x = 0: 20 - 30 + 2(x - 10) + F = 0;$
$F = 30 - 2x$

PROBLEMS

(Show complete free-body diagrams in the solutions to all of the following problems.)

1.1　Consider the straight member *ABCD* subjected to the concentrated loads as shown. Compute the internal axial force in (a) segment *AB* and (b) segment *BC*.

A　　B 8 kips　　C 5 kips　　　　D

20 kips

7 kips

$\leftarrow$ 2 ft $\rightarrow$ $\leftarrow$ 3 ft $\rightarrow$ $\leftarrow$ 5 ft $\rightarrow$

(Problems 1.1 and 1.11)

1.2　A straight member is loaded as shown. Find the internal axial force (a) in segment *AB* and (b) at position 1.

A　　　5 kN B　　2 kN/m　　1　　C

15 kN

4 kN

$\leftarrow$ 2 m $\rightarrow$ $\leftarrow$ 2 m $\rightarrow$ $\leftarrow$1 m$\rightarrow$

(Problems 1.2 and 1.12)

1.3　Find the internal axial force at positions 1 and 2 in the member shown.

50 kips　　　　　　10 kips　　　3 kips/ft　　10 kips

$\leftarrow$ 5 ft $\rightarrow$ $\leftarrow$ 5 ft $\rightarrow$ $\leftarrow$ 5 ft $\rightarrow$ $\leftarrow$ 5 ft $\rightarrow$

1　　　　　　　　2

(Problems 1.3 and 1.13)

1.4　Consider the straight member loaded as shown. Compute the internal axial force at position 1.

A　　4 kN/m　　　　　　B　10 kN　　C

$\leftarrow$ 2 m $\rightarrow$ $\leftarrow$2 m$\rightarrow$ $\leftarrow$ 2 m $\rightarrow$ $\leftarrow$ 1.5 m $\rightarrow$ $\leftarrow$ 1.5 m $\rightarrow$
1　　　　　　　　　　2

$\rightarrow x$

(Problems 1.4, 1.5, 1.6, and 1.14)

1.5   In reference to the figure shown in Problem 1.4, develop a relation between the internal axial force F and the variable x for any position in segment AB.

1.6   In reference to the figure shown in Problem 1.4, find the internal axial force in position 2.

1.7   Find the internal axial force at positions 1 and 2 for the straight member loaded as shown.

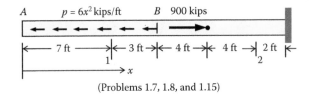

(Problems 1.7, 1.8, and 1.15)

1.8   In reference to the figure shown in Problem 1.7, develop an expression for the internal axial force for any position in segment AB in terms of the variable x.

1.9   Consider the straight member ABC loaded as shown. Express the internal axial force in terms of the variable x_1 for any position in segment AB. What is the magnitude of the internal axial force just to the left of point B?

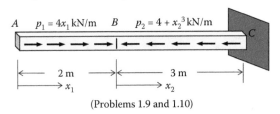

(Problems 1.9 and 1.10)

1.10 In reference to the figure shown in Problem 1.9, express the internal axial force in terms of the variable x_2 for any position in segment BC. What is the magnitude of the internal axial force at $x_2 = 2$ m and $x_2 = 3$ m?

1.11 Construct the internal force diagram for the member of Problem 1.1.

1.12 Construct the internal force diagram for the member of Problem 1.2.

1.13 Construct the internal force diagram for the member of Problem 1.3.

1.14 Construct the internal force diagram for the member of Problem 1.4.

1.15 Construct the internal force diagram for the member of Problem 1.7.

1.3   NORMAL AND SHEARING STRESSES

The concept of *stress* is one of the most important concepts in the study of the *mechanics of materials*. The internal force in an axially loaded member provides very useful but incomplete information regarding the behavior of the member and the question as to whether this member will fail under the action of the applied loads. This question can best be answered by comparing the maximum internal unit force (referred to as the maximum *stress*) in the member to some acceptable design criterion known as the *allowable stress* (to be discussed later) for the material in question.

1.3.1   NORMAL STRESS

By definition, *normal stress* is force per unit of area (unit force). Thus, if the internal axial force at some position in the member is F and the cross-sectional area of the member is A, the *average normal stress* in the member (*tensile stress* if F is tensile and *compressive stress* if F is compressive) is given by the relation

$$\sigma_{AVE} = \frac{F}{A} \qquad\qquad (1.2)$$

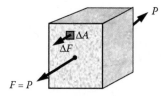

FIGURE 1.7 Member showing the mathematical definition of average stress over area ΔA.

where the Greek letter σ represents the *normal* stress, and the word *normal* implies that the stress is *perpendicular to the plane on which it acts*. The need for the adjective *normal* for the stress σ, lies in the fact that, in general, a second type of stress exists on certain planes in the member. This second type of stress is known as the *shearing stress,* acts parallel to its plane and is given by the Greek symbol τ. We will discuss first the normal stress σ and discuss the shearing stress τ later.

It should be emphasized that the normal stress in an axially loaded member changes from point to point within a given cross-sectional area of the member. The normal stress at a *given point* is defined by the relation

$$\sigma = \operatorname*{Lim}_{\Delta A \to 0} \frac{\Delta F}{\Delta A} = \frac{dF}{dA} \tag{1.3}$$

where ΔA is a small area at the point in the member as shown in Figure 1.7 and ΔF is the small force that acts over this area. Thus, the ratio $\Delta F/\Delta A$ represents the average value of stress over the small area ΔA and, in the limit when ΔA approaches zero, the normal stress σ at the point becomes the derivative of the force F with respect to the area A as indicated by Equation 1.3. To find the normal stress σ at some position along the member, we first pass a cutting plane at the position of interest, perpendicular to the axis of the member, and determine the internal axial force F using the methods discussed in Section 1.1. We then employ Equation 1.2 to determine the average normal stress acting at that position. As we stated in Section 1.1, the internal force F is actually the resultant of a distributed internal axial force intensity (stress). This stress distribution is, in general, not uniform throughout the cross-sectional area of the member. According to *Saint-Venant's* principle, the closer we are to the points of load application, the less uniform is the stress distribution, and the farther away we move from these points, the closer we come to achieving the conditions expressed in Equation 1.2. Thus, consider the member shown in Figure 1.8a, which is subjected to the axial tensile forces P as shown. At a position such as A, sufficiently removed from the point of application of P (at least a distance equal to the diameter d from the ends of the member where the concentrated forces P are

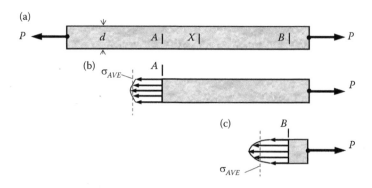

FIGURE 1.8 Member subjected to tensile axial load P showing the stress distributions at section A and at section B.

applied), the stress distribution approaches that given by Equation 1.2 as shown in Figure 1.8b. At position B, on the other hand, sufficiently close to the point of application of P (a distance less than the diameter d), the stress distribution is the type shown in Figure 1.8c, where the maximum stress is much larger, and the minimum stress much smaller than σ_{AVE}. In practice, however, we make the assumption that the stress in a straight member subjected to an axial force is *uniformly distributed* and Equation 1.2 is used along with a *factor of safety* to be discussed later. It should be emphasized, however, that the resultant internal force F of the distribution shown in either Figure 1.8b or Figure 1.8c is exactly equal to the applied force P, thus satisfying the conditions of equilibrium.

1.3.2 STRESS ELEMENT

We now introduce the concept of *stress element*, which is defined as an infinitesimally small block of material, at a given point in a stressed body, showing all the stresses acting on it. Let us construct the free-body diagram of that part of the member of Figure 1.8a to the right of plane X as shown in Figure 1.9a, which also indicates the stress distribution on that plane. Note that we define a plane by the axis to which the plane is normal. Thus, when we speak of the X plane, we are referring to a plane that is perpendicular to the x axis. Let us isolate the small gray square block shown in Figure 1.9a and magnify it in order to show details as depicted in Figure 1.9b. Note that the two vertical planes (X planes), of differential length dy, are subjected to the normal stress σ_x, (equal to σ_{AVE}), while the two horizontal planes (Y planes), of differential length dx, are free of any stress. Note also that the plane whose outward normal points in the positive sense of its axis is designated as a *positive plane*. Thus, for example, the right X plane is a *positive plane*, while the left one is a *negative plane*. The *subscript* in the normal stress σ refers to the plane on which it acts (the X plane), and to the axis along which it is pointed (the x axis). The stress element shown in Figure 1.9b has a constant small depth into the page and represents the simplest stress condition that we encounter in practice and is referred to as a *uniaxial* (along one axis) stress condition. Since, by Equation 1.2, $F = (\sigma_{AVE})A$, it follows that the stress element of Figure 1.9b is in equilibrium because the resultant force (the product of the stress and the infinitesimal area over which it acts) acting to the right is identical to that acting to the left. In a three-dimensional case of stress referred to a right-handed x–y–z coordinate system, two other stresses come into play and are given the symbols σ_y and σ_z.

1.3.3 SHEARING STRESS

We now consider the shearing stress τ which, as stated earlier, acts parallel to its plane. To illustrate the existence of the shearing stress, we consider a horizontal member supported and loaded as shown in Figure 1.10a. Let us isolate the central part of the member and construct its free-body diagram as shown in Figure 1.10b. Equilibrium and symmetry dictate that the resisting forces V on the two exposed planes be each equal to $P/2$. Each of the two forces V represents the resultant of

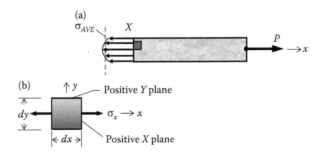

FIGURE 1.9   Member subjected to tensile axial load P showing the simplest stress element of dimensions dx by dy, known as a uniaxial stress condition.

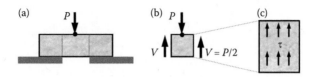

FIGURE 1.10   Member (a) subjected to a load P, (b) producing shearing forces V, (c) which produces a distribution of shearing stresses τ.

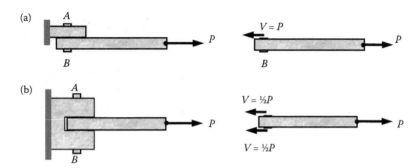

FIGURE 1.11   Diagrams showing (a) pin AB in single shear and (b) Pin AB in double shear.

the shearing stresses τ, which are distributed in some manner over the exposed area, as indicated in Figure 1.10c.

The average value, τ_{AVE}, of the shearing stress acting over an area A is defined by the equation

$$\tau_{AVE} = \frac{V}{A} \qquad (1.4)$$

Consider the practical case of a pin that is holding two members together as shown in Figure 1.11a, and where the concept of the average shearing stress is useful in the design process. The pin AB shown in Figure 1.11a is said to be in *single shear* because only one of its planes is subjected to shearing action. Thus, the average shearing stress in this pin is given by the relation

$$\tau_{AVE} = \frac{V}{A} = \frac{P}{A} \text{ (for a pin in single shear)} \qquad (1.5)$$

The symbol A in Equation 1.5 represents the cross-sectional area of the pin.

In Figure 1.11b, pin AB is said to be in *double shear* because two of its planes are subjected to shearing action. In this case, with the symbol A representing the cross-sectional area of the pin, the average shearing stress is given by

$$\tau_{AVE} = \frac{V}{A} = \frac{P}{2A} \text{ (for a pin in double shear)} \qquad (1.6)$$

1.3.4   STRESSES ON INCLINED PLANES

Let us now consider the axially loaded member shown in Figure 1.12a. Two planes N and T perpendicular to each other are shown such that the N plane is inclined at an angle θ with the longitudinal

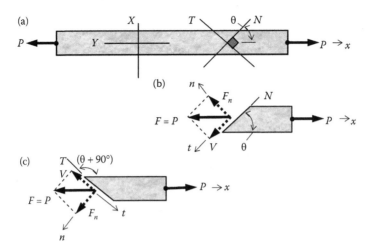

FIGURE 1.12 Diagram showing the development of a stress element on inclined planes.

axis of the member. Also shown is another set of orthogonal X and Y planes, where the X plane is normal to the x axis along the member. Let us construct the free-body diagram of that part of the member to the right of plane N as shown in Figure 1.12b. The internal force F, which of course is equal to P, is decomposed into the two components F_n, normal to the N plane, and $F_t = V$, parallel to it. The force F_n may be viewed as the resultant of a normal stress σ_n distribution over the inclined exposed area of the member and V, as the resultant of a shearing stress τ_{nt} distribution over the same area. Similarly, as shown in Figure 1.12c, we construct the free-body diagram of that part of the member to the right of plane T. The internal force $F = P$, is decomposed into the two components, F_n, normal to the T plane, and V, parallel to this plane. Again, the force F_n represents the resultant of a normal stress σ_n distribution over the inclined exposed area of the member and V, the resultant of a shearing stress τ_{tn} distribution over the same area.

Consider now the stress element shown in gray at the intersection of the N and T planes in Figure 1.12a. This stress element is depicted, magnified, in Figure 1.13, with all of the stresses acting on it. Note that the normal stresses are tensile in this case, as required by the free-body diagrams of Figures 1.12b and 1.12c. The notation used for the normal stress is the same as explained earlier, namely, that the subscript refers to the plane on which it acts (either the N or the T plane), and to the axis along which it points (either the n or the t axis). The notation for the shearing stress requires two subscripts but is similarly based on both its plane and its axis. Thus, the shearing stress τ_{nt} acts on the N plane and is pointed along the t axis and the shearing stress τ_{tn} acts on the T plane and is pointed along the n axis. Assuming the thickness h of the stress element into the page is small and constant, each of the two shearing stresses τ_{nt} creates a shearing force equal to $\tau_{nt}(dt)h$, where the product $(dt)h$ is the area of the N plane. These two shearing forces constitute a counterclockwise couple on the stress element equal to $\tau_{nt}(dt)h(dn)$. Also, each of the two shearing stresses τ_{tn} results in a shearing force equal to $\tau_{tn}(dn)h$, where the product $(dn)h$ is the area of the T plane. The two shearing forces on the T planes produce a clockwise couple equal to $\tau_{tn}(dn)h(dt)$. Rotational equilibrium is assured if these two couples are equal. Thus, $\tau_{tn}(dn)h(dt) = \tau_{nt}(dt)h(dn)$. Simplifying, we conclude that

$$\tau_{tn} = \tau_{nt} \qquad\qquad (1.7)$$

Translational equilibrium of the stress element in Figure 1.13 is assured in the n direction because of the balance provided by the *normal forces* created by the normal stresses σ_n, and in the t direction because of the balance provided by the normal forces created by the normal stresses σ_t.

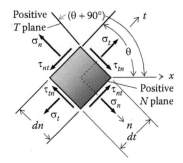

FIGURE 1.13 A stress element *dn* by *dt*, on inclined planes, showing the stresses σ_t and σ_n.

The stress element of Figure 1.13 represents the most complex stress condition that we encounter in two-dimensional stress systems and is referred to as a *biaxial* (along two axes) or a *plane stress* condition.

General symbolic expressions will now be derived for the stresses σ_n and σ_t. Since $F = P$ in Figure 1.12b, its two components acting on plane *N* are $F_n = P \sin \theta$ and $V = F_t = P \cos \theta$. Let the cross-sectional area of the member be *A*. Then the area of inclined plane *N* is $A_n = A/\sin \theta$. Therefore,

$$\sigma_n = \frac{F_n}{A_n} = \left(\frac{P}{A}\right) \sin^2 \theta = \sigma_x \sin^2 \theta \tag{1.8}$$

$$\tau_{nt} = \frac{F_t}{A_n} = \left(\frac{P}{A}\right) \sin \theta \cos \theta = \sigma_x \sin \theta \cos \theta \tag{1.9}$$

Note that in both of the above equations, the quantity *P/A* was replaced by the normal stress σ_x, the normal stress on plane *X*. Note also that the maximum value of σ_n is σ_x, and occurs when $\theta = \pm 90°$. Also, the maximum value of τ_{nt} is $\sigma_x/2$, and occurs when $\theta = \pm 45°$. These conclusions may be arrived at either by inspection or by differentiating each of the two equations and setting the derivative equal to zero.

If the *x* and *y* axes coincide, respectively, with the *n* and *t* axes, then the normal stresses σ_n and σ_t become σ_x and σ_y, respectively, and the shearing stress τ_{nt} becomes τ_{xy}. As discussed in Chapter 7, in a three-dimensional case referred to an *x–y–z* coordinate system, we have a total of six stress components, σ_x, σ_y, σ_z, τ_{xy}, τ_{yz}, and τ_{xz}.

1.3.5 UNITS

Since stress is force per unit area as expressed in Equation 1.2, it follows that the units of stress are those of force (*pounds* or *kilopounds*, *Newtons* or *kiloNewtons*) divided by those of area (*square inches* or *feet, square millimeters* or *meters*). Thus, such units as *pounds per square inch* (psi) and *kilopounds per square inch* (ksi) are common in the U.S. Customary system of measure, and *Newtons per square meter* (N/m²) in the SI (*metric*) system of units. The stress unit, N/m², is called the *Pascal* and denoted by the symbol Pa. It is an extremely small quantity and another stress unit is used known as the *megaPascal* and denoted by the symbol MPa = 10^6 Pa.

1.3.6 SIGN CONVENTION

A normal stress is positive if it points in the direction of the outward normal to its plane. Thus, a positive normal stress produces tension and a negative normal stress produces compression. A shearing stress component is positive if it acts on a positive coordinate plane and points in the positive direction of the axis designated by its second subscript, *or*, if it acts on a negative coordinate plane and points in the negative direction of the axis designated by its second subscript. Thus, for example, referring to Figure 1.13, σ_n on the positive N plane is positive because it points in the direction of the outward normal. Also, τ_{nt} on this plane is positive because it points in the positive sense of the t axis. Furthermore, σ_n on the negative N plane is positive because it points in the direction of the outward normal and τ_{nt} on this plane is positive because it points in the negative sense of the t axis. Note that all of the stresses acting on the stress element of Figure 1.13 are positive.

EXAMPLE 1.3

a. Use the equilibrium method to determine the average normal stress in segment BC of the member loaded as shown in figure below. The cross section of the member is circular with a radius of 2 in.

b. Construct a stress element in segment BC at the intersection of plane X, perpendicular to the axis of the member, and plane Y, parallel to it.

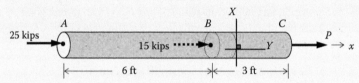

SOLUTION

a. The applied unknown force P at C is determined by considering the equilibrium of the entire member and is found to be $P = -40$ kips. A cutting plane perpendicular to the x axis (X plane) is placed anywhere between B and C and an RFB diagram is constructed as shown in figure below. On the basis of equilibrium, we conclude that the internal force F in the exposed area is compressive with a magnitude of 40 kips.

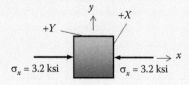

The average normal stress is given by Equation 1.2, where $F = -40$ kips as found above and $A = \pi r^2 = \pi(2^2) = 12.566$ in.^{2} Thus, by Equation 1.2, we obtain

$$\sigma_{AVE} = -\frac{40}{12.566} = -3.18 = -3.2 \text{ ksi}$$
$$= 3.2 \text{ ksi (C)} \qquad\qquad \textbf{ANS.}$$

The symbol (C) in the answer above is used to indicate compression.

b. The gray square block indicated in figure above is the required stress element with two planes parallel and two planes perpendicular to the x axis. This stress element is shown, magnified, in figure below, which indicates that the X planes are subjected to the stress $\sigma_x = \sigma_{AVE} = -3.2$ ksi, while the Y planes are free of stress.

$\sigma_x = 3.2$ ksi $\qquad\qquad$ $\sigma_x = 3.2$ ksi

EXAMPLE 1.4

Construct a complete stress element for a point at the intersection of the *N* and *T* planes as shown. This point, magnified, is represented by the differential gray square in the sketch. The cross section of the member is a 200 mm × 400 mm rectangle. Solve by using the equilibrium method and check the magnitudes of the stresses thus obtained on the *N* plane by using Equations 1.8 and 1.9.

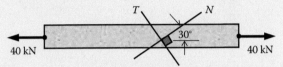

SOLUTION

Construct a free-body diagram of that part of the member to the right of the *N* plane as shown in figure below. By equilibrium, the internal force *F* is tensile with a magnitude of 40 kN. Its two components F_n and *V* are found as follows:

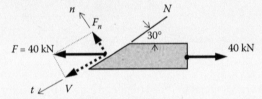

$$F_n = 40 \sin 30° = 20.0 \text{ ksi}; \quad V = 40 \cos 30° = 34.64 \text{ ksi}$$

The inclined cross-sectional area A_n of the member over which F_n and *V* act is

$$A_n = \frac{200 \times 400}{\sin 30°} = 160,000 \text{ mm}^2 = 0.16 \text{ m}^2$$

The normal and shearing stresses on the *N* plane become

$$\sigma_n = \frac{20}{0.16} = 125.0 \text{ kN/m}^2 = 0.125 \text{ MPa} \qquad \textbf{ANS.}$$

$$\tau_{nt} = \frac{34.64}{0.16} = 216.5 \text{ kN/m}^2 = 0.217 \text{ MPa} \qquad \textbf{ANS.}$$

We now construct a free-body diagram of that part of the member to the right of the *T* plane as shown in figure below. Equilibrium dictates that the internal force *F* be tensile with a magnitude of 40 kN. F_n and *V* are computed as follows:

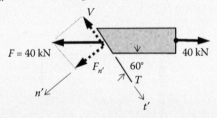

$$F_{n'} = 40 \sin 60° = 34.64 \text{ kN}; \quad V = 40 \cos 60° = 20 \text{ kN}$$

The inclined cross-sectional area over which F_n and V act is

$$A_t = \left(\frac{200 \times 400}{\sin 60°} \right) = 92{,}376 \text{ mm}^2 = 0.092376 \text{ m}^2$$

Therefore, the stresses acting on the T plane are

$$\sigma_t = \left(\frac{34.64}{0.092376} \right) = 375 \text{ kN/m}^2 = 0.375 \text{ MPa} \qquad \textbf{ANS.}$$

$$\tau_{tn} = \left(\frac{20}{0.092376} \right) = 216.5 \text{ kN/m}^2 = 0.217 \text{ MPa} \qquad \textbf{ANS.}$$

Note that the magnitude of τ_{tn} is identical to that of τ_{nt} as expressed in Equation 1.7. Using Equations 1.8 and 1.9, we obtain

$$\sigma_n = \left(\frac{40 \times 10^3}{(0.2)(0.4)} \right) \sin 30° = 0.125 \text{ MPa}$$

$$\tau_{nt} = \left(\frac{40 \times 10^3}{(0.2)(0.4)} \right) \sin 30° \cos 30° = 0.217 \text{ MPa}$$

All of these values are identical to those obtained earlier by equilibrium. The required stress element is shown in figure below.

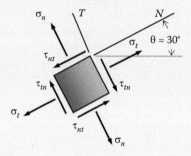

PROBLEMS

1.16 (a) Find the internal axial forces in members AB and CD. Note that ABC is a right triangle. (b) Compute the normal stresses in members AB and CD on planes perpendicular to their respective axes. All members have the same circular cross-sectional area with a diameter of 1.0 in.

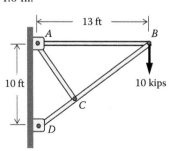

1.17 (a) Use the method of sections to find the forces in members *AB*, *CB*, and *CD*. (b) Find the normal stresses in these members if their diameters are $D_{AB} = 20$ mm, $D_{CB} = 30$ mm, and $D_{CD} = 25$ mm.

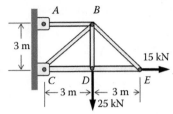

1.18 Determine the force *P* if the normal stress in member *AC* is known to be 75 MPa. Note that *ABC* is a right triangle. The cross-sectional area of member *AC* is 500 mm².

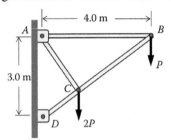

(Problems 1.18 and 1.19)

1.19 In reference to the figure shown in Problem 1.18, find the force *P* if the pin at *D* is known to carry an average shearing stress of 50 MPa. The pin at *D* is in double shear with a diameter of 15 mm, and *ABC* is a right triangle.

1.20 Find the force *P* if the pin at *A* is in single shear and the average shearing stress in it is not to exceed 10 ksi and its diameter is 1.25 in.

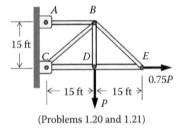

(Problems 1.20 and 1.21)

1.21 In reference to the figure shown in Problem 1.20, determine the force *P* if the normal stress in member *BC* is not to exceed a magnitude of 50 ksi and its cross-sectional area is 2.5 in.²

1.22 Determine the normal stress in rod *AC* if its diameter is 1.5 in.

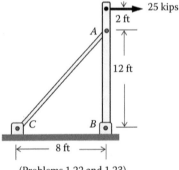

(Problems 1.22 and 1.23)

1.23 In reference to the figure shown in Problem 1.22, find the average shearing stress in the pin at B if it is in single shear and it has a diameter 1.5 in.

1.24 (a) Find the normal stress in member AB if its cross section is a 30 mm × 20 mm rectangle. (b) Construct a stress element in member AB with two planes perpendicular to its axis (X planes) and two planes parallel to its axis (Y planes). Use the equilibrium method.

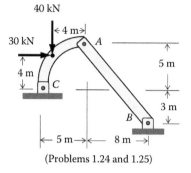

(Problems 1.24 and 1.25)

1.25 In reference to the figure shown in Problem 1.24, determine the average shearing stress in the pin at C if it is in single shear and its diameter is 30 mm.

1.26 Find the magnitude of the force P if the normal stress in rod AB is 30 ksi. The diameter of the rod is 3/8 in.

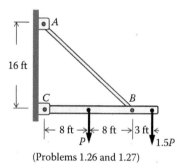

(Problems 1.26 and 1.27)

1.27 In reference to the figure shown in Problem 1.26, determine the magnitude of the force P if the average shearing stress in the pin at C is 10 ksi. The diameter of the pin is 1 in. and the pin is in single shear.

1.28 Find the largest magnitude of the force P if the normal stress in member BC is not to exceed 75 MPa. Member BC has a 25 mm × 20 mm rectangular cross section.

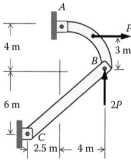

(Problems 1.28 and 1.29)

1.29 In reference to the figure shown in Problem 1.28, determine the largest magnitude of the force P if the average value of the shearing stress in the pin at C cannot exceed 25 MPa. The pin is in single shear and its diameter is 30 mm.

1.30 Construct a stress element in segment AB of the shaft with two planes perpendicular to its axis (X planes) and two planes parallel to it (Y planes). The diameter of segment AB is 5 in. and that of BC is 3 in. Use the equilibrium method.

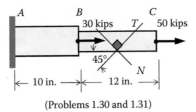

(Problems 1.30 and 1.31)

1.31 In reference to the figure shown in Problem 1.30, construct a stress element in segment BC of the shaft at the intersection of the N and T orthogonal planes as shown. The diameter of segment AB is 4 in. and that of BC is 2 in. Use the equilibrium method.

1.32 Construct a stress element in segment BC of the member with two planes perpendicular to its axis (X planes) and two planes parallel to it (Y planes). The cross section of the member is a 60-mm-diameter circle. Use the equilibrium method.

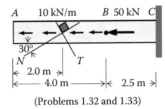

(Problems 1.32 and 1.33)

1.33 In reference to the figure shown in Problem 1.32, construct a stress element in segment AB of the member, at a distance of 2.0 m point A, at the intersection of the N and T orthogonal planes as shown. The cross section of the member is a 50 mm × 30 mm rectangle. Use the equilibrium method and check some of the answers using Equations 1.8 and 1.9.

1.34 Construct the stress element in member DC at the intersection of the two orthogonal N and T planes as shown. Member DC has a hollow rectangular cross section with outside dimensions of 80 mm × 40 mm and inside dimensions of 65 mm × 25 mm. Use the equilibrium method.

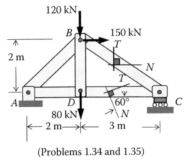

(Problems 1.34 and 1.35)

1.35 In reference to the figure shown in Problem 1.34, construct the stress element in member BC at the intersection of the two orthogonal N and T planes as shown. Member BC has the same cross-sectional area as given for member DC in Problem 1.34. Use the equilibrium method.

1.36 Construct the stress element in member DC at the intersection of the two orthogonal N and T planes as shown. Member DC has a hollow rectangular cross-sectional area with outside dimensions of 8 in. × 6 in. and inside dimensions of 7.5 in. × 5.5 in. Use equilibrium and check the answers on the N plane using Equations 1.8 and 1.9.

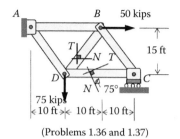

(Problems 1.36 and 1.37)

1.37 In reference to the figure in Problem 1.36, construct the stress element in member *BD* at the intersection of the two orthogonal *N* and *T* planes as shown. Member *BD* has the same cross-sectional dimensions as member *DC* in Problem 1.36. Use the equilibrium method.

1.38 Determine the intensity *p* of the distributed load if the normal stress on plane *N′*, at a distance 0.8 m from *A*, is $\sigma_{n'} = 80$ MPa. The cross section of the member is a 40 mm × 30 mm rectangle.

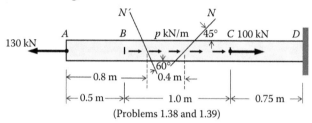

(Problems 1.38 and 1.39)

1.39 In reference to the figure in Problem 1.38, find the intensity *p* of the distributed load if the shearing stress on plane *N*, at a distance of 1.2 m from point *A*, is $\tau_{nt} = 50$ MPa. The cross section of the member is a 40 mm × 30 mm rectangle.

1.4 NORMAL STRAIN AND STRESS–STRAIN DIAGRAMS

When a real body is subjected to stresses, it undergoes *deformations* and *distortions*. The term *deformation* refers to the geometric changes that take place in the dimensions of the body (extensions or contractions), while the term *distortions* represents geometric changes in its shape. We will discuss the term *deformation* in more detail in Section 1.5, and the term *distortion* in Chapter 2. For our purpose here, however, any line element in a body is said to experience *deformation* if its length increases or decreases.

1.4.1 NORMAL STRAIN

The term *strain* of a line element is used to signify its unit deformation (i.e., the deformation of the line element divided by its initial length). The unit deformation of a line element is referred to as *normal* (or *linear*) strain. Therefore, if the initial length of a line element *AB* is *L*, as shown in Figure 1.14, and its deformation due to the applied loads *P* is δ (distance *BC*), an *average* value for the normal strain is given by

$$\varepsilon_{AVE} = \frac{\delta}{L} \tag{1.10}$$

FIGURE 1.14   Diagram showing the definition of the deformation δ.

If the x axis lies along the line element as shown in Figure 1.14, the normal strain of the member may be represented by the symbol ε_x in keeping with the notation used for normal stress. Obviously, if strains occur along the y and z axes, they would be labeled ε_y and ε_z, respectively.

1.4.2 UNITS

Note that by definition, the normal strain ε is a dimensionless quantity. However, it is sometimes expressed in terms of units of length divided by units of length. Thus, for example, in the U.S. Customary system of measure, the unit in./in. and in the SI system, the unit m/m is widely used.

1.4.3 SIGN CONVENTION

The sign convention used in this text for normal strain is such that a *positive normal strain represents an extension* and a *negative normal strain represents a contraction*. This sign convention is, of course, compatible with the one established for normal stress in Section 1.3.

1.4.4 MECHANICAL PROPERTIES

The term *mechanical properties* refers to certain material characteristics that, in general, are different for different materials. These mechanical properties for a given material are obtained in the laboratory by means of a tension or compression test. In general, these mechanical properties are obtained at room temperature under loads that are applied very gradually (i.e., static loads). It should be noted, however, that these properties may be very sensitive to the effects of temperature and to the rate of application of the loads. For example, under impact loads at elevated temperatures, a material would exhibit properties entirely different from those at room temperature under static loads.

The tension or compression test is performed on a small sample of a given material known as the *test specimen*, whose dimensions have been standardized by the American Society for Testing and Materials. The design of three such standardized specimens is shown in Figure 1.15. The load on the test

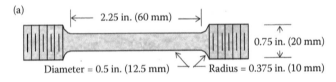

Round tension test specimen

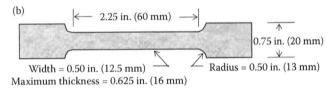

Rectangular tension test specimen

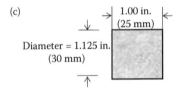

Round compression test specimen

FIGURE 1.15   Standard round and rectangular test specimens.

specimen is applied by means of a machine known as a *universal testing machine*. One type of universal testing machine is shown in Figure 1.16. As the load is gradually increased, the specimen changes in length. This change in length δ is usually measured for a given length L_o, known as the *gage length*, by means of an instrument known as the *extensometer*, a picture of which is shown in Figure 1.17. One type of a compression machine normally used in the testing of concrete specimens is shown in Figure 1.18.

Corresponding values of load and deformation are taken while the load is gradually increased from zero until fracture in a tensile test, or until fracture or some predetermined value of load is reached in a compression test. Values of load are converted into values of normal stress σ by Equation 1.2, and values of deformation into values of normal strain σ by Equation 1.10. In determining the normal stress from Equation 1.2, the original cross-sectional area of the test specimen is used, even though this cross-sectional area decreases as the load is increased. The value of the normal stress thus obtained is known as the *engineering stress*. Also, the value of the normal strain obtained from Equation 1.10 by using the original gage length is referred to as the *engineering strain*. Under certain conditions, it is necessary to account for the changes that occur in the cross-sectional area and in the gage length of the specimen. When such changes are taken into consideration (i.e., when

FIGURE 1.16   MTS Universal Testing Machine (Criterion Model 44). (Photo courtesy of MTS Systems Corp. With permission.)

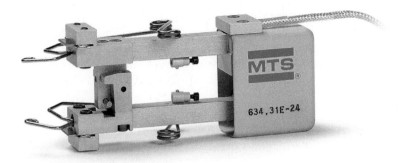

FIGURE 1.17   MTS Extensometer (Model 634.31E-24). (Photo courtesy of MTS Systems Corp. With permission.)

the instantaneous cross-sectional area and the instantaneous gage length are used in Equations 1.2 and 1.10), the resulting normal stress and normal strain are known as the *true stress* and *true strain,* respectively. The following discussion is limited to the engineering stress and engineering strain, since these are the values generally used in the selection of materials for specific purposes, while the true stress and true strains are, in general, useful for conducting research on materials.

The stress–strain diagrams shown in Figures 1.19 and 1.20 define two distinct ranges of material behavior known as the *elastic* and *plastic* (or *inelastic*) *ranges.* In general, the *elastic range* is that

FIGURE 1.18   MC-300PR, 300,000 lb (136,000 kg)-Capacity Compression Testing Machine by Gilson Company, Inc. (Photo by the authors.)

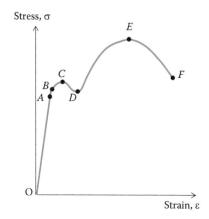

FIGURE 1.19   Stress–strain diagrams for one type of a ductile material, showing both elastic and plastic ranges.

part of the diagram defining a linear relation between the stress and the strain (approximately seg-ment *OA* in Figures 1.19 and 1.20). The upper limit for the stress σ for which the relation between stress and strain is linear is known as the *proportional limit* stress σ_p represented by the value of σ at point *A* in Figures 1.19 and 1.20.

The simple linear relation that exists between the stress and the strain below the proportional limit is expressed by the relation

$$\sigma = E\varepsilon \tag{1.11}$$

The quantity *E* in Equation 1.11 is a factor of proportionality between the normal stress and the normal strain and represents a unique property for a given material known as *Young's modulus of elasticity* (or simply the *modulus of elasticity*) after Thomas Young, who, in 1807, was the first to define it. It can be seen from Equation 1.11, since ε is a dimensionless quantity, that *E* has the same units as the stress, namely, psi or ksi in the U.S. Customary system and MPa or GPa in the SI system. The relation expressed in Equation 1.11 is referred to as *Hooke's law* in honor of Robert Hooke, who, in 1678, was the first to formulate a statement relating force to deformation.

The *plastic* or *inelastic range* is that part of the stress–strain diagram that defines a nonlinear relation between the stress and the strain and is represented by segment BF in Figures 1.19 and 1.20.

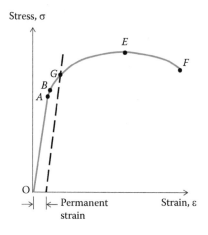

FIGURE 1.20   Stress–strain diagram for another type of a ductile material, showing both elastic and plastic ranges.

Several empirical equations have been proposed to describe the inelastic relation between stress and strain under uniaxial loading conditions, but the most widely used is the one known as the *Ramberg–Osgood* equation, which may be expressed by the relation

$$\varepsilon = \frac{\sigma}{E} + \left(\frac{\sigma}{B}\right)^{n} \tag{1.12}$$

The quantities B and n in Equation 1.12 are constants for a given material.

The most significant mechanical properties will now be stated and defined. Typical values of mechanical properties for some of the most commonly used engineering materials are given in Appendix D.

Proportional limit, σ_p: The proportional limit for a given material represents the largest value of stress beyond which the stress is no longer proportional to the strain. The proportional limit is represented by the ordinate to point A in Figures 1.19 and 1.20.

Elastic limit, σ_e: The elastic limit for a given material is the value of stress beyond which the material experiences permanent deformation after the stress is removed. Thus, if the material is loaded to any level of stress within the elastic limit and the load is then removed, it will regain its original dimensions and shape and is said to behave *elastically*. However, if the load exceeds the elastic limit before it is removed, the material does not fully regain its initial dimensions and shape. In such a case, the material is said to experience *permanent* deformation. The elastic limit is represented by the ordinate to point B in Figures 1.19 and 1.20. Its determination, experimentally, is extremely difficult and, therefore, its exact location on the stress–strain diagram is usually not known, even though it is generally higher than the proportional limit σ_p. For all practical purposes, however, the elastic limit σ_e and the proportional limit σ_p may be assumed to have the same values.

Modulus of elasticity, E: The modulus of elasticity, or Young's modulus of elasticity, is the constant of proportionality between stress and strain in Hooke's law as expressed in Equation 1.11. Physically, it represents the slope of the stress–strain diagram within the proportional range of the material (i.e., the slope of the straight segment OA in Figures 1.19 and 1.20). The term *stiffness* is used to describe the capacity of materials to resist deformation in the elastic range and it is measured by the modulus of elasticity. For example, steels with a modulus of elasticity of about 30×10^6 psi are stiffer than aluminum with a modulus of elasticity of about 10×10^6 psi.

Yield point, σ_y: The yield point is the stress at which the material continues to deform without further increase in the stress. The stress may even decrease slightly as the deformation continues past the yield point. Some materials, notably the plain carbon steels, exhibit a well-defined yield point, as shown by point C in Figure 1.19. If the stress decreases past this point, it is referred to as the *upper yield point*, in contrast to the *lower yield point* represented by point D in Figure 1.19. Beyond the lower yield point, the stress increases with further increase in the strain, a phenomenon known as *strain hardening*.

Yield strength, σ_s: For materials having a stress–strain diagram such as shown in Figure 1.20 (those that do not exhibit a well-defined yield point), a value of stress, known as the *yield strength*, is defined as one producing a certain amount of permanent strain. Although several values of permanent strain may be used for defining the yield strength, the most commonly encountered values range between 0.0020 and 0.0035. To determine the yield strength σ_s, the assigned numerical value of permanent strain is measured from the origin along the strain axis to locate a point through which a line is drawn parallel to the straight portion (segment OA) of the stress–strain diagram. The straight line is then extended until it intersects the stress–strain curve. This construction is shown schematically in Figure 1.20, in which the ordinate of point G represents the value of the yield strength for this material.

Ultimate strength, σ_u: The ultimate strength represents the ordinate to the highest point in the stress–strain diagram and is equal to the maximum load carried by the specimen divided by its original cross-sectional area. The ultimate strength is represented by the ordinate of point *E* in Figures 1.19 and 1.20.

Fracture strength, σ_f: The fracture strength, also known as the *rupture* or *breaking strength*, is the engineering stress at which the specimen fractures and complete separation of the specimen parts occurs. This strength is represented by the ordinate of point *F* in Figures 1.19 and 1.20 and is equal to the load at fracture divided by the original cross-sectional area of the specimen.

Poisson's ratio, μ: When a member is subjected to stress (tension or compression), it exhibits strain not only in the direction of the stress, referred to as *longitudinal strain*, ε_L, but also along the two axes perpendicular to the direction of the stress, known as *transverse strains*, ε_T. For example, if the member is subjected to tension, its longitudinal strain is an extension (positive), while its transverse strains are contractions (negative). Experiments show that the ratio of any transverse to the longitudinal strain is a constant for a given material. This constant is given the symbol μ and is known as Poisson's ratio, in honor of Simon D. Poisson, who, in 1811, was the first to define it. As an example, if the longitudinal strain is ε_x, then the transverse strains would be $\varepsilon_y = \varepsilon_z = -\mu\varepsilon_x$. Since Poisson's ratio is a material property, it is expressed as a pure number without regard to the sign difference between the transverse and the longitudinal strains. Thus, as a pure number, Poisson's ratio is given by

$$\mu = \left| \frac{\varepsilon_T}{\varepsilon_L} \right| \qquad\qquad (1.13)$$

Ductility: The property of materials known as ductility is a measure of their capacity to deform in the plastic or inelastic range. Thus, materials that exhibit large plastic deformations, represented by segment *BEF* in Figures 1.19 and 1.20, are said to be ductile materials. Examples of ductile materials include steel and aluminum alloys as well as the alloys of copper. Specimens of these ductile materials, when subjected to tension, undergo considerable plastic deformation. Not only do they exhibit large extensions, but after the ultimate strength is reached, they also undergo considerable reduction in the cross-sectional dimension, known as *necking*, in the region of fracture as shown in Figure 1.21a. On the other hand, materials that fracture with little or no measurable plastic deformation are known as *brittle materials*. Examples of brittle materials include most cast irons as well as concrete. A tension specimen made of cast iron, for example, would exhibit a very slight extension and no appreciable necking at fracture, as shown in Figure 1.21b. Another characteristic of brittle materials is that they are generally much stronger in compression than they are in tension.

The ductility of a material is usually measured by one or both of the following two properties:

1. *Percent elongation, %EL*: The percent elongation is defined as 100 multiplied by the change in length, ΔL, at fracture divided by the original gage length, L_o, of the tension specimen. Thus,

$$\%EL = \left(\frac{\Delta L}{L_o} \right) \times 100 \qquad\qquad (1.14)$$

2. *Percent reduction of area, %ROA*: The percent reduction of area is defined as 100 multiplied by the change in the cross-sectional area ΔA at fracture divided by the original cross-sectional area A_o. Thus,

$$\%ROA = \left(\frac{\Delta A}{A_o} \right) \times 100 \qquad\qquad (1.15)$$

(a)

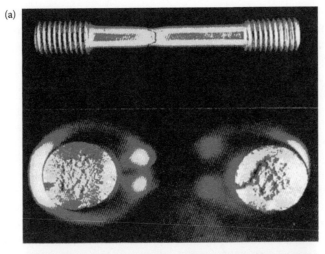

(b)

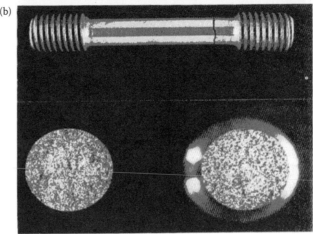

FIGURE 1.21 (a) Photograph showing ductile fracture with appreciable plastic deformation. (b) Photograph showing brittle fracture lacking measurable plastic deformation. (Photo by the authors.)

Energy absorption capacity: Often materials are called upon to resist the action of dynamic and impact loads. Under such conditions, the energy absorption capacity becomes a useful tool in comparing materials for dynamic and impact applications. Two such energy quantities are defined:

1. *Modulus of resilience, u_R*: The modulus of resilience represents the amount of energy per unit volume, absorbed by the material when stressed up to the proportional limit. Thus, the modulus of resilience represents the amount of energy, known as *elastic strain energy*, that the material can absorb elastically.

   When a gradually increasing internal axial force F acts on a member, it produces a deformation δ. The stored elastic strain energy dU, assumed equal to the work of the force F producing an infinitesimal deformation $d\delta$, is given by the relation $dU = F\, d\delta$. Thus, the total stored elastic strain energy due to the force F deforming the member by the amount δ becomes

$$U = \int F\, d\delta \tag{1.16}$$

If deformations are assumed to be within the proportional range of behavior, the force $F = k\delta$, where k is a constant of proportionality. Substituting for F into Equation 1.16 and integrating, we obtain

$$U = k\int \delta \, d\delta = \frac{k\delta^2}{2} = \frac{F\delta}{2} \qquad (1.17)$$

Since $F = \sigma A$ and $\delta = \varepsilon L$, where L and A are the length and cross-sectional area of the member, respectively, the stored elastic strain energy U may be expressed by

$$U = \frac{\sigma\varepsilon}{2}(AL) \qquad (1.18)$$

Note that the product AL represents the volume V of the member and, if both sides of Equation 1.18 are divided by it, the resulting quantity is the elastic strain energy per unit volume of material. Thus,

$$\frac{U}{AL} = \frac{\sigma\varepsilon}{2} \qquad (1.19)$$

The elastic strain energy per unit volume, $u = U/AL$, may be expressed in the form

$$u = \frac{dU}{dV} = \frac{\sigma\varepsilon}{2} \qquad (1.20)$$

We conclude, therefore, that the modulus of resilience for a given material is $\sigma_p\varepsilon_p/2$, where σ_p is the proportional limit and ε_p is the strain at the proportional limit. If ε is replaced by σ/E, Equation 1.20 may be expressed as

$$u = \frac{\sigma^2}{2E} \qquad (1.21)$$

Thus, the modulus of resilience for the material becomes

$$u_R = \frac{\sigma_p^2}{2E} \qquad (1.22)$$

Examination of Equation 1.22 reveals that the modulus of resilience is equal to the area under the stress–strain diagram up to the proportional limit, as shown in Figure 1.22.

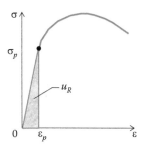

FIGURE 1.22   Stress–strain diagram showing the *modulus of resilience*.

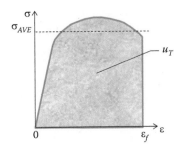

FIGURE 1.23   Stress–strain diagram showing the *modulus of toughness.*

2. *Modulus of toughness, u_T:* The modulus of toughness is the amount of energy per unit volume that the material can absorb before fracture. Thus, the modulus of toughness is a measure not only of the strength of the material, but also of its ductility. By definition, then, the modulus of toughness is given by the area under the entire stress–strain diagram, as indicated in Figure 1.23. An approximate value for this property is obtained by estimating an average stress σ_{AVE} as shown in Figure 1.23, and multiplying it by the strain at fracture ε_f. Thus, in terms of these quantities, we have

$$u_T = \sigma_{AVE}\varepsilon_f \tag{1.23}$$

1.4.5   DESIGN CONSIDERATIONS

The primary purpose for testing materials and obtaining their mechanical properties is to permit the design engineer to properly select a material for a given system (structure or machine) under design consideration. There are many factors during the lifetime of the system that the design engineer cannot foretell accurately. Because of these factors, the design engineer ensures that the magnitude of stress to which the system is subjected during its lifetime is considerably less than the material property obtained in the laboratory. Therefore, a margin of safety is introduced into the design by using what is referred to as the *allowable stress* σ_{ALL}, which is defined by

$$\sigma_{ALL} = \frac{\text{Mechanical property}}{\text{Factor of safety}} \tag{1.24}$$

In Equation 1.24, the term *mechanical property* refers to the yield point, the yield strength, or the ultimate strength; and the *factor of safety* (FS) is a number larger than unity defined as the ratio of the mechanical property to the allowable stress as implied in Equation 1.24. Values of allowable stresses for various materials have been arrived at through vast experience and sound engineering judgment. These values are normally specified in city ordinances and other building, machine, and structural codes.

Terms such as *allowable*, *working*, and *service*, when used as adjectives with terms such as *stress*, *strain*, *deformation*, and *rotation*, refer to permissible values of these quantities associated with satisfactory behavior of structural or machine components under the actual loadings to which they may be safely subjected during their lifetimes. Use of this terminology implies that a factor of safety has already been incorporated in the stated value. For example, an allowable stress stated for a given member implies that the stated stress incorporates a factor of safety, and that it is the stress to which this member might reasonably be subjected during its useful life.

It should be pointed out that other definitions of the factor of safety based upon load and energy, instead of stress, are in use under certain conditions. However, the only definition used for the factor of safety in this text is the one implied by Equation 1.24.

EXAMPLE 1.5

A steel rod 0.5 in. in diameter with a gage length of 4 in. is subjected to a gradually increasing tensile load. At the proportional limit, the value of the load was 20 kips, the change in the gage length was 0.014 in., and the change in diameter was 0.0005 in. Determine (a) the proportional limit σ_p, (b) the modulus of elasticity E, (c) Poisson's ratio μ, and (d) the modulus of resilience u_R.

SOLUTION

a. The proportional limit σ_p may be found from the basic definition of stress, $\sigma = F/A$, where $F = F_p$ is the internal axial force at the proportional limit, and $A = A_o$ is the original cross-sectional area. Thus,

$$\sigma_p = \frac{F_p}{A_o} = \frac{20}{(\pi/4)(0.5)^2}$$
$$= 101.9 \text{ ksi} \qquad \textbf{ANS.}$$

b. The modulus of elasticity is obtained from Hooke's law as expressed in Equation 1.11, $\sigma = E\,\varepsilon$. Thus, $E = \sigma_p/\varepsilon_p$, where

$$\varepsilon_p = \frac{\Delta L_p}{L_o} = \frac{0.014}{4} = 0.0035.$$

Therefore,

$$E = \frac{101.9}{0.0035} = 29.1 \times 10^3 \text{ ksi} \qquad \textbf{ANS.}$$

c. Poisson's ratio μ is given by Equation 1.13 as the ratio between the transverse strain and the longitudinal strain. The longitudinal strain at the proportional limit was found in part (b) to be $\varepsilon_L = 0.0035$. The transverse strain ε_T is obtained by dividing the change in diameter by the original diameter. Thus,

$$\varepsilon_T = \frac{\Delta D_p}{D_o} = \frac{-0.0005}{0.50} = -0.001$$

Thus,

$$\mu = \left| \frac{-0.001}{0.0035} \right| = 0.286 \qquad \textbf{ANS.}$$

d. The modulus of resilience is given by Equation 1.22. Thus,

$$u_R = \frac{101.9^2}{2(29.1 \times 10^3)} = 0.178 \text{ kip} \cdot \text{in./in.}^3 \qquad \textbf{ANS.}$$

PROBLEMS

1.40 A steel bar 3 in. in diameter and a gage length of 5 in. is subjected to a compressive force. The proportional limit for this steel is 45,000 psi, its modulus of elasticity 30×10^6 psi, and its Poisson's ratio is 0.30. Determine (a) the maximum compressive load that may be applied without exceeding the proportional limit and (b) the changes in diameter and length at the proportional limit.

1.41 The modulus of resilience for a given material is known to be $4 \times 10^5 \, \text{N} \cdot \text{m/m}^3$. If the modulus of elasticity is $E = 120 \, \text{GPa}$, determine (a) the proportional limit stress and the proportional limit strain for the material and (b) the axial tensile load required to reach the proportional limit in a specimen whose cross-sectional area is $10 \times 10^{-4} \, \text{m}^2$.

1.42 A steel rod 0.50 in. in diameter and a gage length of 2 in. is subjected to a gradually increasing tensile force. At the proportional limit, the value of the applied load was 8800 lb, the gage length measured 2.003 in., and the diameter measured 0.49975 in. Find (a) the modulus of elasticity, (b) Poisson's ratio, and (c) the modulus of resilience.

1.43 A compressive force of 600 kN is gradually applied to a prismatic rectangular bar whose cross section is 40 mm × 60 mm and whose length is 200 mm. The 60 mm dimension changed to 60.100 mm and the 200 mm length to 198.700 mm. Determine (a) Poisson's ratio, (b) the final value of the 40 mm dimension, and (c) the modulus of elasticity for the material.

1.44 A steel bar ($E = 30 \times 10^3$ ksi) whose diameter is 0.55 in. and whose length is 3.5 in. is subjected to a compressive axial force that produces a shortening of 0.0045 in. at the proportional limit for the material. Determine (a) the proportional limit stress and (b) the modulus of resilience.

1.45 An aluminum bar with a 15 mm × 30 mm rectangular cross section and a gage length of 300 mm is subjected to a tensile load of 130 kN. The 30 mm dimension changed to 29.96 mm and the 300 mm gage length changed to 301.2 mm. Find (a) Poisson's ratio, (b) the final value of the 15 mm dimension, (c) the normal stress in the bar, and (d) the modulus of elasticity.

1.46 A bar of a certain material with a rectangular cross section 25 mm × 35 mm and a gage length of 100 mm is subjected to a compressive load of 600 kN. The 25 mm dimension changed to 25.030 mm and the 100 mm length to 99.40 mm. Determine (a) Poisson's ratio, (b) the final value of the 35 mm dimension, and (c) the modulus of elasticity.

1.47 A tension test on a metallic specimen resulted in the following corresponding values of load and deformation. The initial diameter and gage length of the specimen were 18 and 120 mm, respectively. Construct the engineering stress–strain diagram and determine (a) the ultimate strength, (b) the fracture strength, and (c) the percent elongation and the percent reduction of area if its final diameter is 17.89 mm.

Load (kN)	Deformation (mm)	Load (kN)	Deformation (mm)
0.00	0.00	17.700	29.50×10^{-2}
1.770	1.43×10^{-2}	18.585	36.62×10^{-2}
3.540	2.86×10^{-2}	19.470	46.25×10^{-2}
5.310	4.29×10^{-2}	20.335	58.54×10^{-2}
7.080	5.71×10^{-2}	20.886	75.02×10^{-2}
8.850	7.14×10^{-2}	20.886	90.16×10^{-2}
10.620	8.57×10^{-2}	20.355	125.50×10^{-2}
12.390	10.00×10^{-2}	19.030	150.40×10^{-2}
14.160	13.20×10^{-2}	16.900	170.70×10^{-2}
15.930	18.32×10^{-2}	13.100	$195.00 \times 10^{-2}*$
16.815	24.11×10^{-2}		

*Fracture

1.48 Redraw the initial portion of the data given in Problem 1.47 using a larger scale and find (a) the proportional limit, (b) the 0.20% yield strength (the yield strength corresponding to a strain of 0.0020), and (c) the modulus of elasticity.

1.49 Refer to the data given in Problem 1.47 and determine (a) the modulus of toughness and (b) the modulus of resilience.

1.50 A tension test was performed on a specimen whose initial diameter and gage length were 0.50 and 2.0 in., respectively. Corresponding values of load and deformation are given in the following tabulation. Construct the engineering stress–strain diagram for the material and determine (a) the ultimate strength and (b) the fracture strength.

Load (kips)	Deformation (in.)	Load (kips)	Deformation (in.)
0.0	0.0	12.4	0.0170
1.6	0.0004	12.3	0.0200
3.2	0.0009	13.0	0.0270
4.8	0.0015	14.0	0.0330
6.4	0.0021	15.0	0.0400
8.0	0.0026	16.0	0.0490
9.6	0.0031	17.5	0.0670
11.2	0.0035	19.0	0.1050
12.0	0.0040	19.5	0.1500
12.4	0.0050	20.0	0.2000
12.6	0.0070	20.0	0.2500
12.8	0.0100	18.6	0.3200
12.8	0.0130	17.0	0.4000
12.6	0.0150	16.5	0.4500*

*Fracture

1.51 Redraw the initial portion of the data given in Problem 1.50 using a larger scale and determine (a) the modulus of elasticity, (b) the upper yield point, (c) the lower yield point, and (d) the proportional limit.

1.52 If the final diameter of the specimen in Problem 1.50 was 0.325 in., find (a) the percent elongation and (b) the percent reduction of area.

1.53 Refer to the data in Problem 1.50 and determine (a) the modulus of resilience and (b) the modulus of toughness.

1.54 A plain concrete compression specimen has a diameter of 6 in. and a length of 10 in. When subjected to compressive loads, the specimen fails suddenly by material breakdown. Force in pounds and deformation in inches are as follows: (0.0, 0.0), (28,300; 0.0025), (56,500; 0.0050), (84,800; 0.0080), (99,000; 0.0100), (113,100; 0.0125), (127,200; 0.0150), (141,400; 0.0250), (135,700; 0.0300), (113,100; 0.0350). Construct the stress–strain diagram for this specimen and determine (a) the ultimate strength and (b) the corresponding strain.

1.55 The compression stress σ as a function of the compressive strain ε for concrete is readily approximated by a parabolic function that passes through two points, the origin and the ultimate strength $(\sigma_u, \varepsilon_u)$. This function is given by

$$\sigma = \sigma_u \left[\frac{2\varepsilon}{\varepsilon_u} - \left(\frac{\varepsilon}{\varepsilon_u} \right)^2 \right]$$

In order to determine the behavior of concrete experimentally, a cylindrical specimen is tested to obtain values of the ultimate strength and the corresponding strain. Let $\sigma_u = 5200$ psi and $\varepsilon_u = 0.00210$. Assign values of ε from zero to 0.0035 in increments of 0.0005 and compute the corresponding stresses. Plot an idealized stress–strain curve for this concrete and draw a straight line through the origin and the point having an ordinate equal to $0.5\sigma_u$. Determine the slope of this line, which is known as *the secant modulus of elasticity*, for the concrete.

1.56 High-strength steel wires are used for prestressed concrete construction. Data for wires having an ultimate strength of 250 ksi are given in the tabulation below. Plot the stress–strain curve for this material and determine (a) the modulus of elasticity, (b) the proportional limit, (c) the ultimate strain, and (d) the modulus of toughness.

Stress (ksi)	Strain (in./in.)	Stress (ksi)	Strain (in./in.)
0.0	0.0	235.0	0.02000
50.0	0.00175	240.0	0.03000
100.0	0.00345	247.0	0.04000
210.0	0.00724	250.0	0.05000*
225.0	0.01000		

*Fracture

1.57 A member with a circular cross section is to be designed to carry a tensile axial load of 20 kips. If the allowable tensile stress, $\sigma_{ALL} = 5$ ksi, what should be the diameter of the member.

1.58 A member is to be designed to carry a compressive load P. If the cross-sectional area of the member is a 30 mm × 40 mm rectangle, determine the axial compressive load P if the yield strength is 150 MPa and the factor of safety is 2.5.

1.59 A member is to be designed to carry a tensile axial load of 400 kN. The member is to have a circular cross section with a diameter of 60 mm. If the material of the member has a yield strength of 250 MPa, determine the factor of safety based on the yield strength. What is the allowable stress?

1.60 A member is to be designed to carry a compressive axial load of 100 kips. The member has a 1.5 in. × 2.0 in. rectangular cross section. The ultimate strength for the material $\sigma_U = 60$ ksi. Determine the factor of safety based on the ultimate strength. What is the allowable stress?

1.61 Two pieces of lumber are spliced together along segment BC as shown using wood glue that has an allowable shearing strength of 70 psi. Each piece of lumber has cross-sectional dimensions of $t \times 8$ in., where $t = 4$ in. Determine the maximum force P that the spliced member may carry. Let $L = 12$ in.

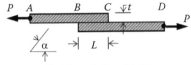

(Problems 1.61 and 1.62)

1.62 In reference to the figure shown in Problem 1.61, two pieces of lumber are spliced together along segment BC as shown using wood glue that has an allowable shearing strength of 0.45 MPa. Each piece of lumber has cross-sectional dimensions of $t \times 0.5$ m and the bonding material between its grains has an allowable strength of 0.3 MPa in tension and 0.25 MPa in shear. The angle defining the orientation of the grains is $\alpha = 45°$. Let $t = 120$ mm, $L = 0.3$ m and determine the maximum permissible force P.

1.63 A joint is fabricated using two eye bolts and a connecting threaded bolt as shown. If the allowable shearing strength in the connecting bolt is 75 ksi and its diameter is 0.375 in., determine the largest safe magnitude of the force P. Assume the two eye bolts to be strong enough to carry the load.

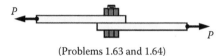

(Problems 1.63 and 1.64)

1.64 In reference to the figure shown in Problem 1.63, a joint is fabricated using two aluminum eye bolts and a threaded steel connecting pin as shown. The diameter of the eye bolt is 15 mm, and that of the steel pin is 10 mm. The allowable strength of the steel pin is 50 MPa in shear and that in the aluminum bolt is 25 MPa in tension. Determine the largest load P that the connection can carry.

1.65 A side view of a spliced connection is shown in the sketch. It consists of four threaded steel bolts (only the front two are shown), each having a diameter of 0.75 in. The dimension $d = 6$ in., and the spacing of the bolts into the page is 8 in. If the allowable shearing stress in the bolts is 25 ksi, determine the maximum permissible force P.

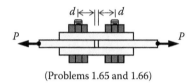

(Problems 1.65 and 1.66)

1.66 In reference to the figure shown in Problem 1.65, a side view of a spliced connection is shown in the sketch. It consists of six threaded aluminum bolts (only the front two are shown), each having a diameter of 30 mm. The dimension $d = 150$ mm and the spacing of the bolts into the page is 150 mm. The main plates are also aluminum with a thickness of 20 mm. The depth of these plates into the page is 0.40 m. The allowable shear stress in the bolts is 150 MPa, and the allowable tensile strength in the main plates is 100 MPa. Determine the maximum permissible load P. Assume that the splice plates are sufficiently strong to carry the load. *Ignore the reduction in the area of the main plates due to bolt holes.*

1.67 Determine the thickness t to be able to simultaneously punch circular holes in plates A and B. Material A is steel ($\tau_Y = 90$ ksi, thickness $= 0.30$ in.) for which $d_A = 1.5$ in., and material B is aluminum ($\tau_Y = 50$ ksi) for which $d_B = 3.0$ in. What is the needed minimum magnitude of P?

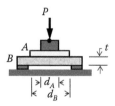

1.5 LOAD-DEFORMATION RELATIONS

If a member is subjected to an axial load, it experiences a deformation. Consider the unloaded member shown in Figure 1.24a, for which the cross-sectional geometry varies along its length in an arbitrary manner. In Figure 1.24b, this same member is subjected to a tensile axial load P. In order to define the strain at a point, we need to consider the deformation in the neighborhood of a point

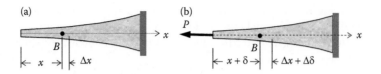

FIGURE 1.24   Definition of *strain at a point.*

such as B. The length x in Figure 1.24a experiences a deformation δ, and the length Δx a deformation $\Delta\delta$, when the load P is applied, as shown in Figure 1.24b. Thus, the *strain at point B* becomes

$$\varepsilon = \underset{\Delta x \to 0}{\text{Lim}}\frac{(\Delta x + \Delta\delta) - \Delta x}{\Delta x} = \underset{\Delta x \to 0}{\text{Lim}}\frac{\Delta\delta}{\Delta x} = \frac{d\delta}{dx} \qquad (1.25)$$

Therefore, the strain at a point is the derivative of the deflection at the point with respect to the axial coordinate x.

From Equation 1.25, we conclude that

$$d\delta = \varepsilon\,dx \qquad (1.26)$$

Two cases are considered here:

1. The first is the case where the member has a constant cross-sectional area A, is made of the same material so that E is constant, and is subjected to a force P at its ends (tension or compression) as shown in Figure 1.25a, where P is tensile. Considering the free-body diagram of the left segment of length x, we conclude that the internal force F on the cut section is equal to P as shown in Figure 1.25b.

    Using Equation 1.26, we obtain

$$\delta = \int_0^L \varepsilon\,dx = \int_0^L \left(\frac{\sigma}{E}\right)dx \qquad (1.27)$$

    where ε was replaced by σ/E. If we now replace σ by F/A, take out of the integral sign the constant quantity F/AE, and integrate the remaining differential dx between the indicated limits, we obtain

$$\delta = \frac{FL}{AE} \qquad (1.28)$$

Note that Equation 1.28 is equally applicable to both tensile and compressive axial forces, and that the deformation δ would be an extension when F is tensile and a contraction when F is compressive.

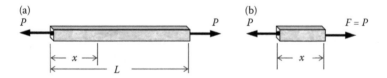

FIGURE 1.25   Definition of *strain at a point* in a member of constant cross-sectional area.

For a member subjected to concentrated forces at several locations along its length, the deformation δ would be obtained by algebraically adding the deformations of the various segments of the member. In such cases, the deformation is obtained from the relation

$$\delta = \sum \frac{FL}{AE} \tag{1.29}$$

2. The second case deals with axially loaded members in which any or all of the three quantities, F, A, or E, varies or vary along the length of the member. In such cases, we substitute F/A for σ in Equation 1.27 to obtain

$$\delta = \int \frac{F\,dx}{AE} \tag{1.30}$$

Before we can make use of Equation 1.30, the variation with x of any or all of the three quantities, F, A, or E, must be known or determined from given information in order to properly perform the indicated integration.

EXAMPLE 1.6

The system shown in figure (a) below consists of a rigid beam ABC hinged at A to a fixed support and at B to an aluminum rod ($E = 70$ GPa) with a diameter of 20 mm. The rod is fastened to a rigid plate D, and a linear spring having a spring constant $k = 5000$ kN/m is placed between this rigid plate and a second plate E, which is fixed in position. The aluminum rod passes freely through a hole in plate E. The rigid beam ABC is horizontal and the rod and the spring are undeformed before the load P is applied. Determine (a) the displacement of plate D and (b) the displacement of point B.

SOLUTION

a. Consider the free-body diagram of beam ABC, shown in figure (b) below, where F_{BD} is the force in rod BD. Using the coordinate system shown and summing moments about point A, we have

$$\sum M_A = 0 \Rightarrow F_{BD}(2) - 25(4) = 0; \quad F_{BD} = 50 \text{ kN}$$

The 50-kN force is transmitted to rigid plate D, which, in turn, transmits it to the spring as a compressive force.

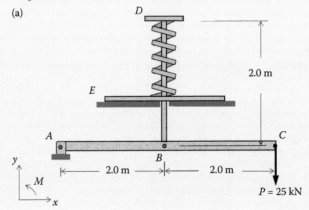

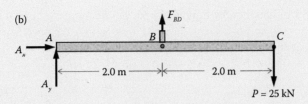

The spring experiences a contraction representing the displacement of plate D. Therefore, since for a spring, $F = k\delta$, it follows that

$$\delta_D = \frac{F_{BD}}{k} = \frac{50}{5000} = 0.010 \text{ m} = 10 \text{ mm} \downarrow \qquad \textbf{ANS.}$$

b. The displacement of point B consists of two parts. The first is the downward displacement of plate D, found above. The second is the deformation δ_R experienced by rod BD, which is subjected to a tensile force of 50 kN. Thus, by Equation 1.28, we have

$$\delta_R = \left(\frac{FL}{AE}\right)_R = \frac{50 \times 10^3 \times 2.0}{(\pi/4)(0.02^2) \times 70 \times 10^9} = 0.004 \text{ m} = 4 \text{ mm}$$

Therefore,

$$\delta_B = \delta_D + \delta_R = 14 \text{ mm} \downarrow \qquad \textbf{ANS.}$$

EXAMPLE 1.7

Member ABC is subjected to the loads as shown in figure (a) below. Determine (a) the deformation of segment BC and (b) the displacement of point A. The origin for the coordinate x is at A. The cross-sectional areas are

$A_{AB} = 0.50$ in.^{2}; $A_{BC} = 0.90$ in.^{2}
The modulus of elasticity for the entire member is $E = 10 \times 10^6$ psi.

SOLUTION

a. A cut is made in any place between B and C and a free-body diagram is constructed as shown in figure (b) above.
Summing forces in the x direction, we have

$$\sum F_x = 0 \implies F_{BC} - 5(x - 3) + 60 - 20 = 0$$

$F_{BC} = 5x - 55$. Therefore, by Equation 1.30, we have

$$\delta_{BC} = \left(\int_3^8 \frac{F\,dx}{AE}\right)_{BC} = \frac{12 \times 10^3}{(0.90)(10 \times 10^6)} \int_3^8 (5x - 55)\,dx$$

$$= -0.183 \text{ in.} \qquad\qquad \textbf{ANS.}$$

b. The displacement of point A is obtained by adding, algebraically, the deformations of segments AB and BC. The deformation of segment AB is obtained by Equation 1.28, which requires the internal force in segment AB. It is obvious, in this simple case, without a free-body diagram, that the internal force $F_{AB} = 20$ kips. Thus,

$$\delta_{AB} = \left(\frac{FL}{AE}\right)_{AB} = \frac{(20 \times 10^3)(3 \times 12)}{(0.50)(10 \times 10^6)} = 0.144 \text{ in.}$$

Therefore,

$$\text{Displacement of } A = \delta_{AB} + \delta_{BC} = 0.144 - 0.183$$

$$= -0.039 \text{ in.} = 0.039 \text{ in.} \rightarrow \qquad\qquad \textbf{ANS.}$$

EXAMPLE 1.8

Member BC is suspended from a support at B as shown in figure (a) below and subjected to the end load of 5 kN. The member is made of a material for which the density is 8800 kg/m^3 and the modulus of elasticity is 180 GPa. Determine the displacement of the member at point C due to both the end load and to its weight. The cross-sectional area of the member is rectangular, 100 mm × 200 mm.

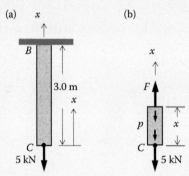

SOLUTION

Since a part of the internal force acting on member BC is variable, we need to use Equation 1.30 to find the deformation δ of member BC, which is equal to its displacement at point C. Before we can apply this equation, however, we need to find the internal force F from the free-body diagram shown in figure (b) above, where the quantity p represents the weight distribution and is found as follows:

$$p = \frac{Weight}{Unit\ length} = (0.1 \times 0.2)(1.0)(8800 \times 9.81) = 1726.6 \text{ N/m} = 1.727 \text{ kN/m}$$

Summing forces in the x direction, we have

$$\sum F_x = 0 \Rightarrow F - 1.727x - 5 = 0; \quad F = 5 + 1.727x$$

Thus, Equation 1.30 becomes

$$\delta = \int_0^3 \frac{(5 + 1.727x)(10^3)\,dx}{(0.1 \times 0.2)(180 \times 10^9)} = \frac{(5 \times 10^3)(3)}{(0.1 \times 0.2)(180 \times 10^9)} + \frac{(1.727 \times 10^3)}{(0.1 \times 0.2)(180 \times 10^9)}\left[\frac{x^2}{2}\right]_0^3$$

$$= (4.167 + 2.159) \times 10^{-6} = 6.326 \times 10^{-6} \text{ m}$$

$$= 6.3 \times 10^{-3} \text{ mm} \qquad \textbf{ANS.}$$

PROBLEMS

1.68 Determine the deformation of member AB if it is made of steel for which $E = 30 \times 10^3$ ksi, and its cross-sectional area is 1.5 in.²

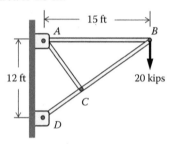

(Problems 1.68 and 1.69)

1.69 In reference to the figure shown in Problem 1.68, find the deformation of member DC if it is made of a certain material for which $E = 15 \times 10^3$ ksi and its cross-sectional area is a 1.5 in. × 3.0 in. rectangle.

1.70 Find the deformation of member BE if $P_1 = 10$ kips, $P_2 = 20$ kips, and $a = 5$ ft. The material is steel for which $E = 30 \times 10^3$ ksi and the cross-sectional area $A = 2.5$ in.²

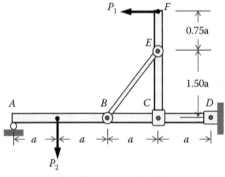

(Problems 1.70 and 1.71)

1.71 In reference to the figure shown in Problem 1.70, determine the magnitude of P_1 if the allowable deflection of member BE is 3.0 mm. Let $P_2 = 1.5P_1$ and $a = 2$ m. The member is aluminum for which $E = 70$ GPa and the cross-sectional area is a 25 mm × 40 mm rectangle.

1.72 Find the maximum permissible force P if the allowable displacement at C is 0.25 in. Assume beam DCB to be rigid. Member ABC is made of steel for which $E = 30 \times 10^6$ psi. The cross-sectional area of segment AB is 0.75 in.^{2} and that of BC is 0.35 in.^{2}

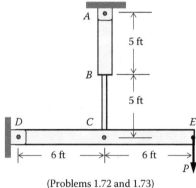

(Problems 1.72 and 1.73)

1.73 In reference to the figure shown in Problem 1.72, let $P = 25$ kips. Segment AB of the vertical member is made of a material for which $E = 16 \times 10^3$ ksi and has a rectangular cross-sectional area 0.80 in. $\times$ 1.00 in. Segment BC is made of a material for which $E = 22 \times 10^3$ ksi and has a cross-sectional area of 0.24 in.^{2} Determine the displacement at point C.

1.74 The mechanism shown in the sketch consists of a rigid beam ABC hinged at A to a steel ($E = 200$ GPa) rod AE, which has a diameter of 15 mm and a length of 3.0 m. The linear spring, which is placed between rigid plates D and E, has a spring constant of 500 kN/m. If $P = 20$ kN, determine (a) the displacement of point A and (b) the deformation of the spring. Rigid plate D is fixed in position and member ABC is horizontal before the load P is applied. Rod AE passes freely through the spring and a hole in plate D.

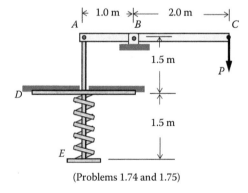

(Problems 1.74 and 1.75)

1.75 In reference to the figure shown in Problem 1.74, the mechanism shown in the sketch consists of a rigid beam ABC hinged at A to an aluminum ($E = 70$ GPa) rod AE, which has a diameter of 25 mm and a length of 3.0 m. The linear spring, which is placed between rigid plates D and E, has a spring constant of 1200 kN/m. Determine (a) the magnitude of P if the allowable displacement at A is 10 mm and (b) the deformation of rod AE.

1.76 Find (a) the displacement at C and (b) the displacement at D. The following information is provided: For segment AB, $P_1 = 50$ kips and the diameter is $D_1 = 2.0$ in. for segment BC, $P_2 = 30$ kips and $D_2 = 1.0$ in. for segment CD, $P_3 = 25$ kips and $D_3 = 0.5$ in. The material for the entire member is steel for which $E = 30 \times 10^3$ ksi.

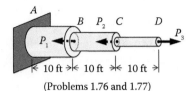

(Problems 1.76 and 1.77)

1.77 In reference to the figure shown in Problem 1.76, find P_2 if the allowable displacement at C is 0.25 in. The following information is given: For segment AB, $P_1 = 60$ kips, and the cross-sectional area is $A_1 = 1.25$ in.²; for segment BC, $A_2 = 0.90$ in.²; for segment CD, $P_3 = 20$ kips and $A_3 = 0.625$ in.² The material for the entire member is aluminum for which $E = 10 \times 10^3$ ksi.

1.78 The allowable displacement at D is 3.5 mm. The cross-sectional areas of segments AB, BC, and CD are, respectively, A_1, A_2, and A_3 such that $A_1 = 0.50$, $A_2 = 0.75$, and A_3. Determine the three cross-sectional areas if $P_1 = 150$ kN, $P_2 = 100$ kN, and $P_3 = 125$ kN. All segments are made of the same material for which $E = 120$ GPa.

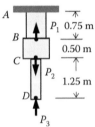

(Problems 1.78 and 1.79)

1.79 In reference to the figure shown in Problem 1.78, segment AB is made of a material for which $E = 75$ GPa, segment BC of a material for which $E = 100$ GPa, and segment CD of a material for which $E = 150$ GPa. The cross-sectional areas of segments AB, BC, and CD are, respectively, 1200, 2000, and 450 mm². If $P_1 = 75$ kN, $P_2 = 150$ kN, and $P_3 = 200$ kN, find the displacements at B, C, and D.

1.80 Let the diameter of segment AB be $D_1 = 1.5$ in., its modulus of elasticity be $E_1 = 15 \times 10^6$ psi; the diameter of segment BC be $D_2 = 0.50$ in., and its modulus of elasticity be $E_2 = 25 \times 10^6$ psi. Also, let $P_1 = 50$ kips and $P_2 = 20$ kips and determine the ratio a/b so that the displacement of point C is zero.

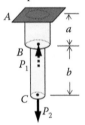

(Problems 1.80 and 1.81)

1.81 In reference to the figure shown in Problem 1.80, let the diameter of segment AB be D_1, its modulus of elasticity be $E_1 = 10 \times 10^3$ ksi; the diameter of segment BC be D_2, its modulus of elasticity be $E_2 = 20 \times 10^3$ ksi. Also, let $a = 18$ in., $b = 10$ in., $P_1 = 50$ kips, and $P_2 = 30$ kips and determine the ratio D_1/D_2 so that the displacement at C is zero.

1.82 The system shown in the sketch consists of an aluminum rod ($E = 70$ GPa) 15 mm in diameter and 0.90 m long welded to a rigid plate A at its top. A linear spring with a spring constant $k = 1200$ kN/m is placed between rigid plates A and B, and a plastic tube ($E = 3.5$ GPa) with outside diameter of 45 mm and inside diameter of 43 mm is placed between rigid plates B and C, the latter being fixed in position as shown. The aluminum rod passes freely through the spring and the tube, and through holes in plates B and C. Determine the displacement of (a) plate A, (b) plate B, and (c) point D. Let $P = 75$ kN and neglect possible localized buckling of the plastic tube.

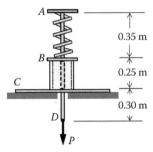

(Problems 1.82 and 1.83)

1.83 Consider the same system described in Problem 1.82. Find the load P if the allowable displacement of point D is 50 mm.

1.84 Member AB is a truncated cone with a circular cross section that changes in diameter linearly from 2.0 in. at A to 4.0 in. at B. The cone is made of a certain material for which $E = 20 \times 10^3$ ksi. Determine the elongation of the cone if $P = 250$ kips.

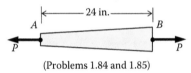

(Problems 1.84 and 1.85)

1.85 In reference to the figure shown in Problem 1.84, member AB is a truncated cone with a circular cross section that changes in diameter linearly from 1.5 in. at A to 3.5 in. at B. The cone is made of a material for which $E = 15 \times 10^3$ ksi. If the allowable deformation of the cone is 0.05 in., determine the largest magnitude of P.

1.86 Member AB is a hollow truncated cone with a square cross section that changes linearly from a 50 mm × 50 mm outside at A to 100 mm × 100 mm outside at B. The thickness is constant throughout and equal to 5 mm. The material is such that it has a modulus of elasticity $E = 30$ GPa. If $P = 200$ kN, determine the shortening of the cone.

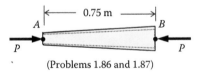

(Problems 1.86 and 1.87)

1.87 In reference to the figure shown in Problem 1.86, member AB is a hollow truncated cone with a rectangular cross section that changed linearly from a 36 mm × 46 mm outside at A to a 36 mm × 76 mm outside at B. The thickness is constant throughout and equal to 6 mm. The material has a modulus of elasticity $E = 110$ GPa. Determine the largest magnitude of P if the allowable shortening of the cone is 3.5 mm.

1.88 The 29-ft steel piling weighing 210 lb/ft was damaged during driving and must be pulled out. A force of 30 kips is needed to start the pulling-out process. In the position shown, the frictional resistance p may be assumed uniform throughout the piling length. The cross-sectional area of the piling is 61.8 in.^{2} and the modulus of elasticity for steel is $E = 30 \times 10^3$ ksi. Determine (a) the intensity of the frictional force p, (b) the total deformation of the piling, and (c) the normal stress in the piling at 15 ft below ground level.

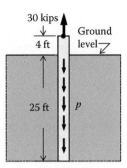

1.89 A solid circular shaft whose cross-sectional area is 500 mm² is being pulled by a force
P out of a sleeve into which it had been press-fitted. In the position shown, the frictional
resistance *p* between the shaft and the sleeve may be assumed uniform and equal to
150 kN/m. The material is aluminum for which $E = 70$ GPa. Determine (a) the force *P*,
(b) the maximum normal stress in the shaft, and (c) the deformation of the shaft.

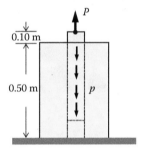

1.90 The soil-sampling device shown is designed for possible use in planetary exploration. A
force *P* is needed to begin the withdrawal process. The soil resistance is $p = 1000x$ N/m,
where *x* is measured from the top of the device as shown. The material is aluminum,
for which $E = 70$ GPa. Determine (a) the force *P*, (b) the normal force in the device at
$x = 0.20$ m, and (c) the deformation of the device if its cross-sectional area is 2×10^{-5} m².

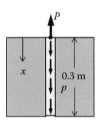

1.6   STATICALLY INDETERMINATE MEMBERS

The type of problems that we have dealt with in *statics* and those in the preceding sections may
be classified as *statically determinate* because we were able to obtain a complete solution for the
unknown forces simply by using the equations of equilibrium. In the most general case of such a
problem (three-dimensional case), six equations of equilibrium are available to solve for a maximum of six unknown quantities, and in the two-dimensional case, we can solve for a maximum of
three unknown quantities, using the three available equations of equilibrium. The term *statically
indeterminate* member refers to one in which the unknown forces cannot be fully ascertained by
using only the equations of equilibrium. In such a case, the number of unknown quantities exceeds
the number of available equilibrium equations, and additional conditions are needed to complete
the solution. These additional conditions are generally deduced from knowledge of the deformation
characteristics of the system under consideration.

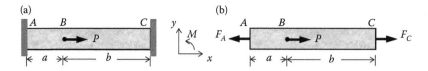

FIGURE 1.26 Diagram showing one case of *statically indeterminate member*.

Consider, for example, member ABC shown in Figure 1.26a, which is fixed between two rigid walls so that its length, $L = a + b$, cannot change when the load P is applied at B. Thus, when the force P is applied, segment AB undergoes an increase in length, which must be exactly the same in magnitude as the decrease in length of segment BC because, as stated above, the total length of the member, $L = a + b$, cannot be changed. The free-body diagram of member ABC is given in Figure 1.26b, showing the two unknown internal forces F_A (in segment AB), and F_C (in segment BC), both assumed to be positive (i.e., tension). There is only one equation of equilibrium that may be used in this case, and, in reference to the coordinate system shown, this equation is, $\Sigma F_x = 0$. Thus, we are faced with a situation in which two unknown quantities exist, and only one equilibrium equation is available for their determination. Therefore, the member in Figure 1.26 is said to be statically indeterminate to the *first degree* because there is one unknown quantity more than the available equations of equilibrium. Thus, one additional condition (equation) is needed to complete the solution. This equation is obtained by considering the deformation characteristics of the member. As stated above, the deformations of the two segments of the member are equal in magnitude, a condition that may be stated by saying that $\delta_{AB} + \delta_{BC} = 0$. This type of solution is illustrated in Example 1.9. There are cases (some of which will be encountered later in this text) in which the number of unknown quantities exceeds the number of available equations by more than one, and these cases are said to be statically indeterminate to the *nth degree*, where n stands for the number of unknown quantities by which the available number of equilibrium equations is exceeded. The n unknown quantities are, sometimes, referred to as *redundants*, which implies that they are not necessary for equilibrium. Example 1.10 makes use of the concept of redundants along with the principle of *superposition* (adding together algebraically) to obtain a solution. Note that the principle of superposition is only valid in the elastic range for the material.

EXAMPLE 1.9

Consider the member shown in Figure 1.26 and let $P = 40$ kips. The entire member is made of the same material (constant E) and the cross-sectional area A is constant throughout. Determine (a) the unknown internal forces F_A and F_C and (b) the displacement of point B if $E = 10 \times 10^3$ ksi, $A = 0.5$ in.^{2}, $a = 3$ ft, and $b = 5$ ft.

SOLUTION

a. Applying the only available equation of equilibrium, we have

$$\sum F_x = 0 \Rightarrow 40 - F_A + F_C = 0 \tag{1.9.1}$$

The second equation follows from the condition stated earlier that

$$\delta_{AB} + \delta_{BC} = 0 \Rightarrow \left[\frac{FL}{AE} \right]_{AB} = - \left[\frac{FL}{AE} \right]_{BC}$$

Substituting the known quantities into the above equation, we obtain

$$\frac{F_A (3 \times 12)}{AE} = - \frac{F_C (5 \times 12)}{AE}$$

Since A and E are the same for both segments, the above equation reduces to

$$3F_A = -5F_C \qquad (1.9.2)$$

Thus, we have expressed the solution to this statically indeterminate problem in terms of two simultaneous equations. Solving them yields

$$F_A = 25 \text{ kips}; \; F_C = -15 \text{ kips} \qquad \textbf{ANS.}$$

b. The displacement of point B may be determined by finding either δ_{AB} or δ_{BC} because they are equal in magnitude. Choosing the former, we have

$$\text{Displacement of point } B = \delta_{AB} = \frac{(25)(3 \times 12)}{(0.5)(10 \times 10^3)} = 0.18 \text{ in.} \rightarrow \qquad \textbf{ANS.}$$

EXAMPLE 1.10

Determine the normal stresses in segments AB and BC of the composite rod shown in figure (a) below. Segment AB is made of steel with a diameter of 40 mm and $E_S = 200$ GPa. Segment BC is made of aluminum with a diameter of 60 mm and $E_A = 70$ GPa. Let $P = 150$ kN.

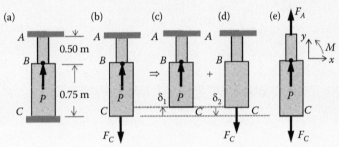

SOLUTION
This example will be solved by two methods as follows:

I. Superposition
Let us assume the reaction at C, F_C, to be the redundant and remove the support at C as shown in figure (b) above. Removing the support at C allows the entire member to contract by the amount δ_1 under the influence of P as shown in figure (c) above. This contraction cannot actually take place and a force F_C, of appropriate magnitude, must be applied as shown in figure (d) above in order to delete it (i.e., to produce an extension δ_2 equal in magnitude to δ_1) and bring the member back to where it ought to be. Therefore, as indicated in the diagrams above, the superposition of the actions in figures (c) and (d) above is equivalent to that in figure (b) above. In other words, $\delta_1 + \delta_2 = 0$. Thus,

$$\delta_1 = \delta_{AB} = -\frac{(150 \times 10^3)(0.5)}{\pi(0.040^2/4)(200 \times 10^9)} = -298.416 \times 10^{-6} \text{ m}$$

$$\delta_2 = \delta_{AB} + \delta_{BC} = \frac{F_C(0.50)}{\pi(0.04^2/4)(200 \times 10^9)} + \frac{F_C(0.75)}{\pi(0.060^2/4)(70 \times 10^9)} = 5.779 \times 10^{-6} F_C \text{ m}$$

Thus, since $\delta_1 + \delta_2 = 0$, we have

$$-298.416 \times 10^{-6} + 5.779 \times 10^{-6} F_C = 0$$

Therefore,

$$F_C = 51.618 \text{ kN}$$

The internal force at A, F_A, is obtained by using the free-body diagram shown in figure (e) above. Thus,

$$\sum F_y = 0 \implies F_A + P - 51.618 = 0$$

from which

$$F_A = 51.618 - P = -98.382 \text{ kN}$$

Therefore, the stresses are

$$\left. \begin{aligned} \sigma_{AB} &= \frac{-98.382 \times 10^3}{\pi(0.06^2/4)} = -34795.5 \times 10^3 \text{ Pa} \approx -34.8 \text{ MPa} \\ \sigma_{BC} &= \frac{51.618 \times 10^3}{\pi(0.04^2/4)} = 41076.3 \times 10^3 \text{ Pa} \approx 41.1 \text{ MPa} \end{aligned} \right\} \qquad \textbf{ANS.}$$

II. Simultaneous Equations

Referring to the free-body diagram of figure (e) above, we write

$$\sum F_y = 0 \implies F_A + P - F_C = 0 \qquad (1.10.1)$$

$$\delta_{AB} + \delta_{BC} = 0 \implies \left[\frac{FL}{AE}\right]_{AB} + \left[\frac{FL}{AE}\right]_{BC} = 0$$

Substituting the given data into the above deflection equation, we have

$$\frac{F_A(0.5)}{\pi(0.04^2/4)(200 \times 10^9)} + \frac{F_C(0.75)}{\pi(0.06^2/4)(70 \times 10^9)} = 0$$

When simplified, this equation leads to

$$F_A = -1.905 F_C \qquad (1.10.2)$$

Solving Equations 1.10.1 and 1.10.2 simultaneously, we obtain

$$F_A = -98.365 \text{ kN} \quad \text{and} \quad F_C = 51.635 \text{ kN}$$

The small differences between these values for F_A and F_C and those obtained earlier are due to truncation of numbers during the solutions. Of course, the stresses in both segments of the member are obtained as was done in Method (I).

EXAMPLE 1.11

Consider the frame shown in figure (a) below, consisting of a rigid beam *ABC*, which is supported by three rods. The two outside rods are made of brass for which the modulus of elasticity $E_B = 15 \times 10^6$ psi and have identical cross-sectional areas $A_B = 0.375$ in.^{2}. The center rod is made of steel and has a cross-sectional area $A_S = 1.0$ in.^{2} and a modulus of elasticity $E_S = 30 \times 10^6$ psi. Determine the normal stresses in the brass and in the steel rods if $P = 13.75$ kips and $a = 3$ in.

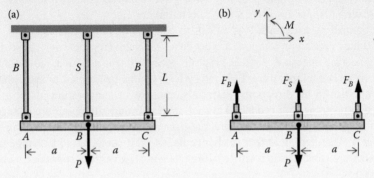

SOLUTION

The free-body diagram of rigid beam *ABC* is constructed as shown in figure (b) above. Note that because of symmetry about a vertical plane through point *B*, the conclusion was reached that both of the brass rods have the same force of magnitude F_B. This same conclusion may be reached by assuming these two forces to be different and summing moments about point *B*. Summing forces in the *y* direction, we obtain

$$\sum F_y = 0 \implies 2F_B + F_S - P = 0 \tag{1.11.1}$$

The above equation contains the two unknown quantities, F_B and F_S. A second equation is needed to solve the problem. This second equation comes from the deformation characteristics of the frame. Because of symmetry and the fact that the beam is rigid, we conclude that it moves parallel to itself and that the deformations of the brass and steel rods are equal (i.e., $\delta_B = \delta_S$). Thus, since all lengths are the same, we obtain

$$\left[\frac{FL}{AE}\right]_B = \left[\frac{FL}{AE}\right]_S \implies \frac{F_B}{(0.375)(15 \times 10^6)} = \frac{F_S}{(1.0)(30 \times 10^6)}$$

$$F_S = (16/3)F_B \tag{1.11.2}$$

Solving Equations 1.11.1 and 1.11.2 simultaneously, we obtain

$$F_B = (3/22)P = 1.875 \text{ kips} \quad \text{and} \quad F_S = (16/22)P = 10.0 \text{ kips}$$

Therefore, the stresses are

$$\sigma_B = \frac{1.875}{0.375} = 5.0 \text{ ksi}; \quad \sigma_S = \frac{10.0}{1.000} = 10.0 \text{ ksi} \qquad \textbf{ANS.}$$

1.6.1   TEMPERATURE EFFECTS

Engineering materials, when not restrained, experience a change in dimensions if subjected to a change in temperature. The change in length per unit length per degree change in temperature is

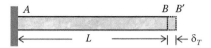

FIGURE 1.27   Change in dimension due to change in temperature in a statically determinate member.

referred to as the *coefficient of thermal expansion* and is denoted by the symbol α. The coefficient of thermal expansion is a unique property for a given material and representative values of this property are given in Appendix D for a few selected materials.

For most engineering materials, α is a positive constant quantity, although over wide ranges in temperature, α may depend on temperature (as do all mechanical properties). Also, there are a few important engineering materials, such as graphite, for which α is negative, indicating a tendency to contract with increased temperature. Laboratory tests can be performed to show that for *isotropic* (i.e., same properties in all directions) materials, changes in temperature affect *only* the normal strains; shearing strains are not produced in such materials under temperature changes.

Consider, for example, the case of the prismatic bar AB of length L shown in Figure 1.27. This bar is fixed only at A. Therefore, it is free to change its length if subjected to a change in temperature. Thus, if the bar were subjected to a temperature change ΔT, it would experience a change in length $\delta_T = BB'$ given by

$$\delta_T = \alpha(\Delta T)L \qquad\qquad (1.31)$$

For example, if ΔT is positive (a temperature increase), δ_T would represent an increase in length as shown in Figure 1.27, and if ΔT is negative (a temperature drop), δ_T would represent a decrease in length.

If the prismatic bar, however, were restrained so that the change δ_T could not take place in part or in whole, forces would be induced in it, giving rise to a system of stresses known as *thermal stresses*. Thus, if member AB were fixed at both ends to unyielding supports as shown in Figure 1.28a, it would not be able to experience the change in length δ_T if subjected to a temperature change ΔT. Thus, if ΔT is a temperature drop, bar AB would have a tendency to shrink in length but it is unable to do so because of the two unyielding supports at A and B which, in effect, apply a restraining tensile force F that would create a system of tensile thermal stresses in bar AB.

Determination of the unknown restraining tensile force F is most conveniently accomplished using superposition. Support B is removed, as shown in Figure 1.28b, and the rod is allowed to experience free contraction by the amount $\delta_T = \alpha(\Delta T)L$. This free contraction is eliminated, as shown in Figure 1.28b, by the application of the restraining tensile force F, which is adjusted in magnitude to produce an extension $\delta = FL/AE$, equal in magnitude to the temperature contraction δ_T. Thus, $\delta + \delta_T = 0$, from which the magnitude of the force F may be determined if all of the other information is known. Substituting FL/AE for δ_F and $\alpha(\Delta T)L$ for δ_T and simplifying, we obtain

$$F = \alpha(\Delta T)AE \qquad\qquad (1.32)$$

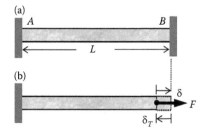

FIGURE 1.28   Creation of tensile stresses due to a decrease in temperature.

Thus, the average thermal stress becomes

$$\sigma_T = \frac{F}{A} = \alpha(\Delta T)E \tag{1.33}$$

EXAMPLE 1.12

Refer to the rod shown in Figure 1.28a and let $L = 24$ in., $A = 1.25$ in.^{2}, $E = 30 \times 10^3$ ksi, and $\alpha = 6.5 \times 10^{-6}$ in./in./°F. If the allowable thermal stress is $\sigma_{ALL} = 25$ ksi, determine the maximum temperature drop to which the member may be subjected.

SOLUTION

Since the allowable stress is 25 ksi and the cross-sectional area of the rod is 1.25 in.^{2}, it follows that the restraining force in the rod is

$$F = (25)(1.25) = 31.25 \text{ kips}$$

Therefore, the deformation that F would produce is

$$\delta = \frac{(31.25 \times 10^3)(24)}{(1.25)(30 \times 10^6)} = 20.0 \times 10^{-3} \text{ in.}$$

The deformation δ_T produced by the temperature drop ΔT may be found in terms of the unknown ΔT. Thus,

$$\delta_T = -6.5 \times 10^{-6}(\Delta T)(24) = -(1.56 \times 10^{-4})\Delta T$$

Therefore, since $\delta + \delta_T = 0$, it follows that

$$20.0 \times 10^{-3} - (1.56 \times !0^{-4})\Delta T = 0$$

from which

$$\Delta T = 128.2°\text{F} \qquad \textbf{ANS.}$$

PROBLEMS

1.91 Repeat Example 1.10 using the following information: segment AB is titanium whose diameter is 60 mm and for which $E_T = 115$ GPa; segment BC is aluminum whose diameter is 80 mm and for which $E_A = 70$ GPa. Let $P = 120$ kN.

1.92 Refer to Example 1.11 and determine the force P if the allowable stresses are $\sigma_B = 20$ MPa and $\sigma_S = 50$ MPa. Use the following data: $A_B = 600$ mm^2, $A_S = 1500$ mm^2, $E_B = 100$ GPa, $E_S = 200$ GPa, $L = 1.5$ m, and $a = 125$ mm.

1.93 A composite member consisting of a concentric steel rod S and an aluminum tube A is shown in the sketch in a cut-out view. Determine the normal stresses in the steel rod and in the aluminum tube if $E_S = 30 \times 10^3$ ksi, $E_A = 10 \times 10^3$ ksi, $A_S = 0.7$ in.^{2}, and $A_A = 0.4$ in.^{2} Let $P = 50$ kips and $h = 10$ in. The plate on top of the member is rigid and used only to apply the load P.

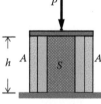

(Problems 1.93 and 1.94)

1.94 In reference to the figure shown in Problem 1.93, a composite member consisting of a concentric steel rod S and an aluminum tube A is shown, in cross section, in the sketch. Determine the largest force P if the allowable stress in the steel rod is 40 ksi and that in aluminum is 15 ksi. The following data is provided: $A_S = 0.5$ in.^{2}, $A_A = 0.35$ in.^{2} Use the moduli of elasticity for the two materials as given in Problem 1.93. Find also the vertical displacement of the top rigid plate.

1.95 Member ABC is made of the same material for which $E = 150$ GPa. Segment AB has a 15 mm × 30 mm rectangular cross-sectional area and segment BC a 50 mm × 80 mm rectangular cross-sectional area. Let $P = 200$ kN, $a = 250$ mm, and $b = 500$ mm and determine (a) the normal stresses in AB and BC and (b) the vertical displacement at B.

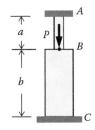

(Problems 1.95 and 1.96)

1.96 In reference to the figure shown in Problem 1.95, member ABC is made of two materials. Segment AB is steel for which $E = 200$ GPa, and has a diameter of 35 mm and segment BC is aluminum for which $E = 70$ GPa and has a diameter of 75 mm. If the allowable stresses in AB and BC are, respectively, 750 and 120 MPa, determine the maximum permissible load P. What will be the vertical displacement at B under those conditions? Let $a = 300$ mm and $b = 600$ mm.

1.97 The frame shown in the sketch consists of two magnesium outside bars marked M and a central aluminum bar marked A. Before P is applied, the aluminum bar is $\Delta = 0.012$ in. shorter than the two magnesium bars. Each bar has a cross-sectional area of 3 in.^{2} Let $E_M = 6.5 \times 10^3$ ksi and $E_A = 10 \times 10^3$ ksi. If $P = 200$ kips, determine (a) the normal stresses in the three bars and (b) the deformation of the two magnesium bars.

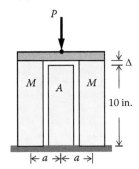

(Problems 1.97 and 1.98)

1.98 In reference to the figure shown in Problem 1.97, the frame shown in the sketch consists of two magnesium outside bars marked M and a central aluminum bar marked A. If the aluminum bar is $\Delta = 0.01$ in. shorter than the two magnesium bars, determine the force P so that the normal stresses in the three bars are identical. Each bar has a cross-sectional area of 2.5 in.^{2} Let $E_M = 6.5 \times 10^3$ ksi and $E_A = 10 \times 10^3$ ksi. What is the shortening of the two magnesium bars?

1.99 A reinforced concrete column with a diameter $D = 400$ mm has five steel reinforcing rods, each with a diameter of 32 mm. The height of the column is 4 m. Compute the normal stresses in the steel rods and in the concrete if $E_S = 200$ GPa and $E_C = 30$ GPa and the centrally applied compressive load on the column is 2000 kN.

(Problems 1.99 and 1.100)

1.100 In reference to the figure shown in Problem 1.99, a reinforced concrete column with a diameter $D = 450$ mm has five steel reinforcing rods, each with a diameter of 35 mm. The height of the column is 3.5 m. Compute the maximum magnitude of a centrally applied compressive force that the column can carry if the allowable stress in concrete is 20 MPa and that in steel is 220 MPa. The moduli of elasticity for the two materials are as given in Problem 1.99.

1.101 The system shown in the sketch consists of two steel outside posts and an aluminum central post, each with a rectangular cross section 2 in. × 2.5 in. Before the load $P = 50$ kips is applied, the two steel posts are shorter than the aluminum post by an amount $e = 0.001$ in. Let $E_S = 30 \times 10^3$ ksi and $E_A = 10 \times 10^3$ ksi and determine the stresses in the three posts.

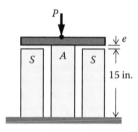

(Problems 1.101 and 1.102)

1.102 In reference to the figure shown in Problem 1.101, the system shown in the sketch consists of two steel outside posts and an aluminum central post, each with a rectangular cross section 1.5 in. × 3.0 in. Before the load $P = 60$ kips is applied, the two steel posts are shorter than the aluminum post by some unknown magnitude e. Determine this quantity e so that the stress in the steel posts is exactly twice that in the aluminum post. The moduli of elasticity are as given in Problem 1.101.

1.103 The rigid horizontal member is supported by the aluminum rod A and brass rod B as shown. The rigid member remains horizontal, and the normal stress in the brass rod becomes $\sigma_B = 15$ MPa, when the load $P = 50$ kN is applied. Let $a = 0.15$ m and $b = 0.10$ m. The moduli of elasticity for aluminum and brass are, respectively, $E_A = 75$ GPa and $E_B = 100$ GPa. Determine the necessary cross-sectional areas of the two rods.

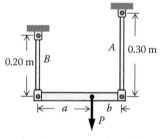

(Problems 1.103 and 1.104)

1.104 In reference to the figure shown in Problem 1.103, the rigid horizontal member is supported by the aluminum rod A and brass rod B as shown. The rigid member remains horizontal when the load $P = 60$ kN is applied. Determine the ratio a/b if the cross-sectional areas of the aluminum and brass rods are, respectively, $A_A = 6500$ mm^2 and $A_B = 13{,}000$ mm^2. The moduli of elasticity are as given in Problem 1.103. What are the normal stresses in both rods?

1.105 A compression member consists of two materials, wood on the left marked W, measuring 4 in. $\times$ 5 in. in cross section and magnesium on the right marked M and measuring 3 in. $\times$ 5 in. in cross section. The force $P = 75$ kips is applied to the top rigid plate as shown. The moduli of elasticity for wood and magnesium are, respectively, $E_W = 2 \times 10^6$ psi and $E_M = 6.5 \times 10^6$ psi. Determine (a) the position of P, measured from the left edge of the rigid plate, so that this rigid plate remains horizontal, and (b) the normal stresses induced in the two materials. Let $h = 10$ in.

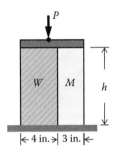

(Problems 1.105 and 1.106)

1.106 In reference to the figure shown in Problem 1.105, a compression member consists of two materials, wood on the left marked W and magnesium on the right marked M. The force $P = 60$ kips is applied to the top rigid plate as shown. The moduli of elasticity are as given in Problem 1.105. If the rigid plate remains horizontal when the load is applied and the allowable stress in the wood is 7.5 ksi and that in magnesium is 30.0 ksi, determine (a) the necessary depth into the page, in inches, of both materials assuming they are the same and (b) the position of P measured from the left edge of the rigid plate. Let $h = 12$ in.

1.107 The rigid horizontal member CD is supported by the two rods A and B as shown. The force $P = 100$ kN is so placed that the rigid member remains horizontal. Let $\phi = 30°$ and assume the following: $A_A = 450$ mm^2, $A_B = 350$ mm^2, $E_A = 120$ GPa, and $E_B = 150$ GPa. Determine (a) the angle θ for equilibrium and (b) the position of the 100-kN force measured from point C on the horizontal rigid member.

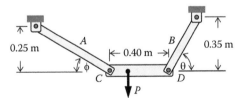

(Problems 1.107 and 1.108)

1.108 In reference to the figure shown in Problem 1.107, the rigid horizontal member CD is supported by the two rods A and B as shown. The force $P = 150$ kN is so placed that the rigid member remains horizontal. Let $\theta = 60°$ and assume the following: $A_A = 500$ mm^2, $A_B = 400$ mm^2, and $E_A = E_B = 150$ GPa. Determine (a) the angle ϕ for equilibrium, (b) the position of the 150-kN force measured from point C on the horizontal rigid member, and (c) the normal stresses in rods A and B.

1.109 The assembly shown in the sketch consists of an outer cylinder of material A, a central solid core of material B, and filling the concentric space between A and B is material C. When the concentric compressive load $P = 200$ kips is applied through the rigid plate, the entire assembly shrinks uniformly by an amount Δ. Determine Δ and the normal stresses in the various components of the assembly. Let $E_A = 20 \times 10^3$ ksi, $E_B = 15 \times 10^3$ ksi, $E_C = 10 \times 10^3$ ksi, $d_1 = 14$ in., $d_2 = 12$ in., and $d_3 = 3$ in.

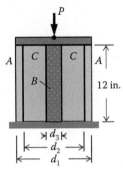

(Problems 1.109 and 1.110)

1.110 In reference to the figure shown in Problem 1.109, the assembly shown in the sketch consists of an outer cylinder of material A, a central solid core of material B, and filling the concentric space between A and B is material C. When the concentric load P is applied through the rigid plate, the entire assembly shrinks uniformly by $\Delta = 0.00225$ in. Determine the load P and the normal stresses in the various components of the assembly. Let $d_1 = 12$ in., $d_2 = 10$ in., and $d_3 = 4$ in. The moduli of elasticity are as given in Problem 1.109.

1.111 The rigid beam BCD is hinged at B and supported at C and D, respectively, by an aluminum rod marked A and a steel rod marked S. Let $P = 125$ kN and determine the stresses and deformations induced in both rods. The following information is given: $E_S = 205$ GPa, $A_S = 200$ mm², $E_A = 75$ GPa, and $A_A = 1200$ mm².

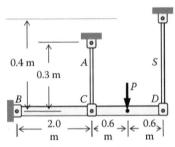

(Problems 1.111 and 1.112)

1.112 In reference to the figure shown in Problem 1.111, the rigid beam BCD is hinged at B and supported at C and D, respectively, by an aluminum rod marked A and a steel rod marked S. Determine the maximum permissible magnitude of P if the allowable stress in aluminum is 25 MPa and that in steel is 100 MPa. The following information is provided: $A_S = 180$ mm² and $A_A = 900$ mm². The moduli of elasticity are as given in Problem 1.111.

1.113 The system shown in the sketch consists of a rigid beam hinged at D and supported at C by a steel ($E_S = 30 \times 10^3$ ksi, $A_S = 0.50$ in.²) flexible cable and at B by a concrete support ($E_C = 4.5 \times 10^3$ ksi, $A_C = 60$ in.²). If the load $P = 200$ kips, determine the normal stresses in the two materials.

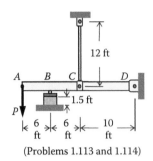

(Problems 1.113 and 1.114)

1.114 In reference to the figure shown in Problem 1.113, the system shown in the sketch consists of a rigid beam hinged at D and supported at C by a steel ($A_S = 0.75$ in.^{2}) flexible cable and at B, by a concrete support ($A_C = 70$ in.^{2}). The moduli of the two materials are as given in Problem 1.113. If the allowable stress in concrete is 5 ksi and that in steel is 15 ksi, determine the maximum permissible load P.

1.115 The system shown in the sketch consists of a rigid beam hinged at A and supported at B by a flexible cable made of material X. Before the load $P = 400$ kN is applied, a gap $\Delta = 1.75$ mm exists between the beam and the support at C, which is made of material Y. Let $E_X = 150$ GPa, $A_X = 175$ mm^2, $E_Y = 60$ GPa, and $A_Y = 6000$ mm^2. Determine the normal stresses in the two materials and the displacement of the rigid beam at C.

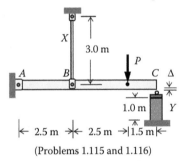

(Problems 1.115 and 1.116)

1.116 In reference to the figure shown in Problem 1.115, the system shown in the sketch consists of a rigid beam hinged at A and supported at B by a flexible cable made of material X. Before the load $P = 450$ kN is applied, a gap $\Delta = 2.0$ mm exists between the beam and the support at C, which is made of material Y. Determine the needed cross-sectional areas for the two materials if the allowable stress in material Y is 90 MPa and that for material X is 120 MPa. The moduli of elasticity are as give in Problem 1.115. Given: $A_Y = 20 A_X$.

1.117 The entire member shown is made of aluminum for which $E = 75$ GPa. If $P = 300$ kN and the diameters are $d_{AB} = 60$ mm and $d_{BC} = 90$ mm, determine the normal stresses in both parts of the member and the displacement at B.

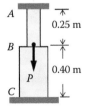

(Problems 1.117 and 1.118)

1.118 In reference to the figure shown in Problem 1.117, if in addition to the force $P = 300$ kN, the member experiences a temperature drop $\Delta T = 30°C$, determine the normal stresses in both parts of the member. Let $\alpha = 22 \times 10^{-6}$ m/m/°C.

1.119 The system shown consists of a rigid containing frame C, two aluminum bars marked A, each with a 0.5 in. × 1.5 in. rectangular cross section, rigid member BD, and a single-threaded steel bolt marked S, with a diameter 2.0 in. and a pitch (the advance of the bolt per complete turn of the nut) of 0.1 in. After the nut at the top of the bolt is turned to a snug fit, it is wrench-tightened one complete turn. Determine the normal stresses in the two materials, assuming that $E_S = 30 \times 10^3$ ksi and $E_A = 10 \times 10^3$ ksi.

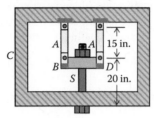

(Problems 1.119 and 1.120)

1.120 In reference to Problem 1.119, in addition to the conditions stated there, the temperature of the two aluminum bars and the steel bolt are dropped by $\Delta T = 35°F$. Let $\alpha_A = 12.5 \times 10^{-6}$ in./in./°F, $\alpha_S = 6.5 \times 10^{-6}$ in./in./°F, and determine the normal stresses developed in the two materials.

1.121 The system shown in the sketch consists of a 0.030°m-diameter titanium rod marked T attached rigidly to a 0.060°m-diameter aluminum rod marked A at its top and to a rigid frame at its bottom. When the temperature is 60°C, the clearance between the head of the aluminum rod and the rigid frame is $\Delta = 0.25$ mm. Determine the normal stresses induced in the two materials when the temperature drops to 15°C. Let $P = 0$, $E_T = 110$ GPa, $\alpha_T = 10 \times 10^{-6}$ m/m/°C, $E_A = 70$ GPa, and $\alpha_A = 24 \times 10^{-6}$ m/m/°C.

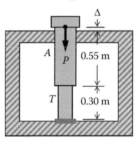

(Problems 1.121 and 1.122)

1.122 In reference to Problem 1.121, determine the largest force P if the allowable stress in the aluminum is 30.0 MPa and that in titanium is 110.0 MPa. All other conditions are as stated in Problem 1.121.

1.123 Segment AB is steel with a diameter of 1.5 in. and a modulus of elasticity $E_S = 30 \times 10^6$ psi. Segment BC is aluminum with a diameter 5.0 in. and a modulus of elasticity $E_A = 10 \times 10^6$ psi. Let $P = 30$ kips. If the entire member experiences a temperature increase $\Delta T = 30°F$, determine the stresses developed in the two materials. Let $\alpha_S = 6.5 \times 10^{-6}$ in./in./°F and $\alpha_A = 12.5 \times 10^{-6}$ in./in./°F.

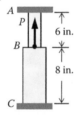

(Problems 1.123 and 1.124)

1.124 In reference to Problem 1.123, determine the largest force P if the allowable stress in aluminum is 10 ksi and that in steel is 65 ksi, both in tension or compression. All other conditions are as stated in Problem 1.123.

*1.7 STRESS CONCENTRATION

The stresses computed from the equations developed earlier (e.g., $\sigma_{AVE} = F/A$) are only average values referred to as *nominal stresses*. These equations do not account for the fact that localized stresses may reach much higher values, under static loading conditions, as a result of the following factors:

1. Loads or reactions delivered to a member over a relatively small area such as concentrated loads
2. Internal or surface flaws in the material of which the member is fabricated
3. Geometric changes in the component under consideration

Item 1 above refers to the condition already discussed in Section 1.3 in connection with *Saint-Venant's* principle, which states that stresses in the neighborhood of the point of application of a concentrated force can reach much higher values than those given by the simple equation $\sigma_{AVE} = F/A$, unless we move far enough away from that point. Condition 2 deals with the fact that engineering materials are not ideal—*homogeneous* (i.e., same properties throughout) and *isotropic* (i.e., same properties in all directions) because they include foreign matter and other impurities as well as internal and surface manufacturing flaws that can give rise to above-average stress values. Finally, Condition 3, refers to design necessities that require the introduction of geometric discontinuities in a member, such as holes and fillets, which result in stresses much higher than those given by the equation $\sigma_{AVE} = F/A$.

We should note here that, in general, the existence of stress concentration is a significant design consideration especially under dynamic loadings such as impact and fatigue. Under static conditions, however, with the exception of brittle materials, the presence of stress concentration does not, in general, play a significant role in the design process. This is so because ductile materials are capable of plastic deformation in the neighborhood of stress concentrations, thus providing some relief from the high stresses caused by these concentrations. Brittle materials, however, are incapable of undergoing this relief process.

To account for the increase in stresses at geometric discontinuities, we introduce a factor k, known as the *stress-concentration factor*. Values of the stress-concentration factors are determined, mainly, experimentally using *photoelasticity*, a method especially suited for this purpose. Therefore, to determine the maximum stress σ_{MAX} due to a given geometric discontinuity, we employ the equation

$$\sigma_{MAX} = k\left(\frac{F}{A_{net}}\right) = k\sigma_{AVE*} \tag{1.34}$$

The symbol A_{net} in Equation 1.34 represents the net, rather than the gross, area of the member and $\sigma_{AVE*} = F/A_{net}$ represents the average normal stress over this net area. For example, in Figure 1.29a

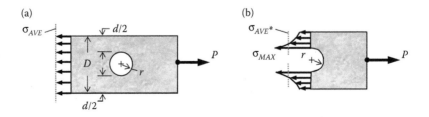

FIGURE 1.29 Stress concentration in a member, with a circular hole at its center, subjected to an axial tensile force.

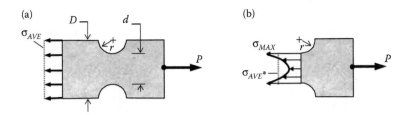

FIGURE 1.30 Stress concentration in a member with symmetrically placed grooves subjected to an axial tensile force.

is shown a member with a circular hole of radius r subjected to a tensile force P, which results in a system of tensile stresses that may be assumed uniformly distributed ($\sigma = \sigma_{AVE}$) if the position considered is far removed from the geometric discontinuity. Figure 1.29b, however, shows the tensile stress distribution at a section through the center of the hole. Note that because of the concentration of the stress in the neighborhood of the hole, the maximum stress is considerably larger than both σ_{AVE} and σ_{AVE^*}, where $\sigma_{AVE^*} > \sigma_{AVE}$. The net area in this case would be obtained by multiplying the dimension d by the thickness of the member t measured into the page.

In Figure 1.30a is shown a member with symmetrically located semicircular grooves of radius r, one on each side, and subjected to a tensile force P. The stress distribution across the section may be assumed uniform if the section considered happens to be far enough from the geometric discontinuity. However, right at the bottom of the grooves (i.e., at the narrowest section where the grooves are located) as shown in Figure 1.30b, the stress distribution is not uniform and the maximum stress is considerably larger than either of the two average stresses.

Finally, Figure 1.31a shows a member with symmetrically located fillets of radius r and subjected to a tensile force P. Again, the stress distribution may be assumed uniform at sections far removed from the geometric discontinuity. At the narrowest section in the neighborhood of the fillets, however, the stress distribution is not uniform, as shown in Figure 1.31b and the maximum stress is much larger than either of the two average stresses.

As stated earlier, the values of the stress-concentration factors for use in Equation 1.34 are obtained, for the most part, experimentally, although some analytical work has been done using computers along with the finite element methods. Such stress-concentration factors are plotted in graphs similar to that shown in Figure 1.32, where the stress-concentration factor k is plotted on the vertical axis against the ratio r/d on the horizontal axis for the three cases discussed above.

The curve marked A provides stress-concentration factors for axially loaded members with a circular hole, the curve marked B gives those for an axially loaded member with semicircular grooves and the one marked C gives stress-concentration factors for axially loaded members with fillets. It should be pointed out that the information shown in Figure 1.32 is but a sample of a large volume of information available in the literature on the subject of stress-concentration factors.

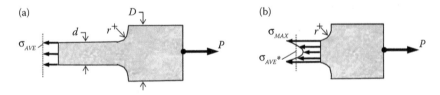

FIGURE 1.31 Stress concentration in a member with symmetrically placed fillets subjected to an axial tensile force.

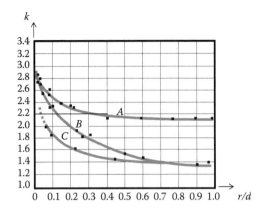

FIGURE 1.32 Stress-concentration factors for axially-loaded members: A = Circular holes; B = Semicircular grooves; C = Fillets. (After Wahl and Beeuwkes, Stress concentration produced by holes and notches, *Trans ASME*, 56, 617–625, 1934.)

EXAMPLE 1.13

An axially loaded plate similar to the one shown in Figure 1.29a is 3.00 in. wide, has a thickness of 0.50 in., and a 1.00-in. diameter hole drilled through it at its center. The plate is subjected to a tensile force $P = 10$ kips. Determine the maximum tensile normal stress and compare it to the average tensile normal stresses σ_{AVE} and $\sigma_{AVE}*$.

SOLUTION

The parameters needed to find the stress-concentration factor are computed as follows:

$$d = 3.00 - 1.00 = 2.00 \text{ in.,} \quad r = \frac{1.00}{2} = 0.50 \text{ in., and} \quad \frac{r}{d} = 0.25$$

Entering Figure 1.32 with a value of $r/d = 0.25$ and reading from the curve marked A; we estimate the stress-concentration factor to be $k = 2.25$. Thus, using Equation 1.34, we have

$$\sigma_{MAX} = 2.25\left(\frac{10}{2(0.50)}\right) = 22.5 \text{ ksi} \qquad \textbf{ANS.}$$

Also

$$\sigma_{AVE}* = \frac{10}{2(0.50)} = 10.0 \text{ ksi}$$

$$\sigma_{AVE} = \frac{10}{(3.00)(0.50)} = 6.667 \text{ ksi}$$

Therefore, the ratio of the maximum to the average normal stresses, in this case, is

$$\sigma_{MAX} = \left(\frac{22.5}{10.0}\right) = 2.25\sigma_{AVE}* = \left(\frac{22.5}{6.667}\right)\sigma_{AVE} = 3.375\sigma_{AVE} \qquad \textbf{ANS.}$$

Thus, for the conditions stated in this problem, the magnitude of the maximum tensile normal stress is over two times $\sigma_{AVE}*$ and over three times σ_{AVE}. As stated in an earlier section, according to Saint-Venant's principle, the effects of the geometric discontinuity would gradually disappear as we move farther and farther away from the location of the hole. It has been shown, both experimentally and analytically, that these effects disappear at distances approximately equal to the largest dimension of the cross section. Thus, we can expect to find practically uniform stress distributions on either side of the hole at distances equal to or larger than the dimension D, which in this case is 3.00 in.

PROBLEMS

1.125 The plate shown in the sketch has a width $D = 100$ mm and a centrally placed hole of radius $r = 15$ mm. It is subjected to a tensile force $P = 200$ kN. Determine the maximum tensile stress and the average tensile stresses σ_{MAX} and σ_{AVE}. Let $t = 20$ mm.

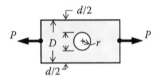

(Problems 1.125, 1.126, and 1.127)

1.126 In reference to the figure shown in Problem 1.125, the plate shown in the sketch has a width $D = 4$ in. and a centrally placed hole of radius $r = 0.75$ in. If the maximum stress is to be limited to 35 ksi, determine the maximum allowable force P. Let $t = 0.5$ in.

1.127 In reference to the figure shown in Problem 1.125, the plate shown in the sketch has a width $D = 85$ mm and a centrally placed hole of radius r. Let $P = 250$ kN and $\sigma_{ALL} = 300$ MPa. Let $t = 30$ mm and $\sigma_{AVE}{}^* = 130$ MPa. Determine the minimum radius r consistent with the given data.

1.128 The plate shown in the sketch has two semicircular grooves of radius r symmetrically placed, one on each side. Let $D = 6$ in., $t = 2$ in., and $d = 4$ in. If the allowable stress is 25 ksi, find the maximum permissible magnitude of the force P.

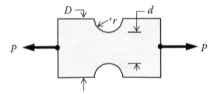

(Problems 1.128, 1.129, and 1.130)

1.129 In reference to the figure shown in Problem 1.128, the plate shown in the sketch has two semicircular grooves of radius r symmetrically placed, one on each side. Let $P = 300$ kN, $d = 75$ mm, $t = 50$ mm, and $\sigma_{ALL} = 150$ MPa. Determine the smallest radius of the groove consistent with the given data.

1.130 In reference to the figure shown in Problem 1.128, the plate shown in the sketch has two semicircular grooves of radius $r = 0.5$ in. symmetrically placed, one on each side. Let $\sigma_{AVE}{}^* = 30$ ksi and $\sigma_{ALL} = 45$ ksi and determine the dimension d and the force P consistent with the given information. Let $t = 1.0$ in.

1.131 The plate shown in the sketch has circular fillets of radius $r = 5$ mm symmetrically placed, one on each side. If the compressive force $P = 150$ kN and $d = 35$ mm, determine the maximum compressive stress as well as the average stress that would occur at distances far removed from the fillets. Let $t = 30$ mm.

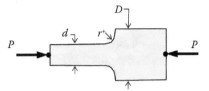

(Problems 1.131, 1.132, and 1.133)

1.132 In reference to the figure shown in Problem 1.131, the plate shown in the sketch has circular fillets of radius r symmetrically placed, one on each side. The compressive force $P = 65$ kips, $D = 3$ in., $\sigma_{AVE}{}^* = 25$ ksi, and $\sigma_{ALL} = 40$ ksi. Determine the minimum radius r consistent with the given data. Let $t = 1.5$ in.

1.133 In reference to the figure shown in Problem 1.131, the plate shown in the sketch has circular fillets of radius $r = 7$ mm symmetrically placed, one on each side. Let $d = 40$ mm, $t = 20$ mm, and $\sigma_{ALL} = 125$ MPa and determine the maximum permissible compressive force P.

1.134 The plate shown in the sketch has two semicircular groves of radius r_1 and a circular hole of radius r_2, all symmetrically placed with respect to its central axis. It is subjected to an axial load $P = 75$ kips. Let $D = 8$ in., $r_1 = 1.5$ in., $r_2 = 0.5$ in., and $t = 1.25$ in. and determine the maximum tensile stress in the plate.

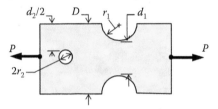

(Problems 1.134 and 1.135)

1.135 In reference to the figure shown in Problem 1.134, the plate shown in the sketch has two semicircular grooves of radius r_1 and a circular hole of radius r_2, all symmetrically placed with respect to its central axis. It is subjected to an axial load P. Let $D = 180$ mm, $r_1 = 40$ mm, $r_2 = 30$ mm, and $t = 50$ mm. If the allowable stress in the plate is 200 MPa, find the maximum permissible force P.

*1.8 IMPACT LOADING

Often in practice, an axially loaded member is called upon to resist the action of impact loading. A very good example of this kind of loading occurs in bridge-truss members when heavy trucks and other traffic pass over a bridge. Consider, for example, the case of member BC, shown in Figure 1.33,

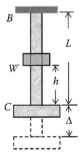

FIGURE 1.33   Member BC subjected to a falling weight W from a height h.

which is fixed to a rigid support at B and subjected to the action of the falling weight W. As the falling weight W contacts the rigid end plate at C, it transmits its kinetic energy to member BC, which undergoes a maximum deformation Δ as shown. As will be discovered shortly, this deformation Δ is considerably larger than the deformation that would be experienced by the member if the weight W were placed gently (statically) on the rigid plate at C.

In analyzing the effects of impact loading, certain assumptions need to be made as follows:

1. All of the potential energy possessed by the weight W is transformed, without losses, to kinetic energy as the member falls through the height h, and then to *elastic strain energy* in the member as it impacts the rigid plate at C. Actually, some of the energy is lost to heat and sound and, during impact, to the process of damping out the oscillations that take place after impact.
2. All impacted members behave linearly elastically during impact so that they do not experience plastic deformations.
3. The dead-load stresses associated with the weights of the impacted bodies are negligible compared to stresses associated with impact.
4. The work done by the weight of the impacted bodies during deformation is ignored.
5. The impacting body does not bounce back carrying with it some of the available energy.

All of the above statements may be summarized mathematically by the relation

$$W(h + \Delta) = U \tag{1.35}$$

The symbol U in Equation 1.35 is the total strain energy stored in the member.

Consider now a differential element of volume dV in member BC of Figure 1.33. At the instant when the weight W has come to rest at the end of its downward travel, the member has been stretched its maximum amount Δ and the stress in the volume dV has been built up from zero to its maximum value σ. This, of course, means that the average stress during the deformation of the member is $1/2\sigma$. The differential strain energy dU stored in the differential volume dV is given by $1/2\sigma \varepsilon$ (see also Equation 1.20 in Section 1.4), where ε is the maximum strain of the member associated with the maximum stress σ when the maximum deformation Δ is reached. Thus,

$$u = \frac{dU}{dV} = \frac{\sigma\varepsilon}{2} \tag{1.20, Repeated}$$

Equation 1.20 is, once again, illustrated in Figure 1.34 by the area under the stress–strain curve within the elastic range for the material.

The total energy U stored in the member is obtained by integrating Equation 1.20. Therefore,

$$U = \int \frac{1}{2}\sigma\varepsilon\, dV = \frac{1}{2}\left(\frac{F}{A}\right)\left(\frac{\Delta}{L}\right)\int_{vol} dV = \frac{1}{2}\left(\frac{F}{A}\right)\left(\frac{\Delta}{L}\right)V \tag{1.36}$$

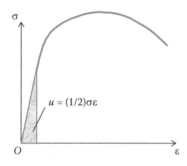

FIGURE 1.34 Stress–strain diagram showing the *elastic strain energy* due to the falling weight in Figure 1.33.

The quantities F/A and Δ/L in Equation 1.36 were substituted for σ and ε, respectively, and taken out of the integral sign under the assumption that they are constant throughout the entire volume of the member. If it is assumed further that the member is prismatic (has a constant cross-sectional area A), the volume V may be replaced by the product AL. When this product is substituted for V in Equation 1.36, and the product AL in the numerator is canceled out by the product AL in the denominator, we obtain

$$U = \frac{1}{2}F\Delta \tag{1.37}$$

Equation 1.37 is now substituted into Equation 1.35 to obtain the basic relation used in the solution of problems dealing with the axial impact. Thus,

$$W(h + \Delta) = \frac{1}{2}F\Delta \tag{1.38}$$

We should note here that, in Equation 1.38, F is the maximum force experienced by the member during impact and that the deformation $\Delta = FL/AE$, as given by Equation 1.28.

The *impact factor*, *IF*, is defined as the maximum force F during impact to the static force W if applied gently without impact. Thus, since $F = 2W(1 + h/\Delta)$ from Equation 1.38, we obtain

$$IF = \frac{F}{W} = 2\left(1 + \frac{h}{\Delta}\right) \tag{1.39}$$

EXAMPLE 1.14

Refer to the member shown in Figure 1.33. The following data is provided: $W = 200$ N, $h = 0.15$ m, $L = 2.5$ m, $A = 200$ mm^2, and $E = 200$ GPa. Assume that 25% of the available energy is lost to heat, noise, etc. and determine (a) the maximum stress in member *BC* due to impact, (b) the impact factor, and (c) the maximum deflection.

SOLUTION

a. Since 25% of the available energy is lost, Equation 1.38 needs to be modified to reflect that loss. Thus,

$$0.75W(h + \Delta) = \frac{1}{2}F\Delta$$

$$\Delta = \frac{FL}{AE} = \frac{F(2.5)}{(2 \times 10^{-4})(200 \times 10^9)} = 6.25 \times 10^{-8} F$$

Substituting into the modified energy equation the given numerical values for W and h and the expression for Δ in terms of F, we obtain, after simplification

$$F^2 - 300F - 7.2 \times 10^8 = 0$$

Solving this quadratic equation yields

$$F = 26983.2\,\text{N} \approx 26.983\,\text{kN}$$

Therefore,

$$\sigma = \frac{F}{A} = \frac{26983.2}{2 \times 10^{-4}} = 134.916 \approx 134.9\,\text{MPa} \qquad \textbf{ANS.}$$

b. The impact factor is found from Equation 1.39. Thus,

$$IF = \frac{F}{W} = \frac{26983.2}{200} = 134.9 \qquad \textbf{ANS.}$$

c. The maximum deflection is obtained by using Equation 1.28. Thus,

$$\Delta = \frac{(26983.2)(2.5)}{(2 \times 10^{-4})(200 \times 10^9)} = 0.001686 \text{ m} \approx 1.7 \text{ mm} \qquad \textbf{ANS.}$$

Note that $h = 0.15$ m is much larger than $\Delta = 0.001686$ m. This implies that the linear term in the quadratic equation could have been neglected without serious error. The resulting impact factor (134.9) means that the values of maximum stress and maximum deflection under impact are 134.9 times those obtained under static application of the load W. However, these impact stresses and deflections occur only for an instant of time, which somewhat mitigates their effect on the material, as compared to long-time loading. It is, of course, assumed the temperature is high enough so that the brittle fracture does not take place.

EXAMPLE 1.15

A block of weight W moves to the left at a speed $v = 60$ ft/s and impacts a linearly elastic spring of spring constant $k = 1$ kip/ft as shown. If the displacement of the spring is limited to 6 in., determine the weight W that would produce this displacement.

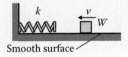

Smooth surface

SOLUTION

If we assume that no energy is lost during impact, then the left-hand side of Equation 1.38 is given by the kinetic energy possessed by the block prior to impact. The force F on the right-hand side of Equation 1.38 is equal to $k\Delta$ for a linearly elastic spring so that the quantity on the right of Equation 1.38 becomes equal to $1/2k\Delta^2$. We recognize this quantity as the strain energy stored in the spring. Thus, assuming no losses, the kinetic energy of the block is transformed in its entirety to strain energy in the spring and Equation 1.38 may be expressed as follows:

$$\frac{1}{2}mv^2 = \frac{1}{2}k\Delta^2$$

Thus,

$$\left(\frac{W}{32.2}\right)(60^2) = (1 \times 10^3)\left(\frac{6}{12}\right)^2$$

Solving for the weight W, we obtain

$$W = 2.236 \approx 2.2 \text{ lb} \qquad \textbf{ANS.}$$

PROBLEMS

1.136 The weight $W = 15$ lb drops from rest from a height $h = 4$ in. and impacts the rigid plate at C. Let $L = 18$ in., $E = 10 \times 10^6$ psi, and the diameter of rod BC is $d = 0.75$ in. Determine (a) the maximum stress induced in the rod, (b) the maximum deflection of rod BC, and (c) the impact factor. Assume no energy losses.

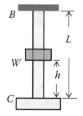

(Problems 1.136 and 1.137)

1.137 In reference to the figure shown in Problem 1.136, the weight $W = 120$ N drops from rest from a height $h = 0.75$ m and impacts the rigid plate at C. The material has a proportional limit $\sigma_p = 250$ MPa and a modulus of elasticity $E = 200$ GPa. Let $L = 1.50$ m and determine the cross-sectional area of member BC to avoid exceeding the proportional limit. Find also the maximum deformation of member BC and the impact factor. Assume that 20% of the available energy is lost.

1.138 Rods AB and CD are identical in all respects, have a diameter of 0.35 in. and are made of a material for which $E = 10 \times 10^3$ ksi. The post marked P serves as a guide for the weight $W = 25$ lb, and may be assumed frictionless and weightless. The plate attached to the two rods at B and D is rigid. Let $h = 5$ in. and $L = 10$ in. Assume that 25% of the available energy is lost, and determine (a) the maximum stress induced in the two rods, (b) the maximum deflection of the rods, and (c) the impact factor.

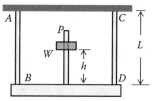

(Problems 1.138 and 1.139)

1.139 In reference to the figure shown in Problem 1.138, rods AB and CD are identical in all respects, have a diameter of 40 mm, and are made of a material for which $E = 75$ GPa and a proportional limit $\sigma_p = 200$ MPa. The post marked P serves as a guide for the weight W, and may be assumed frictionless and weightless. The plate attached to the rods at B and D is rigid. Let $h = 20$ mm and $L = 500$ mm. Assume that there are no energy losses and determine (a) the maximum weight W that may be dropped without exceeding the proportional limit and (b) the maximum deflection of the rods.

1.140 Collar C of weight $W = 5$ lb moves along the frictionless and weightless rod BD at a constant speed $v = 25$ ft/s and impacts the rigid plate attached to the rod at D. Rod BD has the following characteristics: $E = 10 \times 10^3$ ksi, $L = 7$ ft, and a cross-sectional area $A = 1.5$ in.^{2}. Determine (a) the maximum stress induced in the rod and (b) the maximum deflection experienced by the rod.

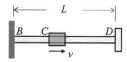

(Problems 1.140 and 1.141)

1.141 In reference to the figure shown in Problem 1.140, collar C of weight W moves along the frictionless and weightless rod BD at a constant speed $v = 7$ m/s and impacts the rigid plate attached to the rod at D. Rod BD has the following characteristics: $E = 70$ GPa, $L = 3$ m, and a cross-sectional area $A = 1.25 \times 10^{-3}$ m^2. Determine the maximum weight W not to exceed the allowable stress of 150 MPa. What is the maximum deflection of the rod?

1.142 A block of weight $W = 10$ lb moves on a frictionless surface at a speed v and impacts a linearly elastic spring with a spring constant $k = 200$ lb/in. The maximum deflection experienced by the spring is 5.5 in. Determine the speed v at which the block impacts the spring.

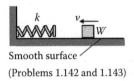

Smooth surface

(Problems 1.142 and 1.143)

1.143 In reference to the figure shown in Problem 1.142, a block of weight $W = 100$ N moves on a frictionless surface at a speed $v = 5$ m/s and impacts a linearly elastic spring with a spring constant k. The maximum deflection of the spring was measured at 250 mm. Determine the spring constant k.

REVIEW PROBLEMS

R1.1 Develop the internal axial force functions for segments AB, BC, and CD and plot the internal axial force diagram for the entire member shown.

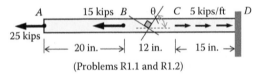

(Problems R1.1 and R1.2)

R1.2 In reference to the figure shown in Problem R1.1, construct the stress element indicated on the member that is located 6 in. to the right of point B if $\theta = 45°$. The member has a rectangular cross-sectional area of 2.0 in.^{2}

R1.3 Let $P = 100$ kN and determine the average shearing stress in the pin at D. This pin is in single shear and has a diameter of 60 mm.

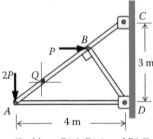

(Problems R1.3, R1.4, and R1.5)

R1.4 In reference to the figure shown in Problem R1.3, if the axial stress in member BD (area = 450 mm^2) is not to exceed 150 MPa in tension or compression, find the largest allowable magnitude of the applied load P.

R1.5 In reference to Problem R1.3, construct the stress element at point Q located anywhere in member AB (area = 500 mm^2).

R1.6   The punch press shown has a capacity $P = 100$ kips. It is used to punch rectangular holes in a steel plate of thickness $t = 1/4$ in. If the ultimate shearing strength of the steel is 15 ksi, determine the largest rectangular hole that can be punched by specifying the dimension a, where a is the depth into the page of the rectangular hole.

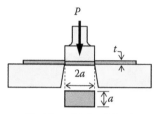

(Problems R1.6 and R1.7)

R1.7   In reference to the figure shown in Problem R1.6, rectangular holes, with dimension $a = 35$ mm as shown, are to be punched in a 15 mm aluminum plate. If the ultimate shearing strength of the aluminum is known to be 75 MPa, determine the minimum capacity of the press by specifying the magnitude of P. Note that a is the depth of the hole.

R1.8   Member *ABCD* was formed by securely joining together rods of three different materials as follows:

>   *AB*: Aluminum ($E = 10 \times 10^3$ ksi), diameter = 0.5 in.
>   *BC*: Steel ($E = 30 \times 10^3$ ksi), diameter = 0.25 in.
>   *CD*: Magnesium ($E = 6.5 \times 10^3$ ksi), diameter = 1.0 in.

Determine the movement of point *A*.

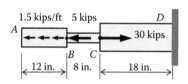

R1.9   Member *ABC* was formed by securely joining together rods of two different materials as follows:

>   *AB*: Aluminum ($E = 70$ GPa), diameter = 15 mm.
>   *BC*: Steel ($E = 30 \times 10^3$ ksi), diameter = 10 mm.

The member is subjected to the forces shown. If the movement of point *C* is to be zero, find the value of *L*.

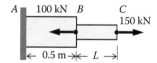

R1.10 The system shown in the sketch consists of brass ($E = 15 \times 10^3$ ksi) rod *AD*, 3/4 in. in diameter and 24 in. long attached securely to a rigid plate *A* at one end. A linear spring with a spring constant $k = 100$ kips/in. is placed between rigid plates *A* and *B*, and a magnesium ($E = 6.5 \times 10^3$ ksi) tube with an outside diameter of 2.0 in. and inside diameter of 1.75 in. is placed between rigid plates *B* and *C*, the latter being fixed in position as shown. The brass rod passes freely through the spring and the tube and through holes in plates B and C and is hinged to rigid beam EDG at D. If $P = 20$ kips, determine the displacement of (a) plate A, (b) plate B, and (c) point D. Neglect possible localized buckling of the magnesium tube.

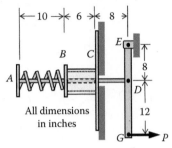

All dimensions
in inches

(Problems R1.10 and R1.11)

R1.11 In reference to the figure shown in Problem R1.10, the system shown in the sketch consists of brass ($E = 15 \times 10^3$ ksi) rod AD, 3/4 in. in diameter and 24 in. long, attached securely to a rigid plate A at one end. A linear spring with a spring constant $k = 100$ kips/ft is placed between rigid plates A and B and a magnesium ($E = 6.5 \times 10^3$ ksi) tube with outside diameter of 2.0 in. and inside diameter of 1.75 in. is placed between rigid plates B and C, the latter being fixed in position as shown. The brass rod passes freely through the spring and the tube and through holes in plates B and C and is hinged to rigid beam EDG at D. If the largest allowable movement of point G is limited to 1.0 in., determine the maximum permissible magnitude of P. Neglect possible localized buckling of the magnesium tube.

R1.12 Rigid beam $BCDG$ is hinged at B and supported at C by a spring of spring constant $k = 2000$ kN/m and at D by a steel ($E = 200$ GPa) rod as shown. If a vertical downward load $P = 150$ kN is applied at G, determine the deformation of the spring and the normal stress in the steel rod if its diameter is 30 mm.

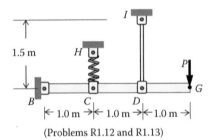

(Problems R1.12 and R1.13)

R1.13 In reference to the figure shown in Problem R1.12, rigid beam $BCDG$ is hinged at B and supported at C by a spring of spring constant $k = 2000$ kN/m and at D by a steel ($E = 200$ GPa) rod as shown. If the allowable deformation of the spring is 3 mm and the allowable normal stress in the steel rod, whose diameter 30 mm, is 150 MPa, determine the largest permissible load P applied vertically downward at G.

R1.14 An aluminum ($E = 70$ GPa) cylinder, 240 mm outside diameter, 220 mm inside diameter, and a height $h = 320$ mm, is placed on a rigid support at A. Rigid plate B is placed on top of the cylinder as shown. The threaded steel ($E = 200$ GPa) bolt marked S, with a diameter of 30 mm and a pitch (advance of bolt per complete turn of nut) of 0.5 mm, is securely fastened to the support at A and passes freely through rigid plate B. After the nut at the top of the threaded bolt is snugly fitted, it is wrench-tightened two full turns. Determine the normal stresses in the steel bolt and in the aluminum cylinder. The thickness of the rigid plate at B is negligibly small.

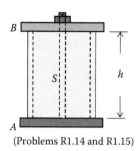

(Problems R1.14 and R1.15)

R1.15 In reference to Problem R1.14, after the nut at the top of the steel bolt is snugly fitted, no further tightening occurs but the temperature of the aluminum cylinder and the steel bolt is increased by $\Delta T = 30°C$. Determine the normal stresses in the steel bolt and in the aluminum cylinder. Let $\alpha_S = 11.7 \times 10^{-6}/°C$ and $\alpha_A = 23.4 \times 10^{-6}/°C$. All other information and data given in Problem R1.14 remains the same.

R1.16 The plate shown has two semicircular grooves of radius r_1, two circular fillets of radius r_1, and a circular hole of radius r_2, all symmetrically placed with respect to its central axis. If the allowable stress is 30 ksi, determine the largest permissible force P. Let $r_1 = 1.5$ in., $r_2 = 1.0$ in., $D = 6.0$ in., and $t = 1.5$ in.

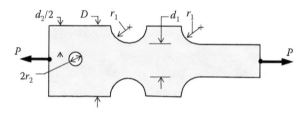

R1.17 A rigid plate AB is placed on top of a steel ($E = 200$ GPa) cylinder of length $L = 0.75$ m, outside diameter of 240 mm, and inside diameter of 180 mm that rests on a rigid surface as shown. A weight W is dropped from a height $h = 30$ mm on top of rigid plate AB. The weightless postmarked P serves as a guide for the dropping weight and is centrally located with respect to the cylinder. If the axial stress in the cylinder cannot exceed 200 MPa, determine the largest permissible weight W.

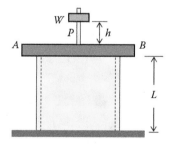

2 Torsional Loads

2.1 INTRODUCTION

In Chapter 1, we focused on the analysis of members subjected to axial loads and learned how to determine the stresses and deformations produced by such loads. In this chapter, we will discuss members subjected to torsional loads, namely, moments (couples) about the centroidal axes of such members. These types of moments or couples are generally referred to as *torques* and the members subjected to torques are generally known as *shafts*. Shafts are encountered often in practice in such engineering applications where it is necessary to transmit power from one location to another. We encounter such applications, for example, in the propeller shafts of ships and aircraft as well as in the drive shafts of automobiles and power tools and other equipment. As an example of such applications, consider the case of an electric motor driving a power tool as shown in Figure 2.1. The connecting shaft AB serves to transmit the power from the electric motor to the power tool. Thus, shaft AB is subjected to couples (torques Q) about its centroidal x axis that may be represented by curved arrows as shown Figure 2.2a. However, in dealing with three-dimensional drawings, it is more convenient, less ambiguous, and easier to interpret if these torques are represented by double-headed vectors along the x axis as shown in Figure 2.2b. In constructing and interpreting these vectors, use is made of the *right-hand rule*. Thus, for example, in the case of the double-headed vector at A, placing the thumb of the right hand along the vector, the curved four fingers point in the direction of rotation that the torque Q would produce. Therefore, in most instances when dealing with three-dimensional drawings, the vector approach, along with the right-hand rule, is used in this text. There are exceptions, of course, especially in certain two-dimensional drawings, where the use of the curved arrow is advantageous.

While different shapes may be used for the cross-sectional areas of shafts, emphasis in this chapter is given to shafts of circular cross sections, both solid and hollow. However, for the sake of completeness, a discussion is given about shafts of noncircular cross sections in Section 2.9. In Section 2.2, we discuss the topic of internal torques and internal torque diagrams; in Section 2.3, we develop the equations for determining the shearing stresses and angles of twist in circular shafts; in Section 2.4, we deal with the problem of statically indeterminate shafts; in Section 2.5, we talk about shafts that are used for the transmission of power; in Section 2.6, we discuss the topic of combined loads; in Section 2.7, we deal with stress concentrations in circular shafts; in Section 2.8, we introduce the topic of impact loading of circular shafts; in Section 2.9, as stated above, we analyze shafts of noncircular cross sections using analytical as well as the experimental membrane analogy procedure, and finally, in Section 2.10, we introduce the concept of elastoplastic behavior of shafts.

2.2 INTERNAL TORQUE

2.2.1 CONCENTRATED TORQUE

A concentrated torque may be defined as one that is applied over a very small area (assumed to be a point) of the member on which it acts. In practice, however, the torque, of necessity, is applied over a finite area, and the assumption of point application is only an idealization used to simplify the analysis.

Consider, for example, the shaft AB subjected to the torques Q as shown in Figure 2.3a. The torques Q in this case are assumed to be concentrated at specific points on the shaft. The vector Q acting on the circular plane at A is assumed to act at the center point of this circular area. The right-hand rule tells us that this double-headed vector implies a torque Q acting in the circular yellow area at A, as indicated by the broken curved arrow.

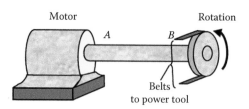

FIGURE 2.1   Example of a motor transmitting power, through shaft AB, to a power tool.

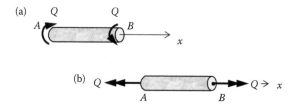

FIGURE 2.2   Representation of a three-dimensional torque Q by a double-headed arrow. Interpretation is made by the right-hand rule.

Let us assume that we were interested in the internal torque T_C at any position such as C in the shaft. To find this internal torque, we do the same thing we did in the case of axially loaded members and cut the shaft at the chosen position in order to expose the cross-sectional area of interest, as well as the resultant torque acting on it. A free-body diagram is, of course needed, and in this particular case, the portion of the shaft to the right of position C (the RFB) is used as shown in Figure 2.3b to determine the internal torque T_C at C. Equilibrium is then applied by summing moment (torques) about the x axis. Thus,

$$\sum T_x = 0 \Rightarrow Q - T_C = 0; \quad T_C = Q$$

Note that, in this case, the internal torque is a constant equal to Q throughout the length of the shaft. This conclusion is reached by selecting any other position and applying the condition of equilibrium to the resulting free-body diagram as was done for position C. It should be noted that the internal torque T at any position such as C, whether indicated as a double-headed vector or a curved arrow, is nothing more than the resultant of a shearing stress distribution acting over the entire circular yellow area at C. The distribution of the shearing stresses on cross sections of shafts of circular shapes will be discussed later in Section 2.3.3.

2.2.2   SIGN CONVENTION

As in the case of axially loaded members, in determining the internal torque at any position in a member, we make use of two distinct sign conventions. The first is a *physical sign convention* that states that a torque is positive, if the two-headed vector representing it points away from the surface

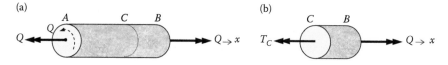

FIGURE 2.3   Representation of a concentrated torque Q at the center point in a circular shaft.

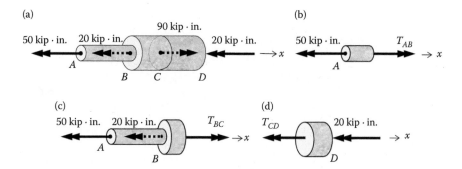

FIGURE 2.4 Shaft subjected to concentrated torques at a number of positions along its length, and determination of internal torques in segments *AB*, *BC*, and *CD* using equilibrium.

on which it acts, and negative, if it points into this surface. Thus, all the torques shown in Figure 2.3 are positive by the physical sign convention. The second sign convention is the arbitrary *equations sign convention*. Once a free-body diagram is constructed using the physical sign convention, and a convenient coordinate system established, the equations sign convention is used to apply the equations of equilibrium. It is strongly recommended that when using the physical sign convention, all of the unknown torques be assumed positive on the free-body diagram. The solution of the equilibrium equations will then tell us if these assumptions are correct.

Consider now the case of a shaft subjected to concentrated torques not only at its ends, but also at intermediate points as shown, for example, in Figure 2.4a. To determine the internal torque in segment *AB*, we need to cut the shaft at any position between *A* and *B* and construct either the LFB or the RFB diagram as shown in Figure 2.4b, where the LFB diagram was chosen. To determine the internal torque in segment *BC*, we cut the shaft at any position between *B* and *C* and construct a free-body diagram as shown in Figure 2.4c where an LFB diagram was, once again, chosen. Finally, to determine the internal torque in segment *CD*, we cut the shaft at any position between *C* and *D* and construct a free-body diagram as shown in Figure 2.4d, where an RFB diagram was selected. Note that in all of the three cases, the unknown torque was assumed to be positive according to the physical sign convention. Now that the free-body diagrams are completed and an arbitrarily selected coordinate *x* axis established with origin at *A* as shown, we proceed to apply the equilibrium equation, $\sum T_x = 0$. This equation is applied to each of the three cases shown in Figures 2.4b through 2.4d, in that order, to determine the internal torques T_{AB}, T_{BC}, and T_{CD}. Thus,

$$\sum T_x = 0 \Rightarrow T_{AB} - 50 = 0; \quad T_{AB} = 50 \text{ kip} \cdot \text{in.}$$

$$\sum T_x = 0 \Rightarrow T_{BC} - 20 - 50 = 0; \quad T_{BC} = 70 \text{ kip} \cdot \text{in.}$$

$$\sum T_x = 0 \Rightarrow -20 - T_{CD} = 0; \quad T_{CD} = -20 \text{ kip} \cdot \text{in.}$$

The minus sign found for the internal torque T_{CD} indicates that this torque does not point away from its surface as assumed, but into it.

2.2.3 DISTRIBUTED TORQUES

Occasionally, we encounter shafts subjected to torques that are distributed in some fashion along their lengths. Such is the case with shaft *AB* shown in Figure 2.5a, where the torque intensity *q*, measured in terms of units of torque per unit of length is, in this case, constant at 10 kip·ft/ft over the entire length of the shaft. To determine the internal torque *T*, at any position a distance *x* from

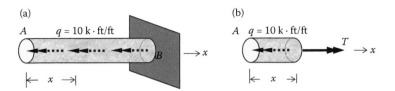

FIGURE 2.5 Example of a constantly distributed torque $q = 10 \, \text{k} \cdot \text{ft/ft}$ over the entire length of the shaft AB and determination, by equilibrium, of internally distributed torque at any position a distance x from end A.

the origin at A, we cut the shaft at the position of interest and construct a convenient free-body diagram as shown in Figure 2.5b. Equilibrium is then applied to determine the required internal torque T. Thus, using the free-body diagram of Figure 2.5b, we have

$$\sum T_x = 0 \Rightarrow T - qx = 0; \quad T = qx = 10x \, \text{kip} \cdot \text{ft}$$

Instead of q being a constant, as was assumed above, let us assume that it varies with x in some manner dictated by the function $q = f(x)$ as shown in Figure 2.6a. In order to find the internal torque T at any position along the shaft, we isolate a small segment dx and construct its free-body diagram as shown in Figure 2.6b. We then apply the condition of equilibrium along the x axis. Thus,

$$\sum F_x = 0 \Rightarrow T + dT - qdx - T = 0; \quad dT = qdx$$

Therefore,

$$\int_1^2 dT = \int_1^2 qdx$$

Integrating and applying the limits, we obtain

$$\Delta T_{1-2} = T_2 - T_1 = \int_1^2 qdx \qquad \textbf{(2.1)}$$

Equation 2.1 allows us to find the change in the internal torque between any two arbitrary positions such as 1 and 2 in Figure 2.6. Obviously, the function $q = f(x)$ must be known in order to perform the indicated integration.

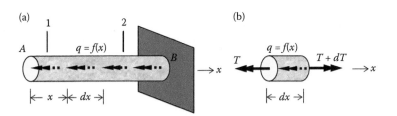

FIGURE 2.6 Example of a variably distributed torque $q = f(x)$ over the length of shaft AB and determination, using equilibrium, of internally distributed torque at any position in the shaft.

It is sometimes convenient to construct what is known as the *internal torque diagram*, which is a diagram that shows, at a glance, how the internal torque changes from point to point along the length of the shaft. Such a diagram is constructed by first determining the internal torque at several positions along the member, including relations that may exist between the torque and the variable x in segments where distributed torques exist. Once this is accomplished, a graph is constructed with the internal torque on a vertical axis and the position along the shaft, measured by the variable x, on a horizontal axis. The resulting points are then connected by straight lines and/or curves to complete the construction. The construction of internal torque diagrams is illustrated in Examples 2.2 and 2.3.

EXAMPLE 2.1

A shaft is subjected to a concentrated as well as a distributed torque as shown in figure (a) below. Find the length L so that the shaft is in equilibrium and develop an expression for the internal torque for any position defined by x.

SOLUTION

Using the entire shaft (figure (a) below) as a free-body diagram and applying the equation of equilibrium, we have

(a)

$q = 0.5x^2$ kN $\cdot$ m/m

5 kN $\cdot$ m

L (m)

x

(b)

$q = 0.5x^2$ kN $\cdot$ m/m

5 kN $\cdot$ m

T

x

$$\sum T_x = 0 \Rightarrow \int_0^L 0.5x^2 dx - 5 = 0; \quad \frac{0.5L^3}{3} = 5; \quad \Rightarrow L = 3.107 \text{ m} \qquad \textbf{ANS.}$$

The free-body diagram shown in figure (b) above is constructed and equilibrium is applied. Thus,

$$\sum T_x = 0 \Rightarrow T + \int_0^x 0.5x^2 dx - 5 = 0; \quad T = 5 - 0.167x^3 \text{ kN} \cdot \text{m} \qquad \textbf{ANS.}$$

EXAMPLE 2.2

Construct the internal torque diagram for the shaft shown in figure (a) below. Let $q = 4$ kip $\cdot$ ft/ft.

SOLUTION

As shown in figure (b) below, the internal torque diagram is conveniently placed directly under the shaft. The unknown torque Q at D is determined by using the entire shaft as a free-body diagram and applying the equation of equilibrium, $\sum T_x = 0$, which leads to $Q = -15$ kip $\cdot$ ft.

The internal torque diagram is obtained by creating three free-body diagrams after cutting the shaft at three locations: the first is any location between A and B (i.e., $0 < x < 2$ ft), the second is any place between B and C (i.e., $2 < x < 5$ ft), and the third is any position between C and D (i.e., $5 < x < 10$ ft). These three free-body diagrams are shown in figure (c) below along with the values of the unknown torques T, which were determined by using the equilibrium equation $\sum T_x = 0$. Note that, in all cases, the origin of the x coordinate is at A. Thus,

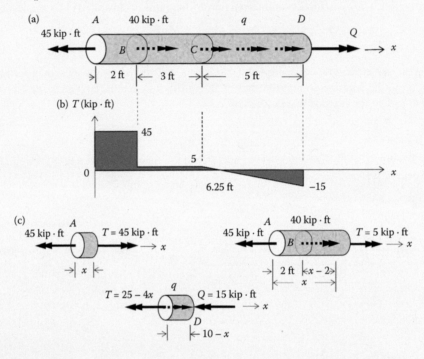

Segment AB ($0 < x < 2$ ft)

$$\sum T_x = 0 \Rightarrow T - 45 = 0; \quad T = 45 \text{ kip} \cdot \text{ft}$$

Segment BC ($2 < x < 5$ ft)

$$\sum T_x = 0 \Rightarrow T + 40 - 45 = 0; \quad T = 5 \text{ kip} \cdot \text{ft}$$

Segment CD ($5 < x < 10$ ft)
(Note that the RFB diagram was chosen because it is easier to construct and to use than the LFB diagram.)

$$\sum T_x = 0 \Rightarrow -T + 4(10 - x) - 15 = 0; \quad T = (25 - 4x) \text{ kip} \cdot \text{ft}$$

The internal torque diagram has a discontinuity at $x = 2$ ft where the torque drops abruptly from 45 kip·ft to 5 kip·ft. This is due to the fact that the 40 kip·ft torque is assumed to act at a point. In practice, however, a torque is applied over a finite area and the transition from 45 kip·ft to 5 kip·ft is a gradual one. A second discontinuity (slope) occurs at $x = 5$ ft where the slope changes abruptly from zero to -4 kip·ft/ft.

EXAMPLE 2.3

The shaft shown in the sketch, in a two-dimensional view, is subjected to several torques as indicated. Let $q = 9x^2$, where q is in $kN \cdot m/m$ and x (origin at A) is in meters, and construct the internal torque diagram.

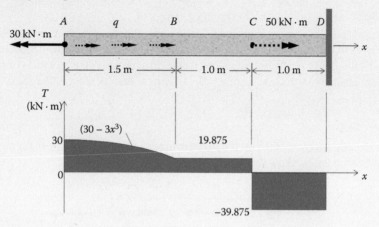

SOLUTION

The solution to this example is obtained in a manner similar to that used in Example 2.2. Cutting planes are passed through three positions along the shaft, and corresponding free-body diagrams are constructed as indicated below. The first free-body diagram corresponds to a cutting plane through any position between A and B ($0 < x < 1.5$ m), the second through any position between B and C ($1.5 < x < 2.5$ m), and the third through any position between C and D ($2.5 < x < 3.5$ m). The equation of equilibrium, $\sum T_x = 0$, is then applied to each free-body diagram to obtain the indicated unknown internal torque T. Thus,

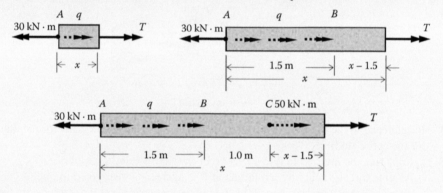

Segment AB ($0 < x < 1.5$ m)

$$\sum T_x = 0 \Rightarrow T + \int_0^x 9x^2 dx - 30 = 0; \quad T = 30 - 3x^3 \ kN \cdot m$$

Segment BC ($1.5 < x < 2.5$ m)

$$\sum T_x = 0 \Rightarrow T + \int_0^{1.5} 9x^2 dx - 30 = 0; \quad T = 19.875 \ kN \cdot m$$

Segment CD ($2.5 < x < 3.5$ m)

$$\sum T_x = 0 \Rightarrow T + 50 + \int_0^{1.5} 9x^2 dx - 30 = 0; \quad T = -39.875 \text{ kN} \cdot \text{m}$$

The internal torque diagram is placed directly under the sketch of the shaft as shown above. Once again, note the discontinuity that exists at $x = 2.5$ m due to the 50-kN·m concentrated torque. A discontinuity of a different nature (slope discontinuity) is observed at $x = 1.5$ m due to the abrupt change in the slope from $q = 9x^2$ to zero at that location in the shaft.

PROBLEMS

2.1   Let the torques $Q_1 = -60$ kip·ft and determine the internal torques in segments AB and BC. In both cases, show free-body diagrams.

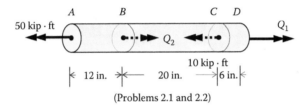

(Problems 2.1 and 2.2)

2.2   In reference to the figure shown in Problem 2.1, let the applied torque $Q_2 = 40$ kip·ft. and determine the internal torques in segments BC and CD. In both cases, show free-body diagrams.

2.3   Determine the reactive torque at the fixed support and the internal torques in segments AB and BC. Show all free-body diagrams.

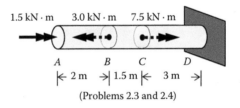

(Problems 2.3 and 2.4)

2.4   In reference to the figure shown in Problem 2.3, construct the internal torque diagram for the entire shaft.

2.5   a. Construct the internal torque diagram for the shaft loaded as shown.

b. Assume that the concentrated torque at B is, instead, a distributed torque of constant intensity q over a central span of 0.5 ft. Find the intensity q of this distributed torque and construct the internal torque diagram for this second case. Compare the two diagrams and comment on the question of reducing the span over which q acts from 0.5 ft to zero.

(Problems 2.5 and 2.6)

2.6   In reference to the figure shown in Problem 2.5, assume that the two concentrated end torques at A and C are, instead, distributed torques of constant intensity q, each over a span of 0.5 ft. Determine this intensity q and construct the corresponding internal torque diagram for the entire shaft. Compare this diagram to that obtained in part (a) of Problem 2.5 and comment on the question of reducing the spans over which q acts, at both A and C, from 0.5 ft to zero.

2.7   a. Find an expression for the internal torque in segment BC of the shaft if $q = 20$ kN · m/m and determine the resisting torque at the fixed support at C.

   b. Draw the internal torque diagram for the entire shaft.

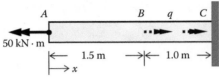

(Problems 2.7 and 2.8)

2.8   In reference to the figure shown in Problem 2.7:

   a. Find an expression for the internal torque in segment BC of the shaft if $q = 100(x - 1.5)^2$ kN · m/m, and determine the resisting torque at the fixed support at C.

   b. Draw the internal torque diagram for the entire shaft.

2.9   The stepped shaft is subjected to a concentrated torque and two distributed torques as shown. Let $Q = 50$ kip · ft, $a = 2.5$ ft, $b = 5.0$ ft, $q_1 = 6x_1$, and $q_2 = 3x_2^2$, both expressed in kip · ft/ft. Develop expressions for the internal torque in segments AB and BC as functions of x_1 and x_2, respectively. Plot the internal torque diagram for the entire shaft.

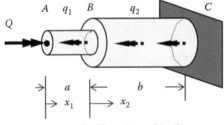

(Problems 2.9 and 2.10)

2.10   In reference to the figure shown in Problem 2.9, the stepped shaft is subjected to a concentrated torque and two distributed torques as shown. Let $Q = 40$ kN · m, $a = 2.0$ m, $q_1 = 6x_1^2$, and $q_2 = 10x_2$, both expressed in kN · m/m. Develop expressions for the internal torque in segments AB and BC as functions of x_1 and x_2, respectively. Find the dimension b if the torque at the fixed support is 101 kN · m. Plot the internal torque diagram for the entire shaft.

2.11   The airplane wing has a wing span $L = 50$ ft from wing tip to the fuselage as shown. In addition to bending and shear, it is subjected to a distributed external torque $q = 250 + 3x^2$ where q is in lb · ft/ft and x is in ft. Determine the internal torque T as a function of x and plot the internal torque diagram.

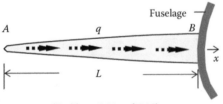

(Problems 2.11 and 2.12)

2.12 In reference to the figure shown in Problem 2.11, the airplane wing has a wing span L from wing tip to the fuselage as shown. In addition to bending and shear, it is subjected to a distributed internal torque $q = 1.2 - 0.2x$, where q is in kN·m/m and x is in m. Determine the internal torque T as a function of x. Determine the maximum permissible length L if the torque at the fuselage cannot exceed 50.0 kN·m. Plot the internal torque diagram for the entire wing.

2.3 STRESSES AND DEFORMATIONS IN CIRCULAR SHAFTS

As noted in the introductory statements, Chapter 2 deals primarily with shafts of circular cross sections. The reason for this is the fact that, mathematically, the torsion of circular shafts represents the simplest of all torsion problems. Later in Section 2.9, some analyses will be provided for shafts of noncircular cross sections.

2.3.1 SHEARING STRAIN

As was stated in Chapter 1, the term *distortion* signifies a change in the initial shape of the body. A more precise definition of distortion, however, is needed, and for this purpose, the change in angle between two initially perpendicular lines is used. The change in the 90° angle between the two line elements n and t, as shown in Figure 2.7, represents a distortion and is given the name *shearing strain* denoted by the symbol γ_{nt}. The rectangular stress element $OABC$ shown in Figure 2.7, becomes the rhombus $OA'B'C$ under the action of the shearing stresses τ_{nt} and τ_{tn}. Thus, the shearing strain γ_{nt} represents the angle between the initial orientation of the t axis and its final orientation t', or the *change in the 90° angle* that existed between the n and t axes. For very small distortions, the angle γ_{nt} may be assumed equal to the tangent of this angle. Thus, referring to Figure 2.7, we have

$$\gamma_{nt} \approx \tan\gamma_{nt} = \frac{AA'}{OA} \tag{2.2}$$

As indicated in Equation 2.2, the shearing strain γ_{nt} is a dimensionless quantity since it represents the ratio of two lengths. However, sometimes it is expressed in such units as in./in. in the U.S. Customary system of units and m/m in the SI system of units.

If the x and y axes coincide with the n and t axes, respectively, then the symbol for the shearing strain becomes γ_{xy}. In a three-dimensional case of strain referred to an x–y–z right-handed coordinate system, two other shearing strains exist symbolized by γ_{yz} and γ_{xz}.

As in the case of tension and compression in the elastic range for the material, a linear relation exists between the shearing strain γ and the shearing stress τ given by what is known as *Hooke's*

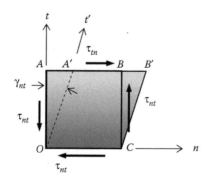

FIGURE 2.7 Definition of shearing strain γ_{nt} as being approximately equal to the tangent of the shearing angle γ_{nt}.

law in shear. A detailed discussion of properties of materials in shear is given later in this section, but for the moment, we state Hooke's law without elaboration. Thus,

$$\tau = G\gamma \tag{2.3}$$

In Equation 2.3, G is the proportionality factor between the shearing stress and the shearing strain known as the *modulus of rigidity*. As seen from Equation 2.3, since γ is dimensionless, G has the same units as τ.

2.3.2 SIGN CONVENTION

The sign convention used in this text for shearing strain is such that *a positive shearing strain represents a decrease in the* 90° *angle* and *a negative shearing strain represents an increase in this angle*. This sign convention is consistent with the sign convention adopted earlier for shearing stresses. Thus, for example, the shearing strain γ_{nt} shown in Figure 2.7 is positive since it represents a decrease in the 90°.

2.3.3 SHEARING STRESS AND SHEARING DEFORMATION

Consider the case of the shaft shown in Figure 2.8, which is subjected to the torque Q at one end and fixed rigidly at the other end. Certain assumptions are made in the development of the governing equations. The most significant assumptions are:

1. Cross sections such as those at D and E in Figure 2.8 are plane prior to twisting and remain plane after twisting of the circular shaft.
2. Straight line elements on the surface of the shaft such as line AB are assumed to remain straight after twisting occurs. Point A moves to A' and line $A'B$ is assumed to be straight for small angles θ even though its true shape is helical.
3. The material is assumed to behave elastically in order to be able to use Hooke's law.

Figure 2.8 shows the distortions that would occur once the torque Q is applied. Consider line element AB on the surface of the shaft. Point A moves to A', and radius OA rotates through the angle θ to OA'. This angle θ is known as the *angle of twist*. However, point B at the fixed end of the shaft is not free to move. Thus, at a radius R (on the outside surface), the shaft experiences a distortion (shearing strain) γ_R, which is the maximum possible distortion. At a radius $\rho < R$, point C moves to C' and the distortion that takes place is $\gamma_\rho < \gamma_R$. Since the angle θ as well as the distortions γ_R and γ_ρ are very small, the length of the curved lines $AA' = L\gamma_R = R\theta$. Thus, $\gamma_R/R = \theta/L$. Also, the length of the curved line $CC' = L\gamma_\rho = \rho\theta$ and $\gamma_\rho/\rho = \theta/L$. Equating these two values of θ/L, we obtain

$$\frac{\gamma_\rho}{\rho} = \frac{\gamma_r}{r} = \frac{\theta}{L} = k \tag{2.4}$$

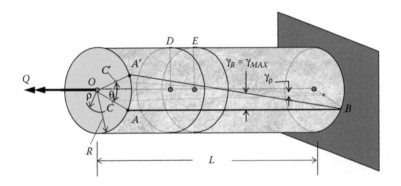

FIGURE 2.8 Shaft showing distortion under the influence of a torque Q and the definition of the angle of twist θ.

Thus, the quantity k in Equation 2.4 is a constant equal to ratio θ/L. Solving for γ_ρ and γ_R in terms of k, we obtain

$$\left.\begin{array}{l} \gamma_\rho = k\rho \\ \gamma_R = kR \end{array}\right\} \tag{2.5}$$

The first of Equations 2.5 show that the shearing strain is proportional to the radial distance ρ and increases from zero at the center of the shaft to its maximum value, kR, at the outside surface.

From Hooke's law, Equation 2.3, we conclude that $\gamma = \tau/G$. Substituting this relation for γ in Equations 2.5, we obtain

$$\left.\begin{array}{l} \tau_\rho = (kG)\rho \\ \tau_R = (kG)R \end{array}\right\} \tag{2.6}$$

where the product $kG = \theta G/L$ (see Equation 2.4) is also a constant. The first of Equations 2.6 shows that the shearing stress is proportional to the radial distance ρ and increases from zero at the center of the shaft to its maximum value, $(kG)R$, at the outside surface. Such a stress distribution is depicted in Figure 2.9, which shows the cross section at some position along a circular shaft of radius R where the resultant internal torque is T. This torque results in the linear shearing stress distribution shown where the maximum shearing stress, τ_{MAX}, according to the second of Equations 2.6, is $\tau_R = (kG)R$.

Let us now relate the internal torque T to the resulting shearing stress in the cross section of Figure 2.9. The annular differential area (shown in gray) $dA = (2\pi\rho)d\rho$. The differential force acting over this area is $dF = \tau_\rho\, dA$ and the corresponding differential torque is $dT = \tau_\rho\rho\, dA$. Thus,

$$\left.\begin{array}{l} T = \int \tau_\rho\rho\, dA = 2\pi \int_0^R \tau_\rho\rho^2 d\rho = 2\pi \int_0^R \left(\frac{\tau_\rho}{\rho}\right)\rho^3\, d\rho \\[4mm] T = 2\pi\left(\frac{\tau_\rho}{\rho}\right)\int_0^R \rho^3 d\rho = \left(\frac{\tau_\rho}{\rho}\right)\left(\frac{\pi R^4}{2}\right) = \left(\frac{\tau_\rho}{\rho}\right)J \end{array}\right\} \tag{2.7}$$

Note that in Equation 2.7, the ratio (τ_ρ/ρ) was taken out of the integral sign because, by the first of Equations 2.6, it is equal to kG, which is a constant. Note also that the quantity $(\pi R^4/2)$ was replaced by the symbol J, which represents the *polar moment of inertia* for the circular cross section.

If we solve the second part of Equation 2.7 for the shearing stress τ_ρ, we obtain

$$\tau_\rho = \frac{T\rho}{J} \tag{2.8}$$

Equation 2.8 may be used to find the shearing stress at any point within a circular cross section of a shaft, if we know the internal torque T, the radial position of the point ρ, and the polar moment

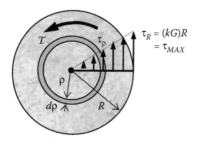

FIGURE 2.9   Circular cross section of a circular shaft showing the stress distribution due to a torque T.

of inertia for the circular cross section J. Of course, the maximum value of the shearing stress, as stated earlier, occurs at the outside surface of the shaft (i.e., at $\rho = R$) and is given by

$$\tau_{MAX} = \tau_R = \frac{TR}{J} \tag{2.9}$$

Let us now refer to Equation 2.6, which states that $\tau_\rho = (kG)\rho$. Since as stated above, $kG = \theta G/L$ and by Equation 2.8, $\tau_\rho = T\rho/J$, we conclude that $T\rho/J = \theta G\rho/L$. Simplifying and solving for the angle of twist, θ, we obtain

$$\theta = \frac{TL}{JG} \tag{2.10}$$

From Equation 2.10, we note that $\theta/L = T/JG = k$ and, therefore, it is, in fact, a constant as stated earlier because for a given set of conditions and at a given position in the shaft, the quantities T, J, and G are all constants.

Equation 2.10 may be used to find the angle of twist θ for a given shaft of length L if the internal torque T and the properties J and G are constant over this length. If, however, as shown in Figure 2.10, the shaft consists of a number of segments in which any of the quantities T, J, or G differs from segment to segment, then the angle of twist must be determined separately for each segment and the results added algebraically to obtain the total angle of twist. This process of algebraic summation is expressed by

$$\theta = \sum \frac{TL}{JG} \tag{2.11}$$

For the conditions stated in Figure 2.10, for example, Equation 2.11 becomes

$$\theta_A = \frac{T_1 L_1}{J_1 G_1} + \frac{T_2 L_2}{J_2 G_2} + \frac{T_3 L_3}{J_3 G_3}$$

The summation in the above equation is understood to be algebraic.

It should also be emphasized that Equation 2.10 can only be used for an entire shaft, or a segment of a shaft, where all of the quantities T, J, and G are constants over the entire shaft, or over the segment thereof. For cases where any or all of these quantities vary with the longitudinal coordinate x, the differential angle of twist $d\theta$ may be written as

$$\left. \begin{aligned} d\theta &= \frac{T\,dx}{JG} \\ \theta &= \int \frac{T\,dx}{JG} \end{aligned} \right\} \tag{2.12}$$

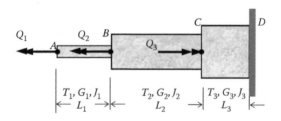

FIGURE 2.10 Shaft consisting of three component parts, each having its own properties T, G, and J, showing that the total angle of twist is the sum of three angles of twist.

2.3.4 Material Properties in Shear

As in the case of tension and compression, materials may be tested in the laboratory to obtain their properties in shear. Mechanical properties in shear for a given material are generally obtained by conducting a torsion test in the laboratory on a cylindrical specimen of that material. During the test, the specimen is subjected to increasing torques and measurements are made of the applied torques and corresponding angles of twist over a specified gage length. Solid circular specimens are often tested, although thin-walled hollow tubes have the advantage of a practically uniform shearing stress across a small wall thickness.

A typical shearing stress–shearing strain diagram is shown in Figure 2.11. The linear portion of such a diagram (segment OP) is obtained by transforming the torque and angle of twist data obtained in the laboratory into shearing stress and shearing strain values, respectively. This is accomplished by using Equation 2.9, $\tau_R = TR/J$, to obtain the shearing stress values from the known quantities T, R, and J. The shearing strains are obtained by using the second of Equations 2.5, $\gamma_R = kR = (\theta/L)R$ from the known quantities θ, R, and L, where L is the specimen gage length.

The initial segment OP of the diagram is linear and represents the elastic behavior of the material. Its slope provides the *modulus of rigidity G* for the material and is described by *Hooke's law* in shear as expressed earlier in Equation 2.3. Note that Equation 2.3 is only valid up to τ_p, the *shearing proportional limit* of the material. Thus,

$$\tau = G\gamma \tag{2.3}$$

As stated earlier, the quantity G is the proportionality factor between the shearing stress and the shearing strain and represents a significant property for the material known as the *modulus of rigidity* or the *modulus of elasticity in shear*. Other mechanical properties similar to those discussed in Section 1.4 for tension and compression, such as yield and ultimate strengths and modulus of resilience are determined for the case of shear using a torsion test. Some of these properties are tabulated in Appendix D.

Segment PU of the diagram in Figure 2.11 shows the general trend of the inelastic or plastic behavior of the material. Point U represents the stage at which the specimen ruptures and the corresponding torque is known as the *rupture* or *ultimate torque* T_U. A property known as the *modulus of rupture*, τ_U, is obtained from Equation 2.9 using the ultimate torque T_U. Thus,

$$\tau_U = \frac{T_U R}{J} \tag{2.13}$$

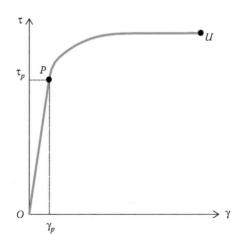

FIGURE 2.11 Shearing stress–strain diagram for a given ductile material, defining the properties: modulus of rigidity G and the modulus of rupture τ_U.

Note that the modulus of rupture, τ_U, is not the true value of the ultimate stress when the specimen ruptures even though the torque used to determine it is the ultimate torque. This is so because Equation 2.9 is valid only within the elastic range for the material. However, the modulus of rupture is a measure of the torsional strength of the material and provides an index for comparing materials for given torsion applications.

Another useful property is the *torsional toughness*, u_T, which is defined as the work per unit volume required to rupture the specimen. The area under curve *OPU* in Figure 2.11 represents the torsional toughness. It is another approximate criterion used in comparing materials for a given torsion application.

It should be noted here that the *design concepts* discussed in Chapter 1 about *allowable stresses* and *factors of safety* are as applicable in the case of members subjected to torsion as they are in the case of members subjected to axial loads. The student is urged to review these concepts, which are discussed in Section 1.4.

2.3.5 STRESS ELEMENT

A segment of length dx is separated from a circular shaft as shown in Figure 2.12a. As a consequence of the torque Q applied to the shaft, a system of shearing stresses is created on the two faces of the segment of differential length dx (i.e., on planes normal to the x axis, namely, the X planes) as shown. This stress system varies from a magnitude τ on the outside surface of the shaft to zero at its center. The stress element, shown in gray, between these two X planes is isolated as shown in Figure 2.12b, where, in addition to the two shearing stresses, τ_{xy}, on the X planes, the two shearing stresses, τ_{yx}, were added on the Y planes (planes normal to the y axis) in order to maintain rotational balance. As proven in Section 1.3, the magnitude of the shearing stress τ_{yx} is equal to that of the shearing stress τ_{xy}. Thus, not only planes perpendicular to the longitudinal axis of a twisted shaft, but also its longitudinal planes (planes that contain this axis) are subjected to shearing stresses of equal magnitude. The stress element shown in Figure 2.12b is a two-dimensional stress condition, and the element is said to be in a state of *pure shear*.

2.3.6 STRESSES ON INCLINED PLANES

It is often useful to be able to determine the stresses in a twisted shaft on planes inclined to those that are perpendicular to the axis of the shaft. Consider the stress element shown in Figure 2.13a, between

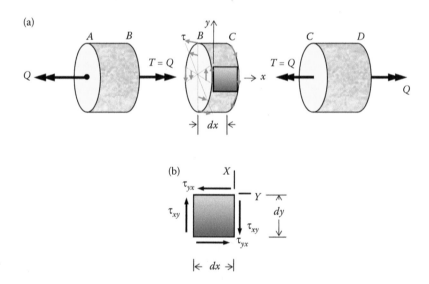

FIGURE 2.12 Diagram showing the state of pure shear in a shaft subjected to a torque Q.

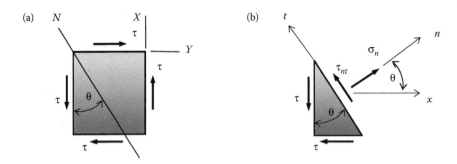

FIGURE 2.13   (a) The state of pure shear and (b) determination of normal and shearing stresses on inclined planes.

the X and Y planes, which is in pure shear. Since $\tau_{yx} = \tau_{xy}$, we will use the simpler notation τ for both of them as shown. It is required to find the normal and shearing stresses on plane N, which is inclined by the counterclockwise angle θ from the X plane. To determine these stresses, we isolate the wedge contained within the X, Y, and N planes and construct its free-body diagram as shown in Figure 2.13b. We then apply the conditions of equilibrium along the n and t axes. In order to apply these equations, we need to transform the stresses acting on the wedge into forces, a process that requires multiplying the stresses by the areas over which they act. If we assume the area of the inclined plane (the N plane) to be A_n, then the area of plane X, $A_x = A_n \cos \theta$, and that of the Y plane, $A_y = A_n \sin \theta$. Thus,

$$\sum F_n = 0 \Rightarrow \sigma_n A_n - \tau(A_n \cos \theta) \sin \theta - \tau(A_n \sin \theta) \cos \theta = 0$$

Therefore,

$$\sigma_n = 2\tau \sin \theta \cos \theta = \tau \sin 2\theta \tag{2.14}$$

$$\sum F_t = 0 \Rightarrow \tau_{nt} A_n - \tau(A_n \cos \theta) \cos \theta + \tau(A_n \sin \theta) \sin \theta = 0$$

Thus,

$$\tau_{nt} = \tau(\cos^2 \theta - \sin^2 \theta) = \tau \cos 2\theta \tag{2.15}$$

Note that in deriving Equations 2.14 and 2.15, we made use of the trigonometric identities $(\sin 2\theta = 2 \sin \theta \cos \theta$ and $\cos 2\theta = \cos^2 \theta - \sin^2 \theta)$. Equation 2.14 shows that σ_n reaches its maximum value $(\sigma_n)_{MAX} = \tau$ when $\theta = 45°$ and a minimum value $(\sigma_n)_{MIN} = -\tau$ when $\theta = -45°$. Equation 2.15 shows that for both of these angles, the shearing stress $\tau_{nt} = 0$. In other words, the shearing stress is zero on planes on which the normal stress is either a maximum or minimum. As we will see in Chapter 7, this statement is true for any plane on which the normal stress is a maximum or a minimum. Such planes are known as *principal planes* and the corresponding stresses as *principal stresses*. Also, Equation 2.15 shows that the shearing stress has its maximum magnitude $(\tau_{nt})_{MAX} = \tau$ when $\theta = 0°$ or $\theta = 90°$. Of course, for either of these two angles, the normal stress $\sigma_n = 0$ as may be ascertained by applying Equation 2.14.

EXAMPLE 2.4

The data tabulated below were obtained in the laboratory during the testing of a hollow steel specimen having an outside diameter of 4.0 in., an inside diameter of 3.6 in., and a gage length of 10.0 in. (a) Plot the torque versus angle of twist diagram and find the modulus of rigidity.

(b) Plot the linear portion of the shearing stress versus shearing strain diagram and determine the modulus of resilience.

Torque (kip·in.)	Angle of Twist (degree)	Torque (kip·in.)	Angle of Twist (degree)	Torque (kip·in.)	Angle of Twist (degree)
0.0	0.0	92.26	0.60	137.0	4.00
15.38	0.10	107.6	0.70	140.0	5.00
30.75	0.20	109.2	0.80	147.5	7.00
46.13	0.30	118.0	1.00	152.3	9.00
61.51	0.40	126.5	2.00	158.0	11.00
76.89	0.50	131.2	3.00		

SOLUTION

- The given data are plotted as shown in figure below. The modulus of rigidity is obtained from Equation 2.10 by using corresponding values of T and θ at one point in the straight segment of the graph. The point with $T = 92.26$ kip·in and $\theta = 0.6°$ is chosen for this purpose. Thus,

$$G = \frac{TL}{\theta J} = \frac{(92.26)(10.0)}{(0.6 \times \pi/180)(\pi/32)(4.0^4 - 3.6^4)}$$

$$G \approx 10.2 \times 10^3 \text{ ksi} \quad \textbf{ANS.}$$

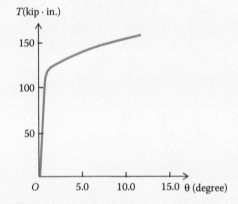

- Since the portion of the graph we have to plot is a straight line, we need only find the coordinates of two points. One of these two points is the origin and the second, any point on the straight segment of the T versus θ graph, preferably the highest point on the straight line segment. The point chosen (highest point, which represents the proportional limit) has the coordinates: $T_p = 107.6$ kip·in., $\theta_p = 0.7°$. Using Equation 2.9, we obtain the maximum shearing stress corresponding to this torque and, since G is now known, we can use Hooke's law, Equation 2.3, to obtain the corresponding maximum shearing strain. Thus,

$$\tau_p = \frac{T_p R}{J} = \frac{(107.6)(2.0)}{(\pi/32)(4.0^4 - 3.6^4)} = 24.898 \text{ ksi}$$

$$\gamma_p = \frac{\tau_p}{G} = \frac{24.898}{10.2 \times 10^3} = 2.441 \times 10^{-3}$$

The required graph is shown in figure below.

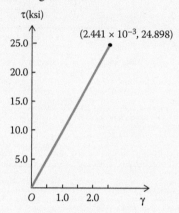

The modulus of resilience is given by the area under the straight line in figure above. Thus,

$$u_R = (1/2)\tau_p\gamma_p = (1/2)(24.898)(2.441 \times 10^{-3})$$
$$u_R = 3.039 \times 10^{-2} \approx 3.0 \times 10^{-2}\,\text{kip}\cdot\text{in./in.}^3 \qquad \textbf{ANS.}$$

EXAMPLE 2.5

A shaft of circular cross section is shown in figure (a) below. Determine (a) the maximum shearing stresses in segments AB (diameter of 1.4 in.) and BC (diameter of 2.0 in.) and (b) the shearing stress distributions in segments AB and BC.

SOLUTION

a. The free-body diagram shown in figure (b) below is now used to determine the internal torque in segment AB. Thus,

$$\sum T_x = 0 \Rightarrow T_{AB} - 10 = 0; \quad T_{AB} = 10\,\text{kip}\cdot\text{in.}$$

Also, for segment AB, $R = 0.7$ in. and $J = \pi(0.7^4)/2 = 0.377148$ in.^{4}. Therefore, using Equation 2.9, we have

$$(\tau_{MAX})_{AB} = \frac{(10)(0.7)}{0.377148} = 18.560 \approx 18.6\,\text{ksi} \qquad \textbf{ANS.}$$

(a)

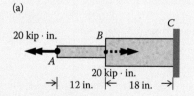

(b)

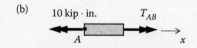

(c)

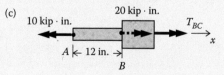

The free-body diagram shown in figure (c) above is used to find the internal torque in segment BC. Thus,

$$\sum T_x = 0 \Rightarrow T_{BC} + 20 - 10 = 0; \quad T_{BC} = -10 \text{ kip} \cdot \text{in.}$$

Also, for segment BC, $R = 1.0$ in. and $J = \pi(1.0^4)/2 = 1.571$ in.[4]. Therefore, using Equation 2.9, we have

$$(\tau_{MAX})_{BC} = \frac{(-10)(1.0)}{1.571} = -6.365 \approx -6.4 \text{ ksi} \qquad \textbf{ANS.}$$

Note that the minus sign on the maximum shearing stress in segment BC simply indicates the sense of that shearing stress. This will be illustrated in part (b) where the stress distributions are shown.

b. The shearing stress distributions for both segments are shown in the diagrams below. These diagrams represent cross sections as viewed from the positive x axis looking toward point A. The proper sense of the shearing stress is determined by applying the right-hand rule to the torque vector and noting the direction in which the four curled fingers are pointed. Thus, since for segment AB (left sketch), the torque vector is positive and points out of the page, the four fingers of the right hand point upward in the first quadrant of the ρ–τ coordinate system. On the other hand, the torque vector is negative and points into the page in segment BC (right sketch) and the four fingers of the right hand point downward in the first quadrant of ρ–τ coordinate system.

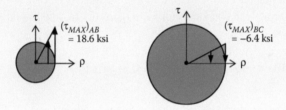

EXAMPLE 2.6

Refer to the shaft of Example 2.5 and determine (a) the angles of twist at A and at B. The material is steel for which $G = 10 \times 10^6$ psi and (b) the normal and shearing stresses on the outside surface of segment AB on a plane that is inclined to cross sections of the shaft at a clockwise angle $\theta = 30°$.

SOLUTION

a. Since the internal torques are constant for the entire length of each segment, the angle of twist is given by Equation 2.11. Thus,

$$\theta_A = \theta_{AB} + \theta_{BC} = \left(\frac{TL}{JG}\right)_{AB} + \left(\frac{TL}{JG}\right)_{BC} = \frac{(10 \times 10^3)(12)}{(0.377148)(10 \times 10^6)} + \frac{(-10 \times 10^3)(18)}{(1.571)(10 \times 10^6)}$$

$$= 0.020 \text{ rad}$$

$$\theta_B = \left(\frac{TL}{JG}\right)_{BC} = \frac{(-10 \times 10^3)(18)}{(1.571)(10 \times 10^6)} = -0.011 \text{ rad} \qquad \textbf{ANS.}$$

The signs on the angles of twist are interpreted by using the right-hand rule. Thus, the plus sign on θ_A indicates that the rotation is clockwise when looking at point A from the positive x axis. The minus sign on θ_B says that the rotation is counterclockwise when looking at B from the positive x axis.

b. Before we embark on the solution of part (b), it will be useful to show the pure-shearing stress element on the outside surface of segment AB. This stress element has two planes parallel and two planes perpendicular to the axis of the shaft as shown in figure (a) below. The shearing stresses τ shown on the element are all of the same magnitude of 18.560 ksi as determined in Example 2.5 and are all negative according to the sign convention stated in Chapter 1. The required plane is also shown and given the symbol N.

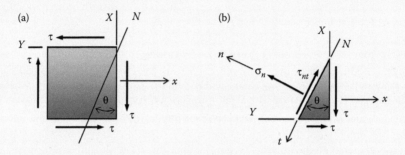

The normal stress on an inclined plane on the outside surface of segment AB is given by Equation 2.14, where $\tau = -18.560$ ksi as found in Example 2.5 and $\theta = 30°$ is negative because it is clockwise. Thus,

$$\sigma_n = -\tau\sin 2\theta \Rightarrow \sigma_n = (-18.560)\sin(-2 \times 30°) = 16.073 \approx 16.1\,\text{ksi} \quad \textbf{ANS.}$$

The shearing stress on this inclined plane is given by Equation 2.15. Thus,

$$\tau_{nt} = -\tau\cos 2\theta \Rightarrow \tau_{nt} = (-18.560)\cos(-2 \times 30°) = -9.28 \approx -9.3\,\text{ksi} \quad \textbf{ANS.}$$

These stresses are shown in figure (b) above on the wedge that is contained within the X, Y, and N planes. Note that the shearing stress on the inclined plane, because it is negative, points in the negative t direction.

EXAMPLE 2.7

- The hollow circular shaft AB (see figure (a) below) is subjected to torques as shown. (a) Construct the internal torque diagram for the entire shaft. (b) Find the maximum shearing stress 0.25 m from point A and sketch the shearing stress distribution across the hollow section. Let the outside and inside diameters be 50 and 30 mm, respectively. (c) Determine the angle of twist at A. The modulus of elasticity for the material is $G = 25$ GPa.

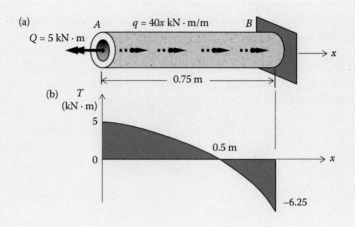

SOLUTION

a. The internal torque diagram is shown in figure (b) above, directly under the sketch of the shaft. This diagram was obtained by first developing the equation of the internal torque as a function of the variable x. To accomplish this, we make use of the free-body diagram shown in figure below. Thus,

$$\sum T_x = 0 \quad \Rightarrow T + \int_0^x 40x \, dx - 5 = 0$$

$$T = 5 - 20x^2$$

This function is plotted as shown in figure (b) above.

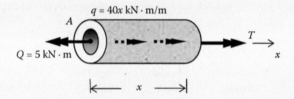

b. The torque at $x = 0.25$ m is found from the internal torque function. Thus,

$$T_{x=0.25} = 5 - 20(0.25^2) = 3.75 \text{ kN} \cdot \text{m}$$

Therefore, by Equation 2.9, we have

$$\tau_{MAX} = \frac{TR}{J} = \frac{3.75(10^3)(0.025)}{(\pi/2)(0.025^4 - 0.015^4)} = 175.539 \approx 175.5 \text{ MPa}$$

$$\tau_{MIN} = \left(\frac{15}{25}\right)(175.539) = 105.323 \approx 105.3 \text{ MPa} \quad \textbf{ANS.}$$

The shearing stress, τ_{MIN}, occurs on the inside surface of the hollow shaft. It was found by taking ratios since the distribution is linear but it could also be found by using Equation 2.9. The required shearing stress distribution is shown in figure below, which was obtained by viewing the cross section from the positive x axis toward point A on the shaft.

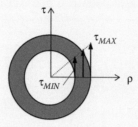

c. The angle of twist at A is found by Equation 2.12. Thus,

$$\theta = \int \frac{T\,dx}{JG} = \int_0^{0.75} \left(\frac{(10^3)(5 - 20x^2)\,dx}{(\pi/2)(0.025^4 - 0.015^4)(25 \times 10^9)} \right)$$

$$= \left(\frac{10^3}{13351} \right) \left[5x - (20/3)x^3 \right]_0^{0.75} = 0.072 \text{ rad} \qquad \textbf{ANS.}$$

PROBLEMS

2.13 The data shown below were obtained during a torsion test of a hollow specimen. The specimen had an outside diameter of 100 mm, an inside diameter of 92 mm, and a gage length of 250 mm. (a) Plot the torque versus angle of twist diagram and find the modulus of rigidity and the modulus of rupture for the material. (b) Plot the linear portion of the shearing stress–shearing strain diagram and determine the proportional limit shearing stress and the corresponding shearing strain.

Torque (kN·m)	Angle of Twist (degree)	Torque (kN·m)	Angle of Twist (degree)	Torque (kN·m)	Angle of Twist (degree)
0.0	0.0	33.32	10.0	43.49	60.0
7.1	0.4	35.58	15.0	44.06	70.0
14.2	0.8	37.28	20.0	44.40	80.0
21.3	1.2	39.54	30.0	46.25	Rupture
24.85	1.4	41.80	40.0		
29.93	5.0	42.93	50.0		

2.14 In a torsion test of a solid alloy steel specimen, measurements indicated that the proportional limit torque was 62.8 kip·in. and the corresponding angle of twist 1.14°. Determine (a) the modulus of rigidity and (b) the modulus of resilience. The specimen had a diameter of 2.0 in. and a gage length of 6.0 in.

2.15 Data from a torsion test on a hollow steel specimen are tabulated below. The specimen had an outside diameter of 4.0 in., an inside diameter of 3.6 in., and a gage length of 10.0 in. (a) Plot the torque versus angle of twist diagram and determine the modulus of rigidity and the modulus of rupture. (b) Plot the linear portion of the shearing stress versus shearing strain diagram and determine the modulus of resilience.

Torque (kip·in.)	Angle of Twist (degree)	Torque (kip·in.)	Angle of Twist (degree)	Torque (kip·in.)	Angle of Twist (degree)
0.0	0.0	293.0	10.0	385.0	60.0
62.9	0.4	315.0	15.0	390.0	70.0
125.7	0.8	330.0	20.0	393.0	80.0
188.6	1.2	350.0	30.0	456.0	Rupture
220.0	1.4	370.0	40.0		
262.0	5.0	380.0	50.0		

2.16 Measurements during a torsion test on a solid specimen with a diameter of 40 mm and a gage length of 150 mm indicated a proportional limit torque of 7.35 kN·m, a corresponding angle twist of 1.2°, and a rupture torque of 18.25 kN·m. Determine (a) the modulus of rigidity, (b) the modulus of resilience, and (c) the modulus of rupture.

2.17 A hollow shaft is subjected to the torques indicated. (a) Find the maximum shearing stress in the shaft if the outside and inside diameters are $d_o = 2.0$ in. and $d_i = 1.25$ in, respectively. (b) Determine the maximum shearing stress in the shaft if $d_o = 2.0$ in. and $d_i = 0$. Compare the results. Let $Q_1 = 5$ kip·in. and $Q_2 = 15$ kip·in.

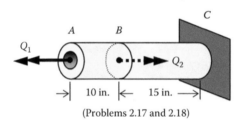

(Problems 2.17 and 2.18)

2.18 In reference to the figure shown in Problem 2.17, a hollow shaft is subjected to the torques indicated. (a) Determine the angle of twist at A if the outside and inside diameters are, $d_o = 1.5$ in. and $d_i = 0.75$ in., respectively. (b) Determine the angle of twist at A if $d_o = 1.5$ in. and $d_i = 0$. Compare the results. The material has a modulus of rigidity $G = 6 \times 10^6$ psi. Let $Q_1 = 5$ kip·in. and $Q_2 = 15$ kip·in.

2.19 The two-diameter shaft consists of segment AB for which $G_{AB} = 40$ GPa, $d_{AB} = 60$ mm, and segment BC for which $G_{BC} = 25$ GPa and $d_{BC} = 80$ mm. Let $Q_1 = 5$ kN·m and $Q_2 = 15$ kN·m and determine (a) the maximum shearing stress in the shaft and (b) the angle of twist of A relative to C.

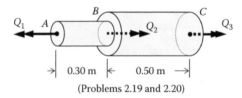

(Problems 2.19 and 2.20)

2.20 Refer to the shaft described in Problem 2.19 and use the data given there. (a) Construct a stress element on the surface of segment AB with two planes perpendicular and two planes parallel to the axis of the shaft. (b) Find the normal and shearing stresses on the surface of segment AB on a plane that is oriented to the cross sections of the shaft by a counterclockwise angle 20°. Show these stresses on a properly oriented stress element.

2.21 A solid circular shaft is made of a material for which the allowable shearing stress, $\tau_{ALL} = 25$ ksi, and the allowable normal stress, $\sigma_{ALL} = 40$ ksi. If the shaft is to transmit a torque of 50 kip·in., determine the smallest possible diameter for this shaft.

2.22 A hollow circular shaft with outside diameter $d_o = 50$ mm and inside diameter $d_i = 30$ mm has a length $L = 0.75$ m. Determine the maximum permissible torque that the shaft may carry if the allowable normal stress $\sigma_{ALL} = 100$ MPa.

2.23 A hollow circular shaft with outside diameter $d_o = 2.0$ in. is to have a length $L = 20$ in. Determine the maximum permissible inside diameter if the shaft is to transmit a torque of 40 kip·in. and the allowable angle of twist $\theta_{ALL} = 10°$. Let $G = 4 \times 10^3$ ksi.

2.24 The two-diameter shaft is subjected to the torques shown. Segment AB has a diameter of 2.0 in. What should be the diameter of segment BC if the allowable shearing stress is 25 ksi for both segments? Let $Q_2 = 20$ kip·in.

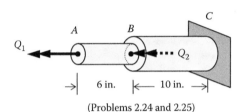

(Problems 2.24 and 2.25)

2.25 In reference to the figure shown in Problem 2.24, the two-diameter shaft is subjected to the torques shown. Segment BC has a diameter of 3.5 in. Find the diameter of segment AB if the angle of twist of A relative to B is identical to that of B relative to C. Let $Q_1 = 25$ kip·in., $Q_2 = 10$ kip·in., $G_{AB} = 10 \times 10^3$ ksi, and $G_{BC} = 4 \times 10^3$ ksi.

2.26 A pipe is constructed of 10-mm steel sheet material by welding along the helix shown where $\theta = 60°$. The torque $Q = 20$ kN·m and the outside diameter of the pipe is 400 mm. Determine the normal and shearing stresses that act on the weld.

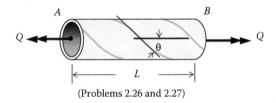

(Problems 2.26 and 2.27)

2.27 In reference to the figure shown in Problem 2.26, a pipe is to be constructed of 1/4-in. steel sheet material by welding along the helix shown, where $\theta = 50°$. If it is to transmit a torque $Q = 50$ kip·ft, and the allowable normal and shearing stresses in the weld are 15 and 10 ksi, respectively, determine the largest inside diameter the pipe may have.

2.28 The three-diameter shaft is made of the same aluminum material for which $G = 25$ GPa. Let $D_1 = 15$ mm, $D_2 = 30$ mm, and $D_3 = 45$ mm. If the allowable shearing stress is 100 MPa, determine the magnitudes of Q_1, Q_2, and Q_3. Let $L_1 = 0.50$ m, $L_2 = 0.30$ m, and $L_3 = 0.20$ m.

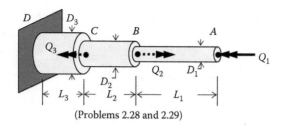

(Problems 2.28 and 2.29)

2.29 In reference to the figure shown in Problem 2.28, the three-diameter shaft is made of three different materials. The moduli of rigidity are $G_{AB} = 10 \times 10^3$ ksi, $G_{BC} = 5 \times 10^3$ ksi, and $G_{CD} = 3 \times 10^3$ ksi, respectively. The allowable angle of twist at A is $15°$. Let $L_1 = 18$ in., $L_2 = 12$ in., and $L_3 = 8$ in. If $D_1 = 2/3D_2 = 1/3D_3$, determine these diameters. Let $Q_1 = 5$ kip·in., $Q_2 = 10$ kip·in., and $Q_3 = 20$ kip·in.

2.30 Let Q_1, Q_2, and Q_3 be 50 kN·m, 25 kN·m, and 125 kN·m., respectively. If $D_1 = 1/2$, $D_2 = 1/3$, $D_3 = 40$ mm, $L_1 = 0.20$ m, $L_2 = 0.10$ m, $L_3 = 0.15$ m, and $L_4 = 0.25$ m, determine the maximum shearing stress in the shaft and specify its location.

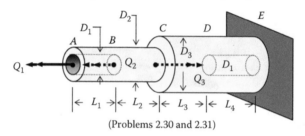

(Problems 2.30 and 2.31)

2.31 Refer to the shaft described in Problem 2.30 and determine the total angle of twist at A. Let $G = 70$ GPa for the entire shaft.

2.32 The electric motor supplies a torque to the shaft at A of 20 kip·in. This torque is distributed such that Q_B, Q_C, and Q_D are, 10, 5, and 5 kip·in., respectively. The diameters are $d_{AB} = 2.5$ in., $d_{BC} = 2.0$ in., $d_{CD} = 1.75$ in., and $d_{DE} = 1.5$ in. Determine the maximum shearing stress in the shaft specifying its location. Let $L = 3$ ft. Assume a frictionless bearing at E.

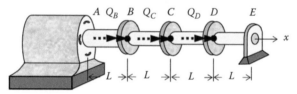

(Problems 2.32, 2.33, and 2.34)

2.33 The electric motor supplies a torque to the shaft at A of 2.5 kN·in. This torque is distributed such that Q_B, Q_C, and Q_D are 1.00, 0.90, and 0.60 kN·m, respectively. The diameters are $d_{AB} = 85$ mm, $d_{BC} = 65$ mm, $d_{CD} = 50$ mm, and $d_{DE} = 40$ mm. Determine the normal and shearing stresses in segment BC on a plane inclined to its cross section by a clockwise angle of $50°$. Let $L = 0.75$ m.

2.34 Refer to Problem 2.32. All shafts are made of steel for which $G = 10 \times 10^3$ ksi. Find the angle of twist of E relative to A.

2.35 Consider the gear-shaft system shown. Shafts AB and CD are both hollow with outside diameter $d_o = 75$ mm and inside diameter $d_i = 45$ mm. If the torque $Q = 10$ kN·m, determine the maximum shearing stress in the shafts. Let $r_1 = 125$ mm and $r_2 = 300$ mm. Assume frictionless bearings.

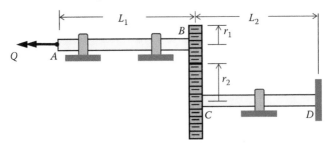

(Problems 2.35 and 2.36)

2.36 In reference to the figure shown in Problem 2.35, consider the gear-shaft system shown. Shafts AB and CD are both solid with diameters $d = 1.75$ in. and made of aluminum for which $G = 4 \times 10^3$ ksi. If the allowable angle of twist of A relative to D is 5°, determine the maximum permissible torque Q. Let $r_1 = 5$ in. and $r_2 = 10$ in. Assume frictionless bearings. Let $L_1 = L_2 = 12$ in.

2.37 The soil-sampling device for possible use in planetary exploration is being withdrawn from the soil by turning the knurled circular handle in a counterclockwise direction and pulling upward simultaneously. Of course, this action is equivalent to a torque Q and an upward axial force (not shown). Ignore the effects of the axial force and consider only the torsional soil resistance q to the applied torque Q. Assume $q = (3/4)x$ lb·in./in., where x is measured from A and determine the maximum shearing stress in the shank of the device. The shank is a hollow tube with outside and inside diameters of 2.0 and 1.5 in., respectively. Let $L_1 = 12$ in. and $L_2 = 24$ in.

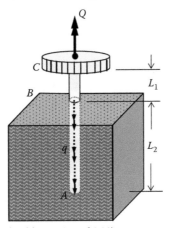

(Problems 2.37 and 2.38)

2.38 In reference to the figure shown in Problem 2.37, the soil sampling device for possible use in planetary exploration is being withdrawn from the soil by turning the knurled circular handle in a counterclockwise direction and pulling upward simultaneously. Of course, this action is equivalent to a torque Q and an upward axial force (not shown). Ignore the effects of the axial force and consider only the effects of the torque Q. If $Q = 48$ N·m, determine (a) the soil torsional resistance q if it is assumed constant between A and B and (b) the angle of twist of A relative to B. The aluminum material of the device has a modulus of rigidity $G = 25$ GPa. The shank of the device is a hollow tube with outside and inside diameters of 55 and 45 mm, respectively. Let $L_1 = 0.35$ m and $L_2 = 0.70$ m.

2.39 A screw is driven into a piece of wood by means of a screwdriver producing a torque Q and a downward axial force (not shown). Ignore the effects of the axial force and consider only the torsional wood resistance q to the applied torque Q. Assume $q = 3x^2$ kN·m/m, where x is measured from A and determine the maximum shearing stress at the root of screw threads. The diameter of the screw at the root of the threads is 6 mm. Ignore stress concentration effects.

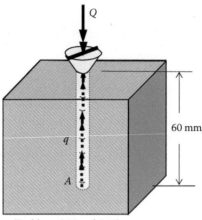

(Problems 2.39 and 2.40)

2.40 In reference to the figure shown in Problem 2.39, a screw is driven into a piece of wood by means of a screwdriver producing a torque Q and a downward axial force (not shown). Ignore the effects of the axial force and consider only the torsional wood resistance q to the applied torque Q. If the allowable shearing stress at the root of the thread is 10 MPa, determine (a) the maximum permissible torque Q that may be applied without failure. Assume the torsional resistance q is constant over the entire embedded length of the screw and find its magnitude. The diameter of the screw at the root of the thread is 8 mm. Ignore the effects of stress concentration.

2.41 The tapered solid shaft has a diameter of 2 in. at A and 4 in. at B. The torque $Q = 30$ kip·in. and the length $L = 20$ in. (a) Assume the torsion equation is applicable and determine the maximum shearing stress in the shaft specifying its location. (b) Find the angle of twist of A relative to B. The material is aluminum for which $G = 4 \times 10^6$ psi.

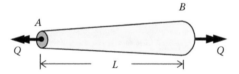

2.42 Determine the maximum permissible torque Q if the allowable shearing stress is 70 MPa and the allowable angle of twist at A is 4°. Assume the torsion equation applies. The diameter at A is 15 mm and at B it is 25 mm. Let $G = 50$ GPa.

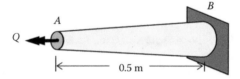

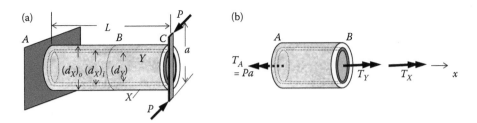

FIGURE 2.14 (a) A two-material shaft subjected to a torque Pa and (b) the free-body diagram of segment AB.

2.4 STATICALLY INDETERMINATE SHAFTS

By definition, a *statically indeterminate shaft* is one in which the number of unknown quantities exceeds the number of available equilibrium equations. Thus, the procedure used in the solution of statically indeterminate shafts is essentially the same as that used for statically indeterminate members under axial loads discussed in Section 1.6. As was done there, the conditions of equilibrium need to be supplemented with relations arising from the deformation characteristics of the shaft under consideration.

Consider, for example, the case of the two-material shaft shown in Figure 2.14a, which consists of a hollow shaft made of material X and a solid shaft made of material Y and placed concentrically inside the hollow shaft. Both shafts are fixed at A and subjected to a torque Pa at C by means of a rigid vertical member welded to both shafts so that the entire assembly acts as a single unit. It is required to determine the maximum shearing stresses in both shafts. Equilibrium dictates that the resisting torque at the fixed support be $T_A = Pa$. Cutting the assembly at B, we construct the free-body diagram of segment AB as shown in Figure 2.14b. The applied torque $Pa = T_A$ is resisted partially by each of the two shafts. The resisting torque in the hollow shaft is T_X, and that in the solid shaft is T_Y as shown, where both of these torques are shown coaxially in Figure 2.14b. Thus, applying the only available equation of equilibrium, we obtain

$$\sum T_x = 0 \Rightarrow T_X + T_Y - Pa = 0 \qquad (2.16)$$

Equation 2.16 contains the two unknown quantities T_X and T_Y, and obviously their determination requires the development of a second relation containing these same two quantities. This second relation is developed by observing that, since the assembly acts as a single unit, the total angle of twist at C for shaft X is the same as that for shaft Y. Thus, over the entire length L of the assembly, we can state that

$$\theta_X = \theta_Y \Rightarrow \left(\frac{TL}{JG}\right)_X = \left(\frac{TL}{JG}\right)_Y \qquad (2.17)$$

As will be illustrated in Example 2.8, Equations 2.16 and 2.17 form the basis for determining T_X and T_Y, which can then be used to find the maximum shearing stresses and any other information desired about the two shafts.

EXAMPLE 2.8

Refer to the system given in Figure 2.14 and use the following numerical values to find (a) the maximum shearing stresses in the two shafts and (b) the rotation of the vertical member at C. Let $L = 20$ in., $P = 15$ kips, $a = 6$ in., $(d_X)_o = 4.0$ in., $(d_X)_i = 3.0$ in., $G_X = 6.0 \times 10^3$ ksi, $d_Y = 2.5$ in., and $G_Y = 3.5 \times 10^3$ ksi.

SOLUTION

a. From Equation 2.16, we obtain

$$T_X + T_Y = Pa = (15)(6) = 90 \text{ kip} \cdot \text{in.}$$

(2.8.1)

Substituting the given numerical values into Equation 2.17 above, and canceling the equal lengths L from both sides of the equation, we have

$$\frac{T_X}{(\pi/32)(4.0^4 - 3.0^4)(5.5 \times 10^3)} = \frac{T_Y}{(\pi/32)(2.5^4)(3.5 \times 10^3)}$$

Simplifying, we conclude that

$$T_X = 7.04 T_Y$$

(2.8.2)

Solving Equations 2.8.1 and 2.8.2 simultaneously, we obtain

$$T_X = 78.806 \text{ kip} \cdot \text{in.} \quad \text{and} \quad T_y = 11.194 \text{ kip} \cdot \text{in.}$$

The maximum values of the shearing stresses are found by Equation 2.9. Thus,

$$\left.\begin{aligned}(\tau_{MAX})_X &= \frac{78.806(2.0)}{(\pi/32)(4.0^4 - 3.0^4)} = 9.174 \approx 9.2 \text{ ksi} \\ (\tau_{MAX})_Y &= \frac{11.194(1.25)}{(\pi/32)(2.5^4)} = 3.649 \approx 3.6 \text{ ksi}\end{aligned}\right\} \quad \textbf{ANS.}$$

b. The rotation of the member at C is found by determining the total angle of twist at this position. Thus, by Equation 2.10, we have

$$\theta_C = \theta_Y = \frac{(11.194)(20)}{(\pi/32)(2.5^4)(3.5 \times 10^3)} = 0.0167 \text{ rad} \approx 0.1° \quad \textbf{ANS.}$$

Instead of using θ_Y, we could have used θ_X and still obtain the same answer because θ_X is, in fact, equal to θ_Y.

EXAMPLE 2.9

The system shown in figure (a) below consists of a 3-in.-diameter steel shaft *DEH* fixed at both ends and welded to a rigid vertical member *BC* of length $2c = 40$ in., which in turn is attached to aluminum rods at B and C as shown. The length of each aluminum rod is $\ell = 20$ in. and each has a cross-sectional area $A = 1.0$ in.^{2}; other needed dimensions are shown in the sketch. The coefficient of thermal expansion for aluminum is $\alpha = 12.5 \times 10^{-6}$ in./in./°F and its modulus of elasticity $E = 10 \times 10^3$ ksi. The modulus of rigidity for steel is $G = 12 \times 10^3$ ksi. If the temperature of the two aluminum rods is dropped by $\Delta T = 200$°F, determine the magnitude of the torque induced in each of the two segments of the steel shaft.

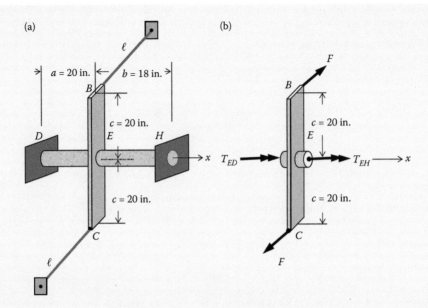

SOLUTION

A free-body diagram of the rigid member BC is shown in figure (b) above. When the temperature of the two aluminum rods drops by $\Delta T = 200°F$, each of these two rods will have a tendency to shorten by an amount $\delta_T = \alpha \Delta T \ell$. However, because of the restraining force F, each of the two rods will shorten only through the amount $\delta = \delta_T - \delta_F = \alpha \Delta T \ell - F\ell/AE$. As a consequence of the deformation δ of the aluminum rods, the rigid member BC rotates through the angle θ, which is also the angle of twist of the two segments of the steel shaft at E. Therefore,

$$\theta = \frac{\delta}{c} = \left(\frac{1}{c}\right)\left[\alpha(\Delta T)\ell - \frac{F\ell}{AE}\right] = \left(\frac{1}{20}\right)\left[12.5 \times 10^{-6}(200)(20) - \frac{20F}{(1)(10 \times 10^6)}\right]$$

$$= (2.5 \times 10^{-3}) - (10^{-7})F \tag{2.9.1}$$

The angles of twist of the two segments of the steel shaft are the same at E (i.e., $\theta_{ED} = \theta_{EH}$). Thus,

$$\left(\frac{TL}{JG}\right)_{ED} = \left(\frac{TL}{JG}\right)_{EH}$$

Since J and G for segments ED and EH are identical, the above relation reduces to

$$aT_{ED} = bT_{EH} \quad \text{and} \quad T_{ED} = \left(\frac{b}{a}\right)T_{EH} = 1.5T_{EH} \tag{2.9.2}$$

Also, since $\theta_{ED} = \theta_{EH} = \theta$, the following relations may be established:

$$\left(\frac{TL}{JG}\right)_{ED} = \theta \Rightarrow T_{ED} = 1.984 \times 10^{-4} - 0.794F \tag{2.9.3}$$

$$\left(\frac{TL}{JG}\right)_{EH} = \theta \Rightarrow T_{EH} = 1.323 \times 10^{-4} - 0.529F \tag{2.9.4}$$

Referring now to the free-body diagram of figure (b) above, we sum torques about the x axis. Thus,

$$\sum T_x = 0 \Rightarrow T_{ED} + T_{EH} - 40F = 0 \qquad (2.9.5)$$

Substituting Equation 2.9.2 into Equation 2.9.5 yields

$$T_{EH} = 16F \qquad (2.9.6)$$

A simultaneous solution of Equations 2.9.4 and 2.9.6 leads to

$$T_{EH} = 12.804 \approx 12.8 \text{ kip} \cdot \text{in.} \qquad \textbf{ANS.}$$

Finally, Equation 2.9.2 yields

$$T_{ED} = 19.206 \approx 19.2 \text{ kip} \cdot \text{in.} \qquad \textbf{ANS.}$$

PROBLEMS

2.43 Shaft ABC is made of aluminum for which $G = 4 \times 10^3$ ksi and has a diameter of 3 in. If $Q = 75$ kip·in., $a = 10$ in., and $b = 20$ in., determine (a) the maximum shearing stress in the shaft and (b) the angle of twist at B.

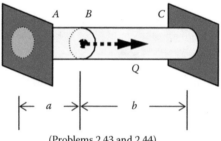

(Problems 2.43 and 2.44)

2.44 In reference to the figure shown in Problem 2.43, segment AB is made of aluminum ($G = 25$ GPa) and segment BC of copper ($G = 40$ GPa). The entire shaft has a diameter of 70 mm, and the two segments are brazed together and act as a unit. The allowable shearing stresses in aluminum and copper are 150 and 90 MPa, respectively. Determine the largest permissible torque Q that may be applied to the shaft. Let $a = 0.5$ m and $b = 1.0$ m.

2.45 The two-diameter shaft is made of the same material for which $G = 40$ GPa and is fixed rigidly at A and C. Segment AB has a diameter of 50 mm and segment BC a diameter of 30 mm. The diameter of the pulley at B is 200 mm, $a = 0.3$ m, and $b = 0.5$ m. Let $F = 100$ kN, determine the maximum shearing stress and the maximum angle of twist in the shaft.

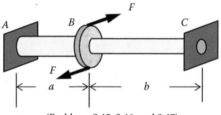

(Problems 2.45, 2.46, and 2.47)

2.46 In reference to the figure shown in Problem 2.45, the two-diameter shaft consists of two segments: AB is aluminum ($G = 4 \times 10^6$ psi, $d_{AB} = 2.0$ in.) and BC is steel ($G = 10 \times 10^6$ psi, $d_{BC} = 1.0$ in.). If the allowable angle of twist is 5°, determine the maximum permissible magnitude of F if the diameter of the pulley at B is 8 in. Let $a = 12$ in. and $b = 20$ in.

2.47 The two-diameter shaft consists of two segments: AB is aluminum ($G = 4 \times 10^3$ ksi, $d_{AB} = 3.0$ in.) and BC is steel ($G = 10 \times 10^3$ ksi, $d_{BC} = 2.0$ in.). If the allowable shearing stresses for aluminum and steel are 10 and 15 ksi, respectively, determine the maximum permissible magnitude of F if the diameter of the pulley at B is 10 in. Let $a = 15$ in. and $b = 25$ in.

2.48 The stepped shaft is made of the same material for which $G = 25$ GPa. Segment AB is hollow with outside diameter of 75 mm and inside diameter of 40 mm while segment BC is solid. Segment CD is solid with a diameter of 40 mm. If the torque $Q = 10$ kN·m, determine (a) the maximum shearing stress in the shaft specifying its location and (b) the maximum angle of twist. Let $L = 0.4$ m.

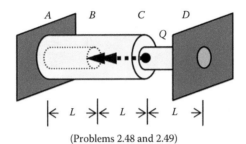

(Problems 2.48 and 2.49)

2.49 In reference to the figure shown in Problem 2.48, the stepped shaft consists of two segments: AB is hollow made of aluminum ($G = 4 \times 10^3$ ksi, outside and inside diameters of 4 and 2 in., respectively) and segment BC is solid, also made of aluminum. Segment CD is solid made of titanium ($G = 6 \times 10^3$ ksi, diameter 2 in.). If the allowable shearing stresses in aluminum and titanium are 30 and 60 ksi, respectively, determine the maximum allowable torque Q. Let $L = 12$ in.

2.50 The sectional view shown is that of a composite system consisting of an aluminum hollow shaft ($G = 4 \times 10^3$ ksi) placed concentrically around a steel solid shaft ($G = 10 \times 10^3$ ksi). The system is attached to a fixed support at its left end and to a rigid plate at its right end, which is used to apply the torque $Q = 50$ kip·in. Let $d_o = 5$ in., $d_i = 3$ in., $d = 2$ in., and $L = 20$ in., determine (a) the maximum shearing stresses in the two shafts and (b) the rotation of the rigid plate.

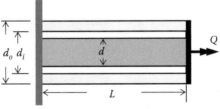

(Problems 2.50, 2.51, and 2.52)

2.51 In reference to the figure shown in Problem 2.50, the sectional view shown is that of a composite system consisting of an aluminum hollow shaft ($G = 25$ GPa) placed concentrically around a steel solid shaft ($G = 75$ GPa). The system is attached to a fixed support at its left end and to a rigid plate at its right end, which is used to apply the torque Q. Let $d_o = 120$ mm, $d_i = 80$ mm, $d = 45$ mm, and $L = 0.5$ m. If the allowable shearing stresses are $(\tau_{ALL})_A = 60$ MPa and $(\tau_{ALL})_S = 100$ MPa, determine the maximum permissible torque Q.

2.52 Solve Problem 2.51 if instead of the allowable shearing stresses, the allowable rotation of the rigid plate is specified at $\theta = 5°$. All other conditions are as stated in Problem 2.51.

2.53 The sectional view shown is that of a composite system consisting of hollow shafts A and B and solid shaft C. The system is attached to fixed supports at both ends and to a rigid plate P, which is used to apply the torque Q. Member A is aluminum ($G = 4 \times 10^6$ psi), B is brass ($G = 6 \times 10^6$ psi), and C is steel ($G = 10 \times 10^6$ psi). Let $d_1 = 8$ in., $d_2 = 6$ in., and $d_3 = 3$ in. If $Q = 750$ kip·in., determine (a) the maximum shearing stress in the system and (b) the rotation of rigid plate P. Let $a = 6$ in. and $b = 10$ in.

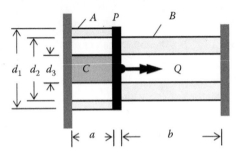

(Problems 2.53, 2.54, and 2.55)

2.54 In reference to the figure shown in Problem 2.53, the sectional view shown is that of a composite system consisting of hollow shafts A and B and solid shaft C. The system is attached to fixed supports at both ends and to a rigid plate P, which is used to apply the torque Q. Member A is aluminum ($G = 25$ GPa), B is brass ($G = 40$ GPa), and C is steel ($G = 75$ GPa). Let $d_1 = 250$ mm, $d_2 = 150$ mm, and $d_3 = 70$ mm. If the allowable shearing stresses for aluminum, brass, and steel are 70, 100, and 150 MPa, respectively, determine the maximum permissible Q. Let $a = 0.2$ m and $b = 0.3$ m.

2.55 Solve Problem 2.54 if instead of the allowable shearing stresses, the allowable rotation of rigid plate P is specified at $\theta = 0.5°$. All other conditions are as stated in Problem 2.54.

2.56 The shaft system shown is attached to rigid supports at both ends. Segment AB is aluminum ($G = 4 \times 10^3$ ksi) with a diameter of 2.5 in., segment BC is brass ($G = 6 \times 10^3$ ksi) with a diameter of 1.5 in., and segment CD is steel ($G = 10 \times 10^3$ ksi) with a diameter of 1.0 in. Let $L = 1.5$ ft, $Q_1 = 10$ kip·ft, and $Q_2 = 15$ kip·ft, determine the maximum shearing stress in the shaft system and the angle of twist at C.

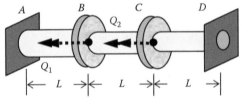

(Problems 2.56 and 2.57)

2.57 In reference to the figure shown in Problem 2.56, the shaft system shown is attached to rigid supports at both ends. Assume all three segments have the same diameter d and are made of the same material for which $G = 30$ GPa. Let $Q_1 = 50$ kN·m, $Q_2 = 75$ kN·m, and $L = 1.0$ m. If the allowable shearing stress is 120 MPa, determine the minimum diameter of the shaft system.

2.58 The shaft-gear system shown consists of shafts AB and BC connected by two gears at one of their ends and to fixed supports at their other ends. Shaft AB is made of steel ($G = 10 \times 10^3$ ksi) and has a diameter $d_{AB} = 1.0$ in. Shaft BC is made of aluminum ($G = 4 \times 10^3$ ksi) and has a diameter $d_{BC} = 2.0$ in. Let $r_1 = 3.0$ in., $r_2 = 6.0$ in., $L_1 = L_2 = 15.0$ in., and $Q = 20$ kip·in., determine the maximum shearing stresses in the two shafts.

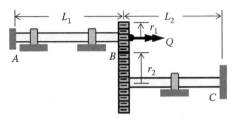

(Problems 2.58 and 2.59)

2.59 In reference to the figure shown in Problem 2.58, the shaft-gear system shown consists of shafts AB and BC of the same diameter d connected by two gears at one of their ends and to fixed supports at their other ends. Both shafts are made of the same material for which $G = 50$ GPa and the allowable shearing stress is 120 MPa. Let $L_1 = L_2 = 0.75$ m, $r_1 = 100$ mm, $r_2 = 150$ mm, and $d = 75$ mm, determine the maximum permissible torque Q.

2.60 Hollow shaft AB with outside and inside diameters of 4.0 and 3.5 in., respectively, is made of brass ($G = 6 \times 10^6$ psi) and is attached to a fixed support at A and to a rigid plate CD at B. The two ends of the rigid plate are attached to identical magnesium ($\alpha = 14.5 \times 10^{-6}$ in./in./°F and $E = 6.5 \times 10^6$ psi) rods at C and D whose cross-sectional areas and lengths are each 1.0 in.² and 18 in., respectively. If the temperature of the two magnesium rods is dropped by $\Delta T = 150$°F, determine the maximum shearing stress in the brass shaft and the rotation of the rigid plate.

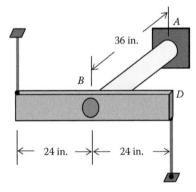

(Problems 2.60 and 2.61)

2.61 In reference to the figure shown in Problem 2.60, hollow shaft AB with outside and inside diameters of 3.0 and 2.5 in., respectively, is made of aluminum ($G = 4 \times 10^6$ psi) and is attached to a fixed support at A and to a rigid plate at B. The two ends of the rigid plate are attached to identical aluminum ($\alpha = 13.0$ in./in./°F, $E = 10 \times 10^6$ psi) rods at C and D whose diameters and lengths are each 0.5 and 15 in., respectively. If the allowable normal and shearing stresses for aluminum are 15 and 7 ksi, respectively, determine the maximum temperature drop to which the two aluminum rods may be subjected.

2.5 DESIGN OF POWER-TRANSMISSION SHAFTS

As stated in the introductory statements, a primary and very practical use of shafts exists in transmitting *power* from one location to another. Numerous applications are found in which power needs to be transmitted from a power source to a power user. Two examples of such applications are the transmission of power from a motor to a power tool in a machine shop and from an internal combustion engine to the rear axle of a car. The shafts used in such applications need to be designed properly in order to avoid failure during service. Two factors, *power*, P, and *angular speed*, ω, play a major role in the *design* of such shafts. These two factors are related by an equation developed in a basic *dynamics* course and is stated here without proof in Equation 2.18. Thus,

$$P = T\omega \qquad (2.18)$$

The quantity T in Equation 2.18 is the internal torque in the transmission shaft and, as stated earlier, P is the power transmitted by the shaft and ω is the angular speed at which the shaft is rotating. We recall that the angular speed ω is expressed in radians per second and that the power P is the rate at which work is performed by the transmission shaft. Thus, in the SI system of units, power may be expressed as $N \cdot m/s = joule/s$ ([watt (W)]) or $kN \cdot m/s$ ([kilowatt (kW)]); in the U.S. Customary system, as $lb \cdot in./s$ or $lb \cdot ft/s$. Note that one *horsepower* (hp) = 6600 $lb \cdot in./s$ = 550 $lb \cdot ft/s$.

Therefore, in designing a shaft, the engineer starts with knowledge of the power to be transmitted and the speed at which this power is to be transmitted. Using Equation 2.18, she determines the internal torque the shaft is to carry. She then selects a material to be used in the fabrication of the shaft, which then determines the allowable stress in shear. Using Equation 2.9, $\tau = TR/J = T/(J/R)$, where the allowable shearing stress is substituted for τ, she determines the ratio J/R from which the safe radius of the shaft can be found. If the angle of twist of the shaft during service is a significant design consideration, the engineer needs to ensure that the allowable angle of twist is not exceeded by using one of Equations 2.10 through 2.12. If this angle of twist is, in fact, exceeded, a redesign becomes necessary in which case the redesign would be based on one of Equations 2.10 through 2.12. In connection with these considerations, it would be very helpful at this point to recall the *design concepts* discussed in Section 1.4 about *allowable stresses* and *factors of safety*.

Example 2.10 illustrates some of the concepts discussed above.

EXAMPLE 2.10

A 10-hp motor is to operate at 1800 rpm. (a) Determine the minimum diameter if a solid aluminum shaft is to be used for which the allowable shearing stress is 10 ksi. Assume the angle of twist is not a significant design consideration. (b) Determine the minimum inside diameter if, instead of a solid aluminum shaft, a hollow steel ($G = 10 \times 10^3$) shaft is to be used with outside diameter $d_o = 1.25$ in. Let $\tau_{ALL} = 15$ ksi and $\theta_{ALL} = 0.10$ rad and the length of the shaft is 18 in.

SOLUTION

Expressing the angular velocity in rad/s, we have

$$\omega = \frac{1800}{60} = 30 \text{ rev./s} = 30 \times 2\pi = 60\pi \text{ rad/s}$$

Thus, using Equation 2.18, we obtain

$$T = \frac{P}{\omega} = \frac{(10)(6600)}{60\pi} = 1100\pi \text{ lb} \cdot \text{in.}$$

Therefore, by Equation 2.9, we have

$$(J/R) = \frac{T}{\tau_{ALL}} = \frac{1100\pi}{10 \times 10^3} = 0.110\pi \text{ in.}^3 \qquad (2.10.1)$$

Thus, since $J = \pi R^4/2$, it follows that $J/R = \pi R^3/2$. Therefore,

$$\frac{\pi R^3}{2} = 0.110\pi$$

$$R = \sqrt[3]{0.220} = 0.604 \text{ in.}; \quad d = 1.207 = 1.2 \text{ in.} \qquad \textbf{ANS.}$$

$$(2J/d_o) = \frac{T}{\tau_{ALL}} = \frac{1100\pi}{15 \times 10^3} = 0.07333\pi \text{ in.}^3 \qquad (2.10.2)$$

Thus, since $J = (\pi/32)(d_o^4 - d_i^4)$, it follows that

$$(2J/d_o) = \left(\frac{2\pi}{32 d_o}\right)(d_o^4 - d_i^4) = \left(\frac{\pi}{(16)(1.25)}\right)(1.25^4 - d_i^4)$$

Therefore,

$$\left(\frac{\pi}{(16)(1.25)}\right)(1.25^4 - d_i^4) = 0.07333\pi; \quad d_i = 0.994 \approx 1.0 \text{ in.} \qquad \textbf{ANS.}$$

A check is now made of the angle of twist to ensure that the allowable angle is not exceeded. Thus,

$$\theta = \frac{TL}{JG} = \frac{(1100\pi)(18)}{(\pi/32)(1.25^4 - 0.994^4)(10 \times 10^6)} = 0.065 \text{ rad} < \text{allowable value of } 0.1 \text{ rad}$$

Since the angle of twist of the shaft is within the allowable limit, the above inside diameter is acceptable. If this had not been the case, a redesign using $\theta = TL/JG$ would become necessary.

2.6   STRESSES UNDER COMBINED LOADS

There are many engineering situations in which a given member is called upon to resist the simultaneous action of axial and torsional loads. Familiar examples of such situations include the case of the screw stem and that of the drill shaft. For example, as a screw is driven by a screwdriver, turning and pushing occur simultaneously. The turning action results in a system of resisting shearing forces between the screw threads and the material into which the screw is driven that give rise to a resisting torque. The pushing action gives rise to a compressive force in the screw stem. Thus, the analysis of the screw stem should account for the action of both the torque and the compressive force.

Analysis of combined axial and torsional loads is most conveniently accomplished by using the principle of *superposition* that was introduced earlier in Section 1.6. It is important to recall that the principle of superposition is only valid if the material behaves elastically. For simplicity and convenience, only members with circular cross sections will be analyzed in this section, although the method of superposition is equally applicable to any cross-sectional form as long as the material is not stressed beyond the elastic limit.

Let us now consider, for example, the solid circular shaft shown in Figure 2.15a, which is subjected to the combined action of the axial force P and the torque Q. Assuming the member to behave

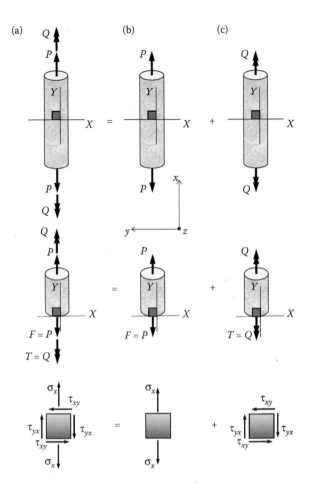

FIGURE 2.15 Analysis of stresses in a shaft subjected to the combined loads, axial force P and a torque Q.

elastically, then by the principle of superposition, we may consider this member to be subjected to two separate and independent loading actions as shown in Figures 2.15b and 2.15c.

The member shown in Figure 2.15b is subjected to the axial force P acting alone. As shown in the free-body diagram of that part of the member above the X plane (the middle part of Figure 2.15b), the internal force on the X plane is $F = P$. As was discussed in Section 1.3, this force produces a normal stress $\sigma_x = F/A$, where A is the cross-sectional area of the member. The stress, σ_x, acts on the stress element contained between two X and two Y planes, shown at the bottom of Figure 2.15b.

The member shown in Figure 2.15c is subjected to the torque Q acting alone. As shown in the free-body diagram of that part of the member above plane X (the middle part of Figure 2.15c), the internal torque in that position is $T = Q$. As was discussed in Section 2.3, this torque produces a system of shearing stresses τ_{xy} on transverse planes and τ_{yx} on longitudinal planes of the member that, according to Equation 2.8, vary from zero at the center of the circular cross section to its maximum at the outside surface. These shearing stresses act on a stress element contained between two X and two Y planes on the outside surface of the circular shaft and shown at the bottom of Figure 2.15c.

Finally, the member shown in Figure 2.15a represents the superposition of the two conditions depicted in Figures 2.15b and 2.15c. The three components shown in Figure 2.15a represent the superposition of the two corresponding components in Figures 2.15b and 2.15c. In particular, note the stress element at the bottom of Figure 2.15a, which combines the stress systems in Figures 2.15b and 2.15c.

All of the stress elements shown in Figure 2.15 represent *plane stress* conditions (two-dimensional stress conditions), which means that all of the stresses acting on the stress elements are in the same plane, namely, the *x–y* plane. This is so because there are no stresses in a direction perpendicular to this plane (i.e., perpendicular to the cylindrical surface of the shaft).

In a given solution using superposition, we may eliminate the steps expressed in Figures 2.15b and 2.15c and go directly to the combined step shown in Figure 2.15a. This process is illustrated in Example 2.11.

2.6.1 HOOKE'S LAW IN TWO DIMENSIONS

Let us consider the stress element shown in Figure 2.16a, which depicts a two-dimensional stress condition referred to as *plane stress*. Let us break it down into three components as shown in Figures 2.16b through 2.16d. This process of breaking down a complex problem into two or more simple components, solving these simple components separately, and then adding them up algebraically to obtain the solution of the original complex problem is known as *superposition*. Superposition is very useful in solving many problems in engineering but we need to remember that it is only valid if the stress is directly proportional to the strain (i.e., in the elastic range).

We will now consider Figure 2.16a in order to reaffirm what was shown earlier by Equation 1.7, namely, that the magnitudes of the shearing stresses on perpendicular planes passing through the same point in a stressed body are equal. To accomplish this purpose, we need to sum moments of the forces created by the given stress system about some point such as *O*. Note that the forces produced by σ_x and σ_y do not contribute to this summation because they pass through point *O*. However, the forces produced by all of the shearing stresses contribute to the summation. In order to transform these shearing stresses into shearing forces, we multiply them by their respective areas. Thus, if we assume that the thickness of the stress element into the page is a small constant equal to *t*, we obtain

$$\sum M_O = 0 \Rightarrow 2(\tau_{xy})(t)(dy)(dx/2) - 2(\tau_{yx})(t)(dx)(dy/2) = 0$$

Simplifying, we conclude that

$$\tau_{yx} = \tau_{xy} \tag{2.19}$$

Having shown that $\tau_{yx} = \tau_{xy}$ in magnitude, it will not be necessary to distinguish between them and henceforth, whenever subscripting is not required, the two shearing stresses will both be labeled τ as shown in Figure 2.16d.

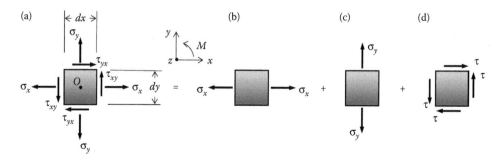

FIGURE 2.16 A general plane-stress condition broken down into three simple components, two uniaxial σx and σy, and one pure shear τ.

Applying Hooke's law to the stress element in Figure 2.16b and using the definition of Poisson's ratio, we obtain

$$\varepsilon_x = \sigma_x/E; \quad \varepsilon_y = -\mu(\sigma_x/E) = \varepsilon_z \tag{2.20}$$

Similarly, applying Hooke's law to the stress element in Figure 2.16c and using the definition of Poisson's ratio, we obtain

$$\varepsilon_x = -\mu(\sigma_y/E) = \varepsilon_z; \quad \varepsilon_y = \sigma_y/E \tag{2.21}$$

At this point in our development, we should note that the pure shear condition shown in Figure 2.16d does not contribute in any way to the normal strains. Superposing (adding algebraically) the strains in the x and y directions as given by Equations 2.20 and 2.21, we obtain

$$\left.\begin{array}{l} \varepsilon_x = \dfrac{\sigma_x}{E} - \mu\left(\dfrac{\sigma_y}{E}\right) = \dfrac{1}{E}(\sigma_x - \mu\sigma_y) \\[2mm] \varepsilon_y = \dfrac{\sigma_y}{E} - \mu\left(\dfrac{\sigma_x}{E}\right) = \dfrac{1}{E}(\sigma_y - \mu\sigma_x) \\[2mm] \varepsilon_z = -\dfrac{\mu}{E}(\sigma_x + \sigma_y); \quad \gamma_{xy} = \dfrac{\tau_{xy}}{G} \end{array}\right\} \tag{2.22}$$

Note that there is a strain ε_z despite the fact that $\sigma_z = 0$ since what we are dealing with is a plane-stress condition. The very last part of Equations 2.22 is the result of the pure shear condition shown in Figure 2.16d and expresses Hooke's law in shear. Equations 2.22 are known as *Hooke's law in two dimensions* or *Hooke's law for plane stress*.

The first two parts of Equations 2.22 may be solved simultaneously for the stresses σ_x and σ_y in order to express them in terms of the strains ε_x and ε_y. Thus,

$$\left.\begin{array}{l} \sigma_x = \dfrac{E}{1-\mu^2}(\varepsilon_x + \mu\varepsilon_y) \\[2mm] \sigma_y = \dfrac{E}{1-\mu^2}(\varepsilon_y + \mu\varepsilon_x) \\[2mm] \sigma_z = 0 \end{array}\right\} \tag{2.23}$$

The third part of Equations 2.23 emphasizes the fact that we are dealing with a plane stress system.

We should observe here that the three material constants, E, μ, and G, that have thus far been defined are not independent of one another. The relationship among these three quantities is developed in Section 8.4 but, for convenience, it is provided here without proof. Thus,

$$G = \dfrac{E}{2(1+\mu)} \tag{2.24}$$

EXAMPLE 2.11

During a drilling operation, the 4-in.-diameter shaft of an oil rig is subjected to the simultaneous action of a push $P = 160$ kips and a torque $Q = 120$ kip·in. as shown schematically in figure (a) below. Consider a point on the outside surface of the shaft at the intersection of the X and Y planes (shown in gray) and construct its stress element.

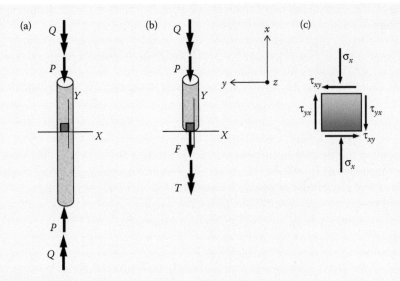

SOLUTION

The free-body diagram of the segment of the shaft above the X plane is shown in figure (b) above. Thus,

$$\sum F_y = 0 \Rightarrow -160 - F = 0; \quad F = -160 \text{ kips}$$

$$\sum T_y = 0 \Rightarrow -120 - T = 0; \quad T = -120 \text{ kip} \cdot \text{in.}$$

Using Equation 1.2, we obtain

$$\sigma_y = \frac{-160}{\pi(2^2)} = -12.732 \approx -12.7 \text{ ksi} \qquad \textbf{ANS.}$$

Using Equation 2.7b, we obtain

$$\tau_{yx} = \tau_{xy} = \left| \frac{-120(2)}{(\pi/32)(4^4)} \right| = -9.549 \approx -9.5 \text{ ksi} \qquad \textbf{ANS.}$$

These stresses are shown properly directed on the stress element of figure (c) above.

EXAMPLE 2.12

The stress element shown in figure (a) below was obtained at a point on the outside surface of a member subjected to the combined actions of a tensile force P and a torque Q as shown in figure (b) below. Using equilibrium, develop equations that give the normal and shearing stresses on any plane such as N inclined to the X plane by the counterclockwise angle θ.

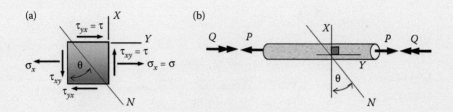

(a)   (b)

SOLUTION

For simplicity, we will use σ and τ for σ_x and $\tau_{yx} = \tau_{xy}$, respectively. We now construct the free-body diagram of the stress-element wedge contained within the X, Y, and N planes as shown in figure below. Let A_n be the area of the inclined plane. Then, $A_x = A_n \cos\theta$ and $A_y = A_n \sin\theta$. Applying the equations of equilibrium, we have

$$\sum F_n = 0 \Rightarrow \sigma_n A_n - \sigma(A_n \cos\theta)\cos\theta - \tau(A_n \cos\theta)\sin\theta - \tau(A_n \sin\theta)\cos\theta = 0$$

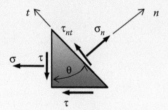

Solving for σ_n, we obtain

$$\sigma_n = \sigma\cos^2\theta + \tau(2\sin\theta\cos\theta)$$

$$\sigma_n = \frac{1}{2}\sigma(1 + \cos 2\theta) + \tau\sin 2\theta \qquad \textbf{ANS.}$$

$$\sum F_t = 0 \Rightarrow \tau_{nt} A_n + \sigma(A_n \cos\theta)\sin\theta - \tau(A_n \cos\theta)\cos\theta + \tau(A_n \sin\theta)\sin\theta = 0$$

Solving for τ_{nt}, we obtain

$$\tau_{nt} = \tau(\cos^2\theta - \sin^2\theta) - \frac{1}{2}\sigma(2\sin\theta\cos\theta)$$

$$\tau_{nt} = \tau\cos 2\theta - \frac{1}{2}\sigma\sin 2\theta \qquad \textbf{ANS.}$$

EXAMPLE 2.13

Measurements of strain in the n and t directions at point O on the outside surface of the shaft shown in figure (a) below, provided the following values: $\varepsilon_n = 750\,\mu$ and $\varepsilon_t = -750\,\mu$. If the diameter of the shaft is 2.0 in. and it is made of aluminum for which $E = 10 \times 10^6$ psi and $\mu = 0.3$, determine the applied torque Q and the shearing strain γ_{nt}.

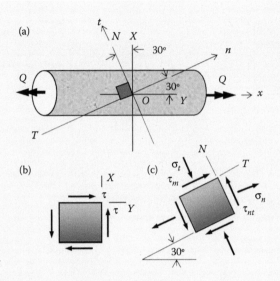

SOLUTION

By Equation 2.23, we have

$$\sigma_n = -\sigma_t = \frac{10 \times 10^6}{(1 - 0.3^2)}\left[750 - 0.3(750)\right] \times 10^{-6} = 5.769 \text{ ksi}$$

Solving Equation 2.14 for the shearing stress τ on the X and Y planes yields

$$\tau = \frac{\sigma_n}{\sin 2\theta} = \frac{5.769}{\sin 60°} = 6.661 \text{ ksi}$$

The shearing stress τ is shown on the stress element of figure (b) above, which is bounded by the X and Y planes as shown.

Equation 2.8 is now solved for the internal torque T in terms of the shearing stress, τ. Thus,

$$T = \frac{\tau J}{R} = \frac{(6.661)(\pi/32)(2.0^4)}{(1.0)} = 10.463 \text{ kip} \cdot \text{in.}$$

In this case, the internal torque T at any point is equal to the applied torque Q. Therefore,

$$Q = T = 10.463 \approx 10.5 \text{ kip} \cdot \text{in.} \qquad \textbf{ANS.}$$

We now use Equation 2.24 to find the modulus of rigidity G and Equation 2.15 to find the magnitude of the shearing stress acting on the N and T planes. Thus,

$$G = \frac{10 \times 10^6}{2(1 + 0.3)} = 3.846 \times 10^6 \text{ psi}; \quad \tau_{nt} = -\tau_{tm} = 6.661 \cos 60° = 3.331 \text{ ksi}$$

Figure (c) above shows the stress element bounded by the N and T planes. Applying Hooke's law in shear, Equation 2.3, we solve for the shearing strain γ_{nt}. Thus,

$$\gamma_{nt} = \frac{3.331 \times 10^3}{3.846 \times 10^6} = 0.8661 \times 10^{-3} \approx 866 \text{ } \mu \qquad \textbf{ANS.}$$

*2.7 STRESS CONCENTRATION

Equation 2.9, $\tau_r = TR/J$, derived in Section 2.3 for the shearing stress in a circular shaft, assumes a number of ideal conditions about the material, the method of torque application, and the geometric conditions of the shaft. In practice, some of these conditions may not be met entirely and, as a consequence, Equation 2.9 does not provide the correct magnitude of the shearing stress.

As in the case of members subjected to axial loads, geometric discontinuities in a shaft of circular cross section give rise to shearing stresses that are much higher than those given by Equation 2.9. To account for the increase in stress at geometric discontinuities, we introduce a factor k, known as the *stress-concentration factor*. As was stated in Section 1.7, stress-concentration factors are, for the most part, determined experimentally using methods such as *photoelasticity* and *electrical analogs*. However, some work has been done using computers along with finite element methods.

To determine the maximum shearing stress due to a geometric discontinuity in shafts, we use the equation

$$\tau_{MAX} = k(\tau_{AVE}) = k\left(\frac{TR}{J_{min}}\right) \tag{2.25}$$

The quantity, J_{min} in Equation 2.25 is the minimum polar moment of inertia of the shaft and, as stated above, k is the *stress-concentration factor*.

Torsional values of stress-concentration factors are shown in Figure 2.17 for a two-diameter shaft with a fillet of radius r. Note that the stress-concentration factor depends not only on the ratio of the diameters of the two shafts but also on the ratio of the radius of the fillet between them to the diameter of the smaller of the two shafts. It should be noted that other kinds of geometric discontinuities, such as keyways and gear teeth, also cause stress concentrations in shafts.

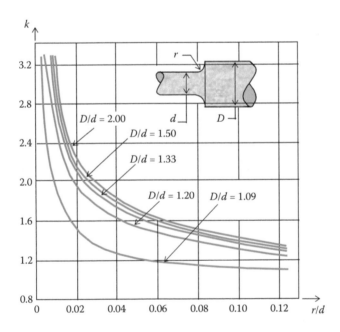

FIGURE 2.17 Stress-concentration factors for torsional loads. (Adapted from the work of L.S. Jacobsen, Torsional stress concentrations in shafts of circular cross section and variable diameter, *Trans ASME*, 47, 619–641, 1925.)

EXAMPLE 2.14

Consider the two-diameter shaft shown in the sketch. Let $r = 4$ mm, $d = 40$ mm, $D = 80$ mm, and $Q = 1.20$ kN·m. Determine the maximum shearing stress in the smaller of the two shafts and sketch the shearing stress distribution across the diameter of the smaller shaft.

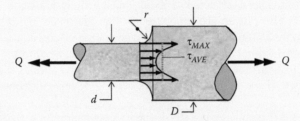

SOLUTION

The ratios needed to determine the stress-concentration factor are computed as follows:

$$\frac{D}{d} = \frac{80}{40} = 2.0$$

$$\frac{r}{d} = \frac{4}{40} = 0.10$$

Entering Figure 2.17 with these values, we estimate the stress-concentration factor to be $k = 1.50$. Thus,

$$\tau_{MAX} = k(\tau_{AVE}) = (1.50)\left(\frac{1.20(10^3)(0.02)}{(\pi/32)(0.04^4)}\right) = 143.239 \approx 143.2 \text{ MPa} \qquad \textbf{ANS.}$$

A sketch of the shearing stress distribution is shown in the drawing of the shaft above.

EXAMPLE 2.15

The two-diameter shaft of Example 2.14 is used to transmit power at 1800 rpm from an electric motor to a machine tool in a shop. Let $D = 6.0$ in., $d = 3.0$, and $r = 0.4$ in., and if the material is steel for which the allowable shearing stress is 10 ksi, determine the maximum power that may be transmitted.

SOLUTION

The ratios needed to determine the stress-concentration factor are obtained as follows:

$$\frac{D}{d} = \frac{6}{4} = 1.5; \quad \frac{r}{d} = \frac{0.4}{4} = 0.10$$

Using Figure 2.17, we estimate the stress-concentration factor to be $k = 2.9$.
Applying Equation 2.25 and solving for the torque, we obtain $T = (\tau_{MAX})J/(kR)$. Thus,

$$T = \frac{(\tau_{MAX})J}{(kR)} = \frac{10(\pi/32)(4^4)}{(2.9)(2)} = 43.332 \text{ kip} \cdot \text{in.}$$

Note that in computing the magnitude of the torque T, we used the values of J and R for the smaller of the two diameters.

Using Equation 2.18, we obtain

$$P = T\omega = (43.332)\left(\frac{(600)(2\pi)}{60}\right) = 2722.63 \text{ kip} \cdot \text{in./s} \approx 413 \text{ hp} \qquad \textbf{ANS.}$$

*2.8 IMPACT LOADING

Occasionally, shafts are subjected to impact loads that impart stresses and strains that are much larger than those experienced during normal (static) loading conditions. It, therefore, becomes desirable to be able to analyze the effects of these impact loads on the behavior of shafts. While the analysis developed here is general enough to apply to shafts of any cross-sectional shape, we will limit our discussion at present to shafts of circular cross sections.

We will show shortly that Equation 1.38 developed for members subjected to impacting axial loads may be used for the analysis of members subjected to impacting torsional loads. Thus, consider the case of the system shown in Figure 2.18, which consists of a shaft of length L, fixed rigidly at one end and supported by a frictionless bearing at the other end where the shaft is attached to a rigid handle of length b. If a weight W is dropped onto the end of the rigid handle from a height h as shown, the right end of the shaft experiences a rotation θ, which represents the maximum angle of twist of the shaft over its length L due to the impact produced by W.

To analyze torsional impact, we make the same assumptions that were made in Section 1.8 concerning the impact loading of members subjected to axial loads. The reader is urged to refer to those assumptions before proceeding with the development that follows. Consider an element of volume dV at a radius ρ taken from the shaft in Figure 2.18 at the instant when W has come to rest at the end Figure 2.18 of its downward travel and the shaft has rotated through its maximum angle of twist $\theta = \Delta/b$, provided that θ is a very small angle. At this instant, the shearing stress in the volume dV has been built up from zero to its maximum value, τ_ρ. This, of course, means that the average shearing stress in dV during the rotation of the shaft is $(1/2)\tau_\rho$. The differential shearing strain energy dU stored in the volume dV becomes $(1/2)\tau_\rho\gamma_\rho$, where γ_ρ is the maximum shearing strain in dV associated with the maximum deflection Δ at the instant when W has completed its downward motion. Thus, the *elastic strain energy, u,* absorbed by the shaft material is

$$u = \frac{dU}{dV} = \left(\frac{1}{2}\right)\tau_\rho\gamma_\rho \qquad (2.26)$$

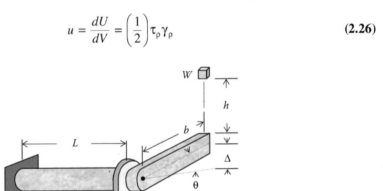

FIGURE 2.18 Diagram showing a shaft of length L subjected to impact loading due to a weight W dropping through a height h.

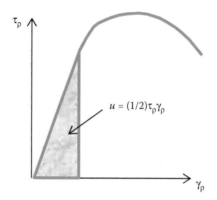

FIGURE 2.19   A shearing stress–strain diagram showing determination of the elastic strain energy u.

A graphical representation of Equation 2.26 is shown in Figure 2.19. From Equation 2.26 and Figure 2.20, we obtain

$$dU = \left(\frac{1}{2}\right)\left(\frac{T\rho}{J}\right)\left(\frac{\tau_\rho}{G}\right)dV = \left(\frac{1}{2G}\right)\left(\frac{T\rho}{J}\right)^2 (dA)(dx) = \left(\frac{1}{2G}\right)\left(\frac{T\rho}{J}\right)^2 (\rho)(d\rho)(d\theta)dx$$

where dx is a differential length along the shaft. Therefore,

$$U = \int dU = \left(\frac{1}{2G}\right)\left(\frac{T}{J}\right)^2 \int_0^L dx \int_0^{2\pi} d\theta \int_0^r \rho^3 d\rho = \left(\frac{1}{2G}\right)\left(\frac{T}{J}\right)^2 (2\pi L)(r^4/4) = \left(\frac{TL}{JG}\right)\left(\frac{\pi r^4}{2\pi r^4}\right)T$$

Replacing TL/JG by θ, T by the product Fb, and simplifying, we obtain

$$U = \left(\frac{1}{2}\right)F(b\theta) = \left(\frac{1}{2}\right)F\Delta \tag{2.27}$$

The product $b\theta$ in Equation 2.27 was replaced by the quantity Δ, which represents the vertical movement of the right end of the rigid handle. Therefore, since the work done by the weight is

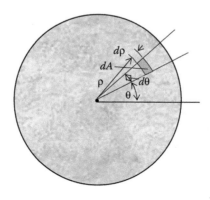

FIGURE 2.20   Cross section of a shaft subjected to a torque T and used to determine the total strain energy U.

$W(h + \Delta)$, and it is assumed that all of the work is transformed into *elastic strain energy*, it follows that

$$W(h + \Delta) = \frac{1}{2} F\Delta \qquad \qquad \textbf{(2.28)}$$

Equation 2.28 is identical to Equation 1.38 developed in Chapter 1 for the case of impacting axial loads.

The use of Equation 2.28 in the solution of problems relating to the impact of torsional members is illustrated in Example 2.16.

EXAMPLE 2.16

Refer to the system shown in Figure 2.18. Let $W = 120$ lb, $h = 0.8$ in., $L = 36$ in., and $b = 10$ in., and assume the shaft to have a diameter of 3 in. and to be steel for which $G = 10 \times 10^6$ psi. If all of the available energy is transformed into elastic strain energy in the shaft, determine (a) the maximum shearing stress, (b) the maximum rotation of the rigid handle, and (c) the impact factor.

SOLUTION

a. Substituting the given numerical values into Equation 2.28, we obtain

$$120(0.8 + \Delta) = \frac{1}{2} F\Delta$$

where $\Delta = \theta b = \left(\dfrac{TL}{JG} \right) b = \dfrac{Fb^2 L}{JG} = \dfrac{F(10^2)(36)}{(\pi/32)(3^4)(10 \times 10^6)} = (4.527 \times 10^{-5})F$

Substituting the above value of Δ into the first equation and simplifying, we obtain

$$F^2 - 240F - 4.241 \times 10^6 = 0$$

Solving this quadratic equation yields

$$F = 2183 \text{ lb.}$$

Therefore,

$$T = Fb = 21{,}830 \text{ lb} \cdot \text{in}$$

The maximum shearing stress in the shaft is given by Equation 2.9. Thus,

$$\tau_{MAX} = \frac{(21{,}830)(1.5)}{(\pi/32)(3^4)} = 4118 \text{ psi} \qquad \textbf{ANS.}$$

b. The maximum rotation of the rigid handle is found from the relation $\theta = \Delta/b$. Thus,

$$\theta = \frac{(4.527 \times 10^{-5})F}{10} = \frac{(4.527 \times 10^{-5})(2{,}183)}{10} = 0.00988 \text{ rad} \approx 0.57° \qquad \textbf{ANS.}$$

c. The impact factor becomes

$$I.F. = \frac{F}{W} = \frac{2183}{120} = 18.2 \qquad \textbf{ANS.}$$

PROBLEMS

2.62 Determine the maximum shearing stress and angle of twist in a solid shaft (diameter $d = 30$ mm) that transmits 120 kW at 900 rpm. The length of the shaft is 1.0 m and $G = 70$ GPa.

2.63 A hollow shaft for which $d_o = 2.0$ in. and $d_i = 1.5$ in. rotates at 600 rpm. If the allowable shearing stress and angle of twist are 8 ksi and 4°, respectively, determine the maximum power that may be transmitted using this shaft. Let $L = 3$ ft and $G = 4 \times 10^3$ ksi.

2.64 A 30-kW motor operates at 1200 rpm. Find the minimum diameter for a solid shaft to transmit this power if the allowable shearing stress and angle of twist are 50 MPa and 0.08 rad, respectively, over a length of 0.50 m. Let $G = 75$ GPa.

2.65 A hollow steel shaft of outside diameter $d_o = 50$ mm is to be designed to transmit 20 kW at 1000 rpm. If the allowable shearing stress and angle of twist are 65 MPa and 0.07 rad over a length of 0.75 m, respectively, determine the maximum inside diameter of the shaft if $G = 75$ GPa.

2.66 A 20-hp motor operates at 600 rpm. Determine the minimum diameter needed for a solid shaft to transmit this power if the allowable shearing stress is 15 ksi. Assume the angle of twist of the shaft to be adequate.

2.67 A 55-mm diameter solid aluminum shaft is to transmit 150 kW to a machine tool. The allowable shearing stress and angle of twist are 25 MPa and 3° per meter, respectively. Find the maximum speed in rpm at which this shaft may operate. Let $G = 26$ GPa.

2.68 The shaft-gear system shown consisting of a motor at A, two gears at B and C and two shafts AB and CD, is used to transmit 20 hp at 300 rpm from the motor to the machine tool at D. The two shafts have equal diameters of 1.5 in. Determine the maximum shearing stresses and the angles of twist in both shafts. Let $r_1 = 2.5$ in., $r_2 = 4.0$ in., $L_1 = 4$ ft, $L_2 = 3$ ft, and $G = 4 \times 10^6$ psi.

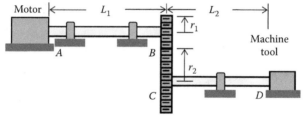

(Problems 2.68 and 2.69)

2.69 In reference to the figure shown in Problem 2.68, the shaft-gear system shown consisting of a motor at A, two gears at B and C, and two shafts AB and CD, is used to transmit 10 kW at 400 rpm from the motor to the machine tool at D. If the allowable shearing stress for the two shafts is 50 MPa, determine the minimum diameter required for each of the two shafts. Assume the angle of twist of both shafts to be adequate for the purpose. Let $r_1 = 40$ mm, $r_2 = 75$ mm, $L_1 = 1.25$ m, $L_2 = 3$ ft, and $G = 75$ GPa.

2.70 An oil-drill shaft with a diameter of 120 mm jams during drilling and has to be retracted. An axial pull $P = 650$ kN and a torque $Q = 20$ kN·m are required for the extraction process. Select a point on the outside surface of the shaft and construct a stress element with two planes parallel and two planes perpendicular to the axis of the shaft.

2.71 An oil-drill shaft with a diameter of 5 in. jams during drilling and has to be retracted. An axial pull $P = 150$ kips and a torque $Q = 150$ kip·in. are required for the extraction process. (a) Select a point on the outside surface of the shaft and construct a stress element with two planes parallel and two planes perpendicular to the axis of the shaft. (b) Use equilibrium to find the normal and shearing stresses on a plane that is oriented at 30° clockwise with the shaft axis. Use Example 2.12 as a guide.

2.72 A hollow steel shaft 5 m long, 150 mm outside diameter, and 6 mm inside diameter is fixed at one end and subjected at the other end to an axial compressive force of 500 kN and a torque of 30 kN·m. Select a point on the outside surface of the shaft and construct a stress element with two planes parallel and two planes perpendicular to the axis of the shaft.

2.73 Refer to the shaft of Problem 2.72. Use equilibrium to determine the normal and shearing stresses on the outside surface of the shaft on a plane making a 40° clockwise angle with the shaft axis. Use Example 2.12 as a guide.

2.74 The shaft shown has a diameter of 3.5 in. and carries an axial tensile force $P = 125$ kips and a torque $Q = 150$ kip·in. Use equilibrium to determine the normal and shearing stresses on plane N. Let $\theta = 45°$.

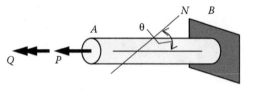

(Problems 2.74 and 2.75)

2.75 In reference to the figure shown in Problem 2.74, the shaft shown has a diameter of 100 mm and sustains an axial tensile force $P = 600$ kN and a torque $Q = 20$ kN·m. Use equilibrium to determine the normal and shearing stresses on plane N. Let $\theta = 60°$.

2.76 Let $P = 150$ kips, $P_1 = 2$ kip, and $P_2 = 3$ kip, pulley diameters at B and C equal 2.0 and 1.5 ft, respectively, $L = 5$ ft, and the diameter of shaft AB equals 4 in. Find the normal and shearing stresses on the outside surface of shaft AB on a plane oriented to the axis of the shaft by a 25° clockwise angle.

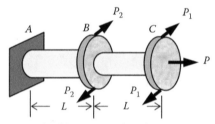

(Problems 2.76 and 2.77)

2.77 In reference to the figure shown in Problem 2.76, let $P = 550$ kN, $P_1 = 8$ kN, and $P_2 = 12$ kN, pulley diameters at B and C equal 0.5 m and 0.75 m, respectively, $L = 2.0$ m, and the diameter of shaft AB equals 120 mm. Find the normal and shearing stresses on the outside surface of shaft AB on a plane oriented to the axis of the shaft by a 30° counterclockwise angle.

2.78 Consider the stress element shown in the sketch. If the normal and shearing stresses on plane N are $\sigma_n = 15$ ksi and $\tau_{nt} = 10$ ksi, respectively, find the stresses τ and σ acting on the stress element if $\theta = 35°$.

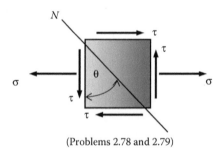

(Problems 2.78 and 2.79)

2.79 In reference to the figure shown in Problem 2.78, consider the stress element shown in the sketch. If the normal and shearing stresses on plane N are $\sigma_n = 150$ MPa and $\tau_{nt} = 100$ MPa, respectively, find the stresses τ and σ acting on the stress element if $\theta = -40°$.

2.80 Let $P = 60$ kips, $Q = 30$ kip·in., the diameter of the shaft $d = 2$ in., $\theta = 30°$, $\mu = 0.33$, and $E = 30 \times 10^3$ ksi. Determine (a) the normal strains along the x and y axes and (b) the shearing strain γ_{xy}.

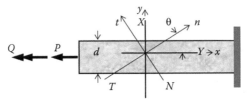

(Problems 2.80, 2.81, 2.82, and 2.83)

2.81 In reference to the figure shown in Problem 2.80, let $P = 150$ kN, $Q = 5$ kN·m, the diameter of the shaft $d = 60$ mm, $\theta = 40°$, $\mu = 0.28$, and $E = 75$ GPa.
Determine (a) the normal strains along the n and t axes and (b) the shearing strain γ_{nt}.

2.82 In reference to the figure shown in Problem 2.80, measurements of the strains along the x and y axes gave the following values: $\varepsilon_x = 400$ μ, $\varepsilon_y = -120$ μ, and $\gamma_{xy} = 800$ μ. Let the diameter of the shaft $D = 2.5$ in., $\mu = 0.3$, and $E = 10 \times 10^6$ psi, and determine the axial force P and the torque Q.

2.83 In reference to the figure shown in Problem 2.80, measurements of the strains along the n and t axes gave the following values: $\varepsilon_n = -336.3$ μ, $\varepsilon_t = 858.6$ μ, $\gamma_{nt} = -1823.7$ μ, and $\theta = 30°$. Let the diameter of the shaft $d = 80$ mm, $\mu = 0.30$, and $E = 200$ GPa, determine the axial force P and the torque Q.

2.84 Let $\sigma_n = 60$ ksi, $\sigma_t = -30$ ksi, $\tau_{nt} = \tau_{tn} = 20$ ksi, $E = 10 \times 10^3$ ksi, $\mu = 0.3$, and $\theta = 30°$, determine (a) the strains along the n and t axes and (b) the shearing strain γ_{nt}.

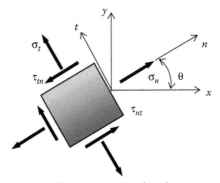

(Problems 2.84, 2.85, and 2.86)

2.85 In reference to the figure shown in Problem 2.84, measurements of the normal strains along the n and t axes gave the following values: $\varepsilon_n = 500$ μ, $\varepsilon_t = 400$ μ, and $\gamma_{nt} = 0$. Find the normal stresses σ_x and σ_y and the shearing stress τ_{xy}. Let $E = 75$ GPa, $\mu = 0.33$, and $\theta = 30°$.

2.86 In reference to the figure shown in Problem 2.84, measurements of the normal and shearing strains along the x and y axes gave the following values: $\varepsilon_x = 450$ μ, $\varepsilon_y = 600$ μ, and $\gamma_{xy} = -500$ μ. If $E = 30 \times 10^3$ ksi, $\mu = 0.3$, and $\theta = 20°$, determine the normal stresses σ_n and σ_t and the shearing stress τ_{nt}.

2.87 The two-diameter shaft is to transmit a torque $Q = 5.0$ kip·in. Find the maximum shearing stress in the shaft if (a) $r = 0.15$ in. and (b) $r = 0.05$ in. Let $D = 3.0$ in. and $d = 2.0$ in.

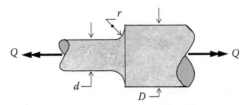

(Problems 2.87, 2.88, and 2.89)

2.88 In reference to the figure shown in Problem 2.87, the stepped shaft is made of a material for which the allowable shearing stress is 45 MPa. Determine the maximum power that may be transmitted at 1200 rpm if (a) $r = 3.0$ mm and (b) $r = 1.0$ mm. Let $D = 100$ mm and $d = 50$ mm.

2.89 In reference to the figure shown in Problem 2.87, the two-diameter shaft transmits 15 hp at 900 rpm. If the allowable shearing stress is 10 ksi, determine the smallest permissible fillet radius r. Let $D = 2.0$ in. and $d = 1.0$ in.

2.90 The two-diameter shaft transmits 50 kW at 180 rpm. If the allowable shearing stress is 60 MPa, $d = 70$ mm, and $r = 4$ mm, determine the largest permissible value of D.

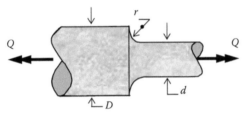

(Problems 2.90, 2.91, and 2.92)

2.91 In reference to the figure shown in Problem 2.90, the stepped shaft transmits 50 hp at 600 rpm. Determine the minimum permissible fillet radius r, if the allowable shearing stress is 6 ksi, $D = 3.0$ in., and $d = 2.0$ in.

2.92 In reference to the figure shown in Problem 2.90, the two-diameter shaft transmits 100 kW. If the allowable shearing stress is 70 MPa and $r = 5$ mm, determine the maximum speed of rotation expressed in rpm. Let $D = 100$ mm and $d = 70$ mm.

2.93 The system shown consists of hollow shaft AB of length $L = 20$ in. ($d_o = 4.0$ in., $d_i = 3.0$ in.), fixed at A and supported at B by a frictionless bearing where it is attached to rigid arm BC. A weight $W = 100$ lb is dropped as shown onto the rigid arm of length $b = 15$ in., from a height $h = 1.2$ in. The shaft is aluminum for which $G = 4 \times 10^6$ psi. Determine (a) the maximum shearing stress in the shaft and (b) the maximum rotation of the rigid arm. Assume a 15% energy loss during impact.

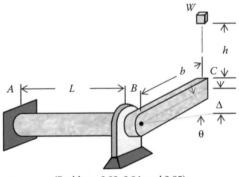

(Problems 2.93, 2.94, and 2.95)

2.94 The system shown consists of solid steel ($G = 75$ GPa) shaft AB of length $L = 0.75$ m and diameter $d = 50$ mm, fixed at A and supported at B by a frictionless bearing where it is attached to rigid arm BC. A weight W is dropped as shown onto the rigid arm of length $b = 1.25$ m, from a height $h = 75$ mm. If the allowable shearing stress in the shaft is 100 MPa and the allowable rotation at B is $5°$, find the largest weight W that may be dropped without exceeding these allowable values. Assume no energy losses during impact.

2.95 The system shown consists of hollow steel ($G = 4 \times 10^3$) shaft AB of length $L = 3$ ft, outside diameter $d_o = 6.0$ in., and inside diameter $d_i = 5.5$ in. fixed at A and supported at B by a frictionless bearing where it is attached to rigid arm BC whose length $b = 5$ ft. A weight $W = 150$ lb is dropped onto the rigid arm as shown from a height h. If the allowable shearing stress in the shaft is 12 ksi, and the allowable angle of twist is 0.2 rad, find the largest height h from which the weight W may be dropped without exceeding these allowable values.

2.96 A proposed diving system for a public swimming pool would consists of a rigid diving board CD attached rigidly at C to the outside surface of a hollow shaft AB ($d_o = 150$ mm, $d_i = 130$ mm), which is fixed at A and supported by a frictionless bearing at B as shown. A diver weighing 800 N jumps up to a height of 125 mm above end D of the board and then drops onto it before diving into the water below. Let $L = 2.0$ m and $a = 2.5$ m, determine the maximum shearing stress induced in the hollow shaft. Also find the angle of twist of the shaft at C if the hollow shaft is aluminum for which $G = 75$ GPa.

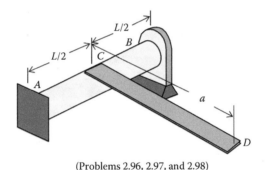

(Problems 2.96, 2.97, and 2.98)

2.97 A proposed diving system for a public swimming pool would consist of a rigid diving board CD attached rigidly to a solid shaft AB, which is fixed at A and supported by a frictionless bearing at B as shown. A 185-lb diver jumps up to a height of 6 in. above end D of the board and then drops onto it before diving into the water below. Let $L = 16$ ft and $a = 10$ ft, determine the smallest permissible diameter if the allowable shearing stress in the shaft is 12 ksi.

2.98 Refer to Problem 2.97 and replace the solid shaft with a hollow shaft ($d_o = 4.0$ in., $d_i = 3.5$ in.) and the frictionless bearing at B with a fixed support identical with that at A. A diver weighing 180 lb jumps up to a height of 5 in. above end D of the board and then drops onto it before diving into the water below. Let $L = 20$ ft and $a = 10$ ft, determine the maximum shearing stress induced in the hollow shaft and the maximum angle of twist of the shaft at C. Let $G = 10 \times 10^3$ ksi.

*2.9   SHAFTS OF NONCIRCULAR CROSS SECTIONS

2.9.1   ANALYTICAL SOLUTIONS

Up to this point, we have focused our attention on the solution of problems relating to shafts of circular cross sections. These solutions are of great practical importance because shafts of circular cross section carry torsional loads efficiently. A primary assumption (verifiable by experiment) made in deriving the fundamental equations is that plane sections of these members before twisting remain plane after twisting as shown in Figure 2.21, which is the photograph of a rubber model of a shaft with circular cross section after twisting. The vertical straight lines represent cross sections of this shaft, which even after twisting, have remained straight. On the other hand, plane sections of members with noncircular cross sections before twisting do not remain plane after twisting as shown in Figure 2.22, which is the photograph of a rubber model of a shaft with a rectangular cross section after twisting. Note the warped "vertical lines," which were straight vertical lines before twisting. Thus, points initially in a plane are displaced in a direction parallel to the axis of twist of the member. Such displacements, which occur in addition to in-plane displacements, are referred to as *warping displacements* and the function describing them analytically is termed the *warping function*. If these displacements were prevented from taking place (e.g., by attaching the ends of the twisted member to rigid supports), normal stresses would arise directed parallel to the axis of twist.

Mathematically, the problem of torsion of noncircular cross sections involves the solution of partial differential equations subject to boundary conditions. Such solutions are beyond the scope of this book. However, it should be pointed out that many such analytical solutions have been obtained, and three of these are given in Figures 2.23 through 2.25.

Figure 2.23 provides the solution for a shaft with an elliptical cross section. Note that the maximum shearing stress occurs at the ends of the minor axis (the two points labeled *A*).

Figure 2.24 gives the solution for a shaft with a rectangular cross section and shows that the maximum shearing stress occurs at the midpoints of the two longest sides (the two points marked *A*).

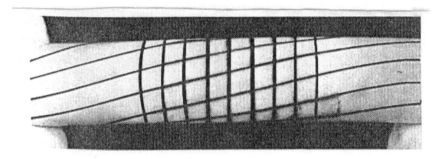

FIGURE 2.21   Diagram showing that, for a shaft of circular cross section, plane sections before twisting remain plane after twisting.

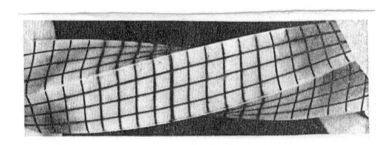

FIGURE 2.22   Diagram showing that, for a shaft of a noncircular cross section, plane sections before twisting do not remain plane after twisting.

$$\tau_{MAX} = \frac{2T}{\pi ab^2}; \quad \text{occurs at } A$$

$$\theta = \frac{TL}{KG}; \quad K = \frac{\pi a^3 b^3}{a^2 + b^2}$$

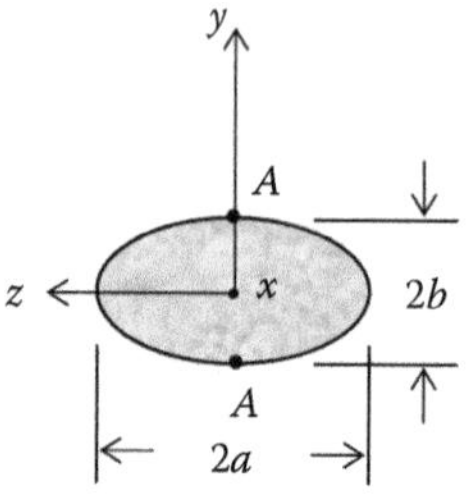

FIGURE 2.23   Mathematically obtained solutions for a shaft of an elliptical cross section.

$$\tau_{MAX} = \frac{T(3a + 1.8b)}{8a^2b^2}; \quad \text{occurs at } A$$

$$\theta = \frac{TL}{KG}; \quad K = ab^3 \left[\left(\frac{16}{3} \right) - 3.36 \left(\frac{b}{a} \right) \left(1 - \frac{b^4}{12a^4} \right) \right]$$

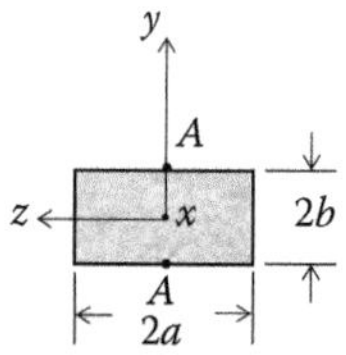

FIGURE 2.24   Mathematically obtained solutions for a shaft of a rectangular cross section.

Finally, Figure 2.25 provides the solution for a shaft with an equilateral triangle and indicates that the maximum shearing stress occurs at the mid-points of the three sides (the three points labeled A).

In each of the above three cases, the angle of twist is given by the equation $\theta = TL/KG$. Note the mathematical similarity of this equation with the one developed for the angle of twist of a circular shaft. The polar moment of inertia J for the circular cross section has been replaced by the

$$\tau_{MAX} = \frac{20T}{a^3}; \quad \text{occurs at } A$$

$$\theta = \frac{TL}{KG}; \quad K = \frac{\left(\sqrt{3} \right) a^4}{80}$$

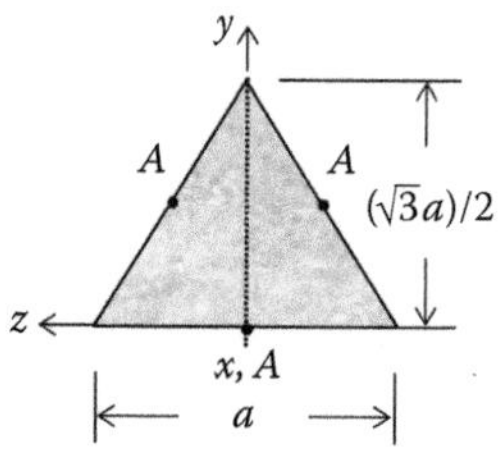

FIGURE 2.25   Mathematically obtained solutions for a shaft of an equilateral triangular cross section.

constant K, which is a unique value for each cross section as indicated in the above three cases. It should be pointed out that the above solutions assume that the two ends of the torsion member are free, and warping may take place freely.

EXAMPLE 2.17

An engineer is confronted with the problem of deciding whether to use a circular or an elliptical cross section for a shaft that will transmit 2000 hp at 1200 rpm. Both shafts have a length 10 ft and meet the geometric clearance requirements. They are both made of the same steel for which $G = 11 \times 10^3$ ksi. The circular shaft is to have a radius of 2.0 in. and the elliptical shaft is to have the same magnitude for its cross-sectional area with a semimajor axis of 2.5 in. Determine the maximum shearing stresses and angles of twist for both shafts and select one on this basis.

SOLUTION

The torque to be transmitted is found from Equation 2.18. Thus,

$$T = \frac{P}{\omega} = \frac{(2000)(550)}{(1200 \times 2\pi/60)} = 8753.522 \text{ lb} \cdot \text{ft}$$

Circular shaft

$$\tau_{MAX} = \frac{TR}{J} = \frac{(8753.522 \times 12)(2)}{(\pi/2)(2^4)} = 8359 \text{ psi}; \theta = \frac{TL}{JG} = \frac{(8753.522 \times 12)(10 \times 12)}{(\pi/2)(2^4)(11 \times 10^6)}$$

$$= 0.046 \text{ rad}$$

Elliptical shaft
Since the two areas are the same, it follows that

$$\pi R^2 = \pi ab; \quad b = \frac{R^2}{a} = \frac{4}{2.5} = 1.6 \text{ in.}$$

$$\tau_{MAX} = \frac{2T}{\pi ab^2} = \frac{2 \times 8753.522 \times 12}{\pi(2.5)(1.6)} = 16718 \text{ psi}; \quad K = \frac{\pi a^3 b^3}{a^2 + b^2} = \frac{\pi(2.5^3)(1.6^3)}{2.5^2 + 1.6^2} = 22.822 \text{ in.}^4$$

$$\theta = \frac{TL}{KG} = \frac{(8753.522 \times 12)(10 \times 12)}{(22.822)(11 \times 10^6)} = 0.050 \text{ rad}$$

The maximum shearing stress in the elliptical shaft is exactly twice that in the circular shaft. Also, the angle of twist in the elliptical shaft is about 9% larger than that in the circular shaft. Since all other conditions are identical, we conclude that the circular shaft is a much better choice for this application.   **ANS.**

2.9.2 EXPERIMENTAL SOLUTIONS

An experimental method for solving the torsion problem, known as the *membrane analogy* (also as the soap-film analogy) was introduced by L. Prandtl in 1903. It is based upon the fact that the mathematical formulation of the torsion problem and the *small* displacements of a homogeneous membrane under pressure give rise to the same boundary-value problems. If a membrane is placed over an opening whose shape is identical to the cross section of a shaft and fastened all around its perimeter, its distended shape when subjected to uniform pressure, can be expressed by an equation that is mathematically similar to the equation defining the stress distribution in the twisted shaft.

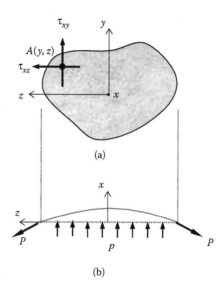

(a)

(b)

FIGURE 2.26   (a,b) The two views of a distended thin membrane subjected to a pressure p used to experimentally solve the torsion problem of noncircular shafts.

Thus, consider the case of a twisted shaft with the arbitrary cross section shown in Figure 2.26a. It can be shown that the stress function $F = (y,z)$ governing the distribution of shearing stresses in the twisted shaft obeys the following partial differential equation:

$$\frac{\partial^2 F}{\partial y^2} + \frac{\partial^2 F}{\partial z^2} = -2G\phi \qquad (2.29)$$

The quantity ϕ in Equation 2.29 is the angle of twist per unit of length and G is the modulus of rigidity. Also, if the stress function F is known, then the components of the shearing stress at any point such as $A(y,z)$ in the cross section may be found from the relations

$$\tau_{xy} = -\frac{\partial F}{\partial z}; \quad \tau_{xz} = \frac{\partial F}{\partial y} \qquad (2.30)$$

where x–y–z is a right-handed coordinate system and the x axis points out of the page. Now, if a thin membrane is stretched over an opening of the same size and shape as that shown in Figure 2.26a, and a uniform pressure p is applied, the thin membrane will be distended. One section of the distended membrane taken in the x–z plane is shown in Figure 2.26b, where P represents the force per unit length around the edge of the membrane along the perimeter of the opening.

It can be shown that the function describing the distended membrane is of the form

$$\frac{\partial^2 x}{\partial y^2} + \frac{\partial^2 x}{\partial z^2} = -\frac{p}{P} \qquad (2.31)$$

If we set the quantity $2G\phi$ numerically equal to the quantity p/P, then Equations 2.29 and 2.31 become mathematically identical. Thus, by examining the properties of the deflected membrane, we can determine the shearing stresses as well as other characteristics of the twisted shaft.

Let us now rewrite Equation 2.29 in the form

$$\frac{\partial}{\partial y}\left(\frac{\partial F}{\partial y}\right) + \frac{\partial}{\partial z}\left(\frac{\partial F}{\partial z}\right) = -2G\phi \tag{2.32}$$

Using Equation 2.30, we can express Equation 2.32 in terms of the shearing stress components. Thus,

$$\frac{\partial}{\partial y}(\tau_{xz}) + \frac{\partial}{\partial z}(-\tau_{xy}) = -2G\phi \tag{2.33}$$

Rewriting Equation 2.31 after replacing the quantity p/P with the quantity $2G\phi$, we obtain

$$\frac{\partial}{\partial y}\left(\frac{\partial x}{\partial y}\right) + \frac{\partial}{\partial z}\left(\frac{\partial x}{\partial z}\right) = -2G\phi \tag{2.34}$$

where $(\partial x/\partial y)$ and $(\partial x/\partial z)$ are the slopes of the deflected membrane in the y and z directions, respectively. Therefore, by comparing Equations 2.32 and 2.34, we conclude that $\tau_{xy} = -(\partial x/\partial z)$ and $\tau_{xz} = \partial x/\partial y$.

It can also be shown that the volume under the deflected membrane is proportional to the torque acting on the twisted shaft. Thus, the following two general principles may be stated:

1. The slope of the distended membrane at any point is proportional to the shearing stress in the twisted shaft. The shearing stress is pointed in a direction perpendicular to the plane in which the slope is measured.
2. Twice the volume under the deflected membrane is proportional to the torque acting on the twisted shaft.

The proof of the second principle above is beyond the scope of this book. However, both principles are shown to be valid when applied to the torsion of a circular shaft as illustrated in Example 2.18 below.

EXAMPLE 2.18

Show that the above two membrane analogy principles are consistent with the solutions obtained previously for a shaft with a circular cross section.

SOLUTION

A circular cross section of radius R and the corresponding deflected membrane are shown in the sketch. Because of the symmetric nature of this cross section, we can deal with ordinary differentials instead of partial differentials. Thus, Equation 2.31 may be written in the following form:

$$2\left(\frac{d^2 x}{d\rho^2}\right) = -\frac{p}{P} = -2G\phi$$

$$\frac{d}{d\rho}\left(\frac{dx}{d\rho}\right) = -G\phi; \quad \frac{dx}{d\rho} = -G\phi\rho + C_1 \tag{2.18.1}$$

The quantity $dx/d\rho$ in Equation 2.18.1 is the slope of the distended membrane and C_1 is a constant of integration that is evaluated from the boundary condition that at $\rho = 0$, $dx/d\rho = 0$ and $C_1 = 0$. Therefore, Equation 2.18.1 becomes

$$\frac{dx}{d\rho} = -G\phi\rho \tag{2.18.2}$$

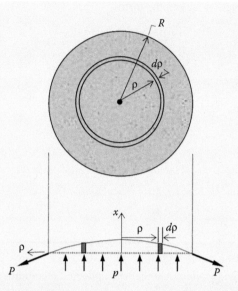

Now, for a circular cross section $\phi = T/JG$ and $T = JG\phi$.
Thus,

$$\tau = \frac{T\rho}{J} = \frac{(JG\phi)\rho}{J} = G\phi\rho \qquad (2.18.3)$$

Setting Equations 2.18.2 and 2.18.3 equal to each other, we obtain

$$\tau = -\frac{dx}{d\rho} \qquad \textbf{ANS.}$$

which shows that the magnitude of the shearing stress is, in fact, given by the slope of the
distended membrane.

Integrating Equation 2.18.2 yields

$$\int dx = -G\phi \int_0^R \rho \, d\rho$$

$$x = -(1/2)G\phi R^2 + C_2$$

where C_2 is a second constant of integration. This constant is evaluated from the boundary
condition that when $\rho = R$, $x = 0$, which leads to $C_2 = (1/2)G\phi R^2$. Therefore,

$$x = (1/2)(G\phi)(R^2 - \rho^2)$$

Since $dV = x(2\pi\rho \, d\rho) = (\pi G\phi)(R^2 - \rho^2)\rho \, d\rho$, the volume under the distended membrane
becomes

$$V = (\pi G\phi)\int_0^R (R^2 - \rho^2)\rho \, d\rho = \left(\frac{G\phi}{2}\right)\left(\frac{\pi R^4}{2}\right) = \frac{G\phi J}{2}$$

Therefore, it follows that

$$2V = G\phi J = T \qquad \textbf{ANS.}$$

which shows that twice the volume under the distended membrane gives the torque in the twisted circular shaft.

2.9.3 APPLICATION TO TORSION OF A LONG, THIN RECTANGLE

A thin long rectangle $2a \times 2b$ is shown in Figure 2.27 in which $2a \ggg 2b$. Also shown is a cross section of the distended membrane taken at a distance far removed from the two ends, that is, close to $y = 0$. Ignoring the distorted conditions at these two ends, Equation 2.31, when applied to this distended membrane, reduces to

$$\frac{d^2 x}{dz^2} + 0 = -\frac{p}{P} = -2G\phi \qquad (2.35)$$

The first term in Equation 2.31, namely, d^2x/dy^2, vanishes under the assumption that, since $2a \ggg 2b$, the distended membrane has a constant height along the y axis (except near the two ends) so that its slope along the y direction is negligibly small.

Integrating Equation 2.35, we obtain

$$\frac{dx}{dz} = -2G\phi z + C_1 \qquad (2.36)$$

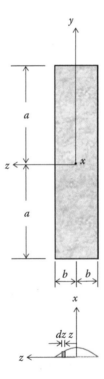

FIGURE 2.27 Experimental solution of the torsion problem of a shaft having a long and thin rectangular cross section.

where C_1 is evaluated from the condition that when $z = 0$, $dx/dz = 0$. This, of course, leads to $C_1 = 0$. Integrating Equation 2.36, we obtain

$$x = -G\phi z^2 + C_2 \tag{2.37}$$

where $C_2 = G\phi b^2$ from the condition that when $z = b$, $x = 0$. It follows, therefore, that

$$x = G\phi(b^2 - z^2) \tag{2.38}$$

A good approximation for the volume under the distended membrane is obtained by assuming that its cross-sectional area along the y axis is constant and given by

$$A = 2\int_0^b x\,dz \tag{2.39}$$

Since the volume under the distended membrane is the product of this area and the length ($2a$) of the thin rectangle, this volume becomes

$$V = (2a)A = (2a)(2)\int_0^b x\,dz = (4G\phi a)\int_0^b (b^2 - z^2)\,dz \tag{2.40}$$

Integrating Equation 2.40 and simplifying, we obtain

$$V = \left(\frac{8}{3}\right)ab^3 G\phi \tag{2.41}$$

Equation 2.41 forms the basis for the torsional analysis of shafts having rectangular cross sections that are long and thin. Thus, according to the second principle of the membrane analogy, we have

$$T = 2V = \left(\frac{16}{3}\right)ab^3 G\phi \tag{2.42}$$

Solving Equation 2.42 for the angle of twist per unit length ϕ leads to

$$\phi = \frac{3T}{16ab^3 G} \tag{2.43}$$

The maximum shearing stress is, according to the first principle of the membrane analogy, given by the maximum slope of the distended membrane that occurs at $z = \pm b$, at the midpoints of the two long sides. Thus, using Equation 2.36 above with $C_1 = 0$, we obtain

$$\tau_{MAX} = \left(\frac{dx}{dz}\right)_{z=0} = -2G\phi b \tag{2.44}$$

Ignoring the negative sign in order to deal only with the stress magnitude, substituting for ϕ from Equation 2.43 and simplifying, we obtain

$$\tau_{MAX} = \frac{3T}{8ab^2} \tag{2.45}$$

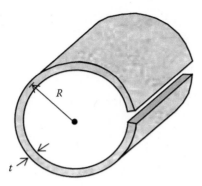

FIGURE 2.28   Solution of the torsion problem of a shaft with a narrow circular section having a thin slit.

2.9.4   SPECIAL CASES

2.9.4.1   Narrow Circular Section with Thin Slit

Equations 2.43 and 2.45 may be used to solve the torsion problem of a shaft whose cross section is thin and circular with a narrow slit along its length as shown in Figure 2.28. In this case, the thickness of the cross section is $t = 2b$ and the circumference is $2\pi R = 2a$ in Equations 2.43 and 2.45. Thus, making these substitutions and simplifying leads to

$$\left. \begin{array}{c} \phi = \dfrac{3T}{2\pi\, Rt^3 G} \\[3mm] \tau_{MAX} = \dfrac{3T}{2\pi\, Rt^2} \end{array} \right\} \tag{2.46}$$

2.9.4.2   Sections Composed of Narrow Rectangles

Figures 2.29a and 2.29b show two examples of structural cross sections of shafts for which the solution of the torsion problem may be obtained by using Equations 2.43 and 2.45. Rewriting Equation 2.43, we have

$$\phi = \frac{T}{(1/3)(2a)(2b)^3\, G} = \frac{T}{(1/3)L\, t^3 G} = \frac{T}{KG} \tag{2.47}$$

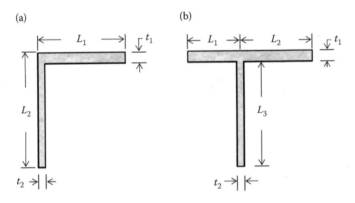

FIGURE 2.29   Diagrams showing cross sections composed of narrow rectangles for which solutions can be obtained.

where K is the *torsion constant* equal to $(1/3)Lt^3$ for a single thin rectangle and, for a section consisting of two or more thin rectangles, such as shown in Figures 2.29a and 2.29b, it would be given by

$$K = \sum (1/3)Lt^3 \tag{2.48}$$

Similarly, Equation 2.45 may be rewritten in the form

$$\tau_{MAX} = \frac{Tt}{(1/3)Lt^3} = \frac{Tt}{K} \tag{2.49}$$

where K is the torsion constant given by Equation 2.48. Note the mathematical similarities between Equations 2.47 and 2.49 on the one hand and Equations 2.8 through 2.12, developed earlier for the circular cross section, on the other. Note also that τ_{MAX} occurs in the thickest rectangle.

EXAMPLE 2.19

A circular shaft with a thin slit, as shown in Figure 2.28 ($R = 1.5$ in., $t = 0.1$ in.), is being considered to carry a torque of 400 lb·in. Compare the performance of this shaft with that of another of the same volume V and same length L, having an angular cross section as shown in Figure 2.29a, where $L_1 = 4$ in., $t_1 = 0.08$ in., and $L_2 = 6$ in. Select one of these two cross sections for the application. The material is steel for which $G = 10 \times 10^6$ ksi.

SOLUTION

Circular section

$$\tau_{MAX} = \frac{3T}{2\pi R t^2} = \frac{3(400)}{2\pi(1.5)(0.1^2)} = 12.732 \text{ ksi}$$

$$\phi = \frac{3T}{2\pi R t^3 G} = \frac{3(400)}{2\pi(1.5)(0.1^3)(10 \times 10^6)} = 0.01273 \text{ rad/in.}$$

Angle section
The volume of the circular section is

$$V = (2\pi R)t = 2\pi(1.5)(0.1)(L) = 0.94248(L) \text{ in.}^3$$

The volume of the angular section is $(4 \times 0.08 + 6t_2)(L)$. Since the two volumes are to be the same, it follows that

$$4 \times 0.08 + 6t_2 = 0.94248; \quad t_2 = 0.10375 \text{ in.}$$

Therefore, the torsion constant becomes

$$K = \sum (1/3)Lt^3 = (1/3)[4 \times 0.08^3 + 6 \times 0.10375^3] = 2.916 \times 10^{-3} \text{ in.}^4$$

$$\tau_{MAX} = \frac{Tt}{K} = \frac{(400)(0.10375)}{2.916 \times 10^{-3}} = 14.232 \text{ ksi}$$

$$\phi = \frac{T}{KG} = \frac{400}{(2.916 \times 10^{-3})(10 \times 10^6)} = 0.01372 \text{ rad/in.}$$

Comparing the two shafts, we conclude that on the basis of both the shearing stress and angle of twist, the shaft with the circular cross section is the better choice.   **ANS.**

2.9.5 Application to Torsion of Thin-Walled Tubes

The cross section of a thin-walled torsional tube is shown in Figure 2.30a and a sectional view *ABCD* of the distended membrane in Figure 2.30b. The distended membrane consists of the weightless horizontal rigid plate *AB* having the same size and shape as the inside of the tube. The membrane (represented by curves *AC* and *BD*) connects this rigid plate to the perimeter of an opening that has the same size and shape as the outside of the tube. Note that the weightless plate is horizontal and that the wall thickness *t* of the tube may vary around the tube perimeter.

In deriving the applicable equations, the assumption is made that the membrane is not curved but straight (i.e., curves *AC* and *BD* are straight lines) making it easier to determine the volume under the distended membrane. This volume can then be approximated by the product of the small height *h* of the distended membrane and A_m, the mean of the two areas of the tube, the outside area contained within the outer boundary and the inside area bounded by the inner perimeter.

Since twice the volume under the distended membrane is proportional to the applied torque, and the shearing stress is proportional to the slope of the distended membrane, we have

$$\left. \begin{array}{l} T = C(2h\,A_m) \\[2mm] \tau = C\left(\dfrac{h}{t}\right) \end{array} \right\}$$
                                                                                                     (2.50)

where h, A_m, and t have already been defined, and C is a constant of proportionality. Note that the second of Equation 2.50 yields an average value of the shearing stress because we assumed the slope of the membrane to be constant. This, of course, implies that the shearing stress is constant across the wall thickness of the tube, a condition that is very close to the actual situation since the wall thickness is assumed to be very small so that the difference among the maximum,

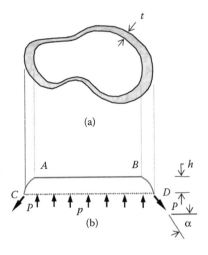

(a)

(b)

FIGURE 2.30   (a,b) The two views of a distended thin membrane subjected to a pressure *p* used to experimentally solve the torsion problem of thin-walled tubes of noncircular shafts.

minimum, and average shearing stresses is insignificantly small. Of course, the central horizontal plate implies no shearing stresses in the opening where there is no material to resist such
stresses.

Solving the first of Equation 2.50 for h, substituting into the second and simplifying, we obtain

$$\tau = \frac{T}{2tA_m} \tag{2.51}$$

Multiplying both sides of Equation 2.51 by the thickness t, we obtain the product $t\tau$, which has
units of force per unit of length. This product is known as the shear flow, in analogy to the flow of
fluid in a pipe or channel and given the symbol q. Thus,

$$q = t\tau = \frac{T}{2A_m} \tag{2.52}$$

Since the torque T and the mean area A_m are constants, it follows that the shear flow q is also
constant around the perimeter of the tube.

The angle of twist per unit of length, ϕ, is obtained by considering the vertical equilibrium of the
distended membrane shown in Figure 2.30b. Thus, since for small distensions and small angles, α,
$\sin \alpha \cong \tan \alpha = h/t = \tau$, we conclude that the vertical component of P is $(h/t)P = \tau P$. Thus,

$$\sum F_{VERT.} = 0 \Rightarrow pA_m - \int_0^s \tau P \, ds = 0 \tag{2.53}$$

where the quantity ds is a differential length around the perimeter of the tube. Since for small distensions, the force per unit length, P, may be assumed constant around the perimeter, it is taken out
of the integral sign, and both sides of the equation divided by it. Also, replacing the ratio p/P by the
quantity $2G\phi$ and τ by $T/2tA_m$, Equation 2.53 may be rewritten in the form

$$2G\phi A_m = \int_0^s \left(\frac{T}{2tA_m} \right) ds \tag{2.54}$$

The quantity $T/2A_m$ is a constant and is taken out of the integral sign, and the resulting equation is
solved for ϕ and simplified to yield

$$\phi = \left(\frac{T}{4GA_m^2} \right) \int_0^s \left(\frac{ds}{t} \right) \tag{2.55}$$

If the hollow tube consists of a number of simple geometric shapes such as, for example, a hollow rectangular tube, the integration in Equation 2.55 reduces to a summation and the angle of twist
becomes

$$\phi = \left(\frac{T}{4GA_m^2} \right) \sum \left(\frac{s}{t} \right) \tag{2.56}$$

EXAMPLE 2.20

A rectangular hollow tube, whose cross section is shown in the sketch, is subjected to an internal torque $T = 60$ kN·m. Determine (a) the maximum shearing stress in the cross section specifying its location and (b) the angle of twist of the tube per meter. Let $G = 28$ GPa.

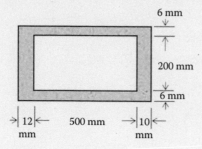

SOLUTION

a. The shearing stress is given by Equation 2.51. However, the maximum shearing stress in the cross section occurs in the middle of the two rectangles where the thickness t is the least, i.e., in the two horizontal rectangles. Thus,

$$A_m = (1/2)[(0.522 \times 0.212) + (0.50 \times 0.20)] = 0.105332 \text{ m}^2$$

$$\tau_{MAX} = \frac{T}{2tA_m} = \frac{60 \times 10^3}{2(0.006)(0.105332)} = 47.469 \approx 47.5 \text{ MPa} \qquad \textbf{ANS.}$$

b. The angle of twist is given by Equation 2.56. Thus,

$$\phi = \left(\frac{T}{4GA_m}\right)\sum\left(\frac{s}{t}\right) = \frac{60 \times 10^3}{4(28 \times 10^9)(0.105332)}\left[2\left(\frac{0.511}{0.006}\right) + \frac{0.206}{0.010} + \frac{0.206}{0.012}\right]$$

$$\phi = 0.011 \text{ rad/m}$$

Note that the lengths s of the four rectangles making up the hollow section were measured from mid-thickness to mid-thickness.

EXAMPLE 2.21

The cross section shown is that for a tube designed to carry a torque. If the allowable shearing stress and angle of twist are 8 ksi and 0.0012 rad/in., respectively, determine the maximum permissible torque. Let $G = 4 \times 10^3$ ksi.

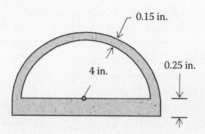

SOLUTION

Since the mean area A_m is needed, we will compute this quantity first. Thus,

$$A_m = \left(\frac{1}{2}\right)\left[\pi(4.15^2) + (0.25 \times 0.83) + \pi(4^2)\right] = 53.223 \text{ in.}^2$$

From Equation 2.51, we obtain

$$T_\tau = 2\tau t A_m = 2(8)(0.15)(53.223) = 127.735 = 127.7 \text{ kip} \cdot \text{in.} \qquad \textbf{ANS.}$$

From Equation 2.56, we obtain

$$T_\phi = (4GA_m^2\phi)\left[\left(\frac{1}{\sum\left(\frac{s}{t}\right)}\right)\right] = 4 \times 4 \times 10^3(53.223^2) \times 0.0012)\left[\frac{1}{\left(\frac{\pi \times 4.075}{0.15}\right) + \left(\frac{8.15}{0.25}\right)}\right]$$

$$T_\phi = 461.125 \text{ kip} \cdot \text{in.}$$

Note that T_τ was selected instead of T_ϕ because exceeding this value would violate the limitation of the allowable shearing stress.

PROBLEMS

2.99 A shaft with an elliptical cross section as shown in the sketch ($a = 2$ in., $b = 1$ in.) is to carry a torque of 30 kip·in. Determine (a) the maximum shearing stress specifying its location and (b) the angle of twist per inch of length. Let $G = 4 \times 10^3$ ksi.

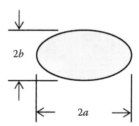

(Problems 2.99, 2.100, and 2.101)

2.100 A shaft of an elliptical cross section as shown in the sketch ($a = 60$ mm, $b = 40$ mm) has a length of 1.5 m and is to carry a torque Q. Determine the maximum permissible torque Q if the allowable angle of twist is $12°$ and the allowable shearing stress is 60 MPa. The shaft material is steel for which $G = 75$ GPa.

2.101 An elliptical cross section is to be used for a shaft whose length is 3.5 ft and is to carry a torque of 40 kip·in. The dimension $a = 3.0$ in. If the allowable shearing stress is 8 ksi, determine the smallest permissible dimension b for this cross section.

2.102 A membrane placed over an elliptical opening for which $2a = 70$ mm and $2b = 50$ mm was subjected to a uniform pressure and the volume under the distended membrane was measured at 3.0×10^{-6} m³. If the calibration ratio is $2G\phi/(p/P) = 80$ MPa, find the maximum shearing stress and the corresponding angle of twist over a length of 1.5 m in a shaft of the same cross section. Let $G = 25$ GPa. Note that your solution may be a combination of both the membrane analogy and analytical methods.

2.103 A shaft is to have a rectangular cross section as shown ($a = 50$ mm, $b = 30$ mm) and is to transmit a torque of 500 N·m. Find (a) the maximum shearing stress specifying its location in the cross section and (b) the angle of a twist over a length of 1.20 m. Let $G = 30$ GPa.

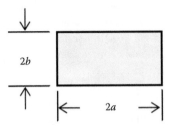

(Problems 2.103, 2.104, and 2.105)

2.104 In reference to the figure shown in Problem 2.103, a shaft is to have a rectangular cross section as shown ($a = 2.5$ in., $b = 1.5$ in.) and a length of 3 ft. Determine the maximum permissible torque that this shaft can carry if the allowable shearing stress and angle of twist are 15 ksi and 0.02 rad/ft, respectively. Let $G = 4 \times 10^3$ ksi.

2.105 In reference to the figure shown in Problem 2.103, a rectangular cross section is to be used for a shaft that is to carry a torque of 15 kN·m. The dimension $a = 45$ mm. If the allowable shearing stress is 35 MPa, determine the smallest permissible dimension b for this cross section.

2.106 A membrane stretched over a rectangular opening for which $2a = 5$ in. and $2b = 3.5$ in. was subjected to uniform pressure and the volume under the distended membrane was measured at 2.5 in.³. If the calibration ratio $2G\phi/(p/P) = 20.5$ ksi, find the maximum shearing stress and the angle of twist over a length of 5 ft. in a shaft of the same cross section. Let $G = 10 \times 10^3$ ksi. Note that your solution may be a combination of both the membrane analogy and the analytical method.

2.107 A shaft has an equilateral triangular cross section as shown ($a = 80$ mm) and is to transmit a torque of 7.5 kN·m. Determine (a) the maximum shearing stress specifying its location in the cross section and (b) the angle of twist over a length of 2.0 m. Let $G = 27$ GPa.

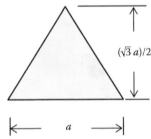

(Problems 2.107, 2.108, and 2.109)

2.108 In reference to the figure shown in Problem 2.107, a shaft is to have an equilateral triangular cross section as shown ($a = 3$ in.) and a length of 5 ft. If the allowable shearing stress and angle of twist are 12 ksi and 0.15 rad over a length of 3 ft, respectively, determine the maximum permissible torque to which this shaft may be subjected. Let $G = 4 \times 10^3$ ksi.

2.109 In reference to the figure shown in Problem 2.107, a shaft is to have an equilateral triangular cross section as shown and is subjected to a torque of 10 kN · m. If the allowable shearing stress and angle of twist over a length of 2.0 m are 20 MPa and 0.2 rad and $G = 27$ GPa, respectively, determine the least permissible value of the dimension a.

2.110 Two shafts (a circular cross section with $R = 1.5$ in. and an elliptical cross section with $b = 1.25$ in.) of the same material for which the allowable shearing stress is 15 ksi are being considered for a given application. If the two shafts are to have the same volume, select the one cross section you would use for the application stating the reasons for your choice. Ignore the rotational characteristics in making your selection.

2.111 Two shafts (an equilateral triangular cross section and a square cross section) of the same material for which the allowable shearing stress is 75 MPa are being considered for a given application. If the two shafts are to have equal volumes and lengths and if the side of the square is 65 mm, select one of the two cross sections for the application stating the reasons for your choice. Do not include the angle of twist in your selection.

2.112 The cross section shown is that for a shaft 4 ft long, which is to carry a torque of 150 kip · in. Find (a) the maximum shearing stress specifying its location and (b) the angle of twist. Let $G = 4 \times 10^3$ ksi and $a = 1.5$ in.

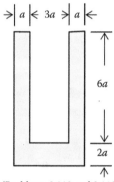

(Problems 2.112 and 2.113)

2.113 In reference to the figure shown in Problem 2.112, the cross section shown is that for a shaft that is 3.0 m long. If the allowable shearing stress and angle of twist are 40 MPa and 6°, respectively, determine the maximum permissible torque Q. Let $G = 25$ GPa and $a = 50$ mm.

2.114 The cross section shown is that for a shaft that is to carry a torque of 100 kip · in. Compute (a) the torsion constant K and find the maximum shearing stress specifying its location and (b) the angle of twist over a length of 3 ft. Let $G = 10 \times 10^6$ psi and $b = 1.25$ in.

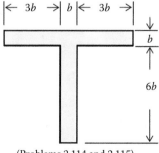

(Problems 2.114 and 2.115)

2.115 In reference to the figure shown in Problem 2.114, the cross section shown is that for a shaft 2 m long. It is made of aluminum for which $G = 27$ GPa. If the allowable shearing stress and angle of twist are 50 MPa and 5°, respectively, determine the maximum permissible torque Q. Let $b = 40$ mm.

2.116 The steel ($G = 10 \times 10^3$ ksi) cross section shown is that for a shaft that is subjected to a torque of 12 kip·in. Let $t = 0.5$ in. and if the allowable shearing stress is 12 ksi, determine the dimension a. Note that the dimensions given in terms of a are measured from mid-thickness to mid-thickness.

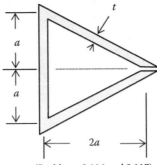

(Problems 2.116 and 2.117)

2.117 In reference to the figure shown in Problem 2.116, the steel ($G = 77$ GPa) cross section shown is that for a shaft that is 1.5 m long. If the allowable shearing stress and angle of twist are 80 MPa and 0.10 rad, respectively, determine the maximum permissible torque Q. Let $a = 100$ mm and $t = 15$ mm. Note that the dimensions given in terms of a are measured from mid-thickness to mid-thickness.

2.118 A W12 × 72 steel section 5 ft long is subjected to a torque of 10 kip·ft. Determine (a) the maximum shearing stress and (b) the angle of twist. Let $G = 10 \times 10^3$ ksi. Note that properties of this section are not in Appendix E but have to be obtained from the *Steel Construction Manual*.

2.119 A C8 × 18.75 American Standard channel (see Appendix E) is used for a member 4 ft long. If the allowable shearing stress and angle of twist are 10 ksi and 10°, respectively, find the maximum permissible torque Q that this member can carry. Let $G = 10 \times 10^3$ ksi.

2.120 The hollow rectangular cross section is that for an aluminum tube 1.0 m long. If a torque of 7.5 kN·m is applied, determine (a) the maximum shearing stress specifying its location, (b) the shear flow, and (c) the angle of twist. Let $G = 27$ GPa and $a = 20$ mm.

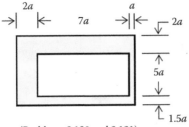

(Problems 2.120 and 2.121)

2.121 In reference to the figure shown in Problem 2.120, the hollow rectangular cross section is that for a steel tube 5 ft long that will carry a torque of 5 kip·in. If the maximum shearing stress and angle of twist are 8 ksi and 0.15 rad, respectively, determine the minimum permissible dimension a. Let $G = 10 \times 10^3$ ksi.

2.122 A tube 6 ft long has the hollow triangular cross section shown. The material is cold-rolled copper for which $G = 6.4 \times 10^3$ ksi. The tube is to transmit a torque of 10 kip·ft. Determine (a) the maximum shearing stress and (b) the angle of twist. Let $b = 1.0$ in.

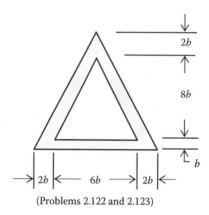

(Problems 2.122 and 2.123)

2.123 In reference to the figure shown in Problem 2.122, a tube 1.5 m long has the hollow triangular cross section shown. It is made of a material for which $G = 50$ GPa. If $b = 20$ mm and the allowable shearing stress and angle of twist are 40 MPa and 1°, respectively, find the maximum permissible torque that may be applied.

2.124 A tube 4 ft long has the hollow cross-sectional shape shown and is to carry a torque of 100 kip·in. The material is aluminum for which $G = 4 \times 10^3$ ksi. If the allowable shearing stress and angle of twist are 12 ksi and 10°, respectively, determine the least permissible dimension a. Note that all dimensions are given from mid-thickness to mid-thickness and that $t = a$ is constant all around the section.

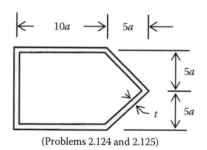

(Problems 2.124 and 2.125)

2.125 In reference to the figure shown in Problem 2.124, a tube 2 m long has the hollow cross-sectional shape shown. The material is such that $G = 27$ GPa. If the allowable shearing stress and angle of twist are 25 MPa and 0.05 rad, respectively, determine the maximum permissible torque that may be applied. Let $t = a = 15$ mm and constant all around the section.

2.126 A tube 3 ft long has the hollow cross-sectional shape shown and is to transmit a torque of 15 kip·in. The material is cold-rolled yellow brass for which $G = 5.5 \times 10^3$ ksi. Let $r = 1.5b$ and $t = 0.25b$ (constant around the cross section). If the allowable shearing stress and angle of twist are 10 ksi and 15°, respectively, determine the least permissible dimension b.

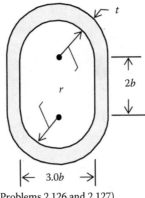

(Problems 2.126 and 2.127)

2.127 A tube 1.5 m long has the hollow cross-sectional shape shown. It is made of aluminum for which $G = 26$ GPa. Let $r = 1.5b$, $t = 0.25b$ (constant around the cross section), and $b = 30$ mm. If the allowable shearing stress and angle of twist are 30 MPa and 4°, respectively, determine the maximum permissible torque Q that may be applied.

*2.10 ELASTOPLASTIC BEHAVIOR

The shearing stress–strain diagram for a mild steel specimen is shown in Figure 2.31. The diagram has been idealized as two straight lines, OY representing elastic behavior up to the shearing yield stress τ_y, and horizontal line YU representing plastic behavior. Such idealized diagrams are referred to as *elastoplastic stress–strain* diagrams and are very useful in obtaining solutions for torsional problems in the plastic range.

Consider the case of a solid circular shaft, made of an elastoplastic material similar to that shown in Figure 2.31, subjected to twisting action. As the torque increases, the maximum shearing stress in the shaft increases until it reaches the yield stress τ_y at its outer surface as shown in Figure 2.32a. The torque corresponding to this condition is the yield torque T_y. If the torque continues to increase above this value, plastic deformation penetrates the outer layers of the circular cross section, where the shearing stress is constant at τ_y, leaving a central elastic core, where the shearing stress τ_p decreases linearly from τ_y at the outer surface of the elastic core to zero at its center as shown in Figure 2.32b. Ultimately, if the torque is increased further, plastic deformation penetrates the entire

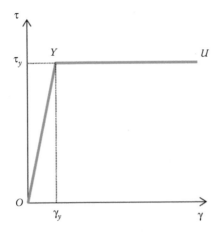

FIGURE 2.31   Stress–strain diagram for an elastoplastic material in which τ_y is the yield stress.

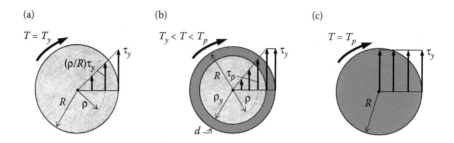

FIGURE 2.32 (a) The shearing stress distribution in a circular shaft when it reaches the yield stress τ_y. (b) The stress distribution when plastic action has reached the outer layers of the circular shaft. (c) The stress distribution when the entire shaft is under plastic action.

circular cross section, in which case the shearing stress is constant at τ_y throughout as shown in Figure 2.32c. The torque corresponding to this condition is T_p, known as the *fully plastic torque*.

The shearing stress in the central elastic core of Figure 2.32b is given by

$$\tau_\rho = \left(\frac{\rho}{\rho_y}\right)\tau_y \tag{2.57}$$

It is useful to determine the torque T needed to produce a certain depth of plastic penetration, $d = (R - \rho_y)$, in the circular shaft as shown in Figure 2.32b. This torque is found by adding the torque corresponding to the plastic zone (gray area) to that corresponding to the elastic central core (orange area). To accomplish this purpose, we need to consider two rings of radius ρ and thickness $d\rho$ (differential area $dA = 2\pi\rho\,d\rho$), one in each of the two zones. The differential force and torque in the elastic core are $dF = (2\pi\rho\,d\rho)\tau_\rho = (2\pi\rho\,d\rho)(\rho/\rho_y)\tau_y$ and $dT = (2\pi\rho^3\,d\rho)(\tau_y/\rho_y)$, respectively. Also, the differential force and torque in the plastic zone are $dF = (2\pi\rho\,d\rho)\tau_y$ and $dT = (2\pi\rho^2\,d\rho)\tau_y$, respectively. Thus, referring to Figure 2.32b, the required torque T is given by

$$T = (2\pi\tau_y/\rho_y)\int_0^{\rho_y}\rho^3\,d\rho + (2\pi\tau_y)\int_{\rho_y}^{R}\rho^2\,d\rho = (2\pi\tau_y)\left[(1/\rho_y)\left(\frac{\rho^4}{4}\right)_0^{\rho_y} + \left(\frac{\rho^3}{3}\right)_{\rho_y}^{R}\right]$$

Simplifying, we obtain

$$T = \frac{\pi\tau_y}{6}(4R^3 - \rho_y^3) \tag{2.58}$$

Equation 2.58 may be specialized to obtain the yield torque T_y and the fully plastic torque T_p. To obtain the yield torque, we set $\rho_y = R$ in Equation 2.58, which, after simplification, reduces to

$$T_y = \frac{\pi\tau_y R^3}{2} \tag{2.59}$$

Note that Equation 2.59 could also be obtained by solving Equation 2.9 for the torque T and setting $\tau = \tau_y$. To obtain the fully plastic torque, we set $\rho_y = 0$ in Equation 2.58, which, after simplification, reduces to

$$T_p = \frac{2\pi\tau_y R^3}{3} \tag{2.60}$$

Equations 2.59 and 2.60 will be reconfirmed in Example 2.22.

Equation 2.58 may be expressed in dimensionless form by dividing it by Equation 2.59. This operation leads to

$$\frac{T}{T_y} = \left(\frac{4}{3}\right)\left[1 - (1/4)\left(\frac{\rho_y}{R}\right)^3\right] \tag{2.61}$$

We should emphasize the fact that the strain distribution across the circular cross section remains linear even after plastic penetration. This, of course, implies that Equation 2.4 is still valid. From this equation, we obtain the following two relations expressed in Equations 2.62 and 2.63:

$$\rho = \frac{L\gamma_\rho}{\theta} \Rightarrow \rho_y = \frac{L\gamma_y}{\theta} \tag{2.62}$$

$$R = \frac{L\gamma_R}{\theta} \Rightarrow R = \frac{L\gamma_y}{\theta_y} \tag{2.63}$$

Note that in Equation 2.62, the radius ρ and the corresponding shearing strain γ_ρ were replaced by ρ_y and γ_y, respectively, in order to express the radius of the elastic core in terms of the angle of twist θ. Note also that in Equation 2.63, we replaced γ_R and θ by γ_y and θ_y, respectively, in order to relate the strain conditions when yielding begins. If we now divide Equation 2.62 by Equation 2.63, we obtain the ratio ρ_y/R. Thus,

$$\frac{\rho_y}{R} = \frac{\theta_y}{\theta} \tag{2.64}$$

Substituting from Equation 2.64 into Equation 2.61 and rearranging, we obtain

$$\frac{T}{T_y} = \left(\frac{4}{3}\right)\left[1 - \frac{1}{4(\theta/\theta_y)^3}\right] \tag{2.65}$$

Note that Equation 2.65 is only applicable in the plastic range, that is, T/T_y and θ/θ_y are both larger than unity. Thus, the plot of Equation 2.65 is represented by curve YB in Figure 2.33. Note

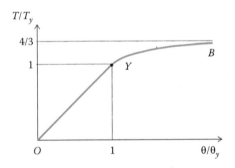

FIGURE 2.33   Diagram showing a plot of Equation 2.65.

also, that, theoretically, the value of the dimensionless ratio T/T_y approaches 4/3 as the dimensionless ratio θ/θ_y approaches infinity. Practically, however, this condition can never be reached, and the fully plastic state can never be attained because the shaft will break before that can happen. The first part in Figure 2.33, segment OY, represents the linear behavior in the elastic range where the torque is directly proportional to the angle of twist as expressed by Equation 2.10. In other words, in the elastic range, $T/T_y = \theta/\theta_y$, which is the equation of the straight line OY in Figure 2.33.

EXAMPLE 2.22

A hollow circular shaft has an inside radius R_i and an outside radius R_o. It is made of an elastoplastic material for which the stress–strain diagram is of the type shown in Figure 2.31. Determine (a) the torque T_y required to bring the shaft to first yield at its outside surface and (b) the torque T_p required to take the shaft to the fully plastic condition. Specialize both answers to the case of a solid circular cross section.

SOLUTION

a. When yielding is initiated, the stress distribution will be as shown in figure below. Since up to first yield the shaft behaves elastically, Equation 2.9 applies and from it, we obtain

$$T = \frac{\tau_{MAX}\,J}{R} \Rightarrow T_y = \frac{\tau_y(\pi/2)(R_o^4 - R_i^4)}{R_o} \qquad \textbf{ANS.}$$

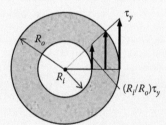

If we let $R_i = 0$, the above equation gives the torque required to initiate yielding in a solid circular cross section. Thus,

$$T_y = \frac{\pi \tau_y R_o^3}{2} \qquad \textbf{ANS.}$$

b. When the fully plastic condition is reached, the stress distribution would be as shown in figure below, where the stress is constant at τ_y across the entire section. The torque T_p producing this stress condition is found by first finding the differential force dF acting on an annular differential ring of radius ρ and thickness $d\rho$. This force is the product of the shearing stress τ_y and the area of the differential ring. Thus, $dF = \tau_y(2\pi\rho\,d\rho)$ and the differential torque dT_p is the product of the differential force dF and the moment arm ρ. Thus, $dT_p = \tau_y(2\pi\rho^2\,d\rho)$. Therefore,

$$T_p = 2\pi\tau_y\int_{R_i}^{R_o}\rho^2 d\rho = \left(\frac{2\pi\tau_y}{3}\right)(R_o^3 - R_i^3) \qquad \textbf{ANS.}$$

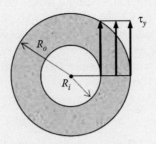

Specializing this to the case of a solid shaft, we obtain

$$T_p = \frac{2\pi\tau_y R_o^3}{3} \quad \textbf{ANS.}$$

Note that the expressions found for T_y and T_p in this example confirm those given by Equations 2.59 and 2.60.

PROBLEMS

2.128 An elastoplastic stress–strain diagram for a hollow torsional specimen is shown in the sketch. The outside and inside diameters are 2.5 in. and 1.5 in., respectively, and the gage length 2 ft. Determine (a) the modulus of rigidity, (b) the yield torque, (c) the angle of twist at yielding, and (d) the modulus of resilience. Let $\tau_y = 12.0$ ksi and $\gamma_y = 2 \times 10^{-3}$.

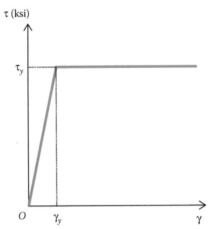

(Problems 2.128, 2.129, 2.130, and 2.131)

2.129 In reference to the figure shown in Problem 2.128, an elastoplastic stress–strain diagram for a solid torsional specimen is shown in the sketch. If the yield torque is 30.0 kN·m and $\tau_y = 350$ MPa, determine (a) the diameter of the specimen, (b) the modulus of rigidity if the angle of twist at yielding is 8° over a gage length of 1.5 m.

2.130 In reference to the figure shown in Problem 2.128, a solid circular shaft 3.5 in. in diameter is made of mild steel that may be assumed to behave in an elastoplastic manner. If $\tau_y = 17$ ksi, determine the depth of plastic penetration d if the applied torque is (a) 90 kip·in. and (b) 80 kip·in.

2.131 In reference to the figure shown in Problem 2.128, a solid circular shaft is made of an elastoplastic material for which $\tau_y = 20$ ksi. If the diameter of the shaft is 4.0 in., determine the torque required to produce a plastic penetration of (a) 0.5 in., (b) 1.0 in., (c) 1.5 in., and (d) 2.0 in.

2.132 A hollow circular shaft, with outside and inside radii of R_o and R_i, respectively, is made of mild steel that may be assumed to behave elastoplastically. If the yield strength is τ_y, show that equation relating the torque T to the radius of the elastic core ρ_y, may be expressed in the form

$$T = \frac{\pi\tau_y}{2}\left(\frac{4R_o^3 - \rho_y^3}{3} - \frac{R_i^4}{\rho_y}\right).$$

2.133 A hollow circular shaft, with outside and inside radii of 100 and 60 mm, respectively, is fabricated of mild steel that is assumed to be elastoplastic. If $\tau_y = 175$ MPa and the applied torque is 32.0 kN·m, determine the depth d of plastic penetration. (*Hint*: Use the equation given in Problem 2.132.)

2.134 A solid circular shaft with a radius of 3.5 in. and a length of 2.5 ft is fabricated of an elastoplastic material for which $\tau_y = 15$ ksi and $G = 10.5 \times 10^3$ ksi. Find the torque required to produce an angle of twist of (a) 0.2 rad and (b) 0.40 rad.

2.135 Refer to the shaft described in Problem 2.134, determine the radius of the elastic core for each of the two given angles of twist.

2.136 A solid circular shaft with a diameter of 90 mm and a gage length of 0.75 m is made of mild steel for which $\tau_y = 150$ MPa and $G = 75$ GPa. If a torque of 25 kN·m is applied, find the radius of the elastic core and the corresponding angle of twist.

2.137 A solid shaft of circular cross section is made of mild steel that may be assumed to be an elastoplastic material for which $\tau_y = 25$ ksi and $G = 10.5 \times 10^3$ ksi. If the diameter $d = 4$ in. and the length $L = 2$ ft, determine (a) using basic fundamentals, the yield torque T_y, the fully plastic torque T_p, and (b) the depth of plastic penetration if the applied torque is $T = 1.15T_y$ and $T = 1.25T_y$.

2.138 A solid shaft of circular cross section is made of mild steel that may be assumed to be an elastoplastic material for which $\tau_y = 160$ MPa and $G = 70$ GPa. If the diameter $D = 80$ mm and the length $L = 0.80$ m, determine (a) using basic fundamentals, the ratio T/T_y if the depth of plastic penetrations are $d = 20$ mm and $d = 40$ mm and (b) the angles of twist corresponding to plastic penetrations of $d = 20$ mm and $d = 40$ mm.

2.139 A hollow circular shaft is made of mild steel that may be assumed to be elastoplastic material for which $\tau_y = 20$ ksi. It is subjected to a torque of 250 kip·in. resulting in plastic penetration with an elastic core of radius $\rho_y = 3.5$ in. If the inside radius of the shaft $R_i = 3.0$ in., determine the outside radius of this hollow shaft.

2.140 A hollow circular shaft is made of mild steel that may be assumed to be an elastoplastic material for which $\tau_y = 10$ ksi. It is subjected to a torque of 200 kip·in. resulting in plastic penetration with an elastic core of radius $\rho_y = 3.3$ in. If the outside radius of the shaft $R_o = 3.5$ in., determine the inside radius of this hollow shaft.

2.141 A hollow shaft is made of an elastoplastic material for which $\tau_y = 150$ MPa and $G = 75$ GPa. If $R_o = 50$ mm, $R_i = 20$ mm, and $L = 0.60$ m, find the torque T and the corresponding angle of twist θ if the depth of plastic penetration is (a) $d = 10$ mm and (b) $d = 20$ mm. Use basic fundamentals in your solution.

2.142 A hollow shaft is made of an elastoplastic material for which $\tau_y = 160$ MPa and $G = 70$ GPa. If $R_o = 70$ mm, $R_i = 40$ mm, and $L = 0.60$ m, find (a) the torques T_y and T_p and the corresponding angles of twist and (b) the depth d of plastic penetration if $T/T_y = 1.20$. What is the angle of twist corresponding to this value of T/T_y? Use basic fundamentals in your solution.

REVIEW PROBLEMS

R2.1 The stepped shaft is fixed at A and subjected to a concentrated torque Q at C and to the distributed torques q_1 and q_2 as shown. Develop expressions for the internal torques in terms of x_1 and x_2 for segments AB and BC and construct the internal torque diagram for the entire shaft. Let $Q = 60$ kip·ft, $q_1 = 3x_1^2$ kip·ft/ft, $q_2 = 12x_2$ kip·ft/ft, $a = 8$ ft, and $b = 3$ ft.

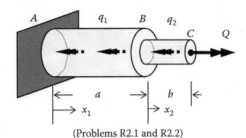

(Problems R2.1 and R2.2)

R2.2 In reference to the figure shown in Problem R2.1, the stepped shaft is fixed at A and subjected to a concentrated torque Q at C and to the distributed torques q_1 and q_2 as shown. Find the internal torque (a) just to the right of A, (b) just to the left of B, and (c) just to the right of B. Let $Q = 80$ kN·m, $q_1 = 3x_1^2$ kN·m/m, $q_2 = 12x_2$ kN·m/m, $a = 3.0$ m, and $b = 1.0$ m.

R2.3 The shaft shown consists of a steel segment AB ($d = 2$ in., $G = 12 \times 10^3$ ksi) and an aluminum segment BC ($d = 3$ in., $G = 4 \times 10^3$ ksi) connected to act as a single unit. Let $Q_1 = 15$ kip·in., $Q_2 = 10$ kip·in., $q = 5$ kip·in./in., $a = 20$ in., $b = 30$ in., determine (a) the angle of twist at A and (b) the maximum shearing stress 10 in. to the right of B.

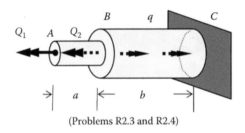

(Problems R2.3 and R2.4)

R2.4 In reference to the figure shown in Problem R2.3, the shaft shown consists of a steel segment AB ($d = 60$ mm, $G = 200$ GPa) and an aluminum segment BC ($d = 120$ mm, $G = 70$ GPa) connected to act as a single unit. If the allowable shearing stresses in steel and aluminum are 120 and 70 MPa, respectively, and the magnitude of $Q_2 = 20$ kN·m, determine the largest permissible magnitude of Q_1. Let $a = 300$ mm and $b = 500$ mm.

R2.5 A pipe is manufactured of 1/4-in. steel sheet material by welding along the helix shown where $\theta = 50°$. If the inside diameter of the pipe is 24 in. and $L = 5$ ft and the allowable normal and shearing stresses on the weld are 20 and 14 ksi, respectively, determine the maximum permissible torque Q that may be applied to the pipe.

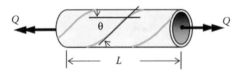

R2.6 The stepped shaft is made of aluminum for which $G = 4 \times 10^3$ ksi. Determine the magnitudes of the torques Q_1 and Q_2 if the allowable shearing stress is 15 ksi and the allowable angle of twist at C is 0.03 rad. The diameter of segment AB is 4 in. and that of BC is 2 in. Also, $a = 12$ in. and $b = 6$ in.

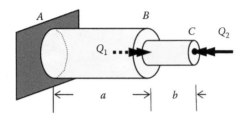

R2.7 The entire shaft shown is made of the same material. Segment AB is solid with a diameter of 4.0 in. and segment BC is hollow with an inside diameter of 3.5 in. Find the ratio Q_1/Q_2 if the maximum shearing stress in segment AB is to be the same as that in segment BC. Determine also the ratio of the angle of twist $\theta_{B/A}$ to the angle of twist $\theta_{C/B}$. Let $a = 20$ in. and $b = 12$ in.

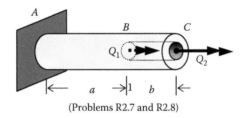

(Problems R2.7 and R2.8)

R2.8 In reference to the figure shown in Problem R2.7, segment AB is solid with a diameter of 120 mm and segment BC is hollow with an inside diameter of 100 mm. Let $Q_1 = 25$ kN·m and $Q_2 = 15$ kN·m. (a) Construct a stress element on the outside surface of segment AB with two planes parallel and two planes perpendicular to the axis of the shaft. (b) Find the normal and shearing stresses on the outside surface of segment BC, first on a plane making an angle of 30° with the shaft axis and then on a plane perpendicular to this axis (i.e., on a plane making an angle of 30° + 90° = 120° with the shaft axis). Construct the corresponding stress element.

R2.9 Shafts AB and CD in the gear-shaft system shown are made of the same material and have the same diameter d. Let $Q = 5$ kip·in., $r_1 = 3$ in., and $r_2 = 7$ in., the allowable shearing stress, $\tau_{ALL} = 20$ ksi, find the least permissible diameter d.

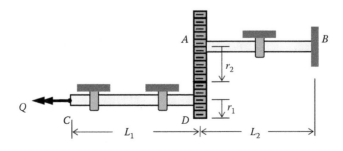

R2.10 Segment AB of the composite shaft shown is solid aluminum for which $G = 4 \times 10^3$ ksi. Segment BC is solid brass for which $G = 6.4 \times 10^3$ ksi with a diameter of 3.0 in. Determine the minimum permissible diameter of the aluminum segment AB if the allowable shearing stress in aluminum is 10 ksi and that in brass is 8 ksi. Let $Q = 50$ kip·in.

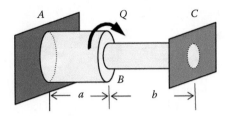

R2.11 The gear system consists of solid shaft AB made of steel ($G = 70$ GPa, $d = 30$ mm, and $\tau_{ALL} = 100$ MPa) and solid shaft CD made of aluminum ($G = 25$ GPa, $d = 60$ mm, and $\tau_{ALL} = 80$ MPa) interconnected by the gears at A and C as shown. Let $r_1 = 180$ mm, $r_2 = 90$ mm, and $L = 300$ mm, determine the maximum permissible torque Q that may be applied.

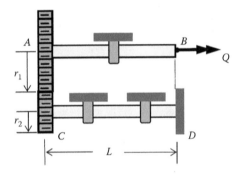

R2.12 A hollow aluminum shaft is to transmit 500 hp at 400 rpm. The allowable shearing stress is 10 ksi and the allowable angle of twist is 0.015 rad/foot of length. If the outside diameter of the hollow shaft is 6.0 in., determine the maximum permissible inside diameter.

R2.13 The system shown consists of an aluminum hollow shaft AB of length $L = 600$ mm ($d_o = 110$ mm, $G = 25$ GPa), fixed at A and supported at B by a frictionless bearing where it is attached to rigid arm BC of length $b = 400$ mm. A weight $W = 500$ N is dropped as shown from a height $h = 50$ mm. Determine the maximum permissible inside diameter of the shaft if the allowable shearing stress is 100 MPa and the allowable rotation of the shaft at B is 0.05 rad. Assume no energy losses.

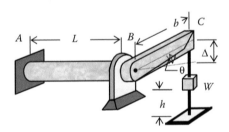

R2.14 Measurements of the strain on the outside surface of a shaft subjected to the combined action of an axial tensile force P and a torque Q resulted in the following values: $\varepsilon_n = 500$ μ and $\varepsilon_t = -300$ μ. If the diameter of the shaft is 100 mm and $\theta = 50°$, and the material is aluminum for which $E = 75$ GPa and $\mu = 0.33$, determine the magnitudes of P and Q.

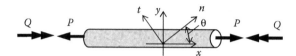

R2.15 A hollow shaft in the form of a regular hexagon as shown is fabricated of yellow brass for which $G = 5 \times 10^3$ ksi and has a length $L = 4$ ft. Determine the maximum permissible torque Q that may be applied to this shaft if the allowable shearing stress is 12 ksi and the allowable angle of twist is $2°$.

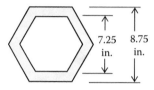

R2.16 Repeat Problem R2.15 if a narrow slit is cut at one of the six corners of the hexagonal tube along the entire length of the shaft.

R2.17 A hollow steel shaft has an outside diameter of 160 mm and an inside diameter of 80 mm. If the material has a yield strength of 200 MPa and the shaft is subjected to a torque of 270 kN·m, determine the depth of plastic penetration.

3 Bending Loads
Stresses

3.1 INTRODUCTION

Beams are very useful components employed in many structural applications such as floor, roof-deck, and bridge-deck systems. By definition, a beam is a long structural member subjected to loads normal to its longitudinal axis (transverse loads) or to pure bending couples. Regardless of the method of loading, the resulting action leads to two effects that are of major significance in the analysis and design of beams. These two effects are, on the one hand, the induced normal and shearing stresses, and on the other, the resulting deformations. Chapters 3 and 4 deal with the first of these two effects, the induced stresses, and Chapters 5 and 6 deal with the subject of deformations (deflections).

In general, there are two distinct types of bending: *symmetric* bending and *unsymmetric* bending. To understand the nature of, and differences between, these two types of bending, we need to know that every shape of cross-sectional area possesses two very important in-plane centroidal axes known as *principal axes of inertia*, which are perpendicular to each other. With respect to one of these two centroidal axes, the moment of inertia of the cross-sectional area is the largest possible, and with respect to the other, the moment of inertia is the least possible. The procedure to determine the principal axes of inertia is summarized in Appendix C.2, and the interested reader is urged to consult this appendix for more details. However, the important matter at present is that we understand that symmetric bending of a beam is the result of bending loads that lie in one of the two *centroidal longitudinal principal planes*. A *centroidal longitudinal principal plane* is a plane that contains the same centroidal principal axis of inertia of all the cross sections along the entire length of the beam. Of course, this statement implies that we are dealing with a prismatic beam (i.e., one that has the same cross-sectional area all along the beam). One observation that should be made here is that an axis of symmetry for a cross section is, in fact, a principal axis of inertia. Since, as stated earlier, the two principal axes of inertia are orthogonal, it follows that the *centroidal axis perpendicular to an axis of symmetry is also a principal axis of inertia regardless of whether it is an axis of symmetry or not*.

Consider, for example, the case of the T section shown in Figure 3.1. The y axis passing through the centroid C of the section is an axis of symmetry and, therefore, it is a centroidal principal axis of inertia. The centroidal z axis, normal to the y axis, is also a principal axis of inertia even though it is *not* an axis of symmetry. A prismatic beam with this cross section will possess two centroidal longitudinal principal planes, one containing all of the y centroidal principal axes, and the other, all of the z centroidal principal axes along the beam. Note that the centroidal x axis points out of the page at C.

Chapter 3 is primarily concerned with the problem of symmetric bending. In Section 3.2, we learn how to determine the internal shear and moment at any cross section in the beam; in Section 3.3, we develop the relations that exist between the shear and the moment; in Section 3.4, we derive the bending stress equation; in Section 3.5, we develop the shearing stress equation; in Section 3.6, we talk about the stresses produced under combined loading; in Section 3.7, we introduce the topic of allowable-stress beam design; and finally, in Section 3.8, we discuss the topic of stress concentration in beams.

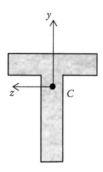

FIGURE 3.1   A T section showing two centroidal principal axes of inertia, namely, y and z.

3.2   INTERNAL SHEAR AND MOMENT

3.2.1   SHEAR AND MOMENT AT SPECIFIED POSITIONS

Consider the simply supported beam shown in Figure 3.2a. Let us suppose that we want to find the *internal shearing force* V_C and the *internal bending moment* M_C at any position such as C along the beam. To this end, we first construct the free-body diagram of the entire beam, shown in Figure 3.2b and apply the conditions of equilibrium in order to find the support reactions A_y and B_y. Once this is done, we cut the beam at C and construct either a left free-body diagram (LFB) or a right free-body diagram (RFB). Figure 3.2c shows the LFB diagram and Figure 3.2d the RFB diagram. It is always a good practice to show a convenient coordinate system as indicated in Figure 3.2e. It is immaterial

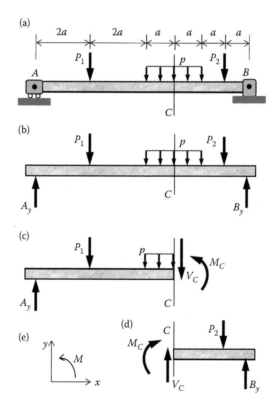

FIGURE 3.2   Determination, using equilibrium, of internal shear V_C and internal moment M_C at any position C along the beam.

which of the two free-body diagrams is used. The equations of equilibrium lead to identical answers for the internal shear V_C and internal moment M_C. In practice, however, one free-body diagram may be more convenient to use than the other depending upon the location on the beam where the internal shear and moment are to be found. For the position C in Figure 3.2, there is no advantage in choosing one free-body diagram over the other. On the other hand, if the position chosen were, for example, between P_1 and the distributed force, p, it would be simpler to use the LFB diagram because fewer forces would be involved in the calculations.

The beam shown in Figure 3.2 is subjected to the concentrated loads P_1 and P_2, assumed to act at specific points, as well as the uniformly distributed load p, which is expressed in terms of units of force per units of length along the beam such as kips/ft and kN/m. Often, the distribution p is not uniform. Regardless of the shape of the distributed load, it is good to recall from *statics* that its *resultant force R is given by the area under the distribution* (in Figure 3.2a, $R = 2pa$), and that *the resultant passes through the centroid of the distribution* (in Figure 3.2a, R *is at 5a to the right of the support at A*).

Let us now consider the LFB diagram (Figure 3.2c). There are only two unknown quantities, V_C and M_C, since A_y was already determined from the free-body diagram of the entire beam (Figure 3.2b). These two unknown quantities may be found by using the coordinate system shown in Figure 3.2e along with the two equilibrium equations $\Sigma F_y = 0$ and $\Sigma M_C = 0$. To check on the correctness of our computations, we could use the RFB diagram (Figure 3.2d), and once again apply the equations of equilibrium to find V_C and M_C. Of course, this second set of values must be identical to the first set obtained from the LFB diagram, or we have incorrectly written or solved the equilibrium equations. In many cases, this method of checking would be impractical. It would be simpler to use the same free-body diagram and write a moment equation about some other point, such as A, to check the results.

3.2.2 SIGN CONVENTION

As in Chapters 1 and 2 where two distinct sign conventions were needed to handle the analysis of internal forces and internal torques, respectively, two distinct sign conventions are also needed to deal with internal shearing forces, V, and internal bending moments, M. The first is the *physical sign convention* and the second is the arbitrary *equations sign convention*. A beam, in general, is used in a horizontal position as illustrated in Figure 3.2. Thus, any cutting plane results in two segments and two corresponding free-body diagrams, an LFB diagram and an RFB diagram. Using such free-body diagrams, the physical sign convention states that *a positive shearing force V is one that points downward on the LFB diagram* and *upward on the RFB diagram*. Also, *a positive bending moment M is one that is counterclockwise in the LFB diagram* and *clockwise in the RFB diagram*. In other words, a positive moment is one that causes a beam to be *concave upward*. Note that the internal shears V_C and the internal moments M_C shown in Figures 3.2c and 3.2d are all positive by the physical sign convention. It is a very good practice to assume the unknown internal shears and moments to be positive when using the *physical sign convention* and to let the application of the *equations sign convention* tell us if the assumption is, in fact, correct. Once the needed free-body diagram is constructed, assuming the unknown internal shear and moment to be positive by the *physical sign convention*, the equations of equilibrium are then applied using the *equations sign convention* along with an arbitrarily chosen coordinate system that indicates the positive sense of the quantities involved. For example, the coordinate system shown in Figure 3.2e dictates that any force pointed upward (i.e., along the positive y axis) is given a positive sign in the equilibrium equation, $\Sigma F_y = 0$, and any moment producing counterclockwise rotation is given a positive sign in the equilibrium equation, $\Sigma M_C = 0$. If the sign of an answer is positive, we conclude that the assumption made about the internal shear or the internal moment on the basis of the physical sign convention is, in fact, correct. If, on the other hand, the answer is negative, the assumed sense of the quantity is incorrect and has to be reversed. The following example illustrates the procedure.

EXAMPLE 3.1

Consider the loaded overhanging beam shown in figure (a) below. Determine the shear and moment at (a) position D and (b) position E, which is just to the right of point B. Let $p = 3$ kips/ft, $P = 25$ kips, $Q = 24$ kip·ft, and $a = 2$ ft.

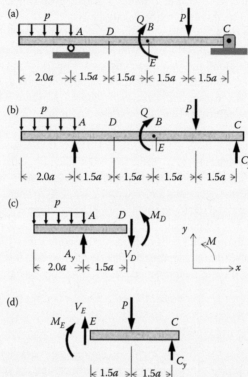

SOLUTION

Before we can answer either of the two questions, we need to find the support reactions at A and C. To accomplish this, we need to construct a free-body diagram of the entire beam as shown in figure (b) above and apply the equations of equilibrium. Thus,

$$\sum M_A = 0 \Rightarrow C_y(6.0a) - P(4.5a) - Q + p(2.0a)(a) = 0;$$

$$C_y = \frac{3P}{4} + \frac{Q}{6a} - \frac{pa}{3}$$

Substituting the given numerical values for P, Q, a, and p, we obtain

$$C_y = 8.75 \text{ kips}$$

$$\sum F_y = 0 \Rightarrow A_y + 8.75 - p(2.0a) - P = 0$$

$$A_y = P + 2pa - 8.75$$

Substituting the given data, we obtain

$$A_y = 28.25 \text{ kips}$$

a. To find the internal shear and moment at position D, we cut the beam at this position and construct an LFB diagram as shown in figure (c) above, and apply the conditions of equilibrium. Note that an RFB diagram could be used instead, but the LFB diagram is, in this case, a little easier to use. Thus,

$$\sum M_D = 0 \Rightarrow M_D + p(2.0a)(2.5a) - A_y(1.5a) = 0$$

$$M_D = 1.5A_y a - 5.0pa^2$$

$$M_D = 24.75 \approx 24.8 \text{ kip} \cdot \text{ft} \qquad \textbf{ANS.}$$

$$\sum F_y = 0 \Rightarrow A_y - 2.0pa - V_D = 0$$

$$V_D = A_y - 2pa = 16.25 \approx 16.3 \text{ kips} \qquad \textbf{ANS.}$$

b. To find the internal shear and moment at E, we cut the beam at that position, construct a free-body diagram as shown in figure (d) above and apply the equilibrium conditions. Note that, in this case, the RFB, instead of the LFB diagram, was used because it is less complicated and easier to use. Thus,

$$\sum M_E = 0 \Rightarrow C_y(3.0a) - P(1.5a) - M_E = 0$$

$$M_E = 3aC_y - 1.5aP = -22.5 \text{ kip} \cdot \text{ft} \qquad \textbf{ANS.}$$

$$\sum F_y = 0 \Rightarrow V_E + C_y - P = 0$$

$$V_E = P - C_y = 16.25 \approx 16.3 \text{ kips} \qquad \textbf{ANS.}$$

The negative answer obtained for the moment, M_E, shows that this moment is not clockwise (positive) as assumed but counterclockwise (negative) according to the physical sign convention.

3.2.3 SHEAR AND MOMENT FUNCTIONS: SHEAR AND MOMENT DIAGRAMS

In general, the internal shear V and the internal moment M change from point to point along the beam. Thus, it is useful to write functions that express these quantities in terms of a longitudinal coordinate x, measuring distances along the beam from some convenient position. Such a position may be selected at a support, a member end or a point of load application. These functions may be used to plot *shear* and *moment* diagrams, which show the variation of these two quantities along the beam. A more direct method of constructing these diagrams will be developed in Section 3.3.

The procedure for determining the internal shear and moment functions is similar to that used in finding the internal shear and moment at a specified position discussed earlier. A free-body diagram is constructed containing the cross section of interest whose location is defined in terms of the variable x. The unknown quantities V and M are assumed positive by our *physical sign convention*, and the equations of equilibrium are applied using a convenient but arbitrary *equations sign convention*. Solving these equations for V and M yields functions in terms of the variable x. Each shear and moment function is valid only for a restricted range of values for x, which must be stated because of the discontinuities that occur in the values of shear and moment at certain locations such as points of load application. Because of such discontinuities, whenever the character of the load changes, or when a support reaction is encountered, we must consider a new free-body diagram to develop the shear and moment functions that apply to the new segment of the beam. In Example 3.2, we develop the shear and moment functions for one beam, and in Example 3.3, we develop these functions for another beam and construct the shear and moment diagrams.

EXAMPLE 3.2

Beam ABC carries a uniform load of intensity p and a concentrated load of magnitude P as shown in figure (a) below. Using the coordinate x with origin at A, develop the internal shear and moment functions for segment AB and segment BC.

SOLUTION
Segment AB

Use the coordinate system shown along with the LFB diagram in figure (b) below to find V and M. Thus,

$$\sum F_z = 0 \Rightarrow -px - V = 0$$

$$V = -px \ldots 0 < x < 3a \quad \textbf{ANS.}$$

$$\sum M_O = 0 \Rightarrow M + px\left(\frac{x}{2}\right) = 0$$

$$M = -\frac{(px^2)}{2} \ldots 0 < x < 3a \quad \textbf{ANS.}$$

Segment BC

We now apply the conditions of equilibrium to the LFB diagram shown in figure (c) above to find V and M in this segment. Thus,

$$\sum F_y = 0 \Rightarrow -px - P - V = 0$$

$$V = -P - px \dots 3a < x < 5a \qquad \textbf{ANS.}$$

$$\sum M_O = 0 \Rightarrow M + P(x - 3a) + px\left(\frac{x}{2}\right) = 0$$

$$M = 3Pa - Px - \frac{px^2}{2} \dots 3a < x < 5a \qquad \textbf{ANS.}$$

EXAMPLE 3.3

Write the internal shear and moment functions and construct the internal shear and moment diagrams for the beam shown in figure (a) below. Let $p = 20.0$ kN/m, $Q = 24.0$ kN·m, and $a = 1.0$ m.

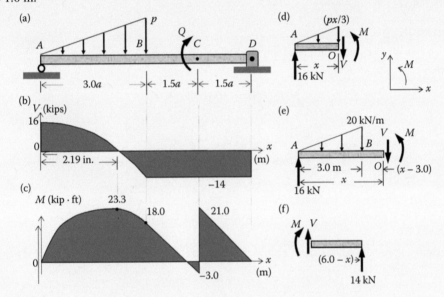

SOLUTION

Using the free-body diagram of the entire beam (not shown), we determine that the support reaction at A is $A_y = 16$ kN and that at D is $D_y = 14$ kN.

Three separate internal shear and moment functions are needed to cover the entire beam, one for segment AB, one for segment BC, and a third for segment CD. This, of course, means that three separate free-body diagrams are needed, one for each segment. These free-body diagrams and the corresponding computations to obtain the required shear and moment functions are shown above. The functions are then plotted to obtain the shear and moment diagrams shown, respectively, in figures (b) and (c) above, directly under the loaded beam. As stated in the case of axial and torsional loads, it is always a good practice to show these diagrams adjacent to the loaded member, either under or to one side of it. Note the moment discontinuity at $x = 4.5$ m. At this point, the internal moment changes abruptly from -3.0 kN·m to

21.0 kN·m. This is due to the existence at this point of the concentrated couple of 24.0 kN·m. The functions used in plotting these two diagrams are developed as follows:

Segment AB

We now apply the equations of equilibrium to the free-body diagram shown in figure (d) above to obtain

$$\sum F_y = 0 \Rightarrow 16 - \left(\frac{1}{2}\right)\left(\frac{px}{3}\right)(x) - V = 0$$

$$V = 16 - \frac{(px^2)}{6} \ldots 0 < x < 3.0 \text{ m} \qquad \textbf{ANS.}$$

$$\sum M_O = 0 \Rightarrow M + \left(\frac{1}{2}\right)\left(\frac{px}{3}\right)(x)\left(\frac{x}{3}\right) - 16x = 0$$

$$M = 16x - \frac{(px^3)}{18} \ldots 0 < x < 3.0 \text{ m} \qquad \textbf{ANS.}$$

Segment BC

The equations of equilibrium are then applied to the free-body diagram of figure (e) above. Thus,

$$\sum F_y = 0 \Rightarrow 16 - 30 - V = 0$$

$$V = -14 \ldots 3.0 < x < 4.5\text{m} \qquad \textbf{ANS.}$$

$$\sum M_O = 0 \Rightarrow M + 30(1.0 + x - 3.0) - 16x = 0$$

$$M = 60 - 14x \ldots 3.0 < x < 4.5\text{m} \qquad \textbf{ANS.}$$

Segment CD

Finally, we apply the equations of equilibrium to the RFB diagram shown in figure (f) above. Thus,

$$\sum F_y = 0 \Rightarrow V + 14 = 0$$

$$V = -14 \ldots 4.5 < x < 6.0 \text{ m} \qquad \textbf{ANS.}$$

$$\sum M_O = 0 \Rightarrow 14(6.0 - x) - M = 0$$

$$M = 84 - 14x \ldots 4.5 < x < 6.0\text{m} \qquad \textbf{ANS.}$$

PROBLEMS

3.1   Beam AB is supported and loaded as shown. Let $P = 30$ kips, $a = 10$ ft, and $b = 5$ ft, find the internal shear and moment at a section (a) 5 ft to the right of support A and (b) 2.5 ft to the left support B.

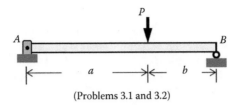

(Problems 3.1 and 3.2)

3.2 In reference to the figure shown in Problem 3.1, beam AB is supported and loaded as shown. Let $P = 60$ kN, $a = 4$ m, and $b = 2$ m, find the internal shear and moment at a section (a) just to the left of the load P and (b) just to the right of this load.

3.3 Let $P = 10$ kips, $p = 2$ kips/ft, and $L = 16$ ft, determine the internal shear and moment at a section (a) 3 ft to the right of support A and (b) 12 ft to the right of support A.

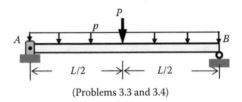

(Problems 3.3 and 3.4)

3.4 In reference to the figure shown in Problem 3.3, let $P = 40$ kN, $p = 5$ kN/m, and $L = 4$ m and determine the internal shear and moment at a section (a) just to the left of the load P and (b) just to the right of the load P.

3.5 The cantilever beam AB is subjected to a distributed load that varies from zero at A to p at B, and to a concentrated force P as shown. Find, in terms of p, P, and L, the internal shear and moment at a section (a) $L/2$ from end A and (b) at the fixed support at B.

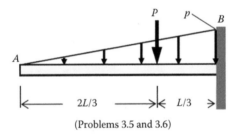

(Problems 3.5 and 3.6)

3.6 In reference to the figure shown in Problem 3.5, the cantilever beam AB is subjected to a distributed load that varies from zero at A to p at B, and to a concentrated force P as shown. Find, in terms of p, P, and L, the internal shear and moment at a section (a) just to the left of the load P and (b) just to the right of the load P.

3.7 Overhanging beam ABC is loaded as shown. Find, in terms of p, P, and L, the internal shear and moment at a section (a) $L/2$ to the right of support A and (b) just to the right of the load P.

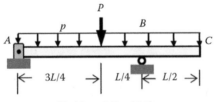

(Problems 3.7 and 3.8)

3.8 In reference to the figure shown in Problem 3.7, overhanging beam *ABC* is loaded as shown. Find, in terms of *p*, *P*, and *L*, the internal shear and moment at a section (a) *just* to the left of support *B* and (b) just to the right of this support.

3.9 Let $Q = 30$ kip·ft, $p = 3$ kips/ft, and $a = 8$ ft. Find the internal shear and moment at a section (a) 12 ft to the right of support *A* and (b) 5 ft to the left of support *B*.

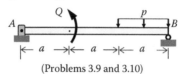

(Problems 3.9 and 3.10)

3.10 In reference to the figure shown in Problem 3.9, let $Q = 50$ kN·m, $p = 5$ kN/m, and $a = 2$ m. Find the internal shear and moment at a section (a) just to the right of the couple *Q* and (b) just to the left of this couple.

3.11 The cantilever beam *AB* supports the simply supported beam *BC* by the built-in hinge at *B*. Let $P = 15$ kips, $p = 2$ kips/ft, and $a = 6$ ft, determine the internal shear and moment at a section (a) 5 ft to the right of the fixed support at *A* and (b) 3 ft to the left of the roller support at C.

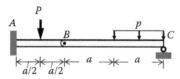

(Problems 3.11 and 3.12)

3.12 In reference to the figure shown in Problem 3.11, the cantilever beam *AB* supports the simply supported beam *BC* by the built-in hinge at *B*. Let $P = 40$ kN, $p = 5$ kN/m, and $a = 2$ m, determine the internal shear and moment at a section (a) just to the right of the built-in hinge at *B* and (b) just to the left of the concentrated load *P*.

3.13 The cantilever beam *AB* supports the simply supported beam *BC* by the built-in hinge at *B*. Let $p_1 = 2$ kips/ft, $p_2 = 4$ kips/ft, and $a = 8$ ft, determine the internal shear and moment at a section (a) 4 ft to the right of the fixed support at *A* and (b) 8 ft to the left of the roller support at *C*.

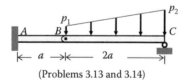

(Problems 3.13 and 3.14)

3.14 In reference to the figure shown in Problem 3.13, the cantilever beam *AB* supports the simply supported beam *BC* by the built-in hinge at *B*. Let $p_1 = 5$ kN/m, $p_2 = 8$ kN/m, and $a = 2$ m, determine the internal shear and moment at a section (a) just to the right of the hinge support at *B* and (b) at the fixed support at *A*.

3.15 Refer to the beam of Problem 3.1. Using an *x* coordinate positive to the right with origin at support *A*, write the shear and moment functions for each of the two segments of the beam.

3.16 Refer to the beam of Problem 3.3. Using an *x* coordinate positive to the right with origin at support *A*, write the shear and moment functions for each of the two segments of the beam.

3.17 Refer to the beam of Problem 3.5. Using an *x* coordinate positive to the right with origin at support *A*, write the shear and moment functions for each of the two segments of the beam.

3.18 Refer to the beam of Problem 3.7. Using an x coordinate positive to the right with origin at support A, write the shear and moment functions for each of the three segments of the beam.

3.19 Refer to the beam of Problem 3.9. Using an x coordinate positive to the right with origin at support A, write the shear and moment functions for each of the three segments of the beam.

3.20 Refer to the beam of Problem 3.11. Using an x coordinate positive, to the right with origin at support A, write the shear and moment functions for each of the four segments of the beam.

3.21 Write the shear and moment functions and use them to construct the shear and moment diagrams for the beam indicated below. Use an x coordinate positive to the right with origin at A.

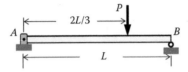

3.22 Write the shear and moment functions and use them to construct the shear and moment diagrams for the beam indicated below. Use an x coordinate positive to the right with origin at A.

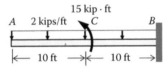

3.23 Write the shear and moment functions and use them to construct the shear and moment diagrams for the beam indicated below. Use an x coordinate positive to the right with origin at A.

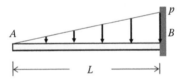

3.24 Write the shear and moment functions and use them to construct the shear and moment diagrams for the beam indicated below. Use an x coordinate positive to the right with origin at A.

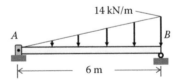

3.25 Write the shear and moment functions and use them to construct the shear and moment diagrams for the beam indicated below. Use an x coordinate positive to the right with origin at A.

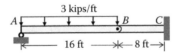

3.26 Write shear and moment functions and use them to construct the shear and moment diagrams for the beam indicated below. Use an x coordinate positive to the right with origin at A.

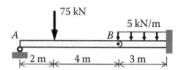

3.3   LOAD, SHEAR, AND MOMENT RELATIONSHIPS

As illustrated in Example 3.3, the shear and moment diagrams for a beam may be constructed by first writing the several shear and moment functions needed, and then plotting these functions. However, if the beam is subjected to a number of concentrated loads or to segmented distributed loads, this procedure becomes lengthy and tedious. In this section, we develop a more direct and less cumbersome method that utilizes the relations that exist among the loads applied to a beam and the internal shears and internal moments induced in the beam.

Let us consider beam AB shown in Figure 3.3a that is subjected to the arbitrary load $p = f(x)$. A differential segment of this beam of length dx at a distance x from the support at A is isolated as a free-body diagram and shown in Figure 3.3b. On the left face of this differential segment, the shear and moment are, respectively, V and M. On the right face, these two quantities become, respectively, $V + dV$ and $M + dM$. The differential quantities dV and dM represent the changes in the shear and moment that occur as we move from left to right a differential length dx. Note that on both faces of the segment, both the shear and the moment are assumed to be positive according to our physical sign convention. Using the arbitrarily chosen coordinate system shown in Figure 3.3, we apply the two equations of equilibrium. Applying the first equation, we have

$$\sum F_y = 0 \Rightarrow V - (V + dV) - p\,dx = 0$$

Thus,

$$dV = -p\,dx \qquad\qquad (3.1)$$

(a) $p = f(x)$

(b)

FIGURE 3.3   Construction of the shear and moment diagrams for a beam by utilizing the relations that exist among the applied loads and the internal shear force V and internal moment M.

Equation 3.1 may be expressed in the form

$$p = -\frac{dV}{dx} \tag{3.2}$$

which states that the load intensity p at any point is equal to minus the derivative with respect to x of the shear function at the same point on the beam. The minus sign is due to the positive sense chosen for p.

If we integrate Equation 3.1 between the limits x_1 and x_2, where the shears are, respectively, V_1 and V_2, we obtain

$$V_2 - V_1 = -\int_{x_1}^{x_2} p\,dx \tag{3.3}$$

The quantity $(V_2 - V_1)$ in Equation 3.3 is the change in shear between x_1 and x_2, and the integral represents the area under the load function (i.e., load diagram) between the same limits. Therefore, Equation 3.3 may be stated as follows:

$$V_2 - V_1 = -(\text{area under load function } p \text{ between } x_1 \text{ and } x_2 \text{ or } (A_p)_{1-2}) \tag{3.4}$$

For purposes of constructing the shear diagram, it is convenient to write Equation 3.4 in a slightly different form as follows:

$$V_2 = V_1 - (A_p)_{1-2} \tag{3.5}$$

Applying the second equation of equilibrium, we have

$$\sum M_O = 0 \Rightarrow (M + dM) + p\,dx\left(\frac{dx}{2}\right) - M - V\,dx = 0$$

The second term in the above equation is ignored because it is a second-order differential, and the remaining terms reduce to

$$dM = V\,dx \tag{3.6}$$

Equation 3.6 may be expressed in the form

$$V = \frac{dM}{dx} \tag{3.7}$$

Equation 3.7 states that the shear V at any point is equal to the derivative of the moment function with respect to x at the same point in the beam.

Let us integrate Equation 3.6 between the limits of x_1 and x_2 at which points the moments are, respectively, M_1 and M_2. Thus,

$$M_2 - M_1 = \int_{x_1}^{x_2} V\,dx \tag{3.8}$$

The quantity $(M_2 - M_1)$ represents the change in moment as we move from x_1 to x_2. The integral on the right-hand side of Equation 3.8 is the area under the shear function (i.e., shear diagram) between x_1 and x_2. Equation 3.8 may be restated in the form

$$M_2 - M_1 = \text{Area under shear function } V$$
$$\text{between } x_1 \text{ and } x_2 \text{ or } (A_V)_{1-2} \tag{3.9}$$

For purposes of constructing the moment diagram, Equation 3.9 is conveniently expressed in the form

$$M_2 = M_1 + (A_V)_{1-2} \tag{3.10}$$

The use of Equations 3.5 and 3.10 requires knowledge of the initial values of the shear and moment to which we can add or subtract whatever changes take place in these quantities as we move from left to right along the beam. These initial values are generally the values of these quantities at the left end of the beam. For example, if the left end of the beam is hinged or supported by a roller, the initial value of the shear is provided by the support reaction at that point. This statement may be confirmed by referring back to the shear function of segment AB in Example 3.3, where if x approaches zero, the value of the shear at the left support approaches the support reaction. Furthermore, assuming no externally applied couples at the left support, the initial value for the moment is zero because an ideal hinge or roller cannot support a moment. We should also keep in mind the discontinuities in the shear caused by concentrated forces and those in the moment produced by concentrated couples. The following two examples illustrate the use of Equations 3.5 and 3.10 in the construction of shear and moment diagrams.

EXAMPLE 3.4

Refer to Example 3.2 and (a) construct the shear and moment diagrams for the beam by plotting the shear and moment functions developed there and (b) check these diagrams by using Equations 3.5 and 3.10.

SOLUTION

a. The free-body diagram of the beam of Example 3.2 is shown in figure (a) below. By the equations of equilibrium, the shear and moment at the fixed support are found to be, respectively, $V_C = -(5pa + P)$ and $M_C = -(12.5pa^2 + 2Pa)$. The shear diagram plotted from the shear functions of Example 3.2 is shown in figure (b) below. The moment diagram plotted from the moment functions developed in Example 3.2 is shown in figure (c) below. Note that the shear diagram consists of straight lines while the moment diagram is made up of two second-degree parabolic functions. Note also the discontinuity at $x = 3a$ caused by the concentrated force P where the shear changes abruptly from $-3pa$ to $-(3pa + P)$. Finally, note that the values of shear and moment at the fixed support are confirmed by the shear and moment functions.

b. The initial value for the shear diagram is the shear at end A, which is zero because this is a free end. By Equation 3.5, we find the shear just to the left of point B where the concentrated force P is applied. Thus, by Equation 3.5,

$$V_B^L = V_A - (A_p)_{A-B} = 0 - p(3a) = -3pa$$

The notation V_B^L represents the shear just to the left of point B. The line connecting the shear values at A and just left of B is a straight line sloping down from zero to $-3pa$, because, by Equation 3.2, the slope of the shear diagram is equal to minus the load intensity p, which is constant. On the other hand, the shear just to the right of point B, $V_B{}^R$, is the algebraic sum of the shear just to the left of B and the concentrated force at this point. Thus,

$$V_B{}^R = -(3pa + P)$$

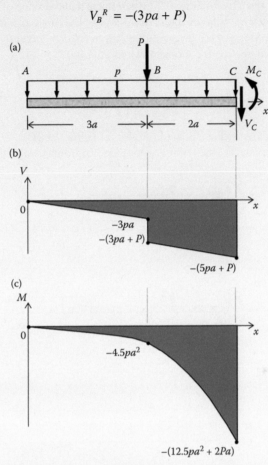

Once again, the shear at C is found by using Equation 3.5. Thus,

$$V_C = V_B{}^R - (A_p)_{B-C} = -(3pa + P) - p(2a) = -(5pa + P)$$

All of the shear values obtained by Equation 3.5 are identical to those given by the shear function in part (a). The line connecting the two shear values just right of B and at C is also a straight line sloping down from $-(3pa + P)$ to $-(5pa + P)$, for the same reason given above for the segment between A and B.

The initial value for moment is that at end A, which is zero because this end is free. Therefore, using Equation 3.10, we have

$$M_B = M_A + (A_V)_{A-B} = 0 + \left(\frac{1}{2}\right)(-3pa)(3a) = -4.5pa^2$$

and

$$M_C = M_B + (A_V)_{B-C} = -4.5pa^2 + \left(\frac{1}{2}\right)\left[-(3pa + P) - (5pa + P)\right](2a) = -(12.5pa^2 + 2P)$$

The moment values obtained by Equation 3.10 are the same as those found by the moment function in part (a). The line connecting the moment values at A and B is a parabolic curve whose slope decreases as we move from a moment of zero at A to a moment of $-4.5pa^2$ at B. The reason for this is given by Equation 3.7, which states that the slope of the moment diagram at any point is equal to the value of the shear at that point. Since the shear decreases as we move from A to B, the slope of the moment diagram must also decrease. The same argument applies to the curve connecting the moment values at B and C.

EXAMPLE 3.5

Consider beam $ABCD$ loaded as shown in figure (a) below. Let $P = 40$ kips, $Q = 80$ kip·ft, and $p = 5$ kips/ft and construct the shear and moment diagrams using Equations 3.5 and 3.10.

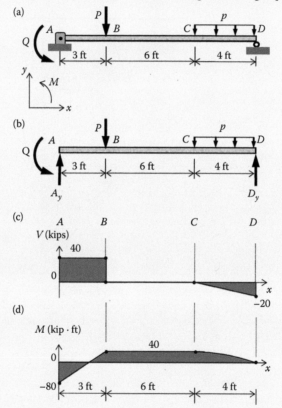

SOLUTION

The free-body diagram of the beam is shown in figure (b) above. Applying the equations of equilibrium, we obtain the support reactions $A_y = 40$ kips and $D_y = 20$ kips.

The starting point for the shear diagram is the value of shear an infinitesimal distance to the right of the support at A. The shear at this point is equal to the support reaction $A_y = 40$ kips and is plotted at $x = 0$ as shown in the shear diagram of figure (c) above.

By Equation 3.5, we get

$$V_B{}^L = V_A - (A_p)_{A-B} = 40 - 0 = 40 \text{ kips}$$

This second point is plotted at $x = 3$ ft as shown. The line connecting the two points at A and B is a horizontal straight line as dictated by Equation 3.2, which states that the slope of the shear at any point is equal to the load intensity p. Since $p = 0$, the slope of the shear diagram must be zero. At point B, there is a downward concentrated force of 40 kips causing an abrupt change (discontinuity) in the shear from 40 kips to zero (i.e., $V_B^R = V_B^L - P = 40 - 40 = 0$). Since there is no load in the segment between B and C, the shear there remains constant at zero (i.e., $V_C = V_B^R - 0 = 0 - 0 = 0$) and the shear diagram in this segment is represented by a horizontal line along the x axis. Finally, the shear at D is determined by Equation 3.5. Thus,

$$V_D = V_C - (A_p)_{C-D} = 0 - 5(4) = -20 \text{ kips}$$

This point is plotted at $x = 13$ ft as shown. The line connecting the two points at C and D is a straight line sloping down from 0 at C to -20 kips at D. The reason for this is provided by Equation 3.4, $p = -dV/dx$. Since p is constant at 5 kips/ft, the slope of this straight line is constant at -5 kips/ft.

The starting point for the moment diagram is the value of moment at support A, which is equal to the applied couple of -80 kip·ft. The moment at B is found by Equation 3.10. Thus,

$$M_B = M_A + (A_V)_{A-B} = -80 + (40)(3) = 40 \text{ kip} \cdot \text{ft}$$

This point is plotted at $x = 3$ ft and connected to the point at $x = 0$ by a straight line because of Equation 3.7, which states that the slope of the moment diagram is equal to the shear at any point in the beam. Since in segment AB, the shear is constant, the slope of the moment diagram in this segment must be constant. The moment at C is found using Equation 3.10. Thus,

$$M_C = M_B + (A_V)_{B-C} = 40 + 0 = 40 \text{ kip} \cdot \text{ft}$$

This point is located at $x = 9$ ft and, by Equation 3.10, a straight line is used to connect the two points at B and C, indicating that the moment is constant at 40 kip·ft throughout segment BC. Finally, using Equation 3.10, we get

$$M_D = M_C + (A_V)_{C-D} = 40 - 1/2(20)(4) = 0$$

This value agrees with the fact that the moment at a roller support must be zero. If this had not been the case, a mistake must have occurred in the computations. This last point is plotted at $x = 13$ ft. It is connected to the point at $x = 9$ ft by a curve the slope of which decreases as we move from C to D. This is so because of Equation 3.7. Since the value of the shear decreases as we move from C to D, so must the slope of the moment diagram.

PROBLEMS

3.27 Refer to the beam of Problem 3.1 and use Equations 3.5 and 3.10 to construct the shear and moment diagrams.

3.28 Refer to the beam of Problem 3.3 and use Equations 3.5 and 3.10 to construct the shear and moment diagrams.

3.29 Refer to the beam of Problem 3.4 and use Equations 3.5 and 3.10 to construct the shear and moment diagrams.

3.30 Refer to the beam of Problem 3.7 and use Equations 3.5 and 3.10 to construct the shear and moment diagrams. Let $P = 30$ kips, $p = 2$ kips/ft, and $L = 12$ ft.

3.31 Refer to the beam of Problem 3.9 and use Equations 3.5 and 3.10 to construct the shear and moment diagrams.

3.32 Refer to the beam of Problem 3.11 and use Equations 3.5 and 3.10 to construct the shear and moment diagrams.

3.33 Refer to the beam of Problem 3.12 and use Equations 3.5 and 3.10 to construct the shear and moment diagrams.

3.34 Refer to the beam of Problem 3.22 and use Equations 3.5 and 3.10 to construct the shear and moment diagrams.

3.35 Refer to the beam of Problem 3.25 and use Equations 3.5 and 3.10 to construct the shear and moment diagrams.

3.36 Refer to the beam of Problem 3.26 and use Equations 3.5 and 3.10 to construct the shear and moment diagrams.

3.37 Use Equations 3.5 and 3.10 to construct the shear and moment diagrams for the beam indicated below.

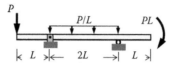

3.38 Use Equations 3.5 and 3.10 to construct the shear and moment diagrams for the beam indicated below.

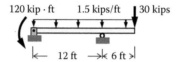

3.39 Use Equations 3.5 and 3.10 to construct the shear and moment diagrams for the beam indicated below.

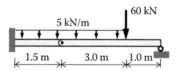

3.40 Use Equations 3.5 and 3.10 to construct the shear and moment diagrams for the beam indicated below.

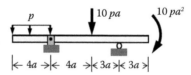

3.4 BENDING STRESSES UNDER SYMMETRIC LOADING

The term *symmetric bending* was defined in Section 3.1. This definition is repeated here for convenience by stating that *symmetric bending* of a beam is the result of bending loads that lie in one of the two *centroidal longitudinal principal planes*. A *centroidal longitudinal principal plane* is a plane that contains the same centroidal principal axis of inertia of all the cross sections along the entire length of the beam. The reader is again urged to consult Appendix C.2 where the *principal centroidal axes of inertia* for a given cross-sectional area are discussed in some detail. Suffice it to say that every cross-sectional area, regardless of shape, possesses two centroidal principal axes of inertia. With respect to one of these two principal axes, the moment of inertia of the area is maximum (*the strong, u axis*), and with respect to the second axis, its moment of inertia is minimum (*the weak, v axis*). It is of interest to note (see Appendix C.2) that an axis of symmetry for an area is a

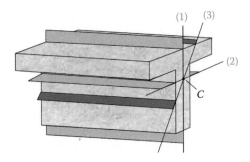

FIGURE 3.4 A T section of a beam showing three centroidal axes: Axis #1 is an axis of symmetry and a principal axis of inertia. Axis #2, perpendicular to Axis #1, is not an axis of symmetry but a principal axis of inertia. Axis #3 is any axis inclined to axes #1 and #2.

principal axis of inertia and that any axis normal to this axis of symmetry (regardless of whether it, itself, is an axis of symmetry) is also a principal axis of inertia.

As an illustration of the statements above, consider the segment of the beam shown in Figure 3.4. The cross section of the beam is a T shape with one axis of symmetry (a principal axis of inertia) given the number (1), which passes through point C, the centroid of the section. Centroidal axis (2), which is perpendicular to the axis (1), is not an axis of symmetry, but it too is a principal axis of inertia. Axes (1) and (2) define two longitudinal principal planes, shown in light gray. Any loads contained within these two principal planes would produce symmetric bending of the beam. Centroidal axis (3), however, inclined to the two centroidal principal axes, defines a longitudinal plane, shown in a teal color, which is not a principal plane. Any loads contained within this plane, or others like it, would produce unsymmetric bending of the beam. Once again, the reader is referred to Appendix C.2 for more details.

By definition, a beam is a long slender member subjected to bending action. This bending action may be due to the application of pure bending moments or to transverse loads. In developing the bending (flexure) equation, we assume that only pure bending moments are applied. As stated above, depending upon the way the beam is loaded relative to its principal axes of inertia, bending is either symmetric or unsymmetric. Symmetric bending is discussed in this section and unsymmetric bending in Section 4.3. In both types of bending, however, the assumption is made that the bending loads are so placed that they produce no twisting action. Furthermore, the beam is assumed to have the same cross-sectional configuration along its entire length (i.e., the beam is prismatic).

The beam shown in Figure 3.5a is subjected to positive bending moments Q that lie in the vertical longitudinal principal plane defined by the y centroidal principal axis. The beam deforms into the configuration shown, in an exaggerated manner, by the broken lines. Longitudinal fibers in the upper part of the beam are shortened, and those in the lower part of the beam are lengthened. At some position between these two sets of fibers, there exists a longitudinal horizontal surface at which the fibers are neither shortened nor lengthened. This surface is known as the *neutral surface* for the beam. The straight line at the intersection of this surface with a cross section of the beam is known as the *neutral axis for the section* and its intersection with the longitudinal principal plane of the loads is known as the *neutral axis for the beam*. As will be shown shortly, when a beam is symmetrically loaded, its neutral axis coincides with one of the two principal axes of inertia of its cross section. In our particular case, symmetric loading is applied in the longitudinal principal plane defined by the y axis. The y axis is an axis of symmetry for the cross section as shown in Figure 3.5b and, therefore, it is one of its two principal axes of inertia. Thus, the neutral axis for bending coincides with the z axis, which is the second principal axis of inertia for this cross section. The shortening of the upper longitudinal fibers in the beam of Figure 3.5a is due to a system of normal compressive stresses, and the lengthening of its lower longitudinal fibers is caused by a system of normal tensile stresses. These stresses are known as *bending* or *flexural* stresses.

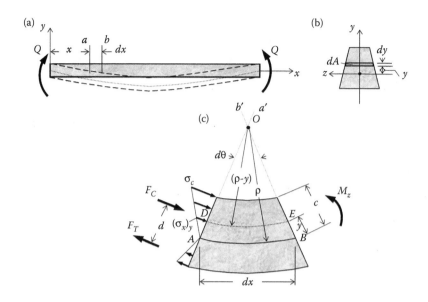

FIGURE 3.5 Beam in (a), with the cross section shown in (b), is subjected to symmetric bending due to applied moment Q and showing the neutral axis after bending in (c).

It should be pointed out that the flexure problem is a rather complex one, and its exact solution is beyond the scope of this book. The simple beam theory presented in this textbook is based upon a number of simplifying assumptions. However, the results obtained are sufficiently accurate for most engineering purposes. In addition to the two assumptions already stated, one dealing with the plane of bending loads for symmetric bending and the other dealing with the prismatic nature of the beam, several other assumptions will be made and stated as we go through the development of the *flexure equation* that enables us to compute the *flexural (bending) stresses* in a given situation.

The most significant assumption made in beam analysis states that *plane cross sections before bending remain plane after bending.* Thus, the two adjacent cross sections a and b in Figure 3.5a, which are an infinitesimal distance dx apart, remain plane but rotate with respect to each other into positions a' and b', respectively, as shown in the magnified view of Figure 3.5c. The line element dx, whose ends are denoted by the symbols A and B, coincides with the neutral surface for the beam and, therefore, it experiences no deformation. Furthermore, the two adjacent sections a and b rotate about their respective neutral axes, which appear in Figure 3.5c as points A and B. The rotated plane sections a' and b' form the angle $d\theta$ between them and, if extended, intersect at point O, known as the *center of curvature* for the deflected beam segment. The radius of curvature ρ, as shown in Figure 3.5c, is the radial distance from the center of curvature to the neutral surface.

Consider the longitudinal beam fiber DE at a distance y above the neutral surface. The initial length of this fiber is $AB = \rho\, d\theta$ and its final length is $DE = (\rho - y)\, d\theta$. Thus, the compressive strain ε_x at a distance y above the neutral surface is given by

$$(\varepsilon_x)_y = \frac{DE - AB}{AB} = \frac{(\rho - y)d\theta - \rho\, d\theta}{\rho\, d\theta} = -\frac{y}{\rho} \qquad (3.11)$$

If the material obeys Hooke's law, we have *elastic behavior* and $\sigma = E\varepsilon$. Thus, the stress σ_x at a distance y above the neutral surface becomes

$$(\sigma_x)_y = -\frac{yE}{\rho} \qquad (3.12)$$

The *magnitude* of the maximum stress occurs at $y = c$, the farthest point from the neutral surface, located on the outside fiber of the beam (see Figure 3.5c) and is given by

$$\sigma_{MAX} = \sigma_c = \frac{cE}{\rho} \tag{3.13}$$

Solving Equation 3.13 for ρ and substituting into Equation 3.12, we obtain

$$(\sigma_x)_y = -\left(\frac{\sigma_c}{c}\right)y = -ky \tag{3.14}$$

The quantity $k = (\sigma_c /c)$ in Equation 3.14 is a constant. Thus, Equation 3.14 states the very important conclusion that the magnitude of the bending (flexural) stress (and assuming Hooke's law applies, the magnitude of the strain as well) is directly proportional to the distance y from the *elastic neutral axis* (*ENA*). For the beam of Figure 3.5, these stresses are compressive above (resultant F_C) and tensile below (resultant F_T) the neutral surface AB as illustrated for plane a' in Figure 5.3c. Figure 5.3c also shows the internal moment M_z acting on plane b'.

Let us return now to the cross section of the beam shown in Figure 3.5b and consider a differential element of area dA at a distance y above the *ENA*. The bending normal stress $(\sigma_x)_y$ at this point produces a differential force $dF = (\sigma_x)_y\, dA$ acting normal to the cross section of the beam (i.e., along the x axis). Equilibrium of forces in the x direction dictates that the algebraic sum of all forces at a given section such as a' be equal to zero. Therefore,

$$\int (\sigma_x)_y\, dA = 0 \tag{3.15}$$

Substituting from Equation 3.14 into Equation 3.15, we conclude that

$$k \int y\, dA = 0 \tag{3.16}$$

Since $k \neq 0$, it follows that

$$\int y\, dA = \bar{y}\, A = 0 \tag{3.17}$$

where A is the cross-sectional area of the section and $\bar{y}$ represents the distance from the *ENA* to the centroidal axis of this section. Since A cannot be zero, it follows that $\bar{y}$ must be zero and the *ENA* coincides with the centroidal axis for the cross section, which, in this case, is the z principal axis of inertia.

In Figure 3.5c, the tensile system of stresses below the *ENA* is equivalent to a tensile force F_T and the compressive system of stresses above this axis is equivalent to a compressive force F_C as shown. Since there are no externally applied forces along the beam axis, equilibrium tells us that $F_C = F_T = F$. It follows, therefore, that these two forces constitute a couple whose magnitude is Fd, where d, as shown in Figure 3.5c, is the perpendicular distance between the two forces. This couple is the internal moment M_z at this position in the beam (see Figure 3.5c).

The differential force $dF = (\sigma_x)_y\, dA$ at a distance y above the *ENA* produces an internal differential moment about this axis whose magnitude is $dM_z = y\, (\sigma_x)_y\, dA$. Thus, the resultant internal moment at the section becomes

$$M_z = \int dM_z = \int y(\sigma_x)_y\, dA \tag{3.18}$$

Substituting $(\sigma_x)_y = -ky$ from Equation 3.14, Equation 3.18 becomes

$$M_z = k \int y^2 dA = k I_z \qquad \textbf{(3.19)}$$

Note that the negative sign was discarded from Equation 3.19 in order to yield a positive bending moment M_z, in agreement with what is applied at the section (see Figure 3.5c). Furthermore, the symbol $I_z = \int y^2\, dA$ (see Appendix C.2) represents the moment of inertia of the cross section about the z axis, which is also the *ENA* for bending. Replacing k by σ_c/c (see Equation 3.14) in Equation 3.19 and solving for σ_c, we obtain

$$\sigma_{MAX} = \sigma_c = \frac{M_z c}{I_z} \qquad \textbf{(3.20)}$$

Substituting the value of σ_c given by Equation 3.20 into Equation 3.14, we obtain

$$(\sigma_x)_y = -\frac{M_z y}{I_z} \qquad \textbf{(3.21)}$$

To simplify the notation, we will dispense with some of the subscripting in Equation 3.21 and write

$$\sigma_x = -\frac{My}{I} \qquad \textbf{(3.22)}$$

Equation 3.22 is known as the *bending* or *flexural* equation and shows that the bending stress increases in magnitude from zero at the *ENA* to a maximum at the farthest point from this axis. It allows us to determine the normal bending stress σ_x at any distance y from the *ENA* for any cross section in the beam where the internal moment M and the moment of inertia I are known or can be found, *both with respect to the ENA*. Determination of the internal moment M was discussed in Sections 3.2 and 3.3. A review of the determination of moments of inertia is provided in Appendix C.2 and values of these quantities are given for selected cross-sectional areas in Appendix C.3.

Note that Equation 3.22 requires the use of proper signs for M and y if it is to produce the proper sign for the stress σ_x. It is possible, however, to rewrite Equation 3.22 in the form

$$\sigma_x = \frac{My}{I} \qquad \textbf{(3.23)}$$

where the negative sign has been omitted and deals exclusively with magnitudes ignoring the signs of the quantities M and y. Of course, the sign of the stress (tension or compression) is significant and will have to be determined. This can be easily accomplished by considering the sign of the internal bending moment M. *If this moment is positive, it creates bending stresses that are compressive above and tensile below the ENA. The reverse is true for a negative internal bending moment.* In the solutions that follow, Equation 3.23 will be given preference over Equation 3.22, which will only be used if there is a compelling reason for keeping track of the signs.

Equations 3.22 and 3.23 were derived for a pure bending moment. In most cases, however, the symmetric bending of beams is not due to pure bending moments but due to *transverse* loads (loads perpendicular to the beam). Under these conditions, the beam is subjected to *internal shearing forces V* in addition to the internal bending moments M as was discussed in Sections 3.2 and 3.3. As will be shown in Section 3.5, the existence of a shearing force is equivalent to a system of shearing stresses in the plane of the section. These shearing stresses produce a certain amount of distortion and the assumption made in deriving Equations 3.22 and 3.23, *of plane sections remaining plane after bending*, is not fully realized. However, more refined solutions obtained for the bending problem, including the effects of shearing stresses, indicate that Equations 3.22 and 3.23 give solutions that are very satisfactory for most practical purposes.

The *magnitude* of the maximum bending stress, σ_{MAX}, at any section of the beam where the bending moment is M, occurs at the farthest point from the *ENA* (i.e., at $y = c$). Thus,

$$\sigma_{MAX} = \frac{Mc}{I} \tag{3.24}$$

It is very convenient in the *allowable-stress design* of beams, discussed in Section 3.7, to rewrite Equation 3.24 in the form shown in Equation 3.25. Thus,

$$\sigma_{MAX} = \frac{|M_{MAX}|}{(I/c)} = \frac{|M_{MAX}|}{S} \tag{3.25}$$

where the quantity $|M_{MAX}|$ is the absolute maximum moment in the beam and $S = I/c$ is a property of structural sections known as the *section modulus*. The use of Equation 3.25 in the design of beams is discussed in Section 3.7.

The following examples illustrate the use of Equations 3.22, 3.23, 3.24, and 3.25 in the solution of problems.

EXAMPLE 3.6

The beam shown in figure (a) below has the rectangular cross section shown in figure (b) below. Consider a position 16 ft from support A and plot the bending stress versus the y coordinate.

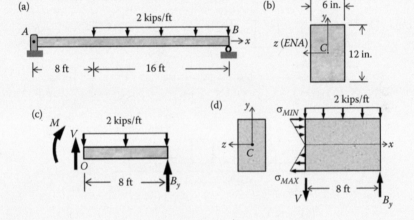

SOLUTION

The reactions at supports A and B are found to be, respectively, $A_y = 10.667$ kips and $B_y = 21.333$ kips. We now cut the beam 16 ft to the right of support A and construct an RFB diagram as shown in figure (c) above. The internal shear and moment are determined by applying the conditions of equilibrium. Thus,

$$\sum F_y = 0 \Rightarrow V + B_y - 2(8) = 0$$
$$V = 16 - B_y = -5.333 = 5.333 \text{ kips} \downarrow$$

$$\sum M_O = 0 \Rightarrow B_y(8) - 2(8)(4) - M = 0$$
$$M = 8B_y - 64 = 106.664 \text{ kip} \cdot \text{ft}$$

The internal shear force V is not needed in the solution of this example but is used to complete the free-body diagrams. However, the internal bending moment M is required in order to use Equation 3.23 for finding the needed bending stresses. Thus,

$$\sigma_x = \frac{My}{I} = \frac{(106.664)(12)y}{(1/12)(6)(12^3)} = 1.481\,y \qquad (3.6.1)$$

Equation 3.6.1 is plotted in figure (d) above, which shows the 8-ft segment of the beam to the left of support B, magnified in order to show details, along with the beam cross section. Since the bending moment is positive, it produces compression above and tension below the ENA as shown. The stresses labeled σ_{MAX} (bottom of the beam) and σ_{MIN} (top of the beam) are, in this case, equal in magnitude and are found by using Equation 3.6.1 above. Thus,

$$\sigma_{MAX} = 1.481(6) = 8.886 \approx 8.9 \text{ ksi}; \quad \sigma_{MIN} = -8.9 \text{ ksi} \qquad \textbf{ANS.}$$

EXAMPLE 3.7

The overhanging beam shown in figure (a) below has the T section depicted in figure (b) below. Determine the maximum tensile and maximum compressive bending stresses in the beam specifying their locations.

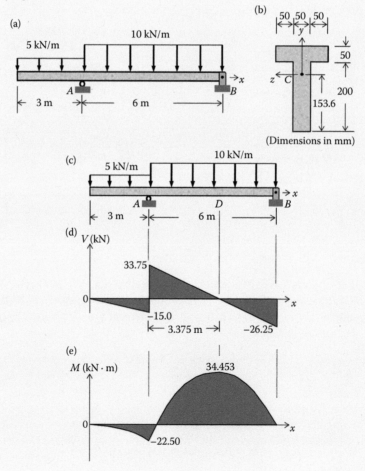

SOLUTION

As shown in figure (b) above, the centroid of the section is located at 153.6 mm above its bottom. The moment of inertia about this axis is found to be $I_z = 1.0186 \times 10^{-4}$ m^4. For convenience, the shear and moment diagrams are drawn directly beneath the beam as shown in figures (c) and (d) above, where the beam of figure (a) above is repeated. There are two positions along the beam where the bending moment is a relative maximum. One occurs at the support at A where the moment is -22.50 kN·m and the other occurs at D, a distance of 3.375 m to the right of A where the moment is 34.453 kN·m.

MAXIMUM TENSION

The bending moment at A is negative producing tension above the *ENA* (i.e., above the z axis). The bending moment at D is positive and produces tension below this axis. The maximum tensile stresses at these two locations are

$$(\sigma_A)_{\text{TEN}} = \frac{(22.50 \times 10^3)(0.250 - 0.1536)}{1.0186 \times 10^{-4}}$$

$$= 21.294 \text{ MPa}$$

$$(\sigma_D)_{\text{TEN}} = \frac{(34.453 \times 10^3)(0.1536)}{1.0186 \times 10^{-4}}$$

$$= 51.953 \text{ MPa}$$

Therefore,

$$(\sigma_{MAX})_{\text{TEN}} = 51.953 = 52.0 \text{ MPa at bottom of section at } D. \qquad \textbf{ANS.}$$

MAXIMUM COMPRESSION

Since the moment at A is negative, it produces compression below the neutral axis. However, the moment at D is positive and produces compression above the neutral axis. The maximum compressive stresses at these two locations are

$$(\sigma_A)_{\text{COMP}} = \frac{(22.50 \times 10^3)(0.1536)}{1.0186 \times 10^{-4}} = 33.929 \text{ MPa}$$

$$(\sigma_D)_{\text{COMP}} = \frac{(34.453 \times 10^3)(0.250 - 0.1536)}{1.0186 \times 10^{-4}} = 32.606 \text{ MPa}$$

Therefore,

$$(\sigma_{MAX})_{\text{COMP}} = 33.929 \approx 33.9 \text{ MPa at the bottom of section at } A \qquad \textbf{ANS.}$$

PROBLEMS

3.41–3.45 At some position along a beam, the bending moment about a horizontal axis was computed as given below for each of the cross sections indicated. In each case, compute the maximum tensile and maximum compressive stresses and sketch the bending stress distribution plotting the bending stress on a horizontal axis versus a vertical centroidal axis.

3.41 $M = 50 \text{ kip} \cdot \text{ft}$

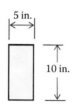

3.42 $M = -15 \text{ kN} \cdot \text{m}$

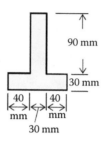

3.43 $M = -600 \text{ kip} \cdot \text{in.}$

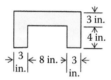

3.44 $M = 20 \text{ kN} \cdot \text{m}$

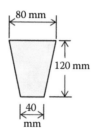

3.45 $M = 15 \text{ kip} \cdot \text{in.}$

3.46 The beam shown in figure (a) below has a rectangular cross section as shown in figure (b) below. If $P = 30$ kN, $L = 6$ m, $b = 60$ mm, and $h = 120$ mm, determine the maximum tensile and maximum compressive bending stresses at 3 m to the right of A.

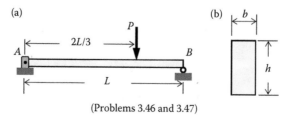

(Problems 3.46 and 3.47)

3.47 In reference to the figure shown in Problem 3.46, the beam shown in figure (a) above has a rectangular cross section as shown in figure (b) above. If $P = 36$ kips, $L = 15$ ft, $b = 4$ in., and $h = 10$ in., determine the maximum tensile and maximum compressive bending stresses right at the load P.

3.48 The beam shown in figure (a) below has the cross section shown in figure (b) below. If $Q = 15$ kip·ft, $p = 2$ kips/ft, $L = 20$ ft, and $a = 3$ in., determine the maximum tensile and maximum compressive bending stresses just to the right of the couple Q.

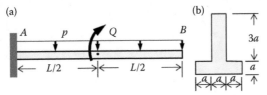

(Problems 3.48 and 3.49)

3.49 In reference to the figure shown in Problem 3.48, the beam shown in figure (a) above has the cross section shown in figure (b) above. If $Q = 30$ kN·m, $p = 5$ kN/m, $L = 8$ m, and $a = 90$ mm, determine the maximum tensile and maximum compressive bending stresses 2 m to the right of A.

3.50 The beam shown in figure (a) below has the cross section shown in figure (b) below. If $p = 4$ kips/ft, $L = 18$ ft, and $a = 2.5$ in., determine the maximum tensile and maximum compressive bending stresses at 9 ft to the right of A.

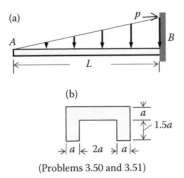

(Problems 3.50 and 3.51)

3.51 In reference to the figure shown in Problem 3.50, the beam shown in figure (b) above has the cross section shown in figure (b) above. If $p = 6$ kN/m, $L = 6$ m, and $a = 80$ mm, determine the maximum tensile and maximum compressive bending stresses 2 m to the left of the fixed support.

3.52 The beam shown in figure (a) below has the cross section shown in figure (a) below. If $p = 6$ kips/ft, $L = 18$ ft, and $a = 3$ in., determine the maximum tensile and maximum compressive bending stresses at 6 ft to the right of A.

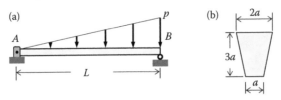

(Problems 3.52 and 3.53)

3.53 In reference to the figure shown in Problem 3.52, the beam shown in figure (a) above has the cross section shown in figure (b) above. If $p = 8$ kN/m, $L = 6$ m, and $a = 70$ mm, determine the maximum tensile and maximum compressive bending stresses at the center of the beam.

3.54 The beam shown in figure (a) below has the cross section shown in figure (b) below. If $P = 12$ kips, $p = 2$ kips/ft, $b = 8$ ft, and $a = 6$ in., determine the maximum tensile and maximum compressive bending stresses at the load P.

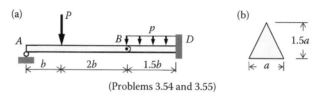

(Problems 3.54 and 3.55)

3.55 In reference to the figure shown in Problem 3.54, the beam shown in figure (a) above, has the cross section shown in figure (b) above. If $P = 30$ kN, $p = 5$ kN/m, $b = 2$ m, and $a = 120$ mm, determine the maximum tensile and maximum compressive bending stresses at the fixed support at D.

3.56 The beam shown in figure (a) below has the cross section shown in figure (b) below. If $p = 3$ kips/ft, $b = 8$ ft, and $a = 2$ in., determine the maximum tensile and maximum compressive bending stresses at 8 ft to the right of the support at A.

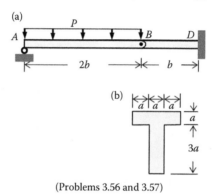

(Problems 3.56 and 3.57)

3.57 In reference to the figure shown in Problem 3.56, the beam shown in figure (a) above has the cross section shown in figure (b) above. If $p = 6$ kN/m, $b = 2$ m, and $a = 60$ mm, determine the maximum tensile and maximum compressive bending stresses at the fixed support at D.

3.58 The beam shown in figure (a) below has the cross section shown in figure (b) below. If $P = 10$ kips, $L = 5$ ft, and $a = 2$ in., determine the maximum tensile and maximum compressive bending stresses 3 ft to the right of the support at A.

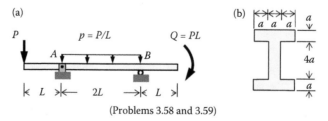

(Problems 3.58 and 3.59)

3.59 In reference to the figure shown in Problem 3.58, the beam shown in figure (a) above, has the cross section shown in figure (b) above. If $P = 120$ kN, $L = 2$ m, and $a = 60$ mm, determine the maximum tensile and maximum compressive bending stresses 2 m left of the support at B.

3.60 Refer to Problem 3.46 and determine the maximum tensile and maximum compressive bending stresses in the beam.

3.61 Refer to Problem 3.47 and determine the maximum tensile and maximum compressive bending stresses in the beam.

3.62 Refer to Problem 3.48 and determine the maximum tensile and maximum compressive bending stresses in the beam.

3.63 Refer to Problem 3.49 and determine the maximum tensile and maximum compressive bending stresses in the beam.

3.64 Refer to Problem 3.51 and determine the maximum tensile and maximum compressive bending stresses in the beam.

3.65 Refer to Problem 3.52 and determine the maximum tensile and maximum compressive bending stresses in the beam.

3.66 Determine the maximum tensile and maximum compressive bending stresses in the beam.

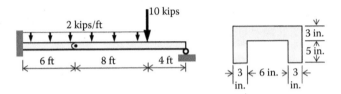

3.67 Determine the maximum tensile and maximum compressive bending stresses in the beam.

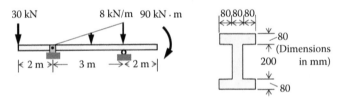

3.68 Determine the maximum tensile and maximum compressive bending stresses in the beam.

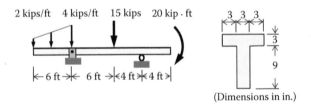

3.69 Determine the maximum tensile and maximum compressive bending stresses in the beam.

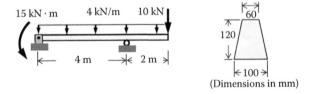

3.70 Refer to the beam of Problem 3.50 and determine the maximum permissible load intensity p if the allowable stress (tension or compression) is 10 ksi. Let $L = 18$ ft and $a = 2.5$ in.

3.71 Refer to the beam of Problem 3.53 and determine the maximum permissible load intensity p if the allowable bending stress (tension or compression) is 100 MPa. Let $L = 6$ m and $a = 70$ mm.

3.72 Refer to the beam of Problem 3.54 and determine the maximum permissible load P if the allowable bending stress (tension or compression) is 130 ksi. Let $p = 2$ kips/ft, $b = 8$ ft, and $a = 6$ in.

3.73 Refer to the beam of Problem 3.57 and determine the maximum load intensity p if the allowable bending stress (tension or compression) is 100 MPa. Let $b = 2$ m and $a = 60$ mm.

3.74 Refer to the beam of Problem 3.58 and determine the maximum permissible load P if the allowable bending stress (tension or compression) is 40 ksi. Let $L = 5$ ft and $a = 2$ in.

3.75 Refer to the beam of Problem 3.59 and determine the maximum permissible load P if the allowable bending stress (tension or compression) is 150 MPa. Let $L = 2$ m and $a = 60$ mm.

3.5 SHEARING STRESSES UNDER SYMMETRIC LOADING

3.5.1 LONGITUDINAL SURFACES NORMAL TO LOADS

In Section 3.2, we discovered that when bending of a beam is produced by transverse loads, all of its cross sections are subjected not only to a bending moment M but also to a shearing force V. Just as the bending moment M represents the resultant of a system of bending stresses whose distribution was established in Section 3.4, the shearing force V represents the resultant of a system of shearing stresses distributed in some manner that will be discussed in the following paragraphs.

In order to derive an approximate distribution for the shearing stresses in a symmetrically loaded beam, consider the beam shown in Figure 3.6a. For simplicity and convenience, the cross-sectional area is assumed to be a rectangle as shown in Figure 3.6b. A differential segment of the beam of length dx between planes (1) and (2) is isolated as shown in Figure 3.6c where the shear force and bending moment on plane (1) are, respectively, V and M, and on plane (2), they are, respectively,

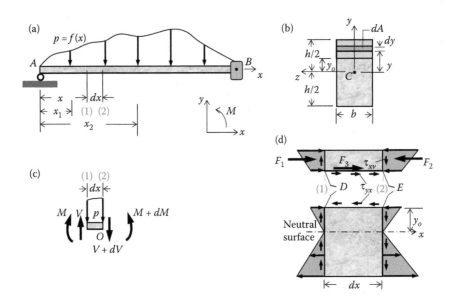

FIGURE 3.6 Beam in (a), with cross section in (b), is subjected to a distributed load $p = f(x)$. (c) A differential element of length dx showing the internal shear V and moment M on the one side and internal $V + dV$ and $M + dM$ on the other side. (d) The shearing and normal stress distributions on the two planes, 1 and 2.

$V + dV$ and $M + dM$. The quantities dV and dM represent the differential changes in the shear force and bending moment that occur, respectively, over the differential length dx.

The moments M and $M + dM$ give rise to bending stresses depicted on planes (1) and (2), respectively, in Figure 3.6d. The shearing forces V and $V + dV$ are the resultants of the shearing stress distributions shown on planes (1) and (2), respectively, in Figure 3.6d. The assumption is made that on any plane such as DE in Figure 3.6d, parallel to the neutral surface, the shearing stress is constant. Inherent in this assumption is the requirement that the width b of the cross-sectional area is relatively small and that its two sides intersecting the neutral axis are perpendicular to this axis. A more refined solution by Saint Venant indicates that the width b of the section cannot be much larger than its height h if this approximate solution is to yield satisfactory answers. Otherwise, the variation of the shearing stress across the width of the section may be sufficiently large to render useless the following development and the resulting equations.

In order to find the *horizontal shearing stress* τ_{yx} on any horizontal surface as the one defined by the line DE a distance y_o above the neutral axis, we isolate the segment above this surface as shown in the upper part of Figure 3.6d. The resultant of these shearing stresses is F_3 as shown in this figure. The three forces, F_1, F_2, and F_3 on this segment act in the x direction and must be in equilibrium. Thus,

$$\sum F_x = 0 \Rightarrow F_1 + F_3 - F_2 = 0; \quad F_3 = F_2 - F_1$$

where

$$F_1 = \int_{y_o}^{h/2} (\sigma_x)_1 dA = (M/I) \int_{y_o}^{h/2} y\, dA; \quad F_2 = \int_{y_o}^{h/2} \sigma_x)_2 dA = (M + dM/I) \int_{y_o}^{h/2} y\, dA;$$

$$F_3 = \tau_{yx} b\, dx$$

Making these substitutions and simplifying, we obtain

$$\tau_{yx} b\, dx = \frac{dM}{I} \int_{y_o}^{h/2} y\, dA$$

Solving for τ_{yx}, we obtain

$$\tau_{yx} = \left(\frac{1}{I\,b}\right)\left(\frac{dM}{dx}\right) \int_{y_o}^{h/2} y\, dA \qquad (3.26)$$

The derivative (dM/dx) is, by Equation 3.7, equal to the shear force V and the integral $\int_{y_o}^{h/2} y\, dA$ represents the first moment of the area between y_o and $h/2$ (shown in orange in Figure 3.6b) about the neutral axis. Alternatively, this integral represents *the first moment, about the neutral axis, of the area contained between an outer edge of the cross section parallel to the neutral axis and the interior surface at which the shearing stress τ_{yx} is to be computed.* Implied in this statement is the *freedom* to choose the area on either side of the surface at which the shearing stress is to be found. This integral is, generally, given the symbol Q. Making these substitutions in Equation 3.26, we obtain the equation for the magnitude of the *horizontal shearing stress* τ_{yx} on any interior longitudinal surface at a distance y from the neutral axis. Thus,

$$\tau_{zx} = \frac{VQ}{I\,b} \qquad (3.27)$$

Based upon the above statements, the quantity Q may be found either from its basic definition, $\int_{y_o}^{h/2} y \, dA$ or from the relation

$$Q = A^* \bar{y} \tag{3.28}$$

where A^* is the area above or below the surface at which the shearing stress is to be found and $\bar{y}$ is the distance of its centroid from the neutral axis of the entire section.

Since Equations 3.22 and 3.23 were used in deriving Equation 3.27, the latter equation is subject to the same limitations dictated by the assumptions made in deriving the first. Also, by Equation 2.19, we come to the conclusion that the *vertical shearing* stress is equal in magnitude to the horizontal shearing stress (i.e., $\tau_{xy} = \tau_{yx}$). Therefore, Equation 3.27 yields not only the magnitude of the shearing stress at any point on a horizontal surface at some distance from the neutral axis but also the magnitude of the shearing stress on a vertical surface passing through the same point. In view of this, we can dispense with the subscripts on the shearing stress and write the equation in a simpler form. Thus,

$$\tau = \frac{VQ}{I \, b} \tag{3.29}$$

Examination of Equation 3.29 reveals that since at a given location along a beam of a certain cross-sectional area, the quantities V and I are constants, the magnitude of τ depends only on the ratio Q/b. Thus, the maximum value of the shearing stress, on a horizontal or a vertical surface, occurs at the location in the cross section, where Q/b is a maximum. In general, for most commonly used structural shapes, this ratio is a maximum at the neutral axis. However, there are shapes for which the maximum shearing stress does not occur at the neutral axis. Therefore, we need to examine each cross-sectional configuration to determine the location for which the ratio Q/b is a maximum.

3.5.2 Shear Flow

The shear force per unit of length on any longitudinal surface is referred to as *shear flow* (in analogy to the flow of fluid in a pipe or channel) and given the symbol q. The shear flow q is obtained by multiplying the shearing stress τ by the dimension b in Equation 3.29. Thus,

$$q = \tau b = \frac{VQ}{I} \tag{3.30}$$

The quantities V, Q, and I in Equation 3.30 have already been defined.

It can be shown that Equations 3.29 and 3.30 apply equally well to any longitudinal surface arbitrarily oriented to the plane of the loads. Therefore, Equations 3.29 and 3.30 are very useful in analyzing built-up sections, as illustrated in Example 3.11, and thin-walled sections with at least one plane of symmetry and subjected to loads in a plane of symmetry, as illustrated in Example 3.12.

EXAMPLE 3.8

A beam has the rectangular cross section shown in figure (a) below. If the shear force is V at some position, develop, for this position, an expression for the shearing stress τ as a function of the coordinate y. Plot this expression.

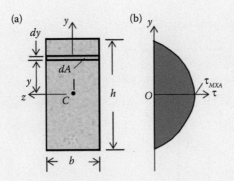

SOLUTION

We begin by finding the quantity Q. By definition, this quantity is

$$Q = \int y\, dA = \int\limits_{y}^{h/2} z\, dA = \frac{b}{2}\left(\frac{h^2}{4} - y^2\right)$$ (3.8.1)

Substituting the value of Q from Equation 3.8.1 into Equation 3.29 and simplifying, we obtain

$$\tau = \frac{6V}{bh^3}\left(\frac{h^2}{4} - y^2\right) \qquad \textbf{ANS.}$$ (3.8.2)

which is the equation of a parabola. At the top and bottom of the rectangular cross section, where y is $+h/2$ and $-h/2$, respectively, the shearing stress τ vanishes. The maximum magnitude of τ occurs when $y = 0$, which is at the neutral axis for the rectangular cross section. Thus, setting $y = 0$ in Equation 3.8.2 and simplifying, we obtain

$$\tau_{MAX} = \left(\frac{3}{2}\right)\left(\frac{V}{bh}\right) = \left(\frac{3}{2}\right)\left(\frac{V}{A}\right)$$ (3.8.3)

The quantity A in Equation 3.8.3 is the area of the rectangular cross section.
The plot of Equation 3.8.2 is shown in figure (b) above.

EXAMPLE 3.9

The T cross section for a beam is shown in figure (a) below. If the shear force at some position along the beam is $V = 33.750$ kN, plot the shearing stress distribution as a function of the coordinate y.

SOLUTION

Two separate shear stress functions are needed to plot the required distribution, one for the flange ($46.4 < y < 96.4$ mm) and the second for the web ($-153.6 < y < 46.4$ mm). The reason for this is the fact that at the junction between the flange and the web, there is a discontinuity in the shearing stress, as will be shown below and illustrated in figure (b) below.

As shown in figure (a) below, the centroid location was computed at 153.6 mm above the base of the T section. Also, the moment of inertia about the neutral axis was found to be $I = 1.0186 \times 10^{-4}$ m^4.

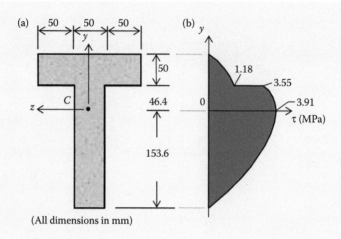

(All dimensions in mm)

FLANGE

$$Q = \int_{y}^{0.0964} y\, dA = 0.15 \int_{y}^{0.964} y\, dy = 7.5 \times 10^{-2}(0.009293 - y^2)$$

$$\tau = \frac{VQ}{Ib} = \frac{33750 \times 7.5 \times 10^{-2}}{(1.0186 \times 10^{-4})(0.15)}(0.009293 - y^2)$$

$$= 165.67(0.009293 - y^2)\, \text{MPa} \cdots 0.0464 < y < 0.0964 \text{ m}$$

WEB

$$Q = \int_{y}^{-0.1536} y\, dA = 0.05 \int_{y}^{-0.1536} y\, dy = 2.5 \times 10^{-2}(0.023593 - y^2)$$

$$\tau = \frac{VQ}{Ib} = \frac{33750 \times 2.5 \times 10^{-2}}{(1.0186 \times 10^{-4})(0.05)}(0.023593 - y^2)$$

$$= 165.67(0.023593 - y^2)\, \text{MPa} \cdots -0.1536 < y < 0.0464 \text{ m}$$

These two functions are plotted above in figure (b). Note the discontinuity in the shearing stress at the junction between the flange and web where the shearing stress changes abruptly from 1.18 to 3.55 MPa. This is due to the sudden change in the dimension b from 0.15 to 0.05 m. Note also that $\tau_{MAX} = 3.91$ MPa at $y = 0$. **ANS.**

EXAMPLE 3.10

The cross-sectional area shown is that for a beam subjected to transverse loads contained in a longitudinal plane represented by the y centroidal axis. At some position along the beam, the shear force is $V = 25$ kips. Find the shearing stress (a) 3 in. below the top of the section and (b) at the neutral axis.

SOLUTION

In Examples 3.8 and 3.9, we were asked to plot the distribution of the shearing stress across the height of the section. To accomplish the purpose, we developed expressions relating the shearing stress to the y coordinate. In this example, we are asked to find the shearing stress at specific points in the cross section and, therefore, we can resort to the alternate definition of Q, which states that it is the first moment about the neutral axis of the area contained between an outer edge of the cross section parallel to the neutral axis and the interior surface at which the shearing stress τ is to be found. Because of the symmetry of the section, the centroid, and hence the neutral axis, are located by inspection as shown in the sketch. Thus, Equation 3.29, $\tau = VQ/Ib$, is used to find the answers for both parts (a) and (b).

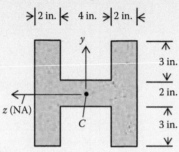

a. At a distance of 3 in. below the top of the section (i.e., at the junction between the horizontal rectangle and the two upper vertical rectangles), the width b may be taken either just above ($b = 4$ in.) or just below ($b = 8$ in.). Therefore, at the junction, we have a shearing stress discontinuity similar to the one encountered in Example 3.9. For either of these two positions, however, Q is the same and determined by finding the first moment about the neutral axis of the *area above* or the *area below* the junction. Using the area above, we get

$$Q = \sum A^* \bar{z} = 2(2 \times 3)(2.5) = 30.0 \text{ in.}^3$$

The moment of inertia about the neutral axis was computed and found to be $I = 173.3$ in.[4]. Therefore,

$$\tau = \frac{25(30.0)}{4(173.3)} = 1.082 \approx 1.1 \text{ ksi}, \quad \text{just above the junction} \qquad \textbf{ANS.}$$

$$\tau = \frac{25(30.0)}{8(173.3)} = 0.541 \approx 0.5 \text{ ksi}, \quad \text{just below the junction} \qquad \textbf{ANS.}$$

b. At the neutral axis, the width b in Equation 3.29 has a unique value of 8 in. The quantity Q becomes

$$Q = \sum A^* \bar{y} = 2(2 \times 3)(2.5) + (8)(1)(0.5) = 34.0 \text{ in.}^3$$

Therefore, at the neutral axis, the shearing stress becomes

$$\tau = \frac{25(34.0)}{8(173.3)} = 0.613 \approx 0.6 \text{ ksi} \qquad \textbf{ANS.}$$

Note that for this particular cross section, the maximum shearing stress does not occur at the neutral axis.

EXAMPLE 3.11

The cross section shown was fabricated by fastening two 12×1 in. steel plates using two $4 \times 4 \times 1/2$ steel angles (see Appendix E). The diameter of the bolts marked D is 5/8 in. and have a longitudinal spacing into the page $p_D = 6$ in. The diameter of the bolts marked E is 7/8 in. and their longitudinal spacing into the page is p_E. The maximum shear force in the beam was found to be $V = 30$ kips. Find (a) the shearing stress in the D bolts and (b) the spacing p_E if the allowable shearing force in each of the E bolts is 15 kips.

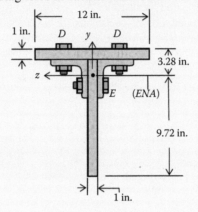

SOLUTION

a. The centroidal y and z axes are located as shown in the sketch and the moment of inertia about the neutral axis is determined to be $I = 431.84$ in.[4]. Note that while more accurate analyses may include the effect of bolt holes, these holes are not included in the present computations. The D bolts are subjected to longitudinal shear flow q at the junction between the horizontal plate and the two angles. This shear flow is given by Equation 3.30, where Q is the first moment of the rectangular area above (or below) the junction about the neutral axis. Using the area above, this quantity becomes $Q = 12(1)(2.78) = 33.36$ in.[3]. Thus,

$$q = \frac{VQ}{I} = \frac{30(33.36)}{431.84} = 2.32 \text{ kips/in.}$$

Therefore, the total force $2F$ carried by the two D bolts in the longitudinal space p_D becomes

$$2F = q(p_D) = 2.32(6) = 13.92 \text{ kips}$$

The shear force carried by each bolt is $F = 13.92/2 = 6.96$ kips. Since all of the D bolts are in single shear, the shearing stress in these bolts is

$$\tau_D = \frac{F}{A_{bolt}} = \frac{6.96}{(\pi/4)(5/8)^2} = 22.686 \approx 22.7 \text{ ksi} \qquad \textbf{ANS.}$$

b. The E bolts are subjected to longitudinal shear flow q at the junction between the vertical plate and the two angles, which is given by Equation 3.30. In this equation, Q is the first moment of the area of the vertical rectangle about the neutral axis, which becomes $Q = 12(1)(3.72) = 44.64$ in.[3]. Thus,

$$q = \frac{VQ}{I} = \frac{30(44.64)}{431.84} = 3.10 \text{ kips/in.}$$

The shear force F carried by each of the E bolts is the shear flow multiplied by the unknown spacing p_E. Thus, since $F = 15$ kips, it follows that $(3.10)p_E = 15$. Therefore,

$$p_E = 4.839 \approx 4.8 \text{ in.} \quad \textbf{ANS.}$$

EXAMPLE 3.12

The inverted channel shown in figure (a) below is to carry a symmetrically applied downward shear force $V = 150$ kN passing through its centroid as shown. (a) Determine the shearing stress at points A and B. (b) Show the shear flow on a sketch of the channel.

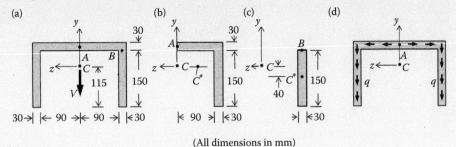

(All dimensions in mm)

SOLUTION

The centroidal z axis of the channel (i.e., its ENA) is located at 115 mm above the base as shown in figure (a) above and its centroidal moment of inertia was computed to be $I = 36.315 \times 10^{-6}$ m⁴.

a. The shearing stress at point A is given by Equation 3.29. The Q in this equation is obtained by Equation 3.28. Since the centroid C^* of the area in figure (b) above is at the same height as that for the entire section, the distance $\bar{y}$ in Equation 3.28 is zero, and thus Q is zero. It follows, therefore, that the shearing stress at point A is zero. **ANS.**

The shearing stress at point B is found from Equation 3.29, where Q is determined by Equation 3.28 using the area to one side of this point as shown in figure (c) above. The distance between the centroid C^* of this area and that of the entire channel is 40 mm (i.e., $\bar{y} = 40$ mm) as shown in figure (c) above. Thus, using Equation 3.28, we have

$$Q = (0.03)(0.15)(0.04) = 1.8 \times 10^{-4} \text{ m}^3$$

and by Equation 3.29, we obtain

$$\tau = \frac{150(10^3)(1.8 \times 10^{-4})}{36.315 \times 10^{-6}(0.03)} = 24.783 \approx 24.8 \text{ MPa} \quad \textbf{ANS.}$$

Note that point B was assumed to be just below the junction where the thickness b in Equation 3.29 is 30 mm.

b. The shear flow q in the channel is shown schematically in figure (d) above. This shear flow starts at zero at point A and increases gradually on both sides of point A as Q increases until it reaches a maximum at the neutral axis for the section. Then it begins to decrease in magnitude until it becomes zero at the two bottom edges of the section. Note that the horizontal rectangle is assumed to carry no vertical shear flow and that the entire vertical shear V is assumed to be carried by the two vertical rectangles. **ANS.**

PROBLEMS

3.76–3.78 At some position along a beam, the shearing force was computed as given below for each of the three cross sections indicated. In each case, develop a relation between the shearing stress and the vertical y axis. Plot the shearing stress on a horizontal axis versus the y axis. What is the maximum shearing stress? Use Examples 3.8 and 3.9 as guides.

3.76 $V = 30$ kips

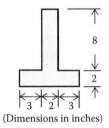

(Dimensions in inches)

3.77 $V = 125$ kN

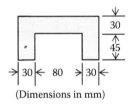

(Dimensions in mm)

3.78 $V = 45$ kips

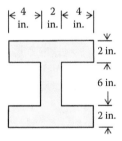

3.79 The beam shown in figure (a) below has a rectangular cross section as shown in figure (b) below. If $P = 60$ kN, $L = 6$ m, $b = 60$ mm, and $h = 120$ mm, determine, at 3 m to the right of A, the shearing stress (a) 40 mm below the top of the section and (b) at its neutral axis.

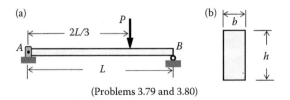

(Problems 3.79 and 3.80)

3.80 In reference to the figure shown in Problem 3.79, the beam shown in figure (a) above has a rectangular cross section as shown in figure (b) above. If $P = 210$ kips, $L = 15$ ft,

$b = 4$ in., and $h = 10$ in., determine, at 9 ft to the right of A, the shearing stress (a) 3 in. above the bottom of the section and (b) 1 in. above its neutral axis.

3.81 The beam shown in figure (a) below has the cross section shown in figure (b) below. If $Q = 15$ kip·ft, $p = 5$ kips/ft, $L = 20$ ft, and $a = 3$ in., determine, at 5 ft to the right of A, the shearing stress (a) just above the junction between the flange and the web and (b) at the neutral axis of the section.

(a)   (b)

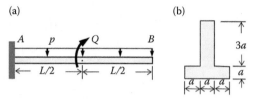

(Problems 3.81 and 3.82)

3.82 In reference to the figure shown in Problem 3.81, the beam shown in figure (a) above has the cross section shown in figure (b) above. If $Q = 20$ kN·m, $p = 7$ kN/m, $L = 8$ m, and $a = 90$ mm, determine, just to the right of the couple Q, the shearing stress (a) just below the junction between the flange and the web and (b) at 90 mm below the top of the section.

3.83 The beam shown in figure (a) below has the cross section shown in figure (b) below. If $p = 20$ kips/ft, $L = 20$ ft, and $a = 2.0$ in., determine, at 5 ft to the right of A, the shearing stress (a) 2 in. above the bottom of the section and (b) at its neutral axis.

(a)   (b)

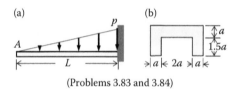

(Problems 3.83 and 3.84)

3.84 In reference to the figure shown in Problem 3.83, the beam shown in figure (a) above has the cross section shown in figure (b) above. If $p = 28$ kN/m, $L = 8$ m, and $a = 80$ mm, determine, just to the left of the fixed support, the shearing stress (a) just above the junction between the horizontal and the two vertical rectangles and (b) just above this junction.

3.85 The beam shown in figure (a) below has the cross section shown in figure (b) below. If $p = 3$ kips/ft, $b = 8$ ft, and $a = 2$ in., determine, at 4 ft to the right of the support at A, the shearing stress (a) just below the junction between the web and the upper flange and (b) at the neutral axis for the cross section.

(a)   (b)

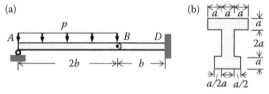

(Problems 3.85 and 3.86)

3.86 In reference to the figure shown in Problem 3.85, the beam shown in figure (a) above has the cross section shown in figure (b) above. If $p = 60$ kN/m, $b = 2$ m, and $a = 60$ mm, determine, at the fixed support at D, the shearing stress (a) just above the junction between the web and the lower flange and (b) at the neutral axis for the cross section.

3.87 The beam shown in figure (a) below has the cross section shown in figure (b) below. If $P = 10$ kips, $L = 5$ ft, and $a = 2$ in., determine, 3 ft to the right of the support at A, the shearing stress (a) at point O on the neutral axis and (b) at point D, which is just above the junction between the right vertical and the horizontal rectangles.

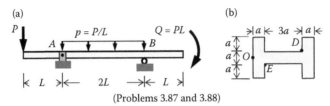

(Problems 3.87 and 3.88)

3.88 In reference to the figure shown in Problem 3.87, the beam shown in figure (a) above has the cross section shown in figure (b) above. If $P = 24$ kN, $L = 2$ m, and $a = 60$ mm, determine, 1.5 m left of the support at B, the shearing stress (a) at point O on the neutral axis and (b) at point E, which is just below the junction between the left vertical and the horizontal rectangle.

3.89 Refer to the beam of Problem 3.79 and determine the maximum shearing stress in the beam specifying its location.

3.90 Refer to the beam of Problem 3.80 and determine the maximum shearing stress in the beam specifying its location.

3.91 Refer to the beam of Problem 3.81 and determine the maximum shearing stress in the beam specifying its location.

3.92 Refer to the beam of Problem 3.82 and determine the maximum shearing stress in the beam specifying its location.

3.93 Refer to the beam of Problem 3.83 and determine the maximum shearing stress in the beam specifying its location.

3.94 Refer to the beam of Problem 3.84 and determine the maximum shearing stress in the beam specifying its location.

3.95 Refer to the beam of Problem 3.85 and determine the maximum shearing stress in the beam specifying its location.

3.96 Refer to the beam of Problem 3.86 and determine the maximum shearing stress in the beam specifying its location.

3.97 Refer to the beam of Problem 3.87 and determine the maximum shearing stress in the beam specifying its location.

3.98 Refer to the beam of Problem 3.88 and determine the maximum shearing stress in the beam specifying its location.

3.99 The built-up section shown is fabricated by nailing four rectangular pieces of wood using steel spikes having a diameter of 3 mm. The shear force that must be carried by the section is $V = 2$ kN. If the allowable shearing stress in the spikes is 30 MPa, determine the required spacing of these spikes. Neglect spike holes in the computations.

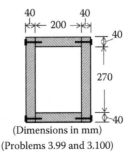

(Dimensions in mm)

(Problems 3.99 and 3.100)

3.100 In reference to the figure shown in Problem 3.99, the built-up section shown is fabricated by nailing four rectangular pieces of wood using steel spikes having a diameter
of 4 mm spaced longitudinally at 200 mm. If the allowable shearing stress in the spikes
is 40 MPa, determine the maximum permissible shear force that the section can carry.
Neglect spike holes in the computations.

3.101 The built-up section shown is fabricated by fastening two 10×1 in. steel plates using
two $4 \times 4 \times 3/8$ in. steel angles (see Appendix E) as shown. The diameter of the vertical
bolts is 1/2 in. and their longitudinal spacing is 5 in. If the allowable shearing stress is
15 ksi per bolt, determine the maximum shear force that the section can carry. Assume
that the horizontal bolts are strong enough for the purpose and neglect the bolt holes in
the computations.

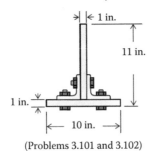

(Problems 3.101 and 3.102)

3.102 In reference to the figure shown in Problem 3.101, the built-up section shown is fabricated by fastening two 10×1 in. steel plates using two $4 \times 4 \times 3/8$ in. steel angles
(see Appendix E) as shown. The diameter of the vertical bolts is 5/16 in. If the shear
force on the section is $V = 15$ kips and the allowable shearing stress in these bolts
is 13 ksi/bolt, determine the required spacing of the vertical bolts. Assume the horizontal bolts are strong enough for the purpose and neglect the bolt holes in the
computations.

3.103 A W21 × 83 section (see Appendix E) is strengthened by fastening a 1 × 8-in. plate to
the bottom flange using two rows of 1/2-in.-diameter bolts as shown. This cross section
is that for a simply supported beam 30 ft long carrying a concentrated load P at midspan. If the longitudinal spacing of the bolts is 8 in. and the allowable shearing stress
in them is 15 ksi/bolt, determine the maximum permissible load P. Neglect bolt holes
in the computations.

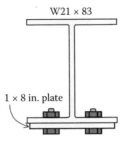

3.104 A W16 × 31 section (see Appendix E) is strengthened by fillet welding a 3/4 × 5 in. plate
to the top of flange as shown. The shear force in the section is known to be 10 kips.
Determine the shear flow in the weld, which is assumed to be continuous along the
length of the beam.

3/4 × 5 in. plate

3.105 Each of the following four thin-walled open sections has at least one plane of symmetry and is subjected to a downward vertical shear force symmetrically placed as shown. For each case, determine (a) the shearing stress at point *A* and (b) the shearing stress at the neutral axis of the section.

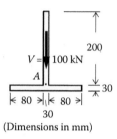

$V =$ 100 kN

A

200

30

|← 80 →|← 80 →|

30

(Dimensions in mm)

3.106

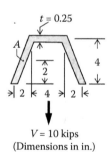

$t = 0.25$

A

2

4

→| 2 |← 4 →| 2 |←

$V =$ 10 kips

(Dimensions in in.)

3.107

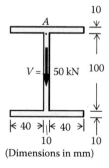

10

A

$V =$ 50 kN   100

|← 40 →|← 40 →|

10   10

(Dimensions in mm)

3.108

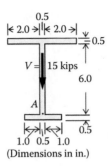

V = 15 kips

(Dimensions in in.)

3.109 Each of the following four thin-walled closed sections has at least one plane of symmetry and is subjected to a downward vertical shear force symmetrically placed as shown. For each case, determine (a) the shearing stresses at points A and B and (b) the shearing stress at the neutral axis.

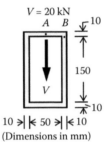

(Dimensions in mm)

3.110

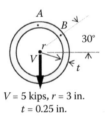

$V = 5$ kips, $r = 3$ in.
$t = 0.25$ in.

3.111

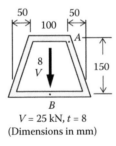

$V = 25$ kN, $t = 8$
(Dimensions in mm)

3.112

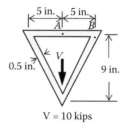

5 in. | 5 in.

0.5 in.

V

9 in.

V = 10 kips

3.6 STRESSES UNDER COMBINED LOADS

When a beam carries transverse loads, it is subjected to the combined action of shear forces and bending moments. This combined action leads to the simultaneous existence of a shearing stress τ and a bending stress σ at any point in the beam. Consider, for example, the segment of a beam of length Δx shown in Figure 3.7a that carries a transverse distributed load p as shown. At any position such as *ABCDE*, there exist a shear force and a bending moment. The shear force gives rise to a shearing stress and the bending moment to a bending normal stress whose magnitudes vary depending upon the location of the point being considered. Five points are considered, two above the neutral surface (points *A* and *B*), two below this surface (points *D* and *E*), and the fifth (point *C*) right at the neutral surface. The stress elements corresponding to these points are depicted in Figure 3.7b. Point *A* at the very top of the beam is subjected only to a compressive bending stress because the shearing stress at this position is zero. Point *B*, somewhere between the top of the beam and its neutral surface, carries both a compressive bending stress and a shearing stress. Point *C*, which is located at the neutral surface, carries only a shearing stress because the bending stress at this position is zero. Point *D*, somewhere between the neutral surface and the bottom of the beam, is subjected to a tensile bending stress as well as a shearing stress. Finally, point *E* at the bottom of the beam is subjected only to a tensile bending stress because the shearing stress at this position is zero. Note that, for all of the five points considered above, the normal and shearing stresses are given, respectively, by Equations 3.22 and 3.23 and Equation 3.29.

Numerous other cases of combined loadings are encountered in practice where the bending action is combined with other types of actions such as axial and or torsional loadings. Some of these cases are illustrated in Examples 3.13, 3.14, and 3.15.

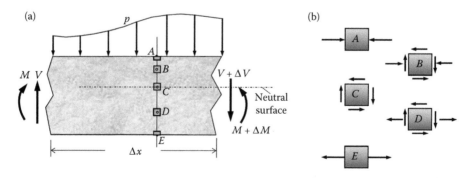

FIGURE 3.7 The beam in (a), under the action of transverse load p, is subjected to the combined action of internal shear V and moment M, resulting in the simultaneous existence of a shearing stress τ and a bending stress σ.

EXAMPLE 3.13

The cantilever beam shown in figure (a) below is subjected to the force P inclined to the horizontal at an angle θ. Select any point at position A, a distance d from the left end, and construct a stress element at this point with two planes parallel to the axis of the beam and two planes perpendicular to it.

SOLUTION

The free-body diagram of the segment of length d is shown in figure (b) below, where point B was selected somewhere between the neutral surface and the top of the beam as shown. The stress element at this point is depicted in figure (c) below, where the normal and shearing stresses are computed as follows:

$$F = P\cos\theta; \quad V = -P\sin\theta; \quad M = -(P\sin\theta)d$$

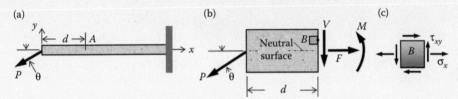

Therefore,

$$\sigma_x = \frac{My}{I} + \frac{F}{A} = \left(\frac{(P\sin\theta)b}{I}\right)y + \frac{P\cos\theta}{A}; \quad \tau_{xy} = \frac{VQ}{Ib} = \frac{(P\sin\theta)Q}{Ib} \qquad \textbf{ANS.}$$

The negative sign on the shearing force indicates that this force is pointed up, not down as shown; hence, the upward sense of the shearing stress on the stress element. The negative sign on the moment tells us that this moment is clockwise, not counterclockwise as indicated; hence, the sense of the normal stress on the stress element. Note that the location of point B (above or below the neutral surface) dictates the sense of the normal stress. Thus, for example, if point B were placed below instead of above the neutral surface, the normal stress could be compressive instead of tensile, depending on the relative magnitudes of the two terms in the equation for σ_x.

EXAMPLE 3.14

Arm AB is attached rigidly to shaft BDE of diameter d and a load P is applied at A as shown in figure (a) below. Construct the stress elements at points G_1 and G_2 on the outside surface of the shaft at section D.

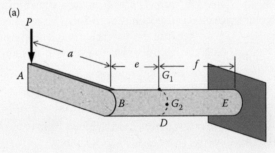

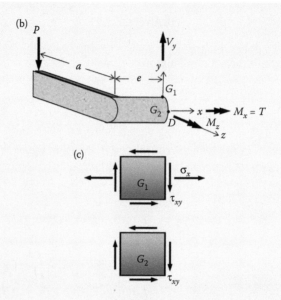

SOLUTION

The free-body diagram of segment *ABD* of the assembly is shown in figure (b) above. The internal reaction components at section *D* consist of a shear force V_y, a torque $M_x = T$, and a bending moment M_z as shown. These internal reactions are found in terms of the applied load *P* as follows:

$$\sum F_y = 0 \Rightarrow V_y - P = 0; \quad V_y = P$$

$$\sum M_x = 0 \Rightarrow M_x - Pa = 0; \quad M_x = Pa$$

$$\sum M_z = 0 \Rightarrow M_z + Pe = 0; \quad M_z = -Pe$$

POINT G_1

The stress element at G_1 (viewing it from above) is shown in figure (c) above. Only two of the three internal reaction components, $M_x = T$ and M_z, produce stresses at this point. This is because the quantity *Q* in the shearing stress equation, *VQ/Ib*, is zero at this point. The stresses σ and τ are found as follows:

$$\sigma_x = \frac{M_z y}{I_z} = \frac{(-Pe)(d/2)}{(\pi/64)d^4}; \quad \tau_{xy} = \frac{Tr}{J} = \frac{M_x r}{J} = \frac{(Pa)(d/2)}{(\pi/32)d^4} \quad \textbf{ANS.}$$

POINT G_2

The stress element at G_2 (viewing it from the front) is also shown in figure (c) above. Here again, only two of the three reaction components, $M_x = T$ and V_y, produce stresses at this point. The third component, M_z, does not because the distance *y* in the bending stress equation, *M y/I*, is zero at this point (point G_2 is at the neutral axis). The stress τ_{xy} is found as follows:

$$\tau_{xy} = \frac{Tr}{J} + \frac{VQ}{Ib} = \frac{(Pa)(d/2)}{(\pi/32)d^4} - \frac{P\left[(\pi/8)d^2)(2d/3\pi)\right]}{\left[(\pi/64)d^4\right]d} \quad \tau_{xy} = \frac{16P}{\pi d^2}\left(\frac{3a-d}{3d}\right) \quad \textbf{ANS.}$$

Note that the two shearing stresses, in this case, subtract from each other. There are cases, however, in which the two shearing stresses add to each other.

EXAMPLE 3.15

The block shown in figure (a) below is subjected to a compressive force P at point $G(y, z)$. Construct stress elements at points B, D, and E.

SOLUTION

The free-body diagram of the segment above plane BDE is shown in figure (b) below. The internal reactive components at this section consist of two moments, M_y and M_z, and the axial force F_x. These reactive components are found as follows:

$$\sum F_x = 0 \Rightarrow F_x + P = 0; \quad F_x = -P$$

$$\sum M_y = 0 \Rightarrow M_y - Pz = 0; \quad M_y = Pz$$

$$\sum M_z = 0 \Rightarrow M_z - Py = 0; \quad M_z = Py$$

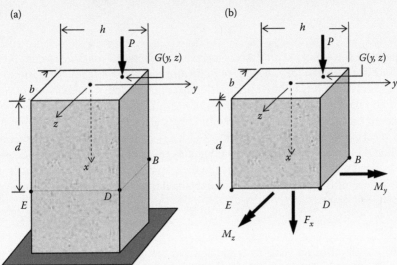

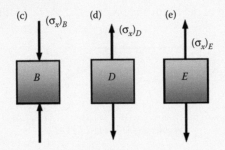

POINT B

The stress element at point B is shown in figure (c) above. The compressive normal stress $(\sigma_x)_B$ at this point consists of three components as follows:

$$(\sigma_x)_B = \frac{F_x}{A} + \frac{M_y z_B}{I_y} + \frac{M_z y_B}{I_z} = -\frac{P}{bh} - \frac{(Pz)(b/2)}{(1/12)hb^3} - \frac{(Py)(h/2)}{(1/12)bh^3} \qquad \textbf{ANS.}$$

POINT D

The stress element at point D is shown in figure (d) above. While the normal stress at this point is indicated as tensile, its character depends upon the relative magnitudes of P, M_y, and M_z. It is found as follows:

$$(\sigma_x)_D = \frac{F_x}{A} + \frac{M_y z_D}{I_y} + \frac{M_z y_D}{I_z} = -\frac{P}{bh} + \frac{(Pz)(b/2)}{(1/12)hb^3} - \frac{(Py)(h/2)}{(1/12)bh^3} \qquad \textbf{ANS.}$$

POINT E

The stress element at point E is shown in figure (e) above. As in the case of point D, the normal stress is shown as tensile. However, its character depends upon the relative magnitudes of P, M_y, and M_z. This stress is found as follows:

$$(\sigma_x)_E = \frac{F_x}{A} + \frac{M_y z_E}{I_y} + \frac{M_z y_E}{I_z} = -\frac{P}{bh} + \frac{(Pz)(b/2)}{(1/12)hb^3} + \frac{(Py)(h/2)}{(1/12)bh^3} \qquad \textbf{ANS.}$$

In summary, *on beginning the analysis of members subjected to combined loads, the above three examples illustrate the need for a good free-body diagram that includes the position of the point at which the stress element is to be constructed. Each of the reactive components is then examined to determine its effect on the stress element and the effects of the several reactive components are then superposed (added algebraically).* Once the stress element at the specific point is drawn, the normal and shearing stresses may be determined on any inclined plane by constructing an appropriate free-body diagram and applying the equations of equilibrium. While the determination of stresses on inclined planes was not performed in the above three examples, the reader is referred to Example 2.12 where this type of analysis was explained thoroughly. A more exhaustive analysis of stress is found in Chapter 7.

PROBLEMS

3.113 The cantilever beam has a rectangular cross section and carries the load P as shown. Let $P = 20$ kips, $\theta = 30°$, $a = 10$ ft, $b = 3$ in., and $h = 5$ in. and construct stress elements at points A and B with two planes perpendicular to the beam axis and two parallel to it. Point A is at the top of the beam and point B 11/2 in. below the top.

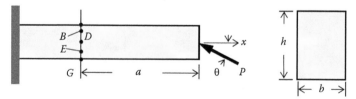

(Problems 3.113, 3.114, and 3.115)

3.114 In reference to the figure shown in Problem 3.113, the cantilever beam has a rectangular cross section and carries the load P as shown. Let $P = 50$ kN, $\theta = 20°$, $a = 3$ m, $b = 60$ mm, and $h = 120$ mm and construct stress elements at points E and G with two planes perpendicular to the beam axis and two parallel to it. Point E is 30 mm above the bottom of the beam and point G is at the bottom.

3.115 In reference to the figure shown in Problem 3.113, the cantilever beam has a rectangular cross section and carries the load P as shown. Let $P = 40$ kips, $\theta = 45°$, $a = 7$ ft, $b = 2$ in., and $h = 4$ in. (a) Construct the stress element at point D, which is at the neutral surface, with two planes perpendicular to the beam axis and two parallel to it. (b) Find the normal and shearing stresses at this point on a plane inclined to the cross section of the beam at 30° cw.

3.116 The frame of a machine press is shown in figure (a) below, and its cross section is illustrated by section B–B in figure (b) below. Construct the stress elements at points D and E with two planes parallel to the vertical axis of the press (i.e., the x axis) and two planes perpendicular to it. Let $P = 200$ kN, $a = 60$ mm, and $b = 750$ mm.

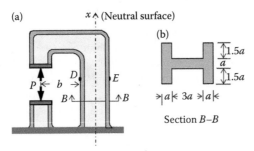

(Problems 3.116, 3.117, and 3.118)

3.117 Refer to Problem 3.116 and determine the normal and shearing stresses at point D on a plane inclined to the vertical axis of the press (the x axis) at 45° ccw.

3.118 In reference to the figure shown in Problem 3.116, the frame of a machine press is shown in figure (a) above, and its cross section is illustrated by section B–B in figure (b) above. If the allowable normal stress (tension or compression) at any section such as B–B is 20 ksi, determine the maximum load P that the press may carry. Let $a = 2$ in. and $b = 2.5$ ft.

3.119 A machine component is shown in figure (a) below and its cross section is illustrated by section B–B in figure (b) below. Construct the stress elements at points D and E with two planes parallel to the vertical axis of the component (i.e., the x axis) and two planes perpendicular to it. Let $P = 100$ kN, $b = 150$ mm, and $a = 30$ mm.

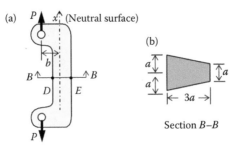

(Problems 3.119, 3.120, and 3.121)

3.120 Refer to Problem 3.119 and determine the normal and shearing stresses at point D on a plane inclined to the vertical axis (i.e., the x axis) of the machine component at 25° cw.

3.121 In reference to the figure shown in Problem 3.119, a machine component is shown in figure (a) above and its cross section is illustrated by section B–B in figure (b) above. If the allowable normal stress (tension or compression) at any section such as B–B is 25 ksi, determine the maximum force P that the component may carry. Let $a = 1$ in. and $b = 6$ in.

3.122 A machine component may be idealized as a 75-mm-diameter circular shaft fixed at one end and attached to a rigid arm at the other. Let $P = 80$ kN, $a = 200$ mm, $b = 150$ mm, and $d = 250$ mm and construct stress elements at points D and E on the outside surface of the shaft with two planes parallel and two planes perpendicular to its axis. Note that point D is at the front end of a horizontal diameter and point *E* at the bottom of a vertical diameter.

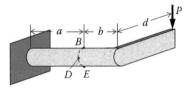

(Problems 3.122 and 3.123)

3.123 Refer to Problem 3.122 and determine the normal and shearing stresses at point *B* on the outside surface of the shaft, on a plane inclined to its axis at 40° cw. Note that point *B* is at the top of a vertical diameter.

3.124 A 1½-in. standard threaded pipe (see Appendix E) is being unscrewed by means of a pipe wrench as shown. Construct stress elements at points *B* and *D* on the outside surface of the pipe with two planes parallel and two planes perpendicular to its axis. Note that point *B* is at the top of a vertical diameter and point *E* at the front end of a horizontal diameter. Let *P* = 100 lb, *a* = 14 in., and *b* = 18 in. and ignore the weight of the pipe and wrench.

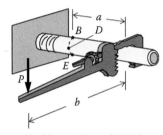

(Problems 3.124 and 3.125)

3.125 Refer to Problem 3.124 and determine the normal and shearing stresses at point *E* on the outside surface of the pipe on a plane inclined to its axis at 25° ccw.

3.126 A 3.0 × 1.5 m sign is supported by a column with a hollow circular cross section whose outside diameter is 100 mm and inside diameter 75 mm. The wind pressure *p* = 0.75 kN/m² on the sign may be assumed evenly distributed over its entire area. Neglect the weight of the sign and that of the column and construct stress elements at points *B* and *D* on the outside surface of the column with two planes parallel and two planes perpendicular to its axis.

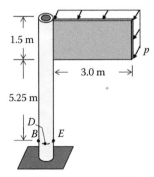

(Problems 3.126 and 3.127)

3.127 Refer to Problem 3.126 and determine the normal and shearing stresses at point *E* on the outside surface of the column, on a plane inclined to its axis at 30° cw.

3.128 A block of rectangular cross section is subjected to a compressive force $P = 200$ kN as shown. Let $b = 80$ mm, $h = 140$ mm, $d = 200$ mm, and $G(y, z) = G(20, 40)$ mm and construct a stress elements at points B and D with two planes parallel to the vertical axis of the block (i.e., the x axis) and two planes perpendicular to it.

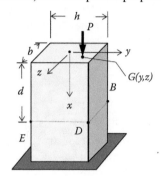

(Problems 3.128, 3.129, and 3.130)

3.129 Refer to Problem 3.128 and determine the normal and shearing stresses at point E on a plane inclined to the vertical axis (i.e., the x axis) of the block at 30° cw.

3.130 In reference to the figure shown in Problem 3.128, a block of rectangular cross section is subjected to a compressive force P at point $G(-0.5, 2.0)$ in. as shown. Determine the magnitude of P if the allowable stress (tension or compression) at point B is 20 ksi. Let $b = 2$ in., $h = 5$ in., and $d = 10$ in.

3.7 ALLOWABLE-STRESS DESIGN

The treatment given in this section for the design of beams is intended to serve only as an introduction to a complex process that requires the application of a number of steps as follows:

1. The loads that act on the beam are computed or estimated and the shear and moment diagrams are constructed to determine the maximum magnitudes of the shear force, $|V_{MAX}|_{MAX}$, and that of the bending moment $|M_{MAX}|$.
2. The allowable stress, σ_{ALL}, for a given application, is obtained from codes that provide design specifications. However, if the codes do not provide values for the allowable stress, the designer consults a table of material properties similar to that provided in Appendix D and divides the appropriate property by a factor of safety, which is usually decided upon on the basis of sound engineering judgment and experience. In this introductory treatment, we will make the assumption that the allowable stress is the same in tension and compression.
3. Assuming that the maximum bending stress in the beam is the controlling design factor, Equation 3.25, $\sigma_{ALL} = |\sigma_{MAX}| = |M_{MAX}|/S$, is used to determine the minimum value of the section modulus, S, required at the position of maximum moment. It should be pointed out here that in certain cases where shear force in the beam is high, the maximum normal stress may not be the largest bending stress in the beam at the farthest point from the neutral axis, but at some other point where the combination of the normal and shearing stresses (see Chapter 7) may lead to a normal stress larger than the maximum bending stress.
4. Reference is now made to tables of design properties for the chosen type of beam, similar to those shown in Appendix E, and only those beams with section moduli larger than the computed minimum section modulus are considered. Since cost is an important design consideration and, the heavier the beam, the more expensive it is, we select the one section with the least weight per unit length. Also, if the assumption of the allowable stress in tension and compression being equal is not acceptable, then both allowable stresses are used in Equation 3.25, and the larger of the two selected section moduli is chosen. *The notation used to*

designate a beam as shown in Appendix E consists of a letter followed by two numbers, as, for example, W10 × 68. The letter in this example indicates that it is a wide flange beam. The first number gives the nominal height, which, in this case, is 10 in. and the second number indicates the weight per unit length, which, in this case, is 68 lb/ft. In the SI system, instead of weight per unit length, mass per unit length is given by the second number.

5. *Under actual design conditions,* the chosen section is checked to ensure that the allowable shearing stress at the position of maximum shear force is not exceeded. Also, if the deflection of the beam is a significant design consideration, it is checked using the methods developed in Chapters 5 and 6.

EXAMPLE 3.16

Select the lightest wide flange section from Appendix E that meets the requirements indicated in figure (a) below if the allowable bending stress is 24 ksi. Assume that the bending stress is the only significant design factor.

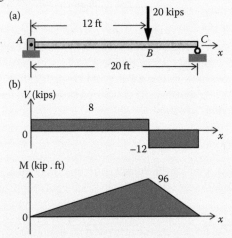

SOLUTION

The shear and moment diagrams are constructed as shown in figure (b) above. The maximum moment occurs at the concentrated load and is found to be 96 kip·ft. Using Equation 3.25, we find the needed section modulus S. Thus,

$$S = \frac{(96)12}{24} = 48 \text{ in.}^3$$

The information given in Appendix E indicates a large number of wide flange sections that have a section modulus larger than the required value of 48 in.^{3}. However, the one section with the smallest weight per foot of length is W14 × 34 ($S = 48.6$ in.^{3}) weighing 34 lb/ft. Unless other design considerations need to be accounted for, the W14 × 34 section is the one selected.   **ANS.**

EXAMPLE 3.17

Select the lightest S shape from Appendix E for the beam shown in figure (a) below. The allowable bending stress in tension or compression is 160 MPa. Assume that the bending stress is the only significant design factor.

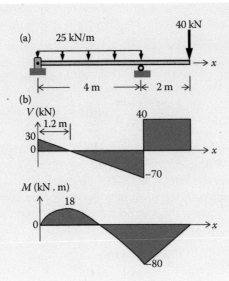

SOLUTION

The shear and moment diagrams are constructed as shown in figure (b) above. The maximum magnitude of the bending moment in the beam is 80 kN·m at the right support as shown. Using Equation 3.25, we find the needed section modulus S. Thus,

$$S = \frac{80 \times 10^3}{160 \times 10^6} = 50 \times 10^{-5} \text{ m}^3$$

In Appendix E, we find many S shapes possessing section moduli larger than the needed value of 50×10^{-5} m³. However, the lightest S shape with a slightly larger section modulus is the S305 × 47.3 ($S = 59.6 \times 10^{-5}$ m³), which has a mass of 47.3 kg/m.   **ANS.**

EXAMPLE 3.18

A structural wood beam is shown in figure (a) below. If the allowable bending stress and the allowable shearing stress for the chosen timber are 1.8 ksi and 0.12 ksi, respectively, select the lightest wood section from Appendix E.

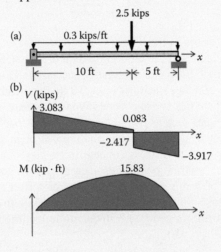

SOLUTION

The shear and moment diagrams are constructed as shown in figure (b) above. The maximum magnitudes of the shear force and bending moment in the beam are, respectively, 3.917 kips and 15.83 kip·ft. Using Equation 3.25, we find the needed section modulus. Thus,

$$S = \frac{(15.83)12}{1.8} = 105.533 \text{ in.}^3$$

Reference to Appendix E indicates that the lightest structural wood section with a section modulus slightly larger than the computed value is a 6×12 section with dressed dimensions of 5.5×11.5 in. and a section modulus of 121.2 in.^{3}. This section, however, needs to be checked to ensure that the maximum shearing stress does not exceed the given allowable value.

Reference is made to Example 3.8 where we discovered that the maximum shearing stress in a rectangular cross section was given by the equation $\tau_{MAX} = (3/2)V/A$. Thus,

$$\tau_{MAX} = \left(\frac{3}{2}\right)\frac{3.917}{(5.5 \times 11.5)} = 0.093 \text{ ksi}$$

This value of the maximum shearing stress is less than the given allowable value and, therefore, the 6×12 structural wood section selected on the basis of the allowable bending stress is acceptable.   **ANS.**

*3.8   STRESS CONCENTRATION

In Section 3.4, we derived Equation 3.24, $\sigma_{MAX} = |M|c/I$, which enables us to find the bending stress at any point in a beam subjected to bending action. This equation, however, does not provide the true value of the bending stress at locations in a beam where geometric discontinuities exist. At such discontinuities, the bending stresses are much larger than those given by Equation 3.25. To account for the effects of these geometric discontinuities in a beam, a factor k, known as the *stress-concentration factor*, is introduced in the equation, which then becomes

$$\sigma_{MAX} = k\left(\frac{|M|c}{I_{net}}\right) \tag{3.31}$$

where I_{net} is the moment of inertia of the reduced section of the member and k, as stated above, is the *stress-concentration factor*. Values of the stress-concentration factor k are given in Figure 3.8 for flat members with fillets of radius r and in Figure 3.9 for flat members with notches (grooves) of radius r. It should be noted here that stress-concentration factors are, for the most part, determined experimentally using techniques such as *photoelasticity* and *electrical analog*. However, in recent years, some work has been done using computers and finite element methods.

As was stated in earlier chapters, stress concentration plays a significant role in the design process, particularly under dynamic conditions such as impact and fatigue loading. However, even under static loading of brittle materials, such as cast iron and concrete, stress concentration is of great importance. In the case of ductile materials that are able to deform plastically, on the other hand, stress concentration under static loading is not as important a design consideration.

The use of information of the type provided in Figures 3.8 and 3.9 is illustrated in Example 3.19.

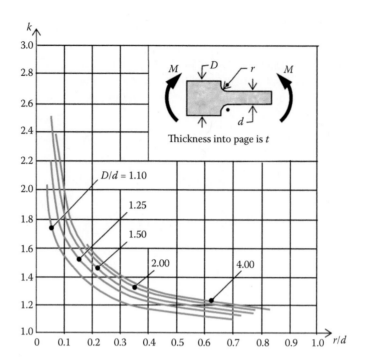

FIGURE 3.8 Stress-concentration factors for flat members with fillets under bending loads. (Adapted from the work of Frocht, M. M. Photoelastic studies in stress concentration, *Mechanical Engineering*, August, 1936, 485–489.)

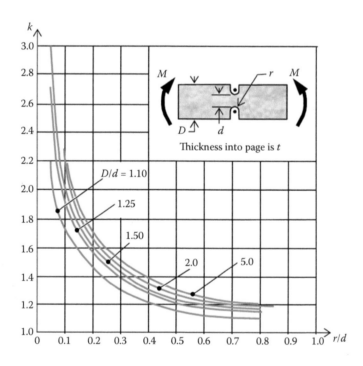

FIGURE 3.9 Stress-concentration factors for flat members with grooves under bending loads. (Adapted from the work of Frocht, M. M. Photoelastic studies in stress concentration, *Mechanical Engineering*, August, 1936, 485–489.)

EXAMPLE 3.19

A flat bending member, of thickness $t = 1.5$ in., is to have fillets of radius r as shown and subjected to a maximum moment $M = 3$ kip·ft. It is made of a material for which the allowable bending stress is 30.0 ksi. If $D = 6$ in. and $d = 3$ in., find the minimum radius r of the fillet consistent with the given information.

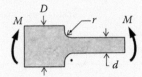

SOLUTION

The nominal stress at the root of the fillet is found by Equations 3.22 and 3.23. Thus,

$$\sigma = \frac{3(12)(1.5)}{(1/12)(1.5)(3^3)} = 16.0 \text{ ksi}$$

Also,

$$\frac{D}{d} = \frac{6}{3} = 2.0; \quad k = \frac{30.0}{16.0} = 1.875$$

Entering Figure 3.8 with these values of D/d and k, we estimate the ratio r/d to be about 0.1. Therefore, the minimum radius r is

$$r = 0.1(d) = 0.1(3) = 0.3 \text{ in.} \quad \textbf{ANS.}$$

PROBLEMS

3.131 Select the lightest wide flange section from among the 18-in sections in Appendix E for the beam shown if the allowable bending stress is 25 ksi. Let $P = 30$ kips and $L = 18$ ft. Assume the bending stress is the only significant design consideration.

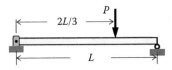

(Problems 3.131, 3.132, and 3.133)

3.132 In reference to the figure shown in Problem 3.131, select the lightest S section from Appendix E for the beam shown if the allowable bending stress is 150 MPa. Let $P = 150$ kN and $L = 9$ m. Assume the bending stress is the only significant design consideration.

3.133 In reference to the figure shown in Problem 3.131, a rectangular structural wood section is to be selected for the beam shown. If the allowable bending stress and that for the shearing stress for the chosen timber is 1.8 ksi and 0.12 ksi, respectively, select the lightest timber section from Appendix E. Let $P = 1.5$ kip and $L = 15$ ft.

3.134 Select the lightest wide flange section from among the 305 sections in Appendix E for the beam shown if the allowable bending stress is 175 MPa. Let $Q = 15$ kN·m, $p = 5$ kN/m, and $L = 6$ m. Assume the bending stress is the only significant design consideration.

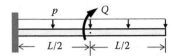

(Problems 3.134, 3.135, and 3.136)

3.135 In reference to the figure shown in Problem 3.134, a rectangular structural wood section is to be selected for the beam shown. If the allowable bending stress and that for the shearing stress for the chosen timber is 1.6 ksi and 0.10 ksi, respectively, select the lightest timber section from Appendix E. Let $Q = 0.75$ kip·ft, $p = 0.2$ kips/ft, and $L = 12$ ft.

3.136 In reference to the figure shown in Problem 3.134, select the lightest C section from Appendix E, two of which will be welded back-to-back, for the beam shown if the allowable bending stress is 250 MPa. Let $Q = 100$ kN·m, $p = 4$ kN/m, and $L = 10$ m. Assume the bending stress is the only significant design consideration.

3.137 Select the lightest wide flange section from Appendix E for the beam shown if the allowable bending stress is 20 ksi. Let $p = 6$ kips/ft and $L = 18$ ft. Assume the bending stress is the only significant design consideration.

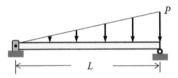

(Problems 3.137, 3.138, and 3.139)

3.138 In reference to the figure shown in Problem 3.137, select the lightest S section from among the 305 group in Appendix E for the beam shown if the allowable bending stress is 150 MPa. Let $p = 12$ kN/m and $L = 12$ m. Assume the bending stress is the only significant design consideration.

3.139 In reference to the figure shown in Problem 3.137, a rectangular structural wood section is to be selected for the beam shown. If the allowable bending stress and that for the shearing stress for the chosen timber is 2.0 ksi and 0.15 ksi, respectively, select the lightest timber section from Appendix E. Let $p = 0.2$ kips/ft and $L = 18$ ft.

3.140 Select the lightest equal-leg angle section from Appendix E, two of which will be welded back-to-back, for the beam shown if the allowable bending stress is 160 MPa. Let $p = 2$ kN/m, $P = 15$ kN, and $b = 2$ m. Assume the bending stress is the only significant design consideration.

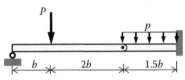

(Problems 3.140, 3.141, and 3.142)

3.141 In reference to the figure shown in Problem 3.140, a rectangular structural wood section is to be selected for the beam shown. If the allowable bending stress and that for the shearing stress for the chosen timber is 1.7 ksi and 0.12 ksi, respectively, select the lightest timber section from Appendix E. Let $P = 0.75$ kips, $p = 0.2$ kips/ft, and $b = 4$ ft.

3.142 In reference to the figure shown in Problem 3.140, select the lightest wide flange section from Appendix E for the beam shown if the allowable bending stress is 24 ksi. Let

$p = 2$ kips/ft, $P = 15$ kips, and $b = 8$ ft. Assume the bending stress is the only significant design consideration.

3.143 Select the lightest wide flange section from Appendix E for the beam shown if the allowable bending stress is 20 ksi. Let $P = 12$ kips and $L = 6$ ft. Assume the bending stress is the only significant design consideration.

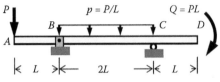

(Problems 3.143, 3.144, and 3.145)

3.144 In reference to the figure shown in Problem 3.143, select the lightest S section from Appendix E for the beam shown if the allowable bending stress is 150 MPa. Let $P = 18$ kN and $L = 2$ m. Assume the bending stress is the only significant design consideration.

3.145 In reference to the figure shown in Problem 3.143, select the lightest C section from Appendix E, two of which will be welded back-to-back, for the beam shown if the allowable bending stress is 20 ksi. Assume the bending stress is the only significant design consideration. Let $P = 8$ kips and $L = 4$ ft.

3.146 The flat bending member of thickness $t = 15$ mm, has two fillets symmetrically placed as shown. Let $M = 0.75$ kN·m, $d = 40$ mm, and $D = 60$ mm. Determine the maximum bending stress at the root of these fillets if (a) $r = 10$ mm and (b) $r = 5$ mm.

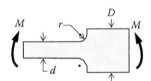

(Problems 3.146, 3.147, and 3.148)

3.147 In reference to the figure shown in Problem 3.146, the flat bending member of thickness $t = 0.75$ in. has two fillets symmetrically placed as shown. It is made of a material for which the allowable bending stress is 12 ksi. Let $D = 7$ in., $d = 5$ in., and $r = 0.3$ in., find the maximum permissible moment M that may be safely applied.

3.148 In reference to the figure shown in Problem 3.146, the flat bending member of thickness $t = 16$ mm has two fillets symmetrically placed as shown. It is made of a material for which the allowable bending stress is 160 MPa. Let $D = 120$ mm, $d = 80$ mm, and $M = 1.8$ kN·m, find the minimum permissible fillet radius r.

3.149 The flat bending member of thickness $t = 0.875$ in. has two grooves symmetrically placed as shown. Let $D = 6$ in., $d = 4$ in., and $M = 2$ kip·ft. Determine the maximum bending stress at the root of the grooves if (a) $r = 0.3$ in. and (b) $r = 0.5$ in.

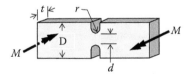

(Problems 3.149, 3.150, and 3.151)

3.150 In reference to the figure shown in Problem 3.149, the flat bending member of thickness $t = 20$ mm has two grooves symmetrically placed as shown. It is made of a material for which the allowable bending stress is 90 MPa. Let $D = 140$ mm, $d = 100$ mm, and $M = 2.0$ kN·m, find the minimum permissible fillet radius r.

3.151 In reference to the figure shown in Problem 3.149, the flat bending member of thickness $t = 0.50$ in. has two grooves symmetrically placed as shown. It is made of a material for which the allowable bending stress is 15 ksi. Let $D = 6$ in., $d = 4$ in., and $r = 0.2$ in., find the maximum permissible moment M that may be safely applied.

3.152 The flat bending member of thickness $t = 30$ mm has two fillets symmetrically placed as shown. It is made of a material for which the allowable bending stress is 120 MPa. Let $D = 120$ mm, $r = 15$ mm, and $M = 0.6$ kN·m, find the stress-concentration factor and the minimum permissible dimension d consistent with the given data. (*Hint:* A trial-and-error solution is needed here.)

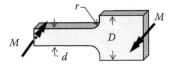

(Problems 3.152 and 3.153)

3.153 In reference to the figure shown in Problem 3.152, the flat bending member of thickness $t = 0.875$ in. has two fillets symmetrically placed as shown. Let $D = 8$ in., $d = 6$ in., and $M = 3$ kip·ft. Determine the maximum bending stress at the root of the grooves if (a) $r = 0.4$ in. and (b) $r = 0.6$ in.

3.154 The flat bending member of thickness $t = 14$ mm has two grooves symmetrically placed as shown. It is made of a material for which the allowable bending stress is 120 MPa. Let $D = 110$ mm, $d = 70$ mm, and $M = 0.8$ kN·m, find the minimum permissible radius r.

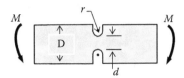

(Problems 3.154 and 3.155)

3.155 In reference to the figure shown in Problem 3.154, the flat bending member of thickness $t = 0.75$ in. has two grooves symmetrically placed as shown. It is made of a material for which the allowable bending stress is 15 ksi. Let $D = 8$ in., $d = 6$ in., find the maximum permissible moment M that may be safely applied if (a) $r = 0.3$ in and (b) $r = 0.6$ in.

REVIEW PROBLEMS

R3.1 Compound beam ABC is supported and loaded as shown. Let $p_1 = 1/2p_2 = 2$ kips/ft, $P = 20$ kips, and $a = 10$ ft, determine the shear and moment 10 ft to the left of the support at C.

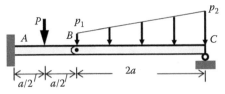

(Problems R3.1 and R3.2)

R3.2   In reference to the figure shown in Problem R3.1, compound beam ABC is supported and loaded as shown. Let $p_1 = 1/2p_2 = 5$ kN/m, $P = 50$ kN, and $a = 3$ m. Establish a coordinate system at B measuring x positive to the right and develop the shear and moment functions in segment BC.

R3.3   Establish an origin at A measuring x positive to the right and write the shear and moment functions for segments AB, BC, and CD. Using these functions, construct the shear and moment diagrams for the entire beam.

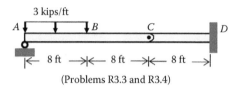

(Problems R3.3 and R3.4)

R3.4   In reference to the figure shown in Problem R3.3, use Equations 3.5 and 3.10 to construct the shear and moment diagrams for the beam shown.

R3.5   Use Equations 3.5 and 3.10 to construct the shear and moment diagrams for the beam shown.

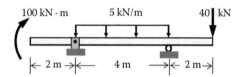

R3.6   The beam shown in figure (a) below has the cross section shown in figure (b) below. Let $a = 3$ in., $b = 10$ ft, and $p = 3$ kips/ft, determine the maximum tensile and maximum compressive stresses just to the left of the support at D.

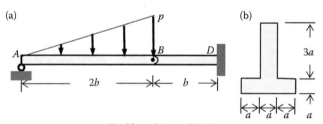

(Problems R3.6 and R3.7)

R3.7   In reference to the figure shown in Problem R3.6, the beam shown in figure (a) above has the cross section shown in figure (b) above. Let $a = 80$ mm, $b = 4$ m, and $p = 10$ kN/m, determine the maximum tensile and maximum compressive stresses in segment AB.

R3.8   The beam shown in figure (a) below has the cross section shown in figure (b) below. If the allowable bending stress (tension or compression) is 20 ksi and the allowable shearing stress is 10 ksi, determine the maximum permissible magnitude of the load P.

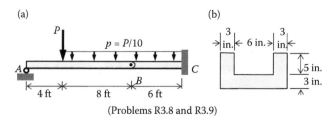

(Problems R3.8 and R3.9)

R3.9 In reference to the figure shown in Problem R3.8, the beam shown in figure (a) above has the cross section shown in figure (b) above. Let $P = 40$ kips and construct the shearing stress distribution at a section just to the left of the hinge at B.

R3.10 The beam shown in figure (a) below has the cross section shown in figure (b) below. This built-up section is fabricated by fastening two 300×30 mm steel plates using two $127 \times 127 \times 19.1$ steel angles (see Appendix E) as shown. The diameter of the vertical bolts is 15 mm, and the allowable shearing stress in these bolts is 120 MPa. Assume that the horizontal bolts are strong enough for the purpose and determine the required spacing for vertical, horizontal bolts. Neglect the bolt holes in the computations.

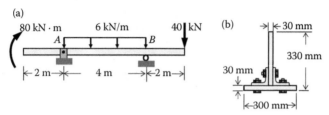

(Problems R3.10 and R3.11)

R3.11 In reference to the figure shown in Problem R3.10, the beam shown in figure (a) above has the built-up cross section given in figure (b) above. This built-up section is fabricated by fastening two 300×30 mm steel plates using two $127 \times 127 \times 19.1$ steel angles (see Appendix E) as shown. The diameter of the horizontal bolts is 12 mm, and the allowable shearing stress in these bolts is 100 MPa. Assume that the vertical bolts are strong enough for the purpose and determine the required spacing for the horizontal bolts. Neglect the bolt holes in the computations.

R3.12 Member $ABCD$, shown in figure (a) below, is supported and loaded as shown. Construct stress elements at points E (top) and H (2 in. below the top) with two planes perpendicular and two planes parallel to the axis of segment AB. The cross section of segment AB is the rectangle shown in figure (b) below. Let $L = 3$ ft, $a = 10$ in., $\theta = 45°$, $b = 4$ in., $h = 8$ in., and $P = 20$ kips. Assume segment BCD of the member to be rigid.

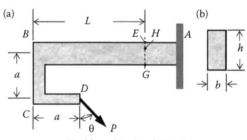

(Problems R3.12 and R3.13)

R3.13 In reference to the figure shown in Problem R3.12, member $ABCD$, shown in figure (a) above, is supported and loaded as indicated. Strain measurements at point G along the longitudinal axis of segment AB gave a reading of $-600\ \mu$. The cross section of segment AB is the rectangle shown in figure (b) above. Let $L = 1$ m, $a = 300$ mm, $\theta = 30°$, $b = 100$ mm, $h = 200$, and $E = 200$ GPa, determine the magnitude of the applied force P. Assume segment BCD of the member to be rigid.

R3.14 A block of rectangular cross section is subjected to a compressive force $P = 50$ kips as shown. Let $b = 3$ in., $h = 6$ in., $d = 12$ in., $y = 2$ in., and $z = 1$ in., determine the normal and shearing stresses at point D on a plane inclined to the vertical axis (i.e., the x axis) of the block at $40°$.

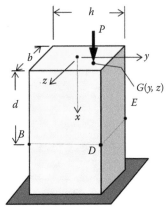

(Problems R3.14 and R3.15)

R3.15 In reference to the figure shown in Problem R3.14, a block of rectangular cross section is subjected to a compressive force P as shown. Let $b = 80$ mm, $h = 180$ mm, $d = 300$ mm, $y = 50$ mm, and $z = 20$ mm, determine the maximum permissible magnitude of P if the allowable stress in tension or compression at any point in section BDE is 100 MPa.

R3.16 Select the lightest wide flange section from Appendix E for the beam shown if the allowable stress is 30 ksi. Let $P = 30$ kips, $p = 2.5$ kips/ft, and $a = 10$ ft. Assume the bending stress is the only significant design consideration.

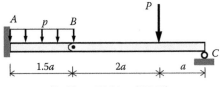

(Problems R3.16 and R3.17)

R3.17 In reference to the figure shown in Problem R3.16, select the lightest S section from Appendix E for the beam shown if the allowable stress is 250 MPa. Let $P = 210$ kN, $p = 20$ kN/m, and $a = 4$ m. Assume the bending stress is the only significant design consideration.

R3.18 A flat bending member, of thickness $t = 0.75$ in., has two grooves and two fillets, both symmetrically placed as shown. Let $r_1 = 1.5$ in., $r_2 = 0.5$ in., $D = 6.0$ in., and $M = 8$ kip·in., determine the magnitude and location of the maximum bending stress.

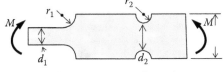

(Problems R3.18 and R3.19)

R3.19 In reference to the figure shown in Problem R3.18, a flat bending member, of thickness $t = 100$ mm, has two grooves and two fillets, both symmetrically placed as shown. Let $r_1 = 50$ mm, $r_2 = 20$ mm, and $D = 180$ mm, determine the magnitude of the maximum permissible moment M if the allowable stress is 200 MPa.

*4 Bending Loads
Additional Stress Topics

4.1 INTRODUCTION

The subject of stresses under bending loads is rather extensive. In addition to the topics already discussed in Chapter 3, there are many other subjects of interest that could be included in a textbook such as this. A few such subjects have been selected for inclusion in this chapter. Thus, in Section 4.2, we discuss the stresses in a member made of two or more materials when subjected to symmetric bending loads; in Section 4.3, we introduce the topic of stresses under *unsymmetric* loading, which results when the plane of the bending loads does not coincide with one of the two centriodal principal axes of inertia of the beam's cross section; in Section 4.4, we deal with the analysis of *thin-walled* open sections and the concept of the *shear center*, at which point the loads must be applied if the beam is to experience only bending with no twisting action; in Section 4.5, we discuss the behavior of curved beams; in Section 4.6, we deal with beams stressed beyond the elastic limit, and with the concepts of *elastoplastic* behavior and the *plastic hinge*; finally, in Section 4.7, we discuss the concept of *fatigue* failure due to repetitive loading.

4.2 BEAMS OF TWO OR THREE MATERIALS LOADED SYMMETRICALLY

4.2.1 GENERAL PRINCIPLES

Beams are often constructed of two or more materials in order to achieve certain desirable behavioral effects. Such is the case, for example, with reinforced concrete beams, which will be discussed in some detail in a later part of this section. In this part of the section, we will develop the general theory underlying the method of solution for beams of two or more materials. The theory is based upon the same fundamental assumptions made in deriving Equation 3.23 for the bending stress in a beam of a homogeneous material. In analyzing a beam of two or more materials, however, the problem is complicated by the fact that we are dealing with two or more different moduli of elasticity. Also, in analyzing the case of a multimaterial beam, the assumption is made that these materials are securely fastened together so that there is no relative displacement among them. Thus, the multimaterial beam can be assumed to act as a single unit. It is assumed further that the bending loads are so placed to produce no twisting action.

We will begin our analysis by considering a beam of two materials as shown in Figure 4.1. Figure 4.1a illustrates a segment of this beam of length Δx subjected to negative bending moments. Figure 4.1b shows the cross section of the beam indicating that it is composed of two materials: material B is used in two parts, one on the left side and the other on the right side to symmetrically sandwich material C between them. These two materials are fastened together such that the composite beam acts as one unit. One basic assumption of beam theory is that plane sections before bending remain plane after bending. According to this assumption, a plane section represented by line a before bending remains plane but rotates into position a' as shown in Figure 4.1a. Thus, at any distance y above the neutral axis (the z axis), the deformations, δ, and hence the strains, ε, of both materials are identical. Therefore,

$$\varepsilon_B = \varepsilon_C \qquad (4.1)$$

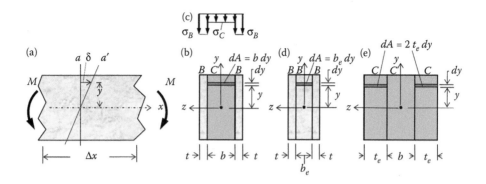

FIGURE 4.1 (a) A segment of beam of two materials of length Δx subjected to bending moment M. (b) The cross section of the beam with material C symmetrically sandwiched between material B. (c) The stress distribution across the section. (d) The equivalent cross section when material C is transformed into material B. (e) The equivalent cross section when material B is transformed into material C.

Assuming that both materials obey Hooke's law, Equation 4.1 may be expressed as $\sigma_B/E_B = \sigma_C/E_C$, where σ_B, E_B and σ_C, E_C represent the stresses and moduli of elasticity for materials B and C, respectively. Therefore,

$$\sigma_C = \left(\frac{E_C}{E_B}\right)\sigma_B = n\sigma_B \qquad (4.2)$$

The ratio $n = (E_C/E_B)$ in Equation 4.2 is a *shrinking* or *magnifying* factor depending on whether $E_C < E_B$ or $E_C > E_B$. In either case, there is an abrupt change in the magnitude of the bending stress at the junction between material B and material C. For purposes of discussion, we will assume that $E_C < E_B$ so that n is less than one and by Equation 4.2, σ_C is less than σ_B for any position defined by the coordinate y as shown by the schematic stress distribution of Figure 4.1c.

In analyzing beams of two materials, it is very convenient to deal with an *equivalent cross section* instead of dealing with the actual cross section of the beam. This equivalent cross section is obtained by transforming either of the two materials into an equivalent amount of the second material. To accomplish this transformation, let us consider the differential force dF_C acting on the differential area $dA = b\,dy$ in material C of the actual cross section shown in Figure 4.1b. This differential force is the product of the stress σ_C and the differential area dA. Thus,

$$dF_C = \sigma_C b\,dy \qquad (4.3)$$

Using Equation 4.2, we replace σ_C in Equation 4.3 by $n\sigma_B$ to obtain

$$dF_C = \sigma_B(nb)dy = \sigma_B b_e\,dy \qquad (4.4)$$

Equation 4.4 shows that that part of the actual cross section made of material C may be transformed into an equivalent amount of material B by changing the actual width b of material C into an equivalent width b_e of material B. It is evident from Equation 4.4 that the relation between the equivalent and the actual widths is

$$b_e = nb \qquad (4.5)$$

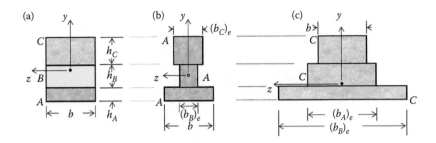

FIGURE 4.2 (a) The cross section of a beam made of three materials, A, B, and C. (b) The equivalent cross section when the three materials are transformed into material A. (c) The equivalent cross section when the three materials are transformed into material C.

Thus, using Equation 4.5, we can transform material C into material B and create the equivalent cross section shown in Figure 4.1d. This equivalent cross section is made entirely of material B and is capable of developing the same resistance to bending as the actual cross section shown in Figure 4.1b. Instead of transforming material C, we could use Equation 4.5 to transform material B into material C and create the equivalent cross section shown in Figure 4.1e, which is made entirely of material C.

Beams of more than two materials may be analyzed in basically the same fashion. Consider, for example, the beam cross section shown in Figure 4.2a, which is made up of the three materials A, B, and C, placed on top of each other. This section is subjected to loads contained in the longitudinal plane defined by the y axis. If it is assumed that $E_A > E_B > E_C$ and that we want to create an equivalent cross section made entirely of material A, we would compute $n_B = E_B/E_A$ and $n_C = E_C/E_A$, which are both less than one, and use Equation 4.5 to find $(b_B)_e$ and $(b_C)_e$ and then construct the equivalent cross section shown in Figure 4.2b, which now consists entirely of the same material A. Note that only the cross-sectional dimensions perpendicular to the loads are changed and that the heights of the individual components as well as the height of the entire cross section (all of which are parallel to the loads) remain unchanged. Two other equivalent cross sections are possible, one consisting entirely of material B and the second entirely of material C. The construction of the equivalent section made up entirely of C is discussed here and that made of material B is left as an exercise for the student.

Assuming as before that $E_A > E_B > E_C$, we compute $n_A = E_A/E_C$ and $n_B = E_B/E_C$, which are both larger than one, and use Equation 4.5 to find $(b_A)_e$ and $(b_B)_e$ and then construct the equivalent cross section shown in Figure 4.2c, which now consists entirely of material C. Again note that only the cross-sectional dimensions perpendicular to the loads are changed and that the heights of the individual components as well as the height of the entire cross section (all of which are parallel to the loads) remain unchanged.

Once an equivalent cross section of a homogeneous material is obtained, we locate its centroid, and hence its neutral axis, and proceed to compute its centroidal moment of inertia and find the bending stresses using the methods developed in Chapter 3. Some of these ideas are illustrated in the solution of Examples 4.1 and 4.2.

EXAMPLE 4.1

Two aluminum (A) plates are used to sandwich an oak (O) block in fabricating the cross section shown in figure (a) below. This cross section is that for a cantilever beam of length $L = 10$ ft used to support a uniform load $p = 0.5$ kip/ft along its entire length acting in the vertical longitudinal plane represented by the y axis. Let $E_A = 10 \times 10^3$ ksi, $E_O = 2 \times 10^3$ ksi and determine the maximum bending stresses in the aluminum and in the oak.

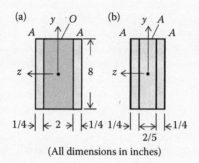

(All dimensions in inches)

SOLUTION

We determine the value of $n = E_O/E_A = 1/5$ in order to find the equivalent width of oak when replaced by aluminum. Thus, using Equation 4.5, we obtain

$$b_e = \left(\frac{1}{5}\right)(2) = \frac{2}{5}\text{ in.}$$

The equivalent cross section, which is made entirely of aluminum, is shown in figure (b) above. The maximum bending stresses occur where the bending moment is maximum at the fixed end of the cantilever beam. Thus, since $M_{MAX} = pL^2/2 = 25$ kip · ft, we use Equation 3.11 to obtain

$$(\sigma_{MAX})_A = \frac{(25)(12)(4)}{(1/12)(0.9)(8^3)} = 31.25 \approx 31.3 \text{ ksi} \qquad \textbf{ANS.}$$

Also, using Equation 4.2, we obtain the maximum bending stress in the oak. Thus,

$$(\sigma_{MAX})_O = \left(\frac{1}{5}\right)(31.25) = 6.25 \approx 6.3 \text{ ksi} \qquad \textbf{ANS.}$$

EXAMPLE 4.2

A segment of a beam of length Δx, subjected to negative bending moments, is shown in figure (a) below. The beam is constructed of two materials such that material B is placed in two parts, one on top and the other on the bottom, sandwiching material C between them as shown in figure (b) below. Transform material C into material B to create an equivalent cross section made entirely of material B and sketch the bending stress distribution showing how this stress varies from the top to the bottom of the cross section.

SOLUTION

Assuming that plane sections remain plane after bending, we conclude the deformations δ and hence the strains ε of the two materials are related by

$$\varepsilon_C = \left(\frac{y_C}{y_B}\right)\varepsilon_B$$

$$(4.2.1)$$

Using Hooke's law, $\varepsilon = \sigma/E$, we transform Equation 4.2.1 into a stress relationship. Thus,

$$\sigma_C = \left(\frac{y_C}{y_B}\right)\left(\frac{E_C}{E_B}\right)\sigma_B = \left(\frac{y_C}{y_B}\right)n\sigma_B \qquad (4.2.2)$$

where $n = E_C/E_B$ is once again a shrinking or magnifying factor, depending upon whether $E_C < E_B$ or $E_C > E_B$. In either case, there is an abrupt change in the magnitude of the bending stress at the junction between material B and material C at the point in the cross section, where $y_B = y_C$. For the purposes of this example, we will assume that $E_C < E_B$ so that n is less than one. It follows from Equation 4.2.2 that at the junction where $y_B = y_C$, σ_C is less than σ_B. A stress distribution (σ vs. y) is shown qualitatively in figure (d) below.

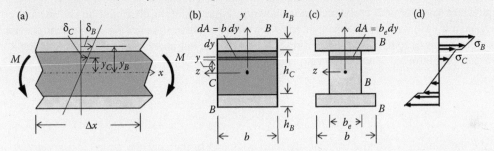

The development of the equivalent cross section made entirely of material B proceeds in essentially the same manner as was followed in the development of Equation 4.5. Since we are transforming material C into material B, we define a differential area $dA = b\,dy$ within material C as shown in figure (b) above. The differential force acting on this differential area becomes

$$dF_C = \sigma_C b\,dy \qquad (4.2.3)$$

Substituting for σ_C from Equation 4.2.2, we obtain

$$dF_C = \left(\frac{y_C}{y_B}\right)\sigma_B(nb)dy = \left(\frac{y_C}{y_B}\right)\sigma_B b_e\,dy \qquad (4.2.4)$$

The quantity $(y_C/y_B)\sigma_B$ in Equation 4.2.4 represents the stress at a distance y_C from the neutral axis in a beam of homogeneous material B. Therefore, Equation 4.2.4 shows that material C may be transformed into an equivalent amount of material B by changing the actual width b of this material into an equivalent width b_e of material B. Note that the equivalent width, by Equation 4.2.4, becomes $b_e = nb$, which is identical to Equation 4.5. The equivalent cross section is shown in figure (c) above.

Note that instead of transforming material C into material B, we could have transformed material B into material C. It is left as an exercise for the student to make this latter transformation and create a new equivalent cross section.

4.2.2   APPLICATION TO REINFORCED CONCRETE

The general principles developed earlier for the analysis of beams of two materials are used here, with some modification, to analyze the case of *reinforced concrete beams*.

The rectangular reinforced concrete beam cross section shown in Figure 4.3a is subjected to negative bending moments. In creating the equivalent or *transformed cross section*, the assumption is made that the portion of concrete subjected to tensile stresses does not exist. This assumption is justified on the basis that concrete, while strong in compression, is very weak in tension. The

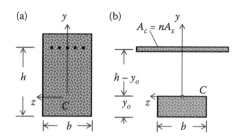

FIGURE 4.3   (a) The rectangular cross section of a reinforced concrete beam. (b) The equivalent cross section when the reinforcing steel is transformed into concrete.

equivalent (transformed) cross section is shown in Figure 4.3b. In obtaining this equivalent cross section, it is customary and convenient to deal with *areas* rather than *widths* as was done previously for other types of composite cross sections. We may begin our analysis by adapting Equation 4.2 to the case of reinforced concrete. Thus,

$$\sigma_S = \left(\frac{E_S}{E_C}\right)\sigma_C = n\sigma_C \tag{4.6}$$

where the subscripts S and C refer to the steel rods and to the concrete material, respectively, and $n = E_S/E_C$ is known as the *modular ratio*. Finding the differential force in a differential steel area we have

$$dF_S = \sigma_S b\, dy = \sigma_C n\, dA_S = \sigma_C\, dA_C \tag{4.7}$$

If we compare the last two terms in Equation 4.7, we conclude that $dA_C = n\, dA_S$. It follows, therefore, that

$$A_C = nA_S \tag{4.8}$$

Note that the equivalent area $A_C = nA_S$ is assumed to be a long, thin horizontal strip as shown in Figure 4.3b.

The location of the centroid C above the bottom of the equivalent cross section (i.e., the distance y_o) is obtained by insuring that the first moment of this cross section with respect to its neutral axis is zero. Thus,

$$\sum A\bar{y} = 0 \Rightarrow nA_s(h - y_o) - (by_o)\left(\frac{y_o}{2}\right) = 0$$

After simplification, this equation reduces to the following quadratic equation:

$$y_o^2 + \left(\frac{2nA_s}{b}\right)y_o - \left(\frac{2nA_s h}{b}\right) = 0 \tag{4.9}$$

The use of the above equations and concepts is illustrated in Example 4.3.

EXAMPLE 4.3

A reinforced concrete cross section is shown in figure (a) below. The total cross-sectional area of the reinforcing steel rods is 3.2 in.^{2} and the moduli of elasticity for the steel and concrete are $E_S = 30 \times 10^3$ ksi and $E_C = 3 \times 10^3$ ksi, respectively. If the applied bending moment is -100 kip · ft, determine the maximum bending stresses induced in the concrete and in the steel reinforcing rods as well as the total force in these rods.

SOLUTION

Since $n = E_s/E_c = 10$, using Equation 4.8, we find the equivalent concrete area to be $10(3.2) = 32$ in.² as shown in figure (b) below. Substituting the numerical values into Equation 4.9 and simplifying, we obtain the following quadratic equation:

$$y_o^2 + 4.571\, y_o - 109.714 = 0$$

Solving this quadratic equation, we obtain $y_o = 8.436$ in. and -13.007 in. The negative answer is discarded on physical grounds and the value of y_o locating the neutral axis from the bottom of the section is $y_o = 8.436$ in. The equivalent cross section made entirely of concrete is shown in figure (b) below.

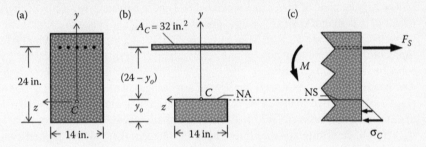

The maximum compressive bending stress in concrete, σ_C, is now computed using Equation 3.23, $\sigma = My/I$. The moment of inertia of the equivalent cross section about its neutral axis is given by

$$I = \left(\frac{1}{3}\right)(14)(8.436^3) + 32(24 - 8.436)^2 = 10{,}553.3 \text{ in}^4$$

$$\sigma_C = -\frac{My}{I} = -\frac{100(12)(8.436)}{10{,}553.3} = -0.959 \approx -1.0 \text{ ksi} \qquad \textbf{ANS.}$$

The tensile bending stress in the steel reinforcing rods is given by

$$\sigma_S = n\left(\frac{M(24 - y_o)}{I}\right) = 10\left(\frac{100(12)(24 - 8.436)}{10{,}553.3}\right) = 17.698 \approx 17.7 \text{ ksi} \qquad \textbf{ANS.}$$

Thus, the total force F_S developed in the steel rods becomes

$$F_S = \sigma_S A_S = 17.698(3.2) = 56.634 \approx 56.6 \text{ kips} \qquad \textbf{ANS.}$$

Figure (c) above shows a small length of the beam in equilibrium under the action of the applied bending moment M on the left side and the concrete stress distribution σ_C plus the total force F_S in the steel rods on the right side.

PROBLEMS

4.1 The beam cross section shown consists of a rectangle made of material 1 sandwiched rigidly between two plates of material 2. Material 1 is aluminum ($E_A = 10 \times 10^3$ ksi) and material 2 is brass ($E_B = 17 \times 10^3$ ksi). At some location in the beam, the bending moment is $M = 500$ kip · in. acting symmetrically about a horizontal axis. Let $h = 8$ in., $b = 4$ in., and $t = 1$ in., determine the maximum bending stresses developed in both materials.

(Problems 4.1 and 4.2)

4.2 In reference to the figure shown in Problem 4.1, the beam cross section shown consists of a rectangle made of material 1 sandwiched rigidly between two plates of material 2. Material 1 is magnesium ($E_M = 45$ GPa) and material 2 is brass ($E_B = 95$ GPa). If the allowable stresses are $\sigma_M = 35$ MPa and $\sigma_B = 70$ MPa, determine the maximum bending moment that may be applied symmetrically about a horizontal axis. Let $h = 250$ mm, $b = 120$ mm, and $t = 20$ mm.

4.3 The beam cross section shown consists of a rectangle made of material 1 sandwiched rigidly between two plates of material 2. Material 1 is aluminum ($E_A = 10 \times 10^3$ ksi) and material 2 is brass ($E_B = 17 \times 10^3$ ksi). If the allowable stresses are $\sigma_A = 30$ ksi and $\sigma_B = 50$ ksi, determine the maximum bending moment that may be applied symmetrically about a horizontal axis. Let $h = 10$ in., $b = 6$ in., and $t = 1$ in.

(Problems 4.3 and 4.4)

4.4 In reference to the figure shown in Problem 4.3, the beam cross section shown consists of a rectangle made of material 1 sandwiched rigidly between two plates of material 2. Material 1 is magnesium ($E_M = 45$ GPa) and material 2 is brass ($E_B = 95$ GPa). At some location in the beam, the bending moment is $M = 50$ kN · m acting symmetrically about a horizontal axis. Let $h = 240$ mm, $b = 160$ mm, and $t = 30$ mm, determine the maximum bending stresses developed in both materials.

4.5 A simply supported beam 15 ft long carries a uniform load of intensity $p = 2$ kips/ft acting downward in the vertical plane of symmetry. Its cross section consists of a rectangle of material 1 sandwiched rigidly between two plates of material 2. Material 1 is wood ($E_W = 2 \times 10^3$ ksi) and material 2 is steel ($E_S = 30 \times 10^3$ ksi). Let $b = 6$ in., $h = 10$ in., and $t = 1/2$ in., determine the maximum bending stresses developed in both of these materials.

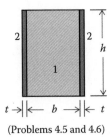

(Problems 4.5 and 4.6)

4.6 In reference to the figure shown in Problem 4.5, a cantilever beam 4 m long carries a concentrated load P at its free end acting downward in the vertical plane of symmetry. Its cross section consists of a rectangle of material 1 sandwiched rigidly between two plates of material 2.

   Material 1 is wood ($E_W = 15$ GPa) and material 2 is aluminum ($E_A = 75$ GPa). Let $b = 200$ mm, $h = 300$ mm, and $t = 20$ mm, determine the largest permissible value of P if the allowable stresses are $\sigma_W = 35$ MPa and $\sigma_A = 150$ MPa.

4.7 A cantilever beam 18 ft long carries a linearly distributed load that varies from p at the fixed end to zero at the free end acting downward in the vertical plane of symmetry. Its cross section consists of a rectangle of material 1 sandwiched rigidly between two plates of material 2. Material 1 is wood ($E_W = 1.5 \times 10^3$ ksi) and material 2 is steel ($E_S = 29 \times 10^3$ ksi). Let $b = 6$ in., h = 8 in., and $t = 3/8$ in., determine the largest permissible value of p if the allowable stresses are $\sigma_S = 40$ ksi and $\sigma_W = 2$ ksi.

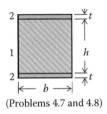

(Problems 4.7 and 4.8)

4.8 In reference to the figure shown in Problem 4.7, a simply supported beam 6 m long carries a concentrated load $P = 60$ kN at midspan acting downward in the vertical plane of symmetry. Its cross section consists of a rectangle of material 1 sandwiched rigidly between two plates of material 2. Material 1 is wood ($E_W = 12$ GPa) and material 2 is steel ($E_S = 200$ GPa). Let $b = 180$ mm, $h = 260$ mm, and $t = 15$ mm, determine the maximum bending stresses in both materials.

4.9 The beam of figure (a) below has the cross section of figure (b) below, which consists of two plates of material 1 and two rectangular pieces of material 2. Material 1 is steel ($E_S = 30 \times 10^3$ ksi) and material 2 is wood ($E_W = 2 \times 10^3$ ksi). Let $P = 15$ kips, $L = 5$ ft, $a = 4$ in., $b = 6$ in., $t = 3/4$ in., and $h = 12$ in., determine the maximum bending stresses in both materials.

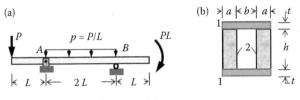

(Problems 4.9 and 4.10)

4.10 In reference to the figure shown in Problem 4.9, the beam of figure (a) above has the cross section of figure (b) above, which consists of two plates of material 1 and two rectangular pieces of material 2. Material 1 is brass ($E_B = 120$ GPa) and material 2 is aluminum ($E_A = 70$ GPa). Let $L = 2$ m, $a = 100$ mm, $b = 160$ mm, $t = 25$ mm, and $h = 300$ mm, determine the maximum permissible value of P if the allowable stresses are $\sigma_B = 150$ MPa and $\sigma_A = 90$ MPa.

4.11 The beam of figure (a) below has the cross section of figure (b) below, which consists of two plates of material 1 and two rectangular pieces of material 2. Material 1 is steel ($E_S = 30 \times 10^3$ ksi) and material 2 is wood ($E_W = 2 \times 10^3$ ksi) Let $L = 6$ ft, $a = 4$ in., $b = 10$ in., $t = 1/2$ in., and $h = 6$ in., determine the maximum permissible value of p if the allowable stresses are $\sigma_S = 12$ ksi and $\sigma_W = 1$ ksi.

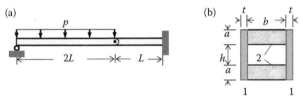

(Problems 4.11 and 4.12)

4.12 In reference to the figure shown in Problem 4.11, the beam of figure (a) above has the cross section of figure (b) above, which consists of two plates of material 1 and two rectangular pieces of material 2. Material 1 is brass ($E_B = 120$ GPa) and material 2 is aluminum ($E_A = 70$ GPa). Let $p = 15$ kN/m, $L = 3$ m, $a = 140$ mm, $b = 240$ mm, $t = 30$ mm, and $h = 200$ mm, determine the maximum stresses induced in both materials.

4.13 The cross section for a beam consists of three rectangular components made of materials 1, 2, and 3 as shown. Material 1 is aluminum ($E_A = 10 \times 10^3$ ksi), material 2 is brass ($E_B = 15 \times 10^3$ ksi), and material 3 is steel ($E_S = 30 \times 10^3$ ksi). If the beam carries a negative bending moment of 500 kip · in. acting symmetrically about a horizontal axis, determine the maximum bending stresses in the three materials. Let $b = 6$ in., $h_1 = 4$ in., $h_2 = 3$ in., and $h_3 = 1$ in.

(Problems 4.13 and 4.14)

4.14 In reference to the figure shown in Problem 4.13, the cross section for a beam consists of three rectangular components made of materials 1, 2, and 3 as shown. Material 1 is wood ($E_W = 12$ GPa), material 2 is aluminum ($E_A = 70$ GPa), and material 3 is steel ($E_S = 200$ GPa). If the beam carries a positive bending moment of 5 kN · m acting symmetrically about a horizontal axis, determine the maximum bending stresses in the three materials. Let $b = 50$ mm, $h_1 = 60$ mm, $h_2 = 40$ mm, and $h_3 = 20$ mm.

4.15 The cross section for a beam consists of three rectangular components made of materials 1, 2, and 3 as shown. Material 1 is aluminum ($E_A = 10 \times 10^3$ ksi), material 2 is wood ($E_W = 2 \times 10^3$ ksi), and material 3 is steel ($E_S = 30 \times 10^3$ ksi). If the allowable stresses in the three materials are $\sigma_A = 15$ ksi, $\sigma_W = 2$ ksi, and $\sigma_S = 24$ ksi, find the largest permissible bending moment that may be applied symmetrically about a horizontal axis. Let $b = 6$ in., $h_1 = 1$ in., $h_2 = 8$ in., and $h_3 = 1/2$ in.

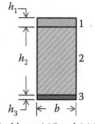

(Problems 4.15 and 4.16)

4.16 In reference to the figure shown in Problem 4.15, the cross section for a beam consists of three rectangular components made of materials 1, 2, and 3 as shown. Material 1 is aluminum (E_A = 70 GPa), material 2 is wood (E_W = 15 GPa), and material 3 is steel (E_S = 205 GPa). If the allowable stresses in the three materials are σ_A = 100 MPa, σ_W = 30 MPa, and σ_S = 165 MPa, find the largest permissible bending moment that may be applied symmetrically about a horizontal axis. Let b = 160 mm, h_1 = 30 mm, h_2 = 220 mm, and h_3 = 20 mm.

4.17 The reinforced concrete cross section shown is that for a cantilever beam carrying a uniform load of intensity p = 1.5 kips/ft over the entire length of 12 ft acting downward in the vertical plane of symmetry. Each of the five reinforcing steel rods has a diameter of 5/8 in. The moduli of elasticity are E_S = 30 × 10³ ksi and E_C = 3 × 10³ ksi. Let b = 12 in. and h = 18 in., find the maximum bending stresses developed in the concrete and in the steel.

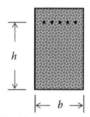

(Problems 4.17 and 4.18)

4.18 In reference to the figure shown in Problem 4.17, the reinforced concrete cross section shown is that for a cantilever beam of length L = 6 m carrying a concentrated load P at its free end acting downward in the vertical plane of symmetry. Each of the five reinforcing steel rods has a diameter of 20 mm. Let E_S = 200 GPa, E_C = 20 GPa and the allowable stresses σ_S = 140 MPa and σ_C = 12 MPa. If b = 180 mm and h = 360 mm, determine the maximum permissible load P.

4.19 The reinforced concrete cross section shown is that for a simply supported beam with a span of 8 m carrying a downward uniform load of intensity p = 6 kN/m over the entire length acting downward in the vertical plane of symmetry. The total cross-sectional area of the steel reinforcing rods is A_S = 14 × 10⁻⁴ m². The moduli of elasticity are E_S = 200 GPa and E_C = 20 GPa. Let b = 160 mm and h = 380 mm, determine the maximum bending stresses developed in the concrete and in the steel.

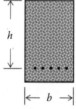

(Problems 4.19 and 4.20)

4.20 In reference to the figure shown in Problem 4.19, the reinforced concrete cross section shown is that for a simply supported beam with a span of 18 ft carrying a concentrated load P at a distance of 6 ft from the left support acting downward in the vertical plane of symmetry. The total area of the steel reinforcing rods is 1.0 in.^{2}. The moduli of elasticity are $E_S = 30 \times 10^3$ ksi and $E_C = 3 \times 10^3$ ksi and the allowable stresses are $\sigma_S = 30$ ksi and $\sigma_C = 2$ ksi. Let $b = 8$ in. and $h = 15$ in., find the largest permissible load P.

4.21 Compute the maximum permissible moment that may be applied symmetrically about a horizontal axis to the reinforced concrete section shown if the allowable bending stresses are $\sigma_C = 1.6$ ksi and $\sigma_S = 20$ ksi and the total cross-sectional area of the reinforcing rods is 5.0 in.^{2}. Let $b = 6$ in., $h_1 = 6$ in., $h_2 = 8$ in., $h_3 = 16$ in., $h_4 = 2.5$ in., and $n = 9$. (*Hint:* Assume that the neutral axis intersects the cutout.)

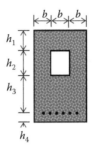

(Problems 4.21 and 4.22)

4.22 In reference to the figure shown in Problem 4.21, the reinforced concrete section shown is subjected to a bending moment $M = 250$ kN · m acting symmetrically about a horizontal axis. Let $b = 150$ mm, $h_1 = 150$ mm, $h_2 = 400$ mm, $h_3 = 400$ mm, $h_4 = 70$ mm, $n = 9$, and the total cross-sectional area of the reinforcing rods is 3.0×10^{-3} m^2. Find the maximum bending stresses developed in both materials. (*Hint:* Assume that the neutral axis intersects the cutout.)

4.23 A reinforced concrete section is said to be under *balanced conditions* if the maximum bending stresses in both materials reach their respective allowable values simultaneously. If the section shown is under balanced conditions and the allowable stresses are $\sigma_S = 20$ ksi and $\sigma_C = 1.6$ ksi, determine (a) the distance y_o locating the neutral axis from the top of the section when it is subjected to a bending moment about a horizontal axis and (b) the maximum bending moment that can be applied. Let $b = 8$ in., $h = 20$ in., $n = 10$, and the total cross-sectional area of the reinforcing rods is 0.6 in.^{2}. Let $E_S = 30 \times 10^3$ ksi and $E_C = 3 \times 10^3$ ksi.

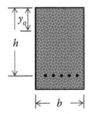

(Problems 4.23 and 4.24)

4.24 In reference to the figure shown in Problem 4.23, assume the reinforced concrete section shown is under balanced conditions (see Problem 4.23). If the allowable stresses are σ_S and σ_C, the ratio $E_S/E_C = n$ and the section is subjected to a bending moment about a horizontal axis, find the distance y_o locating the neutral axis from the top of the section. Express your answer in terms of the two allowable stresses, the dimension h and the ratio n. What is y_o if $n = 9$, $h = 720$ mm, $\sigma_S = 160$ MPa, and $\sigma_C = 12$ MPa?

4.3 BENDING STRESSES UNDER UNSYMMETRIC LOADING

As stated in Chapter 3 and discussed in detail in Appendix C.2, every shape of cross-sectional area possesses two in-plane centroidal principal axes of inertia, which are perpendicular to each other. In Chapter 3, we discussed the topic of symmetric bending of beams, which deals with bending loads that act in one of the two centroidal longitudinal principal planes. If, however, the bending loads *do not* act in one of the two centroidal longitudinal principal planes, the resulting action is known as *unsymmetric* bending.

4.3.1 ARBITRARY CENTROIDAL AXES

Consider the arbitrary cross-sectional area shown in Figure 4.4, for which the arbitrary y and z centroidal axes have been established. This cross-sectional area is that of a beam for which the plane of the loads, L, makes the counterclockwise angle ϕ with the longitudinal centroidal plane defined by the y axis. For any position along the beam, the bending moment M may be represented by the right-hand rule using a double-headed vector perpendicular to the load axis L as shown in Figure 4.4. The sense of the moment M would depend upon the way the beam is loaded and supported. Without loss of generality, we can assume a cantilever beam with the free end out of the page, loaded at the free end downward and to the right in the plane of the loads resulting in a moment M whose vector, by the right-hand rule, points upward and to the right. This vector may be decomposed into two perpendicular components as shown. These two components are

$$\left.\begin{array}{l} M_y = M \sin \phi \\ M_z = M \cos \phi \end{array}\right\} \tag{4.10}$$

The sign convention for these moments states that M_y *is positive if it produces tension where z is positive and M_z is positive if it produces tension where y is positive.* Thus, M_y and M_z in Figure 4.4 are both positive.

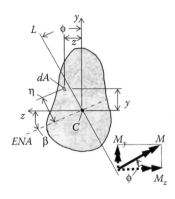

FIGURE 4.4 Arbitrarily chosen cross section showing the plane of the load L, the determined centroid C, and arbitrarily chosen y and z axes.

As in the case of symmetric bending, we assume that bending is produced by pure bending moments or by transverse loads so that no axially applied forces exist. Under these conditions (see the development of Equation 3.17 and subsequent results), the conclusion is reached that the elastic neutral axis (*ENA*) for bending must pass through the centroid of the section as shown in Figure 4.4. Let the neutral axis for bending make the counterclockwise angle β from the z axis as shown. We further make the same assumptions that were made earlier under symmetric bending (particularly that plane sections remain plane) and conclude, as was done earlier from Equation 3.14, that the bending stress is directly proportional to the distance from the neutral axis. Thus, for an element of area dA defined by the coordinates y and z, at a distance η from the neutral axis, the bending stress becomes

$$\sigma_\eta = K\eta \tag{4.11}$$

The quantity K in Equation 4.11 is a constant of proportionality. The distance η can be expressed in terms of the coordinates y and z. Thus, from the geometry in Figure 4.4, we conclude that

$$\eta = y\cos\beta + z\sin\beta \tag{4.12}$$

Substitution of Equation 4.12 into Equation 4.11 yields

$$\sigma_\eta = K(y\cos\beta + z\sin\beta) \tag{4.13}$$

Equilibrium of moments about the y axis leads to

$$M_y = \int \sigma_\eta z\,dA = K\int(y\cos\beta + z\sin\beta)z\,dA = K(\cos\beta\int yz\,dA + \sin\beta\int z^2 dA) \tag{4.14}$$

The first integral in the extreme right component of Equation 4.14 represents the product of inertia I_{yz} with respect to the centroidal y and z axes. The second integral of this component represents I_y, the moment of inertia of the cross-sectional area about the centroidal y axis. Thus,

$$M_y = K(I_{yz}\cos\beta + I_y\sin\beta) \tag{4.15}$$

Solving for K, we obtain

$$K = \frac{M_y}{(I_y\sin\beta + I_{yz}\cos\beta)} \tag{4.16}$$

Similarly, by considering equilibrium of moments about the z axis, we conclude that

$$M_z = K(I_z\cos\beta + I_{yz}\sin\beta) \tag{4.17}$$

Again, solving for K yields

$$K = \frac{M_z}{(I_z\cos\beta + I_{yz}\sin\beta)} \tag{4.18}$$

The quantity I_z is the moment of inertia about the centroidal z axis. Using the geometry in Figure 4.4, we obtain

$$\tan\phi = \frac{M_y}{M_z} = \frac{I_y \sin\beta + I_{yz}\cos\beta}{I_z\cos\beta + I_{yz}\sin\beta} = \frac{I_y(\tan\beta) + I_{yz}}{I_z + I_{yz}(\tan\beta)} \tag{4.19}$$

The solution of Equation 4.19 for $\tan\beta$ yields

$$\tan\beta = \frac{I_{yz} - I_z(\tan\phi)}{I_{yz}(\tan\phi) - I_y} \tag{4.20}$$

Equation 4.20 allows us to find the angle β that the neutral axis makes with the centroidal z axis, provided we know the centroidal moments of inertia I_y, I_z, the centroidal product of inertia I_{yz}, and the angle ϕ that the plane of the loads makes with the centroidal y axis. *It is important to remember that while the angle ϕ is measured from the y axis, the angle β is measured from the z axis, both in the same sense, either clockwise or counterclockwise.* To one side of the neutral axis, fibers are stretched (tension zone), and to the other, fibers are compressed (compression zone).

Substituting Equation 4.16 into Equation 4.13, we obtain

$$\left.\begin{aligned}\sigma_\eta &= \frac{M_y(y\cos\beta + z\sin\beta)}{I_y\sin\beta + I_{yz}\cos\beta} \\ &= \frac{M_y(y + z\tan\beta)}{I_y\tan\beta + I_{yz}}\end{aligned}\right\} \tag{4.21}$$

Similarly, substituting Equation 4.18 into Equation 4.13, we obtain

$$\left.\begin{aligned}\sigma_\eta &= \frac{M_z(y\cos\beta + z\sin\beta)}{I_z\cos\beta + I_{yz}\sin\beta} \\ &= \frac{M_z(y + z\tan\beta)}{I_z + I_{yz}\tan\beta}\end{aligned}\right\} \tag{4.22}$$

Elimination of the quantity, $\tan\beta$, between Equations 4.21 and 4.22 yields

$$\sigma = \frac{M_y(yI_{yz} - zI_z) + M_z(zI_{yz} - yI_y)}{I_{yz}^2 - I_yI_z} \tag{4.23}$$

For the sake of simplicity, the subscript η on the stress σ was discarded with the understanding that σ represents the bending stress along the beam axis at any point in the cross section defined by the coordinates y and z.

4.3.2 Principal Centroidal Axes

If we let the centroidal y and z axes coincide, respectively, with the centroidal v (weak) principal axis and centroidal u (strong) principal axis (see Appendix C.2), Equations 4.20 and 4.23 may be expressed in terms of principal axes and principal moments of inertia. Making these changes and realizing that the product of inertia with respect to principal axes of inertia vanishes, Equation 4.20 reduces to

$$\tan\beta = \left(\frac{I_u}{I_v}\right)\tan\phi \tag{4.24}$$

Also, Equation 4.23 reduces to

$$\sigma = \left(\frac{M_u}{I_u}\right)v + \left(\frac{M_v}{I_v}\right)u \tag{4.25}$$

Note that Equation 4.25 indicates that the unsymmetric bending problem can be looked upon as the superposition of two separate symmetric bending problems. The first problem is contained in the first term of Equation 4.25, which describes symmetric bending about the u axis that serves as the neutral axis for bending. The second problem is embodied in the second term of Equation 4.25, which describes symmetric bending about the v axis that serves as the neutral axis for bending. While the sign convention stated above for M_y (M_v) and M_z (M_u) could be used, *it is usually more convenient to determine the sign of each term by inspection as was done in Chapter 3 for symmetric bending.*

Equations 4.24 and 4.25 make it possible to locate the neutral axis for bending and find the bending stress at any point in the cross section if the principal centroidal axes u and v and the principal moments of inertia I_u and I_v are known. Thus, these two equations are very useful in solving bending problems in which the beam cross section possesses at least one axis of symmetry in which case the principal centroidal axes and principal centroidal moments of inertia can be easily determined without extensive computations. On the other hand, Equations 4.20 and 4.23 are more advantageous for cases of cross sections that do not possess any axes of symmetry. They eliminate the need for finding the principal centroidal axes and principal centroidal moments of inertia required by Equations 4.24 and 4.25.

The following two examples show how the above concepts are used in the solution of unsymmetric bending problems. Example 4.4 illustrates the use of Equations 4.24 and 4.25 and Example 4.5 the use of Equations 4.20 and 4.23.

EXAMPLE 4.4

A cantilever beam 15 ft long has a rectangular cross-sectional area as shown in the sketch. The beam carries a concentrated load $P = 10$ kips at the free end acting at a counterclockwise angle of 30° from the v principal centroidal axis as shown. Locate the neutral axis and find the maximum tensile bending stress in the beam.

SOLUTION

The rectangular cross section possesses two axes of symmetry, which, of course, coincide with the two principal centroidal axes of inertia as shown in the sketch. The principal centroidal moments of inertia are found to be $I_u = 170.7$ in.^{4} and $I_v = 42.7$ in.^{4} (see Appendix C.2) and the angle $\phi = 30$ in Equation 4.24, as shown in the sketch. Therefore, by Equation 4.24, we obtain

$$\beta = \tan^{-1}\left(\frac{170.7}{42.7}\right)\tan 30° = 66.6°$$

Since the angle $\beta = 66.6°$ is positive, it is counterclockwise from the u principal axis as shown in the sketch. Fibers in the beam upward and to the left of the *ENA* (gray area) are in the tension zone, and those downward and to the right (marble area) are in the compression zone.

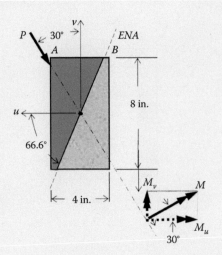

The maximum tensile bending stress occurs at the position along the beam where the bending moment is a maximum, namely, at the fixed end of the beam, where $M = 150$ kip · ft. This moment and its two perpendicular components $M_u = M \cos 30° = 129.9$ kip · ft and $M_v = M \sin 30° = 75.0$ kip · ft are shown in the sketch of the cross section. Since the bending stress is directly proportional to the distance from the neutral axis, the maximum tensile stress is found in the tensile zone at the point farthest from this axis. If the cross-sectional area and the orientation of the neutral axis are drawn approximately to scale, it is usually possible to determine, by inspection, the point in either zone that is farthest from the neutral axis. In some cases, however, the determination of such points may require a little more effort. For the case at hand, the maximum tensile stress occurs at point A. The tensile bending stress at this point is found by Equation 4.25. Thus,

$$\sigma_A = \left(\frac{129.9 \times 12}{170.7}\right)(4) + \left(\frac{75 \times 12}{42.7}\right)(2) = 36.527 + 42.155 = 78.682 \approx 78.7 \text{ ksi}$$

ANS.

The first term in the above equation represents the bending stress at point A due to the moment M_u, which produces symmetric bending about the u axis. Note that the u axis is the neutral axis for bending for this first case. Above the u axis, fibers are stretched and, therefore, point A is in tension. The second term represents the bending stress at point A due to M_v, which produces symmetric bending about the v axis. Again, note that the v axis is the neutral axis for bending for this second case. To the left of the v axis, fibers are stretched and therefore point A is again in tension. Thus, for point A, both terms in the above equation are tensile and hence positive. On the other hand, if the stress at some other point such as B were required instead of, or in addition to, that at point A, M_u would still produce tension, while M_u would produce compression.

EXAMPLE 4.5

A simply supported beam of length $L = 6$ m carries a downward uniform $p = 4$ kN/m along its entire length acting vertically downward. The cross-sectional area for the beam is the angle shown. Assume the load is so placed that no twisting of the section occurs and determine the maximum tensile and maximum compressive bending stresses in the beam.

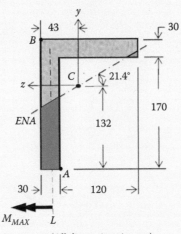

(All dimensions in mm)

SOLUTION

The centroid C of the section is located as shown and the centroidal moments of inertia determined to be $I_y = 17.43 \times 10^{-6}$ m^4 and $I_z = 36.53 \times 10^{-6}$ m^4. Also, the product of inertia is found to be $I_{yz} = -14.34 \times 10^{-6}$ m^4 (see Appendix C.2). While the centroidal principal moments of inertia may be determined using the methods of Appendix C.2, and Equations 4.24 and 4.25 used for the solution, it is much more convenient in this case to use Equations 4.20 and 4.23. Thus, by Equation 4.20, knowing that $\phi = 0$, we obtain

$$\beta = \tan^{-1}\left(\frac{-14.34 \times 10^{-6} - 0}{0 - 36.53 \times 10^{-6}}\right) = 21.4°$$

The positive sign implies that the angle is β is counterclockwise from the z axis because in the development of the equations, a counterclockwise angle β was assumed positive. The neutral axis is positioned through the centroid as shown. Below and to the right of the neutral axis, we have the tension zone (the gray area), and above and to the left of this axis, we have the compression zone (marble area). Thus, the maximum tensile stress occurs at point A(13, 132 mm) and the maximum compressive stress at point B (43, −68 mm).

These maximum stress values occur at midspan where the bending moment has its maximum value $M_{MAX} = pL^2/8 = 18$ kN · m acting along the z axis. Therefore, $M_y = 0$ and $M_z = M_{MAX} = -18$ kN · m. The negative sign on M_z is due to the fact that it produces compression where y is positive. Thus, using Equation 4.23, we obtain

$$\sigma_A = \frac{0 - 18 \times 10^3((0.013)(-14.34 \times 10^{-6}) - (-0.132)(17.43 \times 10^{-6}))}{(-14.34 \times 10^{-6})^2 - (36.53)(17.43) \times 10^{-12}} = 88.3 \text{ MPa} \qquad \textbf{ANS.}$$

Note that Equation 4.23 provides not only the magnitude but also the sign of the stress. Thus, as surmised by inspection, the answer indicates that point A is in tension. Also,

$$\sigma_B = \frac{0 - 18 \times 10^3((0.043)(-14.34 \times 10^{-6}) - (0.068)(17.43 \times 10^{-6}))}{(-14.34 \times 10^{-6})^2 - (36.53)(17.43) \times 10^{-12}} = -75.3 \text{ MPa} \qquad \textbf{ANS.}$$

Once again note that the sign of the answer for the stress at point B agrees with the sign deduced by inspection.

PROBLEMS

4.25 This cross section is that for a cantilever beam 5 m long subjected to a concentrated force $P = 50$ kN acting at the free end as shown. The plane of the load L is such that $\phi = -25°$. Let $h = 400$ mm and $b = 250$ mm, determine the maximum tensile and maximum compressive bending stresses in the beam specifying their locations.

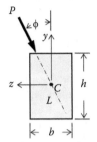

(Problems 4.25 and 4.26)

4.26 In reference to the figure shown in Problem 4.25, this cross section is that for a simply supported beam 16 ft long subjected to a concentrated force $P = 25$ kips acting at midspan as shown. The plane of the load L is such that $\phi = -40°$. Let $h = 10$ in. and $b = 5$ in., determine the maximum tensile and maximum compressive bending stresses in the beam specifying their locations.

4.27 This cross section is that for a simply supported beam 8 m long subjected to a uniform load of intensity $p = 6$ kN/m acting as shown over the entire length. Let $\phi = 30°$, $h = 340$ mm, and $b = 260$ mm, determine the maximum tensile and maximum compressive bending stresses in the beam specifying their locations.

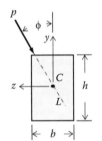

(Problems 4.27 and 4.28)

4.28 In reference to the figure shown in Problem 4.27, this cross section is that for a cantilever beam 18 ft long subjected to a uniform load of intensity $p = 2$ kips/ft acting as shown along the entire length. Let $\phi = 50°$, $h = 10$ in., $b = 8$ in., determine the maximum tensile and maximum compressive bending stresses in the beam specifying their locations.

4.29 The cross section shown is that for a cantilever beam 3 m long carrying a concentrated load of 1.5 kN at the free end acting vertically downward such that no twisting action is produced. Determine the maximum tensile and maximum compressive bending stresses specifying their locations. Let $a = 20$ mm, $b = 80$ mm, and $h = 110$ mm. (*Hint:* Use Equations 4.20 and 4.23 for your solution.)

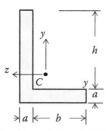

(Problems 4.29 and 4.30)

4.30 In reference to the figure shown in Problem 4.29, the cross section shown is that for a simply supported beam 12 ft long carrying two concentrated loads P, each with a magnitude of 3 kips placed at the third points, acting vertically downward such that no twisting action is produced. Determine the maximum tensile and maximum compressive bending stresses specifying their locations. Let $a = 1$ in., $b = 6$ in., and $h = 10$ in. (*Hint:* Use Equations 4.20 and 4.23 for your solution.)

4.31 The cross section shown is that for a simply supported beam 8 ft long carrying a uniform load of intensity $p = 1.5$ kip/ft acting vertically downward over the right half of the beam such that no twisting action is produced. Determine the maximum tensile and maximum compressive bending stresses specifying their locations. Let $a = 0.75$ in., $b = 6.0$ in., and $h = 4.0$ in. (*Hint:* Use Equations 4.20 and 4.23 for your solution.)

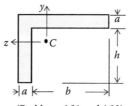

(Problems 4.31 and 4.32)

4.32 In reference to the figure shown in Problem 4.31, the cross section shown is that for a cantilever 4 m long carrying a uniform load of intensity $p = 2$ kN/m acting vertically downward over the entire length such that no twisting action is produced. Determine the maximum tensile and maximum compressive bending stresses specifying their locations. Let $a = 25$ mm, $b = 120$ mm, and $h = 80$ mm. (*Hint:* Use Equations 4.20 and 4.23 for your solution.)

4.33 This cross section is that for a cantilever 6 m long carrying a uniform load of intensity $p = 5$ kN/m acting as shown over the entire length. Let $a = 40$ mm and $\phi = 20°$, find the maximum tensile and maximum compressive bending stresses specifying their locations. Solve the problem using Equations 4.24 and 4.25 and check your answers using Equations 4.20, 4.21, and 4.22.

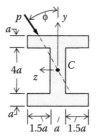

(Problems 4.33 and 4.34)

4.34 In reference to the figure shown in Problem 4.33, this cross section is that for a simply supported beam, 16 ft long, subjected to a uniform load of intensity $p = 3$ kips/ft acting as shown over the entire length. Let $a = 2$ in., $\phi = 30°$ and determine the maximum tensile and maximum compressive bending stresses specifying their locations. Solve the problem using Equations 4.24 and 4.25 and check your answers using Equations 4.20, 4.21, and 4.22.

4.35 The cross section shown is that for a simply supported beam 20 ft long subjected to a concentrated load $P = 15$ kips acting as shown at 8 ft from the left support. Let $a = 2$ in. and $\phi = -35°$, determine the maximum tensile and maximum compressive bending stresses specifying their locations. Solve the problem using Equations 4.24 and 4.25 and check your answers using Equations 4.20, 4.21, and 4.22.

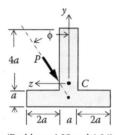

(Problems 4.35 and 4.36)

4.36 In reference to the figure shown in Problem 4.35, the cross section shown is that for a cantilever beam 6 m long carrying a concentrated load $P = 25$ kN acting as shown at the free end. Let $a = 50$ mm and $\phi = -25°$, find the maximum tensile and maximum compressive bending stresses specifying their locations. Solve the problem using Equations 4.24 and 4.25 and check your answers using Equations 4.20, 4.21, and 4.22.

4.37 The cross section shown is that for a cantilever beam 8 m long carrying a concentrated load $P = 30$ kN acting vertically downward through its centroid at the free end. Let $a = 60$ mm, $h = 220$ mm, and $b = 160$ mm, find the maximum tensile and maximum compressive bending stresses specifying their locations.

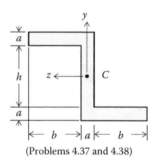

(Problems 4.37 and 4.38)

4.38 In reference to the figure shown in Problem 4.37, the cross section shown is that for a simply supported beam 18 ft long subjected to a uniform load of intensity $p = 3$ kips/ft acting vertically downward through its centroid over the left two-thirds of the beam. Let $a = 2$ in., $h = 8$ in., and $b = 6$ in., determine the maximum tensile and maximum compressive bending stresses specifying their locations.

4.39 The cross section shown is that for a simply supported beam 8 m long subjected to a uniform load of intensity $p = 6$ kN/m acting vertically downward over the entire beam, so placed that no twisting action occurs. Let $a = 40$ mm, $h = 100$ mm, $b = 80$ mm, and $c = 60$ mm, determine the maximum tensile and maximum compressive bending stresses specifying their locations.

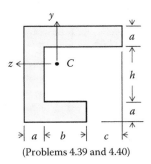

(Problems 4.39 and 4.40)

4.40 In reference to the figure shown in Problem 4.39, the cross section shown is that for a cantilever beam 14 ft long carrying a concentrated load $P = 20$ kips acting vertically downward at the free end, so placed that no twisting action occurs. Let $a = 1.5$ in., $h = 6$ in., $b = 4$ in., and $c = 4$ in., find the maximum tensile and maximum compressive bending stresses specifying their locations.

4.41 The cross section shown in figure (a) below is that for the beam shown in figure (b) below. Let $a = 10$ in., $h = 12$ in., $b = 6$ in., $p = 2$ kips/ft, $P = 20$ kips, $L = 10$ ft, and $\phi = -60°$, determine the maximum tensile and maximum compressive bending stresses specifying their locations.

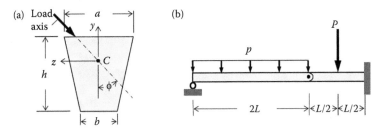

(Problems 4.41 and 4.42)

4.42 In reference to the figure shown in Problem 4.41, the cross section shown in figure (a) above is that for the beam shown in figure (b) above. Let $a = 260$ mm, $h = 270$ mm, $b = 100$ mm, $p = 6$ kN/m, $P = 50$ kN, $L = 4$ m, and $\phi = -40°$, determine the maximum tensile and maximum compressive bending stresses specifying their locations.

4.43 The cross section shown in figure (a) below is that for the beam shown in figure (b) below. Let $a = 40$ mm, $h = 80$ mm, $b = 60$ mm, $p = 8$ kN/m, $P = 60$ kN, $L = 4$ m, and $\phi = 0°$, determine the maximum tensile and maximum compressive bending stresses specifying their locations.

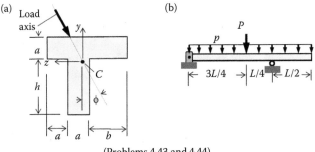

(Problems 4.43 and 4.44)

4.44 In reference to the figure shown in Problem 4.43, the cross section shown in figure (a) above is that for the beam shown in figure (b) above. Let $a = 2$ in., $h = 6$ in., $b = 4$ in.,

$p = 2$ kips/ft, $P = 40$ kips, $L = 16$ ft, and $\phi = 20°$, determine the maximum tensile and maximum compressive bending stresses specifying their locations.

4.45 A W12 × 58 (see Appendix E) is used for a cantilever beam 16 ft long carrying a concentrated load P at the free end as shown where $\phi = -10°$. If the allowable bending stress is 20 ksi, determine the maximum permissible load P.

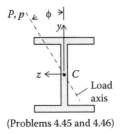

(Problems 4.45 and 4.46)

4.46 In reference to the figure shown in Problem 4.45, a W356 × 56.5 (see Appendix E) is used for a simply supported beam 8 m long carrying a uniform load of intensity $p = 5$ kN/m over the entire length where $\phi = -15°$. If the allowable bending stress is 150 MPa, determine the maximum permissible load intensity p.

4.47 A C178 × 21.9 (see Appendix E) is used for a simply supported beam 4 m long carrying a uniform load of intensity p over the entire length where $\phi = 5°$. If the allowable bending stress is 180 MPa, determine the maximum permissible load intensity p.

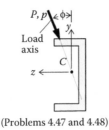

(Problems 4.47 and 4.48)

4.48 In reference to the figure shown in Problem 4.47, a C10 × 25 (see Appendix E) is used for a cantilever beam 12 ft long carrying a concentrated load P at the free end as shown where $\phi = 8°$. If the allowable bending stress is 25 ksi, determine the maximum permissible load P.

4.49 A L8 × 8 × 1 (see Appendix E) is used for a cantilever beam 8 ft long carrying a concentrated load Q at the free end acting vertically downward such that no twisting occurs. If the allowable bending stress is 18 ksi, determine the maximum permissible load P.

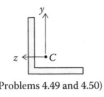

(Problems 4.49 and 4.50)

4.50 In reference to the figure shown in Problem 4.49, an L152 × 152 × 25.4 (see Appendix E) is used for a simply supported beam 4 m long carrying a uniform load of intensity p acting vertically downward over the entire length such that no twisting action occurs. If the allowable bending stress is 150 MPa, determine the maximum permissible load intensity p.

4.4   THIN-WALLED OPEN SECTIONS: SHEAR CENTER

In Section 3.5, we learned how to determine the shear flow in thin-walled sections that have at least one plane of symmetry and subjected to loads in the plane of symmetry. Under those conditions, a beam having such a cross section bends without twisting, and the bending stress can then be found using Equation 3.23 and the shear flow using Equation 3.30. However, in the case of thin-walled open sections with planes of symmetry subjected to loads not in a plane of symmetry, the beam experiences twisting in addition to bending even if the loads are applied through the centroid of the section. Also, if the thin-walled open section does not possess planes of symmetry, loading in any plane, even in those that pass through the centroid, would lead to unsymmetric bending as well as twisting of the section. In either of these two cases, we may avoid the twisting action and produce only pure bending of the beam (symmetric or unsymmetric), if the plane of the loads is made to pass through a point known as the *shear center* for the cross section.

4.4.1   SYMMETRIC BENDING

Consider, for example, a beam whose cross section is the thin-walled channel shown in Figure 4.5. In Figure 4.5a, the applied load P acts in the only longitudinal plane of symmetry for the section defined by the z axis. In such a case, bending is symmetric and the neutral surface coincides with the longitudinal plane defined by the y axis. This case is no different from the cases discussed previously; the bending stress is given by Equation 3.23 and the shear flow by Equation 3.30. However, if this cross section were loaded as shown in Figure 4.5b, the loading would be symmetric, but the beam would not only bend but also twist at the same time, creating a complex stress system that includes torsional shearing stresses in addition to the bending stresses and the shear flow due to the shear force V. Eliminating the torsional shearing stresses requires that the load P be applied through the shear-center point for the cross section—a point that lies on the z axis of symmetry for the section that also serves as the neutral axis for bending.

Locating the shear center for such a section requires finding the shearing stresses in all of the thin rectangles. These shearing stresses result in shearing forces in the rectangles that exert a twisting moment that has to be balanced in order to avoid twisting of the section.

Let us assume that the channel cross section discussed above is that for a cantilever beam subjected to a concentrated force P as shown in Figure 4.6a. The centroidal z axis is a principal axis of inertia because it is an axis of symmetry (see Appendix C.2). It follows, therefore, that the y axis is also a principal axis of inertia. Assume that the thickness t is constant around the entire section and very small in comparison to the dimensions h and w. Since the force P is applied parallel to the y principal centroidal axes, it follows that the z principal centroidal axis is the neutral axis for bending. Consider a small segment of the upper flange at a distance x from the origin of the coordinate system having a differential length dx and a finite width b. This small segment is isolated and its free-body diagram shown magnified in Figure 4.6b. The three forces, F_1, F_2, and F_3 on this segment all act in the x direction and must be in equilibrium. Thus,

$$\sum F_x = 0 \Rightarrow F_2 + F_3 - F_1 = 0; \quad F_3 = F_1 - F_2 \tag{4.26}$$

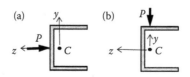

FIGURE 4.5   (a) A thin-walled channel with the applied load P acting along the longitudinal plane of symmetry. (b) The same channel with load P producing both symmetric bending and twisting.

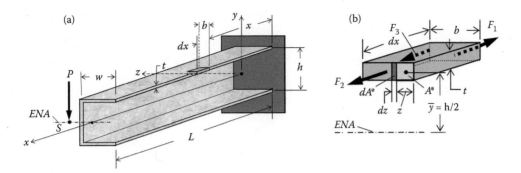

FIGURE 4.6 (a) A beam with a channel cross-subjected to a load P applied through the shear center. (b) A small segment of the beam in the upper flange, at distance x from the origin of the coordinate system, of dimensions dx by b.

The force F_1 is the resultant of the bending stresses acting over the area tb of the flange segment at a distance x from the origin where the moment is M. Therefore, $F_1 = \int (M\,y/I)\,dA$. Also, the force F_2 represents the resultant of the bending stresses acting over the area tb of the flange segment at a distance $x + dx$ from the origin where the moment is $M + dM$. Thus, $F_2 = \int ((M + dM)y/I)\,dA$. Finally, the force F_3 is the resultant of the shearing stresses τ_{zx} acting over the inner longitudinal area $t\,dx$, which must exist in order to balance forces in the x direction. Since the thickness t is very small, the assumption is made that the shearing stress τ_{zx} is contant across the thickness and, therefore, $F_3 = \tau_{zx}t\,dx$. Substituting these values into Equation 4.26 and simplifying yields

$$\tau_{zx}\,t\,dx = \left(\frac{dM}{I}\right)\int y\,dA \tag{4.27}$$

Solving Equation 4.27 for the shearing stress τ_{zx}, we obtain

$$\tau_{zx} = \left[\frac{(dM/dx)}{It}\right]\int y\,dA = \left(\frac{V}{It}\right)\int y\,dA \tag{4.28}$$

where the internal shear force V was substituted for the derivative dM/dx. Note that, in this case, the internal vertical shear force at any position in the beam is equal to the applied force P. The existence of the shearing stress τ_{zx} on longitudinal internal flange planes requires that a shearing stress τ_{xz}, *pointed to the left*, and of equal magnitude, exists on transverse internal flange planes. The integral $\int y\,dA$ is now replaced by the symbol $Q = A^*\bar{y}$, which represents the first moment of the orange-colored area shown in Figure 4.6b about the neutral axis of the section. Thus, Equation 4.28 becomes

$$\tau_{xz} = \tau_{zx} = \frac{VQ}{It} \tag{4.29}$$

Equation 4.29 may be simplified by dropping the subscripting on the shearing stresses. Thus,

$$\tau = \frac{VQ}{It} = \left(\frac{V}{It}\right)A^*\bar{y} \tag{4.30}$$

Note that, in obtaining Equation 4.30, the quantity Q in Equation 4.29 was replaced by the product $A^*\bar{y}$.

Note also that Equation 3.30, developed in Chapter 3 for the horizontal and vertical shearing stresses due to transverse loads in beams, is the same as Equation 4.30 except for the fact that the symbol b has been replaced by the symbol t. Since $Q = \int y\, dA = A^* \overline{y} = (h/2)(t\, z)$, it follows, from Equation 4.30, that the shearing stress in the upper flange is given by

$$\tau = \left(\frac{Vh}{2I} \right) z \tag{4.31}$$

Equation 4.31 shows that the shearing stress in the upper flange increases linearly from zero at its free edge to a maximum value of $(Vhw)/2I$ at the junction between it and the web of the channel where $z = w$. In general, the internal shear force V in any thin rectangular component of the section is the resultant of, and acts in the same sense as, the shearing stresses in that rectangular component. Thus, using Equation 4.30, we obtain

$$V = \int \tau\, dA = \left(\frac{V}{It} \right) \int A^* \overline{y}\, dA^* \tag{4.32}$$

We now make reference to Figures 4.7a and 4.7b. Using Equation 4.32, and the information contained in Figure 4.6b, the internal shearing force V_1 in the upper flange, for example, becomes

$$V_1 = \left(\frac{V}{It} \right) \int_0^w (tz)\left(\frac{h}{2} \right) t\, dA = \left(\frac{Vht}{2I} \right) \int_0^w z\, dz = \frac{Vhtw^2}{4I} \tag{4.33}$$

Similarly, by considering the lower flange (see Figures 4.7a and 4.7b), we conclude that the internal shearing force there has the same magnitude V_1 as was found for the upper flange, except that it points to the right. Also, by assuming that the entire vertical internal shearing force V at, any the section, is carried by the web of the channel, we conclude, on the basis of equilibrium, that the internal shearing force in the web $V_2 = V$. This assumption is valid as long as the thickness t is very small compared to the other dimensions.

The beam of Figure 4.6 is cut at a distance x from the fixed end and the free-body diagram of the part of length L–x is constructed as shown in Figure 4.7a where the internal resisting shearing forces have been properly placed and shown hidden with broken lines on the back surfaces. While both vertical and horizontal equilibrium of this part are assured, rotational equilinbrium requires that the applied force P be placed at point S, known as *the shear center for the section*, a distance e from the centerline of the web, in order to balance the clockwise torque produced by the internal shear forces V_1. Thus, summing torques about point O, we obtain

$$Pe = V_1 h \tag{4.34}$$

Solving for the distance e, Equation 4.34 leads to

$$e = \frac{V_1 h}{P} = \frac{V_1 h}{V} \tag{4.35}$$

Note that the quantity P in Equation 4.35 was replaced by the vertical internal shearing force V because the internal vertical shearing force at any position in the cantilever beam is equal to the applied force P. Substituting for V_1 from Equation 4.33 into Equation 4.35, we obtain

$$e = \frac{th^2 w^2}{4I} \tag{4.36}$$

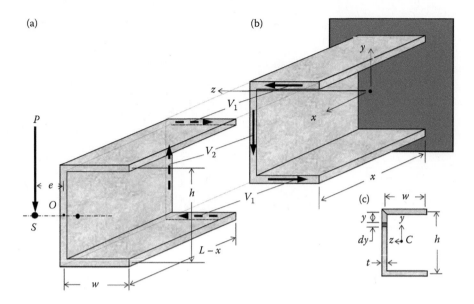

FIGURE 4.7 (a) and (b) A cantilever beam with a channel cross section, subjected to a bending load P acting through its shear center. (c) The beam cross section.

Equation 4.36 shows that the location of the shear center, point S, is strictly a function of the cross-sectional dimensions of the section and, therefore, *the shear center for a given cross section is a geometric property of the section.*

The fact that V_2 is equal to the entire vertical shear force V may also be established by the use of Equation 4.32. Thus, referring to Figure 4.7c, we obtain

$$V_2 = \left(\frac{V}{It}\right)\int_0^h \left[(A^*\bar{y})_{\text{flange}} + (A^*\bar{y})_{\text{web}}\right] dA^* \tag{4.37}$$

where the product $(A^*\bar{y})_{\text{flange}}$ represents the quantity Q for the entire upper flange and is equal to $twh/2$, and the quantity $(A^*\bar{y})_{\text{web}}$ represents the quantity Q for that part of the web between this flange and the location defined by the coordinate y at which the shearing stress is desired. Thus, the sum of these two quantities represents the first moment of the area between the right free surface of the upper flange and the location in the web at which the shearing stress is to be determined. Again, referring to Figure 4.7c, Equation 4.37 becomes

$$V_2 = \left(\frac{V}{It}\right)\int_0^h \left[\frac{twh}{2} + \frac{ty(h-y)}{2}\right] t\,dy = \frac{V}{I}\left(\frac{twh^2}{2} + \frac{th^3}{12}\right) = \frac{V}{I}\left(\frac{th^2}{2}\right)\left(w + \frac{h}{6}\right) \tag{4.38}$$

Neglecting terms containing the quantity t^3 in comparison to other terms, we can show that the moment of inertia of the channel about the z axis of symmetry is

$$I = \frac{twh^2}{2} + \frac{th^3}{12} = \frac{th^2}{2}\left(w + \frac{h}{6}\right) \tag{4.39}$$

Thus, as before, we conclude that the shear force V_2 in the web of the channel is equal to the entire vertical shear force V in the section, which in this case is equal to the applied force P.

EXAMPLE 4.6

1. Determine the location of the shear center, point S, for the channel shown in figure (a) below.
2. Sketch the distribution of the shearing stresses if a downward vertical load $P = 75$ kN is applied at the shear center. Let $h = 160$ mm, $w = 100$ mm, and $t = 5$ mm.

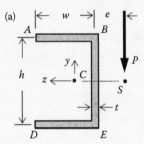

SOLUTION

a. We first determine the moment of inertia of the channel about the z axis (elastic neutral axis) by using Equation 4.39 of the previous development. Thus,

$$I = \frac{th^2}{2}\left(w + \frac{h}{6}\right) = \frac{(0.05)(0.16^2)}{2}\left(0.10 + \frac{0.16}{6}\right)$$

$$= 81.0 \times 10^{-6}\,\text{m}^4$$

The location of the shear center, point S, is now determined using Equation 4.36 developed above. Thus,

$$e = \frac{th^2w^2}{4I} = \frac{(0.05)(0.16^2)(0.10^2)}{4(81.0 \times 10^{-6})}$$

$$= 0.039\,\text{m} = 39\,\text{mm} \quad \textbf{ANS.}$$

b. The shearing stress distribution in the upper flange is given by Equation 4.31 in the above development, which shows that it varies linearly from zero at A to a maximum value at B equal to

$$\tau_B = \frac{Vhw}{2I} = \frac{(75 \times 10^3)(0.16)(0.10)}{2(81.0 \times 10^{-6})} = 7.407 \approx 7.4\,\text{MPa} \quad \textbf{ANS.}$$

Because of symmetry, the shearing stresses in the lower flange are identical to those in the upper flange.

The shearing stress at any point in the web is obtained by adding τ_B to the shearing stress given by Equation 4.30. Therefore,

$$\tau = \tau_B + \frac{V}{It}[A^*(\overline{y})] = \tau_B + \frac{V}{It}\left[ty\left(\frac{h-y}{2}\right)\right] = \tau_B + \frac{V}{2I}(hy - y^2)$$

which represents a second-degree parabolic function and its maximum value occurs at the neutral axis. Thus,

$$\tau_{MAX} = \tau_{ENA} = 7.407 + \frac{75 \times 10^3}{2(81.0 \times 10^{-6})}\left[(0.16)\left(\frac{0.16}{2}\right) - \left(\frac{0.16}{2}\right)^2\right] = 10.4 \text{ MPa} \qquad \textbf{ANS.}$$

The shearing stress distributions in the flanges and in the web of the channel, as well as a qualitative resisting shear flow, are shown in figure (b) below.

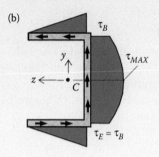

(b)

EXAMPLE 4.7

a. Find the shearing forces in rectangles AB and BD for the section shown in figure (a) below, expressing your answers in terms of the vertical downward shearing force V, the moment of inertia I of the section about the z axis (ENA), and the quantity b. (b) Find the value of e locating the shear center S.

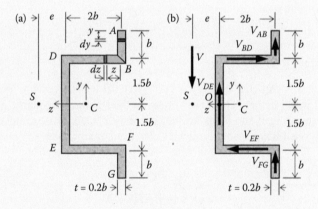

SOLUTION

a. The shearing force in any of the rectangles making up the given cross section may be obtained using Equation 4.32. Thus,

$$V_{AB} = \frac{V}{It}\int [A^*\overline{y}]dA^* = \frac{V}{I(0.2b)}\int_0^b (0.2b)y\left(\frac{2.5b - y}{2}\right)(0.2b)dy$$

$$= (0.217b^4)\frac{V}{I} \quad \textbf{ANS.}$$

$$V_{BD} = \frac{V}{I(0.2b)}\int_0^{2b} [(0.2b)b(2b) + (0.2b)z(1.5b)](0.2b)dz$$

$$= (1.400b^4)\frac{V}{I} \quad \textbf{ANS.}$$

The shear forces in rectangles *FG* and *EF* are identical in magnitude to those in *AB* and *BD*, respectively. All of these resisting shearing forces are indicated in figure (b) above, along with the vertical shearing force V_{DE} in rectangle *DE*.

b. The location of the shear center is found by summing torques about point *O* at the center of rectangle *DE*. Thus,

$$eV + 2V_{AB}(2b) - 2V_{BD}(1.5b) = 0$$

Solving for *e*, we obtain

$$e = \frac{(3b)V_{BD} - (4b)V_{AB}}{V} = \frac{(3b)(1.400b^4)(V/I) - (4b)(0.217b^4)(V/I)}{V} = \frac{3.332b^4}{I}$$

The moment of inertia of the section about the *z* axis (*ENA*) is found to be $I = 3.886b^4$. Making this substitution in the above value for *e*, we obtain

$$e = 0.854 \approx 0.9 \text{ in.} \quad \textbf{ANS.}$$

Note that the location of the shear center is independent of the magnitude of the quantity *b*.

4.4.2 UNSYMMETRIC BENDING

As discussed earlier, unsymmetric bending occurs when the loads are not in one of the two centroidal longitudinal principal planes. Thus, even cross sections possessing planes of symmetry experience unsymmetric bending if the loads are not properly placed. For example, the channel cross section shown in Figure 4.8 is *not* bent symmetrically, even though the load passes through the shear

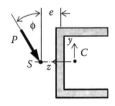

FIGURE 4.8 Unsymmetric bending produced in a beam when the load is not in one of the two centroidal longitudinal principal planes, even though the load acts through the shear center.

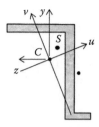

FIGURE 4.9   Locating the shear center for a section possessing no axis of symmetry.

center, because the applied load P does not act in one of the two centroidal longitudinal principal planes. Since the applied load P passes through the shear center, the cross section would not be subject to twisting action but the bending stresses created in the cross section need to be determined using the methods developed in Section 4.3.

Note that the shear center S for sections possessing one axis of symmetry (a principal axis of inertia) lies along this axis and locating it requires finding only one distance along this axis. However, cross sections, such as the one shown in Figure 4.9, that possess no axes of symmetry experience unsymmetric bending and locating the shear center S for this type of section requires finding two perpendicular distances along the orthogonal u (strong) and v (weak) principal axes of inertia. Example 4.8 illustrates the method used to find the shear center for this type of cross section.

EXAMPLE 4.8

Locate the shear center for the Z section shown in figure (a) below. Figure (a) below shows the location of the centriodal y and z axes as well as the orientation of the principal centroidal u and v axes. Given: $I_u = 95.415$ in.^{4} and $I_v = 8.777$ in.^{4}.

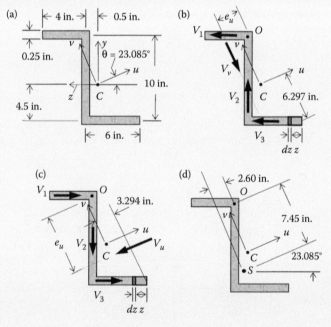

SOLUTION

We begin the solution by applying a force P_v parallel to the v principal centroidal axis which results in an internal shearing force V_v acting as shown in figure (b) above. Figure (b) above also shows the shearing forces V_1, V_2, and V_3 induced in the three rectangular components of the Z section. If we summed torques about point O, V_1, and V_2 would be eliminated from the torque equation because they pass through this point. Thus, the only internal shearing force that needs to be found in terms of V_v is V_3. This is done using Equation 4.32. Thus,

$$V_3 = \frac{V_v}{95.415(0.25)}\int_0^6 0.25z\left(6.297 - \frac{z}{2}\sin 23.085°\right)(0.25dz) = 0.260V_v$$

Referring to figure (b) above, we sum torques about point O. Thus,

$$\sum T_O = 0 \Rightarrow V_v e_u - 0.260V_v(10) = 0$$

Therefore,

$$e_u = 2.60 \text{ in.} \quad \textbf{ANS.}$$

The next step is to apply a force P_u parallel to the u principal centroidal axis, which results in an internal shearing force V_u acting as shown in figure (c) above. Figure (c) above also shows the shearing forces V_1, V_2, and V_3 developed in the three rectangular components of the Z section. Here again, we sum torque about point O, thus eliminating the need to find V_1 and V_2. The force V_3 is, once again, found using Equation 4.32. Thus,

$$V_3 = \frac{V_u}{8.777(0.25)}\int_0^6 0.25z\left(3.294 - \frac{z}{2}\cos 23.085°\right)(0.25dz) = 0.745V_u$$

Referring to figure (c) above, we sum torques about point O. Thus,

$$\sum T_O = 0 \Rightarrow V_u e_v - 0.745V_u(10) = 0$$

Thus,

$$e_v = 7.45 \text{ in.} \quad \textbf{ANS.}$$

The location of the shear center S with respect to point O is shown in figure (d) above.

PROBLEMS

4.51 Locate the shear centers for the two cross sections shown, which possess one axis of symmetry. (*Hint:* The shear center may be located by inspection.)

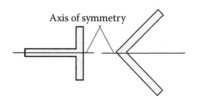

Axis of symmetry

4.52 Use Equation 4.32 to determine the shearing force V_{DE} in rectangle DE of the cross section in Example 4.7. Check your answer using equilibrium. Given: $I = 3.883b^4$.

4.53 (a) Assume loading normal to the horizontal plane of symmetry and that the slit at A is insignificantly small. Determine the shearing forces induced in rectangles AB and BD. (b) Locate the shear center for the cross section. Given: $I = h^4/3$.

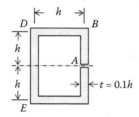

(Problems 4.53, 4.54, and 4.55)

4.54 Refer to Problem 4.53 and, using Equation 4.32, determine the shearing force developed in rectangle DE of the section. Check your answer using equilibrium.

4.55 In reference to the figure shown in Problem 4.53, the cross section shown is that for a cantilever beam carrying a downward vertical load of 10 kN at its free end placed so that it acts along the centerline of rectangle DE. Determine the torque that this loading produces in the beam. Let $h = 120$ mm.

4.56 A hollow equilateral tube is slit at the midpoint of one of its sides as shown. Assume the width of the cut to be insignificantly small. The slit tube is used as a cantilever beam to carry loads normal to the axis of symmetry. (a) Find the shearing force in the two vertical rectangles and (b) locate the shear center for this cross section. Given: $I = 0.025a^4$.

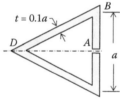

(Problems 4.56 and 4.57)

4.57 Refer to Problem 4.56 and, using Equation 4.32, determine the shearing force induced in rectangle BD of the slit section. Check your answer using equilibrium.

4.58 (a) Assume loading normal to the horizontal plane of symmetry of the section and determine the shearing forces in rectangles AB and BC. (b) Locate the shear center for the cross section. Given: $I = 119.9$ in.[4]

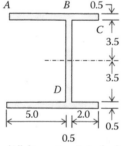

(All dimensions in inches)

(Problems 4.58, 4.59, and 4.60)

4.59 Refer to Problem 4.58 and, using Equation 4.32, determine the shearing force developed in rectangle *BD* of the section. Check your answer using equilibrium.

4.60 In reference to the figure shown in Problem 4.58, the cross section shown is that for a simply supported beam carrying a downward vertical load of 15 kips at midspan placed so that it acts along the centerline of the web (rectangle *BD*). Determine the torque that this loading produces in the beam.

4.61 (a) Assume loading normal to the horizontal plane of symmetry of the section and determine the shearing force in rectangles *AB*. (b) Locate the shear center for the cross section. Given: $I = 0.24714a^4$.

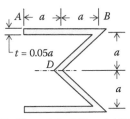

(Problems 4.61, 4.62, and 4.63)

4.62 Refer to Problem 4.61 and, using Equation 4.32, determine the shearing force developed in rectangle *BD* of the section. Check your answer using equilibrium.

4.63 In reference to the figure shown in Problem 4.61, the cross section shown is that for a cantilever beam carrying a downward vertical load of 10 kN at corner *B*. Determine the torque that this loading produces in the beam. Let $a = 100$ mm.

4.64 (a) Assume loading normal to the horizontal plane of symmetry of the section and determine the shearing force in rectangle *AB*. (b) Locate the shear center for the cross section. Given: $I = 0.10a^4$.

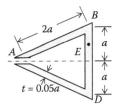

(Problems 4.64, 4.65, and 4.66)

4.65 Refer to Problem 4.64 and, using Equation 4.32, determine the shearing force developed in rectangle *BD* of the section. Check your answer using equilibrium.

4.66 In reference to the figure shown in Problem 4.64, the cross section shown is that for a simply supported beam, 10 ft long, subjected to a uniform load of 100 lb/ft, over the entire length, acting vertically downward through the shear center. Consider a section 2.5 ft from the left support and construct a stress element at point *E*, 2 in. above the axis of symmetry, with two planes parallel to the axis of the beam and two perpendicular to it. Let $a = 4$ in. and $t = 0.25$ in.

4.67 Assume loading normal to the horizontal plane of symmetry and determine the shearing stress distribution around the thin circular open cross section. Express your answer in terms of *R*, the applied vertical shearing force *V*, the moment of inertia *I*, and the variable angle ϕ. Assume the slit to be insignificantly small.

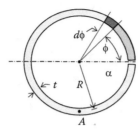

(Problems 4.67, 4.68, and 4.69)

4.68 Refer to Problem 4.67 and find the shear center for this thin cross section. Express the answer in terms of R. Given: $I = \pi R^3 t$.

4.69 In reference to the figure shown in Problem 4.67, the cross section shown is that for a simply supported beam, 4 m long, subjected to a uniform load of 0.75 kN/m, over the entire length, acting vertically downward through the shear center. Consider a section 1 m from the left support and construct a stress element at point A, at the bottom of the cross section, with two planes parallel to the axis of the beam and two perpendicular to it. Let $R = 100$ mm and $t = 10$ mm.

4.70 (a) Assume loading normal to the horizontal plane of symmetry of the section and determine the shearing forces in rectangles AB and BD. (b) Locate the shear center for the cross section. Given: $I = 1.84130a^4$.

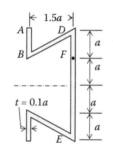

(Problems 4.70, 4.71, and 4.72)

4.71 Refer to Problem 4.70 and, using Equation 4.32, determine the shearing force developed in vertical rectangle DE of the section. Check your answer using equilibrium.

4.72 In reference to the figure shown in Problem 4.70, the cross section shown is that for a cantilever beam, 12 ft long, subjected to a uniform load of 200 lb/ft, over the entire length, acting vertically downward through the shear center. Consider a section at the fixed end and construct a stress element at point F with two planes parallel to the axis of the beam and two perpendicular to it. Let $a = 4$ in.

4.73 (a) Locate the shear center for the L203 × 152 × 12.7 angle shown. (b) Assume a downward vertical shear $V = 2$ kN through the shear center and develop an expression for the shearing stress in the vertical rectangle. Let $a = 80$ mm and $t = 10$ mm. Note that the shearing stress is the sum of two shearing stresses produced by two shearing forces, each parallel to one of the two centroidal principal axes of inertia. Given: $A = 4350 \times 10^{-6}$ m², $I_y = 9.03 \times 10^{-6}$ m⁴, $I_z = 18.44 \times 10^{-6}$ m⁴, $r_y = 0.33$ m, $\alpha = 29.162°$, $h = 0.203$ m, $b = 0.152$ m, $t = 0.127$ m, $z = 0.0373$ m, and $y = 0.0627$ m.

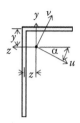

4.74 Repeat Problem 4.73 for an L6 × 4 × 3/8 unequal leg angle and $V = 3.0$ kips. Given: $A = 3.61$ in.^{2}, $I_y = 4.90$ in.^{4}, $I_z = 13.50$ in.^{4}, $r_v = 0.877$ in., $\alpha = 24.037°$, $h = 6$ in., $b = 4$ in., $t = 0.375$ in., $z = 0.941$ in., and $y = 1.940$ in.

4.75 (a) Locate the shear center for the unsymmetric T section shown. (*Hint:* The shear center may be located by inspection.) (b) Assume a downward vertical shear $V = 3$ kips through the shear center and develop an expression for the shearing stress in the left part of the horizontal rectangle. Let $a = 3$ in. and $t = 0.25$ in. Given: $I_y = 17.730$ in.^{4}, $I_z = 37.980$ in.^{4}, and $I_{yz} = 7.594$ in.^{4} (*Hint:* The shearing stress is the sum of two shearing stresses produced by two shearing forces, each parallel to one of the two centroidal principal axes of inertia.)

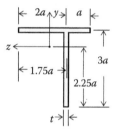

4.76 Compute the location of the shear center for the following unsymmetric cross section. Note the useful information provided in the sketch.

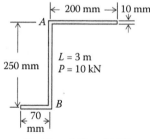

(Problems 4.76 and 4.79)

4.77 Compute the location of the shear center for the following unsymmetric cross section. Note the useful information provided in the sketch.

(Problems 4.77 and 4.80)

4.78 Compute the location of the shear center for the following unsymmetric cross section. Note the useful information provided in the sketch.

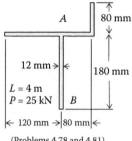

(Problems 4.78 and 4.81)

4.79 In reference to the figure shown in Problem 4.76, the unsymmetric cross section is for a cantilever beam of length L. If the downward load P is placed at the free end so that it acts along rectangle AB, determine the torque induced in the beam. The magnitudes of L and P for the beam are indicated in the figure.

4.80 In reference to the figure shown in Problem 4.77, the unsymmetric cross section is for a cantilever beam of length L. If the downward load P is placed at the free end so that it acts along rectangle AB, determine the torque induced in the beam. The magnitudes of L and P for the beam are indicated in the figure.

4.81 In reference to the figure shown in Problem 4.78, the unsymmetric cross section is for a cantilever beam of length L. If the downward load P is placed at the free end so that it acts along rectangle AB, determine the torque induced in the beam. The magnitudes of L and P for the beam are indicated in the figure.

4.5 CURVED BEAMS

When an initially curved beam is subjected to bending action, two types of normal stresses are induced at every point in the member. One of these normal stresses acts on radial planes and points in the circumferential direction. These stresses, known as *circumferential stresses*, are, in general, the most significant stresses in a curved beam and are the only stresses discussed here. The second normal stress acts on circumferential planes pointing in the radial direction and is referred to as a *radial stress*. Although of interest in many situations, the radial stress will not be discussed in this text.

Consider the curved beam shown in Figure 4.10a. This curved beam has the arbitrary cross-sectional area shown in Figure 4.10b in which the y centroidal axis is an axis of symmetry and, therefore, a

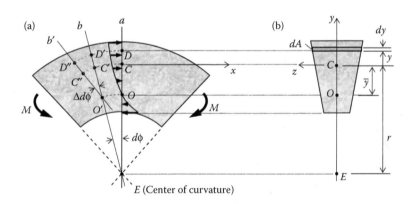

FIGURE 4.10 Curved beam subjected to bending moments M shown in (a) with cross section shown in (b).

principal axis of inertia. It follows, therefore, that the z centroidal axis is also a principal axis of inertia. The initial *radius of curvature* for the curved beam, measured from the centroidal axis of the cross section is denoted by the symbol r. The center of curvature for the curved beam is located at point E.

The curved beam is subjected to the pure bending moments M, which are assumed to act in the plane of symmetry. *Note that bending moments that tend to decrease the radius of curvature are, by convention, positive moments.* Thus, the moments in Figure 4.10 are positive. Except for the fact that the beam is not initially straight, the fundamental assumptions used in deriving the equation for the circumferential stress in the curved beam are the same as those used in deriving the bending equation (Equation 3.23) for straight beams in Chapter 3. Thus, consider two adjacent cross sections (radial planes) a and b in Figure 4.10a separated by the differential angle $d\phi$. Using the assumption that plane sections before bending remain plane after bending, plane section b remains plane but rotates about the axis of *zero stress* represented by point O, into position b' through the small angle $\Delta d\phi$. The circumferential deformation $C'C''$ at the centroidal axis may be expressed in terms of ε_c, the circumferential strain at this axis by the relation

$$C'C'' = \varepsilon_c(CC') = \varepsilon_c(rd\phi) \tag{4.40}$$

The circumferential strain ε at any point D a distance y from the centroidal axis then becomes

$$\varepsilon = \frac{D'D''}{DD'} = \frac{C'C'' + y(\Delta d\phi)}{DD'} = \frac{\varepsilon_c(rd\phi) + y(\Delta d\phi)}{(r + y)d\phi} = \frac{\varepsilon_c r + y(\Delta d\phi/d\phi)}{(r + y)} \tag{4.41}$$

Equation 4.41 may be written in a more convenient form by performing some mathematical manipulations. Thus,

$$\varepsilon = \frac{\varepsilon_c r}{(r + y)} + \frac{y}{(r + y)}\left(\frac{\Delta d\phi}{d\phi}\right) = \frac{\varepsilon_c(r + y - y)}{(r + y)} + \frac{y}{(r + y)}\left(\frac{\Delta d\phi}{d\phi}\right)$$

$$= \frac{\varepsilon_c(r + y)}{(r + y)} - \frac{\varepsilon_c y}{(r + y)} + \frac{y}{(r + y)}\left(\frac{\Delta d\phi}{d\phi}\right)$$

Rearranging and combining terms, we obtain

$$\varepsilon = \varepsilon_c + \left(\frac{y}{r + y}\right)\left(\frac{\Delta d\phi}{d\phi} - \varepsilon_c\right) \tag{4.42}$$

If the material obeys Hooke's law, the circumferential stress at any point D a distance y from the centroidal axis may be expressed in terms of the circumferential strain given in Equation 4.42. Hence,

$$\sigma = E\left[\varepsilon_c + \left(\frac{y}{r + y}\right)\left(\frac{\Delta d\phi}{d\phi} - \varepsilon_c\right)\right] \tag{4.43}$$

Equation 4.43 reveals that the circumferential stress in a curved beam is not directly proportional to the distance y measured from the centroidal axis, as is the case with straight beams. Its exact variation with the coordinate y will be determined after finding the two unknown quantities ε_c and $\Delta d\phi/d\phi$. These two unknown quantities are determined using two conditions of equilibrium. The first condition follows from the fact that the curved beam is subjected to pure bending moments. Therefore, the circumferential stresses over a given cross section result in circumferential forces that are self-equilibrating. Thus,

$$\int \sigma \, dA = 0 \tag{4.44}$$

Substituting from Equation 4.43 and rearranging terms, we obtain

$$
\left.
\begin{aligned}
\varepsilon_c A &= -\left(\frac{\Delta d\phi}{d\phi} - \varepsilon_c\right)\int\left(\frac{y}{r+y}\right)dA; \\
\varepsilon_c &= \left(\frac{\Delta d\phi}{d\phi} - \varepsilon_c\right)\left(-\frac{1}{A}\int\left(\frac{y}{r+y}\right)dA\right) = \left(\frac{\Delta d\phi}{d\phi} - \varepsilon_c\right)K
\end{aligned}
\right\}
\tag{4.45}
$$

where

$$
K = -\frac{1}{A}\int\left(\frac{y}{r+y}\right)dA
\tag{4.46}
$$

and represents a dimensionless property of the cross-sectional area of the curved beam.

The second condition of equilibrium is given by the fact that the resisting moment is equal to the applied moment. Therefore,

$$
M = \int \sigma y\, dA
\tag{4.47}
$$

Substituting for σ from Equation 4.43 yields

$$
M = E\left[\varepsilon_c \int y\, dA + \left(\frac{\Delta d\phi}{d\phi} - \varepsilon_c\right)\int\left(\frac{y^2}{r+y}\right)dA\right]
\tag{4.48}
$$

Since y is measured from the centroidal axis, we conclude that $\int y\, dA = 0$. Also, the quantity $\int [y^2/(r+y)]dA$ may be expressed in terms of the property K as follows:

$$
\int\left(\frac{y^2}{r+y}\right)dA = \int\left(\frac{y^2 + ry - ry}{r+y}\right)dA = \int\left(\frac{y(r+y) - ry}{r+y}\right)dA = \int y\, dA - r\left(\frac{A}{A}\right)\int\left(\frac{y}{r+y}\right)dA = KAr
\tag{4.49}
$$

Substituting from Equation 4.49 into Equation 4.48, we obtain

$$
M = \left(\frac{\Delta d\phi}{d\phi} - \varepsilon_c\right)EKAr
\tag{4.50}
$$

A simultaneous solution of Equations 4.45 and (4.50) yields the values of the unknown quantities $\Delta d\phi/d\phi$ and ε_c. Thus,

$$
\frac{\Delta d\phi}{d\phi} = \frac{M}{EAr}\left(1 + \frac{1}{K}\right); \quad \varepsilon_c = \frac{M}{EAr}
\tag{4.51}
$$

Returning now to Equation (4.43) and replacing $\Delta d\phi/d\phi$ and ε_c by their values from Equations 4.51, we obtain the equation for the circumferential stress in the curved beam. Thus,

$$
\sigma = \frac{M}{Ar}\left[1 + \frac{y}{K(r+y)}\right]
\tag{4.52}
$$

A sketch of the stress distribution given by Equation 4.52 is shown in Figure 4.10. Note that point O, representing the axis of zero stress, is located at a distance $\bar{y}$ from the centroidal axis. This distance may be determined, for a pure bending condition, from Equation 4.46 by setting $\sigma = 0$. Doing this leads to

$$\bar{y} = -\left(\frac{Kr}{K+1}\right) \tag{4.53}$$

The negative sign in Equation 4.53 signifies that the location of point O is toward (not away from) the center of curvature of the curved beam from the centroid of the cross section.

Determination of the factor K is illustrated in Example 4.9 for a rectangular cross section. Numerical values of K for use in the solution of problems are provided within the problem statements. Example 4.9 illustrates how Equation 4.46 may be used to determine the quantity K and Example 4.10 shows how to find the circumferential stresses at various locations in a curved beam subjected to the combined action of axial and bending loads.

EXAMPLE 4.9

A curved beam has a rectangular cross-sectional area as shown. Use the basic definition of K (Equation 4.46) to find its value in terms of the given dimensional quantities.

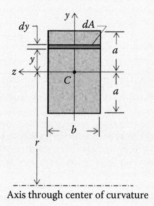

Axis through center of curvature

SOLUTION

A differential element of area $dA = b\,dy$, as shown in the sketch, needs to be defined so that the integral in Equation 4.46 can be evaluated. Thus,

$$K = -\frac{1}{A}\int\left(\frac{y}{r+y}\right)dA = -\frac{1}{2ba}\int_{-a}^{+a}\left(\frac{y}{r+y}\right)b\,dy$$

$$= -\frac{1}{2a}\int_{-a}^{+a}\left(\frac{y}{r+y}\right)dy = -\frac{1}{2a}\left[\int_{-a}^{+a}\left(\frac{r+y-r}{r+y}\right)dy\right] = -\frac{1}{2a}\left[\int_{-a}^{+a}dy - \int_{-a}^{+a}\left(\frac{r}{r+y}\right)dy\right]$$

$$K = -\frac{1}{2a}[y - r\ln(r+y)]_{-a}^{+a}$$

$$K = -1 + \left(\frac{r}{2a}\right)\ln\left[\frac{r+a}{r-a}\right] \qquad \textbf{ANS.}$$

EXAMPLE 4.10

A curved beam is fixed at one end and subjected to a load P at the other end as shown. The cross-sectional area of the beam is a 2 in. × 4 in. rectangle. Determine the value of the dimension d, measured from the centroid of the rectangular section, so that the maximum tensile and maximum compressive circumferential stresses on section n–n do not exceed 5 and 15 ksi, respectively. Determine also the location of the point of zero stress at section n–n for the value found for the distance d. Let $P = 8$ kips, $r = 4$ in., $a = 2$ in., and $b = 2$ in.

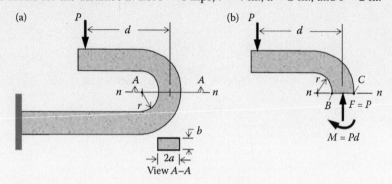

SOLUTION

The value of K is found using the equation developed in Example 4.9. Thus,

$$K = -1 + \left(\frac{r}{2a}\right)\ln\left[\frac{r+a}{r-a}\right] = -1 + \left(\frac{6}{4}\right)\ln\left[\frac{6+2}{6-2}\right] = 0.03972$$

Note that the value of K is, in general, a very small quantity, and care must be exercised to obtain an accurate value.

The free-body diagram of that portion of the curved beam above section n–n is shown in figure (b) above. Note that at this section, the curved beam is subjected not only to a positive bending moment, $M = Pd$, resulting in circumferential stresses given by Equation 4.52, but also to a compressive axial force, $F = P$, leading to a compressive stress system that, for all practical purposes, may be assumed uniform. Therefore, for any position at this section of the curved beam, the normal stress is obtained by the superposition of the two stresses. Thus,

$$\sigma = \frac{F}{A} + \frac{M}{Ar}\left[1 + \frac{y}{K(r+y)}\right]$$

Consider point B on the inner surface of the curved beam. This point is subjected to compressive stresses due to both F and M. Therefore,

$$\sigma_B = -15 = -\frac{8}{2(4)} - \frac{8d}{2(4)\times 6}\left[1 + \frac{-2}{0.03972(6-2)}\right]$$

Solution of this equation yields $d = 7.25$ in.

Now let us consider point C on the outer surface of the curved beam. This point is subjected to a compressive stress due to F and to a tensile stress due to M. Thus,

$$\sigma_C = 5 = -\frac{8}{2(4)} + \frac{8d}{2(4)\times 6}\left[1 + \frac{2}{0.03972(6+2)}\right]$$

Solution of this equation yields $d = 4.94$ in.

Therefore, the tensile stress is the controlling factor and the maximum permissible distance d is

$$d = 4.94 \text{ in.} \quad \textbf{ANS.}$$

The point of zero stress in section $n-n$ is located by finding the distance y at which the algebraic sum of the axial and bending stresses vanishes. Thus,

$$-\frac{8}{2(4)} - \frac{8(4.94)}{2(4) \times 6}\left[1 + \frac{-y}{0.03972(6 - y)}\right] = 0$$

The solution of this equation for y yields

$$y = 0.052 \text{ in.} \quad \textbf{ANS.}$$

PROBLEMS

4.82 A curved beam with a rectangular cross section is subjected to a 5-kip load as shown. Determine the maximum tensile and maximum compressive circumferential stresses in the member specifying their locations. Use the equation derived in Example 4.9 to find the value of K.

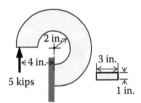

(Problems 4.82 and 4.83)

4.83 At the location in the curved beam considered in the solution of Problem 4.82, construct a complete circumferential stress distribution, computing the stresses at 1/2-in. intervals and accurately locating the point of zero stress.

4.84 Consider the curved member shown. If the allowable tensile and compressive stresses are, respectively, 150 and 200 MPa, determine the largest magnitude of the load P that the member may safely carry. The member has a solid circular cross section (not shown) 100 mm in diameter for which the value of $K = 0.0718$.

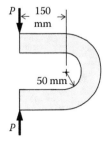

4.85 Repeat Problem 4.84 if the member has a trapezoidal cross section, as shown in the figure below, with $K = 0.1588$.

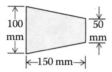

4.86 A curved beam has an isosceles triangular cross section as shown. Compute the maximum tensile and maximum compressive stresses (a) at an infinitesimal distance above section a–a and (b) at any section such as b–b in the straight part of the member. Use $K = 0.0583$.

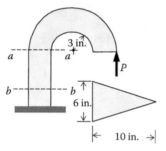

4.87 The hook has a rated capacity of 20 kN and has the cross section shown in figure (a) below for which $K = 0.0886$. Compute the maximum tensile and maximum compressive stresses when the hook is subjected to its maximum rated capacity load W. Let $a = 100$ mm, $b = 50$ mm, $c = 30$ mm, $d = 60$ mm, and $R = 80$ mm.

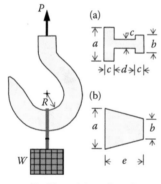

(Problems 4.87 and 4.88)

4.88 Repeat Problem 4.87 if the cross section is the trapezoid shown in figure (b) above. Let $a = 100$ mm, $b = 50$ mm, $e = 120$ mm, and $K = 0.1306$.

4.89 A clamp, shown schematically, has the given T cross section. The material is such that the allowable tensile and compressive stresses are, respectively, 5 and 15 ksi. Determine the maximum capacity (i.e., maximum permissible P) of the clamp. Use $K = 0.1102$.

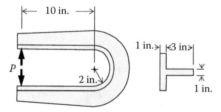

4.6   ELASTOPLASTIC BEHAVIOR: PLASTIC HINGE

4.6.1   SHAPE FACTOR

There are four basic assumptions made concerning the *inelastic behavior* of bending members. These assumptions are stated below:

1. The stress–strain diagram for the material is idealized to consist of two straight lines as shown in Figure 4.11. Members fabricated of such a material would, upon loading, have internal normal stresses defined by

$$\sigma = E\varepsilon \ldots (0 \leq \varepsilon \leq \varepsilon_y) \tag{4.54}$$

$$\sigma = \sigma_y \ldots (\varepsilon \geq \varepsilon_y) \tag{4.55}$$

It is important to note that regardless of the value of the strain, ε, after it reaches ε_y, the stress σ remains constant at σ_y. In other words, there is an infinite number of strains corresponding to the yield strength of the material. This is a convenient simplification of ductile material behavior that ignores *strain hardening.*

2. The tensile and compressive properties of the material are assumed to be identical.
3. The cross section of the beam has a vertical axis of symmetry defining the longitudinal plane of the loads.
4. The beam is subjected to pure bending moments.

With these four assumptions, the behavior of a bending member in the inelastic range is greatly simplified. Thus, consider the segment of a beam of length Δx shown in Figure 4.12. Figure 4.12a shows the bending stress distribution for the case where the bending moment prduces a maximum

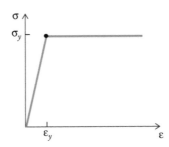

FIGURE 4.11   Stress–strain diagram for a material consisting of two straight lines where σ_y remains at a constant value after yielding.

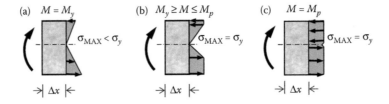

FIGURE 4.12   (a) The stress distribution when the maximum stress is equal to σ_y. (b) The stress distribution when the top and bottom fibers of the section have yielded through some depth. (c) The stress distribution when the entire cross section has yielded.

bending stress on the extreme beam fibers equal to the yield stress σ_y. This bending moment is known as the *yield moment* and given the symbol M_y, whose magnitude may be determined from Equation 3.22 and 3.23, $M_y = \sigma_y\,(I/c)$, where I is the moment of inertia of the cross section about its neutral axis and c is the distance from this axis to the farthest point where yielding first occurs. Figure 4.12b indicates the bending stress distribution when the bending moment has exceeded M_y but is still less than M_p, which represents the moment needed to produce yielding in the entire cross section. The distribution shown in Figure 4.12b indicates that the fibers on both top and bottom of the beam have yielded and a certain amount of plastic penetration has taken place on both sides (top and bottom) of the beam. Finally, Figure 4.12c shows the bending stress distribution corresponding to the condition when the entire cross section has yielded and the stresses over the entire cross section are at the yield value σ_y. At this stage, the bending moment acting on the beam is M_p, known as the *fully plastic moment*. This same distribution is repeated in Figure 4.13a together with the forces F_c and F_t, which represent the resultants of the compressive and tensile stress distributions, respectively. Figure 4.13b shows the cross section of the beam that possesses a vertical axis of symmetry. Note that because of this vertical symmetry, the cross section is divided into two areas, A_c and A_t, by a horizontal neutral axis known as the *plastic neutral axis* (PNA). Note also, that since the beam is subjected to a pure bending moment, the two forces $F_c = \sigma_y A_c$ and $F_t = \sigma_y A_t$ must be equal and opposite in order to have equilibrium in the x direction. Thus, we conclude that

$$A_c = A_t \tag{4.56}$$

Equation 4.56 emphasizes the fact that the *PNA* divides the cross-sectional area into two equal parts. In general, the *PNA* need not pass through the centroid of the area, which is the case for elastic bending, and thus there is a shifting of the neutral axis as the bending moment is increased. Equilibrium of moments about the *PNA* yields

$$\sum M_{PNA} = 0 \Rightarrow \sigma_y A_c z_c + \sigma_y A_t z_t - M_p = 0$$

Solving for M_p, we obtain

$$M_p = \sigma_y (A_c \bar{y}_c + A_t \bar{y}_t) \tag{4.57}$$

The *shape factor f*, which is a cross-sectional property, is defined as the ratio of M_p to M_y. Thus,

$$f = \frac{M_p}{M_y} = \frac{A_c \bar{y}_c + A_t \bar{y}_t}{I/c} \tag{4.58}$$

Thus, the shape factor f is a purely geometric property, which may be determined for any cross section.

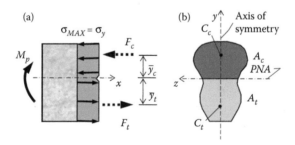

FIGURE 4.13   (a) A segment of a beam under the action of the fully plastic moment M_p. (b) The cross section of this beam.

4.6.2 Plastic Hinge

Let us consider a segment of a beam of unit length fabricated of a ductile material. If we subject this segment to increasing bending moments and plot the actual variation of the bending moment versus the curvature per unit length, $\phi = 1/\rho$, we obtain the curve shown by the broken lines in Figure 4.14. In the simplified theory, the idealized curve consisting of two straight lines is used to represent what is known as *plastic hinge* behavior. This behavior is best understood by reflecting on the behavior of a frictionless real hinge, shown by the red horizontal line in Figure 4.14 coincident with the horizontal ϕ axis. This, of course, indicates that a frictionless real hinge rotates freely with zero resistance to applied moments. On the other hand, the ideal plastic hinge, once created, allows the beam to rotate freely under a constant moment equal to the fully plastic moment M_p.

As an example of the above behavior, let us examine what happens to the simply supported beam shown in Figure 4.15, which is subjected to the uniform load p over the entire length. The maximum internal moment for such a beam occurs at midspan and, as the load increases, this moment is the first to reach the fully plastic value M_p under the action of the fully plastic load p_p. A larger load cannot be applied because, once p_p is reached, the beam will have hinges at three locations, two real hinges at its ends and a plastic hinge at its center. Under these conditions, the beam acts like a *mechanism* and will continue to deflect under the constant load, p_p. Since the third hinge required for the mechanism to form is not a real hinge, the beam mechanism is referred to as a *pseudo-mechanism* or a fictitious mechanism. Note that while the simple plastic theory assumes the plastic hinge to form at a single point, in reality, it extends over a finite length as shown by the green area in Figure 4.15.

A statically indeterminate beam has more unknown reaction components than can be determined by the available equations of equilibrium and knowledge of its deflection characteristics is needed to effect a solution. This type of solution will be discussed in detail in Chapter 5. However, in this chapter, we will deal with some of these beams and discuss their behavior under the idealized conditions introduced above. Consider, for example, the case of the beam fixed at both ends, loaded as shown in Figure 4.16. In such beams, the needed number of plastic hinges for a pseudo-mechanism, in general, do not form simultaneously but form first at sections where the internal

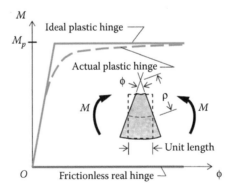

FIGURE 4.14 Diagram showing an idealized M_p versus ϕ curve and defining the plastic hinge.

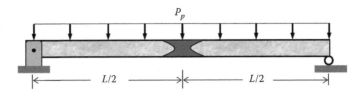

FIGURE 4.15 Simply-supported beam when the applied load reaches p_p and showing a real plastic hinge at its center.

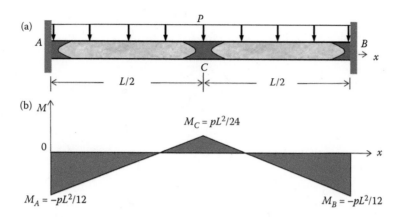

FIGURE 4.16 (a) A statically indeterminate beam with three plastic hinges before a pseudo-mechanism can form. (b) The corresponding moment diagram.

bending moments reach their maximum absolute values, and subsequently at other sections where the internal bending moments reach the fully plastic moment M_p. For any value of the uniform load $p < p_p$, the beam in Figure 4.16a has maximum moments at A and B as shown by the moment diagram of Figure 4.16b. The values given for the moments at A, B, and C are obtained by the methods discussed in Chapter 5. Thus, the first sections in the beam to reach the fully plastic condition are at the fixed supports at A and B. At this stage, however, the beam has only two plastic hinges that are not sufficient for a pseudo-mechanism. As the uniform load p is increased beyond that required for the two plastic hinges at A and B, moment resistance to increased loading must come from sections other than those at the two fixed ends where the moments have already reached the limiting value of M_p. The last section to reach the fully plastic condition, of course, is at C when the uniform load p reaches its ultimate value or the fully plastic load p_p. Once this condition is reached, there are three plastic hinges in the beam as shown in Figure 4.16a, a pseudo-mechanism has formed and no further increase in the load is possible. The progressive increase in moment resistance after the end plastic hinges have formed is referred to as *moment redistribution*.

EXAMPLE 4.11

Consider the cross section shown in the sketch and determine its shape factor for bending about a horizontal axis.

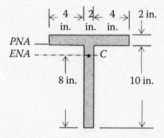

SOLUTION

The centroid C of the section is determined and the *ENA* is positioned at 8 in. above the base as shown. Also, the moment of inertia about this axis is found to be $I_{ENA} = 533.333$ in.[4].

Since the *PNA* divides the area of the cross section into two equal parts, it may be located by inspection at the junction between the flange and the web as shown. Assuming a positive bending moment, the area above the *PNA* is in compression and that below it is in tension. Thus, $A_c = A_t = 2(10) = 20$ in.^{2}. The shape factor is then determined by Equation 4.53 as follows:

$$f = \frac{M_p}{M_y} = \frac{A_c \bar{y}_c + A_t \bar{y}_t}{I/c} = \frac{20(1) + 20(5)}{533.333/8} = 1.80 \qquad \textbf{ANS.}$$

EXAMPLE 4.12

A fixed-ended beam is subjected to a uniform load p over the entire span as shown in figure (a) below. Find the load intensity p_y required for initiation of yielding and the load intensity p_p needed for complete collapse of the beam. The cross section of the beam is that of Example 4.11. Let $\sigma_y = 50$ ksi and $L = 30$ ft.

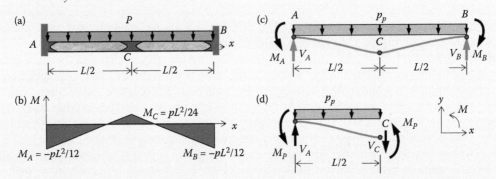

SOLUTION

The elastic moment diagram for the beam is determined using the methods in Chapter 5 and constructed as shown in figure (b) above. It is evident from this diagram that the maximum moments exist at the fixed ends and, therefore, first yielding takes place at these two locations. Thus, using the section properties found in Example 4.11, we conclude, by Equation 3.11, that

$$M_y = \frac{\sigma_y I_{ENA}}{c} = \frac{50(533.333)}{8} = 3333.3 \text{ kip} \cdot \text{in.} = 277.8 \text{ kip} \cdot \text{ft}$$

Therefore, using the given moment expression at the fixed ends, we conclude that at first yielding

$$\frac{p_y L^2}{12} = 277.8 \Rightarrow p_y = \frac{12(277.8)}{30^2} = 3.7 \text{ kips/ft} \qquad \textbf{ANS.}$$

Also,

$$M_p = f M_y = 1.80 M_y = 500 \text{ kip} \cdot \text{ft}$$

Consider now the free-body diagram of the entire beam shown in figure (c) above where plastic hinges are idealized as points. Symmetry tells us that $V_A = V_B$. Thus, summing forces in the y direction, we obtain

$$\sum F_y = 0 \Rightarrow 2V_A - p_pL = 0 \tag{4.12.1}$$

The free-body diagram for the left half of the beam is shown in figure (d) above. Summing moments about point C yields

$$\sum M_C = 0 \Rightarrow 2M_p + p_p\left(\frac{L}{2}\right)\left(\frac{L}{4}\right)$$

$$-V_A\left(\frac{L}{2}\right) = 0 \tag{4.12.2}$$

Solving Equation 4.12.1 for V_A, we obtain

$$V_A = \frac{p_pL}{2} \tag{4.12.3}$$

Substituting Equation 4.12.3 into Equation 4.12.2 and solving yields

$$p_p = \frac{16M_p}{L^2} = \frac{16(500)}{30^2} = 8.8 \text{ kips/ft} \qquad \textbf{ANS.}$$

EXAMPLE 4.13

A two-span continuous beam is subjected to proportional loading as shown in figure (a) below. Find the value of p_p and the constant of proportionality K such that pseudo-mechanisms ABC and CDE form simultaneously. Let $M_p = 50$ kN · m and $L = 4$ m.

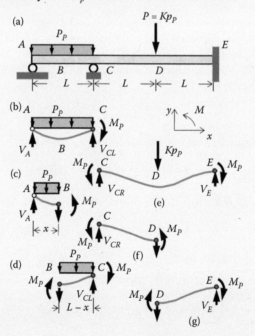

SOLUTION

Span ABC

Free-body diagram shown in figure (b) above:

$$\sum F_y = 0 \Rightarrow V_A + V_{CL} - p_p L = 0... \tag{4.13.1}$$

Free-body diagram shown in figure (c) above:

$$\sum M_B = 0 \Rightarrow M_p + p_p\left(\frac{x^2}{2}\right) - V_A x = 0$$

$$V_A = \left(\frac{M_p}{x}\right) + p_p\left(\frac{x}{2}\right)... \tag{4.13.2}$$

Free-body diagram shown in figure (d) above:

$$\sum M_B = 0 \Rightarrow 2M_p + p_p\frac{(L-x)^2}{2}$$

$$-V_{CL}(L-x) = 0$$

$$V_{CL} = \frac{2M_p}{(L-x)} + \frac{p_p(L-x)}{2}... \tag{4.13.3}$$

Substituting Equations 4.13.2 and 4.13.3 into Equation 4.13.1 and solving for p_p, we obtain

$$p_p = \frac{2M_p}{L}\left[\frac{L+x}{x(L-x)}\right]... \tag{4.13.4}$$

For each value of x in the interval between zero and L, the load intensity p_p will have a different value as shown by Equation 4.13.4. A mechanism will form first under the smallest of this infinite set of p_p values. To determine this value, we differentiate Equation 4.13.4 with respect to x and set the resulting function equal to zero. After simplification, this operation yields a quadratic equation in x. Thus,

$$x^2 + 2Lx - L^2 = 0$$

Solving this quadratic equation, we obtain

$$x = 0.414L = 0.414(4) = 1.656 \text{ m}$$

Substituting this value of x into Equation 4.13.4 with $M_p = 50$ kN $\cdot$ m and $L = 4$ m, we obtain

$$p_p = 36.428 \approx 36.4 \text{ kN/m} \qquad \textbf{ANS.}$$

Span CDE

Free-body diagram shown in figure (e) above:

$$\sum F_y = 0 \Rightarrow V_{CR} + V_E - Kp_p = 0... \tag{4.13.5}$$

Free-body diagram shown in figure (f) above:

$$\sum M_D = 0 \Rightarrow 2M_p - V_{CR}L = 0$$

$$V_{CR} = \frac{2M_p}{L}\cdots$$

(4.13.6)

Free-body diagram shown in figure (g) above:

$$\sum M_D = 0 \Rightarrow V_E L - 2M_p = 0$$

$$V_E = \frac{2M_p}{L}\cdots$$

(4.13.7)

Substituting Equations 4.13.6 and 4.13.7 into Equation 4.13.5 and solving for K, we obtain

$$K = \frac{4M_P}{Lp_p} = \frac{4(50)}{4(36.428)} = 1.37 \approx 1.4 \qquad \textbf{ANS.}$$

PROBLEMS

4.90 Find the shape factor for the cross section shown if bending is about a horizontal axis. If the cross section is that for a beam made of a material for which $\sigma_y = 36$ ksi, determine the yield moment M_y and the fully plastic moment M_p. Let $b = 10.0$ in.

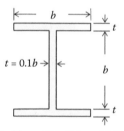

(Problems 4.90, 4.91, and 4.92)

4.91 The cross section of Problem 4.90 is that for a simply supported beam 18 ft long subjected to a concentrated load P at midspan. Find the yield load P_y and the fully plastic collapse load P_p.

4.92 The cross section of Problem 4.90 is that for a fixed-ended beam 24 ft long subjected to a concentrated load P at midspan. Find the yield load P_y and the fully plastic collapse load P_p. Given that, up to first yield, the bending moment and shearing force at one of the fixed supports are $PL/8$ and $P/2$, respectively.

4.93 Find the shape factor for the cross section shown if bending is about a horizontal axis. If the cross section is that for a beam made of a material for which $\sigma_y = 100$ MPa, determine the yield moment M_y and the fully plastic moment M_p. Let $b = 160$ mm.

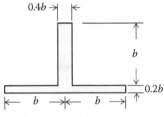

(Problems 4.93, 4.94, and 4.95)

4.94 The cross section of Problem 4.93 is that for a simply supported beam 6 m long subjected to a uniform load p over the entire span. Determine the yield load intensity p_y and the fully plastic collapse load intensity p_p.

4.95 The cross section of Problem 4.93 is that for a fixed-ended beam 8 m long subjected to a uniform load p over the entire span. Find the yield load intensity p_y and the fully plastic collapse load intensity p_p. Use the solution of Example 4.12 as a guide.

4.96 Find the shape factor for the cross section shown for bending about a horizontal axis. If the cross section is that for a beam made of a material for which $\sigma_y = 20$ ksi, determine the yield moment M_y and the fully plastic moment M_p. Let $b = 2.0$ in.

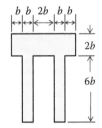

(Problems 4.96, 4.97, and 4.98)

4.97 The cross section of Problem 4.96 is that for a simply supported beam 21 ft long subjected to a concentrated load P at its third point. Find the yield load P_y and the fully plastic collapse load P_p.

4.98 The cross section of Problem 4.96 is that for a fixed-ended beam 27 ft long subjected to a concentrated load P at a third point. Find the yield load P_y and the fully plastic collapse load P_p. Given that, up to first yield, the bending moment and shearing force at the fixed support closest to the load are $4PL/27$ and $20P/27$, respectively.

4.99 Find the shape factor for the hollow square cross section shown for bending about a horizontal axis. If the cross section is that for a beam made of a material for which $\sigma_y = 100$ MPa, determine the yield moment M_y and the fully plastic moment M_p. Let $b = 20$ mm.

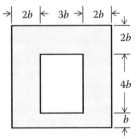

(Problems 4.99, 4.100, and 4.101)

4.100 The cross section of Problem 4.99 is that for a simply supported beam 7 m long subjected to a uniform load p over the entire span. Determine the yield load intensity p_y and the fully plastic collapse load intensity p_p.

4.101 The cross section of Problem 4.99 is that for a fixed-ended beam 9 m long subjected to a uniform load p over the entire span. Find the yield load intensity p_y and the fully plastic collapse load intensity p_p. Use the solution of Example 4.12 as a guide.

4.102 Find the shape factor for the cross section shown if bending is about a horizontal axis. If the cross section is that for a beam made of a material for which $\sigma_y = 20$ ksi, determine the yield moment M_y and the fully plastic moment M_p. Let $b = 3.0$ in.

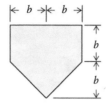

(Problems 4.102, 4.103, and 4.104)

4.103 The cross section of Problem 4.102 is that for a simply supported beam 16 ft long subjected to a concentrated load P at its third point. Find the yield load P_y and the fully plastic collapse load P_p.

4.104 The cross section of Problem 4.102 is that for a fixed-ended beam 21 ft long subjected to a concentrated load P at its third point. Find the yield load P_y and the fully plastic collapse load P_p. Given that, up to the first yield, the bending moment and shearing force at the fixed support closet to the load are $4PL/27$ and $20P/27$, respectively.

4.105 The fixed-ended beam will form a pseudo-mechanism when plastic hinges form at A, B, and C. Determine the collapse load, P_p, if $M_p = 700$ kip $\cdot$ in. and $L = 30$ ft.

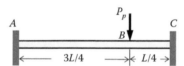

4.106 The two-span continuous beam is subjected to proportional loading as shown. If $M_p = 900$ kip $\cdot$ in., find the constant of proportionality K such that pseudo-mechanisms ABC and CDE form simultaneously.

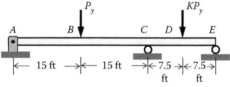

(Problems 4.106 and 4.107)

4.107 In reference to the figure shown in Problem 4.106, if $K = 1.5$, determine which of the two pseudo-mechanisms (ABC or CDE) forms first and the corresponding value of P_p.

4.108 Determine the magnitude of P_p required for the formation of pseudo-mechanism ABCD if My = 75 kN $\cdot$ m and $f = 1.10$. Let $a = b = 3$ m.

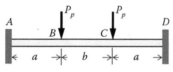

(Problems 4.108 and 4.109)

4.109 Repeat Problem 4.108 if $M_y = 1000$ kip · in., $f = 1.35$, $a = 12$ ft, and $b = 8$ ft.

4.110 The three-span continuous beam is subjected to proportional loading. Determine the proportionality constants K_1 and K_2 such that the three pseudo-mechanisms, ABC, CDE, and EFG, form simultaneously. Let $M_p = 60$ kN · m.

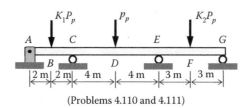

(Problems 4.110 and 4.111)

4.111 In reference to the figure shown in Problem 4.110, the three-span continuous beam is subjected to proportional loading. Let $K_1 = 2$ and $K_2 = 1.25$, determine which pseudo-mechanism (ABC, CDE, or EFG) forms first and the corresponding value of P_p. Let $M_p = 50$ kN · m.

4.112 The fixed-ended beam is subjected to proportional loading as shown. Find the proportionality constant K such that the pseudo-mechanisms ABC and CDE form simultaneously. Let $M_y = 1000$ kip · in., $f = 1.4$ and $L = 20$ ft.

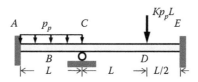

(Problems 4.112 and 4.113)

4.113 In reference to the figure shown in Problem 4.112, the fixed-ended beam is subjected to proportional loading as shown. Let $K = 0.5$, $M_p = 900$ kip · in., and $L = 15$ ft, determine which pseudo-mechanism forms first and the corresponding value of p_p.

4.114 A two-span continuous beam is subjected to uniform loads as shown. Find the ratio p_{p2}/p_{p1} such that pseudo-mechanisms ABC and CDE form simultaneously. Let $M_y = 600$ kip · in. and $f = 1.32$. Note that the plastic hinge at B does not form at the center of span ABC. Use Example 4.13 as guide.

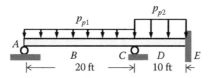

4.115 A three-span continuous beam is loaded as shown. Determine the ratio of L_2 to L_1 such that pseudo-mechanisms ABC, CDE, and EFG form simultaneously. Let $M_y = 50$ kN · m and $f = 1.28$. Note that the plastic hinges at B and F do not form at the centers of their respective spans. Use Example 4.13 as a guide.

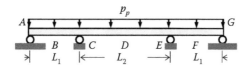

4.7 FATIGUE

Engineers have long recognized that subjecting a metallic member to a large number of stress cycles will result in fracture of the member at stresses much lower than those required for failure under static conditions. Poncelet in 1839 was probably the first to introduce the term *fatigue* and to discuss the property by which materials resist repeated cycles of stress. Modern investigators would probably use a term such as *progressive fracture* instead of the term *fatigue*.

Between 1852 and 1869, August Wohler, a German engineer, designed the first repeated-load testing machine. On the basis of his testing, Wohler came to two very interesting conclusions:

1. The number of cycles of stress rather than elapsed time is significant.
2. Ferrous materials stressed below a certain limiting value can withstand an indefinite number of stress cycles without failure.

The literature of fatigue studies is voluminous. These studies have proceeded along two lines: fundamental research seeking to explain the fatigue phenomenon and empirical investigations to provide information for practical analysis and design. Recent progress in *crack propagation* studies and the use of the electron microscope have enhanced fundamental research efforts, but empirical methods will continue to be widely employed for engineering purposes.

Fatigue fractures originate at points of stress concentrations, such as fillets, keyways, holes, and screw threads, or at points of internal inclusions or defects in the material. Cracks begin at these locations and then propagate through the cross section until the remaining uncracked regions are insufficient to resist the applied forces and fracture occurs suddenly.

The design of structural and machine components, where the loading is either fluctuating or repeated, must take fatigue action into consideration. For example, when traffic passes over a bridge, the structural components of the bridge experience loads that fluctuate above and below the dead loads that already exist. Also, the crankshaft of an automobile is subjected to repeated loading every time the automobile is driven.

An illustration of the type of fracture surface experienced in fatigue failures is shown in Figure 4.17, which represents the fatigue failure of a steering sector shaft of a late-model

FIGURE 4.17   Crack propagation in a fatigue failure of a steering sector shaft.

automobile. The fatigue fracture originated on the outside surface of the shaft (see bottom of figure), probably as a result of stress concentration at the root of the notch, and extended slowly almost through the entire cross section before final failure occurred toward the top of the figure. Note the *beach marks* that are more or less concentric with the final rupture zone. These beach marks represent the locations where the propagating crack was briefly arrested and then began to propagate again.

Depending upon the specific need, fatigue testing may be performed on specimens subjected to axial, torsional, bending, or combined loads. Space does not permit a discussion of all these tests but we will focus, in some detail, on the fatigue of members subjected to bending loads. *R. R. Moore* devised the rotating beam fatigue testing machine shown schematically in Figure 4.18a. The rotating specimen is subjected to a constant bending moment over the gage length as illustrated by the moment diagram of Figure 4.18b. There are several important variables related to fatigue test specimens that include size, shape, method of fabrication, and surface finish. Thus, to minimize the effects of stress concentration, the surface of the specimen is polished and the geometry is chosen to provide a gradual change in cross-sectional dimension on either side of the *critical section* located at its center. Care is also taken to remove, by annealing, any residual stresses induced in the specimen during fabrication.

In a given test, the choice of the weight W and the spacing, a, of the machine bearings determines the bending moment applied to the specimen as it rotates at a constant angular speed, ω. The critical cross section of the specimen is shown in 4.19a. At any point P on its circumference, the distance y perpendicular to the neutral axis (the z axis) varies with time according to the relation $y = R \sin \omega t$. Therefore, as the specimen rotates with the angular speed, ω, the bending stress at any point P varies according to the relation

$$\sigma = \frac{My}{I} = \frac{(Wa/2)R\sin \omega t}{\pi R^4/4} = \left(\frac{2Wa}{\pi R^3}\right)\sin \omega t$$

$$\sigma = \sigma_R \sin \omega t \tag{4.59}$$

where $\sigma_R = (2Wa/\pi R^3)$ is the amplitude of the stress that varies sinusoidally with time according the relation expressed in Equation 4.59. A sketch of this equation is given in Figure 4.19b. This variation between maximum tension $(+\sigma_R)$ and maximum compression $(-\sigma_R)$ is known as *complete stress reversal* and is the most severe form of stress variation. Provided that the stress σ_R is

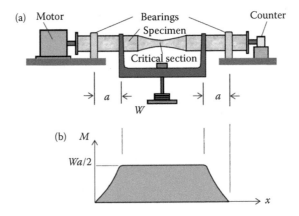

FIGURE 4.18 (a) An R. R. Moore fatigue testing machine. (b) The bending moment to which the rotating specimen is subjected throughout the test.

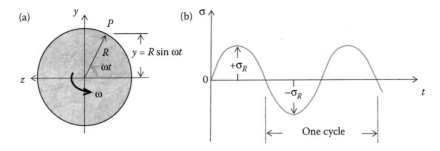

FIGURE 4.19 (a) The critical cross section of a fatigue specimen under test. (b) The variation of stress with time during a fatigue test.

not set too low, the specimen will experience fatigue failure after a certain number of cycles N that depends upon the applied stress level of σ_R. The testing machine stops automatically when the specimen fractures and the number of cycles N is read at the counter. This type of test is repeated on a large number of specimens with different levels of applied stress and the results are plotted on a $\sigma_R - N$ diagram. Actually, because of the large numbers involved, the test results are plotted on log–log paper to create a $\log\sigma_R - \log N$ diagram as shown for a cast alloy steel material in Figure 4.20.

Information obtained from diagrams such as that shown in Figure 4.20 forms the basis for the analysis and design of fatigue loadings. Fatigue testing machines are also available for axial, torsional, and combined loadings, and appropriate fatigue diagrams may be constructed for these other loadings. Let us refer now to the diagram in Figure 4.20 in order to introduce two definitions related to the fatigue phenomenon.

Endurance limit, σ_E: The *endurance limit* is the maximum stress that can be completely reversed an indefinite number of times without producing fatigue failure. For the material of Figure 4.20, the endurance limit is 40 ksi.

Endurance strength, σ_S: The *endurance strength* is the fatigue strength corresponding to a specified number of cycles. For example, for the material in Figure 4.20, if the specified number of cycles is 10^5, the fatigue strength is about 56.6 ksi.

Experimental results from fatigue tests do not lie perfectly on a straight line as may be erroneously surmised from Figure 4.20 but rather they scatter within a band indicated by the dashed lines

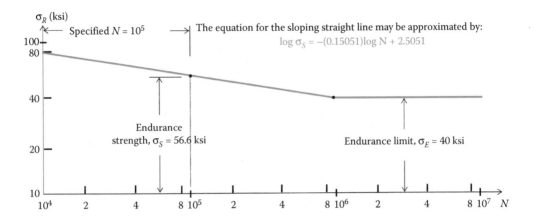

FIGURE 4.20 Log σ_R – log N diagram for a cast alloy steel under completely reversed bending.

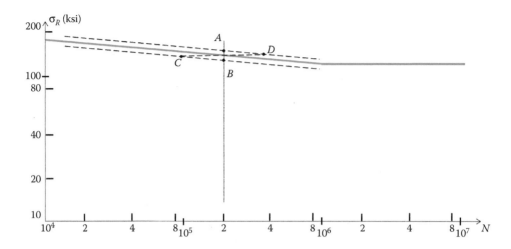

FIGURE 4.21   Log σ_R – log N diagram for an iron-base superalloy steel under completely reversed bending.

above and below the red line of Figure 4.21. This red line represents the best fit for the experimental data obtained for an iron-base superalloy for which the endurance limit is 130 ksi. An illustration of the effect of scatter is given in Figure 4.21. Stress levels for a given number of cycles can be stated with reasonable accuracy. For example, at $N = 2 \times 10^6$ cycles (average endurance strength of about 150 ksi), the scatter band is intersected at points A (endurance strength of about 160 ksi) and B (endurance strength of about 140 ksi), which are in the ratio of 8 to 7. However, for the same stress level of 150 ksi, we intersect the scatter band at points C (9×10^4 cycles) and D (3.7×10^5 cycles), which are in the ratio of about 1 to 4. These effects of fatigue data scatter are usually covered by a factor of safety, which is generally considerably larger than the factor of safety for static loadings.

As seen in Figures 4.20 and 4.21, the fatigue curves for steel level off at a large number of cycles, generally around 10^6 cycles. However, there are many nonferrous materials for which a continual decline in stress levels occurs with an increasing number of cycles. In such cases, engineers estimate the number of stress cycles that a given member will encounter during its lifetime and determine the endurance strength from the fatigue curves. Estimating the number of cycles during the useful life of a given member is not always a simple task, which is further complicated when a member is subjected to high-frequency vibrations during which the number of cycles can increase significantly in a very short time. Fatigue failures under such condition have been blamed for a number of jet and helicopter crashes.

In certain applications, the stress level varies sinusoidally with time between a maximum tensile value σ_{MAX} and a minimum tensile value σ_{MIN} as shown in Figure 4.22a. This stress variation may be decomposed into a static component σ_{AVE}, shown in Figure 4.22b, and a completely reversed sinusoidal variation σ_R, shown in Figure 4.22c. This type of decomposition is desirable because a large portion of the experimental fatigue data has been obtained for completely reversed cycling.

Three relationships between the static component and the completely reversed component are shown in Figure 4.23. These relationships form the basis for the three theories of fatigue failure shown in Figure 4.23: the *Gerber parabola*, the *Goodman straight line*, and the *Soderberg straight line* are drawn in an attempt to describe experimental fatigue behavior under combined static and completely reversed stresses. Note that when the static component is absent, all three theories converge at one point. Following are mathematical expressions for the three fatigue theories just stated:

Gerber parabola: The Gerber parabola may be stated in the following dimensionless mathematical form:

$$\frac{\sigma_R}{\sigma_E} = 1 - \left(\frac{\sigma_{AVE}}{\sigma_u} \right)^2 \tag{4.60}$$

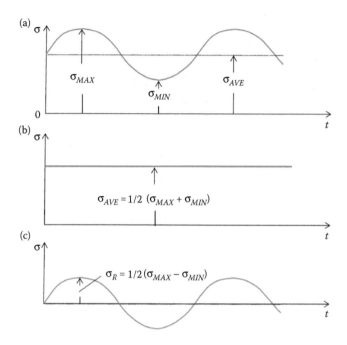

FIGURE 4.22 Decomposition of the stress variation in a fatigue test, into a static component, σ_{AVE}, and a completely reversed sinusoidal stress σ_R.

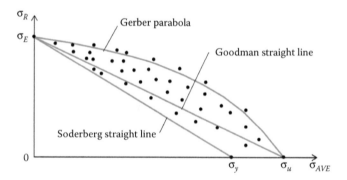

FIGURE 4.23 Relationship between σ_R and σ_{AVE} forming the basis for the three theories of fatigue failure, the Gerber parabola, the Goodman straight line, and the Soderberg straight line.

In practice, for design purposes, a factor of safety (*F.S.*) is added to Equation 4.60. This addition leads to

$$\frac{\sigma_R}{(\sigma_E/F.S.)} = 1 - \left(\frac{\sigma_{AVE}}{(\sigma_u/F.S.)}\right)^2 \qquad (4.61)$$

It should be noted here that the Gerber parabola has not been widely used in American engineering practice because it is not conservative with respect to a great deal of the experimental data.

Goodman straight line: A dimensionless mathematical expression for the Goodman straight line may be stated as follows:

$$\frac{\sigma_{AVE}}{\sigma_u} + \frac{\sigma_R}{\sigma_E} = 1 \qquad (4.62)$$

As in the case of the Gerber parabola, a factor of safety may be included for purposes of design. This yields

$$\frac{\sigma_{AVE}}{(\sigma_u/F.S.)} + \frac{\sigma_R}{(\sigma_E/F.S.)} = 1 \tag{4.63}$$

It should be observed that the Goodman straight line theory has been widely used in American engineering practice.

Soderberg straight line: The mathematical expression for the Soderberg straight line is similar to that of the Goodman straight line. Thus,

$$\frac{\sigma_{AVE}}{\sigma_y} + \frac{\sigma_R}{\sigma_E} = 1 \tag{4.64}$$

When a factor of safety is added, Equation (4.64) becomes

$$\frac{\sigma_{AVE}}{(\sigma_y/F.S.)} + \frac{\sigma_R}{(\sigma_E/F.S.)} = 1 \tag{4.65}$$

This fatigue theory provides a very conservative design approach because, unlike the Goodman straight line theory, it is based on the yield strength σ_y, instead of the ultimate strength σ_u for the static component of stress.

EXAMPLE 4.14

A cantilever machine component, of solid circular cross section with a diameter d and a length of 25 in., is to be fabricated of an iron-base superalloy (AISI grade 635) for which the endurance limit is 54 ksi. The component is to be subjected at the free end to a completely reversed loading of 2.0 ksi maximum value. Ignore stress concentrations and use a factor of safety of 3 to determine the required diameter of the member.

SOLUTION

Since the member is subjected only to a completely reversed loading (i.e., no static component), all three theories yield the same answer. Thus, choosing the Goodman straight line theory, Equation 4.63, we have

$$\frac{\sigma_R}{(\sigma_E/F.S.)} = 1; \quad \sigma_R = \frac{\sigma_E}{F.S.} = \frac{54}{3} = 18\,\text{ksi}$$

Now,

$$\sigma_R = \frac{My}{I}; \quad 18 = \frac{2(25)d/2}{\pi/64(d^4)}$$

Solving for d yields

$$d = 3.05 \approx 3.1\,\text{in.} \quad \textbf{ANS.}$$

EXAMPLE 4.15

The beam shown is fabricated of a nickel alloy (Monel 400) for which the endurance limit in bending is 289 MPa, the yield strength is 300 MPa, and the ultimate strength is 570 MPa. The thickness of the beam is uniform and equal to 50 mm. The stress-concentration factor due to the fillet at B is 1.3. The applied load ranges from a downward load P to an upward load 1/2 P as shown. Use factors of safety of 4 with respect to fatigue, 3 with respect to the ultimate strength, and 2 with respect to yielding and determine the maximum permissible load P using (a) the Gerber parabola, (b) the Goodman straight line, and (c) the Soderberg straight line.

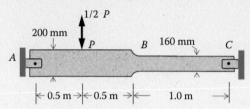

SOLUTION

The critical section is at B where the bending moment is 0.25P. The maximum stress that occurs during the downward part of the cycle at this location is

$$\sigma_{MAX} = k\left(\frac{My}{I}\right) = 1.3\left[\frac{0.25P(0.08)}{(1/12)(0.05)(0.16^3)}\right] = 1520P$$

Therefore, since the upward stroke is 1/2P, it follows that

$$\sigma_{MIN} = -760P$$

The static component and the completely reversed component of the stress become

$$\sigma_{AVE} = \frac{1}{2}(\sigma_{MAX} + \sigma_{MIN}) = 380P; \quad \sigma_R = \frac{1}{2}(\sigma_{MAX} + \sigma_{MIN}) = 1140P$$

a. The Gerber parabola containing a factor of safety is expressed by Equation 4.61. Thus,

$$\frac{\sigma_R}{(\sigma_E/F.S.)} = 1 - \left(\frac{\sigma_{AVE}}{(\sigma_u/F.S.)}\right)^2; \quad \frac{1140P}{(289 \times 10^6/4)} = 1 - \left(\frac{380P}{570 \times 10^6/3}\right)^2$$

Solving this quadratic equation yields

$$P = 62.4 \text{ kN} \qquad \textbf{ANS.}$$

b. The Goodman straight line containing a factor of safety is given by Equation 4.62. Thus,

$$\frac{\sigma_{AVE}}{(\sigma_u/F.S.)} + \frac{\sigma_R}{(\sigma_E/F.S.)} = 1; \quad \frac{380P}{(570 \times 10^6/3)} + \frac{1140P}{(289 \times 10^6/4)} = 1$$

The solution of the above equation leads to

$$P = 56.2 \text{ kN} \qquad \textbf{ANS.}$$

c. The Soderberg straight line with a factor of safety is expressed by Equation 4.65. Thus,

$$\frac{\sigma_{AVE}}{(\sigma_y/F.S.)} + \frac{\sigma_R}{(\sigma_E/F.S.)} = 1; \quad \frac{380P}{(300 \times 10^6/2)} + \frac{1140P}{(289 \times 10^6/4)} = 1$$

Solving this linear equation yields

$$P = 54.6 \text{ kN} \qquad \textbf{ANS.}$$

EXAMPLE 4.16

A shaft of solid circular cross section is to be fabricated of a material for which the torsion ultimate strength is 90 ksi and the torsion endurance limit is 24 ksi. It is to carry an alternating torque varying between a maximum of 4 kip · ft and a minimum of 1 kip · ft. A small keyway in the shaft provides for a stress-concentration factor of 1.3 but is sufficiently small to permit the use of the gross cross-sectional properties in the calculations. Use a factor of safety of 4 relative to the endurance limit and 3 relative to the ultimate strength and compute the least acceptable diameter of the shaft by the Goodman straight line theory.

SOLUTION

Since we are dealing with a torsion problem, the equations developed earlier are modified by replacing σ by τ. Thus, the maximum and minimum shearing stresses are computed as follows:

$$\tau_{MAX} = k\frac{T_{MAX}R}{J} = 1.3\frac{4000(12)(d/2)}{(\pi/32)d^4} = \frac{3.178 \times 10^5}{d^3}$$

$$\tau_{MIN} = \left(\frac{1}{4}\right)\tau_{MAX} = \frac{7.945 \times 10^4}{d^3}$$

Therefore,

$$\tau_{AVE} = \frac{1}{2}(\tau_{MAX} + \tau_{MIN}) = \frac{1.986 \times 10^5}{d^3}; \quad \tau_R = \frac{1}{2}(\tau_{MAX} - \tau_{MIN}) = \frac{1.192 \times 10^5}{d^3}$$

We now modify the Goodman straight line theory given by Equation 4.63. Thus,

$$\frac{\tau_{AVE}}{(\tau_u/F.S.)} + \frac{\tau_R}{(\tau_E/F.S.)} = 1; \quad \frac{1.986 \times 10^5}{d^3(90,000/3)} + \frac{1.192 \times 10^5}{d^3(24,000/4)} = 1$$

The solution of this equation for the diameter d yields

$$d = 2.98 \approx 3.0 \text{ in.} \qquad \textbf{ANS.}$$

EXAMPLE 4.17

A rod of circular of circular cross section, 65 mm in diameter, is to be subjected to alternating tensile forces that vary from a minimum of 160 kN to P_{MAX}. It is to be fabricated of a material with an ultimate tensile strength of 700 MPa and an endurance limit for complete stress reversal of 560 MPa. Use the Goodman straight line theory to find P_{MAX} if the factors of safety with respect to the ultimate strength is 3.0 and with respect to the endurance limit is 3.5. A small hole in the rod provides for a stress-concentration factor of 1.5 but is sufficiently small so that the gross cross-sectional area may be used in the calculations.

SOLUTION

The maximum and minimum tensile stresses are found as follows:

$$\sigma_{MAX} = k\frac{P_{MAX}}{A} = 1.5\frac{P_{MAX} \times 10^3}{(\pi/4)(0.065^2)} = 452038 P_{MAX}$$

$$\sigma_{MIN} = k\frac{P_{MIN}}{A} = 1.5\frac{160 \times 10^3}{(\pi/4)(0.065^2)} = 72.326 \text{ MPa}$$

Therefore,

$$\sigma_{AVE} = \frac{1}{2}(\sigma_{MAX} + \sigma_{MIN}) = 226019 P_{MAX} + 36.163 \times 10^6$$

$$\sigma_R = \frac{1}{2}(\sigma_{MAX} - \sigma_{MIN}) = 226019 P_{MAX} - 36.163 \times 10^6$$

We now apply the Goodman straight line theory given by Equation 4.63. Thus,

$$\frac{\sigma_{AVE}}{(\sigma_u/F.S.)} + \frac{\sigma_R}{(\sigma_E/F.S.)} = 1; \quad \frac{226019 P_{MAX} + 36.163 \times 10^6}{(700 \times 10^6/3.0)} + \frac{226019 P_{MAX} - 36.163 \times 10^6}{(560 \times 10^6/3.5)} = 1$$

Solving, we obtain

$$P_{MAX} = 450 \text{ kN} \qquad \textbf{ANS.}$$

PROBLEMS

4.116 A machine component, with a square cross section ($b \times b$), is to be subjected to alternating axial loading varying from a minimum of 80 kips to a maximum 380 kips. Using the Goodman straight line theory with a factor of safety of 4 with respect to fatigue and 3 with respect to the ultimate, determine the least acceptable value of b. The endurance limit for completely reversed axial fatigue is 60 ksi and the static ultimate strength is 130 ksi. Construct a scaled plot of the Goodman straight line showing the point that represents your solution. Ignore effects of stress concentration.

4.117 A simply-supported beam, 12 ft long, is to carry a downward alternating load, at 4 ft from one support, which varies from 4 kips to 14 kips. It is to have a rectangular cross section with a depth, d, four times its width, w. The following material properties are provided: endurance limit = 20 ksi and yield strength = 40 ksi. Using the Soderberg straight line theory with a factor of safety of 3 with respect fatigue and 2 with respect to yielding, determine the least acceptable cross-sectional dimensions w and d. Ignore effects of stress concentration.

4.118 A hollow shaft of circular cross section (inside diameter is 0.8 of outside diameter) is to be fabricated of a material for which the endurance limit is 165 MPa and the static ultimate strength is 620 MPa. It is to be subjected to an alternating torque that varies from a minimum of 1 kN · m to a maximum of 5 kN · m. Using the Goodman straight line theory with a factor of safety of 4 with respect fatigue and 3 with respect to the ultimate, determine the least acceptable diameters of the hollow section. A small keyway provides for a stress-concentration factor of 1.25 but is sufficiently small to allow use of the gross sectional properties.

4.119 A hollow shaft of circular cross section (inside diameter is 0.7 of outside diameter) is to be designed for a lifespan of 10^5 for which the material has an endurance strength of 34.63 ksi and a static torsional ultimate strength of 88 ksi. It is to be subjected to an alternating torque that varies from a maximum of 3 kip · ft to a minimum of −1 kip · ft. A small hole in the shaft provides for a stress-concentration factor of 1.35 but is small enough to allow the use of gross dimensions in the calculations. Assume the Goodman straight line theory is valid and use it with a factor of safety of 3.5 with respect to fatigue and 2.8 with respect to the static ultimate strength to compute the least acceptable diameters of the hollow section.

4.120 A 2-m-long cantilever beam is to be subjected, at its free end, to a downward alternating load that varies 1 to 12 kN. Use the Goodman straight line theory with a factor of safety of 4 relative to fatigue (endurance limit = 160 MPa) and 3 relative to the ultimate strength (ultimate strength = 440 MPa) to find the least acceptable dimension of a rectangular cross section whose depth, d, is four times its width, w. Ignore effects of stress concentration.

4.121 An 8-ft-long simply-supported beam, of hollow circular cross section (inside diameter is 0.7 of outside diameter) is to be designed for a lifespan of 4×10^5 cycles and is to be made of the material described in Figure 4.20 for which the static bending ultimate strength is 88 ksi. It is to be subjected at midspan to a downward alternating force that varies from a maximum of 10 kips to a minimum of 2 kips. Assume the Gerber parabolic theory is valid and use it with a factor of safety of 3.0 with respect to fatigue and 2.0 with respect to the static ultimate strength to compute the least acceptable diameters of the hollow section. Ignore effects of stress concentration.

4.122 A 10-ft-long cantilever beam with a rectangular cross section (depth, d three times width, w) is to be designed for a lifespan of 5×10^5 cycles and is to be made of the material described in Figure 4.20 for which the static bending ultimate strength is 95 ksi. It is to be subjected at the free end to a downward alternating force that varies from a maximum of 5 kips to a minimum of 2 kips. A small hole in the beam provides for a stress-concentration factor of 1.3 but is small enough to allow the use of gross dimensions in the calculations. Assume the Goodman straight line theory is valid and use it with a factor of safety of 4 with respect to fatigue and 3 with respect to the static ultimate strength to compute the least acceptable dimensions of the section.

4.123 A machine component with a rectangular cross section (depth, d twice the width, w) is to carry an alternating axial load that varies from a maximum of 400 kips to a minimum of 60 kips. The material of which the component is to be fabricated has an endurance limit of 40 ksi for completely reversed axial loading and a static ultimate strength of 90 ksi. Use the Goodman straight line theory with factors of safety of 4 relative to fatigue and 3 relative to the static ultimate and determine the least acceptable dimensions of the section. Apply a stress-concentration factor of 1.2.

4.124 Solve Problem 4.123 using the Soderberg straight line theory, if the yield strength is 60 ksi. Use a factor of safety of 2. Construct a scaled plot of the Soderberg straight line theory showing a point corresponding to your solution.

4.125 Solve Problem 4.123 using the Gerber parabolic theory. Construct a scaled plot of the Gerber parabolic theory showing the point corresponding to your solution theory showing a point corresponding to your solution.

4.126 A 49-mm-diameter steel rod is subjected to an alternating axial load that varies from a maximum of P_{MAX} to a minimum of 90 kN. The steel of which the rod is fabricated has a static ultimate strength of 560 MPa and an endurance limit for complete reversal of tensile loading of 300 MPa. Use the Goodman straight line theory with factors of safety of 3 relative to fatigue and 2 relative to the static ultimate and determine P_{MAX}. Neglect effects of stress concentration.

4.127 A 10-ft simply supported beam having a rectangular cross section (depth, d three times width, w) is subjected at its center to an alternating downward load that varies from zero to a maximum of 10 kips. The material of which the beam is made has an endurance limit of 20 ksi for complete bending reversal and a static ultimate strength of 60 ksi. Use the Goodman straight line theory with a factor of safety of 3 relative to fatigue and 2 relative to the static ultimate to find the least acceptable dimensions of the section. Neglect effects of stress concentration.

4.128 Redesign the beam of Problem 4.127 for a center alternating load that varies from 10 kips to 20 kips.

4.129 A rod of circular cross section with a diameter of 70 mm is to carry an alternating tensile axial load that varies from a minimum of $2P$ to a maximum of $5P$. It is to be fabricated of a material for which the ultimate strength is 900 MPa and the endurance limit is 700 MPa. Use the Goodman straight line theory with a factor of safety of 4 relative to fatigue and 3.5 relative to the ultimate strength and determine the largest permissible load P that may be applied. Use a stress-concentration factor of 1.65 and the gross cross-sectional area in the computations.

4.130 Repeat Problem 4.129 if the alternating tensile axial load varies from P to $31P$.

4.131 A hollow shaft of circular cross section (outside diameter, D_o is 1.6 times inside diameter, D_i) is to be fabricated of a material for which the endurance limit is 24 ksi and a static torsional ultimate strength of 88 ksi. It is to be subjected to an alternating torque that varies from a maximum of 2.5 kip · ft to a minimum of 0.5 kip · ft. A small hole in the shaft provides for a stress-concentration factor of 1.4 but is small enough to allow the use of gross dimensions in the calculations. Use the Goodman straight line theory with a factor of safety of 3.8 with respect to fatigue and 2.9 with respect to the static ultimate strength to compute the least acceptable diameters of the hollow section.

4.132 Repeat Problem 4.131 using the Gerber parabolic theory.

REVIEW PROBLEMS

R4.1 The beam shown in figure (a) below has the cross section shown in figure (b) bellow. This cross section consists of three rectangular components made of materials 1, 2, and 3 as shown. Material 1 is aluminum ($E_A = 10 \times 10^3$ ksi), material 2 is wood ($E_W = 2 \times 10^3$ ksi), and material 3 is steel ($E_S = 30 \times 10^3$ ksi). If $p = 3$ kips/ft, determine the maximum bending stresses in the three materials. Let $b = 5$ in., $h_1 = 1$ in., $h_2 = 12$ in., and $h_3 = 1/2$ in.

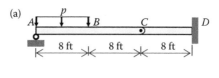

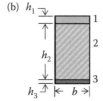

(Problems R4.1 and R4.2)

R4.2 In reference to the figure shown in Problem R4.1, the beam shown in figure (a) above has the cross section shown in figure (b) above. This cross section consists of three rectangular components made of materials 1, 2, and 3 as shown. Material 1 is aluminum ($E_A = 10 \times 10^3$ ksi), material 2 is wood ($E_W = 2 \times 10^3$ ksi), and material 3 is steel ($E_S = 30 \times 10^3$ ksi). If the allowable stresses in the three materials are $\sigma_A = 15$ ksi, $\sigma_W = 2$ ksi, and $\sigma_S = 24$ ksi, find the largest permissible magnitude of the load intensity p. Let $b = 5$ in., $h_1 = 1$ in., $h_2 = 12$ in., and $h_3 = 1/2$ in.

R4.3 The reinforced concrete beam shown in figure (a) below has the cross section shown in figure (b) below. Let, $n = 9$, $b = 150$ mm, $h_1 = 150$ mm, $h_2 = 200$ mm, $h_3 = 400$ mm, and $h_4 = 70$ mm and the total cross-sectional area of the reinforcing rods is 3.0×10^{-3} m². Find the maximum bending stresses in both materials. (*Hint:* Assume that the neutral axis intersects the cutout.)

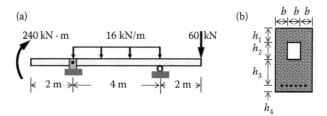

R4.4 The beam shown in figure (a) below has the cross section given in figure (b) below. Let $P = 30$ kips, $a = 2$ in., $h = 8$ in., and $b = 6$ in., determine the maximum tensile and maximum compressive bending stresses specifying their locations in the beam and in the cross section.

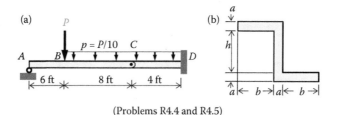

(Problems R4.4 and R4.5)

R4.5 In reference to the figure shown in Problem R4.4, the beam shown in figure (a) above has the cross section shown in figure (b) above. If the allowable bending stress (tension

or compression) is 40 ksi, determine the maximum permissible magnitude of the load
P. Let $a = 2$ in., $h = 8$ in., and $b = 6$ in.

R4.6 The beam shown in figure (a) below has the S20 × 75 section depicted in figure (b)
below where the loading is applied as indicated with $\phi = 7°$. If the distributed load $p = 3$
kips/ft and $a = 5$ ft, determine the maximum tensile and maximum compressive bending stresses in the beam specifying their locations in the beam and in the cross section.

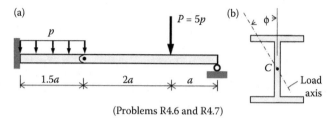

(Problems R4.6 and R4.7)

R4.7 In reference to the figure shown in Problem R4.6, the beam shown in figure (a) above
has the W457 × 68.4 section depicted in figure (b) above where the loading is applied
as indicated with $\phi = 5°$. If the allowable bending stress (tension or compression) is
150 MPa, determine the maximum permissible magnitude of the distributed load p. Let
$a = 2$ m.

R4.8 Assume loading normal to the horizontal plane of symmetry of the cross section and
determine the location of the shear center. Express your answer in terms of the dimension a.

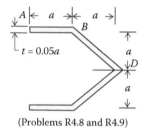

(Problems R4.8 and R4.9)

R4.9 In reference to the figure shown in Problem R4.8, the cross section shown is subjected
to a downward load of 2.5 kips at point B acting normal to the horizontal plane of symmetry of the section. Let $a = 5$ in., determine the maximum shearing stress produced
by the induced torque.

R4.10 Assume loading normal to the plane of symmetry of the cross section. (a) Locate
the shear center for the cross section. (b) If a downward load of 10 kN is applied
at point B, find the maximum shearing stress produced by the induced torque. Let
$a = 120$ mm.

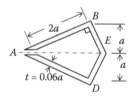

R4.11 A ring has a narrow slit and is subjected to the loads $P = 1.5$ kips as shown applied at
the centroid of its cross section, which is a solid square 2.0 in. on each side. If the inner
radius of the ring is 3.0 in., determine the stress at (a) point A and (b) point B.

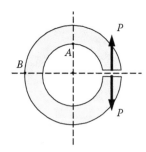

R4.12 Find the shape factor for the given cross section if bending is about a horizontal axis. If the cross section is that for a beam made of a material for which $\sigma_y = 35$ ksi, determine the yield moment M_y and the fully plastic moment M_p. Let $b = 10$ in.

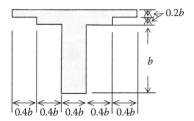

R4.13 The cross section of Problem R4.12 is that for a beam, 30 ft long, fixed at one end simply supported at the other and carrying a concentrated load P at midspan. Find the yield load P_y and the fully plastic load P_p. Given that, up to first yield, the bending moment and shearing force at the fixed support are $3PL/16$ and $11P/16$, respectively.

R4.14 The three-span continuous beam is subjected to proportional loading as shown. Let $K_1 = 1.5$ and $K_2 = 1.75$, determine which pseudo-mechanism (*ABC*, *CDE*, or *EFG*) forms first and the corresponding value of p_p. Note that the plastic hinge at B does not form at the center of span *ABC*. Use Example 4.13 as a guide. Let $M_p = 75$ kN · m and $L = 2$ m.

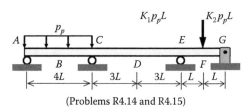

(Problems R4.14 and R4.15)

R4.15 In reference to the figure shown in Problem R4.14, the three-span continuous beam is subjected to proportional as shown. Find the proportionality constants K_1 and K_2 such that pseudo-mechanisms *ABC*, *CD*, and *DEF* form simultaneously. Note that the plastic hinge at B does not form at the center of span *ABC*. Use Example 4.13 as a guide. Let $M_p = 2000$ kip · in. and $L = 3$ ft.

R4.16 A 2.5-m-long cantilever beam is subjected at its free end to a downward alternating load that varies from P to $5P$. It is made of a material for which $\sigma_E = 180$ MPa and $\sigma_u = 360$ MPa. The cross section for the beam is a 50 mm × 170 mm rectangle. Use the Goodman straight line theory with a factor of safety of 3 relative to fatigue and 2 relative to the static ultimate strength and determine the maximum permissible magnitude of P. Ignore effects of stress concentration. The beam is bent about the strong axis.

R4.17 Repeat Problem R4.16 using the Gerber parabolic theory.

5 Bending Loads
Deflections under Symmetric Loading

5.1 INTRODUCTION

Beams are used extensively in many different types of structures, and their deflections become a significant design consideration in many applications. The general trend toward fabricating and constructing lighter and more flexible components that are safe from the standpoint of load-carrying capacity has led to a number of problems attributable, at least in part, to deflections that are too large to be tolerated either on physical or psychological grounds. Consider, for example, the case of a high-rise building. If deflections are not kept within acceptable limits, there may be breakage of windows and cracking of plaster in addition to possible psychological stress experienced by the occupants of the upper floors of such buildings even though failure is unlikely. As a second example, consider the case of shafts subjected to additional bending action. Because of large deflections, these shafts may become misaligned in their bearings, resulting in excessive wear and possible malfunction. Numerous other examples may be cited for maintaining beam deflections within certain limits.

In addition to the practical reasons stated above, knowledge of beam deflections forms the basis for the analysis of indeterminate beams to be discussed in Sections 5.7, 5.8, and 5.9. Modern machine and structural designs are more likely to contain indeterminate beams. As in the case of axial and torsional loads discussed in Chapters 1 and 2, the analysis of statically indeterminate beams requires that the available equations of equilibrium be supplemented by compatibility functions, which involve deflection calculations.

This chapter deals with a number of topics related to the deflection of beams. In Section 5.2, we develop the relationship between the bending moment at any point along the beam and the curvature of the beam at that point. This *moment–curvature relationship* forms the basis for the various methods used to determine beam deflections and slopes. Four such methods are used in this chapter. Each of these methods has advantages and disadvantages as will be seen when these methods are discussed. Thus, in Section 5.3, we deal with the method known as the *two successive* or the *double integration* method. Section 5.3 starts out with a second-order differential equation and by performing two integrations, we obtain both the slope and the deflection functions. Section 5.4, in addition to establishing a second method for beam deflections, which starts out at a fourth-order differential equation, deals with the derivatives of the deflection function and their physical meanings. In Section 5.5, we develop the third method of beam deflections using superposition that has already been used several times in preceding chapters. The fourth deflection method, the *area–moment* method, makes use of the moment diagram of the beam in determining its deflections and slopes and is discussed in Section 5.6.

The last three sections in this chapter deal with the solution of statically indeterminate beams. Thus, Section 5.7 makes use of the method of two successive integrations for this purpose; Section 5.8 employs the method of superposition in solving the statically indeterminate beam problem; finally, in Section 5.9, we make use of the area–moment method to find the solution for the statically indeterminate beam problem. Each of these methods has advantages and disadvantages, as will become clear in these last three sections.

5.2   MOMENT–CURVATURE RELATIONSHIP

The beam shown in Figure 5.1a is subjected to positive bending moments Q that lie in the vertical longitudinal principal plane of symmetry defined by the weak (y, v) centroidal principal axis, thus resulting in symmetric bending about the *strong axis* (z, u) of the beam. The beam deforms into the configuration shown by the broken lines. The trace of the deformed neutral surface is known as the *elastic curve*. The cross section of the beam, showing the two orthogonal principal centroidal axes of inertia (see Appendix C.2), is shown in Figure 5.1b.

Reference is now made to Equation 3.11, developed in Chapter 3 to relate the strain at any point in the beam to its position defined by the coordinate y and to the radius of curvature ρ of the elastic curve. This equation is repeated here for convenience. Thus,

$$(\varepsilon_x)_y = -\frac{y}{\rho}$$

(3.11, Repeated)

The *magnitude* of the maximum strain, ε_{MAX}, occurs at the farthest point from the neutral axis (i.e., at $y = c$). Thus,

$$\varepsilon_{MAX} = \frac{c}{\rho}$$

(5.1)

Note that the negative sign has been omitted in Equation 5.1 in order to deal exclusively with magnitudes of the quantities involved. Solving for the beam curvature, $1/\rho$, we obtain

$$\frac{1}{\rho} = \frac{\varepsilon_{MAX}}{c}$$

(5.2)

If we assume that the material obeys Hooke's law, then by Equation 1.11, we can write that $\varepsilon_{MAX} = \sigma_{MAX}/E$. Therefore, since by Equation 3.24 the *magnitude* of the maximum bending stress is $\sigma_{MAX} = M\,c/I$, we conclude that

$$\varepsilon_{MAX} = \frac{\sigma_{MAX}}{E} = \frac{Mc}{EI}$$

(5.3)

Solving for the quantity ε_{MAX}/c, we obtain

$$\frac{\varepsilon_{MAX}}{c} = \frac{M}{EI}$$

(5.4)

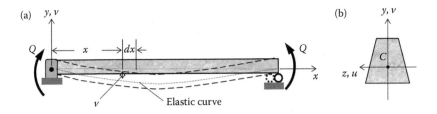

FIGURE 5.1   (a) A beam subjected to symmetric bending about the principal centroidal $z(u)$ axis. (b) The cross section of the beam with the two orthogonal principal centroidal $y(v)$ and $z(u)$ axes.

Combining Equations 5.2 and 5.4 leads to

$$\frac{1}{\rho} = \frac{M}{EI} \tag{5.5}$$

It is important to recall that, in Equation 5.5, both the moment M and the moment of inertia I must be found about the neutral axis for bending, which in this case is the $z(u)$ principal centroidal axis for the beam. Also, strictly speaking, Equation 5.5 is valid only for beams subjected to pure bending moments (i.e., for cases where shearing forces are not present). However, for most cases of practical interest, the effect of shearing forces on the curvature and deflection is relatively small in comparison to the effect of the bending moment, even for beams subjected to transverse loads. Thus, Equation 5.5 is used in the solution of deflection problems, particularly for beams that are very long compared to their cross-sectional dimensions. However, it should be pointed out that short, deep beams, for which shearing forces are significant, may require special analysis.

EXAMPLE 5.1

The steel ($E = 30 \times 10^3$ ksi) strap AB is bent by the couples Q into an arc of a circle $A'B'$ as shown. The cross-sectional dimensions of the strap are $t = 0.10$ in and $b = 0.75$ in. The allowable stress for the steel is 30 ksi. (a) Find the minimum radius ρ of the arc into which the strap may be bent. (b) Find the magnitude of the couples Q needed.

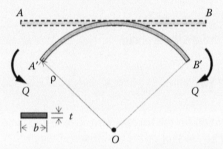

SOLUTION

a. The magnitude of the allowable strain is found from Hooke's law. Thus,

$$\varepsilon_{ALL} = \varepsilon_{MAX} = \frac{\sigma_{ALL}}{E} = \frac{30}{30 \times 10^3} = 0.001$$

Solving Equation 5.2 for ρ, we obtain

$$\rho = \frac{c}{\varepsilon_{MAX}} = \frac{(0.10/2)}{0.001} = 50.0 \text{ in.} \qquad \textbf{ANS.}$$

b. We now solve Equation 5.5 for Q to obtain

$$Q = M = \frac{EI}{\rho} = \frac{(30 \times 10^3)(1/12)(0.75 \times 0.1^3)}{50.0}$$

$$= 0.0375 \text{ kip} \cdot \text{in.} = 37.5 \text{ lb} \cdot \text{in.} \qquad \textbf{ANS.}$$

5.3   DEFLECTION: TWO SUCCESSIVE INTEGRATIONS

As the title of this chapter indicates, we are interested here in the deflections of beams under symmetric bending conditions. The assumption is made, unless otherwise stated, that bending of the beam is about its strong axis (the (z, u) principal centroidal axis) so that its deflection occurs along its weak axis (the (y, v) principal centroidal axis). Thus, in this section, we will develop functions of the form $v = f(x)$, where v represents the deflection of the beam at any position along its length defined by the variable x.

Equation 5.5 $1/\rho = M/EI$, was developed for the case of pure bending of beams, but, as stated above, it can be used without discernible sacrifice in accuracy for beams subjected to transverse loads as long as the shearing forces are relatively small. In order to derive the deflection function, $v = f(x)$, we make use of a relation developed in calculus courses relating the curvature of a plane curve at any point $O(x, v)$ to the derivatives of v with respect to x. Thus,

$$\frac{1}{\rho} = \frac{(d^2v/dx^2)}{[1 + (dv/dx)^2]^{3/2}} \tag{5.6}$$

The quantities dv/dx and d^2v/dx^2 in Equation 5.6 are, respectively, the first and second derivatives of the deflection function with respect to x. However, for most beams of practical interest, the slope of the deflected curve, dv/dx, is relatively small, and when squared this quantity becomes negligible when compared to unity. Making this approximation, the beam curvature becomes

$$\frac{1}{\rho} = \frac{d^2v}{dx^2} \tag{5.7}$$

Combining Equations 5.5 and 5.7, we obtain

$$\frac{d^2v}{dx^2} = v'' = \frac{M}{EI} \tag{5.8}$$

where v'' is a simplified notation for the second derivative of the deflection function with respect to x. This primed notation will be used for other derivatives of the deflection function. Thus, $v' = dv/dx$, $v'' = dv^2/dx^2$, $v''' = dv^3/dx^3$, etc.

Equation 5.8 is a second-order linear differential equation that defines the shape of the elastic curve. To obtain the elastic curve function, $v = f(x)$, from this differential equation, we need to integrate it twice. In order to perform these two integrations, we need to express the moment M and the product EI, known as the *flexural rigidity*, as functions of the variable x. Expressing the moment M as a function of x requires a free-body diagram (see Section 5.3.2). For most beams of practical interest, which are prismatic and are made of one material, the flexural rigidity EI is constant. This, of course, means that the quantity $1/EI$ can be taken out of the integral sign. With this in mind, the first integration of Equation 5.8 yields

$$v' = \frac{1}{EI} \int M \, dx + C_1 \tag{5.9}$$

where v' represents the slope of the elastic curve and C_1 is a constant of integration. The second integration yields

$$v = \frac{1}{EI} \int \left(\int M \, dx + C_1 \right) dx + C_2 \tag{5.10}$$

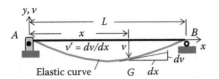

FIGURE 5.2   A schematically deflected elastic curve for a beam under symmetric bending.

where v is the beam deflection and C_2 is a second constant of integration. The two constants of integration, C_1 and C_2, need to be determined in order to have fully meaningful functions for both the slope v' and the deflection v. This determination requires knowledge of what is termed *boundary conditions* (BC), which are conditions dictated by the beam supports.

Figure 5.2 shows schematically the elastic curve for a beam. At some point G, at a distance x from the origin of the coordinate system (support A), the deflection v is indicated representing the vertical position of the elastic curve measured from its original position. Also indicated at this point is the slope v' of the elastic curve.

It should be pointed out that Equations 5.9 and 5.10 need to be determined for every segment of a beam requiring a moment function different from those of other segments in the beam. Thus, for example, a simply supported beam subjected to a concentrated load at midspan requires two applications of Equations 5.9 and 5.10, one for the segment to the left of the concentrated load and another for the segment to the right of this load. The four needed integrations lead to four constants of integration requiring four conditions for their determination. Two of these conditions are derived from the support system (boundary conditions) and two from the conditions that exist at the point on the elastic curve where the concentrated load is applied known as *continuity conditions* or *matching conditions* (MC). This, of course, leads to the solution of four simultaneous equations in order to find the four constants of integration.

The boundary conditions deal with knowledge of the slope and deflection at the supports of a given beam. Thus, for example, the supports of the beam shown in Figure 5.2 dictate that the deflection at A be zero (i.e., at $x = 0$, $v = 0$) and that the deflection at B also be zero (i.e., at $x = L$, $v = 0$). If we assume that the deflection of the beam of Figure 5.2 is caused by a downward load at some position in the beam, say at $L/3$ from support A, the continuity or matching conditions would then be at $x = L/3$, $v_L = v_R$ and at $x = L/3$, $v'_L = v'_R$, where the symbols v_L and v_R, v'_L and v'_R refer, respectively, to the deflections and slopes at $L/3$ obtained from the deflection and slope functions to the left and to the right of the concentrated load, respectively.

EXAMPLE 5.2

The cantilever beam shown in figure (a) below has a constant flexural rigidity EI and is subjected to a downward uniformly distributed load of intensity p. Develop the slope and deflection equations and find these quantities at the free end at A.

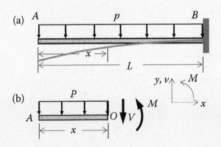

SOLUTION

We begin our solution with the differential equation stated in Equation 5.8. In order to use this equation, we need to express the moment M as a function of the variable x. This is accomplished by using the free-body diagram in figure (b) above and summing moments about point O. Thus,

$$\sum M_O = 0 \Rightarrow M + px\left(\frac{x}{2}\right) = 0; \quad M = -\frac{px^2}{2}$$

Substituting this value of M into Equation 5.8, we obtain

$$v'' = \frac{1}{EI}\left(-\frac{px^2}{2}\right)$$

Performing two consecutive integrations with respect to x yields

$$v' = \frac{1}{EI}\left[-\frac{px^3}{6} + C_1\right] \tag{5.2.1}$$

$$v = \frac{1}{EI}\left[-\frac{px^4}{24} + C_1 x + C_2\right] \tag{5.2.2}$$

The boundary conditions (BC) are now used to obtain the constants C_1 and C_2. These are obtained from the constraints imposed on the deflected curve (see red curve in figure (a) above) at the fixed support at B. If this support is unyielding, then the two boundary conditions are BC1: $x = L$, $v' = 0$ and BC2: $x = L$, $v = 0$. Substituting BC1 into Equation 5.2.1 yields the value of $C_1 = pL^3/6$. Substituting BC2 into Equation 5.2.2 yields $C_2 = -pL^4/8$. Thus,

$$v' = \frac{1}{EI}\left[-\frac{px^3}{6} + \frac{pL^3}{6}\right] \quad \textbf{ANS.} \tag{5.2.3}$$

$$v = \frac{1}{EI}\left[-\frac{px^4}{24} + \frac{pL^3 x}{6} - \frac{pL^4}{8}\right] \quad \textbf{ANS.} \tag{5.2.4}$$

The slope and deflection at A are found, respectively, from Equations 5.2.3 and 5.2.4 by letting $x = 0$. This yields

$$v_A = -\frac{pL^4}{8EI} \quad \textbf{ANS.}$$

$$v_A' = \frac{pL^3}{6EI} \quad \textbf{ANS.}$$

The positive sign on v_A' indicates a counterclockwise rotation and the negative sign on v_A indicates a downward deflection.

EXAMPLE 5.3

A simply-supported beam is loaded as shown in figure (a) below. (a) Develop the slope and deflection equations for the entire beam. Assume that the beam has a constant modulus of rigidity EI. (b) Find the slope and deflection at $x = L/2$.

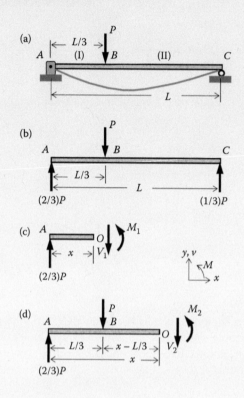

SOLUTION

a. The support reactions are found using the free-body diagram shown in figure (b) above. Two separate moment functions are needed, one for the segment to the left of the load marked (I) and a another for the segment to the right of the load marked (II).

Segment (I) $(x < L/3)$: Using the free-body diagram of figure (c) above, we have

$$\sum M_O = 0 \Rightarrow M_I - \left(\frac{2}{3}\right)Px = 0; \quad M_I = \left(\frac{2}{3}\right)Px$$

Substituting M_I into Equation 5.8 and integrating twice, we obtain

$$v_I'' = \frac{1}{EI}\left[\left(\frac{2}{3}\right)Px\right]$$

$$v_I' = \frac{1}{EI}\left[\left(\frac{1}{3}\right)Px^2 + C_1\right] \qquad (5.3.1)$$

$$v_I = \frac{1}{EI}\left[\left(\frac{1}{9}\right)Px^3 + C_1 x + C_2\right]$$ (5.3.2)

Segment (II) ($x > L/3$): Using the free-body diagram of figure (d) above, we have

$$\sum M_O = 0 \Rightarrow M_{II} + P\left(\frac{x-L}{3}\right) - \left(\frac{2}{3}\right)Px = 0; \quad M_{II} = \left(\frac{1}{3}\right)PL - \left(\frac{1}{3}\right)Px$$

Substituting M_{II} into Equation 5.8 and integrating twice, we obtain

$$v_{II}'' = \frac{1}{EI}\left[\left(\frac{1}{3}\right)PL - \left(\frac{1}{3}\right)Px\right]$$

$$v_{II}' = \frac{1}{EI}\left[\left(\frac{1}{3}\right)PLx - \left(\frac{1}{6}\right)Px^2 + C_3\right]$$ (5.3.3)

$$v_{II} = \frac{1}{EI}\left[\left(\frac{1}{6}\right)PLx^2 - \left(\frac{1}{18}\right)Px^3 + C_3 x + C_4\right]$$ (5.3.4)

The boundary conditions for this beam are BC1: $x = 0$, $v_I = 0$, and BC2: $x = L$, $v_{II} = 0$. The matching conditions are MC1: $x = L/3$, $v_I = v_{II}$, and MC2: $x = L/3$, $v_I' = v_{II}'$. BC1 is substituted into Equation 5.3.2 yielding

$$C_2 = 0$$

When BC2 is substituted into Equation 5.3.4, it leads to

$$C_3 L + C_4 = -\left(\frac{1}{9}\right)PL^3$$ (5.3.5)

According to MC1, we need to equate Equations 5.3.2 and 5.3.4 when $x = L/3$. After simplification, this yields

$$(C_1 - C_3)L - 3C_3 = \left(\frac{1}{27}\right)PL^3$$ (5.3.6)

Finally, by MC2, we equate Equations 5.3.1 and 5.3.3 when $x = L/3$. After simplifying, we obtain

$$C_1 - C_3 = \left(\frac{1}{18}\right)PL^2$$ (5.3.7)

Equations 5.3.5 through 5.3.7 are now solved simultaneously for the remaining constants of integration to obtain

$$C_1 = -\left(\frac{5}{81}\right)PL^2, \quad C_3 = -\left(\frac{19}{162}\right)PL^2, \quad C_4 = \left(\frac{1}{162}\right)PL^3$$

Substituting these constants into the slope and deflection functions, we obtain
Segment I $(x < L/3)$

$$v_I' = \frac{1}{EI}\left[\left(\frac{1}{3}\right)Px^2 - \left(\frac{5}{81}\right)PL^2\right] \qquad \textbf{ANS.} \qquad (5.3.8)$$

$$v_I = \frac{1}{EI}\left[\left(\frac{1}{9}\right)Px^3 - \left(\frac{5}{81}\right)PL^2x\right] \qquad \textbf{ANS.} \qquad (5.3.9)$$

Segment II $(x > L/3)$

$$v_{II}' = \frac{1}{EI}\left[\left(\frac{1}{3}\right)PLx - \left(\frac{1}{6}\right)Px^2 - \left(\frac{19}{162}\right)PL^2\right] \qquad \textbf{ANS.} \qquad (5.3.10)$$

$$v_{II} = \frac{1}{EI}\left[\left(\frac{1}{6}\right)PLx^2 - \left(\frac{1}{18}\right)Px^3 - \left(\frac{19}{162}\right)PL^2x + \left(\frac{1}{162}\right)PL^3\right] \qquad \textbf{ANS.} \quad (5.3.11)$$

b. Since the point at which the slope and displacement are needed is to the right of the load, these quantities are found by substituting $x = L/2$ into Equations 5.3.10 and 5.3.11, respectively. After simplification, this leads to

$$\left.\begin{array}{l} v_{x=L/2}' = v_{II}' = \left(\dfrac{77}{648}\right)\dfrac{PL^2}{EI} \\[4mm] v_{x=L/2} = v_{II} = -\left(\dfrac{23}{1296}\right)\dfrac{PL^3}{EI} = -\dfrac{0.017747PL^3}{EI} \end{array}\right\} \qquad \textbf{ANS.}$$

The positive sign on the slope indicates a counterclockwise rotation and the negative sign on the deflection tells us that it is downward.

EXAMPLE 5.4

Refer to the beam of Example 5.3 and determine the location and magnitude of the maximum deflection.

SOLUTION

The location of the maximum deflection is obtained by differentiating the deflection function with respect to x, setting the result equal to zero and solving for x. Since, in this particular case, two deflection functions are needed for the entire beam, a decision has to be made as to which one is the correct deflection function to use. If the concentrated load were placed at the center of the beam, the maximum deflection would occur at the load, and either deflection function would provide the magnitude of the maximum deflection by setting $x = L/2$. In our case, however, the load is to the left of center, and since the beam deflection is a continuous curve, the maximum deflection occurs to the right of the load. Thus, the deflection function v_{II} is the correct function to use. Differentiation of this function with respect to x yields v'_{II}, which is given by Equation 5.3.10 in Example 5.3. Thus,

$$v'_{II} = \frac{1}{EI}\left[\left(\frac{1}{3}\right)PLx - \left(\frac{1}{6}\right)Px^2 - \left(\frac{19}{162}\right)PL^2\right] = 0 \qquad (5.4.1)$$

Simplifying Equation 5.4.1 and rearranging terms leads to

$$x^2 - 2Lx + \left(\frac{19}{27}\right)L^2 = 0 \qquad (5.4.2)$$

Equation 5.4.2 is a quadratic function, which when solved for x yields

$$x = 1.544L, 0.456L \qquad \textbf{ANS.}$$

The first answer is unacceptable on physical grounds and the correct location for the maximum deflection is $x = 0.456L$. When this value of x is substituted into Equation 5.3.11 of Example 5.3, we obtain the magnitude of the maximum deflection. Thus,

$$v_{MAX} = (v_{II})_{x=0.456L} = -\frac{0.017929PL^3}{EI} \qquad \textbf{ANS.}$$

As expected, this value is a little larger in magnitude than that obtained for the deflection at the center of the beam obtained in Example 5.3.

It should be observed that, if instead of using the correct deflection function v_{II} as was done, a mistake was made and the function v_I was used, the first derivative of this function would be that given by Equation 5.3.8 of Example 5.3. When this is set equal to zero and the resulting equation solved for x, a value would be obtained that is unrealistic or physically unacceptable. For example, if Equation 5.3.8 of Example 5.3 were set equal to zero and the equation solved for x, we would obtain $x = \pm 0.430L$. The negative sign is physically unacceptable, and the positive one is a point to the right of the load, outside the limits of applicability of the function given by Equation 5.3.8 of Example 5.3.

PROBLEMS

5.1   A steel rod having an equilateral triangle for the cross section, 0.75 in. on each side, is bent into an arc of a circle of radius $\rho = 20$ ft as shown. Find the maximum tensile and maximum compressive bending stresses that are induced. Assume elastic behavior and let $E = 30 \times 10^6$ psi.

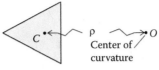

(Problems 5.1 and 5.2)

5.2 In reference to the figure shown in Problem 5.1, an aluminum ($E = 75$ GPa) rod having an equilateral triangle for the cross section, 25 mm on each side, is bent by couples applied at its ends, into an arc of a circle of radius ρ as shown. If the allowable stress is 150 MPa, determine (a) the minimum radius of curvature and (b) the magnitude of the couples producing this curvature.

5.3 A steel ($E = 30 \times 10^3$ ksi) saw blade passes over a 10-in.-diameter pulley as shown. Ignoring the blade teeth, the cross section of the blade may be approximated by a rectangle as shown by section A–A. If $t = 0.02$ in. and $w = 0.75$ in., determine the maximum stress in the blade.

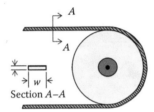

(Problems 5.3, 5.4, and 5.5)

5.4 In reference to the figure shown in Problem 5.3, a steel ($E = 205$ GPa) saw blade passes over a 300-mm-diameter pulley as shown. Ignoring the blade teeth, the cross section of the blade may be approximated by a rectangle as shown by section A–A. If $w = 16$ mm and the allowable stress is 270 MPa, determine the maximum permissible thickness t of the blade.

5.5 In reference to the figure shown in Problem 5.3, a steel ($E = 30 \times 10^3$ ksi) saw blade passes over a pulley as shown. Ignoring the blade teeth, the cross section of the blade may be approximated by a rectangle as shown by section A–A. If the allowable stress is 50 ksi, $t = 0.04$ in., and $w = 0.80$ in., determine the minimum diameter of the pulley.

5.6 A 1/2-in.-diameter brass rod is coiled around a spool, as shown, for transporting purposes. Let $E = 17 \times 10^3$ ksi and $d = 14$ ft. Assume that the material behaves elastically, determine (a) the maximum bending stress in the rod and (b) the couple required to coil the rod around the spool.

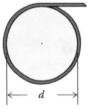

(Problems 5.6 and 5.7)

5.7 A 10-mm-diameter aluminum rod is coiled around a spool, as shown, for transporting purposes. Let $E = 75$ GPa, determine (a) the minimum diameter of the spool if the allowable bending stress in the rod is 150 MPa and (b) the couple required to coil the rod around the spool.

5.8 Assume EI to be a constant and develop the slope and deflection equations for the cantilever beam shown. Find the maximum values of the slope and deflection expressing your answer in terms of the applied load, the length L, and the flexural rigidity EI. Use the given coordinate system with origin at A.

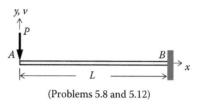

(Problems 5.8 and 5.12)

5.9   Assume EI to be a constant and develop the slope and deflection equations for the cantilever beam shown. Find the maximum values of the slope and deflection expressing your answer in terms of the applied load, the length L, and the flexural rigidity EI. Use the given coordinate system with origin at A.

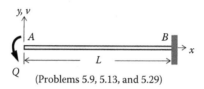

(Problems 5.9, 5.13, and 5.29)

5.10   Assume EI to be a constant and develop the slope and deflection equations for the cantilever beam shown. Find the maximum values of the slope and deflection expressing your answer in terms of the applied load, the length L, and the flexural rigidity EI. Use the given coordinate system with origin at A.

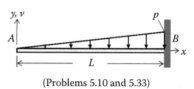

(Problems 5.10 and 5.33)

5.11   Assume EI to be a constant and develop the slope and deflection equations for the cantilever beam shown. Find the maximum values of the slope and deflection expressing your answer in terms of the applied load, the length L, and the flexural rigidity EI. Use the given coordinate system with origin at A.

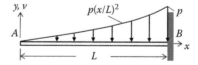

5.12   Ignore the given coordinate system and establish a new x, (y, v) system with origin at B, measuring x positive to the left, and solve Problem 5.8. Note the difference in the values of the integration constants.

5.13   Ignore the given coordinate system and establish a new x, (y, v) system with origin at B, measuring x positive to the left, and solve Problem 5.9. Note the difference in the values of the integration constants.

5.14   Assume EI to be a constant and develop the slope and deflection equations for the simply supported beam shown. Find the slopes at A and B and the deflection at midspan expressing your answers in terms of the applied load, the length of the beam L, and the constant EI.

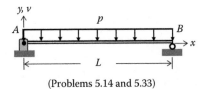

(Problems 5.14 and 5.33)

5.15 Assume EI to be a constant and develop the slope and deflection equations for the simply supported beam shown. Find the slopes at A and B and the deflection at midspan expressing your answers in terms of the applied load, the length of the beam L, and the constant EI.

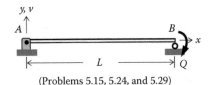

(Problems 5.15, 5.24, and 5.29)

5.16 Assume EI to be a constant and develop the slope and deflection equations for the simply supported beam shown. Find the slopes at A and B and the deflection at midspan expressing your answers in terms of the applied load, the length of the beam L, and the constant EI.

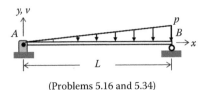

(Problems 5.16 and 5.34)

5.17 Assume EI to be a constant and develop the slope and deflection equations for the simply supported beam shown. Find the slopes at A and B and the deflection at midspan expressing your answers in terms of the applied load, the length of the beam L, and the constant EI.

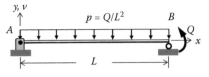

(Problems 5.17, 5.25, and 5.30)

5.18 Assume EI to be a constant and develop the slope and deflection equations for the simply supported beam shown. Find the slopes at A and B and the deflection at midspan expressing your answers in terms of the applied load, the length of the beam L, and the constant EI.

5.19 Assume EI to be a constant and develop the slope and deflection equations for the simply supported beam shown. Find the slopes at A and B and the deflection at midspan expressing your answers in terms of the applied load, the length of the beam L, and the constant EI.

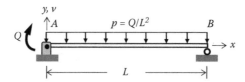

5.20 Assume EI to be a constant and develop the slope and deflection equations for the simply supported beam shown. Find the slopes at A and B and the deflection at midspan expressing your answers in terms of the applied load, the length of the beam L, and the constant EI.

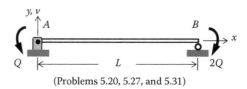

(Problems 5.20, 5.27, and 5.31)

5.21 Assume EI to be a constant and develop the slope and deflection equations for the simply supported beam shown. Find the slopes at A and B and the deflection at midspan expressing your answers in terms of the applied load, the length of the beam L, and the constant EI.

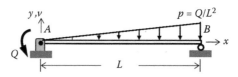

(Problems 5.21, 5.26, 5.32, and 5.35)

5.22 Assume EI to be a constant and develop the slope and deflection equations for the simply supported beam shown. Find the slopes at A and B and the deflection at midspan expressing your answers in terms of the applied load, the length of the beam L, and the constant EI.

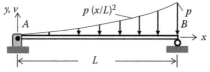

(Problems 5.22, 5.28, and 5.36)

5.23 Assume EI to be a constant and develop the slope and deflection equations for the simply supported beam shown. Find the slopes at A and B and the deflection at midspan expressing your answers in terms of the applied load, the length of the beam L, and the constant EI.

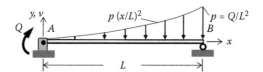

5.24 In reference to the figure shown in Problem 5.15, assume EI to be a constant and find the location and magnitude of the maximum deflection. Express your answers in terms of the applied load, the length of the beam L, and the constant EI.

5.25 In reference to the figure shown in Problem 5.17, assume EI to be a constant and find the location and magnitude of the maximum deflection. Express your answers in terms of the applied load, the length of the beam L, and the constant EI.

5.26 In reference to the figure shown in Problem 5.21, assume EI to be a constant and find the location and magnitude of the maximum deflection. Express your answers in terms of the applied load, the length of the beam L, and the constant EI.

5.27 In reference to the figure shown in Problem 5.20, assume EI to be a constant and find the location and magnitude of the maximum deflection. Express your answers in terms of the applied load, the length of the beam L, and the constant EI.

5.28 In reference to the figure shown in Problem 5.22, assume EI to be a constant and find the location and magnitude of the maximum deflection. Express your answers in terms of the applied load, the length of the beam L, and the constant EI.

5.29 In reference to the figure shown in Problem 5.15, select the lightest wide flange section from Appendix E if the maximum deflection of the beam is to be limited to 0.5 in. Let $Q = 30$ kip · ft, $L = 20$ ft, and $E = 30 \times 10^3$ ksi.

5.30 In reference to the figure shown in Problem 5.17, select the lightest wide flange section from Appendix E if the maximum deflection of the beam is to be limited to 0.5 in. Let $Q = 30$ kip · ft, $L = 20$ ft, and $E = 30 \times 10^3$ ksi.

5.31 In reference to the figure shown in Problem 5.20, select the lightest wide flange section from Appendix E if the maximum deflection of the beam is to be limited to 0.5 in. Let $Q = 30$ kip · ft, $L = 20$ ft, and $E = 30 \times 10^3$ ksi.

5.32 In reference to the figure shown in Problem 5.21, select the lightest wide flange section from Appendix E if the maximum deflection of the beam is to be limited to 0.5 in. Let $Q = 30$ kip · ft, $L = 20$ ft, and $E = 30 \times 10^3$ ksi.

5.33 In reference to the figure shown in Problem 5.10, if the maximum deflection of the beam is to be limited to a value of 10 mm, determine the magnitude of the maximum permissible load intensity p. Let $L = 10$ m and $EI = 400 \times 10^3$ kN · m². Repeat using the Figure in problem 5.14.

5.34 In reference to the figure shown in Problem 5.16, if the maximum deflection of the beam is to be limited to a value of 10 mm, determine the magnitude of the maximum permissible load intensity p. Let $L = 10$ m and $EI = 400 \times 10^3$ kN · m².

5.35 In reference to the figure shown in Problem 5.21, if the maximum deflection of the beam is to be limited to a value of 10 mm, determine the magnitude of the maximum permissible load intensity p. Let $L = 10$ m and $EI = 400 \times 10^3$ kN · m².

5.36 In reference to the figure shown in Problem 5.22, if the maximum deflection of the beam is to be limited to a value of 10 mm, determine the magnitude of the maximum permissible load intensity p. Let $L = 10$ m and $EI = 400 \times 10^3$ kN · m².

5.37 Assume a constant EI and develop the slope and deflection functions for segment AB of the beam. Use a coordinate system with origin at A, measuring x positive to the right and y,v positive upward. Also find the deflection at the midpoint between A and B. Express your answers in terms of the given length, the applied load, and the constant EI.

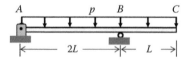

(Problems 5.37 and 5.41)

5.38 Assume a constant *EI* and develop the slope and deflection functions for segment *AB* of the beam. Use a coordinate system with origin at *A*, measuring *x* positive to the right and *y,v* positive upward. Also find the deflection at the midpoint between *A* and *B*. Express your answers in terms of the given length, the applied load, and the constant *EI*.

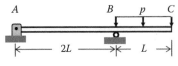

(Problems 5.38 and 5.42)

5.39 Assume a constant *EI* and develop the slope and deflection functions for segment *AB* of the beam. Use a coordinate system with origin at *A*, measuring *x* positive to the right and *y,v* positive upward. Also find the deflection at the midpoint between *A* and *B*. Express your answers in terms of the given length, the applied load, and the constant *EI*.

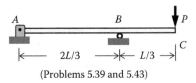

(Problems 5.39 and 5.43)

5.40 Assume a constant *EI* and develop the slope and deflection functions for segment *AB* of the beam. Use a coordinate system with origin at *A*, measuring *x* positive to the right and *y*, *v* positive upward. Also find the deflection at the midpoint between *A* and *B*. Express your answers in terms of the given length, the applied load, and the constant *EI*.

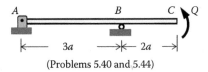

(Problems 5.40 and 5.44)

5.41 In reference to the figure shown in Problem 5.37, assume *EI* to be a constant and find the location and magnitude of the maximum deflection in segment *AB* of the beam. Use a coordinate system with origin at *A*, measuring *x* positive to the right and *y*, *v* positive upward. Express your answers in terms of the given length, the applied load, and the constant *EI*.

5.42 In reference to the figure shown in Problem 5.38, assume *EI* to be a constant and find the location and magnitude of the maximum deflection in segment *AB* of the beam. Use a coordinate system with origin at *A*, measuring *x* positive to the right and *y,v* positive upward. Express your answers in terms of the given length, the applied load, and the constant *EI*.

5.43 In reference to the figure shown in Problem 5.39, assume *EI* to be a constant and find the location and magnitude of the maximum deflection in segment *AB* of the beam. Use a coordinate system with origin at *A*, measuring *x* positive to the right and *y,v* positive upward. Express your answers in terms of the given length, the applied load, and the constant *EI*.

5.44 In reference to the figure shown in Problem 5.40, assume *EI* to be a constant and find the location and magnitude of the maximum deflection in segment *AB* of the beam. Use a coordinate system with origin at *A*, measuring *x* positive to the right, and *y,v* positive upward. Express your answers in terms of the given length, the applied load, and the constant *EI*.

5.45 Assume *EI* to be a constant and develop the slope and deflection functions for the entire beam. Use a coordinate system with origin at *A*, measuring *x* positive to the right and *y*, *v* positive upward. Express your answers in terms of the given length, the applied load, and the constant *EI*.

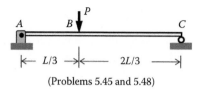

(Problems 5.45 and 5.48)

5.46 Assume *EI* to be a constant and develop the slope and deflection functions for the entire beam. Use a coordinate system with origin at *A*, measuring *x* positive to the right and *y,v* positive upward. Express your answers in terms of the given length, the applied load, and the constant *EI*.

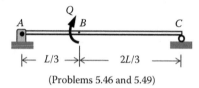

(Problems 5.46 and 5.49)

5.47 Assume *EI* to be a constant and develop the slope and deflection functions for the entire beam. Use a coordinate system with origin at *A*, measuring *x* positive to the right and *y,v* positive upward. Express your answers in terms of the given length, the applied load, and the constant *EI*.

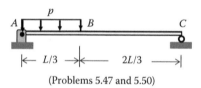

(Problems 5.47 and 5.50)

5.48 In reference to the figure shown in Problem 5.45, assume *EI* to be a constant and find the location and magnitude of the maximum deflection in the beam. Use a coordinate system with origin at *A*, measuring *x* positive to the right and *y*, *v* positive upward. Express your answers in terms of the given length, the applied load, and the constant *EI*.

5.49 In reference to the figure shown in Problem 5.46, assume *EI* to be a constant and find the location and magnitude of the maximum deflection in the beam. Use a coordinate system with origin at *A*, measuring *x* positive to the right and *y,v* positive upward. Express your answers in terms of the given length, the applied load, and the constant *EI*.

5.50 In reference to the figure shown in Problem 5.47, assume *EI* to be a constant and find the location and magnitude of the maximum deflection in the beam. Use a coordinate system with origin at *A*, measuring *x* positive to the right and *y,v* positive upward. Express your answers in terms of the given length, the applied load, and the constant *EI*.

5.4 DERIVATIVES OF THE DEFLECTION FUNCTION

The equation of the elastic curve for a beam, $v = f(x)$, and its first four derivatives are very useful in understanding the physical characteristics of the loaded beam. As we discovered in Section 5.3, the deflection function, *v*, and its first derivative, v', provide, respectively, the deflection and the slope of the deflected beam at any position along its length. Also, Equation 5.8 states that the second derivative of the deflection function, at any position of the loaded beam, is equal to the internal moment at this section divided by the flexural rigidity, *EI*, of the beam. Differentiation of Equation 5.8 yields

$$v''' = \left(\frac{1}{EI}\right)\frac{dM}{dx} = \frac{V}{EI} \tag{5.11}$$

Note that, in Equation 5.11, we used Equation 3.7 to replace dM/dx by the shearing force V. Thus, Equation 5.11 states that the third derivative of the deflection function at any section of the loaded beam is equal to the internal shearing force at this section divided by the flexural rigidity, EI, of the beam.

Differentiating Equation 5.11, we obtain

$$v'''' = \left(\frac{1}{EI}\right)\frac{d^2M}{dx^2} = \left(\frac{1}{EI}\right)\frac{dV}{dx} = -\frac{p}{EI} \tag{5.12}$$

Note that, using Equation 3.2, we replaced dV/dx by the negative of the applied load intensity p. Therefore, Equation 5.12 states that the fourth derivative of the deflection function at any section of the loaded beam is equal to the negative of the load intensity p divided by the flexural rigidity, EI, of the beam.

Equation 5.12 makes it possible to begin the solution of the deflection problem from knowledge of the applied load. Of course, this type of solution requires four integrations resulting in four unknown constants. Determination of these four constants necessitates knowledge of four boundary conditions that include not only those relating to the slopes and deflections but also conditions relating to the shears and moments in the beam. Thus, for example, in the case of a beam simply supported at both ends and subjected to a continuous load over the entire span, two boundary conditions would come from the knowledge that the deflection is zero at both supports and two from the fact that the bending moment is zero at these two supports as well. As a second example, in the case of a cantilever beam subjected to a continuous load over the entire span, two conditions follow from the fact that at the free end, both the shearing force and bending moment are zero and two from the fixed support where both the slope and deflection are zero. Of course, other support and loading conditions different from these two would require a little more effort. As an example, a simply supported beam subjected to one concentrated load at any position would require two sets of equations, each leading to four constants of integration for a total of eight unknown constants needing eight conditions for their determination. Six of these are obtained from the boundary conditions at the two simple supports where the two deflections and the two bending moments are zero, and the two shearing forces can be found by statics. The remaining two conditions are obtained from the matching (continuity) conditions at the point where the concentrated load is applied. These two matching conditions result from the fact that, at this point, the slope and deflection from one set of equations must be equal, respectively, to the slope and deflection from the second set. As these examples illustrate and as shown in Example 5.3, the integration method for finding deflections of beams (whether we start from the second or the fourth derivative) is cumbersome except for the simplest cases of supports and loading. More suitable methods for complex cases are discussed in Sections 5.5 and 5.6 as well as in Chapter 6.

EXAMPLE 5.5

A simply supported beam is loaded as shown. Determine the shear, moment, slope, and deflection functions. Show sketches of these four functions directly under the beam. Start the solution using Equation 5.12.

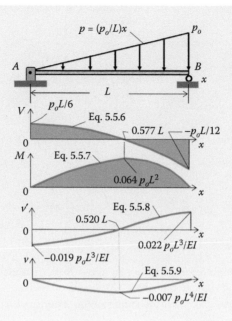

SOLUTION

As shown in the sketch, the load function is $p = (p_o/L)x$. Thus, according to Equation 5.12, we have

$$v'''' = -\frac{1}{EI}\left[\left(\frac{p_o}{L}\right)x\right] \tag{5.5.1}$$

$$v''' = -\frac{1}{EI}\left[\left(\frac{p_o}{2L}\right)x^2 + C_1\right] \tag{5.5.2}$$

$$v'' = -\frac{1}{EI}\left[\left(\frac{p_o}{6L}\right)x^3 + C_1 x + C_2\right] \tag{5.5.3}$$

$$v' = -\frac{1}{EI}\left[\left(\frac{p_o}{24L}\right)x^4 + \frac{C_1}{2}x^2 + C_2 x + C_3\right] \tag{5.5.4}$$

$$v = -\frac{1}{EI}\left[\left(\frac{p_o}{120L}\right)x^5 + \frac{C_1}{6}x^3 + \frac{C_2}{2}x^2 + C_3 x + C_4\right] \tag{5.5.5}$$

The constants C_1 through C_4 are found using the following boundary conditions: BC1: $x = 0$, $v = 0$; BC2: $x = 0$, $v'' = 0$ because $M = 0$. BC3: $x = L$, $v = 0$; BC4: $x = L$, $v'' = 0$ because $M = 0$.

When BC1 is substituted into Equation 5.5.5, it yields $C_4 = 0$. Now, BC2 is substituted into Equation 5.5.3 to obtain $C_2 = 0$. If BC4 is used in Equation 5.5.3, we obtain $C_1 = -p_oL/6$. Finally, if BC3 is used in Equation 5.5.5, we obtain $C_3 = 7p_oL^3/360$. Thus, we conclude that

$$V = EIv''' = -\frac{p_o}{2}\left[\frac{x^2}{L} - \frac{L}{3}\right] \qquad \text{ANS.} \qquad (5.5.6)$$

$$M = EIv'' = -\frac{p_o}{6}\left[\frac{x^3}{L} - Lx\right] \qquad \text{ANS.} \qquad (5.5.7)$$

$$v' = -\frac{p_o}{12EI}\left[\frac{x^4}{2L} - Lx^2 + \frac{7L^3}{30}\right] \qquad \text{ANS.} \qquad (5.5.8)$$

$$v = -\frac{p_o}{36EI}\left[\frac{3x^5}{10L} - Lx^3 + \frac{7L^3 x}{10}\right] \qquad \text{ANS.} \qquad (5.5.9)$$

Equations 5.5.6 through 5.5.9 are sketched above directly under the sketch of the loaded beam. Note that some significant quantities have been computed and shown on these sketches to three significant figures except for the values of the shearing forces at the two supports that are given as fractions.

PROBLEMS

5.51 The equation of the elastic curve for a simply supported beam of length L and constant flexural rigidity is given by

$$v = \frac{p_o}{EI}\left[-\frac{x^5}{120L} + \frac{Lx^3}{36} - \frac{7L^3 x}{360}\right]$$

where p_o is the load intensity at the right end of the beam, x is measured positive to the right from an origin at the left end of the beam and the deflection v is measured positive upward. Determine expressions, in terms of x, for the load intensity, shear force, bending moment, and slope.

5.52 The equation of the elastic curve for a cantilever beam of length L and constant flexural rigidity is given by

$$v = \frac{p_o}{EI}\left[-\frac{x^6}{360L^2} + \frac{Lx^3}{18} - \frac{L^2 x^2}{8}\right]$$

where p_o is the load intensity at the right end of the beam, x is measured positive to the right from an origin at the fixed end (left end) of the beam, and the deflection v is measured positive upward. Determine expressions, in terms of x, for the load intensity, shear force, bending moment, and the slope.

5.53 In reference to the figure shown in Problem 5.8, use the method of Section 5.4 to develop expressions for the shear force, bending moment, slope, and deflection as functions of x.

5.54 In reference to the figure shown in Problem 5.9, use the method of Section 5.4 to develop expressions for the shear force, bending moment, slope, and deflection as functions of x.

5.55 In reference to the figure shown in Problem 5.10, use the method of Section 5.4 to develop expressions for the shear force, bending moment, slope, and deflection as functions of x.

5.56 In reference to the figure shown in Problem 5.11, use the method of Section 5.4 to develop expressions for the shear force, bending moment, slope, and deflection as functions of x.

5.57 In reference to the figure shown in Problem 5.14, use the method of Section 5.4 to develop expressions for the shear force, bending moment, slope, and deflection as functions of x.

5.58 In reference to the figure shown in Problem 5.15, use the method of Section 5.4 to develop expressions for the shear force, bending moment, slope, and deflection as functions of x.

5.59 In reference to the figure shown in Problem 5.22, use the method of Section 5.4 to develop expressions for the shear force, bending moment, slope, and deflection as functions of x.

5.60 In reference to the figure shown in Problem 5.23, use the method of Section 5.4 to develop expressions for the shear force, bending moment, slope, and deflection as functions of x.

5.61 Use the method of this section to develop, in terms of x, expressions for the shear force, bending moment, slope, and deflection for the given beam.

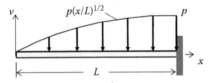

5.62 Use the method of this section to develop, in terms of x, expressions for the shear force, bending moment, slope, and deflection for the given beam.

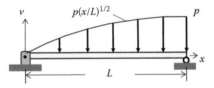

5.63 Use the method of this section to develop, in terms of x, expressions for the shear force, bending moment, slope, and deflection for the given beam.

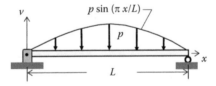

5.5   DEFLECTION: SUPERPOSITION

The method of superposition was used previously in the solution of some complex problems for cases where members behave elastically, that is, deformations are linear functions of the applied loads. This method allows us to break down a complex problem into two or more simple components, find the solution for each of these components, and then superpose (add algebraically) the various simple solutions to obtain the solution of the complex problem.

Consider, for example, the case of a beam subjected to two or more separate loadings denoted by the subscripts 1, 2, etc. Equation 5.8 is now applied separately to each of the loadings. Thus,

$$\frac{d^2 v_1}{dx^2} = \frac{M_1}{EI} \Rightarrow v_1; \quad \frac{d^2 v_2}{dx^2} = \frac{M_2}{EI} \Rightarrow v_2; \quad \text{etc.} \tag{5.13}$$

Since d^2/dx^2 is a linear operator and the flexural rigidity EI is assumed constant, we may write

$$\frac{d^2(v_1 + v_2 + \cdots)}{dx^2} = \frac{M_1 + M_2 + \cdots}{EI} \Rightarrow v_1 + v_2 + \cdots \tag{5.14}$$

Comparing Equations 5.13 and 5.14, we conclude that, if a beam is subjected to two or more loadings, it is possible to find the deflections corresponding to each of these loadings separately and use superposition (algebraic addition) to obtain the deflection of the composite beam. A similar argument may be made for the slope of the composite beam.

In using the method of superposition, it is convenient to have available the solutions for the slopes and deflections for simple cases of beam loadings. A selection of such solutions has been compiled and placed in Appendix G. *Since for very small angles the slope is approximately numerically equal to the angle,* Appendix G *shows angles* θ *instead of slopes* v'. Of course, the selection is not exhaustive. Therefore, in a given solution, we may encounter cases of beams not found in the selections of Appendix G. In such cases, it becomes necessary to consult the literature or develop the needed solution using the method of two successive integrations or, any of the other methods introduced later in this chapter or in Chapter 6.

EXAMPLE 5.6

Using the method of superposition, determine the deflection and slope at A for the beam shown in figure (a) below. Let $L = 12$ ft, $P = 2$ kips, $Q = 20$ kip · ft, $E = 30 \times 10^3$ ksi, and $I = 160$ in.^{4}.

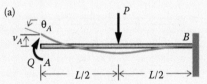

(a)

SOLUTION

Decompose the given beam into the two simpler beams shown in figure (b) and (c) below. As stated above, the deflection at A is equal to the superposition of the deflections found separately for cases (a) and (b). The deflection and slope at A for the beam of figure (b) below are found from case 1 of Appendix G. Thus, since $a = L/2$, we have

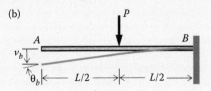

(b)

$$v_b = -\frac{P(L/2)^2}{6EI}\left(3L - \frac{L}{2}\right)$$

$$= -\frac{2(6 \times 12)^2}{6(30 \times 10^3)(160)}[3(12 \times 12) - 6 \times 12] = -0.1296 \text{ in.}$$

$$\theta_b = \frac{P(L/2)^2}{2EI} = \frac{2(6 \times 12)^2}{2(30 \times 10^3)(160)} = 0.00108 \text{ rad}$$

Similarly, the deflection and slope at A for the beam of figure (c) below are found from case 2 of Appendix G. Thus,

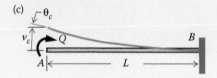

$$v_c = \frac{QL^2}{2EI} = \frac{20(12)(12 \times 12)^2}{2(30 \times 10^3)(160)} = 0.5184 \text{ in.}$$

$$\theta_c = -\frac{QL}{EI} = -\frac{20(12)(12 \times 12)}{(30 \times 10^3)(160)} = -0.0072 \text{ rad}$$

Therefore, the deflection and slope at A for the beam of figure (a) above are

$$\left.\begin{array}{l} v_A = v_b + v_c = -0.1296 + 0.5184 = 0.389 \approx 0.4 \text{ in.} \\ \theta_A = \theta_b + \theta_c = 0.00108 - 0.0072 = -0.00612 \approx 0.006 \text{ rad} \end{array}\right\} \quad \textbf{ANS.}$$

The positive sign on the deflection indicates this deflection to be upward, and the negative sign on the slope shows that this slope is clockwise. A sketch of the elastic curve for the given beam is shown in figure (a) above.

PROBLEMS

5.64 Use the information in Appendix G along with *superposition* to find the slope and deflection at point A for the beam shown below. Express your answers in terms of the load P, the length L, and the constant EI.

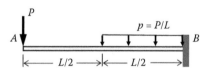

5.65 Use the information in Appendix G along with *superposition* to find the slope and deflection at point A for the beam shown below. Express your answers in terms of the load P, the length L, and the constant EI.

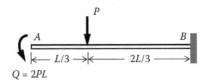

5.66 Use the information in Appendix G along with *superposition* to find the slope and deflection at point *A* for the beam shown below. Express your answers in terms of the couple *Q*, the length *L*, and the constant *EI*.

5.67 Use the information in Appendix G along with *superposition* to find the slope and deflection at point *A* for the beam shown below. Express your answers in terms of the load *P*, the length *L*, and the constant *EI*.

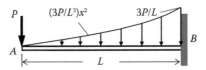

5.68 Find the deflection at point *B* and the slope at *A*. Express your answers in terms of the load *P*, the length *L*, and the constant *EI*.

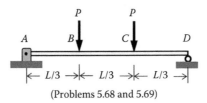

(Problems 5.68 and 5.69)

5.69 In reference to the figure shown in Problem 5.68, find the maximum allowable load *P* if the maximum deflection of the beam is to be limited to 0.75 in. Let $L = 18$ ft and $EI = 100 \times 10^6$ kip $\cdot$ in.^{2}.

5.70 Determine the slope at *A* and the deflection at *B*. Express your answers in terms of the load *P*, the length *L*, and the constant *EI*.

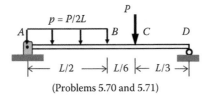

(Problems 5.70 and 5.71)

5.71 In reference to the figure shown in Problem 5.70, determine the maximum allowable load *P* if the deflection of the beam at *B* is to be limited to 30 mm. Let $L = 6$ m and $EI = 10 \times 10^3$ kN $\cdot$ m^2.

5.72 a. Use superposition to determine p in terms of Q so that the deflection of the cantilever beam at A is zero.

  b. Using the result of part (a) and the method of two successive integrations, find the location and magnitude of the maximum deflection, when both Q and p are applied, expressing the answers in term of Q, L, and EI.

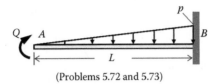

(Problems 5.72 and 5.73)

5.73 Repeat parts (a) and (b) of Problem 5.72 if the deflection at A is upward with a magnitude of $v_A = QL^2/50EI$ instead of zero.

5.74 Select the lightest wide flange section from Appendix E such that the deflection of the beam at A does not exceed a magnitude of 3 in. Let $P = 15$ kips, $p = 2$ kips/ft, and $L = 15$ ft. Ignore the weight of the beam in the computations. Let $E = 30 \times 10^3$ ksi.

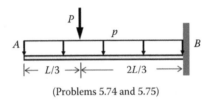

(Problems 5.74 and 5.75)

5.75 In reference to the figure shown in Problem 5.74, a W14 $\times$ 48 section is to be used for the beam shown. If the deflection at A is to be limited to 4 in., determine the maximum permissible value of P. Let $p = P/2L$ and $L = 18$ ft. Ignore the weight of the beam in the computations. Let $E = 30 \times 10^3$ ksi.

5.76 Select the lightest wide flange ($E = 200$ GPa) section from Appendix E such that the maximum deflection of the beam does not exceed a magnitude of 70 mm. Let $P = 15$ kN, $p = 5$ kN/m, $L = 6$ m, and $E = 200$ GPa. Ignore the weight of the beam in the computations.

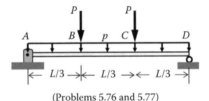

(Problems 5.76 and 5.77)

5.77 In reference to the figure shown in Problem 5.76, a W250 $\times$ 167 ($E = 200$ GPa) section is to be used for the beam shown. If the maximum deflection of the beam is to be limited to 50 mm, determine the maximum permissible value of P. Let $p = P/L$, $L = 8$ m, and $E = 200$ GPa. Ignore the weight of the beam in your computations.

5.6 DEFLECTION: AREA–MOMENT

Underlying the *area–moment* method are two theorems that are based upon the relation that exists between the curvature of a beam at any position along its length and the internal bending moment at this position. This relation was derived in Section 5.2, expressed in Equation 5.5, and reexpressed in terms of the second derivative of the elastic curve in Equation 5.8. Equation 5.8 may be rewritten in the form

$$\frac{d^2v}{dx^2} = \frac{d}{dx}\left(\frac{dv}{dx}\right) = \frac{d\theta}{dx} = \frac{M}{EI} \tag{5.15}$$

Note that the slope, dv/dx, was replaced by angle θ. This is made possible by the fact, stated earlier in Section 5.5, that for very small angles measured in radians, the tangent of the angle is essentially equal to the angle itself. For our present purposes, Equation 5.15 may be expressed as

$$d\theta = \frac{M}{EI}\,dx \tag{5.16}$$

Theorem I

The first theorem of the area–moment method follows from Equation 5.16. Thus, consider a beam subjected to some general loading, $p = f(x)$, as shown in Figure 5.3a. Figure 5.3a also shows the corresponding elastic curve. Let us consider any two arbitrarily selected points A and B on the elastic curve of the beam as shown. Equation 5.16 states that the differential change in slope between points C and D on the elastic curve, a differential distance dx apart, is given by the differential area under the M/EI diagram between these two points as shown in Figure 5.3b. Therefore, integrating Equation 5.16 from A to B yields

$$\int_{\theta_A}^{\theta_B} d\theta = \theta_B - \theta_A = \int_{x_A}^{x_B} \frac{M}{EI}\,dx \tag{5.17}$$

The quantity $(\theta_B - \theta_A)$ in Equation 5.17 represents the change in slope between points A and B on the elastic curve, and the integral on the right-hand side of this equation represents the total area

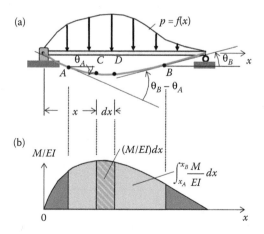

FIGURE 5.3   (a) A beam under a symmetric load $p = f(x)$ showing the resulting elastic curve. (b) The corresponding M/EI diagram.

under the M/EI diagram between the same two points. Thus, the first theorem of the *area–moment* method may be stated as follows:

The change in slope between any two points on the elastic curve of a beam is given by the area under the M/EI diagram between the same two points.

Note that the sign of the slope change (clockwise or counterclockwise) is the same as the sign of the area under the M/EI diagram. Thus, as shown in Figure 5.3, a positive area under the M/EI diagram corresponds to a counterclockwise change in the slope and a negative area under this diagram corresponds to a clockwise change in the slope.

Theorem II

Consider, again, the two points C and D of Figure 5.3a, repeated in Figure 5.4a. The change in slope $d\theta$ between these two points is given by Equation 5.16. We now construct a line perpendicular to the beam axis in Figure 5.4a (i.e., a vertical line because the beam is horizontal in this case) through point B on the elastic curve. Multiplying the differential angle $d\theta$ by the distance x_1 yields a differential arc length that is approximated by the straight segment dt along the vertical line through point B. Thus,

$$dt = x_1 d\theta = x_1 \left(\frac{M}{EI} \right) dx \tag{5.18}$$

Let us integrate Equation 5.18 between points A and B. Thus,

$$\int_0^{t_{B/A}} dt = t_{B/A} = \int_{x_A}^{x_B} x_1 \left(\frac{M}{EI} \right) dx \tag{5.19}$$

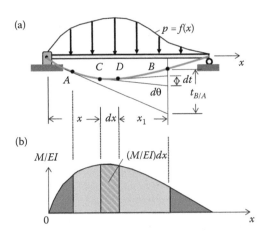

FIGURE 5.4   Repetition of points C and D from Figure 5.3a in order to find the change of angle $d\theta$ between these two points using Equations 5.15 and 5.16. (a) A beam under a symmetric load $p = f(x)$ showing the resulting elastic curve; (b) the corresponding M/EI diagram.

Equation 5.19 is the mathematical form of the second theorem of the *area–moment* method. The quantities dt and t are known as *tangential deviations* in which the subscripts provide a specific meaning. Thus, for example, the quantity $t_{B/A}$ signifies the tangential deviation (the distance perpendicular to the beam axis) of point B on the elastic curve with respect to the tangent drawn at point A, which is also on the elastic curve. The quantity $x_1(M/EI)dx$, which by Equation 5.18 is equal to the differential tangential deviation dt, represents the first moment of the differential area $(M/EI)dx$ about point B from which x_1 is measured. Therefore, the integral of $x_1(M/EI)dx$ between the limits x_A and x_B, which by Equation 5.19 is equal to the tangential deviation $t_{B/A}$, represents the first moment about point B of the total area under the M/EI diagram between points A and B. Thus, an alternate way of expressing Equation 5.19 is

$$t_{B/A} = \bar{x}(Area)_{A-B} \qquad (5.20)$$

The quantity $(Area)_{A-B}$ in Equation 5.20 represents the area under the M/EI diagram between the points A and B and $\bar{x}$ is the distance from the centroid of this area to point B at which the tangential deviation is desired. *Note that the first symbol in the subscript of the tangential deviation represents the point at which the tangential deviation is desired and about which the first moment is taken and, consequently $t_{B/A} \neq t_{A/B}$.* Therefore, the second theorem of the *area–moment* method may be stated as follows:

> The tangential deviation $t_{B/A}$ at any point B on the elastic curve with respect to the tangent at any other point A on this curve is equal to the first moment about point B of the area under the M/EI diagram between A and B.

In general, tangential deviations are not equal to beam deflections. However, a given beam deflection may be expressed geometrically in terms of tangential deviations, which may then be found using Equation 5.20. Also, note that *a positive M/EI diagram yields a positive tangential deviation, which implies that the point in question is above the corresponding tangent. A negative M/EI diagram, on the other hand, yields a negative tangential deviation implying that the point in question is below the corresponding tangent.* Examples of both a positive and negative tangential deviation are shown in Figure 5.5 where beam ABC is subjected to a concentrated couple at B. The elastic curve is shown in red in an exaggerated fashion. Since the M/EI diagram in segment AB is positive, it follows that $t_{B/A}$ is also positive, meaning that point B is above the tangent drawn at point A as shown in Figure 5.5. On the other hand, the M/EI diagram in segment BC is negative, yielding a negative tangential deviation $t_{C/B}$, which means that point C is below the tangent drawn at point B as indicated in Figure 5.5.

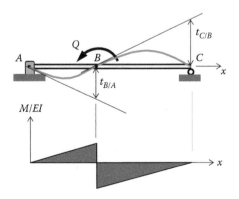

FIGURE 5.5   Beam subjected to a concentrated couple at B showing the M/EI diagram.

5.6.1 MOMENT DIAGRAMS BY CANTILEVER PARTS

When using the *area–moment* method, it becomes necessary to find the areas of the M/EI diagrams and the location of their centroids in order to determine changes in slopes, as well as tangential deviations. This process is greatly simplified if we construct the M/EI for each of the loads acting on the beam and find, separately, the changes in slopes and tangential deviations due to each of the loads and superpose the various components to obtain the required answers. One convenient way of accomplishing this purpose is the method of *cantilever parts*. This method breaks down the original beam into several cantilever beams, each loaded with one of the several loads (applied and reactive) acting on the original beam. The location of the fixed end of these cantilevers may be selected arbitrarily, because the bending moments in a beam are physical quantities that are independent of the choice of an origin for the variable x in terms of which these moments may be expressed. Example 5.7 illustrates the procedure.

EXAMPLE 5.7

Consider the beam shown in figure (a) below and construct the moment diagram using cantilever parts assuming the fixed end of the cantilevers to be (a) at B and (b) at C.

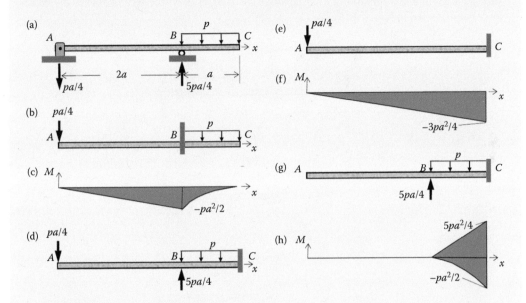

SOLUTION

a. When the fixed end of the cantilevers is placed at B, two cantilevers, BA, and BC, are created as shown in figure (b) above. Cantilever BA, shown in figure (b) above, carries a concentrated load at A (the given support reaction at A), which produces a negative triangular moment diagram with a maximum negative value of $-pa^2/2$. Cantilever BC, also shown in figure (b) above, on the other hand, is subjected to the applied uniform load p producing the parabolic negative moment with a maximum negative value of $-pa^2/2$. Thus, the total moment diagram consists of two simple geometric shapes, as shown in figure (c) above, whose areas and centroid locations may be easily determined.

b. When the fixed end of the cantilevers is placed at C, as shown in figure (d) above, three cantilevers, one AC and two BCs, are created. Cantilever AC, shown in figure (e) above, carries a concentrated load at A (the given support reaction at A), which produces a negative triangular moment diagram, shown in figure (f) above, with a maximum negative value of $-3pa^2/4$. As shown in figure (g) above, the two cantilevers labeled BC are combined into one carrying an upward concentrated load (the given support reaction at B) and the downward applied uniform load p. The upward concentrated load produces a positive triangular moment diagram with a maximum value of $5pa^2/4$ and, the downward uniform load, a negative parabolic moment with a maximum negative value of $-pa^2/2$. Both of these moment diagrams are shown in figure (h) above. Note that the algebraic sum of the moments of the cantilevers at any point along the beam must yield the actual moment in the given beam. Thus, for example, the algebraic sum of the three moments at A is zero, as it should be for a free end.

EXAMPLE 5.8

Use the area–moment method to find the slope and deflection at the free end A for the cantilever beam shown in figure (a) below.

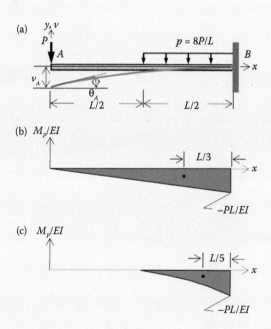

SOLUTION

A tangent constructed at B is a horizontal line as shown in figure (a) above. Therefore, the slope to the elastic curve at B, θ_B, is zero. Also, in this case, the vertical distance from the elastic curve at A to the horizontal tangent at B, $t_{A/B}$, is equal to the deflection v_A. Thus, by Equation 5.17, we have

$$0 - \theta_A = -\left[\frac{PL^2}{2EI} + \frac{PL^2}{6EI}\right] \qquad (5.8.1)$$

$$\theta_A = \frac{2PL^2}{3EI} \qquad \textbf{ANS.}$$

The two terms inside the square brackets of Equation 5.8.1 represent the areas of the M/EI diagrams of the triangle (figure (b) above) and the parabola (figure (c) above), respectively, the values of which were determined using Appendix C.3.

Also, by Equation 5.20, we obtain

$$v_A = t_{A/B} = -\left[\left(\frac{PL^2}{2EI}\right)\left(\frac{2L}{3}\right) + \left(\frac{PL^2}{6EI}\right)\left(\frac{4L}{5}\right)\right] = -\frac{7PL^3}{15EI} \qquad \textbf{ANS.}$$

Note that while the slope is positive, indicating a counterclockwise angle, the deflection is negative or downward.

EXAMPLE 5.9

Use the area–moment method to find the slope at A, θ_A, and the deflection at C, v_C for the beam shown in figure (a) below, where the support reactions at A and B are given as indicated.

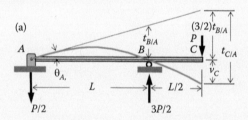

SOLUTION

The elastic curve is sketched and shown in red in figure (a) above. A tangent to the elastic curve is drawn at A and the quantities θ_A, $t_{B/A}$, and $t_{C/A}$ indicated in figure (a) above.

Using cantilever parts, we set the fixed end of the cantilevers at C. This leads to two cantilevers, one carrying a downward concentrated load of $P/2$ (the support reaction at A) and another carrying an upward concentrated load of $3P/2$ (the support reaction at B). Both of these cantilevers are shown in figure (b) below. The M/EI diagram corresponding to the load $P/2$ is shown in figure (c) below, and that corresponding to the load $3P/2$ in figure (d) below.

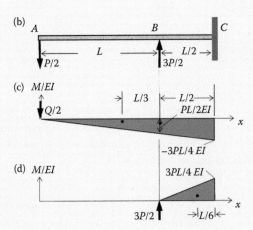

The slope at A may be expressed in terms of tangential deviation $t_{B/A}$. Thus,

$$\theta_A = \frac{|t_{B/A}|}{L} = \left| \frac{1}{2}\left(\frac{-PL}{2EI}\right)(L)\frac{(L/3)}{L} \right|$$

$$= \frac{(PL^3/12EI)}{L} = \frac{PL^2}{12EI} \quad \text{ANS.}$$

The deflection at C may also be expressed in terms of tangential deviations. Thus, as shown in figure (a) above, we may write

$$v_C = t_{C/A} - \frac{3}{2}t_{B/A} = \frac{1}{2}\left(\frac{-3PL}{4EI}\right)\left(\frac{3L}{2}\right)\left(\frac{L}{2}\right) + \left(\frac{1}{2}\right)\left(\frac{3PL}{4EI}\right)\left(\frac{L}{2}\right)\left(\frac{L}{6}\right)$$

$$- \left(\frac{3}{2}\right)\left(\frac{-PL^3}{12EI}\right) = -\frac{PL^3}{8EI} \quad \text{ANS.}$$

Note that the indicated tangential deviations were computed by Equation 5.20 using Appendix C.3 along with the M/EI diagrams shown in figures (c) and (d) above.

EXAMPLE 5.10

Refer to the beam of Example 5.9 and determine the location and magnitude of the maximum deflection between supports A and B.

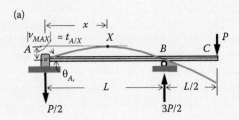

SOLUTION

For convenience, the beam of Example 5.9 is repeated here and shown in figure (a) above. It is obvious that, in this case, the maximum deflection between A and B does not occur at the center between the two supports and, therefore, its exact location must be determined. If we assume this location to be at point X, a distance x from point A, then the tangent to the elastic curve at this point is horizontal ($\theta_X = 0$) as shown in figure (a) above. In order to determine x, we need to use the first theorem of the area–moment method and relate the slope at X to the slope at one of the two supports in the beam. Therefore, the first step in this type of solution is to find the slope at one of the two supports on either side of point X. In our case, the slope at A was determined in Example 5.9 to be $\theta_A = PL^2/12EI$.

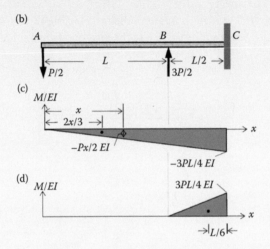

As in Example 5.9, we assume the fixed end of the cantilevers to be at C and construct the M/EI diagrams by cantilever parts as shown in figures (b), (c), and (d) above. Of course, other points, such as B, may be chosen for the fixed end of these cantilevers, but the final answer will be the same.

Thus, by the first theorem, we have

$$0 - \frac{PL^2}{12EI} = \frac{1}{2}\left(\frac{-Px}{2EI}\right)(x) \qquad\qquad (5.10.1)$$

Simplifying Equation 5.10.1 and solving for x yields

$$x = \pm\frac{L}{\sqrt{3}} \qquad \textbf{ANS.}$$

Of course, the negative answer is rejected on physical grounds and only the positive answer is accepted, which places the maximum deflection at a point to the right of the midpoint between supports A and B.

Having the value of x, the maximum deflection may now be determined. As shown in figure (a) above, the maximum deflection is equal in magnitude to the tangential deviation of A

with respect to X (the maximum deflection may also be equated to the tangential deviation of B with respect to X). Thus,

$$v_{MAX} = |t_{A/X}| = \left|\left(\frac{1}{2}\right)\left(\frac{-Px}{2EI}\right)(x)\left(\frac{2x}{3}\right)\right| = \frac{Px^3}{6EI} = \frac{P\left(L/\sqrt{3}\right)^3}{6EI} = \frac{PL^3}{18\sqrt{3}EI} \qquad \textbf{ANS.}$$

EXAMPLE 5.11

The cantilever beam ABC, shown in figure (a) below, consists of two segments AB (flexural rigidity EI) and CD (flexural rigidity $2EI$). Use the area–moment method to find the slope and deflection at A.

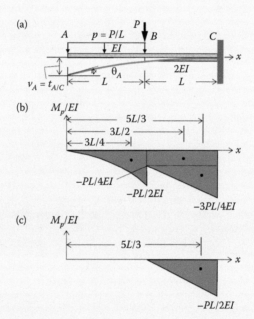

SOLUTION

The moment diagrams are constructed by cantilever parts placing the fixed end of these cantilevers at C. Since the flexural rigidity is not the same for the entire beam, the ordinates to the moment diagrams have to be divided by the corresponding flexural rigidity in order to obtain the correct M/EI diagrams. Thus, the moment diagram produced by the uniform load p is divided by EI in segment AB and by $2EI$ in segment BC to obtain the M/EI diagram shown in figure (b) above. Also, the moment diagram produced by the concentrated load P, which is entirely in segment BC, is divided by $2EI$ to obtain the M/EI diagram shown in figure (c) above.

The slope at A is found by the use of the first theorem of the area–moment method. In applying this theorem, it is convenient to subdivide the area in figure (b) above into three simple geometric shapes: a second-degree parabola, a rectangle, and a triangle as shown. Thus, since the slope at C is zero ($\theta_C = 0$), we obtain

$$0 - \theta_A = -\left[\frac{1}{3}\left(\frac{PL}{2EI}\right)(L) + \left(\frac{PL}{4EI}\right)(L) + \frac{1}{2}\left(\frac{PL}{2EI}\right)(L) + \frac{1}{2}\left(\frac{PL}{2EI}\right)(L)\right] = -\frac{11PL^2}{12EI}$$

$$\theta_A = \frac{11PL^2}{12EI} \quad \textbf{ANS.}$$

As noted in figure (a) above, $v_A = t_{A/C}$. Thus, using the second theorem of the area–moment method, we obtain

$$v_A = t_{A/C} = -\left[\frac{1}{3}\left(\frac{PL}{2EI}\right)(L)\left(\frac{3L}{4}\right) + \left(\frac{PL}{4EI}\right)(L)\left(\frac{3L}{2}\right) + \frac{1}{2}\left(\frac{PL}{2EI}\right)(L)\left(\frac{5L}{3}\right)\right.$$

$$\left. + \frac{1}{2}\left(\frac{PL}{2EI}\right)(L)\left(\frac{5L}{3}\right)\right]$$

$$v_A = -\frac{4PL^3}{3EI} \quad \textbf{ANS.}$$

Note that while the slope at A is positive (counterclockwise), the deflection is negative (downward).

PROBLEMS

5.78 Use the area–moment method to determine the slope and deflection at A. Assume a constant EI for the entire beam.

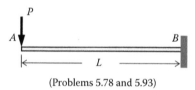

(Problems 5.78 and 5.93)

5.79 Use the area–moment method to determine the slope and deflection at A. Assume a constant EI for the entire beam.

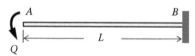

5.80 Use the area–moment method to determine the slope and deflection at A. Assume a constant EI for the entire beam.

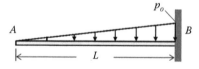

5.81 Use the area–moment method to determine the slope and deflection at A. Assume a constant EI for the entire beam.

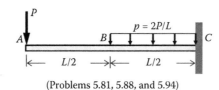

(Problems 5.81, 5.88, and 5.94)

5.82 Use the area–moment method to determine the slope and deflection at A. Assume a constant EI for the entire beam.

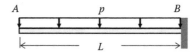

5.83 Use the area–moment method to determine the slope and deflection at A. Assume a constant EI for the entire beam.

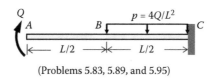

(Problems 5.83, 5.89, and 5.95)

5.84 Use the area–moment method to determine the slope and deflection at A. Assume a constant EI for the entire beam.

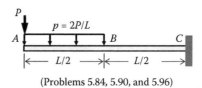

(Problems 5.84, 5.90, and 5.96)

5.85 Use the area–moment method to determine the slope and deflection at A. Assume a constant EI for the entire beam.

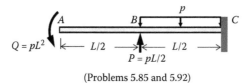

(Problems 5.85 and 5.92)

5.86 Use the area–moment method to determine the slope and deflection at A. Assume a constant EI for the entire beam.

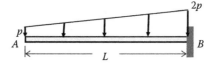

5.87 Use the area–moment method to determine the slope and deflection at A. Assume a constant EI for the entire beam.

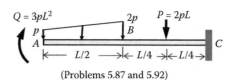

(Problems 5.87 and 5.92)

5.88 In reference to the figure shown in Problem 5.81, the flexural rigidity in segment *AB* of the cantilever beam is *EI* and that in segment *BC* is 2*EI*. Use the area–moment theorems to find the slope and deflection at *A*.

5.89 In reference to the figure shown in Problem 5.83, the flexural rigidity in segment *AB* of the cantilever beam is *EI*, and that in segment *BC* is 2*EI*. Use the area–moment theorems to find the slope and deflection at *A*.

5.90 In reference to the figure shown in Problem 5.84, the flexural rigidity in segment *AB* of the cantilever beam is *EI* and that in segment *BC* is 2*EI*. Use the area–moment theorems to find the slope and deflection at *A*.

5.91 In reference to the figure shown in Problem 5.85, the flexural rigidity in segment *AB* of the cantilever beam is *EI* and that in segment *BC* is 2*EI*. Use the area–moment theorems to find the slope and deflection at *A*.

5.92 In reference to the figure shown in Problem 5.87, the flexural rigidity in segment *AB* of the cantilever beam is *EI* and that in segment *BC* is 2*EI*. Use the area–moment theorems to find the slope and deflection at *A*.

5.93 A W14 × 74 steel ($E = 30 \times 10^3$ ksi) section is used for this beam. Let $P = 20$ kips and $L = 18$, determine the slope and deflection at *A*. Assume bending about the strong axis and a constant *EI*. Refer to the figure shown in Problem 5.78.

5.94 A W406 × 84.8 steel ($E = 205$ GPa) section is used for this beam. Let $p = 2$ kN/m and $L = 6$ m, determine the slope and deflection at *A*. Assume bending about the strong axis and a constant *EI*. Refer to the beam shown in Problem 5.81.

5.95 A W12 × 50 steel ($E = 29 \times 10^3$ ksi) section is used for this beam. However, segment *BC* is reinforced by welding two cover plates as shown in the sketch. Let $Q = 32$ kip · ft and $L = 8$ ft, determine the slope and deflection at *A*. Refer to the beam shown in Problem 5.83.

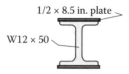

5.96 A W457 × 157.7 steel ($E = 200$ GPa) section is used for this beam. However, segment *BC* is reinforced by welding two cover plates as shown in the sketch. Let $P = 100$ kN and $L = 4$ m, determine the slope and deflection at *A*. Refer to the beam shown in Problem 5.87.

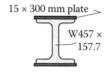

5.97 Use the area–moment to find the slope at *A* and the deflection at *C*. Assume a constant *EI*.

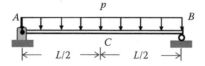

5.98 Use the area–moment to find the slope at *A* and the deflection at *C*. Assume a constant *EI*.

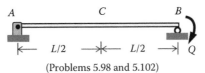

(Problems 5.98 and 5.102)

5.99 Use the area–moment to find the slope at *A* and the deflection at *C*. Assume a constant *EI*.

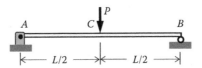

(Problems 5.99, 5.103, and 5.106)

5.100 Use the area–moment to find the slope at *A* and the deflection at *C*. Assume a constant *EI*.

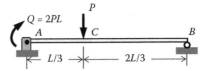

(Problems 5.100, 5.104, and 5.107)

5.101 Use the area–moment to find the slope at *A* and the deflection at *C*. Assume a constant *EI*.

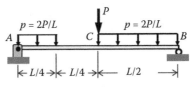

(Problems 5.101 and 5.108)

5.102 In reference to the figure shown in Problem 5.98, assume a constant *EI* and find the location and magnitude of the maximum deflection by the area–moment method.

5.103 In reference to the figure shown in Problem 5.99, assume a constant *EI* and find the location and magnitude of the maximum deflection by the area–moment method.

5.104 In reference to the figure shown in Problem 5.100, assume a constant *EI* and find the location and magnitude of the maximum deflection by the area–moment method.

5.105 Assume a constant *EI* and find the location and magnitude of the maximum deflection by the area–moment method.

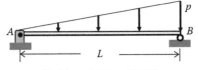

(Problems 5.105 and 5.109)

5.106 In reference to the figure shown in Problem 5.99, segment *AC* has a constant flexural rigidity *EI* and segment *CB* a constant flexural rigidity 2*EI*. Find the slope at *B* and the deflection at *C*.

5.107 In reference to the figure shown in Problem 5.100, segment *AC* has a constant flexural rigidity *EI* and segment *CB* a constant flexural rigidity 2*EI*. Find the slope at *B* and the deflection at *C*.

5.108 In reference to the figure shown in Problem 5.101, segment AC has a constant flexural rigidity EI and segment CB a constant flexural rigidity $2EI$. Find the slope at B and the deflection at C.

5.109 In reference to the figure shown in Problem 5.105, an S20 × 75 steel ($E = 29 \times 10^3$ ksi) section is used for this beam. Let $p = 2.5$ kips/ft and $L = 20$ ft, determine the slope at A and the location and magnitude of the maximum deflection. Assume bending about the strong axis and a constant EI.

5.110 An S457 × 104.1 steel ($E = 200$ GPa) section is used for this beam. Let $P = 50$ kN and $L = 2$ m, determine the slope at A and location and magnitude of the maximum deflection. Assume bending about the strong axis and a constant EI.

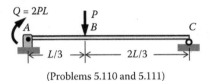

(Problems 5.110 and 5.111)

5.111 In reference to the figure shown in Problem 5.110, a W12 × 30 steel ($E = 30 \times 10^3$ ksi) is used for this beam. However, segment BC is strengthened by welding two 1/2 × 7 in. plates, one on top and one on the bottom, similar to that shown for Problems 5.95 and 5.96. Let $P = 1$ kip and $L = 6$ ft, find the slope at A and the deflection at B.

5.112 A W356 × 101.2 steel ($E = 205$ GPa) is used for this beam. However, segment BC is strengthened by welding two 15 × 270 mm plates, one on top and one on the bottom, similar to that shown for Problems 5.95 and 5.96. Let $P = 75$ kN and $L = 3$ m, determine the slope at C and the deflection at B.

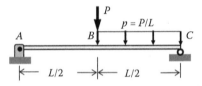

5.113 Assume a constant EI value for the entire beam and determine the slope and deflection at A.

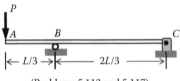

(Problems 5.113 and 5.117)

5.114 Assume a constant EI value for the entire beam and determine the slope and deflection at A.

(Problems 5.114 and 5.118)

5.115 Assume a constant EI value for the entire beam and determine the slope and deflection at A.

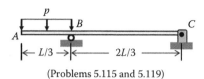

(Problems 5.115 and 5.119)

5.116 Assume a constant EI value for the entire beam and determine the slope and deflection at A.

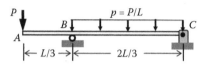

(Problems 5.116 and 5.120)

5.117 In reference to the figure shown in Problem 5.113, find the location and magnitude of the maximum deflection between supports B and C. Assume a constant EI value for the entire beam.

5.118 In reference to the figure shown in Problem 5.114, find the location and magnitude of the maximum deflection between supports B and C. Assume a constant EI value for the entire beam.

5.119 In reference to the figure shown in Problem 5.115, find the location and magnitude of the maximum deflection between supports B and C. Assume a constant EI value for the entire beam.

5.120 In reference to the figure shown in Problem 5.116, find the location and magnitude of the maximum deflection between supports B and C. Assume a constant EI value for the entire beam.

5.121 A W12 × 30 steel ($E = 29 \times 10^3$ ksi) is used for this beam. However, segment AB is strengthened by welding two 1/2 × 7 in. plates, one on top and one on the bottom, similar to that shown for Problems 5.95 and 5.96. Let $p = 0.5$ kips/ft and $L = 12$ ft, find the slope at A and the deflection at C.

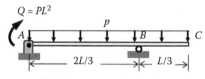

(Problems 5.121 and 5.123)

5.122 A W356 × 101.2 steel ($E = 200$ GPa) is used for this beam. However, segment AB is strengthened by welding two 15 × 270 mm plates, one on top and one on the bottom, similar to that shown for Problems 5.95 and 5.96. Let $p = 6$ kN/m and $L = 4$ m, determine the slope at A and the deflection at C.

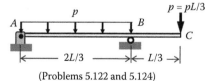

(Problems 5.122 and 5.124)

5.123 Refer to Problem 5.121 and find the location and magnitude of the maximum deflection between supports A and B.

5.124 Refer to Problem 5.122 and find the location and magnitude of the maximum deflection between supports A and B.

5.7 STATICALLY INDETERMINATE BEAMS: TWO SUCCESSIVE INTEGRATIONS

Statically indeterminate beams are analyzed using the same basic procedure discussed in Section 1.6 for statically indeterminate members under axial loads and in Section 2.4 for statically indeterminate members under torsional loads. The applicable equations of equilibrium need to be supplemented by additional relations satisfying the deformation characteristics of the beam. When using the method of two successive integrations, the moment equation for the beam (i.e., the differential equation for the elastic curve) is written in terms of unknown forces and/or moments. Thus, in addition to the two constants of integration, the slope and deflection functions would contain one or more unknown quantities that are evaluated from the boundary conditions of the problem. The following examples illustrate the method of solution.

Examples 5.12 shows the solution of a statically indeterminate beam with one *redundant support reaction*. In other words, this beam is statically indeterminate to the first degree. The term *redundant* was first discussed in Section 1.6, and the reader is urged to review this section. Example 5.13 illustrates the solution of a statically indeterminate beam with two redundant support reactions. Finally, Example 5.14 shows the solution for a statically indeterminate beam with one redundant support that is deformable and subjected to a temperature change. In all of these examples, the available equations of equilibrium are not enough to complete the solution and, therefore, additional conditions related to the deformation of the beam and/or to deformable supports are required to complete the solution.

EXAMPLE 5.12

The beam shown in figure (a) below is simply supported at A and fixed at B. It carries a load that varies linearly in intensity from zero at A to p_o at B. Determine the unknown reactive forces and moments at the two supports.

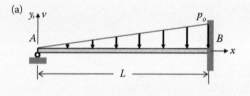

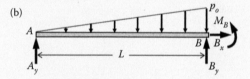

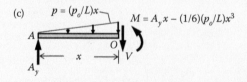

SOLUTION

The free-body diagram of the beam, given in figure (b) above, shows that there are four unknown reactive components, A_y, B_x, B_y, and M_B. Since there are only three applicable equations of equilibrium, the beam is statically indeterminate to the first degree (i.e., it has one redundant support reaction). These three equations are applied as follows:

$$\sum F_x = 0 \Rightarrow B_x = 0 \qquad \textbf{ANS.}$$

$$\sum F_y = 0 \Rightarrow A_y + B_y - \frac{p_o L}{2} = 0 \qquad (5.12.1)$$

$$\sum M_B = 0 \Rightarrow \frac{p_o L^2}{6} + M_B - A_y L = 0 \qquad (5.12.2)$$

Equations 5.12.1 and 5.12.2 contain the three unknown quantities A_y, B_y, and M_B. Thus, one more equation is needed to complete the solution. This third equation is obtained from knowledge of the deflection characteristics of the beam along with the method of two successive integrations. Referring to figure (c) above, we have

$$EIv'' = M = A_y x - \frac{p_o}{6L} x^3$$

Performing two integrations leads to

$$EIv' = \frac{A_y}{2} x^2 - \frac{p_o}{24L} x^4 + C_1 \qquad (5.12.3)$$

$$EIv = \frac{A_y}{6} x^3 - \frac{p_o}{120L} x^5 + C_1 x + C_2 \qquad (5.12.4)$$

The symbols C_1, C_2, and A_y represent three unknown quantities that can be determined from three boundary conditions that exist at the supports in the given beam. These three boundary conditions are BC1: $x = 0$, $v = 0$; BC2: $x = L$, $v = 0$; and BC3: $x = L$, $v' = 0$.

Substitution of these boundary conditions into Equations 5.12.3 and 5.12.4 leads to three simultaneous equations that yield

$$C_1 = -\frac{p_o L^3}{120}; \quad C_2 = 0$$

and

$$A_y = \frac{p_o L}{10} \qquad \textbf{ANS.}$$

Substitution of A_y into Equations 5.12.1 and 5.12.2 leads to two simultaneous equations in B_y and M_B. Their solution yields

$$B_y = \frac{4 p_o L}{10} \qquad \textbf{ANS.}$$

and

$$M_B = -\frac{p_o L^2}{15} \qquad \textbf{ANS.}$$

EXAMPLE 5.13

Beam AB is fixed at both ends and carries a load that varies in intensity from zero at A to p_o at B according to the relation $p = (x/L)^2 p_o$ as shown in figure (a) below. Find the reactive forces and moments at the two supports.

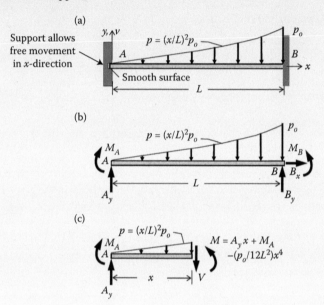

SOLUTION

The free-body diagram of the entire beam is given in figure (b) above, showing that we have a total of five unknown reactive components: A_y, B_x, B_y, M_A, and M_B. Note that the support at A is assumed to allow displacement in the x direction but no rotation about the z axis. This assumption, of course, is a simplification that allows us to eliminate the reactive force A_x. *We will continue to make such an assumption in the solution of problems related to beams fixed at both ends, even though the drawing may not indicate the feature shown in figure (a) above at support A.*

Only three equations of equilibrium are available for the solution (i.e., second-degree redundancy). Thus,

$$\sum F_x = 0 \Rightarrow B_x = 0 \qquad \textbf{ANS.}$$

$$\sum M_B = 0 \Rightarrow M_B + \frac{p_o}{(12L^2)} - A_y L - M_A = 0 \qquad (5.13.1)$$

$$\sum F_y = 0 \Rightarrow A_y + B_y - \frac{(p_o L)}{3} = 0 \tag{5.13.2}$$

Equations 5.13.1 and 5.13.2 contain four unknown quantities: A_y, B_y, M_A, and M_B. Thus, two more equations are needed to complete the solution. These two additional equations are obtained from knowledge of the deflection characteristics of the beam along with the method of two successive integrations. Referring to figure (c) above, we have

$$EIv'' = M = A_y x + M_A - \left(\frac{p_o}{12L^2}\right) x^4$$

Performing two integrations leads to

$$EIv' = \left(\frac{A_y}{2}\right) x^2 + M_A x - \left(\frac{p_o}{60L^2}\right) x^5 + C_1 \tag{5.13.3}$$

$$EIv = \left(\frac{A_y}{6}\right) x^3 + \left(\frac{M_A}{2}\right) x^2 - \left(\frac{p_o}{360L^2}\right) x^6 + C_1 x + C_2 \tag{5.13.4}$$

The quantities C_1, C_2, A_y, and M_A represent four unknown quantities that can be determined from four boundary conditions that are available at the supports in the given beam. These four boundary conditions are: BC1: $x = 0$, $v = 0$; BC2: $x = 0$, $v' = 0$; BC3: $x = L$, $v = 0$; and BC4: $x = L$, $v' = 0$.

Substitution of these boundary conditions into Equations 5.13.3 and 5.13.4 leads to four simultaneous equations that yield $C_1 = C_2 = 0$ and

$$A_y = \frac{p_o L}{15} \qquad \textbf{ANS.}$$

and

$$M_A = -\frac{p_o L^2}{60} \qquad \textbf{ANS.}$$

Substituting the values of A_y and M_A into Equations 5.13.1 and 5.13.2 leads to two simultaneous equations in B_y and M_B. Their solution yields

$$B_y = \frac{4 p_o L}{15} \qquad \textbf{ANS.}$$

and

$$M_B = -\frac{p_o L^2}{30} \qquad \textbf{ANS.}$$

EXAMPLE 5.14

The steel beam AB is supported at A, as shown in figure (a) below, by the aluminum rod AC of length ℓ and cross-sectional area A. The beam is fixed at B and carries a uniform load of intensity p. The coefficient of thermal expansion for aluminum is α and its modulus of elasticity is E_A. The moment of inertia for the beam is I and its modulus of elasticity is E_S. The rod is unstressed before the load q is applied. If the temperature of the rod is dropped by ΔT, find the force in the rod and the reaction components at support B.

(a)

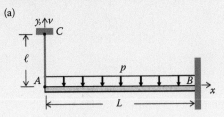

SOLUTION

The free-body diagram of beam AB is shown in figure (b) below, which indicates three unknown reactive forces:

The force in the rod F and the reactive components at the fixed support, B_y and M_B. There are only two available equations of equilibrium. Thus,

$$\sum F_y = 0 \Rightarrow F + B_y - pL = 0 \tag{5.14.1}$$

$$\sum M_B = 0 \Rightarrow M_B + \frac{pL^2}{2} - FL = 0 \tag{5.14.2}$$

Equations 5.14.1 and 5.14.2 contain three unknown quantities, F, B_y, and M_B. An additional equation is, therefore, needed to obtain a solution. This additional equation is obtained from the deflection characteristics of the beam and the rod.

As the temperature drops through ΔT, the rod tends to shorten by the amount $\delta_T = \alpha\,(\Delta T)\ell$. However, the force F, which represents the restraining effect of the beam, reduces this shortening tendency by the amount $\delta_F = F\ell/AE_A$. Thus, if we assume that $\delta_T < \delta_F$, the net deflection of the rod, $\delta = \delta_F - \delta_T = F\ell/AE_A - \alpha\,(\Delta T)\ell$.

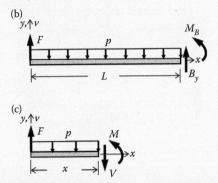

The free-body diagram of a segment of the beam of length x is shown in figure (c) above, from which the differential equation of the elastic curve of the beam becomes

$$E_S I v'' = M = Fx - \left(\frac{p}{2}\right)x^2$$

Two successive integrations yield

$$E_S I v' = \left(\frac{F}{2}\right)x^2 - \left(\frac{p}{6}\right)x^3 + C_1 \tag{5.14.3}$$

$$E_S I v = \left(\frac{F}{6}\right)x^3 - \left(\frac{p}{24}\right)x^4 + C_1 x + C_2 \tag{5.14.4}$$

The three unknown quantities, F, C_1, and C_2, are found from three boundary conditions, which are BC1: $x = 0$, $v = -(F\ell/AE_A - \alpha (\Delta T)\ell)$; BC2: $x = L$, $v = 0$; and BC3: $x = L$, $v' = 0$.

Substitution of these boundary conditions into Equations 5.14.3 and 5.14.4 leads to the following value of F:

$$F = \left(\frac{3AE_A}{8}\right)\left[\frac{pL^4 + 8E_S I\alpha(\Delta T)\ell}{AE_A L^3 + 3IE_S \ell}\right] \qquad \textbf{ANS.}$$

If this value of F is used in Equations 5.14.1 and 5.14.2, we obtain the unknown quantities B_y and M_B. Thus,

$$B_y = qL - \left(\frac{3AE_A}{8}\right)\left[\frac{pL^4 + 8E_S I\alpha(\Delta T)\ell}{AE_A L^3 + 3IE_S \ell}\right] \qquad \textbf{ANS.}$$

$$M_B = \left(\frac{3LAE_A}{8}\right)\left[\frac{pL^4 + 8E_S I\alpha(\Delta T)\ell}{AE_A L^3 + 3IE_S \ell}\right] - \frac{pL^2}{2} \qquad \textbf{ANS.}$$

PROBLEMS

5.125 Use two successive integrations to find the support reactions at A and B.

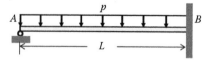

5.126 Use two successive integrations to find the support reactions at A and B.

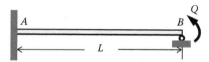

5.127 Use two successive integrations to find the support reactions at A and B.

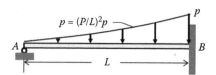

5.128 Use two successive integrations to find the support reactions at A and B.

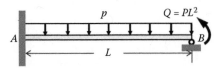

5.129 Use two successive integrations to find the support reactions at A and C.

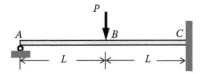

5.130 Use two successive integrations to find the support reactions at A and C.

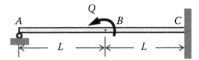

5.131 Use two successive integrations to find the support reactions at A and C.

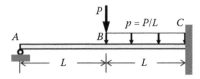

5.132 Use two successive integrations to find the support reactions at A and C.

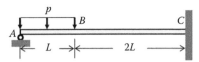

5.133 Determine the support reactions at B and C. Use two successive integrations along with the given xv coordinate system with origin at B.

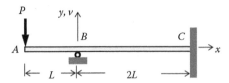

5.134 Determine the support reactions at B and C. Use two successive integrations along with the given xv coordinate system with origin at B.

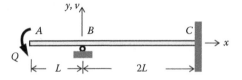

5.135 Determine the support reactions at *B* and *C*. Use two successive integrations along with the given *xv* coordinate system with origin at *B*.

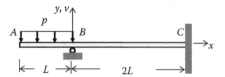

5.136 Determine the support reactions at *B* and *C*. Use two successive integrations along with the given *xv* coordinate system with origin at *B*.

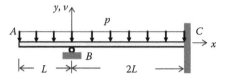

5.137 The overhanging beam *ABC* carries a uniform load of intensity *p* over the span between *B* and *C*. A couple Q_A is applied at *A* as shown. Let $Q_A = pL^2$, determine the reaction components at *B* and *C*. Express your answers in terms of *p* and *L*.

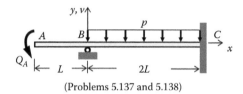

(Problems 5.137 and 5.138)

5.138 In reference to the figure shown in Problem 5.137, the overhanging beam *ABC* carries a uniform load of intensity *p* over the span between *B* and *C*. A couple Q_A is applied at *A* as shown. Determine the magnitude of Q_A if the slope at *B* is zero.

5.139 The aluminum ($E_A = 10 \times 10^3$ ksi) cantilever beam *ABC* is supported at *B* by a steel ($E_S = 30 \times 10^3$ ksi) rod *BD* of length $\ell = 15$ ft and cross-sectional area $A = 0.5$ in.². It carries a uniform load of intensity $p = 0.8$ kips/ft over the entire length. If $L = 12$ ft and the moment of inertia for the aluminum beam is 600 in.⁴, determine (a) the stress in the steel rod, (b) the deflection of the beam at *B*, and (c) the reaction components at *C*.

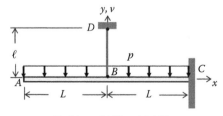

(Problems 5.139 and 5.140)

5.140 In reference to the figure shown in Problem 5.139, the steel ($E = 200$ GPa) cantilever beam *ABC* is supported at *B* by a steel rod *BD* of length $\ell = 4$ m and cross-sectional area $A = 8.0 \times 10^{-4}$ m². It carries a uniform load of intensity $p = 10$ kN/m over the entire length. If $L = 4$ m and the moment of inertia for the aluminum beam is 200×10^{-6} m⁴, determine (a) the stress in the steel rod, (b) the deflection of the beam at *B*, and (c) the reaction components at *C*.

5.8   STATICALLY INDETERMINATE BEAMS: SUPERPOSITION

The concept of superposition has already been used in several cases in previous chapters. In finding the deflections and slopes produced by several loads acting on a beam, we determine the deflection and slope due to each load acting separately and then find the resultant deflection and slope by superposing (i.e., combining algebraically) these separate effects. The basic concept underlying the method of superposition requires the decomposition of the given statically indeterminate beam into two or more statically determinate ones for which solutions may be easily determined, or for which solutions already exist similar to those shown in Appendix G. In a given statically indeterminate beam, the method of superposition is used to satisfy the boundary conditions that exist at the beam supports. This procedure leads to one or more relations among the unknown quantities that supplement the available equations of equilibrium. The use of this method is illustrated in Examples 5.15, 5.16, and 5.17.

EXAMPLE 5.15

Repeat Example 5.12 using the method of superposition.

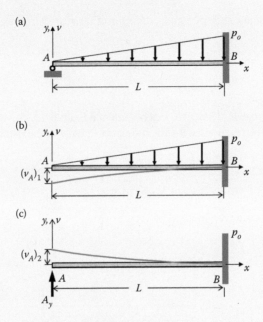

SOLUTION

For convenience, the beam of Example 5.12 is repeated in figure (a) above. The effect of the simple support at A is to provide a vertical reactive force A_y that keeps the beam from deflecting downward at that location. If this redundant support were removed momentarily, the remaining structure would be a statically determinate cantilever beam fixed at B and subjected to the applied linearly varying load as shown in figure (b) above. This cantilever beam would deflect downward as shown by the red curve resulting in the downward deflection $(v_A)_1$ at A as shown in figure (b) above. Obviously, the deflection $(v_A)_1$ cannot take place because of the reactive force A_y. Thus, this reactive force acting at the free end of the cantilever beam would have to be of a magnitude just sufficient to produce an upward deflection $(v_A)_2$ as shown in figure (c) above, equal in magnitude to the downward deflection $(v_A)_1$ in order for the resultant deflection at A to be zero. In other words, the algebraic sum of $(v_A)_1$ and $(v_A)_2$ must be zero. Therefore

$$(v_A)_1 + (v_A)_2 = 0 \qquad (5.15.1)$$

Thus, the whole process of analyzing this problem may be summed up by stating that the given statically indeterminate beam shown in figure (a) above is equivalent to the superposition (algebraic sum) of the two statically determinate beams shown in figures (b) and (c) above.

From the information contained in Appendix G, we determine the deflections $(v_A)_1$ and $(v_A)_2$. Thus,

$$(v_A)_1 = -\frac{p_o L^4}{30EI} \qquad (5.15.2)$$

$$(v_A)_2 = \frac{A_y L^3}{3EI} \qquad (5.15.3)$$

Substitution of Equations 5.15.2 and 5.15.3 into Equation 5.15.1 yields an equation in A_y, which may be solved to yield

$$A_y = \frac{p_o L}{10} \qquad \textbf{ANS.}$$

This answer, of course, is identical to that obtained in Example 5.12 by the method of two successive integrations. The reactions at the fixed support B may now be determined by equilibrium.

EXAMPLE 5.16

Repeat Example 5.14 using the method of superposition.

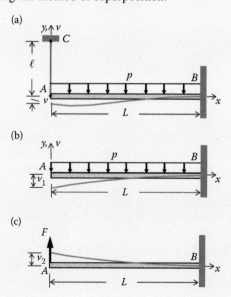

SOLUTION

For convenience, the beam of Example 5.14 is repeated in figure (a) above. As in Example 5.15, the beam is decomposed into two statically determinate beams as shown in figures (b) and (c) above. The condition that needs to be satisfied is that the algebraic sum of the deflections v_1 and v_2 of beams (b) and (c), respectively, be equal to the deflection v of beam (a). Thus,

$$v_1 + v_2 = v \tag{5.16.1}$$

From Appendix G

$$v_1 = -\frac{pL^4}{8E_S I} \tag{5.16.2}$$

$$v_2 = \frac{FL^3}{3E_S I} \tag{5.16.3}$$

The deflection v is also the deflection of rod AC and, as explained in Example 5.14, it is given by

$$v = -\left[\frac{FL}{AE_A} - \alpha(\Delta T)\ell\right] \tag{5.16.4}$$

The negative sign in Equation 5.16.4 is due to the fact that this deflection is downward.

Substituting Equations 5.16.2, 5.16.3, and 5.16.4 into Equation 5.16.1 and solving for F, we obtain

$$F = \left(\frac{3AE_A}{8}\right)\left[\frac{pL^4 + 8E_S I\alpha(\Delta T)\ell}{AE_A L^3 + 3IE_S \ell}\right] \qquad \textbf{ANS.}$$

This answer, of course, is identical to the answer found by the method of two successive integrations in Example 5.14. The support reactions at B may now be found by equilibrium.

EXAMPLE 5.17

The continuous beam ABC is supported and loaded as shown in figure (a) below. Before the load P is applied, there is a small gap δ between the beam and the center support at B. After the load P is applied, however, contact is made between the support at B and the beam so that the support at B carries part of the load. Determine the support reactions at A, B, and C.

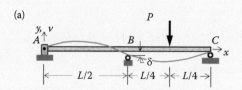

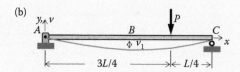

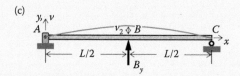

SOLUTION

The given statically indeterminate beam shown in figure (a) above is decomposed into two statically determinate beams as shown in figures (b) and (c) above. As indicated in figure (b) above, the applied load P produces a deflection v_1 at B. The unknown support reaction at B, B_y, produces a deflection v_2 as shown in figure (c) above. The magnitude of the deflection v_2 is less than that of v_1 by the amount δ. In other words, the algebraic sum of v_1 and v_2 must be equal to the negative of the gap δ. Therefore

$$v_1 + v_2 = -\delta \tag{5.17.1}$$

The deflections v_1 and v_2 are obtained from Appendix G. Thus,

$$v_1 = -\frac{11PL^3}{768EI} \tag{5.17.2}$$

$$v_2 = \frac{B_y L^3}{48EI} \tag{5.17.3}$$

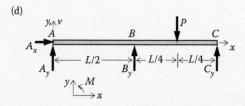

Substituting Equations 5.17.2 and 5.17.3 into Equation 5.17.1 and solving for B_y, we obtain

$$B_y = \frac{11P}{16} - \frac{48EI\delta}{L^3} \qquad \textbf{ANS.}$$

The free-body diagram shown in figure (d) above is now used to find the remaining support reactions by applying the three available equations of equilibrium. Thus,

$$\sum F_x = 0 \Rightarrow A_x = 0 \qquad \textbf{ANS.}$$

$$\sum M_C = 0 \Rightarrow A_y = \frac{24EI\delta}{L^3} - \frac{3P}{32} \qquad \textbf{ANS.}$$

$$\sum F_y = 0 \Rightarrow C_y = \frac{13P}{32} + \frac{24EI\delta}{L^3} \qquad \textbf{ANS.}$$

PROBLEMS

5.141 Repeat Problem 5.125 using the method of superposition.
5.142 Repeat Problem 5.128 using the method of superposition.
5.143 Repeat Problem 5.129 using the method of superposition.
5.144 Repeat Problem 5.131 using the method of superposition.
5.145 Repeat Problem 5.133 using the method of superposition.
5.146 Repeat Problem 5.134 using the method of superposition.
5.147 Repeat Problem 5.136 using the method of superposition.
5.148 Repeat Problem 5.137 using the method of superposition.
5.149 Repeat Problem 5.138 using the method of superposition.
5.150 Repeat Problem 5.139 using the method of superposition.
5.151 Beam AB is fixed at A and supported at B at the midpoint of beam CD, which is simply supported and perpendicular to beam AB. Both beams are steel for which $E = 30 \times 10^3$ ksi and have the same moment of inertia $I = 900$ in.^{4}. Using superposition, determine the force at B acting on beam CD and the reaction components at A. Also find the maximum deflection of beam CD. Let $P = 80$ kips, $L = 24$ ft, and $l = 21$ ft.

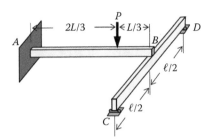

(Problems 5.151 and 5.152)

5.152 In reference to the figure shown in Problem 5.151, beam AB is fixed at A and supported at B at the midpoint of beam CD, which is simply supported and perpendicular to beam AB. Both beams are steel for which $E = 10 \times 10^3$ ksi and have the same moment of inertia $I = 1000$ in.^{4}. Determine the maximum permissible load P if the allowable deflection at B is 0.6 in. Let $L = 24$ ft and $\ell = 18$ ft.

5.153 Beam AB, carrying two concentrated loads P, is simply supported at A and B and a roller is placed at G that in turn rests on simply supported beam CD as shown. Determine the reaction components at A, B, C, and D. Express your answers in terms of P.

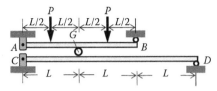

(Problems 5.153 and 5.154)

5.154 In reference to the figure shown in Problem 5.153, beam AB, carrying two concentrated loads P, is simply supported at A and B and a roller is placed at E, which rests on simply supported beam CD as shown. Let $L = 4$ m, determine the maximum permissible load P if the allowable deflection at G is 20 mm. Both beams are steel for which $E = 200$ GPa and $I = 300 \times 10^6$ mm^4.

5.155 The steel beam is fixed at C and supported by a steel rod at B. The modulus of elasticity for steel is E, the moment of inertia for the beam is I, and the cross-sectional area of the rod is A. Use superposition to find, in terms of p, L, ℓ, E, I, and A, the force in the steel rod and the reaction components at C.

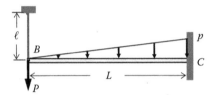

5.156 Before the load P is applied, there is a gap δ between the ends of the two cantilever beams. However, the two beams make contact after the load is applied. Assume that both beams have the identical value of E and I. Using superposition find, in terms of P, L, E, I, and δ, the reaction components at A, B, and C.

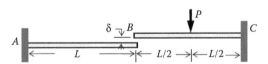

5.157 Two identical cantilever beams are joined at B by a frictionless hinge and have W14 $\times$ 82 sections. Let $p = 3$ kips/ft, $L = 10$ ft, and $E = 30 \times 10^3$ ksi, determine the reaction components at A, B, and C.

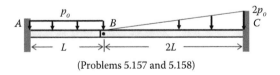

(Problems 5.157 and 5.158)

5.158 In reference to the figure shown in Problem 5.157, two identical cantilever beams are joined at B by a frictionless hinge and have W410 $\times$ 85 sections. Let $L = 4$ m and $E = 205$ GPa. If the allowable shear force on the pin at B is 50 kN, determine the maximum permissible load intensity p.

5.9 STATICALLY INDETERMINATE BEAMS: AREA–MOMENT

In solving statically indeterminate beam problems, the applicable equations of equilibrium may be supplemented by satisfying the boundary conditions through the use of the *area–moment* method. This procedure requires the development of geometric relations, in terms of tangential deviations and/or slope changes that are consistent with the boundary conditions existing in a given support system. It is, therefore, important that an approximate but neat representation be made of the elastic curve of the beam so that the necessary geometric relations are correctly determined. Determining tangential deviations and slope changes is greatly facilitated if the *M/EI* diagrams are constructed using cantilever parts as was discussed earlier in Section 5.6.

EXAMPLE 5.18

Solve Example 5.12 using the area–moment method.

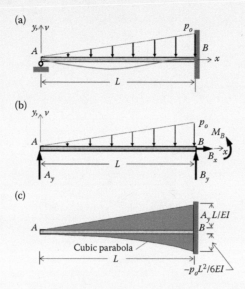

SOLUTION

For convenience, the beam of Example 5.12 is repeated in figure (a) above, which also shows an approximation for the elastic curve of the beam. The free-body diagram for the beam is shown in figure (b) above. Using point B as the fixed end of the cantilevers, we construct the *M/EI* diagram by cantilever parts as shown in figure (c) above.

The nature of the support at B is such that a tangent drawn to the elastic curve at this point passes through point A. Therefore, by the second theorem of the area–moment method, we conclude that the tangential deviation of point A with respect to point B is zero. Thus,

$$t_{A/B} = \frac{1}{2}\frac{A_y L}{EI}(L)\frac{2L}{3} - \frac{1}{4}\frac{p_o L^2}{6EI}(L)\frac{4L}{5} = 0$$

Solving the above equation for A_y yields

$$A_y = \frac{p_o L}{10} \qquad \textbf{ANS.}$$

This answer, of course, is identical to that obtained in Example 5.12. The support reactions at B may now be obtained by applying the equations of equilibrium to the free-body diagram shown in figure (b) above.

EXAMPLE 5.19

Use the area–moment method to solve Example 5.17.

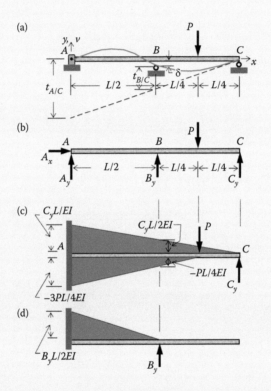

SOLUTION

For convenience, the beam of Example 5.17 is repeated in figure (a) above, which also shows in red, an approximate elastic curve. The free-body diagram for the beam is given in figure (b) above. Using point A as the fixed end of the cantilevers, we construct the M/EI diagrams by cantilever parts as shown in figures (c) and (d) above. A tangent to the elastic curve is drawn at C as shown in figure (a) above, which also indicates the two tangential deviations $t_{A/C}$ and $t_{B/C}$. The geometric relations of figure (a) above, leads us to conclude that

$$\frac{1}{2} |t_{A/C}| = |t_{B/C}| + |\delta| \tag{5.19.1}$$

The tangential deviations are now found from the M/EI diagrams of figures (c) and (d) above. Thus,

$$t_{A/C} = \frac{1}{2}\frac{C_y L}{EI}(L)\frac{L}{3} - \frac{1}{2}\frac{3PL}{4EI}\left(\frac{3L}{4}\right)\frac{L}{4} + \frac{1}{2}\frac{B_y L}{2EI}\left(\frac{L}{2}\right)\frac{L}{6}$$

$$= \frac{C_y L^3}{6EI} + \frac{B_y L^3}{48EI} - \frac{9PL^3}{128EI} \tag{5.19.2}$$

$$t_{B/C} = \frac{1}{2}\frac{C_y L}{2EI}\left(\frac{L}{2}\right)\frac{L}{6} - \frac{1}{2}\frac{PL}{4EI}\left(\frac{L}{4}\right)\frac{L}{12} = \frac{C_y L^3}{48EI} - \frac{PL^3}{384EI} \tag{5.19.3}$$

Substituting Equations 5.19.2 and 5.19.3 into Equation 5.19.1 yields

$$16C_y + 8B_y - P = \frac{768EI\delta}{L^3} \tag{5.19.4}$$

Equation 5.19.4 contains the two unknown quantities C_y and B_y. A second equation containing the same two unknowns can be obtained by summing moments about point A in figure (b) above. Thus,

$$\sum M_A = 0 \Rightarrow C_y + \left(\frac{B_y}{2}\right) - \left(\frac{3P}{4}\right) = 0 \tag{5.19.5}$$

A simultaneous solution of Equations 5.19.4 and 5.19.5 yields

$$B_y = \frac{11P}{16} - \frac{48EI\delta}{L^3} \qquad \textbf{ANS.}$$

$$C_y = \frac{13P}{32} + \frac{24EI\delta}{L^3} \qquad \textbf{ANS.}$$

Of course, these values are identical to those obtained in Example 5.17. The remaining unknown quantities may now be obtained by equilibrium.

EXAMPLE 5.20

The beam shown in figure (a) below is fixed at A and B and subjected to a concentrated load at C. Use the area–moment method to find the reaction components at supports A and B.

SOLUTION

An approximate deflection curve is shown in red in figure (a) below. The free-body diagram of the beam is shown in figure (b) below. Note that the horizontal support reactions at A and B have been ignored on the basis of the assumptions discussed in Example 5.13. Using point B as the fixed end of the cantilevers, we construct the M/EI diagram by cantilever parts as shown in figures (c) and (d) below.

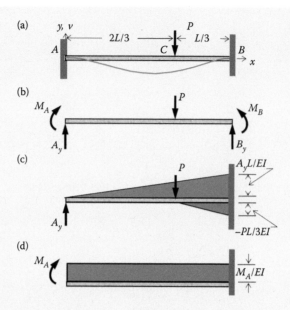

The nature of the supports leads us to conclude that a tangent drawn at point A passes through point B and that the change in slope between points A and B is zero. Thus,

$$t_{B/A} = \frac{1}{2}\frac{A_y L}{EI}(L)\frac{L}{3} - \frac{1}{2}\frac{PL}{3EI}\left(\frac{L}{3}\right)\frac{L}{9} + \frac{M_A}{EI}(L)\frac{L}{2} = 0$$

This equation reduces to

$$A_y\left(\frac{L}{3}\right) + M_A - \left(\frac{PL}{81}\right) = 0 \qquad (5.20.1)$$

Also

$$\theta_B - \theta_A = \frac{1}{2}\frac{A_y L}{EI}(L) - \frac{1}{2}\frac{PL}{3EI}\left(\frac{L}{3}\right) + \frac{M_A}{EI}(L) = 0$$

This equation leads to

$$A_y\left(\frac{L}{2}\right) + M_A - \left(\frac{PL}{18}\right) = 0 \qquad (5.20.2)$$

A simultaneous solution of Equations 5.20.1 and 5.20.2 yields

$$A_y = \frac{7P}{27} \qquad \textbf{ANS.}$$

$$M_A = -\frac{2PL}{27} \qquad \textbf{ANS.}$$

The remaining two unknown quantities are found by applying equilibrium to the free-body diagram of figure (b) above. This application yields

$$B_y = \frac{20P}{27} \qquad \textbf{ANS.}$$

$$M_B = -\frac{4PL}{27} \qquad \textbf{ANS.}$$

PROBLEMS

5.159 Repeat Problem 5.125 using the area–moment method.

5.160 Repeat Problem 5.129 using the area–moment method.

5.161 Repeat Problem 5.133 using the area–moment method.

5.162 Repeat Problem 5.135 using the area–moment method.

5.163 Repeat Problem 5.137 using the area–moment method.

5.164 Repeat Problem 5.139 using the area–moment method.

5.165 Repeat Problem 5.151 using the area–moment method.

5.166 Repeat Problem 5.153 using the area–moment method.

5.167 Repeat Problem 5.155 using the area–moment method.

5.168 Repeat Problem 5.157 using the area–moment method.

5.169 Use the area–moment method to find the support reactions at B and C. Express your answers in terms of p and L.

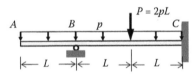

(Problems 5.169 and 5.170)

5.170 In reference to the figure shown in Problem 5.169, before the loads are applied, the support at B is below the beam such that there is a gap between them of magnitude δ. Find the support reactions at B and C. Express your answers in terms of p, δ, and L.

5.171 The steel cantilever beam is fixed at B and restrained by two steel rods of length ℓ connected to the ends of rigid arms securely fastened to the beam at midspan. Let the cross-sectional area of the rods be A, the modulus of elasticity for steel be E and the moment of inertia for the beam be I, find the force in each rod and the reactions at B.

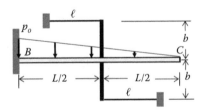

5.172 Determine the reaction components at supports A and B. Use the area–moment method and express your answer in terms of P.

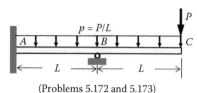

(Problems 5.172 and 5.173)

5.173 In reference to the figure shown in Problem 5.172, before the loads are applied, there is a misalignment so that the support at B is above the bottom of the beam such that the beam is deflected upward at B by the quantity δ. Determine the reaction components at A and B expressing your answers in terms of P and δ.

5.174 The system shown consists of two identical W16 × 89 cantilever beams with a rigid roller between them at B. Let $p = 4$ kips/ft, $L = 15$ ft, and $E = 30 \times 10^3$ ksi, determine the reaction components at A and C and the deflection of the two beams at B. Use the area–moment method.

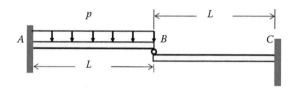

5.175 Beam $BCDE$ is supported at C and D by rigid rollers and at E by a flexible cable. The beam and cable are made of the same material for which the modulus of elasticity is E. The moment of inertia for the beam about the bending axis is I and the cross-sectional area of the cable is A. Determine the support reactions at C and D and the force in the cable. Let $L = 2$ m, $P = 20$ kN, $E = 70$ GPa, $I = 200 \times 10^{-6}$ m^4, and $A = 40 \times 10^{-4}$ m^2.

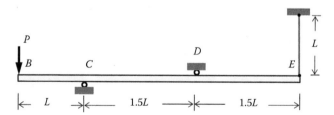

REVIEW PROBLEMS

R5.1 For shipping purposes, a steel ($E = 30 \times 10^3$ ksi) rod of diameter d is coiled around a spool of diameter D, as shown. The allowable stress in the steel is 40 ksi. (a) If $d = 1/2$ in., determine the least spool diameter D. (b) If $D = 150$ in., determine the maximum permissible rod diameter d.

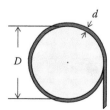

R5.2   Assume *EI* is a constant and use the method of two successive integrations to develop the slope and deflection equations for the cantilever beam shown. Use the given coordinate system and express your answers in terms of p, L, and EI.

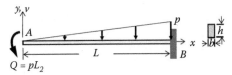

(Problems R5.2 and R5.3)

R5.3   In reference to the figure shown in Problem R5.2, let $L = 3$ m, $E = 200$ GPa, $I = 21 \times 10^{-6}$ m^4, and $h = 0.150$ m and using the method of two successive integrations, determine the largest permissible load intensity p if the maximum deflection cannot exceed 120 mm and the allowable stress is 50 MPa.

R5.4   If the maximum deflection is to be limited to 20 mm and the allowable bending stress is 270 MPa, find the maximum permissible magnitude of the applied load intensity p. Let $EI = 420 \times 10^3$ kN · m^2, $I/c = 10 \times 10^6$ m^3, and $L = 6$ m. Use the method of two successive integrations.

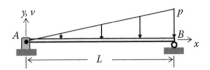

(Problems R5.4 and R5.5)

R5.5   In reference to the figure shown in Problem R5.4, using the method of two successive integrations, select the lightest S section from Appendix E if the maximum deflection is not to exceed 0.75 in. and the allowable bending stress in the beam is 20 ksi. Let $p = 5$ kips/ft, $L = 24$ ft, and $E = 30 \times 10^3$ ksi.

R5.6   Assume *EI* to be a constant and using the method of two successive integrations, determine the magnitude and location of the maximum deflection in the beam. Express your answers in terms of P, L, and EI.

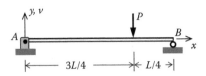

R5.7   Use the method of Section 5.4 to find the maximum permissible magnitude of the intensity p_o if the maximum deflection cannot exceed 10 mm. Let $L = 6$ m, $E = 70$ GPa, and $I = 200 \times 10^{-6}$ m^4.

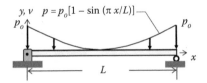

R5.8 Select the lightest S section from Appendix E such that the maximum deflection of the beam does not exceed 0.75 in. and the allowable stress is 20 ksi. Let $P = 10$ kips and $p = 6$ kips/ft. Use superposition for the analysis. Let $E = 30 \times 10^6$ ksi and $L = 18$ ft.

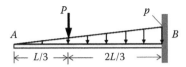

R5.9 The beam shown has a W356 × 122 section. Determine the maximum permissible magnitude of P if the center beam deflection cannot exceed 15 mm and the allowable bending stress is 160 MPa. Let $E = 200$ GPa and $L = 9$ m and use superposition for the analysis.

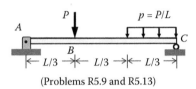

(Problems R5.9 and R5.13)

R5.10 Assume a constant EI for the entire beam and using the area–moment method, find the slope and deflection at A. Express your answers in terms of P, L, and EI.

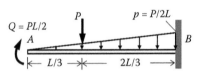

R5.11 Let EI be constant for the entire beam. Using the area–moment method, determine the location and magnitude of the maximum deflection.

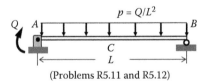

(Problems R5.11 and R5.12)

R5.12 In reference to the figure shown in Problem R5.11, the flexural rigidity in segment AC of the beam is EI and that in segment CB is $2EI$. Use the area–moment method to find the slope and deflection at point C, which is at the center of the beam.

R5.13 Refer the beam of Problem R5.9 and select the lightest wide flange section from Appendix E such that the maximum deflection of the beam does not exceed 1.5 in. and the allowable stress is 22 ksi. Let $E = 30 \times 10^3$ ksi and $L = 24$ ft and ignore the weight of the beam. Use the area–moment method and assume bending about the strong axis.

R5.14 Use the method of two successive integrations along with the given coordinate system to find the reaction components at B and C. Select the lightest wide flange section from

Appendix E if the maximum deflection in segment BC cannot exceed 1.5 in. and the allowable bending stress in this segment is 25 ksi. Let $p = 2$ kips/ft and $a = 5$ ft.

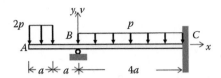

R5.15 The steel ($E = 200$ GPa) beam ABC has a W14 × 48 section and is supported at B by a linear spring of spring constant k. It carries a uniform load of intensity $p = 12$ kN/m over the entire length. If the deflection of the beam at B is to be limited to 20 mm, determine the minimum required spring constant k. Use the method of two successive integrations along with the given coordinate system. Let $L = 3$ m.

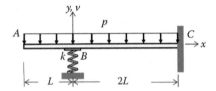

R5.16 Solve Problem R5.10 using the method of superposition.

R5.17 Beam ABC is fixed at A and supported at B at the midpoint of beam DE, which is simply supported and perpendicular to beam ABC. The two beams are identical in every respect with a modulus of elasticity E and a moment of inertia I. Using superposition, determine the reactive force between the two beams at B and the reaction components at A. Also find the maximum deflection of beam DE.

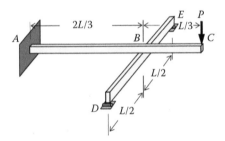

R5.18 Solve Problem R5.17 using the area–moment method.

R5.19 The flexural rigidity of segment AB of the beam is EI, and that of segment BC is $2EI$. Use the area–moment method to find the reaction components at B and C. Express answers in terms of the load intensity p, the length L, and the flexural rigidity EI.

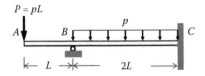

*6 Bending Loads
Additional Deflection Topics

6.1 INTRODUCTION

In Chapter 5, we discussed three different methods of obtaining the deflection of a beam: two successive integrations, superposition, and area–moment. Other deflection methods are available in the literature that, in some ways, may present advantages over the methods developed in Chapter 5. Two such methods are discussed in this chapter, namely, the method of *singularity functions* and the energy method known as *Castigliano's second theorem*. The problem of determining beam deflections under *unsymmetric* loading conditions and the impact loading of beams are also discussed.

Singularity functions, discussed in Section 6.2, make use of and eliminate some of the shortcomings of the method of two successive integrations. Admittedly, the method of two successive integrations is a relatively simple technique for determining beam deflections when the loading is continuous and expressible by a single continuous mathematical function. This method, however, becomes extremely cumbersome for cases in which the beam is subjected to concentrated forces and moments or to distributed loads that change in intensity abruptly. This type of problem can be handled much more conveniently by the use of singularity functions specifically designed to deal with discontinuous functions. This method is used in Section 6.5 to solve statically indeterminate beam problems.

In Chapter 1, we derived Equation 1.17, $U = F\delta/2$, which enables us to find the strain energy stored in the body experiencing a deflection δ due to an internal axial force F. In Section 6.3, we develop the relation $dU = 1/2M\,d\theta$ that provides the strain energy dU due to an internal moment M producing a differential rotation $d\theta$. We will further show that the partial derivative of the strain energy U with respect to any external force P or external couple Q acting on a body gives, respectively, the deflection or rotation of that body at the point of application and in the direction of force P or couple Q. This last statement represents what is known as *Castigliano's second theorem*. This theorem is used in the solution of statically indeterminate beam problems in Section 6.6. Finally, in Section 6.7, we explore the behavior of beams under impact-loading conditions.

6.2 DEFLECTION: SINGULARITY FUNCTIONS

The method of two successive integrations discussed in Chapter 5, while simple for certain types of loading conditions, becomes extremely laborious for other, more complex discontinuous loading systems. The use of *singularity functions* along with the method of two successive integrations enables us to solve deflection problems under discontinuous load systems without much expenditure of time and effort. Singularity functions are special mathematical tools specifically designed to handle discontinuous functions. Only those properties of singularity functions needed for the beam deflection problem will be given in this section.

A singularity function $f(x)$ is written using angle brackets to distinguish it from an ordinary function. Thus,

$$f(x) = \langle x - a \rangle^n \tag{6.1}$$

In the singularity function given in Equation 6.1, the quantity x is any distance measured from some origin in the beam, a is a specific value of x, and n is any positive or negative integer, including zero. For cases where $n \geq 0$, the singularity function expressed in Equation 6.1 obeys the following four operational rules:

1. When $x < a$, the quantity $\langle x - a \rangle^n$ vanishes.
2. When $x \geq a$, the quantity $\langle x - a \rangle^n$ becomes the ordinary function $(x - a)^n$ and obeys the same rules as any other ordinary function.
3. $\int \langle x - a \rangle^n \, dx = (\langle x - a \rangle^{n+1}/n + 1) + C.$
4. $(d/dx)\langle x - a \rangle^n = n\langle x - a \rangle^{n-1}.$

In the solution of deflection problems using the method of two successive integrations, we need to express the internal moment M in the beam in terms of the variable x. When using singularity functions, this process is facilitated if we know the singularity function corresponding to a specific loading condition. Figure 6.1 shows four basic loading conditions and the corresponding singularity function that expresses the internal moment M as a function of x at any position, such as C, on the beam. While a cantilever beam was used in Figure 6.1, any other type of beam could have been used for the purpose. Moreover, while only four basic loads are shown in Figure 6.1, there are other basic loads that are encountered in practice and their singularity functions need to be determined. One way to find these functions is to begin the process by constructing a free-body diagram that includes the moment to be determined, expressing this moment by ordinary functions. We then transform the ordinary function into a singularity function making sure that it meets all requirements of the problem. Consider, for example, the basic case shown in Figure 6.1d. The free-body diagram of that part of the beam to the left of position C is constructed as shown in Figure 6.2. We then apply equilibrium to determine the bending moment at C in terms of an ordinary function as shown in Equation 6.1.1 of Figure 6.2. Next, we transform the ordinary function into a singularity function as given in Equation 6.1.2 of Figure 6.2. Note that the ordinary function for the moment requires a statement of the limits of applicability of the function. However, the singularity function does not need a statement of limits because the limits are built into the function by the rules of singularity functions stated above. Thus, the singularity moment function of Equation 6.1.2 in Figure 6.2 is zero for $x < a$ according to Rule 1 and becomes the ordinary function $-(p_o/6b)(x - a)^3$ for $x \geq a$ according to Rule 2.

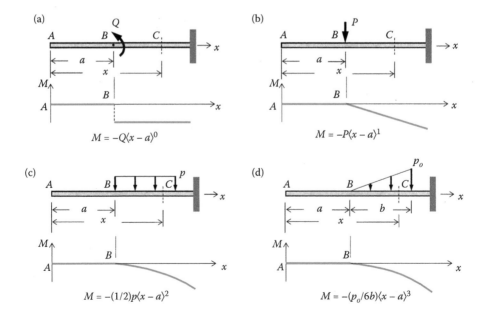

FIGURE 6.1 A diagram showing four basic loading conditions and the corresponding singularity function that expresses the internal moment M as a function of x at any position, such as C, on the beam.

$$\Sigma M_C = 0 \Rightarrow M + \frac{1}{2}\left(\frac{p_o}{b}\right)(x-a)(x-a)\left(\frac{1}{3}\right)(x-a) = 0$$

$$M = -\left(\frac{p_o}{6b}\right)\langle x - a \rangle^3 \text{......}(x \le a) \qquad \text{(a)}$$

$$M = -\left(\frac{p_o}{6b}\right)\langle x - a \rangle^3 \text{.....Singularity function for moment.} \quad \text{(b)}$$

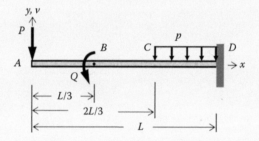

FIGURE 6.2   The free-body diagram of that part of the beam to the left of position C in Figure 6.1d, showing the transformation of an ordinary function into a singularity function.

In writing the singularity function for a beam with a complex loading system, we can proceed in one of two ways. In the first, we need to decompose the given loading system into two or more basic loads and, using information such as given in Figure 6.1, write the singularity function corresponding to each basic load and superpose all those functions to obtain the one that describes the complex loading. In the second, we construct a free-body diagram that includes the moment in question, and applying equilibrium we find this moment using ordinary functions. Next, we transform these ordinary functions to singularity functions making sure that all requirements of the problem are met. Example 6.1 illustrates the first method and Example 6.2 the second.

EXAMPLE 6.1

Using singularity functions and the method of two successive integrations, derive the slope and deflection equations for the cantilever beam shown in the sketch.

SOLUTION

Without the use of singularity functions, we would need to write three separate moment functions, one for each of the three segments $0 < x < L/3$, $L/3 < x < 2L/3$, and $2L/3 < x < L$.

We would then apply the method of two successive integrations to each of the three segments resulting in three slope and three deflection equations. These six equations would contain six constants of integration requiring a total of six boundaries and matching conditions yielding six simultaneous equations the solution for which would be cumbersome and tedious. The use of a singularity function to write a single moment equation serves to minimize the labor needed since the two integrations yield only one slope and one deflection equation containing only two constants of integrations instead of six.

The loading system applied to the cantilever beam may be obtained by superposing basic loading conditions (b), (a), and (c) of Figure 6.1. Thus, using the given coordinate system, the singularity function for the internal moment in the beam becomes

$$M = -Px - Q\left\langle x - \frac{L}{3} \right\rangle^0 - \left(\frac{P}{2}\right)\left\langle x - \frac{2L}{3} \right\rangle^2 \qquad (6.1.1)$$

Equation 6.1.1 expresses the internal moment at any position in the cantilever beam. Thus, for example, in the interval, $L/3 < x < 2L/3$, Equation 6.1.1 reduces to $M = -Px - Q(x - L/3)^0 = -Px - Q$, since $(x - L/3)^0 = 1$ and $\langle x - 2L/3 \rangle^2$ vanishes according to the first operational rule of singularity functions because $x < 2L/3$. Therefore, the differential equation for the elastic curve becomes

$$EIv'' = -Px - Q\left\langle x - \frac{L}{3} \right\rangle^0 - \left(\frac{p}{2}\right)\left\langle x - \frac{2L}{3} \right\rangle^2 \qquad (6.1.2)$$

Using the third rule, we integrate Equation 6.1.2 to obtain the slope function. Thus,

$$EIv' = \frac{-Px^2}{2} - Q\left\langle x - \frac{L}{3} \right\rangle^1 - \left(\frac{p}{6}\right)\left\langle x - \frac{2L}{3} \right\rangle^3 + C_1 \qquad (6.1.3)$$

Substituting the first boundary condition, $v' = 0$ at $x = L$, both of the singularity functions in Equation 6.1.3 reduce to ordinary functions and the constant C_1 becomes

$$C_1 = \left(\frac{PL^2}{2}\right) + \left(\frac{2QL}{3}\right) + \left(\frac{pL^3}{162}\right) \qquad (6.1.4)$$

Therefore, the slope equation may be expressed in the form

$$EIv' = -\frac{Px^2}{2} - Q\left\langle x - \frac{L}{3} \right\rangle^1 - \frac{p}{6}\left\langle x - \frac{2L}{3} \right\rangle^2 + \frac{PL^2}{2} + \frac{2QL}{3} + \frac{pL^3}{162} \qquad \textbf{ANS.} \quad (6.1.5)$$

Again, using the third rule, we integrate Equation 6.1.5 to obtain the deflection function. Thus,

$$EIv = -\frac{Px^3}{6} - \left(\frac{Q}{2}\right)\left\langle x - \frac{L}{3} \right\rangle^2 - \left(\frac{p}{24}\right)\left\langle x - \frac{2L}{3} \right\rangle^4$$
$$+ \left(\frac{PL^2}{2} + \frac{2QL}{3} + \frac{pL^3}{162}\right)x + C_2 \qquad (6.1.6)$$

Substituting the second boundary condition, $v = 0$ at $x = L$, both of the singularity functions in Equation 6.1.6 reduce to ordinary functions and the constant C_2 becomes

$$C_2 = -\left(\frac{PL^3}{3}\right) - \left(\frac{4QL^2}{9}\right) - \left(\frac{11pL^4}{1944}\right)$$

Therefore, the deflection equation becomes

$$EIv = -\frac{Px^3}{6} - \frac{Q}{2}\left\langle x - \frac{L}{3} \right\rangle^2 - \frac{p}{24}\left\langle x - \frac{2L}{3} \right\rangle^4 + \left(\frac{PL^2}{2} + \frac{2QL}{3} + \frac{qL^3}{162}\right)x$$
$$- \left(\frac{PL^3}{3} + \frac{4QL^2}{9} + \frac{11pL^4}{1944}\right) \qquad \textbf{ANS.}$$

EXAMPLE 6.2

Use singularity functions to develop the slope and deflection equations for the beam shown below.

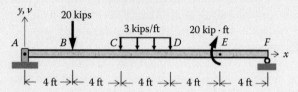

SOLUTION

Instead of using the moment singularity functions for basic loadings given in Figure 6.1, we will develop the needed moment singularity function by constructing an appropriate free-body diagram. The support reactions at A and F are first found to be $A_y = 21$ kips and $F_y = 11$ kips. Next, we cut the beam at any point between E and F and construct the free-body diagram shown below. Note that the truncated distributed load is extended to the cut because singularity functions are open-ended. To ensure that we retain the given loading, we need to subtract what was added by applying an upward distributed load of equal magnitude and span as was added. Summing moments about point O, we obtain the internal moment M expressing it in terms of ordinary functions. Thus,

$$\sum M_O = 0 \Rightarrow M - 21x + 20(x-4) + \left(\frac{3}{2}\right)(x-8)^2 - \left(\frac{3}{2}\right)(x-12)^2 - 20 = 0$$

$$M = 21x - 20(x-4) - \left(\frac{3}{2}\right)(x-8)^2 + \left(\frac{3}{2}\right)(x-12)^2 + 20 \tag{6.2.1}$$

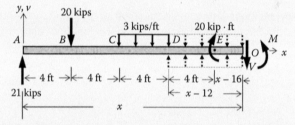

Equation 6.2.1 expresses the moment M in terms of ordinary functions and we need to transform them to singularity functions before solving the deflection problem. This transformation yields

$$M = 21x - 20\langle x-4 \rangle^1 - \frac{3}{2}\langle x-8 \rangle^2 + \frac{3}{2}\langle x-12 \rangle^2 + 20\langle x-16 \rangle^0 \tag{6.2.2}$$

Now, using the first two rules of singularity functions, we check to make certain that Equation 6.2.2 yields the correct internal moment for each of the five segments of the beam.

Segment AB ($x < 4$ ft): All of the singularity functions vanish yielding $M = 21x$, which is correct.

Segment BC ($4 < x < 8$ ft): All singularity functions vanish except $-20\langle x-4 \rangle^1$, which becomes $-20(x-4)$. Thus, the internal moment in this segment becomes $M = 21x - 20(x-4)$, which is correct.

Segment CD ($8 < x < 12$ ft): The last two singularity functions vanish. The first and second become, respectively, $-20(x-4)$ and $-(3/2)(x-8)^2$. The internal moment becomes $M = 21x - 20(x - 4) - (3/2)(x - 8)^2$, which is the correct expression for the internal moment in this segment.

Segment DE ($12 < x < 16$ ft): Only the last singularity function vanishes and the first three reduce to, respectively, $-20(x - 4)$, $-(3/2)(x - 8)^2$, and $(3/2)(x - 12)^2$. The internal moment becomes $M = 21x - 20(x - 4) - (3/2)(x - 8)^2 + (3/2)(x - 12)^2$, which is correct for this segment.

Segment EF ($16 > x < 20$ ft): All of the singularity functions become ordinary functions and since the term $(x - 16)^0$ is unity, the correct internal moment in this segment is $M = 21x - 20(x - 4) - (3/2)(x - 8)^2 + (3/2)(x - 12)^2 + 20$.

Now that we have the correct internal moment expressed in terms of singularity functions, we can apply the method of two successive integrations to obtain the slope and deflection equations. Thus,

$$EIv'' = 21x - 20\langle x - 4\rangle^1 - \frac{3}{2}\langle x - 8\rangle^2 + \frac{3}{2}\langle x - 12\rangle^2 + 20\langle x - 16\rangle^0$$

Using the third rule for singularity functions, we integrate to obtain the slope function. Thus,

$$EIv' = \frac{21}{2}x^2 - 10\langle x - 4\rangle^2 - \frac{1}{2}\langle x - 8\rangle^3 + \frac{1}{2}\langle x - 12\rangle^3 + 20\langle x - 16\rangle^1 + C_1$$

The first boundary condition, $v = 0$ at $x = 0$, enables us to conclude that $C_1 = 0$.
A second integration yields

$$EIv = \frac{7}{2}x^3 - \frac{10}{3}\langle x - 4\rangle^3 - \frac{1}{8}\langle x - 8\rangle^4 + \frac{1}{8}\langle x - 12\rangle^4 + 10\langle x - 16\rangle^2 + C_2$$

Applying the second boundary condition, $v = 0$ at $x = 20$ ft, we obtain $C_2 = -12{,}426.7$ kip·in.[3] Therefore, the slope and deflection equations for this beam may be expressed as follows:

$$EIv' = \frac{21}{2}x^2 - 10\langle x - 4\rangle^2 - \frac{1}{2}\langle x - 8\rangle^3 + \frac{1}{2}\langle x - 12\rangle^3 + 20\langle x - 16\rangle^1 \qquad \textbf{ANS.}$$

$$EIv = \frac{7}{2}x^3 - \frac{10}{3}\langle x - 4\rangle^3 - \frac{1}{8}\langle x - 8\rangle^4 + \frac{1}{8}\langle x - 12\rangle^4 + 10\langle x - 16\rangle^2 - 12{,}426.7 \qquad \textbf{ANS.}$$

EXAMPLE 6.3

Use singularity functions and determine the slope and deflection at points C and E of the beam shown in figure below. Let $E = 200$ GPa and $I = 400 \times 10^6$ mm^4.

SOLUTION

The support reactions at A and B are found to be $A_y = 12.143$ kN and $D_y = 12.857$ kN. Using the free-body diagram shown in figure below, the singularity moment function for the entire beam becomes

$$M = 12.143x - 10\langle x - 2\rangle^1 - \frac{5}{9}\langle x - 4\rangle^3 + 12.857\langle x - 7\rangle^1 + 5\langle x - 7\rangle^2 + \frac{5}{9}\langle x - 7\rangle^3 \quad (6.3.1)$$

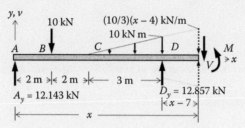

Note that because of the open-ended nature of singularity functions, the distributed load was extended to where the beam was cut and what was added was then taken away by adding an equal upward distributed load as shown. Using the first two rules of singularity functions, we now check Equation 6.3.1 to make sure it applies to each and every segment of the beam.

Segment AB ($0 < x < 2$ m): All terms of Equation 6.3.1 vanish except the first, leading to $M = 12.143x$, which is correct.

Segment BC ($2 < x < 4$ m): Only the first two terms in Equation 6.3.1 remain, yielding $M = 12.143x - 10(x - 2)$, which is the correct moment in this segment.

Segment CD ($4 < x < 7$ m): Only the first three terms in Equation 6.3.1 remain, yielding $M = 12.143x - 10(x - 2) - (5/9)(x - 4)^3$, which is the correct moment for this segment.

Segment DE ($7 < x < 10$ m): All terms of Equation 6.3.1 are applicable, resulting in $M = 12.143x - 10(x - 2) - (5/9)(x - 4)^3 + 12.857(x - 7) + 5(x - 7)^2 + (5/9)(x - 7)^3 = 20$ kN·m, which is the correct answer for this segment.

Applying the method of two successive integrations, we obtain

$$EIv'' = 12.143x - 10\langle x - 2\rangle^1 - \frac{5}{9}\langle x - 4\rangle^3 + 12.857\langle x - 7\rangle^1 + 5\langle x - 7\rangle^2 + \frac{5}{9}\langle x - 7\rangle^3$$

$$EIv' = \frac{12.143}{2}x^2 - 5\langle x - 2\rangle^2 - \frac{5}{36}\langle x - 4\rangle^4 + \frac{12.857}{2}\langle x - 7\rangle^2 + \frac{5}{3}\langle x - 7\rangle^3 + \frac{5}{36}\langle x - 7\rangle^4 + C_1$$

$$EIv = \frac{12.143}{6}x^3 - \frac{5}{3}\langle x - 2\rangle^3 - \frac{5}{180}\langle x - 4\rangle^5 + \frac{12.857}{6}\langle x - 7\rangle^3 + \frac{5}{12}\langle x - 7\rangle^4$$
$$+ \frac{5}{180}\langle x - 7\rangle^5 + C_1 x + C_2$$

The first boundary condition, $v = 0$ at $x = 0$, substituted into the deflection equation yields $C_2 = 0$. The second boundary condition, $v = 0$ at $x = 7$ m, substituted into the same equation yields $C_1 = -68.442$ kN·m².

Applying the rules of singularity functions to the slope and deflection equations developed above, we find the slope and deflection at point C ($x = 4$ m) and at point E ($x = 10$ m) as follows:

Point C ($x = 4$ ft)

$$\theta_C = v'_C = -1.1 \times 10^{-4} \text{ rad}; \quad v_C = -2.0 \text{ mm} \qquad \textbf{ANS.}$$

Point E ($x = 10$ ft)

$$\theta_E = v'_E = 1.9 \times 10^{-3} \text{ rad}; \quad v_E = 4.6 \text{ mm} \qquad \textbf{ANS.}$$

PROBLEMS

(Use singularity functions in the solution of the following problems.)

6.1 Write the slope and deflection equations and determine the slope at A and the deflection at B.

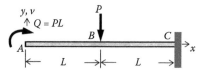

6.2 Write the slope and deflection equations and determine the slope at A and the deflection at B.

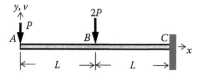

6.3 Write the slope and deflection equations and determine the slope at A and the deflection at B.

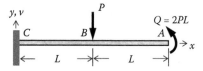

6.4 Write the slope and deflection equations and determine the slope at A and the deflection at B.

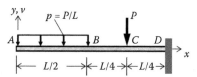

6.5 Let $P = 32$ kips and $L = 16$ ft, determine the slope and deflection at A. The cross section for the beam is a W16 × 100 (see Appendix E) bent about the strong axis. Assume $E = 30 \times 10^3$ ksi.

6.6 Let $P = 32$ kips and $L = 16$ ft, determine the slope and deflection at A. The cross section for the beam is a W16 × 100 (see Appendix E) bent about the strong axis. Assume $E = 30 \times 10^3$ ksi.

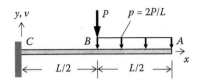

6.7 Let $P = 50$ kN, $Q = 40$ kN·m, and $L = 8$ m, determine the slope and deflection at A. The cross section for the beam is a W356 × 101.2 (see Appendix E) bent about the strong axis. Assume $E = 200$ GPa.

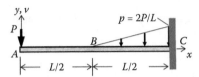

6.8 Let $P = 50$ kN, $Q = 40$ kN·m, and $L = 8$ m, determine the slope and deflection at A. The cross section for the beam is a W356 × 101.2 (see Appendix E) bent about the strong axis. Assume $E = 200$ GPa.

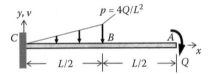

6.9 Develop the slope and deflection equations and determine the slope at A and the deflection at B.

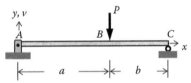

6.10 Develop the slope and deflection equations and determine the slope at A and the deflection at B.

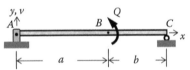

6.11 Develop the slope and deflection equations and determine the slope at A and the deflection at B.

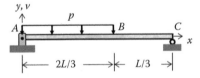

6.12 Develop the slope and deflection equations and determine the slope at A and the deflection at B.

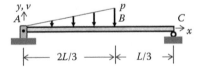

6.13 Write the slope and deflection equations and determine the slope at *A* and the deflection at the center of the beam.

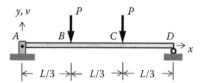

6.14 Write the slope and deflection equations and determine the slope at *A* and the deflection at the center of the beam.

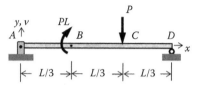

6.15 Write the slope and deflection equations and find the slope at *A* and the deflection at *B*.

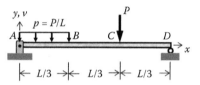

6.16 Write the slope and deflection equations and find the slope at *A* and the deflection at *B*.

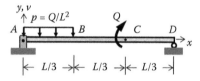

6.17 Find the slope at *B* and *C* and the maximum deflection indicating its location. Express your answers in terms of p, L, and EI.

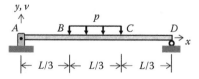

6.18 Find the slope at *B* and *C* and the maximum deflection indicating its location. Express your answers in terms of p, L, and EI.

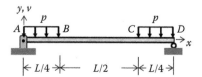

6.19 Find the slope at A and the deflection at C. Let $p = 10$ kN/m, $L = 6$ m, and $E = 200$ GPa. The cross section for the beam is an S152 × 25.7 (see Appendix E) bent about its strong axis.

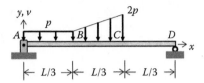

6.20 Find the slope at A and the deflection at C. Let $p = 10$ kN/m, $L = 6$ m, and $E = 200$ GPa. The cross section for the beam is an S152 × 25.7 (see Appendix E) bent about its strong axis.

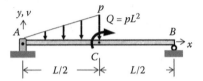

6.21 Develop the slope and deflection equations and find the slope at A and the deflection at D.

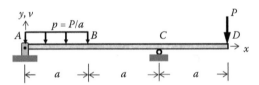

6.22 Develop the slope and deflection equations and find the slope at C and the deflection at D.

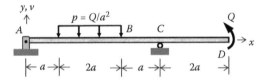

6.23 The beam shown in figure (a) below has the cross section shown in figure (b) below, which is fabricated by welding two C15 × 50 channels (see Appendix E). Let $p = 3$ kips/ft, $a = 5$ ft, and $E = 30 \times 10^3$ ksi, find the slope at A and the deflection at C.

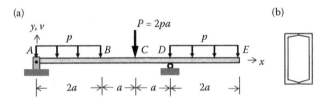

6.3 DEFLECTION: CASTIGLIANO'S SECOND THEOREM

When he was 26 years old, Alberto Castigliano (1847–1884) presented his thesis for an engineer's degree to the Turin Polytechnic Institute in Italy. It contained a statement of his two famous theorems. His first theorem, although applicable to *nonlinearly* elastic material and, therefore, more

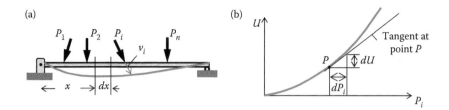

FIGURE 6.3 (a) A beam subjected to a load P_i and the corresponding deflection curve. (b) A curve illustrating the variation of the energy U with load P_i.

general than his second theorem, is not as widely used and hence will not be covered in this text. His second theorem, however, is widely used and will be derived and used to find beam deflections in this section and to solve statically indeterminate beam problems in Section 6.5.

Consider the simply supported beam subjected to the system of forces $P_i(i = 1,2,3,...,n)$ as shown in Figure 6.3a. The beam deforms as indicated by the red elastic curve and the system of forces P_i does external work, which is stored as internal *elastic strain energy* in the beam. By the law of conservation of energy, the stored *elastic strain energy* U is equal to the external work done since all energy losses are ignored. Note that no external work is done at the supports by the reactive forces since the supports are assumed rigid and no displacement can occur.

The curve of Figure 6.3b shows the variation of the strain energy U with the force P_i. Figure 6.3b also shows that when the force P_i is increased by the amount dP_i, the strain energy is also increased by the amount $dU = (\partial U/\partial P_i)dP_i$. Therefore, if after applying the system of loads to the beam, any one of the n forces is increased, the total stored strain energy due to this first sequence of force application becomes

$$(U_{TOTAL})_1 = U + dU = U + \frac{\partial U}{\partial P_i}dP_i \qquad (6.2)$$

Let us now reverse the order in which the loads are applied. We first apply the incremental force dP_i and then apply the system of forces P_i. Since the force dP_i builds up linearly from zero to its full value of dP_i, its average value is $(1/2)dP_i$. Thus, the strain energy stored in the beam due to dP_i is equal to $(1/2)dP_i\,dv_i$, where dv_i is the incremental deflection produced by dP_i *in the direction of this force*. When the system of forces P_i is next applied, it stores an additional amount of energy U but, at the same time, the force dP_i, at its full value, is carried along through the deflection v_i, thus producing an added strain energy of dP_iv_i. Therefore, the total stored energy for the second sequence of force application becomes

$$(U_{TOTAL})_2 = dU + U = \frac{1}{2}dP_idv_i + dP_iv_i + U \qquad (6.3)$$

Since elastic deformations are reversible and energy losses are neglected, it follows that the total elastic strain energy stored in the beam is independent of the sequence in which the forces are applied. Therefore, the total energies expressed by the right-hand sides of Equations 6.2 and 6.3 are identical and may be equated. Thus,

$$U + \frac{\partial U}{\partial Q_i}dP_i = \frac{1}{2}dP_idv_i + dP_iv_i + U \qquad (6.4)$$

The term $(1/2)dP_i\,dv_i$ is a second-order differential and is ignored. Simplifying the remaining terms, we conclude that

$$v_i = \frac{\partial U}{\partial P_i}$$ (6.5)

Equation 6.5 is the mathematical form of *Castigliano's second theorem*. It states that the deflection v_i of a beam at the point of application and in the direction of the force P_i is given by the partial derivative of the total elastic strain energy U stored in the beam with respect to the force P_i.

Retracing the steps followed in the development of Equation 6.5 for deflection, we may develop Castigliano's second theorem for the slope of a beam at the point of application of a couple Q_i. Thus,

$$\theta_i = \frac{\partial U}{\partial Q_i}$$ (6.6)

To make use of Equations 6.5 and 6.6 in solving for deflections and slopes in a given beam, we need an expression relating the strain energy stored in a beam to the internal bending moments. To this end, we isolate a differential length dx of the beam shown in Figure 6.3a and construct its free-body diagram as shown, magnified, in Figure 6.4a. Note that only the internal moments acting on the two faces of this differential segment are shown because the effect of the internal shearing forces on the strain energy is negligibly small when compared with that of the moments. Figure 6.4b shows a curve relating the internal moment M to angle $d\theta$ within the elastic range of material behavior. Since the moment builds up from zero to its final value of M with an average value of $1/2M$ as the angle increases from zero to its final value of $d\theta$, it follows that the work done by M is $1/2M(d\theta/2)$. Thus, since there are two couples, one on each face of the differential segment, we conclude that the total work performed, and hence the total differential strain energy dU stored in the beam is

$$dU = 2\left(\frac{1}{2}M\right)\left(\frac{d\theta}{2}\right) = \frac{1}{2}Md\theta$$ (6.7)

From Equation 5.16, the differential angle $d\theta = (M/EI)dx$. Substituting into Equation 6.7 and simplifying, the differential strain energy dU becomes

$$dU = \frac{1}{2}\frac{M^2}{EI}dx$$ (6.8)

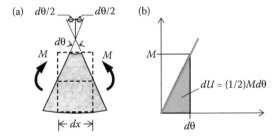

FIGURE 6.4 (a) The free-body diagram of a small segment of length dx of the beam in Figure 6.3a after deformation due to moment M. (b) How the differential angle $d\theta$ changes with the moment M and showing that $dU = (1/2)M\,d\theta$.

The total strain energy in the beam U can now be obtained by integrating Equation 6.8 over the entire length of the beam. Thus,

$$U = \frac{1}{2}\int_0^L \frac{M^2}{EI}\, dx \tag{6.9}$$

In the solution of a beam deflection and slope problem using Castigliano's second theorem, we could first determine the total energy in the beam by Equation 6.9 and then perform the partial differentiation indicated in Equations 6.5 and 6.6. However, it is much more convenient to perform the partial differentiation first and then the integration. Thus, the deflection v_i given by Equation 6.5 and the slope θ_i given by Equation 6.6 become

$$v_i = \frac{\partial U}{\partial P_i} = \int_0^L \frac{M}{EI}\frac{\partial M}{\partial P_i}\, dx \tag{6.10}$$

and

$$\theta_i = \frac{\partial U}{\partial Q_i} = \int_0^L \frac{M}{EI}\frac{\partial M}{\partial Q_i}\, dx \tag{6.11}$$

Occasionally, we may need to find the deflection or slope at a point in a beam where no actual force or couple is acting. In such a case, we apply a fictitious force P or a fictitious couple Q at the point in question and proceed to solve the problem as though the fictitious load or couple is real. After performing the partial differentiation indicated in Equation 6.10 or 6.11, we set the fictitious force P or the fictitious couple Q equal to zero in order to obtain the solution due to the real loads.

EXAMPLE 6.4

The cantilever beam carries a concentrated load P as shown below. Find the deflection and the slope at A.

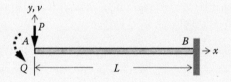

SOLUTION

Since the slope is needed at A, we apply a fictitious couple Q there as indicated. The free-body diagram shown below is used to find the internal moment at x. Thus,

$$M = -Q - Px$$

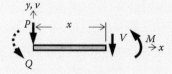

The partial derivatives of the moment with respect to P and Q for use in Equations 6.10 and 6.20, respectively, are

$$\frac{\partial M}{\partial P} = -x; \quad \frac{\partial M}{\partial Q} = -1$$

Using Equation 6.10, and assuming a constant EI, we find the beam deflection at A. Thus,

$$v_i = \frac{\partial U}{\partial P_i} = \int_0^L \frac{M}{EI} \frac{\partial M}{\partial P_i} dx; \quad v_A = \frac{1}{EI} \int_0^L (-Q - Px)(-x)dx = \frac{1}{EI} \int_0^L (0 + Px^2)dx = \frac{PL^3}{3EI} \quad \textbf{ANS.}$$

Using Equation 6.11, and again assuming a constant EI, we find the slope of the beam at A. Thus,

$$\theta_i = \frac{\partial U}{\partial Q_i} = \int_0^L \frac{M}{EI} \frac{\partial M}{\partial Q_i} dx; \quad \theta_A = \frac{1}{EI} \int_0^L (-Q - Px)(-1)dx = \frac{1}{EI} \int_0^L (0 + Px)dx = \frac{PL^2}{2EI} \quad \textbf{ANS.}$$

Note that in both cases, the fictitious couple Q was set equal to zero after the moment function was partially differentiated with respect to P and Q. Note also that in both cases, the answers are positive, indicating that the deflection is in the same sense as the load P (downward) and that the slope is in the same sense as that assumed for the fictitious couple Q (counterclockwise).

EXAMPLE 6.5

The simply supported beam is subjected to a linearly varying load of maximum intensity p as shown below. Determine the slope at A.

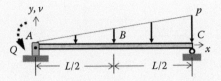

SOLUTION

Since the slope at A is needed, we apply a fictitious couple Q at this point as shown. Using the free-body diagram shown below, the support reactions are found to be

$$A_y = \frac{pL}{2} + \frac{Q}{L}; \quad C_y = \frac{pL}{3} - \frac{Q}{L}$$

The bending moment is now found from the free-body diagram shown below at any position defined by x. Thus,

$$M = \left(\frac{pL}{6} + \frac{Q}{L}\right)x - Q - \frac{p}{6L}x^3$$

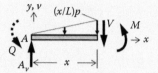

The partial derivative of this moment with respect to Q for use in Equation 6.11 is

$$\frac{\partial M}{\partial Q} = \frac{x}{L} - 1$$

Assuming a constant EI and substituting into Equation 6.11 after setting $Q = 0$, we obtain

$$\theta_A = \frac{1}{EI}\int_0^L \left(\frac{pL}{6}x - \frac{p}{6L}x^3\right)\left(\frac{x}{L} - 1\right)dx = \frac{p}{6EI}\int_0^L \left(x^2 - Lx - \frac{x^4}{L^2} + \frac{x^3}{L}\right)dx = -\frac{7pL^3}{360} \quad \textbf{ANS.}$$

The negative sign of the answer indicates that the slope at A does not have the same sense as the assumed Q but opposite to it. In other words, the slope of the beam at A is clockwise.

EXAMPLE 6.6

A simply supported beam is subjected to a distributed load of intensity p over a segment of length a as shown. Find the deflection at point B. Assume EI is constant.

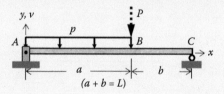

SOLUTION

Since the deflection is to be found at B, a fictitious force P is applied as shown. Using the free-body diagram shown below, the support reactions are found to be

$$A_y = \frac{Pb}{L} + \frac{pa}{2L}(a + 2b); \quad C_y = \frac{Pa}{L} + \frac{pa^2}{2L}$$

The internal moments in segments AB and CD are found by using the free-body diagrams of figures (a) and (b) below, respectively. Thus,

$$M_{AB} = \left(\frac{Pb}{L} + \frac{pa}{2L}(a + 2b) \right)x - \frac{p}{2}x^2$$

$$M_{BC} = \left(\frac{Pa}{L} + \frac{pa^2}{2L} \right)(L - x)$$

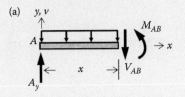

The partial derivatives of these moments with respect to P are

$$\frac{\partial M_{AB}}{\partial P} = \frac{b}{L}x; \quad \frac{\partial M_{BC}}{\partial P} = \frac{a}{L}(L - x)$$

Since EI is constant, when we substitute into Equation 6.10 and after we set $P = 0$, we obtain

$$v_B = \frac{b}{EIL}\int_0^a \left(\frac{pa}{2L}(a + 2b)x - \frac{p}{2}x^2 \right)dx + \frac{a}{EIL}\int_a^L \frac{pa^2}{2L}(L - x)^2 dx$$

$$= \frac{pba^4}{24EIL^2}(a + 5b) + \frac{pa^3b^3}{6EIL^2} = \frac{pba^3}{24EIL^2}(a^2 + 5ba + 4b^2)$$

$$= \frac{pba^3}{24EIL^2}((a^2 + 2ab + b^2) + 3b(a + b))$$

$$v_B = \frac{pba^3}{24EIL^2}(L^2 + 3bL) = \frac{pba^3}{24EIL}(L + 3b) = \frac{pba^3}{24EIL}(a + 4b) \quad \text{ANS.}$$

PROBLEMS

(Use Castigliano's second theorem in the solution of the following problems.)

6.24 The cantilever beam has a constant EI and carries a load as shown. Find the slope and deflection and slope at A.

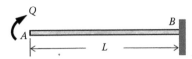

6.25 The cantilever beam has a constant *EI* and carries a load as shown. Find the slope and deflection and slope at *A*.

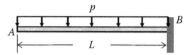

6.26 The cantilever beam has a constant *EI* and carries a load as shown. Find the slope and deflection and slope at *A*.

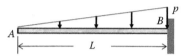

6.27 The cantilever beam has a constant *EI* and carries a load as shown. Find the slope and deflection and slope at *A*.

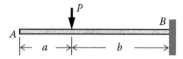

6.28 Assume a constant *EI* and determine the slope at *A* and the deflection at *B*.

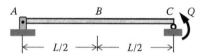

6.29 Assume a constant *EI* and determine the slope at *A* and the deflection at *B*.

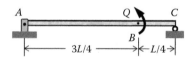

6.30 Assume a constant *EI* and determine the slope at *A* and the deflection at *B*.

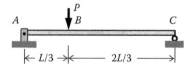

6.31 Assume a constant *EI* and determine the slope at *A* and the deflection at *B*.

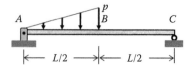

6.32 Find the slope at A and the deflection at C. Assume a constant EI.

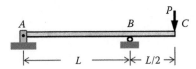

6.33 Find the slope at A and the deflection at C. Assume a constant EI.

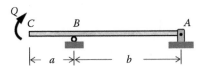

6.34 Assume a constant EI and determine the slope at A and the deflection at B.

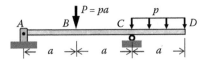

6.35 Assume a constant EI and determine the slope at A and the deflection at B.

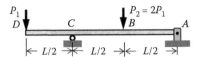

6.36 The cantilever beam ABC has a constant $EI = 29 \times 10^3$ ksi and a W16 × 57 section (see Appendix E) bent about its strong axis. Determine the deflection at A. (*Hint:* Solve symbolically and then substitute numerical values.)

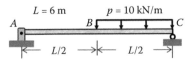

6.37 The simply supported beam ABC has a constant $EI = 200$ GPa and a W406 × 46.1 section (see Appendix E) bent about the strong axis. Find the deflection at B. (*Hint:* Solve symbolically and then substitute numerical values.)

6.4 DEFLECTION: UNSYMMETRIC BENDING LOADS

In Chapter 5, we learned how to determine beam deflections under symmetric loading. In this section, we will develop a method to find the deflection of beams when subjected to unsymmetric bending loads. Symmetric loading was originally discussed in Section 3.4 and later used in Section 4.3 to determine the stresses developed in a beam under conditions of unsymmetric loading.

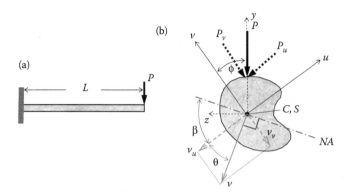

FIGURE 6.5   (a) The unsymmetric bending of a beam subjected to load P at its end. (b) The arbitrarily chosen cross section in which the shear center S coincides with the centroid C and, the two perpendicular deflections v_u, due to load component P_u and v_v, due to load component P_v.

Consider, as an example, the case of finding the end deflection of the cantilever beam shown in Figure 6.5a. This beam is subjected at its free end to a vertical downward load P, applied through the shear center S of the cross section (see Section 4.4), insuring that the beam bends without twisting. In the arbitrarily chosen cross section shown in Figure 6.5b, the shear center S and the centroid of the section C are assumed to coincide. In cases where this is not true, the bending loads must be applied through the shear center.

The u and v principal axes of inertia are found using the methods discussed in Appendix C.2. The deflection of a beam under unsymmetric loading may be determined by superposing two separate deflections, each of which is produced under symmetric loading conditions. Thus, the applied load P is resolved into the two components $P_u = P \sin \phi$ and $P_v = P \cos \phi$ as shown in Figure 6.5b. The component P_u produces symmetric bending about the v principal axis, resulting in the deflection v_u. Besides, the component P_v produces symmetric bending about the u principal axis resulting in the deflection v_v.

The end deflection of a cantilever beam under a symmetrically applied load is $v = PL^3/3EI$ as may be ascertained from Appendix G. Thus, the component deflections v_u and v_v may be stated as follows:

$$v_u = \frac{(P \sin \phi)L^3}{3EI_v}; \quad v_v = \frac{(P \cos \phi)L^3}{3EI_u}$$

To determine the resultant end deflection v of the cantilever beam, we find the vector sum of the two component deflections as shown in Figure 6.5b. Thus,

$$v = \sqrt{v_u^2 + v_v^2}$$

As indicated in Figure 6.5b, the resultant deflection is perpendicular to the neutral axis of the cross section. This fact may be proven by showing that the sum of the two angles β and θ is 90°. The angle β positioning the neutral axis from the u principal axis is given by Equation 4.24 and repeated here for convenience:

$$\tan \beta = \left(\frac{I_u}{I_v}\right) \tan \phi \qquad \textbf{(4.24, Repeated)}$$

As indicated in Figure 6.5b, the angle ϕ in Equation 4.24 defines the orientation of the plane of the load relative to the v principal axis.

The angle θ in Figure 6.5b, between the u principal axis and the resultant deflection, may be determined by finding its tangent from the two components of the deflection. Thus,

$$\tan \theta = \frac{v_v}{v_u} = \frac{(P\cos\phi)L^3/EI_u}{(P\sin\phi)L^3/EI_v} = \frac{I_v\cos\phi}{I_u\sin\phi} = \frac{1}{(I_u/I_v)\tan\phi}$$

We conclude, therefore, that

$$\tan \theta = \frac{1}{\tan \beta} \tag{6.12}$$

Since the angles θ and β have tangents that are reciprocals of each other, they must be complementary angles and their sum is 90°.

Although our discussion has focused on an end-loaded cantilever beam, the approach presented is quite general. All we need to do is select the appropriate deflection equation from such tabulations as given in Appendix G, or develop the needed deflection equation if it cannot be found in those tabulations. These equations are then used to determine the component deflections along the two principal axes of inertia of the cross section under consideration. The following example further illustrates the procedure.

EXAMPLE 6.7

A simply supported steel ($E = 30 \times 10^3$ ksi) beam, 16 ft long, is subjected to a concentrated downward load $P = 4$ kips at midspan. The cross section for the beam is a $9 \times 4 \times 1$ in. angle the centroid of which, located by the methods of Appendix C.3, is shown in the sketch. Determine the deflection of the beam at midspan. The following area properties are given: $I_z = 97$ in.^{4}, $I_y = 12$ in.^{4}, and $I_{zy} = 18$ in.^{4}.

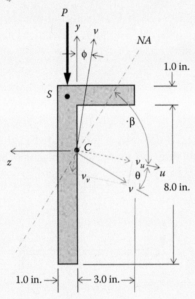

SOLUTION

Using the methods of Appendix C.2, we find the orientation of the principal axes and the magnitudes of the principal moments of inertia to be

$$\phi = 11.48°; \quad I_u = 100.65 \text{ in.}^4; \quad I_v = 8.35 \text{ in.}^4$$

The deflection at the center of a simply supported beam is found from Appendix G to be $PL^3/48EI$. Therefore,

$$v_u = \frac{P_u L^3}{48EI_v} = \frac{4 \sin 11.48°(16 \times 12)^3}{48(30 \times 10^3)8.35} = 0.469 \text{ in.}$$

$$v_v = \frac{P_v L^3}{48EI_u} = \frac{4 \cos 11.48°(16 \times 12)^3}{48(30 \times 10^3)100.65} = 0.191 \text{ in.}$$

$$v = \sqrt{v_u^2 + v_v^2} = 0.506 \text{ in.} \qquad \textbf{ANS.}$$

The orientation of the deflection vector relative to the u principal axis is defined by the angle θ, where

$$\theta = \tan^{-1} \frac{v_v}{v_u} = 22.16° \qquad \textbf{ANS.}$$

Moreover, the angle β defining the orientation of the neutral axis from the u principal axis is given by Equation 4.24. Thus,

$$\beta = \tan^{-1}\left[\frac{I_u}{I_v} \tan \phi\right] = \tan^{-1}\left[\frac{100.65}{8.35}(\tan 11.48°)\right] = 67.78°$$

Note that the sum $\theta + \beta = 89.94° \approx 90.0°$, which confirms the fact that the resultant deflection v is normal to the neutral axis, NA. The slight difference between the two numbers (89.94 and 90.0) is due to the truncations used in obtaining all of the above quantities.

PROBLEMS

6.38 A steel cantilever beam, of length $L = 10$ ft, has a $9 \times 4 \times 1$ in. angle section for its cross section. It carries a downward concentrated load $P = 1$ kip through the shear center at its free end directed along the 9-in. angle leg. Determine the resultant end deflection for the beam. Use the section properties found in Example 6.7. Let $E = 30 \times 10^3$ ksi.

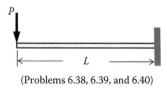

(Problems 6.38, 6.39, and 6.40)

6.39 Repeat Problem 6.38 after replacing the concentrated end load by a uniform load $p = 0.5$ kip/ft over its entire length.

6.40 Repeat Problem 6.38 after replacing the cross section by a $6 \times 6 \times 1$ in. angle (see Appendix E).

6.41 The aluminum ($E = 75$ GPa) beam of figure (a) below has the rectangular cross section shown in figure (b) below. Let $P = 2$ kN, $L = 6$ m, $b = 100$ mm, $h = 200$ mm, and $\phi = 30°$, determine the resultant center beam deflection.

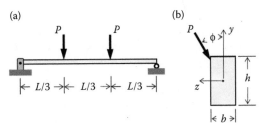

(Problems 6.41 and 6.42)

6.42 In reference to the figure shown in Problem 6.41, consider the steel ($E = 30 \times 10^3$ ksi) beam shown in figure (a) above and replace the rectangular cross section by a W16 × 100 (see Appendix E). Let $P = 100$ kips, $L = 15$ ft, and $\phi = 20°$, determine the resultant center beam deflection.

6.43 The aluminum ($E = 70$ GPa) beam of figure (a) below has the angle section shown in figure (b) below. The distributed load $p = 2$ kN/m acts vertically downward through the shear center of the section. Let $L = 6$ m, determine the resultant center beam deflection.

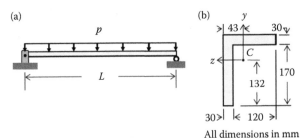

All dimensions in mm

(Problems 6.43 and 6.44)

6.44 In reference to the figure shown in Problem 6.43, consider the steel ($E = 30 \times 10^3$ ksi) beam shown in figure (a) above and replace the angle cross section by an 8 × 8 × 1 in. angle (see Appendix E). The distributed load $p = 2$ kips/ft acts vertically downward through the shear center of the section. Let $L = 15$ ft, determine the resultant center beam deflection.

6.45 The steel ($E = 30 \times 10^3$ ksi) overhanging beam of figure (a) below has the inverted T section shown in figure (b) below. The load $P = 10$ kips acts through the shear center of the section. Let $a = 2$ in., $\phi = 20°$, and $L = 12$ ft, determine the resultant deflection at C.

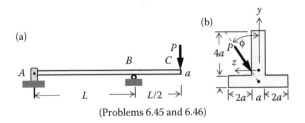

(Problems 6.45 and 6.46)

6.46 In reference to the figure shown in Problem 6.45, consider the steel ($E = 30 \times 10^3$ ksi) beam shown in figure (a) above and replace the inverted T section with a W16 × 26 (see Appendix E). The load, $P = 12$ kips, acts through the centroid, which is also the shear center of the section. Let $L = 16$ ft and $\phi = 10°$, determine the resultant deflection at the center between A and B.

6.47 The aluminum ($E = 75$ GPa) cantilever beam of figure (a) below has the Z section shown in figure (b) below. Let $a = 50$ mm, $h = 200$ mm, and $b = 150$ mm. If $P = 20$ kN and $L = 5$ m, find the resultant end deflection of the beam.

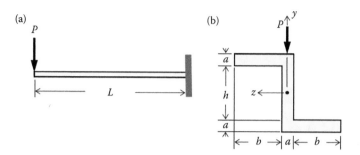

6.48 The steel ($E = 30 \times 10^3$ ksi) simply supported beam of figure (a) below has the section shown in figure (b) below. Let $a = 2$ in., $h = 6$ in., and $b = 4$ in. If $P = 15$ kips and $L = 20$ ft, find the resultant deflection of the beam at midspan.

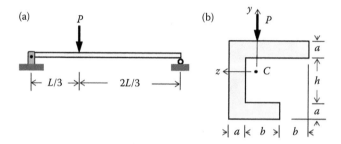

6.5   STATICALLY INDETERMINATE BEAMS: SINGULARITY FUNCTIONS

Singularity functions were introduced in Section 6.2 and used in conjunction with the method of two successive integrations to solve the problem of beam deflection. This same method may be used very effectively in the solution of statically indeterminate beams. This is so because singularity functions simplify the writing of the moment function *for complex load conditions* and make it possible to complete the solution of a statically indeterminate beam problem in a much easier and more compact manner. As a matter of fact, the complexity or simplicity of the loading condition does not make much difference in the application of this method as will be seen in the examples that follow.

EXAMPLE 6.8

Use singularity functions along with the method of two successive integrations to find the support reaction at A for the beam shown below.

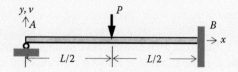

SOLUTION

Using the free-body diagram of shown below as an aid, we express the moment function in the beam in terms of singularity functions. Thus,

$$M = A_y x - P \left\langle x - \frac{L}{2} \right\rangle^1$$

Therefore, the differential equation needed for the solution becomes

$$EIv'' = A_y x - P \left\langle x - \frac{L}{2} \right\rangle^1$$

Assuming a constant EI and performing two successive integrations yields

$$EIv' = A_y \frac{x^2}{2} - \frac{P}{2} \left\langle x - \frac{L}{2} \right\rangle^2 + C_1 \qquad (6.8.1)$$

$$EIv = A_y \frac{x^3}{6} - \frac{P}{6} \left\langle x - \frac{L}{2} \right\rangle^3 + C_1 x + C_2 \qquad (6.8.2)$$

Equations 6.8.1 and 6.8.2 contain the three unknown quantities A_y, C_1, and C_2. Therefore, three boundary conditions are needed to solve for the three quantities. These boundary conditions are as follows:

BC1: $x = 0$, $v = 0$. Substituting into Equation 6.8.2 yields $C_2 = 0$.
BC2: $x = L$, $v = 0$. When substituted into Equation 6.8.2, this boundary condition leads to

$$A_y \frac{L^2}{6} - P \frac{L^2}{48} + C_1 = 0 \qquad (6.8.3)$$

BC3: $x = L$, $v' = 0$. Substituting this boundary condition into Equation 6.8.1 yields

$$A_y \frac{L^2}{2} - P \frac{L^2}{8} + C_1 = 0 \qquad (6.8.4)$$

Solving Equations 6.8.3 and 6.8.4 simultaneously, we obtain

$$A_y = \frac{5P}{16} \qquad \textbf{ANS.}$$

If desired, the support reactions at B may now be obtained by equilibrium.

EXAMPLE 6.9

The steel beam BC, shown in figure below, is supported at B by a steel cable BD of length ℓ and cross-sectional area A. Before the beam is loaded, the cable is unstressed. Develop expressions for the axial stress in the cable and for the deformation of the beam at B after the load is applied.

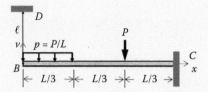

SOLUTION

Using the free-body diagram shown below as an aid, we express the bending moment in the beam in terms of singularity functions. Note that F represents the unknown axial force in the cable. Note also that the distributed load was extended to the cut and that, what was added, was simultaneously deleted. Thus,

$$M = Fx + \left(\frac{P}{2L}\right)\left\langle x - \frac{L}{3}\right\rangle^2 - \left(\frac{P}{2L}\right)x^2 - P\left\langle x - \frac{2L}{3}\right\rangle^1$$

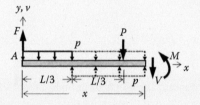

Therefore, the differential equation needed to develop the slope and deflection functions becomes

$$EIv'' = Fx + \left(\frac{P}{2L}\right)\left\langle x - \frac{L}{3}\right\rangle^2 - \left(\frac{P}{2L}\right)x^2 - P\left\langle x - \frac{2L}{3}\right\rangle^1$$

Assuming EI is constant, we integrate twice to obtain the slope and deflection functions. Thus,

$$EIv' = \frac{F}{2}x^2 + \frac{P}{6L}\left\langle x - \frac{L}{3}\right\rangle^3 - \frac{P}{6L}x^3 - \frac{P}{2}\left\langle x - \frac{2L}{3}\right\rangle^2 + C_1 \qquad (6.9.1)$$

$$EIv = \frac{F}{6}x^3 + \frac{P}{24L}\left\langle x - \frac{L}{3}\right\rangle^4 - \frac{P}{24L}x^4 - \frac{P}{6}\left\langle x - \frac{2L}{3}\right\rangle^3 + C_1x + C_2 \qquad (6.9.2)$$

Equations 6.9.1 and 6.9.2 contain the three unknown quantities F, C_1, and C_2. Three boundary conditions are, therefore, needed to evaluate these quantities. Thus,

BC1: $x = 0$, $v = -(F\ell/AE)$. Substituting into Equation 6.9.2 yields $C_2 = -(F\ell I/A)$.
BC2: $x = L$, $v = 0$. After substituting into Equation 6.9.2, and solving for C_1, we obtain

$$C_1 = F\left(\frac{\ell I}{AL} - \frac{L^2}{6}\right) + \frac{77PL^2}{1944} \tag{6.9.3}$$

BC3: $x = L$, $v' = 0$. Substituting into Equation 6.9.1 and solving for C_1, we obtain

$$C_1 = \frac{14PL^2}{81} - \frac{FL^2}{2} \tag{6.9.4}$$

Equating the values of C_1 given by Equations 6.9.3 and 6.9.4 and solving for F, we obtain

$$F = \frac{259APL^3}{648(3\ell I + AL^3)}$$

Therefore, the axial stress in the cable becomes

$$\sigma = \frac{F}{A} = \frac{259PL^3}{648(3\ell I + AL^3)} \qquad \textbf{ANS.}$$

Of course, the deflection of the beam at B, v_B, is identical to the axial deformation of the cable. Therefore,

$$v_B = -\frac{F\ell}{AE} = -\frac{259PL^3\ell}{648(3\ell I + AL^3)E} \qquad \textbf{ANS.}$$

EXAMPLE 6.10

Beam AB is fixed at both ends and loaded as shown below. Determine, in terms of P, Q, and L, the shear and moment at support A. Assume a constant EI.

SOLUTION

Using the free-body diagram shown below as an aid, we write the moment function in the beam in terms of singularity functions. Thus,

$$M = V_A x + M_A - Q\left\langle x - \frac{L}{4}\right\rangle^0 - P\left\langle x - \frac{3L}{4}\right\rangle^1$$

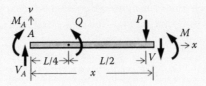

Therefore, the needed differential equation becomes

$$EIv'' = V_A x + M_A - Q\left\langle x - \frac{L}{4}\right\rangle^0 - P\left\langle x - \frac{3L}{4}\right\rangle^1$$

Performing two successive integrations, we obtain the slope and deflection functions. Thus,

$$EIv' = \frac{V_A}{2} x^2 + M_A x - Q\left\langle x - \frac{L}{4}\right\rangle^1 - \frac{P}{2}\left\langle x - \frac{3L}{4}\right\rangle^2 + C_1 \qquad (6.10.1)$$

$$EIv = \frac{V_A}{6} x^3 + \frac{M_A}{2} x^2 - \frac{Q}{2}\left\langle x - \frac{L}{4}\right\rangle^2 - \frac{P}{6}\left\langle x - \frac{3L}{4}\right\rangle^3 + C_1 x + C_2 \qquad (6.10.2)$$

Equations 6.10.1 and 6.10.2 contain the four unknown quantities, $V_A, M_A, C_1,$ and C_2. Therefore, four boundary conditions are needed as follows:

BC1: $x = 0$, $v = 0$. When substituted into Equation 6.10.2, this condition yields $C_2 = 0$.
BC2: $x = 0$, $v' = 0$. Substituted into Equation 6.10.1, this condition leads to $C_1 = 0$.
BC3: $x = L$, $v = 0$. Substituting this condition into Equation 6.10.2 and simplifying, we obtain

$$64V_A L + 192M_A - 108Q - PL = 0 \qquad (6.10.3)$$

BC4: $x = L$, $v' = 0$. When substituted into Equation 6.10.1, this condition, after simplification, leads to

$$16V_A L + 32M_A - 24Q - PL = 0 \qquad (6.10.4)$$

The simultaneous solution of Equations 6.10.3 and 6.10.4 yields

$$M_A = \frac{12Q - 3PL}{64}; \quad V_A = \frac{72Q + 10PL}{64L} \qquad \textbf{ANS.}$$

PROBLEMS

(Use singularity functions in the solution of the following problems.)

6.49 Find the support reactions at A and B. Express your answers in terms of L and the given loading.

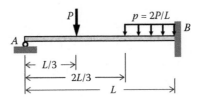

6.50 Find the support reactions at A and B. Express your answers in terms of L and the given loading.

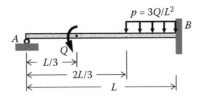

6.51 Find the support reactions at A and B. Express your answers in terms of L and the given loading.

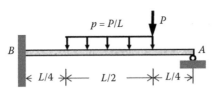

6.52 Find the support reactions at A and B. Express your answers in terms of L and the given loading.

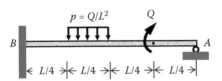

6.53 Beam BC is fixed at C and supported at B by cable BD of length l and cross-sectional area A, made of the same material as the beam for which the modulus of elasticity is E and the moment of inertia is I. Compute the force in the cable expressing your answer in terms of P, L, E, ℓ, A, and I.

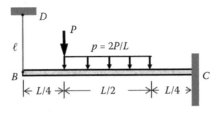

6.54 Beam BC is fixed at C and supported at B by a spring with a spring constant k. The beam is made of a material for which the modulus of elasticity is E and the moment of inertia is I. Compute the force in the spring expressing your answer in terms of Q, L, E, k, and I.

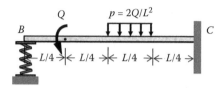

6.55 Beam AB is fixed at B and supported at A by a roller support. However, before the loading is applied, there is a gap Δ between the beam and the roller as shown. The beam has a moment of inertia I and is made of a material for which the modulus of elasticity is E. Compute the reaction at support A expressing your answer in terms of P, L, E, Δ, and I.

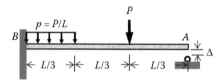

6.56 Beam BC is fixed at B and supported at C by a compression block of length ℓ and cross-sectional area A, made of the same material as the beam. The beam has a moment of inertia I and a modulus of elasticity E. Compute the reaction at C expressing your answer in terms of p, L, E, ℓ, A, and I.

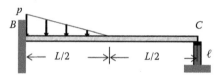

6.57 Beam AB is fixed at both ends and subjected to loads as shown. Assume a constant EI and find the reaction components at A. Express your answers in terms of P and L.

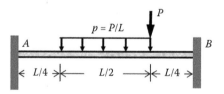

6.58 Beam AB is fixed at both ends and subjected to loads as shown. Assume a constant EI and find the reaction components at A. Express your answers in terms of Q and L.

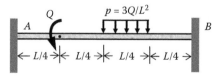

6.59 Beam AB is fixed at both ends and subjected to loads as shown. After the loads are applied, the support at A settles downward by an amount Δ but no rotation of the beam occurs at that point. Assume a constant EI and find the reaction components at A. Express your answers in terms of P, L, Δ, E, and I.

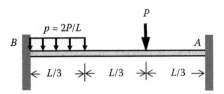

6.60 Beam AB is fixed at both ends and subjected to loads as shown. After the loads are applied, the support at A rotates clockwise through an angle θ but no settlement occurs at this point. Assume a constant EI and find the reaction components at A. Express your answers in terms of p, L, θ, E, and I.

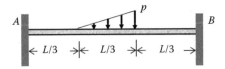

6.61 The steel beam of Problem 6.40 has a W14 × 34 section bent about the strong axis. If $L = 18$ ft and $P = 36$ kips, find the deflection at midspan.

6.62 The steel beam of Problem 6.49 has a W200 × 19.3 section bent about the strong axis. If $L = 6$ m, $\theta = 30°$, and $p = 30$ kN/m, find the deflection at midspan.

6.6 STATICALLY INDETERMINATE BEAMS: CASTIGLIANO'S SECOND THEOREM

Castigliano's second theorem developed in Section 6.3 may be used to supplement the equations of equilibrium in the solution of a given statically indeterminate beam problem. The procedure requires the determination of the internal moment in the beam in terms of one or more unknown reaction components. This moment is then differentiated partially with respect to one or more of the unknown reaction components and Equations 6.10 and/or 6.11 are applied to obtain expressions for the deflection and/or slope. These expressions are then equated to known deflection and/or slope quantities as dictated by the boundary conditions of the problem. This procedure leads to simultaneous equations that are solved to determine some of the unknown reaction components. Note that the coordinate system must be chosen such that the number of equations resulting from the application of Castigliano's second theorem is the same as the number of redundants in the problem. The remaining unknown reaction components may then be found by equilibrium. This method of solution is illustrated in the following examples.

EXAMPLE 6.11

Beam AB is simply supported at A and fixed at B as shown below. It carries a load that varies linearly from zero at A to p_o at B. Assume a constant EI and determine the reaction at support A.

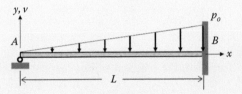

SOLUTION

The free-body diagram shown in figure (a) below indicates all of the unknown support reactions. The free-body diagram shown in figure (b) below is used to determine the internal moment in the beam at any position defined by the coordinate x. The expression for this moment is indicated in figure (b) below.

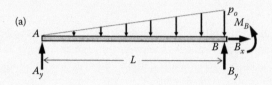

(a)

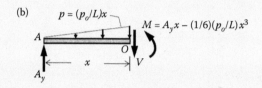

(b)

We now make use of Castigliano's second theorem, Equation 6.10, to satisfy the condition that the deflection at A is zero (i.e., $\partial U/\partial A_y = 0$). Thus,

$$\frac{\partial M}{\partial A_y} = x \qquad\qquad (6.11.1)$$

Substituting Equation 6.11.1 into Equation 6.10, we obtain

$$\int_0^L \frac{M}{EI}\frac{\partial M}{\partial A_y}\,dx = \frac{1}{EI}\int_0^L \left(A_y x^2 - \frac{1}{6}\frac{p_o}{L}x^4\right)dx = 0 \qquad\qquad (6.11.2)$$

Note that we were able to take EI out of the integral sign because it is a constant. Performing the integration indicated in Equation 6.11.2 and substituting the limits, we obtain

$$\left(\frac{1}{EI}\right)\left(\frac{A_y L^3}{3} - \frac{p_o L^4}{30}\right) = 0 \qquad\qquad (6.11.3)$$

Solving Equation 6.11.3 for A_y yields

$$A_y = \frac{p_o L}{10} \qquad \textbf{ANS.}$$

Note that this answer was obtained by other methods for the same problem in Chapter 5. Of course, if needed, the reaction components at support B can now be found by equilibrium.

EXAMPLE 6.12

The steel beam BC is supported at B, as shown below, by the aluminum rod BD of length l and cross section A. The beam is fixed at C and carries a uniform load of intensity p. The coefficient of thermal expansion for aluminum is α and its modulus of elasticity is E_A. The moment of inertia for the beam is I and its modulus of elasticity is E_S. The rod is unstressed before the load p is applied. If the temperature of the rod is dropped by ΔT, find the force in the aluminum rod.

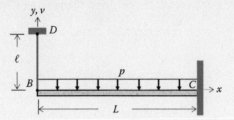

SOLUTION

The free-body diagram of the beam is shown below, where F is the force in the aluminum rod. As the temperature drops through ΔT, the aluminum rod tends to shorten by $\delta_T = \alpha(\Delta T)\ell$. However, the force F in the rod reduces this shortening tendency by the amount $\delta_F = Fl/AE_A$. Thus, if we assume that $\delta_T < \delta_F$, the net deflection of the beam at B is downward and equal to

$$v_B = \delta_F - \delta_T = \frac{F\ell}{AE_A} - \alpha(\Delta T)\ell \qquad (6.12.1)$$

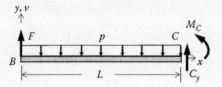

The internal moment in the beam at any position defined by the coordinate x may be found using the free-body diagram is shown below. As indicated in this diagram, the internal moment is $M = Fx - (p/2)x^2$ and its partial derivative with respect to F becomes

$$\frac{\partial M}{\partial F} = x \qquad (6.12.2)$$

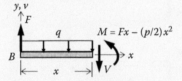

Introducing Equations 6.12.1 and 6.12.2 into Castigliano's second theorem as expressed in Equation 6.10 and assuming the product $E_S I$ for the beam to be a constant, we obtain

$$-\left[\frac{F\ell}{AE_A} - \alpha(\Delta T)\ell\right] = \frac{1}{E_S I}\int_0^L \left(Fx - \frac{p}{2}x^2\right)x\,dx = \frac{1}{E_S I}\left[\frac{FL^3}{3} - \frac{pL^4}{8}\right] \qquad (6.12.3)$$

Solving Equation 6.12.3 for F and simplifying, we obtain

$$F = \frac{3AE_A}{8}\left[\frac{pL^4 + 8E_SI\alpha(\Delta T)\ell}{AE_AL^3 + 3E_SI\ell}\right]\qquad\textbf{ANS.}$$

Note that this answer was obtained by other methods for the same problem in Chapter 5. If needed, the reaction components at the fixed support at C may now be determined by equilibrium.

EXAMPLE 6.13

Beam ACB is fixed at both ends and subjected to the concentrated force P as shown below. Assume a constant EI and determine the reaction components at A and B.

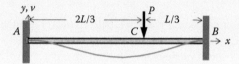

SOLUTION

The free-body diagram of the entire beam is shown in figure (a) below a total of four unknown quantities, M_A, A_y, M_B, and B_y. The internal moments in segments AC and CB are found using the free-body diagrams of figures (b) and (c) below, respectively. These moments and their partial derivatives are as follows:

$$M_{AC} = A_yx + M_A;\quad \frac{\partial M_{AC}}{\partial A_y} = x;\quad \frac{\partial M_{AC}}{\partial M_A} = 1 \qquad (6.13.1)$$

$$M_{CB} = A_yx + M_A - P\left(x - \frac{2L}{3}\right);\quad \frac{\partial M_{CB}}{\partial A_y} = x;\quad \frac{\partial M_{CB}}{\partial M_A} = 1 \qquad (6.13.2)$$

(a)

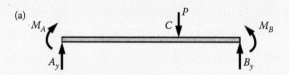

(b)

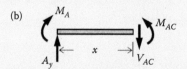

(c)

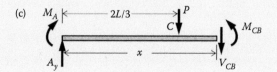

Since the beam is fixed at A, both the deflection and the slope of the beam at this point are zero. Substituting these two conditions into Equations 6.10 and 6.11, respectively, we obtain

$$0 = \frac{\partial U}{\partial A_y} = \int_0^{2L/3} (A_y x + M_A)x\,dx + \int_{2L/3}^{L} \left[A_y x + M_A - P\left(x - \frac{2L}{3}\right) \right] x\,dx$$

$$0 = \frac{\partial U}{\partial M_A} = \int_0^{2L/3} (A_y x + M_A)(1)\,dx + \int_{2L/3}^{L} \left[A_y x + M_A - P\left(x - \frac{2L}{3}\right) \right](1)\,dx$$

Performing the above integrations, applying the limits, and simplifying leads to the following two simultaneous equations containing the two unknown quantities A_y and M_A:

$$54A_y L + 81M_A - 81PL = 0 \qquad\qquad (6.13.3)$$

$$-9A_y L - 18M_A + PL = 0 \qquad\qquad (6.13.4)$$

The simultaneous solution of Equations 6.13.3 and 6.13.4 yields

$$A_y = \frac{7P}{27}; \quad \textbf{ANS.} \quad M_A = -\frac{2PL}{27} \quad \textbf{ANS.}$$

The negative sign on M_A indicates that this moment is counterclockwise and not clockwise as assumed. The reaction components at support B are found by applying the conditions of equilibrium to the free-body diagram of figure (b) above. Thus,

$$B_y = \frac{20P}{27}; \quad \textbf{ANS.} \quad M_B = -\frac{4PL}{27} \quad \textbf{ANS.}$$

Again, the negative sign on M_B tells us that this moment is not counterclockwise as assumed but clockwise. Note that these answers are the same as those obtained in Chapter 5 for the same problem by other methods.

PROBLEMS

(Use Castigliano's second theorem in the solution of the following problems.)

6.63 Beam AB is supported and loaded as shown. Assume a constant EI and find the reaction components at supports A and B.

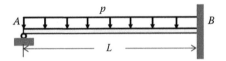

6.64 Beam AB is supported and loaded as shown. Assume a constant EI and find the reaction components at supports A and B.

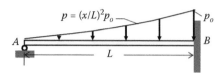

6.65 Beam AB is supported and loaded as shown. Assume a constant EI and find the reaction components at supports A and B.

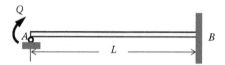

6.66 Beam AB is supported and loaded as shown. Assume a constant EI and find the reaction components at supports A and B.

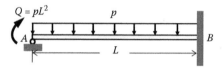

6.67 Beam ABC is supported and loaded as shown. Assume a constant EI and find the reaction components at supports A and C.

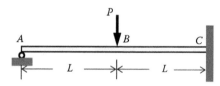

6.68 Beam ABC is supported and loaded as shown. Assume a constant EI and find the reaction components at supports A and C.

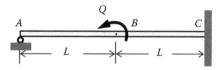

6.69 Beam ABC is supported and loaded as shown. Assume a constant EI and find the reaction components at supports A and C.

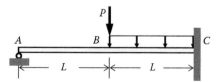

6.70 Beam ABC is supported and loaded as shown. Assume a constant EI and find the reaction components at supports A and C.

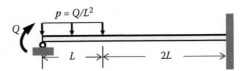

6.71 The steel cantilever beam BC is fixed at B and restrained by two steel rods of length l connected to the ends of rigid arms securely fastened to the beam at midspan. Let the cross-sectional area of the rods be A, the modulus of elasticity for steel be E, and the moment of inertia for the beam be I, find the force in each rod and the reactions at B.

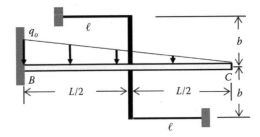

6.72 The system shown consists of two identical W16 × 89 cantilever beams with a rigid roller between them at B. Let $p = 4$ kips/ft, $L = 15$ ft, and $E = 30 \times 103$ ksi, determine the reaction between the two beams at B and the deflection of the two beams at this point.

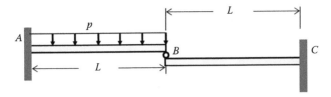

6.73 Refer to the beam of Problem 6.66. Let $L = 20$ ft and $p = 3$ kips/ft, find the reaction components at both supports. Construct the shear and moment diagrams for the beam.

6.74 Beam AB of length L is fixed at both ends and is subjected to loading as shown. Assume a constant EI and determine the reaction components at fixed support at A.

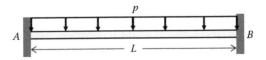

6.75 Beam AB of length L is fixed at both ends and is subjected to loading as shown. Assume a constant EI and determine the reaction components at fixed support at A.

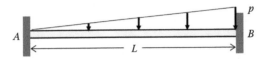

6.76 Beam AB of length L is fixed at both ends and is subjected to loading as shown. Assume a constant EI and determine the reaction components at fixed support at A.

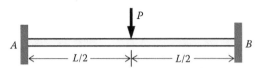

6.77 Beam AB of length L is fixed at both ends and is subjected to loading as shown. Assume a constant EI and determine the reaction components at fixed support at A.

6.78 Beam AB of length L is fixed at both ends and has a standard $S20 \times 75$ section. Let $p = 5$ kips/ft, $L = 30$ ft, and $E = 30 \times 103$ ksi, determine the reaction components at support A.

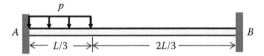

6.79 Beam AB of length L is fixed at both ends and has a standard $W14 \times 120$ section. Let $p = 12$ kips/ft, $L = 24$ ft, and $E = 30 \times 103$ ksi, determine the reaction components at support A.

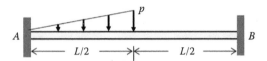

6.80 Beam AB of length L is fixed at both ends and has a standard $W360 \times 101$ section. Let $Q = 40$ kN·m, $L = 9$ m, and $E = 205$ GPa, determine the reaction components at support A.

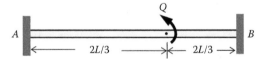

6.7 IMPACT LOADING

Occasionally, beams are subjected to impact loads imparting stresses and strains that are much larger than those experienced during normal (static) loading conditions. It, therefore, becomes desirable to be able to analyze the effects of these impact loads on beams. Examples of beams subjected to impact loads are found in bridges under vehicular traffic conditions, especially when large and heavy trucks are involved.

Consider the simply supported beam shown in Figure 6.6, which is subjected to the impact produced by the weight W dropping through a height h as shown. As a consequence of the impact, the beam deflects through the maximum deflection Δ at the point of impact. Therefore, the total

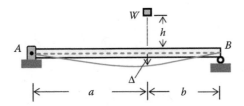

FIGURE 6.6 Simply supported beam subjected to impact loading produced by a load W dropping through a height h.

distance traveled by the dropping weight W is $(h + \Delta)$ resulting in a total work performed by W equal to $W(h + \Delta)$. If we make the assumption that no energy is lost during impact, then the entire work done by W is transformed to elastic strain energy, U, in the beam. The strain energy U in the beam is given by Equation 6.9. We, therefore, conclude that

$$W(h + \Delta) = U = \frac{1}{2} \int_0^L \frac{M^2}{EI} \, dx \tag{6.13}$$

Another way to view the relationship between the work done by the weight W and the strain energy induced in the beam is to deal with the force at the point of impact the moment W strikes the beam. This force and the corresponding beam deflection increase linearly from zero to their maximum values at the end of travel when the force is F and the corresponding deflection is Δ. We, therefore, conclude that the strain energy developed in the beam is $U = 1/2F\Delta$. Thus, the relationship between the work done and the strain energy in the beam, assuming no losses, becomes

$$W(h + \Delta) = U = \frac{1}{2} F\Delta \tag{6.14}$$

We note that Equation 6.14 is identical with Equation 1.38 for axially loaded members and Equation 2.28 for torsionally loaded members. Equations 6.13 and 6.14 imply that no losses occur during impact. Actually, energy losses do occur because some of the work done is transformed into heat and sound as the weight W strikes the beam and in damping out its vibrations after the weight W comes to the end of its travel. Besides, we should keep in mind that Equations 6.13 and 6.14 are valid only within the elastic range for the material. Furthermore, while these equations were developed for the case of a beam with simple supports, they are valid for beams with any other type of support systems.

The question arises as to which of the two equations to use in a given situation. Of course, both equations lead to the same answer and the use of either Equation 6.13 or 6.14 becomes a question of personal choice. Both equations are used in Example 6.14 to illustrate the difference between them.

EXAMPLE 6.14

A 20-lb weight W is dropped from a height $h = 4$ in. onto an simply supported aluminum beam as shown below, producing the maximum deflection Δ at the point of impact. The beam has a rectangular cross section with a height of 2 in. and a depth into the page of 6 in. Let $E = 10 \times 10^3$ ksi and assume that 10% of the work produced by W is lost to heat, sound, and damping and determine the maximum bending stress and the maximum deflection produced in the beam.

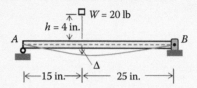

SOLUTION

I. *Solution using Equation 6.13:* The free-body diagram shown below shows beam AB at the end of travel of the weight W when the impact force F at the point of contact has reached its maximum value. In terms of F, the reaction at A is $(5/8)F$ and that at B is $(3/8)F$. The maximum deflection is obtained from the information in Appendix G and, in terms of F, it is $\Delta = 29.297 \times 10^{-6}F$. Moreover, the moment of inertia of the rectangular cross section is found to be $I = 4$ in.^{4} Thus, substituting numerical values into Equation 6.13 and recalling that only 90% of the work done by W is transformed into strain energy, we obtain

$$0.9(20)(4 + 29.297 \times 10^{-6}F) = \left(\frac{1}{2(10 \times 10^6)(4)}\right)\left\{\int_0^{15}\left[\left(\frac{5}{8}\right)Fx\right]^2 dx + \int_0^{25}\left[\left(\frac{3}{8}\right)Fx\right]^2 dx\right\}$$

Simplifying, the above equation reduces to

$$F^2 - 36F - 4.915 \times 10^6 = 0 \qquad\qquad (6.14.1)$$

The solution of this quadratic equation yields $F = 2235$ lb. Therefore,

$$\sigma_{MAX} = \frac{My}{I} = \frac{(5/8)F)(15)(1)}{4} = 5238.3 \text{ psi} \approx 5.2 \text{ ksi} \qquad \textbf{ANS.}$$

$$\Delta = 29.297 \times 10^{-6}F = 65,478.8 \times 10^{-6} \text{ in.} \approx 0.0655 \text{ in.} \qquad \textbf{ANS.}$$

II. *Solution using Equation 6.14:* All of the ingredients needed for this solution are already available. Substituting them into Equation 6.14 yields

$$0.9(20)(4 + 29.297 \times 10^{-6}F) = \left(\frac{1}{2}\right)F(29.297 \times 10^{-6}F)$$

Simplifying, we obtain the same quadratic equation given in Equation 6.14.1 above. The remainder of the solution will obviously yield the same values for F, σ_{MAX}, and Δ as obtained above.

EXAMPLE 6.15

Consider the simply supported beam shown. Assume an arbitrary cross section and no energy losses and develop general expressions in terms of W, h, L, I, and E for the maximum bending stress, maximum deflection, and impact factor.

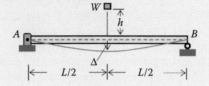

SOLUTION

Equation 6.14 will be used in the solution of this example. From the information given in Appendix G, we conclude that $\Delta = FL^3/48EI$, where F is the maximum impact force corresponding to the maximum deflection Δ. We now substitute this value of Δ into Equation 6.14. Thus,

$$W\left(h + \frac{FL^3}{48EI}\right) = \frac{1}{2}F\left(\frac{FL^3}{48EI}\right) \Rightarrow F^2 - 2WF - 96\left(\frac{WEIh}{L^3}\right) = 0$$

Solving the above quadratic equation, we obtain

$$F = W + \sqrt{W^2 + 96\left(\frac{WEIh}{L^3}\right)}$$

$$\sigma_{MAX} = \frac{Mc}{I} = \frac{(FL/4)c}{I} = \left[W + \sqrt{W^2 + 96\left(\frac{WEIh}{L^3}\right)}\right]\left(\frac{Lc}{4I}\right) \qquad \textbf{ANS.}$$

$$\Delta = \frac{FL^3}{48EI} = \left[W + \sqrt{W^2 + 96\left(\frac{WEIh}{L^3}\right)}\right]\left(\frac{L^3}{48EI}\right) \qquad \textbf{ANS.}$$

$$IF = \frac{F}{W} = 1 + \sqrt{1 + 96\left(\frac{EIh}{WL^3}\right)} \qquad \textbf{ANS.}$$

We recall that the quantity c is the maximum distance from the neutral axis in the cross section.

PROBLEMS

6.81 The simply supported steel beam has a rectangular cross section with a depth of 2 in. and a width into the page of 4 in. Let $W = 20$ lb, $h = 2$ in., $E = 30 \times 10^3$ ksi, and $L = 5$ ft. If 10% of the work performed by W is lost during the process of impact, determine the maximum stress in the beam, the maximum deflection, and the impact factor.

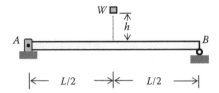

6.82 Solve Problem 6.81 if the rectangular cross section is rotated through 90° so that the 4-in. dimension is the depth of the beam and the 2-in. dimension is the width into the page.

6.83 A weight $W = 40$ lb is dropped from a height $h = 6$ in. onto a fixed-ended steel beam as shown. The beam has rectangular cross section with a depth of 4.0 in. and a width of 8.0 in. into the page. Let $L = 5$ ft and $E = 30 \times 10^3$ ksi, determine the maximum bending stress and maximum deflection due to impact assuming that 15% of the work done by W is lost to heat, sound, and damping.

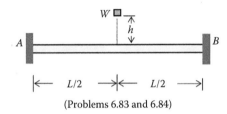

(Problems 6.83 and 6.84)

6.84 In reference to the figure shown in Problem 6.83, a weight W is dropped from a height $h = 150$ mm onto a fixed-ended steel beam as shown. The beam has a rectangular cross section with a depth of 200 mm and a width of 300 mm into the page. The maximum stress at the point of impact was found to be 25 MPa. Determine the weight W that was dropped and the corresponding maximum deflection. Let $L = 2.0$ m and $E = 210$ GPa. Assume no energy losses.

6.85 Two weights $W = 40$ lb are dropped simultaneously onto a simply supported steel beam as shown. The moment of inertia with respect to the horizontal centroidal axis is $I = 150$ in.[4], the depth of the beam is 20 in. and its modulus of elasticity $E = 30 \times 10^3$ ksi. The maximum stress in the beam was found to be 10 ksi. Determine the height h from which the weights W were dropped and the corresponding maximum deflection of the beam. Let $L = 20$ ft and assume 20% energy loss.

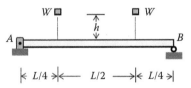

6.86 A weight $W = 80$ N is dropped from a height $h = 100$ mm onto a simply supported aluminum beam as shown. The beam has rectangular cross section with a depth of 50 mm and a width into the page of 150 mm. Let $a = 400$ mm, $b = 600$ mm, and $E = 70$ GPa. Assume that 15% of the work done by W is lost during impact and determine the bending stress at midspan and the corresponding deflection.

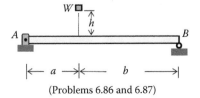

(Problems 6.86 and 6.87)

6.87 In reference to the figure shown in Problem 6.86, a weight $W = 20$ lb is dropped from a height h onto a simply supported aluminum beam as shown. The beam has a rectangular cross section with a depth of 2 in. and a width into the page of 6 in. Let $a = 30$ in., $b = 20$ in., and $E = 10 \times 10^3$ ksi. Assume that 10% of the work done by W is lost during impact and determine the maximum height h if the maximum bending stress in the beam is not to exceed 6.0 ksi. Also determine the corresponding maximum deflection.

6.88 A weight $W = 50$ lb is dropped from a height $h = 4$ in. onto the free end of a steel cantilever beam bent about its strong axis. The cross section of the beam is $W6 \times 20$ (see Appendix E). Let $L = 12$ ft and $E = 29 \times 10^3$ ksi, determine the maximum bending stress and the maximum deflection due to impact. Assume no energy losses.

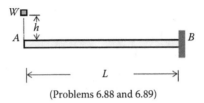

(Problems 6.88 and 6.89)

6.89 In reference to the figure shown in Problem 6.88, a steel cantilever beam is to be used in an application that would subject it to a dropping weight as shown. Let $W = 150$ N, $h = 100$ mm, $L = 4$ m, and $E = 200$ GPa. If the allowable deflection is 20 mm, select the lightest wide flange section from Appendix E suitable for the application. Assume no energy losses.

6.90 A diver represented by the weight $W = 650$ N drops into the diving board through the height $h = 50$ mm. The cross section of the board is rectangular with a depth of 50 mm and a width into the page of 260 mm. Let $a = 1.8$ m, $b = 1.0$ m, and $E = 14.0$ GPa, determine the maximum bending stress and the maximum deflection. Assume no energy losses.

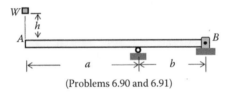

(Problems 6.90 and 6.91)

6.91 In reference to the figure shown in Problem 6.90, a diver represented by the weight W drops into the diving board through the height $h = 10$ in. The cross section of the board is rectangular with a depth of 1.875 in. and a width into the page of 10.5 in. Let $a = 8$ ft, $b = 4$ ft, $E = 2 \times 10^3$ ksi, $\sigma_{ALL} = 30.0$ ksi, and $\Delta_{ALL} = 6.5$ in., determine the maximum diver weight W permissible on this diving board. Assume no energy losses.

6.92 Let $W = 40$ N, $h = 154$ mm, $k = 1000$ N/m, $E = 100$ GPa, $I = 18 \times 10^{-6}$ m^4, $L = 6$ m, and a depth of 0.06 m, determine the maximum bending stress and the maximum deflection experienced by the beam. Assume no energy losses.

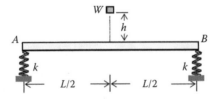

REVIEW PROBLEMS

R6.1   Use singularity functions to develop the slope and deflection equations for the given beam. Find the deflection at B and the slope at C.

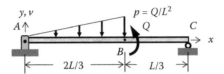

R6.2   Use singularity functions to develop the slope and deflection equations for the given beam. Find the deflection at A and the slope at D.

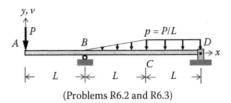

(Problems R6.2 and R6.3)

R6.3   In reference to the figure shown in Problem R6.2, select the lightest wide flange section for the given beam if the deflection at A is not to exceed 5.0 in. Let $P = 10$ kips, $L = 5$ ft, and $E = 30 \times 10^3$ ksi and use singularity functions for the solution.

R6.4   Use Casigliano's second theorem to find the slope at A and the deflection at C. Express the answers in terms of p, L, and the constant EI.

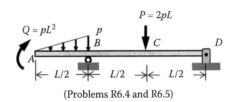

(Problems R6.4 and R6.5)

R6.5   In reference to the figure shown in Problem R6.4, if the deflection at C is to be limited to 20 mm, determine the maximum permissible magnitude of the load intensity p. Let $L = 10$ m and $EI = 200 \times 103$ kN·m². Use Castigliano's second theorem.

R6.6   The aluminum ($E = 10.5 \times 10^3$ ksi) beam of figure (a) below has the cross section shown in figure (b) below. Let $p = 3$ kips/ft, $L = 10$ ft, and $a = 3$ in., determine the resultant deflection under the load P.

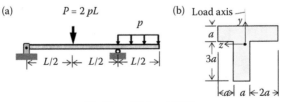

(Problems R6.6 and R6.7)

R6.7   Refer to Problem R6.6 and find the resultant deflection at the right end of the overhang.

R6.8   Beam BC is fixed at C and supported at B by cable BD of length ℓ and cross-sectional area A, made of the same material as the beam for which the modulus of elasticity is

E and the moment of inertia is I. Compute the reaction components at C expressing your answers in terms of P, L, E, ℓ, A, and I. Use singularity functions.

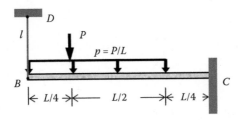

R6.9 Use Castigliano's second theorem to find the reaction components at A and B. Assume a constant EI.

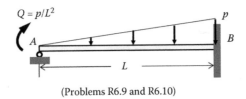

(Problems R6.9 and R6.10)

R6.10 In reference to the figure shown in Problem R6.9, use Castigliano's second theorem to find the deflection at midspan. Assume a constant EI and express your answer in terms of p, L, and EI.

R6.11 Let $k = 40$ kips/ft, $E = 30 \times 10^3$ ksi, $I = 2$ in.^{4}, $L = 5$ ft, and $h = 10$ in. Determine the maximum weight W that may be dropped onto the beam if the maximum permissible stress in the beam is not to exceed 10 ksi. Assume no energy losses.

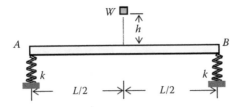

7 Analysis of Stress

7.1 INTRODUCTION

We first introduced the concept of stress in Chapter 1 in connection with members subjected to axial loads. We also discussed and elaborated further on it in Chapter 2 dealing with torsional members and in Chapter 3 dealing with bending members. Furthermore, in Chapters 2 and 3, we learnt how to deal with the stresses resulting from simple cases of combined loads.

In this chapter, we expand our previous knowledge of stress. Thus, in Section 7.2, we generalize the concept of *stress at a point* in a stressed body. In Section 7.3, we define the three components of normal stress and the six components of shearing stress (a total of nine stress components reduced to six) and generalize the sign convention for both normal and shearing stresses. In Section 7.4, we deal with the subject of *plane stress*, consisting of two normal and one shearing stress because three of the six stress components reduce to zero. Moreover, in the case of plane stress, we develop the mathematical relations needed to determine the normal and shearing stresses on inclined planes, as well as the maximum in-plane normal and maximum in-plane shearing stresses existing at the point in question. In Section 7.5, we introduce *Mohr's circle*, which is an alternate semi-graphical solution for the plane stress problem. In Section 7.6, we analyze a three-dimensional stress system and show how Mohr's circle may be used to find the maximum normal and maximum shearing stress in such a stress system. Finally, in Section 7.7, we discuss some of the stress theories of failure that have been developed to attempt to predict at what stress level a member fails if subjected to combined loading.

7.2 STRESS AT A POINT

When a body is subjected to external forces, a system of internal forces is developed. These internal forces tend to separate (tensile forces) or bring closer together (compressive forces) the material particles that make up the body. Let us consider, for example, the body shown in Figure 7.1a, which is subjected to the external forces $P_1, P_2, \ldots, P_i$. Let us imagine a plane that cuts the body into two parts as shown. We then delete the upper part in order to expose the internal forces acting on the lower part at the cut surface as shown in the free-body diagram of Figure 7.1b. The external forces P_1, P_2, and P_3 acting on the lower part of the body are held in equilibrium by an internal system of forces distributed in some manner over the surface area of the imaginary cutting plane and are transmitted from one part of the body to the other through this plane. This internal system of forces may be represented by a single resultant force R and/or by a couple. For the sake of simplicity and without loss of generality, only the resultant force R is assumed to exist. This force R is now decomposed into the two components F_n, perpendicular to the plane and known as the *normal force*, and F_t, parallel to the plane, known as the *shearing force*.

If the area of the imaginary plane is A, the quantities F_n/A and F_t/A represent, respectively, average values of the normal and shearing forces per unit area. As stated in earlier chapters, the quantity F_n/A is known as the normal stress and given the Greek symbol, σ, and the quantity F_t/A as the shearing stress and assigned the Greek symbol, τ. Thus, from these definitions, we conclude that the internal forces F_n and F_t may be obtained, respectively, by multiplying the average values of σ and τ by the corresponding areas. In general, however, these stresses are not uniformly distributed throughout their respective areas and it becomes necessary to determine these stresses at any point within the area. If the normal and shearing forces acting over a differential element of area ΔA in

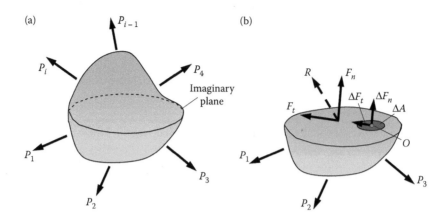

FIGURE 7.1 (a) A body subjected to external forces P_1 to P_i. (b) The internal forces, represented by the resultant R, decomposed into normal and shearing components F_n and F_t, respectively, acting on an imaginary plane leading to the definition of normal and shearing stresses at a point as given by Equation 7.1.

the neighborhood of point O are ΔF_n and ΔF_t, respectively, as shown in Figure 7.1b, then the normal stress σ and the shearing stress τ are given by the following expressions:

$$\sigma = \lim_{\Delta A \to 0} \frac{\Delta F_n}{\Delta A} = \frac{dF_n}{dA}; \quad \tau = \lim_{\Delta A \to 0} \frac{\Delta F_t}{\Delta A} = \frac{dF_t}{dA} \tag{7.1}$$

Special cases of Equations 7.1, where the normal and shearing stresses may be assumed uniformly distributed, were discussed in Chapter 1 dealing with members subjected to axial loads. Moreover, a special case of the second part of Equations 7.1, where the shearing stress is linearly distributed, was discussed in Chapter 2 in connection with a circular shaft subjected to a pure torque. Finally, a special case of the first part of Equations 7.1, where the normal stress is linearly distributed was discussed in Chapter 3 dealing with a beam under a pure bending moment. The reader is urged to refer to these chapters to review some of these special cases, before proceeding further in this chapter.

It is useful to reiterate the units of stress which were first discussed in Chapter 1. From the definition of stress as force per unit area we conclude that stress is measured as units of force divided by units of area. Thus, in the U.S. Customary system of measure, such units as pounds per square inch (psi) and kilopounds per square inch (ksi) are common. In the metric (SI) system of measure, the unit used for stress is the *pascal* denoted by Pa, which represents a Newton per square meter. The pascal, however, is a very small quantity and another SI unit that is widely used is the megapascal (10^6 pascals) and is denoted by the symbol MPa.

7.3 COMPONENTS OF STRESS

Let us consider the most general stress system depicted by the three-dimensional stress element shown in Figure 7.2. This general state of stress is referred to as a *triaxial* stress system. It is convenient to select planes that are normal to the three coordinate x, y, and z axes and designate them as the X, Y, and Z planes, respectively. As indicated in Figure 7.2, we consider a cube bounded by six mutually perpendicular planes, two X planes, two Y planes, and two Z planes.

This cube may be thought of as enclosing a differential volume ($dxdydz$) of material in the neighborhood of some point in a stressed body. Such a differential volume of material showing all of the stresses acting on it is known as a *three-dimensional or triaxial stress element*. On each of the

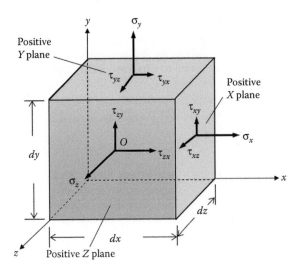

FIGURE 7.2 Sketch showing *a triaxial* stress system in which the three X, Y, and Z planes are normal to the three coordinate x, y, and z axes, respectively. Each of these stresses is decomposed into three components, one along the x, a second along the y, and the third along the z axes.

six mutually perpendicular planes, there exists a normal stress, σ, and a shearing stress, τ, which is represented by its two perpendicular components. Note, however, that while Figure 7.2 shows only the stress components on one X plane (the positive X plane or the one we can see), the stress components on the hidden X plane (negative X plane) are identical to those shown. The same statement applies to the Y and Z planes. Thus, there exist a total of nine stress components that must be specified in order to define the state of stress at any point in a stressed body. However, consideration of rotational equilibrium of the stress element leads us to conclude that $\tau_{xy} = \tau_{yx}$, $\tau_{xz} = \tau_{zx}$ and $\tau_{yz} = \tau_{zy}$ so that this number is reduced to six. For example, summing moments about an z axis through point O eliminates the forces produced by σ_z, τ_{xz}, and τ_{yz} because they are parallel to this axis. The forces created by the two σ_x stresses balance each other or we can eliminate them because they intersect the z axis. A similar statement applies to the two σ_y stresses. Therefore,

$$\sum M_{Oz} = 0 \Rightarrow \tau_{xy}(dz\,dy)(dx/2) - \tau_{yx}(dz\,dx)(dy/2) = 0$$

Note that the products $dzdy$ and $dzdx$ represent the areas of the X and Y planes, respectively. Simplifying, this equilibrium equation reduces to

$$\tau_{yx} = \tau_{xy} \tag{7.2}$$

Let us recall that the concept expressed in Equation 7.2 was also encountered for the two-dimensional case in Chapters 1 and 2. Similarly, we can show that

$$\tau_{xz} = \tau_{zx} \quad \text{and} \quad \tau_{yz} = \tau_{zy} \tag{7.3}$$

The notation for normal and shearing stresses was introduced in Chapter 1 but is repeated here for convenience. It consists of affixing one subscript to a normal stress, indicating the plane on which it acts and the associated coordinate axis, and two subscripts to a shearing stress, the first of which indicates the plane on which it acts and the second the coordinate axis along which it is pointed. Thus, for example, σ_x is the normal stress on the X plane acting along the x axis; τ_{xy} is the component of the shearing stress that acts in the X plane pointing in the y direction; and τ_{xz} is the component of the shearing stress that acts in the X plane pointing in the z direction.

The sign convention for stress was stated in Chapter 1 but is also repeated here for convenience. Thus:

A normal stress is positive if it points in the direction of the outward normal to its plane. Thus, a positive normal stress produces tension and a negative normal stress produces compression. A shearing stress component is positive if it acts on a positive coordinate plane and points in the positive direction of the axis designated by its second subscript, or if it acts on a negative coordinate plane and points in the negative direction of the axis designated by its second subscript. Note that all of the stresses acting on the stress element of Figure 7.2 (shown or hidden) are positive.

7.4 PLANE-STRESS TRANSFORMATION EQUATIONS

We will begin our analysis of stress by focusing on the relatively simple two-dimensional stress system known as *plane stress*. This special case of stress exists when all of the stress components along one of the three coordinate axes are zero. Thus, for example, if $\sigma_z = \tau_{xz} = \tau_{yz} = 0$ in a given stress system, the result is a two-dimensional stress system (plane stress) in the x–y plane (i.e., in the Z plane). Such a plane stress element, with a very thin constant depth dz into the page, is shown in Figure 7.3a. Since $\tau_{yx} = \tau_{xy}$ as established above, the stress element of Figure 7.3a may be simplified as shown in Figure 7.3b where both of the shearing stresses are identified as τ_{xy}. We encounter such stress elements in many different cases on the outside surface of stressed members. For example, we came across this type of stress element on the outside surface of a member subjected to axial loads in Chapter 1 (see Example 1.4). We also encountered this kind of stress element on the outside surface of members subjected to torques and to bending moments in Chapters 2 and 3, respectively (see Examples 2.11 and 3.14).

7.4.1 STRESSES ON INCLINED PLANES

It is useful to relate the stresses on the X and Y planes to those acting on any inclined plane defined by the counterclockwise angle θ measured from the vertical X plane. This inclined plane is shown in a light gray color in Figure 7.3b. To this inclined plane we attach a right-handed n–t–z coordinate system as shown in Figures 7.3b, where the n axis is normal to the inclined N plane. Since θ is the angle between the X and N planes, it is also the angle between the x and n axes. The wedge contained within the X, Y, and N planes is isolated as shown in Figures 7.4. Figure 7.4a shows both the x–y–z and n–t–z coordinate systems as well as the differential cross-sectional areas of the X and Y planes in terms of the differential area dA of the inclined N plane. These simple relations among the three

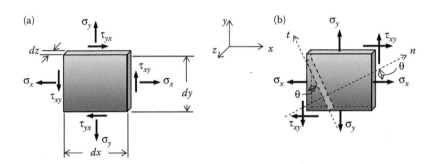

FIGURE 7.3   (a) A two-dimensional stress system known a *plane stress* system. (b) The same plane stress system in a simplified form showing an inclined plane at an angle θ with the X plane.

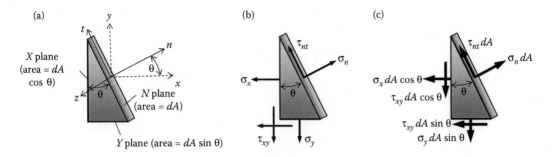

FIGURE 7.4   Diagrams showing how to determine stresses on inclined planes in the two-dimensional case.

differential areas are possible because the depth of the stress element dz is constant. Figure 7.4b shows the stresses on all three planes including the normal stress σ_n and the shearing stress τ_{nt} on the inclined N plane. Finally, in Figure 7.4c, these stresses are transformed into differential forces by multiplying the areas of Figure 7.4a by the corresponding stresses of Figure 7.4b.

To find σ_n and τ_{nt} we apply the force equations of equilibrium in n and t directions. Thus,

$$\sum F_n = 0 \Rightarrow \sigma_n dA - \sigma_x (dA \cos \theta) \cos \theta - \tau_{xy} (dA \cos \theta) \sin \theta$$
$$- \sigma_y (dA \sin \theta) \sin \theta - \tau_{xy} (dA \sin \theta) \cos \theta = 0 \tag{7.4}$$

Solving Equation 7.4 for σ_n we obtain

$$\sigma_n = \sigma_x \cos^2 \theta + \sigma_y \sin^2 \theta + 2\tau_{xy} \sin \theta \cos \theta \tag{7.5}$$

Using the identities $[\sin^2 \theta = (1 - \cos 2\theta)/2]$, $[\cos^2 2\theta = (1 + \cos 2\theta)/2]$ and $[2 \sin \theta \cos \theta = \sin 2\theta]$, Equation 7.5 becomes

$$\sigma_n = \frac{1}{2}(\sigma_x + \sigma_y) + \frac{1}{2}(\sigma_x - \sigma_y) \cos 2\theta + \tau_{xy} \sin 2\theta \tag{7.6}$$

Also,

$$\sum F_t = 0 \Rightarrow \tau_{nt} dA + \sigma_x (dA \cos \theta) \sin \theta - \tau_{xy} (dA \cos \theta) \cos \theta$$
$$- \sigma_y (dA \sin \theta) \cos \theta + \tau_{xy} (dA \sin \theta) \sin \theta = 0 \tag{7.7}$$

Solving Equation 7.7 for τ_{nt} we obtain

$$\tau_{nt} = -(\sigma_x - \sigma_y) \sin \theta \cos \theta + \tau_{xy} (\cos^2 \theta - \sin^2 \theta) \tag{7.8}$$

Using the trigonometric identities given above, plus the identity: $[\cos^2 \theta - \sin^2 \theta = \cos 2\theta]$, Equation 7.8 becomes

$$\tau_{nt} = -\frac{1}{2}(\sigma_x - \sigma_y) \sin 2\theta + \tau_{xy} \cos 2\theta \tag{7.9}$$

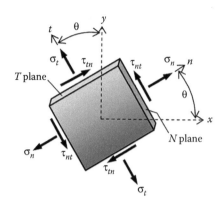

FIGURE 7.5 Diagram showing a rotated, two-dimensional, stress element between the two orthogonal planes N and T.

The normal stress σ_t on the T plane (on the plane perpendicular to the N plane) may be obtained from Equation 7.6 by replacing angle (2θ) by angle $(2\theta + \pi)$. Since $\sin(2\theta + \pi) = -\sin 2\theta$ and $\cos(2\theta + \pi) = -\cos 2\theta$, Equation 7.6 yields

$$\sigma_t = \frac{1}{2}(\sigma_x + \sigma_y) - \frac{1}{2}(\sigma_x - \sigma_y)\cos 2\theta - \tau_{xy} \sin 2\theta \qquad (7.10)$$

Thus, for a plane-stress condition, Equations 7.6 and 7.9 yield the normal and shearing stresses on any plane N inclined to the X plane by angle θ (positive if counterclockwise and negative if clockwise) if the stresses σ_x, σ_y, and τ_{xy} are known. Equation 7.10 yields the normal stress on the T plane perpendicular to the N plane. These equations are known as the *plane-stress transformation equations*. Since the X plane is perpendicular to the x axis, and the N plane perpendicular to the n axis, angle θ is also the angle between the x and the n axes. These geometric relations are given in Figure 7.5 which shows a rotated stress element contained between the N and T planes.

A very useful relation among the normal stresses acting on perpendicular planes is obtained when Equations 7.6 and 7.10 are added. This addition leads to

$$\sigma_n + \sigma_t = \sigma_x + \sigma_y \qquad (7.11)$$

Equation 7.11 states that, for a plane-stress condition, the algebraic sum of the two in-plane normal stresses on any two perpendicular planes is constant regardless of their orientations. In other words, the algebraic sum of the two in-plane normal stresses on any two perpendicular planes is independent of the orientations of these planes.

While a stress element for a plane-stress condition represents a small body of material with some thickness into the page, for convenience, *it is generally shown in a two-dimensional view* and the stresses labeled on only one of the two identical planes (i.e., on one X and one Y plane) as will be done in drawing plane stress elements in subsequent work.

EXAMPLE 7.1

The plane-stress condition at a point in a stressed body is as shown figure (a) below. Use a free-body diagram and the equations of equilibrium to find the normal and shearing stresses on the N plane inclined to the X plane by clockwise angle $\theta = 40°$ as shown.

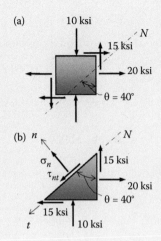

SOLUTION

The triangular wedge contained within the X, Y, and N planes is isolated as shown magnified in figure (b) above which also shows the n–t coordinate system attached to the N plane. Assuming a constant thickness into the page and letting the area of the N plane be A, we determine the areas of the X and Y planes to be $A \cos 40°$ and $A \sin 40°$, respectively. This enables us to transform the stresses shown in figure (b) above into forces and apply the equations of equilibrium.

Thus,

$$\sum F_n = 0 \Rightarrow \sigma_n A - 20(A \cos 40°)\cos 40° + 15(A \cos 40°)\sin 40°$$
$$+ 10(A \sin 40°)\sin 40° + 15(A \sin 40°)\cos 40° = 0$$

Solving for σ_n we obtain

$$\sigma_n = 20 \cos^2 0° - 10 \sin^2 0° - 2(15)\sin 40° \cos 40° = -7.167 \approx -7.2 \text{ ksi} \qquad \textbf{ANS.}$$

$$\sum F_t = 0 \Rightarrow \tau_{nt} A - 20(A \cos 40°)\sin 40° - 15 \cos 40° \cos 40°$$
$$- 10(A \sin 40°)\cos 40° + 15(A \sin 40°)\sin 40° = 0$$

Solving for τ_{nt} yields

$$\tau_{nt} = 30 \sin 40° \cos 40° + 15(\cos^2 0° - \sin^2 0°) = 17.377 \approx 17.4 \text{ ksi} \qquad \textbf{ANS.}$$

The minus sign on the normal stress σ_n indicates that it points in the negative sense of the n axis and not in its positive sense as assumed. In other words, it is compression and *not* tension as was assumed.

EXAMPLE 7.2

Solve Example 7.1 using the plane-stress transformation equations (i.e., Equations 7.6 and 7.9).

SOLUTION

The normal stress on plane N is given by Equation 7.6. Noting that $\theta = -40°$, we obtain,

$$\sigma_n = \frac{1}{2}(\sigma_x + \sigma_y) + \frac{1}{2}(\sigma_x - \sigma_y)\cos 2\theta + \tau_{xy} \sin 2\theta$$

$$= \frac{1}{2}(20 - 10) + \frac{1}{2}(20 + 10)\cos 2(-40°) + 15 \sin 2(-40°) = -7.167 \approx -7.2 \text{ ksi} \qquad \textbf{ANS.}$$

The shearing stress on plane N is given by Equation 7.9. Again, θ must be entered as a negative quantity. Thus,

$$\tau_{nt} = -\frac{1}{2}(\sigma_x - \sigma_y)\sin 2\theta + \tau_{xy} \cos 2\theta$$

$$= -\frac{1}{2}(20 + 10)\sin 2(-40°) + 15 \cos 2(-40°) = 17.377 \approx 17.4 \text{ ksi} \qquad \textbf{ANS.}$$

7.4.2 PRINCIPAL STRESSES

It is important to keep in mind that the dimensions dx and dy of a stress element are infinitesimally small and that, in the limit, they reduce to zero. *Therefore, any stress element, such as that shown in Figure 7.3, is really a point in a stressed body. It is magnified, in order to show details, and contained between two X and two Y planes that are separated by differential distances dx and dy, respectively. In the limit, when dx and dy approach zero, the two X planes merge into one and so do the two Y planes, thus reducing the stress element to a point. However, the X and Y planes are not the only two planes that exist at such a point. As a matter of fact, an infinite number of planes may be imagined to pass through the point, each defined by one of the infinite number of angles θ between zero and 360°.*

An examination of Equations 7.6 and 7.10 reveals that the normal stress varies with changes in the angle θ. Since at any point in a stressed body, there exists an infinite number of planes as stated above, it becomes desirable and useful for design purposes, to find those planes on which the normal stress, at any given point, is either a maximum or a minimum and to determine the magnitudes of these normal stresses. These maximum and minimum normal stresses are known as *principal stresses*. To determine these principal stresses, we need to maximize Equation 7.6 by setting the derivative of σ_n with respect to θ equal to zero. Thus, differentiating Equation 7.6 we obtain

$$\frac{d\sigma_n}{d\theta} = -(\sigma_x - \sigma_y)\sin 2\theta + 2\tau_{xy} \cos 2\theta = 0 \qquad (7.12)$$

Dividing both terms on the left-hand side by $\cos 2\theta$, Equation 7.12 reduces to

$$-(\sigma_x - \sigma_y)\frac{\sin 2\theta}{\cos 2\theta} + 2\tau_{xy} = 0; \Rightarrow -(\sigma_x - \sigma_y)\tan 2\theta + 2\tau_{xy} = 0 \qquad (7.13)$$

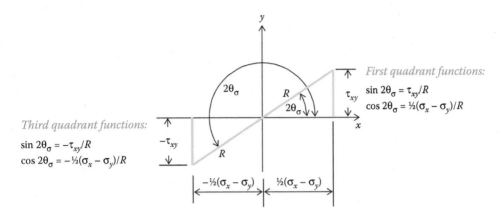

FIGURE 7.6 A sketch showing a graphical interpretation of Equation 7.14 with two values of the angle $2\theta_\sigma$, where the stress σ_n in one is a maximum and on the second, it is a minimum.

Solving Equation 7.13 for the angle 2θ we obtain

$$2\theta_\sigma = \tan^{-1}\left[\frac{\tau_{xy}}{1/2(\sigma_x - \sigma_y)}\right] \tag{7.14}$$

The subscript σ is attached to angle θ to emphasize the fact that Equation 7.14 yields the planes on which the normal stress σ is a maximum or minimum. A graphical interpretation of Equation 7.14 is provided in Figure 7.6. Note that in the range between zero and 360° there are two values of angle $2\theta_\sigma$ differing by 180°, one in the first quadrant and the second in the third quadrant. Of course, this means that there are two values of angle θ_σ differing by 90°. On the plane defined by one of these two values of θ_σ, the normal stress σ_n is a maximum (one *principal stress*) and on the plane defined by the second, the normal stress σ_n is a minimum (a second *principal stress*). The two planes on which these principal stresses act are *perpendicular to each other* and are known as *principal planes*.

The quantity R in Figure 7.6 may be determined in terms of σ_x, σ_y, and τ_{xy} by the Pythagorean theorem. Thus,

$$R = \sqrt{\left(\frac{\sigma_x - \sigma_y}{2}\right)^2 + (\tau_{xy})^2} \tag{7.15}$$

If we now substitute into Equation 7.6 the values of $\sin 2\theta_\sigma$ and $\cos 2\theta_\sigma$ given for the first quadrant in Figure 7.6, after simplification, we obtain the equation for the first principal stress, σ_{MAX}. Thus,

$$\sigma_{MAX} = \frac{1}{2}(\sigma_x + \sigma_y) + \sqrt{\left(\frac{\sigma_x - \sigma_y}{2}\right)^2 + (\tau_{xy})^2} \tag{7.16}$$

Similarly, if the values of $\sin 2\theta_\sigma$ and $\cos 2\theta_\sigma$ given for the third quadrant in Figure 7.6 are substituted into Equation 7.6, we obtain the equation for the second principal stress, σ_{MAX}. Thus,

$$\sigma_{MIN} = \frac{1}{2}(\sigma_x + \sigma_y) - \sqrt{\left(\frac{\sigma_x - \sigma_y}{2}\right)^2 + (\tau_{xy})^2} \tag{7.17}$$

The first term in Equations 7.16 and 7.17, is generally known as the *average normal stress* and given the symbol σ_{AVE}. Thus,

$$\sigma_{AVE} = \frac{1}{2}(\sigma_x + \sigma_y) \qquad (7.18)$$

In general, at any point in a stressed body, we identify three principal stresses and denote them by the symbols σ_1, σ_2, and σ_3. It is convenient and useful, especially in the three-dimensional case, to label these three principal stresses in such a way that, algebraically, $\sigma_1 \geq \sigma_2 \geq \sigma_3$. Thus, σ_1 is the largest algebraic principal stress; σ_3 is the least; and σ_2 is the intermediate algebraic principal stress. It should be emphasized, however, *that in the two-dimensional (plane stress) case, one of the three principal stresses is zero.* Identification of the zero principal stress is not, in general, possible until all three principal stresses are found.

Note that *the shearing stress on a principal plane is zero.* This fact may be confirmed by substituting sin $2\theta_\sigma$ and cos $2\theta_\sigma$ values given in Figure 7.6 (either first or third quadrant) into Equation 7.9. These substitutions lead to

$$\tau_p = 0 \qquad (7.19)$$

The notation τ_p in Equation 7.19 is used to signify the shearing stress on *principal* planes. Therefore, an alternate definition for a principal plane may be stated as *a plane on which the shearing stress is zero.*

7.4.3 Maximum In-Plane Shearing Stress

Since, as stated earlier, an infinite number of planes exist at a point in a stressed body, oriented from the X plane by the infinite values that angle 2θ may assume in the range between zero and 360°, Equation 7.9 must be examined in order to find the planes on which the shearing stress τ_{nt} attains a maximum or a minimum value. To this end, setting the derivative of τ_{nt} with respect to θ in Equation 7.9 equal to zero we have

$$\frac{d\tau_{nt}}{d\theta} = -(\sigma_x - \sigma_y)\cos 2\theta - 2\tau_{xy}\sin 2\theta = 0 \qquad (7.20)$$

Dividing both terms by cos 2θ, Equation 7.20 reduces to

$$-(\sigma_x - \sigma_y) - 2\tau_{xy}\frac{\sin 2\theta}{\cos 2\theta} = 0; \quad \Rightarrow (\sigma_x - \sigma_y) + 2\tau_{xy}\tan 2\theta = 0 \qquad (7.21)$$

Solving Equation 7.21 for angle 2θ we obtain

$$2\theta_\tau = \tan^{-1}\left[\frac{-1/2(\sigma_x - \sigma_y)}{\tau_{xy}}\right] \qquad (7.22)$$

The subscript τ attached to angle θ emphasizes the fact that Equation 7.22 provides the orientation of planes on which the shearing stress τ is a maximum or a minimum. A graphical interpretation of Equation 7.22 is given in Figure 7.7.

Note that in the range between zero and 360° there are two values of angle $2\theta_\tau$ differing by 180°, one in the second quadrant and the other in the fourth quadrant. Of course, this means that there

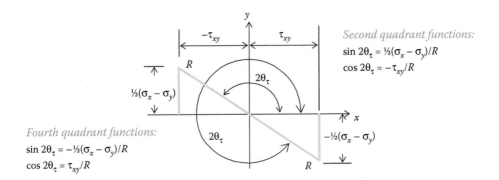

FIGURE 7.7 A sketch showing a graphical interpretation of Equation 7.22 showing two values of $2\theta_\tau$ with the two *maximum in-plane shearing stresses in which* τ_{MAX} is positive in one and negative in the other.

are two values of angle θ_τ differing by 90°. On the plane defined by one of these two values of θ_τ, the shearing stress τ_{nt} is positive, and on the plane defined by the second, the stress τ_{nt} is negative. These two planes are perpendicular to each other and are known simply as the planes of *maximum in-plane shearing stresses*. Since shearing stresses on any two perpendicular planes are equal in magnitude, both of these stresses are labeled in Equation 7.23 as τ_{MAX}, namely the *maximum shearing stresses*. Note that the quantity R in Figure 7.7 is, once again, given by Equation 7.15.

If we substitute into Equation 7.9 the fourth quadrant values for $\sin 2\theta_\tau$ and $\cos 2\theta_\tau$ given in Figure 7.7, we obtain

$$\left| \tau_{MAX} \right| = R = \sqrt{\left(\frac{\sigma_x - \sigma_y}{2} \right)^2 + (\tau_{xy})^2} \tag{7.23}$$

Similarly, if we substitute into Equation 7.9 the values for $\sin 2\theta_\tau$ and $\cos 2\theta_\tau$ given in Figure 7.7 for the second quadrant we obtain a negative quantity of the same magnitude as that given in Equation 7.23.

We now compare the values of $2\theta_\sigma$ and $2\theta_\tau$ given, respectively, by Equations 7.14 and 7.22. We note that $\tan 2\theta_\sigma$ is the negative reciprocal of $\tan 2\theta_\tau$. We recall from trigonometry that such angles are separated by 90°. Thus, the angular separation between $2\theta_\tau$ and $2\theta_\sigma$ is 90°. In other words, the separation between θ_τ and θ_σ is 45°. Obviously, this means that, *for a state of plane stress, the planes of maximum in-plane shearing stress are ±45° from the principal planes.*

It should be noted here that the planes of maximum in-plane shearing stress are not free of normal stresses. The normal stress on a plane of maximum in-plane shearing stress may be found from Equation 7.6 by substituting the values given for $\sin 2\theta$ and $\cos 2\theta$ in Figure 7.7 (either second or fourth quadrant values) corresponding to these planes. These substitutions yield

$$\sigma_{MS} = \frac{1}{2}(\sigma_x + \sigma_y) = \sigma_{AVE} \tag{7.24}$$

The notation σ_{MS} is used to signify the normal stress on planes of *maximum* in-plane *shearing* stress.

Two special cases of plane stress may be identified. The first is the *uniaxial stress* system first encountered in Chapter 1 (see, e.g., Figure 1.9b). In this case, the uniaxial stress element is defined by two planes perpendicular to the axis of a member subjected only to an axial load and two planes parallel to this axis as shown in Figure 7.8a. Examining this case using the information developed above, we conclude that the $\sigma_1 = \sigma_x$, $\sigma_2 = \sigma_3 = 0$, where the plane of σ_1 is the X plane, those of σ_2

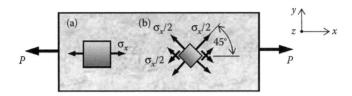

FIGURE 7.8   A sketch showing the stresses under a *uniaxial stress* system.

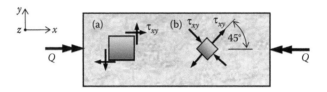

FIGURE 7.9   A sketch showing the stresses under a *pure shear* stress system.

and σ_3 are the Y and Z planes. Besides, the maximum in-plane shearing stress is equal to $\sigma_x/2$ and occurs on planes that make angles of $\pm 45°$ with the X plane as indicated in Figure 7.8b. On the plane of maximum shear, the normal stress is $\sigma_x/2$, which is also indicated in Figure 7.8b. The second is the *pure shear* condition introduced in Chapter 2 (see, e.g., Figure 2.12b). In this case, the pure shear stress element is defined by two planes perpendicular to the axis of a member subjected only to torsion and two planes parallel to this axis as shown in Figure 7.9a. The principal stresses here are $\sigma_1 = \tau_{xy}$, $\sigma_2 = 0$ and $\sigma_3 = -\tau_{xy}$ on planes that make angles of $\pm 45°$ with the X plane as shown in Figure 7.9b. The maximum in-plane shearing stress is τ_{xy} acting on the X plane. The student is urged to confirm these results by applying the equations developed above for plane-stress conditions.

EXAMPLE 7.3

Find the three principal stresses and their planes for the plane-stress condition shown in figure (a) below. Show the in-plane principal stresses on a stress element properly oriented relative to the X plane or the x axis.

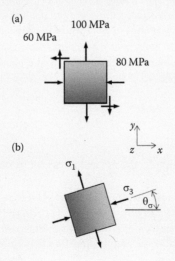

SOLUTION

The principal stresses are obtained by applying Equations 7.16 and 7.17. Thus, by Equation 7.16,

$$\sigma_1 = \frac{1}{2}(\sigma_x + \sigma_y) + \sqrt{\left(\frac{\sigma_x - \sigma_y}{2}\right)^2 + (\tau_{xy})^2}$$

$$= \frac{1}{2}(-80 + 100) + \sqrt{\left(\frac{-80 - 100}{2}\right)^2 + (-60)^2}$$

$$= 10 + 108,167 = 118.167 \approx 118.2 \text{ MPa} \qquad \textbf{ANS.}$$

Also, by Equation 7.17,

$$\sigma_3 = 10 - 108.167 = -98.167 \approx -98.2 \text{ MPa} \qquad \textbf{ANS.}$$

Finally,

$$\sigma_2 = 0 \qquad \textbf{ANS.}$$

Note that the labeling of the three principal stresses is done according to the algebraic scheme, $\sigma_1 \geq \sigma_2 \geq \sigma_3$, stated previously. Note also that this labeling had to wait until all three principal stresses were determined.

One simple check on the accuracy of the computations may be made by noting that the algebraic sum of the two in-plane principal stresses (σ_1 and σ_3) is equal to the algebraic sum of the two given in-plane normal stresses, thus satisfying Equation 7.11.

The plane on which σ_3 acts is found by Equation 7.14. Thus,

$$2\theta_\sigma = \tan^{-1}\left[\frac{\tau_{xy}}{1/2(\sigma_x - \sigma_y)}\right] = \tan^{-1}\left[\frac{-60}{1/2(-80 - 100)}\right] = \tan^{-1}\left(\frac{2}{3}\right) = 33.69°$$

This, of course means that $\theta_\sigma \approx 16.8°$, and since it is positive, it is a counterclockwise angle. Thus, in this case, the plane of σ_3 is 16.8° counterclockwise from the X plane or the principal stress σ_3 is 16.8° counterclockwise from the x axis. As previously stated, the plane of σ_1 is perpendicular to the plane of σ_3. The stress element indicating these two principal stresses and their planes is shown properly oriented in figure (b) above. Note that the plane of $\sigma_2 = 0$ is the Z plane, namely the plane of the page.

EXAMPLE 7.4

Determine the maximum in-plane shearing stress for the stress element shown in figure (a) below and the plane on which this shearing stress acts. Find also the normal stress on the plane of maximum shear and construct a stress element properly oriented relative to the X plane or the x axis.

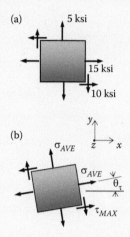

SOLUTION

The maximum in-plane shearing stress is given by Equation 7.23. Thus,

$$\tau_{MAX} = \sqrt{\left(\frac{\sigma_x - \sigma_y}{2}\right)^2 + (\tau_{xy})^2} = \sqrt{\left(\frac{15 - 5}{2}\right)^2 + (-10)^2}$$

$$= 11.18 \approx 11.2 \text{ ksi} \quad \textbf{ANS.}$$

The plane on which this shearing stress acts is found by Equation 7.22. Thus,

$$2\theta_\tau = \tan^{-1}\left[\frac{-1/2(\sigma_x - \sigma_y)}{\tau_{xy}}\right] = \tan^{-1}\left[\frac{-1/2(15 - 5)}{-10}\right] = \tan^{-1}\frac{1}{2} = 26.565°$$

Therefore, $\theta_\tau \approx 13.3°$. This, of course, means that the plane of maximum shearing stress is 13.3° counterclockwise from the X plane. The normal stress on this plane, $\sigma_{MS} = \sigma_{AVE} = 10$ ksi as given by Equation 7.24. A stress element with all of the stresses acting on the planes of maximum shearing stress is shown properly oriented in figure (b) above.

EXAMPLE 7.5

Arm AB is attached rigidly to shaft BDE of diameter $d = 80$ mm and a load $P = 50$ kN is applied at A as shown in figure (a) below. Let $a = 0.3$ m, $e = 0.2$ m, $f = 0.5$ m and construct a stress element at point G with two planes perpendicular to the axis of the shaft (the x axis) and two planes parallel to it. Then determine the principal stresses and their planes and the maximum in-plane shearing stresses and their planes and show them on stress elements properly oriented relative to the X plane or x axis.

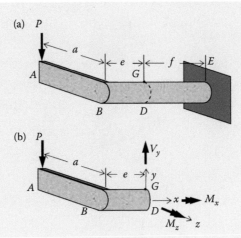

SOLUTION

The free-body diagram of segment ABD is shown in figure (b) above along with the internal reaction components at section DG. These reaction components consist of a shearing force V_y, a torque $T = M_x$, and a bending moment M_z directed as shown. These internal reactions are found by applying the conditions of equilibrium as follows:

$$\sum F_y = 0 \Rightarrow V_y = P = 50 \text{ kN}; \quad \sum M_x = 0 \Rightarrow T = M_x = Pa = 15 \text{ kN} \cdot \text{m};$$

$$\sum M_z = 0 \Rightarrow M_z = -Pe = -10 \text{ kN} \cdot \text{m}$$

The stress element at point G, viewing it from the top looking down, is shown in figure (c) below, where

$$\sigma_x = \frac{M_z y}{I_z} = \frac{(10 \times 10^3)(0.040)}{(\pi/64)(0.080^4)} = 198.9 \text{ MPa}$$

$$\tau_{xy} = \frac{Tr}{J} = \frac{(15 \times 10^3)(0.040)}{(\pi/32)(0.080^4)} = 149.2 \text{ MPa}$$

$$\sigma_1 = \frac{1}{2}(\sigma_x + \sigma_y) + \sqrt{\left(\frac{\sigma_x - \sigma_y}{2}\right)^2 + (\tau_{xy})^2} = \frac{1}{2}(198.9 + 0) + \sqrt{\left(\frac{198.9 - 0}{2}\right)^2 + (-149.2)^2}$$

$$= 99.5 + 179.3 = 278.8 \text{ MPa} \qquad \textbf{ANS.}$$

$$\sigma_3 = \frac{1}{2}(\sigma_x + \sigma_y) - \sqrt{\left(\frac{\sigma_x - \sigma_y}{2}\right)^2 + (\tau_{xy})^2} = 99.5 - 179.3 = -79.8 \text{ MPa} \qquad \textbf{ANS.}$$

In this case, $\sigma_2 = 0$ $\quad$ **ANS.**

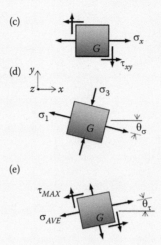

The orientation of the plane of σ_1 is given by Equation 7.14. Thus,

$$2\theta_\sigma = \tan^{-1}\left[\frac{\tau_{xy}}{(1/2)(\sigma_x - \sigma_y)}\right] = \tan^{-1}\left[\frac{-149.2}{(1/2)(198.9 - 0)}\right]$$
$$= \tan^{-1}(-1.50) = -56.31°$$

Therefore, $\theta_\sigma \approx 28.2°$, clockwise. A stress element with the two in-plane principal stresses is shown properly oriented in figure (d) above. The plane of $\sigma_2 = 0$ is the Z plane (the plane of the page).

The maximum in-plane shearing stress is given by Equation 7.23. Thus,

$$\tau_{MAX} = \sqrt{\left(\frac{\sigma_x - \sigma_y}{2}\right)^2 + (\tau_{xy})^2} = 179.3 \text{ MPa}$$

$$2\theta_\tau = \tan^{-1}\left[\frac{-1/2(\sigma_x - \sigma_y)}{\tau_{xy}}\right] = \tan^{-1}\left[\frac{-1/2(198.9 - 0)}{-149.2}\right] = 33.69°$$

Thus, $\theta_\tau \approx 16.8°$, counterclockwise. All planes on which the maximum shearing stresses act are also subjected to normal stresses equal to $\sigma_{MS} = \sigma_{AVE} = (\sigma_1 + \sigma_3)/2 = 99.5$ MPa. A stress element with all of these stresses is shown properly oriented in figure (e) above. Note that the magnitudes of θ_σ and θ_τ add up to 45° indicating, as it should, that the planes of maximum shearing stress are ±45° from the principal planes.

PROBLEMS

7.1   Use free-body diagrams and the equilibrium conditions to find the normal and shearing stresses on the N plane whose orientation with the X plane is $\theta = 50°$.

Assume a standard coordinate system with x positive to the right, y positive upward, and z positive out of the page.

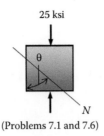

(Problems 7.1 and 7.6)

7.2   Use free-body diagrams and the equilibrium conditions to find the normal and shearing stresses on the N plane whose orientation with the X plane is $\theta = 35°$.
    Assume a standard coordinate system with x positive to the right, y positive upward, and z positive out of the page.

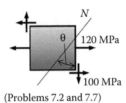

(Problems 7.2 and 7.7)

7.3   Use free-body diagrams and the equilibrium conditions to find the normal and shearing stresses on the N plane whose orientation with the X plane is $\theta = 45°$.
    Assume a standard coordinate system with x positive to the right, y positive upward, and z positive out of the page.

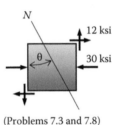

(Problems 7.3 and 7.8)

7.4   Use free-body diagrams and the equilibrium conditions to find the normal and shearing stresses on the N plane whose orientation with the X plane is $\theta = 60°$.
    Assume a standard coordinate system with x positive to the right, y positive upward, and z positive out of the page.

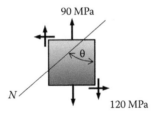

(Problems 7.4 and 7.9)

7.5   Use free-body diagrams and the equilibrium conditions to find the normal and shearing stresses on the N plane whose orientation with the X plane is $\theta = 30°$.

Assume a standard coordinate system with x positive to the right, y positive upward, and z positive out of the page.

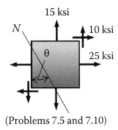

(Problems 7.5 and 7.10)

7.6 Solve Problem 7.1 using the plane-stress transformation Equations 7.6 and 7.9. Assume a standard coordinate system with x positive to the right, y positive upward, and z positive out of the page.

7.7 Solve Problem 7.2 using the plane-stress transformation Equations 7.6 and 7.9. Assume a standard coordinate system with x positive to the right, y positive upward, and z positive out of the page.

7.8 Solve Problem 7.3 using the plane-stress transformation Equations 7.6 and 7.9. Assume a standard coordinate system with x positive to the right, y positive upward, and z positive out of the page.

7.9 Solve Problem 7.4 using the plane-stress transformation Equations 7.6 and 7.9. Assume a standard coordinate system with x positive to the right, y positive upward and z positive out of the page.

7.10 Solve Problem 7.5 using the plane-stress transformation Equations 7.6 and 7.9. Assume a standard coordinate system with x positive to the right, y positive upward, and z positive out of the page.

7.11 Use appropriate equations to find the three principal stresses for the given plane stress system. Show the in-plane principal stresses on a stress element properly oriented relative to the X plane or x axis. Assume a standard coordinate system with x positive to the right, y positive upward, and z positive out of the page.

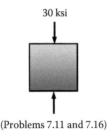

(Problems 7.11 and 7.16)

7.12 Use appropriate equations to find the three principal stresses for the given plane stress system. Show the in-plane principal stresses on a stress element properly oriented relative to the X plane or x axis. Assume a standard coordinate system with x positive to the right, y positive upward, and z positive out of the page.

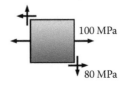

(Problems 7.12 and 7.17)

7.13 Use appropriate equations to find the three principal stresses for the given plane stress system. Show the in-plane principal stresses on a stress element properly oriented relative to the X plane or x axis. Assume a standard coordinate system with x positive to the right, y positive upward, and z positive out of the page.

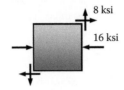

(Problems 7.13 and 7.18)

7.14 Use appropriate equations to find the three principal stresses for the given plane stress system. Show the in-plane principal stresses on a stress element properly oriented relative to the X plane or x axis. Assume a standard coordinate system with x positive to the right, y positive upward and z positive out of the page.

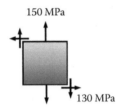

(Problems 7.14 and 7.19)

7.15 Use appropriate equations to find the three principal stresses for the given plane stress system. Show the in-plane principal stresses on a stress element properly oriented relative to the X plane or x axis. Assume a standard coordinate system with x positive to the right, y positive upward, and z positive out of the page.

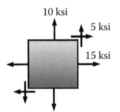

(Problems 7.15 and 7.20)

7.16 In reference to the figure shown in Problem 7.11, use appropriate equations to determine the maximum in-plane shearing stress for the given plane stress element and the plane on which this shearing stress acts. Find also the normal stress on the plane of maximum shear and construct a stress element properly oriented relative to the X plane or the x axis. Assume a standard coordinate system with x positive to the right, y positive upward and z positive out of the page.

7.17 In reference to the figure shown in Problem 7.12, use appropriate equations to determine the maximum in-plane shearing stress for the given plane stress element and the plane on which this shearing stress acts. Find also the normal stress on the plane of maximum shear and construct a stress element properly oriented relative to the X plane or the x axis. Assume a standard coordinate system with x positive to the right, y positive upward, and z positive out of the page.

7.18 In reference to the figure shown in Problem 7.13, use appropriate equations to determine the maximum in-plane shearing stress for the given plane stress element and the
plane on which this shearing stress acts. Find also the normal stress on the plane of
maximum shear and construct a stress element properly oriented relative to the X plane
or the x axis. Assume a standard coordinate system with x positive to the right, y positive upward, and z positive out of the page.

7.19 In reference to the figure shown in Problem 7.14, use appropriate equations to determine the maximum in-plane shearing stress for the given plane stress element and the
plane on which this shearing stress acts. Find also the normal stress on the plane of
maximum shear and construct a stress element properly oriented relative to the X plane
or the x axis. Assume a standard coordinate system with x positive to the right, y positive upward, and z positive out of the page.

7.20 In reference to the figure shown in Problem 7.15, use appropriate equations to determine the maximum in-plane shearing stress for the given plane stress element and the
plane on which this shearing stress acts. Find also the normal stress on the plane of
maximum shear and construct a stress element properly oriented relative to the X plane
or the x axis. Assume a standard coordinate system with x positive to the right, y positive upward, and z positive out of the page.

7.21 Let $P = 200$ kN, $b = 80$ mm, $h = 140$ mm, $d = 200$ mm, and $G(y,z) = G(20$ mm,
40 mm), find the three principal stresses at point B in the compression block shown.
Find also the maximum in-plane shearing stress. Show the in-plane principal stresses
and the maximum in-plane shearing stresses on stress elements properly oriented
relative to the X plane or x axis.

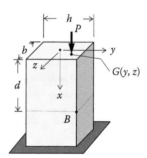

7.22 A 1 1/2-in. standard threaded pipe (see Appendix E) is being unscrewed by means of
a pipe wrench as shown. Let $P = 100$ lb, $a = 14$ in., and $b = 18$ in., determine the three
principal stresses at point B on the outside surface of the pipe. Determine also the maximum in-plane shearing stress. Show the in-plane principal stresses and the maximum
in-plane shearing stresses on stress elements properly oriented relative to the X plane or
x axis.

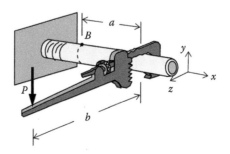

7.23 A machine component may be idealized as a 75-mm-diameter circular shaft fixed at one end and attached to a rigid arm at the other. Let $P = 8$ kN, $a = 200$ mm, $b = 150$ mm, and $d = 250$ mm, find the three principal stresses at point B on the shaft. Find also the maximum in-plane shearing stress. Show the in-plane principal stresses and the maximum in-plane shearing stresses on stress elements properly oriented relative to the X plane or x axis.

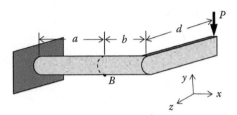

7.24 A 3×1.5 m sign is supported by a column with a hollow circular cross section the outside and inside diameters of which are 100 and 75 mm. The uniform wind pressure on it is $p = 0.75$ kN/m^2. Determine the three principal stresses at point B on the outside surface of the pipe. Determine also the maximum in-plane shearing stress. Show the in-plane principal stresses and the maximum in-plane shearing stresses on stress elements properly oriented relative to the X plane or x axis.

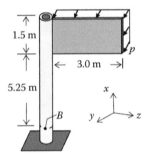

7.25 The stress condition at point B (10, 4, 2) in. on the front face of the compression block of figure (a) below is shown in figure (b) below. Let $b = 4$ in., $h = 8$ in., and $d = 10$ in., determine the magnitude of the compressive load P which is located at $G(0, 3, 0)$ in. *Hint:* Equation 7.11 *is very useful in the solution of this problem.*

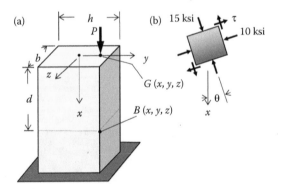

7.5   MOHR'S CIRCLE FOR PLANE STRESS

The plane-stress transformation equation developed in the preceding section may be used to solve a variety of plane stress problems as was done there. However, the use of these equations can become a little cumbersome and an alternate semi-graphical procedure, based upon these equations, is generally more convenient to use in the solution of plane stress problems. This alternative procedure is known as the *Mohr's circle method*, named after Otto Mohr (1835–1918), the German engineer who first introduced it. This method will be developed and discussed in the following paragraphs.

We begin by moving the first term on the right-hand side of Equation 7.6 to the left-hand side and squaring the resulting equation and adding it to the square of Equation 7.9 after expanding the square of the binomials on the right-hand side of both equations. We then simplify the resulting function to obtain

$$\left[\sigma_n - \frac{\sigma_x + \sigma_y}{2} \right]^2 + [\tau_{nt} - 0]^2 = \left[\frac{\sigma_x - \sigma_y}{2} \right]^2 + [\tau_{xy}]^2 \tag{7.25}$$

According to Equation 7.15, the right-hand side of Equation 7.25 is equal to the quantity R^2. Examination of Equation 7.25 reveals that if we replaced the quantity $[\sigma_n - (\sigma_x + \sigma_y)/2]^2$ by the quantity $[x - a]^2$, the quantity $[\tau_{nt} - 0]^2$ by the quantity $[y - b]^2$, we would obtain the familiar equation $[x - a]^2 + [y - b]^2 = r^2$, which is the equation of a circle in the $x - y$ plane with a radius r and a center located at (a, b). Therefore, if we establish a two-dimensional coordinate system in which the horizontal axis is σ_n and the vertical axis is τ_{nt}, Equation 7.25 plots as a circle of radius R given by

$$R = \sqrt{\left(\frac{\sigma_x - \sigma_y}{2} \right)^2 + (\tau_{xy})^2} \tag{7.26}$$

Moreover, this circle is centered at $[(\sigma_x + \sigma_y)/2, 0]$. In other words, the center C of the circle lies on the σ_n axis at a distance from the origin O given by

$$OC = \sigma_{AVE} = \frac{1}{2}(\sigma_x + \sigma_y) \tag{7.27}$$

Such a circle is shown in Figure 7.10, which as stated earlier is known as Mohr's circle. It is important to remember that Mohr's circle represents, parametrically, Equations 7.6 and 7.9. Thus, we must conclude that any plane stress problem solvable by these equations can also be solved by Mohr's circle. Moreover, the coordinates of any point on the circle represents the normal stress (given by the abscissa) and the shearing stress (given by the ordinate) acting on some plane defined by the angle θ in the range between zero and 180°. Thus, for example, point N on the circle represents plane N at some point in a stressed body on which the normal stress is σ_n and the shearing stress is τ_{nt}.

7.5.1   Construction of Mohr's Circle

The construction of Mohr's circle may be accomplished using Equations 7.26 and 7.27 to find, respectively, the radius and center location as was carried out in Figure 7.10. However, it is much more meaningful to construct the circle by locating two diametrically opposite points on its circumference. Two diametrically opposite points on the circle represent two perpendicular planes, or two perpendicular axes, in the stress element. This is so because the transformation equations,

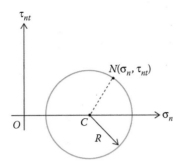

FIGURE 7.10   Mohr's circle for a given *plane stress* system showing the radius R and the location of the center C.

from which Mohr's circle is derived, contain the trigonometric functions of twice the angle (i.e., 2θ). Therefore, two planes (or two axes) in the stress element perpendicular to each other (90° apart) such as the X and Y planes (or the x and y axes) are 180° apart on the circle (diametrically opposite). Since, as we will see shortly, the circle is constructed from knowledge of the *stress conditions* on two perpendicular *planes* in the stress element, it is convenient to think of points on the circle as representing *planes* rather than *axes*. Thus, referring to the stress element of Figure 7.11a, we locate point $X(\sigma_x, -\tau_{xy})$, representing the X plane of the stress element and point $Y(\sigma_y, \tau_{xy})$ representing the Y plane of this element, as shown in Figure 7.11b. Once these two points are plotted, they are connected by diameter XY that intersects the σ_n axis at C, the center of the circle. The reason for changing the sign of the shearing stress in constructing Mohr's circle is to make the rotation of planes (clockwise or counterclockwise) on the stress element the same as that in Mohr's circle. Thus, as shown in Figure 7.11, if we choose to get from the X plane to the N plane in the stress element by a *counterclockwise* rotation through θ, we can also get from the X point to the N point by rotating *counterclockwise* through 2θ on the circle. Of course, another choice would be to rotate *clockwise* through $(180° - \theta)$ in the stress element and *clockwise* through $(360° - 2\theta)$ on the circle.

It should be pointed out, however, that the usual way to construct Mohr's circle is to use a *special sign convention* for the shearing stress instead of changing its sign. This special sign convention consists of plotting the shearing stress as a positive quantity (above the σ_n axis) if it tends to produce clockwise rotation of the stress element, and plotting it as negative (below the σ_n axis) if it tends to produce counterclockwise rotation of the stress element. This special sign convention is consistent with the method used to construct the circle of Figure 7.11b, where the sign of the shearing stress

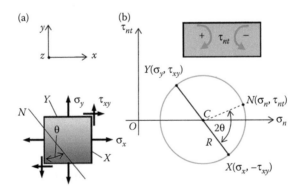

FIGURE 7.11   (a) A plane stress system and (b) construction of the corresponding Mohr's circle.

was changed before plotting. For convenience, this special shearing-stress sign convention is summarized in Figure 7.11b by the curved red arrows inside a gray box. If we use this special sign convention to construct the circle corresponding to the stress element of Figure 7.11a, we obtain the same circle shown in Figure 7.11b. The student is urged to confirm this last statement. *This special shearing-stress sign convention is the only method employed throughout this text in the construction of Mohr's circles.*

7.5.2 Principal Stresses and Maximum In-Plane Shearing Stress

Let us suppose that we were asked to find the two in-plane principal stresses at the point where the stress system is as shown in Figure 7.12a. We begin by establishing a $\sigma_n - \tau_{nt}$ coordinate system and plotting $X(\sigma_x, \tau_{xy})$ below the σ_n axis (because the shearing stresses on the X planes tend to produce a *counterclockwise* rotation of the stress element) and plotting point $Y(\sigma_y, \tau_{xy})$ above the σ_n axis (because the shearing stresses on the Y planes tend to produce *clockwise* rotation of the stress element). These two points are connected by diameter XY which intersects the σ_n axis at the center of the circle, point C. The construction is then completed by drawing a circle with center at C and passing through points X and Y.

There are two properties of Mohr's circle that need to be determined in the solution of any plane stress problem. The first is the distance OC locating the center and the second is the radius R. Both of these properties may be found by analyzing the geometry of the circle. Thus,

$$OC = \frac{1}{2}(\sigma_x + \sigma_y) = \sigma_{AVE} \qquad (7.28)$$

Note that Equation 7.28 is identical to Equations 7.18 and 7.27 derived earlier. The radius of the circle is found by applying the Pythagorean theorem to triangle CDX, where, as shown in Figure 7.12b, side $DX = \tau_{xy}$ and side $CD = 1/2(\sigma_x - \sigma_y)$. Thus,

$$R = \sqrt{\left(\frac{\sigma_x - \sigma_y}{2}\right)^2 + (\tau_{xy})^2} \qquad (7.29)$$

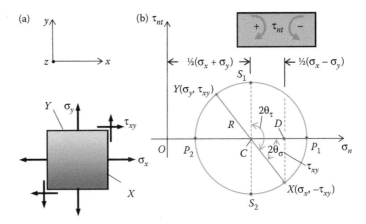

FIGURE 7.12 Diagram showing the given *plane stress* system and the corresponding Mohr's circle needed to determine the two principal stresses given by points P_1 and P_2 and the maximum in-plane shearing stresses given by points S_1 and S_2.

The two in-plane principal stresses, σ_1 and σ_2, are signified on the circle by points P_1 and P_2 which represent the largest and the least normal stresses on the circle. The geometry of the circle is again examined to find these two stresses. Thus,

$$\sigma_1 = \sigma_{MAX} = OC + R = \frac{1}{2}(\sigma_x + \sigma_y) + \sqrt{\left(\frac{\sigma_x - \sigma_y}{2}\right)^2 + (\tau_{xy})^2} \qquad (7.30)$$

$$\sigma_2 = \sigma_{MIN} = OC - R = \frac{1}{2}(\sigma_x + \sigma_y) - \sqrt{\left(\frac{\sigma_x - \sigma_y}{2}\right)^2 + (\tau_{xy})^2} \qquad (7.31)$$

Note that Equations 7.30 and 7.31 are identical to Equations 7.16 and 7.17. Note also that the shearing stress at both points P_1 and P_2 is zero which agrees with what we discovered earlier that the shearing stress vanishes on the principal planes. Moreover, we conclude from the circle (see points P_1 and P_2) that the planes of the two in-plane principal stresses are perpendicular to each other. For the stress element shown in Figure 7.12a, the third principal stress, σ_3, is zero.

We now find the planes on which these principal stresses act relative to some known plane such as the X or Y plane. If we select the X plane as a reference, we note that to get from this reference plane to the plane of σ_1, we have to rotate counterclockwise through angle $2\theta_\sigma$. This angle may be found from triangle CDX in Figure 7.12b. Thus,

$$2\theta_\sigma = \tan^{-1}\left[\frac{\tau_{xy}}{1/2(\sigma_x - \sigma_y)}\right] \qquad (7.32)$$

Again, note that Equation 7.32 is identical to Equation 7.14 developed earlier.

The maximum in-plane shearing stress is identified by two points, labeled as S_1 and S_2, on Mohr's circle in Figure 7.12b. Both of these points provide the same magnitude equal to the radius of Mohr's circle. Therefore,

$$|\tau_{MAX}| = \pm R = \pm\sqrt{\left(\frac{\sigma_x - \sigma_y}{2}\right)^2 + (\tau_{xy})^2} \qquad (7.33)$$

The only difference between points S_1 and S_2 on the circle is the fact that while S_1 is above the σ_n axis, signifying that τ_{MAX} is positive and tends to produce clockwise rotation of the stress element on which it acts, S_2 is below the σ_n axis and τ_{MAX} is negative and tends to produce counterclockwise rotation of the stress element on which it acts.

It is evident from the circle of Figure 7.12b that the orientation of the plane of τ_{MAX} corresponding to point S_1 from the X plane is given by

$$2\theta_\tau = 2\theta_\sigma + 90° \qquad (7.34)$$

Equation 7.34 shows that the angle ($2\theta_\tau$ is 90°) is 90° away from the angle $2\theta_\sigma$ in a counterclockwise sense. In other words, the plane of maximum shearing stress (represented by point S_1) tending to produce clockwise rotation of the element is +90° from the principal plane of σ_1. Moreover, the plane of maximum shearing stress (represented by point S_2) tending to produce counterclockwise of the element is −90° from the plane of σ_1. This, of course, means that, in the *stress element*, the plane of τ_{MAX} tending to produce counterclockwise rotation of the element is −45° from the plane of σ_1 and the plane of τ_{MAX} tending to produce clockwise rotation of the element is $+45°$ from this plane. Thus, we conclude that, in the stress element, the two planes of maximum shearing stress are perpendicular to each other. We should note here that the planes of maximum shearing stress are not free of normal stress. As seen from Mohr's circle in Figure 7.12b, both points, S_1 and S_2 possess not only ordinates representing τ_{MAX}, but also abscissas that represent the normal stress on these planes, given by the distance $OC = \sigma_{AVE}$.

All of the above conclusions are summarized in Figure 7.13. Figure 7.13a is the given plane stress element shown originally in Figure 7.12a, but is repeated here to show the interrelations among the various planes. Figure 7.13b shows the two in-plane principal stresses and the planes on which they act. Note that the plane of σ_1 (labeled plane P_1) is counterclockwise from the X plane by angle θ_σ as required by Mohr's circle of Figure 7.12b because the plane of σ_1 represented by point P_1 is $2\theta_\sigma$ counterclockwise from point X on the circle. Of course, the plane of σ_2 is perpendicular to the plane of σ_1, as required by the position of point P_2 on the circle which is $180°$ from point P_1. Besides, the third principal stress, σ_3, which is zero in this case, acts on the Z plane which coincides with the plane of the page. Finally, Figure 7.13c shows the maximum shearing stress and the planes on which it acts. Note that on the plane that is counterclockwise from the X plane by the angle $(\theta_\sigma + 45°)$, designated as plane S_1, the maximum shearing stress tends to rotate the stress element clockwise. This is so because this plane is represented by point S_1 which is above the σ_n axis, where the shearing stress is positive, implying clockwise rotation of the element according to our special sign convention for shearing stress. The second plane of

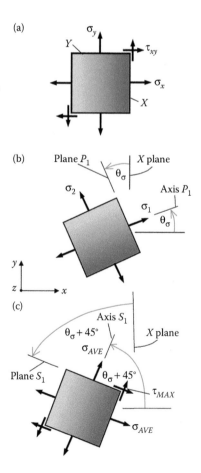

FIGURE 7.13 Summary of conclusions of results obtained from Mohr's circle for a *given plane* stress system.

maximum shearing stress, perpendicular to the first plane just discussed, is represented in Mohr's circle by point S_2 which is below the σ_n axis where the shearing stress is negative, implying counterclockwise rotation of the element according to the special shearing stress sign convention adopted for the plotting of Mohr's circle.

It is convenient to construct a 45° triangular stress element that is contained within the two principal planes and one of the two planes of maximum in-plane shearing stress. Such a stress element has the advantage of showing the important information about the in-plane principal stresses and the maximum in-plane shearing stress, all in one package. Examples of triangular stress elements are shown in Figures 7.14. Here again, for convenience and reference, the original stress element first shown in Figure 7.12a is repeated in Figure 7.14a. Figure 7.14b shows one triangular stress element contained within the two principal planes and one of the two planes of maximum in-plane shearing stress. In reality, this stress element represents the upper half of the stress element of Figure 7.13b. Of course, the lower half of the stress element of Figure 7.13b could have been used instead. Figure 7.14c shows another triangular stress element contained within the two principal planes and the second plane of maximum in-plane shearing stress. In reality, this stress element represents the right half of the stress element of Figure 7.13b. Of course, once again, the left half of the stress element of Figure 7.13b could have been used instead.

The sense of the shearing stress on a plane can be established by locating the point representing this plane on Mohr's circle. For example, let us focus for a moment on the sense of the shearing stress in Figure 7.14b. The plane on which it acts is 45° counterclockwise from the principal plane of σ_1. Rotating in the same sense on the circle from point P_1 through $2(45°) = 90°$, we end up at point S_1. Since this point is above the σ_n axis where the shearing stress is positive, it follows that the shearing stress must tend to produce clockwise rotation of the element on which it acts. Hence, the sense chosen for the shearing stress on this plane. A similar analysis leads us to the sense of the shearing stress in Figure 7.14c.

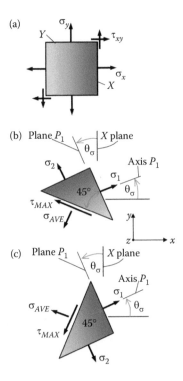

FIGURE 7.14 The 45° triangular stress elements showing σ_1 and σ_2 as well as τ_{MAX} and σ_{AVE}.

EXAMPLE 7.6

Use Mohr's circle to find the two in-plane principal stresses and the maximum in-plane shearing stress for the plane stress element shown in figure below. Use a 45° triangular stress element oriented with respect to the X plane and show all of the above stresses. What is the third principal stress and its plane?

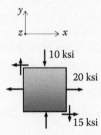

SOLUTION

We locate points $X(20 \text{ ksi}, 15 \text{ ksi})$ and $Y(-10 \text{ ksi}, -15 \text{ ksi})$, which are diametrically opposite, on a $\sigma_n - \tau_{nt}$ coordinate system as shown in figure (a) below. The coordinates of point X are both positive because, on this plane, the normal stress is tension and the shearing stress tends to rotate the stress element clockwise. The coordinates of point Y are both negative because, on this plane, the normal stress is compressive and the shearing stress tends to rotate the element counterclockwise. Once these two points are located, the diameter XY is drawn connecting them and intersecting the σ_n axis at point C, the center of the circle. The circle is then drawn as shown in figure (b) below. We now find the location of the center and radius of this circle as shown below. Note that all computations are carried out using the geometry of the circle and not using the transformation equations developed earlier:

$$OC = \frac{20 + (-10)}{2} = 5 \text{ ksi}$$

$$R = \sqrt{\left(\frac{20 - (-10)}{2}\right)^2 + 15^2} = 21.213 \text{ ksi}$$

(a)

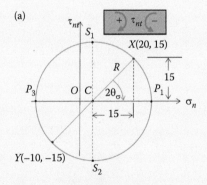

(b)

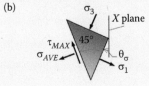

To find the two in-plane principal stresses, we need to determine the ordinates of points P_1 and P_3. Thus,

$$\sigma_1 = OC + R = 26.213 \approx 26.2 \text{ ksi} \quad \textbf{ANS.}$$

$$\sigma_3 = OC - R = -16.213 \approx -16.2 \text{ ksi} \quad \textbf{ANS.}$$

In this case, *the third principal stress, σ_2, is zero.* The plane of this zero principal stress is the Z plane which is coincident with the plane of the page. Note that the labeling of the three principal stresses is based on the scheme introduced earlier that, algebraically, $\sigma_1 \geq \sigma_2 \geq \sigma_3$.

The plane on which σ_1 acts is located by finding the angle $2\theta_\sigma$. Examining the geometry of the circle, we conclude that

$$2\theta_\sigma = \tan^{-1} \frac{15}{15} = 45°; \quad \theta_\sigma = 22.5°$$

We note that point S_1, which locates the plane of maximum shearing stress where this shearing stress must produce clockwise rotation of the element, is 90° counterclockwise from the plane of σ_1 on the circle. This, of course, means that this plane of maximum shearing stress is 45° counterclockwise from the plane of σ_1 in the stress element. The stresses on this plane of maximum shearing stress are

$$\left|\tau_{MAX}\right| = R = 21.213 \approx 21.2 \text{ ksi} \quad \textbf{ANS.}$$

$$\sigma_{MS} = \sigma_{AVE} = OC = 5.0 \text{ ksi} \quad \textbf{ANS.}$$

The required 45° triangular stress element is shown in figure (b) above, properly oriented relative to the X plane. It should be emphasized, however, that the same stress element would be obtained if, instead of the X plane, we chose the Y plane as a reference to measure angles from. It is left as an exercise for the student to confirm this last statement.

EXAMPLE 7.7

The machine component shown in figure (a) below consists of a hollow shaft ($d_o = 100$ mm, $d_i = 75$ mm) attached rigidly to a vertical arm. The component is subjected to loads $L_1 = 25$ kN and $L_2 = 100$ kN as shown. If $a = 350$ mm and $b = 300$ mm, determine the three principal stresses and the maximum in-plane shearing stress at point G, which is on the outside surface of the shaft.

SOLUTION

The free-body diagram shown in figure (b) below is constructed to find the reactive component at the section in the shaft where point G is located. Note that the load L_1 creates the components M_y, V_z, and $M_x = T$, which is the torque acting on the shaft. The load L_2 creates the components M_z and F. These components are found by applying the applicable equations of equilibrium and are stated as follows:

$$F = L_2 = 100 \text{ kN}; \quad V_z = L_1 = 25 \text{ kN}; \quad T = M_x = L_1 a = 8.75 \text{ kN} \cdot \text{m}$$
$$M_y = L_1 b = 7.50 \text{ kN} \cdot \text{m}; \quad M_z = -L_2 a = -35 \text{ kN} \cdot \text{m}$$

The components F, $T = M_x$, and M_y are the only ones to produce stresses at point G. A stress element at point G is drawn with two planes perpendicular to the axis of the shaft and two planes parallel to it. See figure (c) below.

The stresses at point G are computed as follows:

$$\sigma_x = \frac{F}{A} + \frac{M_y z}{I_y} = 140.877 \text{ MPa}; \quad \tau_{xy} = \frac{Tr}{J} = 65.191 \text{ MPa}.$$

Note that $\sigma_y = 0$.

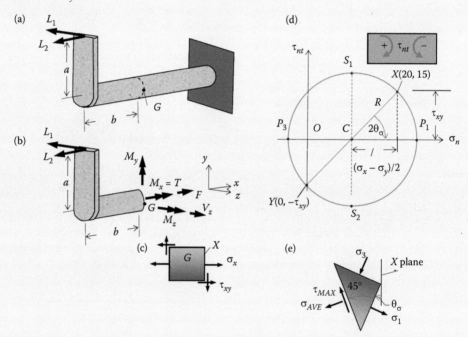

A Mohr's circle is now drawn for this stress element as shown in figure (d) above, having the following properties:

$$OC = \frac{\sigma_x + \sigma_y}{2} = 70.439 \text{ MPa}; \quad R = \sqrt{\left(\frac{\sigma_x - \sigma_y}{2}\right)^2 + (\tau_{xy})^2} = 95.976 \text{ MPa}$$

The two in-plane principal stresses are obtained from the abscissas of points P_1 and P_2. Thus,

$$\sigma_1 = OC + R = 166.415 \approx 166.4 \text{ MPa} \quad \textbf{ANS.}$$

$$\sigma_3 = OC - R = -25.537 \approx -25.5 \text{ MPa} \quad \textbf{ANS.}$$

The plane of σ_1 is located by finding the angle $2\theta_\sigma$. From the geometry of Mohr's circle we obtain

$$2\theta_\sigma = \tan^{-1} \frac{\tau_{xy}}{1/2(\sigma_x - \sigma_y)} = 42.786 \approx 42.8°; \quad \theta_\sigma = 21.4°$$

Therefore, the plane of σ_1 is 21.4° clockwise from the X plane. In this case, as stated above, the third principal stress, σ_2, is zero and acts on the Z plane.

Focusing on point S_1 to obtain the maximum in-plane shearing stress, we note that this point is 90° counterclockwise from point P_1 on the circle. It follows, therefore, that this plane of maximum in-plane shearing stress is 45° counterclockwise from the plane of σ_1 in the stress element. The maximum in-plane shearing stress and the normal stress on its plane are

$$\tau_{MAX} = R = 95.976 \approx 96.0 \text{ MPa} \quad \textbf{ANS.}$$

$$\sigma_{MS} = \sigma_{AVE} = 70.439 \approx 70.4 \text{ MPa} \quad \textbf{ANS.}$$

All of these stresses and their planes are shown in the 45° triangular stress element of figure (e) above.

7.5.3 STRESSES ON INCLINED PLANES

Mohr's circle may also be used to find the normal and shearing stresses on planes inclined to the X plane or to the Y plane. It is important to remember that every point on the circumference of the circle defines one of the infinite number of planes passing through the stressed point represented by the circle. Thus, for example, consider the stress element shown in Figure 7.15a, where the stresses σ_x, σ_y, and τ_{xy} are known and it is necessary to find the normal and shearing stresses on some inclined plane, such as the one labeled N, whose orientation from the X plane is defined by the counterclockwise angle

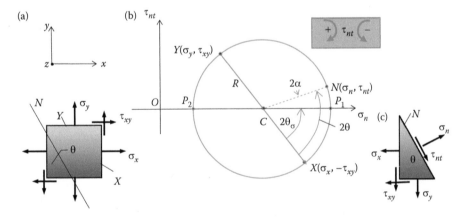

FIGURE 7.15 Construction of Mohr's circle for a given plane stress system and using it to find the stresses σ_n and τ_{nt} on any inclined plane N defined by the angle θ from the X plane.

θ. We begin the solution by constructing Mohr's circle, as shown in Figure 7.15b, for the stress element given in Figure 7.15a. For convenience, the special sign convention for shearing stress is repeated in Figure 7.15b. We then locate plane N on this circle by rotating from point X in the same sense (counterclockwise) through 2θ. The normal and shearing stresses on plane N are determined by using the geometry of Mohr's circle and finding the coordinates (σ_n and τ_{nt}) of point N on this circle. Thus,

$$OC = \sigma_{AVE} = \frac{\sigma_x + \sigma_y}{2}; \quad R = \sqrt{\left(\frac{\sigma_x - \sigma_y}{2}\right)^2 + (\tau_{xy})^2} \tag{7.35}$$

$$\sigma_n = OC + R\cos 2\alpha; \tag{7.36}$$

$$\tau_{nt} = R\sin 2\alpha \tag{7.37}$$

Referring to Mohr's circle, the angle 2α is found from the relation

$$2\alpha = 2\theta - 2\theta_\sigma \tag{7.38}$$

The angle $2\theta_\sigma$ in Equation 7.38 is determined from the geometry of Mohr's circle. Thus,

$$2\theta_\sigma = \tan^{-1}\left|\frac{\tau_{xy}}{1/2(\sigma_x - \sigma_y)}\right| \tag{7.39}$$

Note that only the magnitude of angle $2\theta_\sigma$ is needed for use in Equation 7.38 above. For the position of point N shown on the circle, σ_n is positive, thus producing tension on plane N and, τ_{nt} is above the σ_n axis (positive) and, therefore, it must produce clockwise rotation of the element of which plane N is a part. These stresses are placed on the triangular stress element contained within the X, Y, and N planes as shown in Figure 7.15c.

EXAMPLE 7.8

Determine the normal and shearing stresses on plane N for the stress system shown in figure (a) below. Show these stresses on a triangular stress element contained within the X, Y, and N planes.

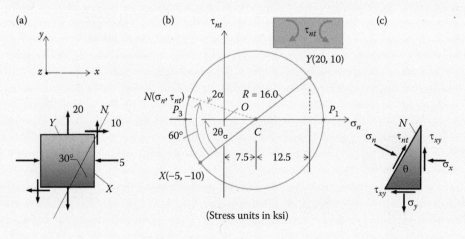

(Stress units in ksi)

SOLUTION

After establishing a $\sigma_n - \tau_{nt}$ coordinate system, we locate the $X(-5$ ksi, -10 ksi) and the point $Y(20$ ksi, 10 ksi) points, and proceed to construct Mohr's circle as shown in figure (b) above. This circle has the following properties:

$$OC = \frac{-5 + 20}{2} = 7.5 \text{ ksi}; \quad R = \sqrt{(12.5)^2 + (10)^2} = 16.0 \text{ ksi}$$

Since plane N is $30°$ clockwise from the X plane, point N on the circle is located by rotating clockwise through $2(30°) = 60°$ from point X as shown. The needed angles are computed as follows:

$$2\theta_\sigma = \tan^{-1} \frac{10}{12.5} = 38.66°$$
$$2\alpha = 60° - 38.66° = 21.34°$$

The normal and shearing stresses on plane N are found using the geometry of the circle. Thus,

$$\sigma_n = OC - R \cos 2\alpha = -7.403 \approx -7.4 \text{ ksi} \qquad \textbf{ANS.}$$

$$\tau_{nt} = R \sin 2\alpha = 5.822 \approx 5.8 \text{ ksi} \qquad \textbf{ANS.}$$

Both of these stresses are shown on plane N of the triangular stress element of figure (c) above.

PROBLEMS

7.26 Solve Problem 7.1 using Mohr's circle.

7.27 Solve Problem 7.2 using Mohr's circle.

7.28 Solve Problem 7.5 using Mohr's circle.

7.29 Solve Problem 7.12 using Mohr's circle.

7.30 Solve Problem 7.13 using Mohr's circle.

7.31 Solve Problem 7.14 using Mohr's circle.

7.32 Solve Problem 7.18 using Mohr's circle.

7.33 Solve Problem 7.19 using Mohr's circle.

7.34 Solve Problem 7.21 using Mohr's circle.

7.35 Solve Problem 7.22 using Mohr's circle.

7.36 Solve Problem 7.23 using Mohr's circle.

7.37 Solve Problem 7.24 using Mohr's circle.

7.38 The plane-stress condition shown is known to exist at some point in a stressed body. Use Mohr's circle to find the normal and shearing stresses on the X and Y planes. Show these stresses on a stress element properly oriented relative to the N plane.

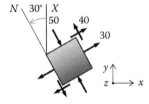

(Stress units in MPa)

(Problems 7.38 and 7.39)

7.39 In reference to the figure shown in Problem 7.38, the plane-stress condition shown is known to exist at some point in a stressed body. Use Mohr's circle to find the two in-plane principal stresses and the maximum in-plane shearing stress. Show these stresses on a stress element properly oriented relative to the N plane. State the third principal stress and its plane.

7.40 The plane-stress condition shown is known to exist at some point in a stressed body. Use Mohr's circle to find the normal and shearing stresses on the X and Y planes. Show these stresses on a stress element properly oriented relative to the N plane.

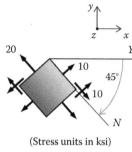

(Stress units in ksi)

(Problems 7.40 and 7.41)

7.41 In reference to the figure shown in Problem 7.40, the plane-stress condition shown is known to exist at some point in a stressed body. Use Mohr's circle to find the two in-plane principal stresses and the maximum in-plane shearing stress. Show these stresses on a stress element properly oriented relative to the Y plane. State the third principal stress and its plane.

7.42 Planes B and D pass through some point in a stressed body. The normal and shearing stresses on plane B are $\sigma_B = 30$ ksi, $\tau_B = 10$ ksi. The shearing stress on plane D is $\tau_D = 15$ ksi. If the angle $\beta = 15°$, determine the normal stress on plane D and the three principal stresses at the point in the stressed body. Show all the stresses on a triangular stress element contained within plane D and the planes of the two in-plane principal stresses.

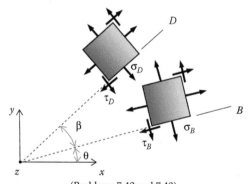

(Problems 7.42 and 7.43)

7.43 In reference to the figure shown in Problem 7.42, planes B and D pass through some point in a stressed body. The normal stress on plane B is $\sigma_B = 50$ MPa and that on plane D is $\sigma_D = 20$ MPa. If the angle $\theta = 15°$ and the angle $\beta = 45°$ determine the shearing stresses on planes B and D. Show these stresses on planes B and D on stress elements that contain the two principal planes along with the two in-plane principal stresses.

7.44 The plane-stress condition at a point in a stressed body is shown in the sketch. Determine the orientation of plane(s), relative plane N, on which the shearing stress has a magnitude of 5 ksi and find the normal stress on this plane(s). Show these stresses on a triangular stress element(s) contained within this plane(s) and the N and T planes.

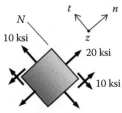

(Problems 7.44 and 7.45)

7.45 In reference to the figure shown in Problem 7.44, the plane-stress condition at a point in a stressed body is shown in the sketch. Determine the orientation of plane(s), relative plane N, on which the normal stress has a magnitude of 13 ksi and find the shearing stress on this plane(s). Show these stresses on a triangular stress element(s) contained within this plane(s) and the N and T planes.

7.46 The plane-stress condition at a point in a stressed body is defined by the two in-plane principal stresses $\sigma_1 = 20$ ksi and $\sigma_2 = 10$ ksi as shown. If $\theta = 45°$, find the normal and shearing stresses on the X and Y planes. Show these stresses on a stress element properly oriented relative to one of the two given principal planes.

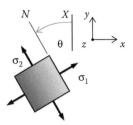

(Problems 7.46 and 7.47)

7.47 In reference to the figure shown in Problem 7.46, the plane-stress condition at a point in a stressed body is defined by the two in-plane principal stresses $\sigma_1 \Rightarrow \sigma_3 = -200$ MPa and $\sigma_2 = -30$ MPa as shown. If $\theta = 30°$, find the normal and shearing stresses on the X and Y planes. Show these stresses on a stress element properly oriented relative to one of the two given principal planes. Since in this case, $\sigma_1 = 0$, the stress σ_1 was changed to σ_3 to satisfy the requirement that $\sigma_1 \geq \sigma_2 \geq \sigma_3$.

7.48 Find the three principal stresses and their planes for the plane stress system obtained by superposing the stress elements shown in figures (a) and (b) below. Draw a rectangular stress element showing the two in-plane principal stresses properly oriented relative to the X plane.

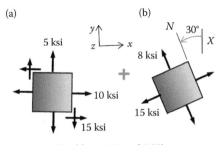

(Problems 7.48 and 7.52)

7.49 Find the three principal stresses and their planes for the plane stress system obtained by superposing the stress elements shown in figures (a) and (b) below. Draw a rectangular stress element showing the two in-plane principal stresses properly oriented relative to the X plane.

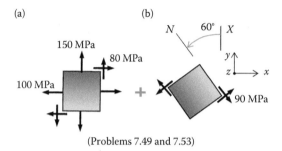

(Problems 7.49 and 7.53)

7.50 Find the three principal stresses and their planes for the plane stress system obtained by superposing the stress elements shown in figures (a) and (b) below. Draw a rectangular stress element showing the two in-plane principal stresses properly oriented relative to the X plane.

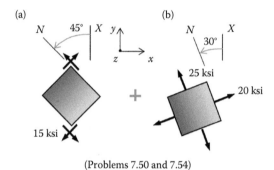

(Problems 7.50 and 7.54)

7.51 Find the three principal stresses and their planes for the plane stress system obtained by superposing the stress elements shown in figures (a) and (b) below. Draw a rectangular stress element showing the two in-plane principal stresses properly oriented relative to the X plane.

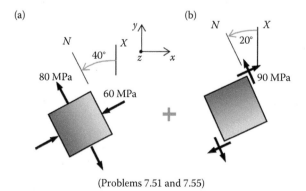

(Problems 7.51 and 7.55)

7.52 In reference to the figure shown in Problem 7.48, find the maximum in-plane shearing stresses and their planes for the plane stress system obtained by superposing the stress elements shown in Problem 7.48. Draw a rectangular stress element properly oriented relative to the X plane and showing the maximum in-plane shearing stress and the normal stress on its plane.

7.53 In reference to the figure shown in Problem 7.49, find the maximum in-plane shearing stresses and their planes for the plane stress system obtained by superposing the stress elements shown in Problem 7.49. Draw a rectangular stress element properly oriented relative to the X plane and showing the maximum in-plane shearing stress and the normal stress on its plane.

7.54 In reference to the figure shown in Problem 7.50, find the maximum in-plane shearing stresses and their planes for the plane stress system obtained by superposing the stress elements shown in Problem 7.50. Draw a rectangular stress element properly oriented relative to the X plane and showing the maximum in-plane shearing stress and the normal stress on its plane.

7.55 In reference to the figure shown in Problem 7.51, find the maximum in-plane shearing stresses and their planes for the plane stress system obtained by superposing the stress elements shown in Problem 7.51. Draw a rectangular stress element properly oriented relative to the X plane and showing the maximum in-plane shearing stress and the normal stress on its plane.

7.56 Find the two in-plane principal stresses acting on the stress element of figure (b) below if when superposed on the stress element of figure (a) below yields the stress element shown in figure (c) below. What is the magnitude of the angle θ_σ?

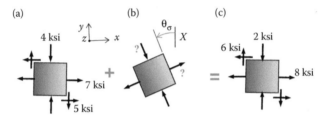

7.57 Find the shearing stress acting on the stress element of figure (b) below if when superposed on the stress element of figure (a) below yields the stress element shown in figure (c) below. What is the magnitude of the angle θ_τ?

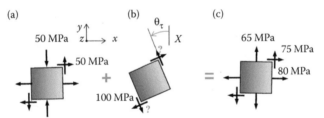

7.58 During operation, the stem of the drill bit (diameter = 1 in.) in a machine shop is subjected to the combined action of the internal force F and internal torque T. If $F = 20$ kips and $T = 5$ kip·in. determine the maximum tensile and compressive stresses on the outside surface, to which the stem is subjected, and the planes on which they act. Locate these planes relative to the X plane.

(Problems 7.58 and 7.59)

7.59 In reference to the figure shown in Problem 7.58, during operation, the stem of the drill bit (diameter = 15 mm) in a machine shop is subjected to the combined action of the internal force F and internal torque T. If $F = 10$ kN and $T = 0.05$ kN·m., determine the maximum in-plane shearing stress on the outside surface, to which the stem is subjected, and the plane on which it acts. Find also the normal stress on this plane. Orient this plane relative to the X plane.

7.60 A reinforced concrete building column of rectangular cross section is subjected to a compressive load P_1 and to a bending load P_2 as shown. Let $P_1 = 100$ kips, $P_2 = 2$ kips, $b = 10$ in., $w = 8$ in., and $h = 5$ ft, determine the maximum compressive and the maximum shearing stresses developed at point B and the planes on which they act. Orient these planes relative to the X plane.

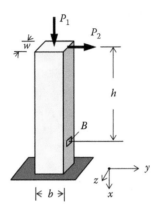

7.61 Arm BC is attached rigidly to shaft AB (diameter = 100 mm) as shown. Cable CD is tensioned to a force of 75 kN. Find the two in-plane principal stresses and the maximum in-plane shearing stress as well as the associated normal stresses at point E located as shown, on the outside surface of the shaft, at the top end of a vertical diameter. Show all of these stresses on a 45° triangular stress element properly oriented relative to the X plane.

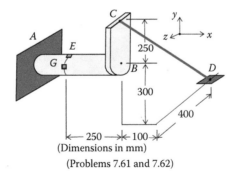

(Dimensions in mm)
(Problems 7.61 and 7.62)

7.62 Repeat Problem 7.61 for point G located as shown, on the outside surface of the shaft, at the front end of a horizontal diameter.

7.63 Cable DE is attached to the 2 in. × 4 in. rectangular compression member as shown. The cable is tensioned to 15 kips. Determine the two in-plane principal stresses and the maximum in-plane shearing stress at point B located at the left edge of the front face of the compression member. Show all of these stresses on a 45° triangular stress element properly oriented relative to the X plane.

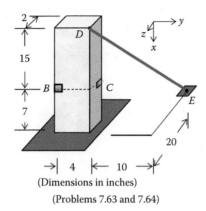

(Dimensions in inches)

(Problems 7.63 and 7.64)

7.64 Repeat Problem 7.63 for point C located at the far edge of the right face of the compression member.

7.65 Arm BC is fastened rigidly to beam AB (W16 × 89) and the assembly is subjected a downward force of 15 kips at C as shown. Determine the two in-plane principal stresses and the maximum in-plane shearing stress at point D located at the center of the top flange of the beam. Show all of these stresses on a 45° triangular stress element properly oriented relative to the X plane.

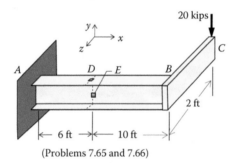

(Problems 7.65 and 7.66)

7.66 Repeat Problem 7.65 for point E located at mid-height of the beam web.

*7.6 THREE-DIMENSIONAL STRESS SYSTEMS

The preceding discussion has focused on the analysis of two-dimensional or plane stress systems. Fortunately, the great majority of the stress problems encountered in engineering practice are tractable using the plane stress solutions developed thus far. Occasionally, however, we encounter stress conditions whose solution requires a three-dimensional analysis. Such stress conditions, in which all of the six stress components (σ_x, σ_y, σ_z, $\tau_{xy} = \tau_{yx}$, $\tau_{xz} = \tau_{zx}$, and $\tau_{yz} = \tau_{zy}$) exist, are referred to as *triaxial stress* systems.

A triaxial stress element in the neighborhood of point O, similar to that of Figure 7.2, is shown in Figure 7.16. To develop the required relations, we consider plane BCD whose normal, n, is inclined to the x, y, and z axes by angles α, β, and γ, respectively. In keeping with the notation established earlier, this plane is labeled as plane N. Note that the normal to this plane, the n axis, intersects it at point E and intersects the positive Z plane at point G.

We now isolate tetrahedron $OBCD$ and construct its free-body diagram as shown in Figure 7.17. Our focus is to determine the three principal stresses at point O in the stressed body and the orientation of the corresponding principal planes.

In general, as a consequence of the normal and shearing stresses on the X, Y, and Z planes, any plane such as BCD is subjected to normal and shearing stresses which lead to the differential

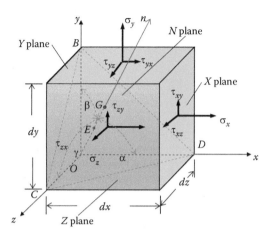

FIGURE 7.16 A general three-dimensional stress system defined by the three perpendicular X, Y, and Z planes and showing three normal and six shearing stresses.

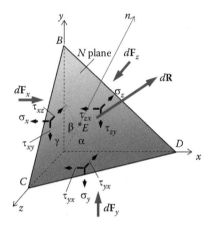

FIGURE 7.17 The free-body diagram of tetrahedron $OBCD$ and determination of the three principal stresses and their orientations at point O. The resultant acting on plane BCD is shown as $d\mathbf{R}$.

resultant force $d\mathbf{R}$ as shown in Figure 7.17. Of course, in the general case, this resultant is *not* perpendicular to plane BCD but has two components, one normal to the plane (due to the normal stresses) and the second parallel to it (due to the shearing stresses), unless plane BCD happens to be a principal plane. Obviously, in such a case, the shearing stresses vanish and only the component of $d\mathbf{R}$ perpendicular to plane BCD remains. The differential forces on the X, Y, and Z planes of the tetrahedron are indicated in Figure 7.17, respectively, as $d\mathbf{F}_x$, $d\mathbf{F}_y$, and $d\mathbf{F}_z$. Since the tetrahedron is in equilibrium, the four forces acting on it must be in balance. Thus,

$$\sum \mathbf{F} = d\mathbf{R} + d\mathbf{F}_x + d\mathbf{F}_y + d\mathbf{F}_z = 0; \quad d\mathbf{R} = d\mathbf{R}_x + d\mathbf{R}_y + d\mathbf{R}_z = -(d\mathbf{F}_x + d\mathbf{F}_y + d\mathbf{F}_z);$$

Therefore,

$$\left. \begin{aligned} d\mathbf{R}_x &= -d\mathbf{F}_x = -(\sigma_x dA_x + \tau_{yx} dA_y + \tau_{zx} dA_z)\mathbf{i} \\ d\mathbf{R}_y &= -d\mathbf{F}_y = -(\tau_{xy} dA_x + \sigma_y dA_y + \tau_{zy} dA_z)\mathbf{j} \\ d\mathbf{R}_z &= -d\mathbf{F}_z = -(\tau_{xz} dA_x + \tau_{yz} dA_y + \sigma_z dA_z)\mathbf{k} \end{aligned} \right\} \tag{7.40}$$

The quantities **i**, **j**, and **k** in Equations 7.40 are the unit vectors along the x, y, and z axes, respectively and the quantities dA_x, dA_y, and dA_z represent the differential areas of the triangular X, Y, and Z planes, respectively.

The unit vector, λ, along the n axis (normal to plane BCD), may be defined in terms of **i**, **j**, and **k**. Thus,

$$\lambda = (\cos\alpha)\mathbf{i} + (\cos\beta)\mathbf{j} + (\cos\gamma)\mathbf{k} = \ell\mathbf{i} + m\mathbf{j} + n\mathbf{k} \tag{7.41}$$

For convenience and simplicity, the quantities $\cos\alpha$, $\cos\beta$, and $\cos\gamma$, known as the *direction cosines* for the n axis, have been replaced, respectively, by the symbols, ℓ, m, and n.

Letting the differential area of inclined plane, BCD, be dA and referring back to Equations 7.40 and to the geometry of Figure 7.17, we conclude that

$$dA_x = dA\cos\alpha = \ell dA \Rightarrow \ell = \frac{dA_x}{dA} \tag{7.42}$$

$$dA_y = dA\cos\beta = mdA \Rightarrow m = \frac{dA_y}{dA} \tag{7.43}$$

$$dA_z = dA\cos\gamma = ndA \Rightarrow n = \frac{dA_z}{dA} \tag{7.44}$$

We now define the three components, T_x, T_y, and T_z, of what is known as the *surface traction T* on plane BCD. By definition, the component T_x, is

$$T_x = \frac{dR_x}{dA} = \sigma_x\left(\frac{dA_x}{dA}\right) + \tau_{yx}\left(\frac{dA_y}{dA}\right) + \tau_{zx}\left(\frac{dA_z}{dA}\right) = \ell\sigma_x + m\tau_{yx} + n\tau_{zx} \tag{7.45}$$

Similarly,

$$T_y = \frac{dR_y}{dA} = \ell\tau_{xy} + m\sigma_y + n\tau_{zy} \tag{7.46}$$

$$T_z = \frac{dR_z}{dA} = \ell\tau_{xz} + m\tau_{yz} + n\sigma_z \tag{7.47}$$

If plane BCD is a principal plane, then the shearing stresses on it vanish and only the normal (principal) stress σ remains. This principal stress is directed along the n axis and its three components along x, y, and z are, $\ell\sigma$, $m\sigma$, and $n\sigma$, respectively and are equal to the corresponding components of surface traction. Therefore,

$$T_x = \ell\sigma = \ell\sigma_x + m\tau_{yx} + n\tau_{zx} \Rightarrow \ell(\sigma_x - \sigma) + m\tau_{yx} + n\tau_{zx} = 0 \tag{7.48}$$

$$T_y = m\sigma = \ell\tau_{xy} + m\sigma_y + n\tau_{zy} \Rightarrow \ell\tau_{xy} + m(\sigma_y - \sigma) + n\tau_{zy} = 0 \tag{7.49}$$

$$T_z = n\sigma = \ell\tau_{xz} + m\tau_{yz} + n\sigma_z \Rightarrow \ell\tau_{xz} + m\tau_{yz} + n(\sigma_z - \sigma) = 0 \tag{7.50}$$

Equations 7.48 through 7.50 are three simultaneous equations in ℓ, m, and n and their trivial solution, $\ell = m = n = 0$, is not admissible because ℓ, m, and n must satisfy the requirement that

$l^2 + m^2 + n^2 = 1$. A nontrivial solution for these equations is obtained from the theory of algebraic equations which states that a solution is possible *only* if the determinant of the coefficients of ℓ, m, and n vanishes. Accordingly,

$$(\sigma_x - \sigma) + \tau_{yx} + \tau_{zx} = 0$$
$$\tau_{xy} + (\sigma_y - \sigma) + \tau_{zy} = 0 \qquad\qquad (7.51)$$
$$\tau_{xz} + \tau_{yz} + (\sigma_z - \sigma) = 0$$

Expanding the determinant in Equation 7.51 and simplifying we conclude that the principal stress σ must satisfy the following cubic equation:

$$\sigma^3 - I_1\sigma^2 + I_2\sigma - I_3 = 0 \qquad\qquad \textbf{(7.52)}$$

The quantities I_1, I_2, and I_3 in Equation 7.52 are constants known as the stress *invariants* because they do not vary with the orientation of the coordinate axes. These invariants are given by

$$\left.\begin{array}{l} I_1 = \sigma_x + \sigma_y + \sigma_z \\[4pt] I_2 = \sigma_x\sigma_y + \sigma_y\sigma_z + \sigma_z\sigma_x - \tau_{xy}^2 - \tau_{yz}^2 - \tau_{zx}^2 \\[4pt] I_3 = \sigma_x\sigma_y\sigma_z + 2\tau_{xy}\tau_{xz}\tau_{yz} - \sigma_x\tau_{yz}^2 - \sigma_y\tau_{xz}^2 - \sigma_z\tau_{xy}^2 \end{array}\right\} \qquad \textbf{(7.53)}$$

The solution of Equation 7.52 yields three values of σ which are the three principal stresses, σ_1, σ_2, and σ_3, acting at the point under consideration. The direction cosines, ℓ, m, and n, corresponding to each of the three principal stresses can then be found by the use of Equations 7.48 through 7.50 along with the relation $l^2 + m^2 + n^2 = 1$. Example 7.9 illustrates this procedure.

In view of the fact that the quantities I_1, I_2, and I_3 are invariants, they may be expressed more compactly in terms of the principal stresses. This, of course, means that, in Equation 7.53, we can replace the axes x, y, and z by the principal axes 1, 2, and 3, respectively, and eliminate all of the shearing stresses. Doing so we obtain

$$\left.\begin{array}{l} I_1 = \sigma_1 + \sigma_2 + \sigma_3 \\[4pt] I_2 = \sigma_1\sigma_2 + \sigma_2\sigma_3 + \sigma_3\sigma_1 \\[4pt] I_3 = \sigma_1\sigma_2\sigma_3 \end{array}\right\} \qquad \textbf{(7.54)}$$

The first of Equations 7.53 and that of Equations 7.54 tell us that the sum of the three normal stresses at a point in a stressed body, acting on three mutually perpendicular planes, is constant. The two-dimensional counterpart of this statement is expressed in Equation 7.11 discussed earlier.

EXAMPLE 7.9

At a point in a stressed body, the three-dimensional state of stress is given as shown below:
$\sigma_x = -\,80\text{ MPa}$, $\sigma_y = 40\text{ MPa}$, $\sigma_z = -\,40\text{ MPa}$, $\tau_{xy} = -\,40\text{ MPa}$, $\tau_{yz} = 120\text{ MPa}$, and $\tau_{xz} = 80\text{ MPa}$.

Find the principal stresses and the direction cosines corresponding to the maximum principal stress.

SOLUTION

The three invariants for the given stress system are found by substituting into Equations 7.53. Thus,

$$I_1 = -80 \text{ MPa}$$
$$I_2 = -24,000 \text{ MPa}^2$$
$$I_3 = 320,000 \text{ MPa}^3$$

Therefore, Equation 7.52 becomes

$$\sigma^3 + 80\sigma^2 - 24,000\sigma - 320,000 = 0 \tag{7.9.1}$$

The solution of Equation 7.9.1 may be obtained by trial-and-error or by other methods for solving polynomials. In this particular case, Newton's method was used to obtain

$$\sigma_1 = 127.6 \text{ MPa}; \quad \sigma_2 = -12.9 \text{ MPa}; \quad \sigma_3 = -194.8 \text{ MPa} \qquad \textbf{ANS.}$$

To find the direction cosines for σ_1, we use any two of Equations 7.48 through 7.50 plus the relation $\ell^2 + m^2 + n^2 = 1$. Employing Equations 7.48 and 7.49 we obtain

$$-207.6\,\ell_1 - 40\,m_1 + 80\,n_1 = 0 \tag{7.9.2}$$

$$-40\,\ell_1 - 87.6\,m_1 + 120\,n_1 = 0 \tag{7.9.3}$$

$$\ell_1^2 + m_1^2 + n_1^2 = 1 \tag{7.9.4}$$

One way to solve Equations 7.9.2 through 7.9.4 is to arbitrarily assume $n_1 = 1.000$ in Equations 7.9.2 and 7.9.3 and rearrange to obtain

$$207.6\,\ell_1 + 40\,m_1 = 80 \tag{7.9.5}$$

$$40\,\ell_1 + 87.6\,m_1 = 120 \tag{7.9.6}$$

Solving Equations 7.9.5 and 7.9.6 simultaneously we obtain the following 'fictitious' values:

$$\ell_1 = 0.133; \quad m_1 = 1.310; \quad n_1 = 1.000$$

Substituting these 'fictitious' values into Equation 7.9.4 yields

$$(0.133)^2 + (1.310)^2 + (1.000)^2 = 2.7337 \tag{7.9.7}$$

Since the right-hand side of Equation 7.9.7 must be unity, we need to reduce the 'fictitious' values by using the scaling factor $(1/2.7337)^{1/2} = 0.605$. Therefore, the true values of the direction cosines of σ_1 are

$$\ell_1 = 0.605(0.133) = 0.081; \quad m_1 = 0.605(1.310) = 0.792; \quad n_1 = 0.605(1.000) = 0.605 \quad \textbf{ANS.}$$

One check on the accuracy of the computations is to substitute these values into Equation 7.9.4 to see if the right-hand side is unity or close enough to it. In this case, the sum in Equation 7.9.4 yields 0.99985 which is acceptable.

7.6.1 Mohr's Circle for Triaxial Stress Systems: Absolute Maximum Shearing Stress

A three-dimensional stress condition possesses three principal stresses, σ_1, σ_2, and σ_3, labeled to satisfy the requirement that, algebraically, $\sigma_1 \geq \sigma_2 \geq \sigma_3$. The two-dimensional (plane) stress system may be considered a special case of the three-dimensional stress system in which one of the three principal stresses is zero.

In analyzing the plane-stress case, we determined the maximum in-plane shearing stress, τ_{MAX}, given by Equation 7.23, or by the radius of Mohr's circle. However, this maximum in-plane shearing stress may not be the largest shearing stress at the point in a stressed body under examination. In general, there is a shearing stress, known as the *absolute maximum shearing stress*, τ_{ABS}, which may be out of plane and larger in magnitude than the maximum in-plane shearing stress. Consider, for example, the three-dimensional stress element shown in Figure 7.18 in which $\sigma_3 = 0$. In reality, therefore, the stress element of Figure 7.18, is a two-dimensional stress system but may be viewed as consisting of three plane-stress systems as follows:

1. *Plane-stress state in the plane of σ_1 and σ_2*: Since this is a plane-stress state in the plane of σ_1 and σ_2 and $\sigma_3 = 0$, as shown in Figure 7.19a, we construct a Mohr's circle for which σ_1 and σ_2 are both positive as shown in Figure 7.19b. The maximum in-plane shearing stress, of course, is equal in magnitude to the radius of the circle. Thus,

$$(\tau_{MAX})_A = R = \frac{\sigma_1 - \sigma_2}{2} \tag{7.55}$$

The stresses σ_1 and σ_2 in Equation 7.55 are the values of the principal stresses corresponding to points P_1 and P_2, respectively, and $(\tau_{MAX})_A$ is the maximum in-plane shearing stress corresponding to either point S_1 or point S_2. As deduced from the circle, point S_1 defines a plane located 45° counterclockwise from the plane of σ_1 on which $(\tau_{MAX})_A$ must tend to produce

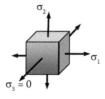

FIGURE 7.18 A three-dimensional stress system showing the stresses $\sigma_1 \geq \sigma_2 \geq \sigma_3$ in which $\sigma_3 = 0$. The following three two-dimensional stress systems, Figures 7.19 through 7.21 are used to determine the *absolute maximum shearing stress, τ_{ABS}*.

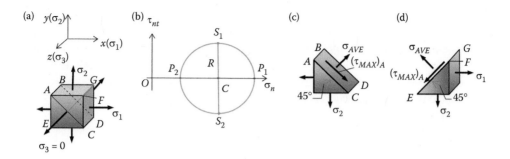

FIGURE 7.19 The three-dimensional stress system of Figure 7.18 is viewed as three, two-dimensional stress systems. This figure shows one of those three including its Mohr's circle in order to determine $(\tau_{MAX})_A$.

clockwise rotation of the element when viewing it as we face the positive plane of zero principal stress (i.e., the plane of σ_3 that we can see). This is the diagonal plane $ABCD$ in Figure 7.19a. A triangular stress element containing this plane is shown in Figure 7.19c. Point S_2, on the other hand, defines a plane 45° clockwise from the plane of σ_1 on which $(\tau_{MAX})_A$ must tend to produce counterclockwise rotation of the element when viewing it as we face the positive plane of zero principal stress. This is the diagonal plane identified by the red line EFG in Figure 7.19a. A triangular stress element containing this plane is shown in Figure 7.19d.

2. *Plane-stress state in the plane of σ_2 and σ_3.* Since this is a plane-stress condition in the plane of σ_2 and σ_3 in which $\sigma_3 = 0$ as shown in Figure 7.20a, Mohr's circle is constructed as shown in Figure 7.20b. The maximum in-plane shearing stress, of course, is equal in magnitude to the radius of the circle. Thus,

$$(\tau_{MAX})_B = R = \frac{\sigma_2 - \sigma_3}{2} \tag{7.56}$$

The stresses σ_2 and σ_3 in Equation 7.56 are the values of the principal stresses corresponding to points P_2 and P_3, respectively, and $(\tau_{MAX})_B$ is the maximum in-plane shearing stress corresponding to either point S_1 or point S_2. As observed from the circle, point S_1 defines a plane located 45° counterclockwise from the plane of σ_2 on which $(\tau_{MAX})_B$ must tend to produce clockwise rotation of the element when viewing it as we face the positive plane of zero principal stress (i.e., the plane of σ_1 that we can see). This is the diagonal plane $ABCD$ in Figure 7.20a. A triangular stress element containing this plane is shown in Figure 7.20c. Point S_2, on the other hand, defines a plane 45° clockwise from the plane of σ_2 on which $(\tau_{MAX})_B$ must tend to produce counterclockwise rotation of the element when viewing it as we face the positive

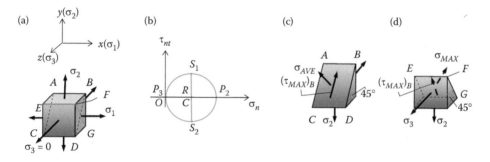

FIGURE 7.20 As in Figure 7.19, this diagram shows the second two-dimensional stress system showing the corresponding Mohr's circle used to determine $(\tau_{MAX})_B$.

plane of zero principal stress. This is the diagonal plane identified by the red line *EFG* in Figure 7.20a. A triangular stress element containing this plane is shown in Figure 7.20d.

3. *Plane-stress state in the plane of σ_1 and σ_3.* Since this is a plane-stress condition in the plane of σ_1 and σ_3 in which $\sigma_3 = 0$ as shown in Figure 7.21a, Mohr's circle is drawn as shown in Figure 7.21b. The maximum in-plane shearing stress, of course, is equal in magnitude to the radius of the circle. Thus,

$$(\tau_{MAX})_C = R = \frac{\sigma_1 - \sigma_3}{2} \tag{7.57}$$

The stresses σ_1 and σ_3 in Equation 7.57 are the values of the principal stresses corresponding to points P_1 and P_3, respectively, and $(\tau_{MAX})_C$ is the maximum in-plane shearing stress corresponding to either point S_1 or point S_2. As deduced from the circle, point S_1 defines a plane located 45° counterclockwise from the plane of σ_1 on which $(\tau_{MAX})_C$ must tend to produce clockwise rotation of the element when viewing it as we face the positive plane of zero principal stress (i.e., the plane of σ_3 that we can see). This is the diagonal plane *ABCD* in Figure 7.21a. A triangular stress element containing this plane is shown in Figure 7.21c. Point S_2, on the other hand, defines a plane 45° clockwise from the plane of σ_1 on which $(\tau_{MAX})_C$ must tend to produce counterclockwise rotation of the element when viewing it as we face the positive plane of zero principal stress. This is the diagonal plane identified by the red line *EFGH* in Figure 7.21a. A triangular stress element containing this plane is shown in Figure 7.21d.

Comparing the magnitudes of the maximum in-plane shearing stresses given by Equations 7.55 through 7.57, we must conclude that the absolute maximum shearing stress, τ_{ABS}, is given by Equation 7.57 because it yields one-half the difference between the largest and the smallest principal stresses as long as these stresses are labeled to satisfy the requirement that, algebraically, $\sigma_1 \geq \sigma_2 \geq \sigma_3$. In other words,

$$\tau_{ABS} = \frac{\sigma_1 - \sigma_3}{2} \tag{7.58}$$

Of course, in the two-dimensional case, $\sigma_3 = 0$ and $\tau_{ABS} = \sigma_1/2$. However, as we shall see shortly, Equation 7.58 is also applicable in the three-dimensional case (as long as the three principal stresses are labeled to satisfy the requirement that, algebraically, $\sigma_1 \geq \sigma_2 \geq \sigma_3$) even though *none* of the three principal stresses is zero. Furthermore, we note that the plane on which τ_{ABS} acts, always bisects the 90° angle between the plane of σ_1 and the plane of σ_3.

The three separate Mohr's circles shown in Figures 7.19 through 7.21 may be combined into one diagram as shown in Figure 7.22. Such a diagram allows us to see at a glance the entire stress

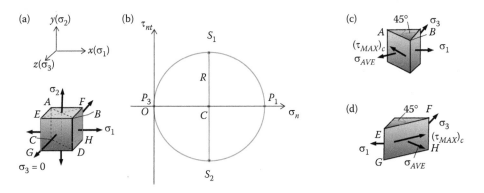

FIGURE 7.21 Again, as in Figure 7.19, this diagram shows the third two-dimensional stress system showing the corresponding Mohr's circle used to determine $(\tau_{MAX})_c$. The *absolute maximum shearing stress*, τ_{ABS} will be the largest of the three shearing stresses found in Figures 7.19 through 7.21.

system and to determine τ_{ABS} as the radius of the largest of the three circles (Equation 7.58). Let us consider the largest circle in Figure 7.22. Points S_1 and S_2, represent the absolute maximum shearing stress. Point S_1 represents a plane located 45° counterclockwise from the plane of σ_1 (point P_1) and 45° clockwise from the plane of σ_3 (point P_3). In other words, this plane must bisect the 90° angle between the principal planes of σ_1 and σ_3. On this plane, τ_{ABS} must be so directed that it tends to produce clockwise rotation of the stress element when viewing it as we face the positive plane of σ_2. A similar analysis of point S_2 leads us to conclude that the plane it represents also bisects the 90° angle between the principal planes of σ_1 and σ_3 and, of course, is perpendicular to the plane represented by point S_1. In general, therefore, we can state *that the plane on which τ_{ABS} acts bisects the 90° angle between the plane of σ_1 and the plane of σ_3.* Note that the normal stress acting on these two planes is $\sigma_{AVE} = (\sigma_1 - \sigma_3)/2 = \sigma_1/2$.

Let us now consider the three-dimensional stress condition in which *none* of the three principal stresses is zero. The case examined here is one in which two principal stresses (σ_1 and σ_2) are positive (tension) and one (σ_3) is negative (compression) as shown in Figure 7.23a. The results obtained for this case are general and apply equally well to other cases including that where all three principal stresses are positive and that where all are negative. Three two-dimensional Mohr's circles are constructed as shown in Figure 7.23b. Each of the three circles is drawn as though the third principal stress did not exist. Of course, τ_{ABS} is given by the radius of the largest circle and, therefore, Equation 7.58 developed for the two-dimensional case is still valid. Points S_1 and S_2, represent the absolute maximum shearing stress. Point S_1 represents a plane located 45° counterclockwise

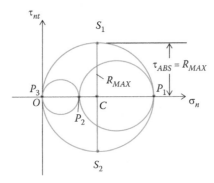

FIGURE 7.22 The three separate Mohr's circles shown in Figures 7.19 through 7.21 are combined into one diagram as shown in this figure.

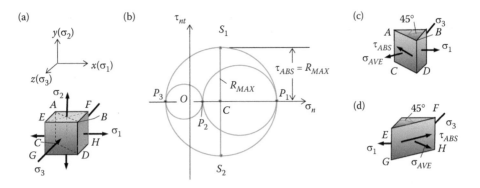

FIGURE 7.23 A three-dimensional stress condition in which *none* of the three principal stresses is zero and σ_3 is compressive. The corresponding Mohr's circle is shown and determining τ_{ABS} and the three principal stresses σ_1, σ_2, and σ_3.

from the plane of σ_1 (point P_1) and 45° clockwise from the plane of σ_3 (point P_3). In other words, this plane (*ABCD* in Figure 7.23a) must bisect the 90° angle between the principal planes of σ_1 and σ_3. On this plane, τ_{ABS} must be so directed that it tends to produce clockwise rotation of the stress element when viewing it as we face the positive plane of σ_2. A triangular stress element containing this plane and τ_{ABS} is shown in Figure 7.23c. A similar analysis of point S_2 leads us to conclude that the plane it represents also bisects the 90° angle between the principal planes of σ_1 and σ_3 (*EFGH* in Figure 7.23a) and is perpendicular to the plane defined by point S_1. A triangular stress element containing this plane and τ_{ABS}, properly directed, is shown in Figure 7.23d. Note that the normal stress on these two planes is $\sigma_{AVE} = (\sigma_1 + \sigma_3)/2$.

EXAMPLE 7.10

A plane-stress condition is shown in figure below. Use Mohr's circles to determine the absolute maximum shearing stress and its plane(s) as well as the normal stress on this plane(s). Show 45° triangular stress element(s) containing this plane(s).

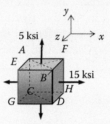

SOLUTION

Since no shearing stresses act on the *X*, *Y*, and *Z* planes of the element, the given normal stresses are principal stresses such that $\sigma_1 = 15$ ksi, $\sigma_2 = 5$ ksi, and $\sigma_3 = 0$. A Mohr's circle for the plane stress state defined by σ_1 and σ_2 is drawn as shown by the solid circle in figure (a) below, where P_1 represents the plane and magnitude of σ_1 and P_2 those of σ_2. Two other circles are drawn in broken lines in figure (a) below. The small one is the circle in the plane of σ_2 and σ_3 and the large one is in the plane of σ_1 and σ_3.

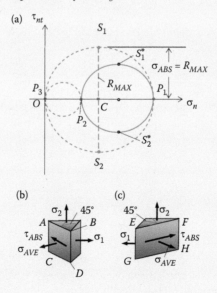

As indicated in figure (a) above, the absolute maximum out-of-plane shearing stress, τ_{ABS}, is given by the radius of the largest circle, which is in the plane of σ_1 and σ_3. Therefore, the magnitude of τ_{ABS} is given by Equation 7.58. Thus,

$$\tau_{ABS} = \frac{15 - 0}{2} = 7.5 \text{ ksi} \qquad \textbf{ANS.}$$

This absolute maximum shearing stress is represented on the circle by points S_1 and S_2. Point S_1 defines a plane located 45° counterclockwise from the plane of σ_1 on which τ_{ABS} must tend to produce clockwise rotation of the element when viewing it as we face the positive plane of σ_2. This is the diagonal plane $ABCD$ in figure above. A triangular stress element containing this plane is shown in figure (b) above. Point S_2 defines a plane located 45° clockwise from the plane of σ_1 on which τ_{ABS} must tend to produce counterclockwise rotation of the element when viewing it as we face the positive plane of σ_2. This is the diagonal plane $EFGH$ in figure above. A triangular stress element containing this plane is shown in figure (c) above. Of course, each of these two planes is subjected to a normal stress which is represented by the distance OC of the large circle and, as stated earlier, is equal to σ_{AVE} given by Equation 7.18. Thus,

$$\sigma_{AVE} = \frac{15 + 0}{2} = 7.5 \text{ ksi} \qquad \textbf{ANS.}$$

Note that in this particular case, σ_{AVE} and τ_{ABS} have identical magnitudes.

For comparison purposes, the maximum in-plane shearing stress relative to the stress element of σ_1 and σ_2, is given by the radius of the solid circle which has a magnitude of 5 ksi. This shearing stress is represented by the two points labeled S_1^* and S_2^*. Point S_1^* defines plane $EADH$, 45° counterclockwise from the plane of σ_1, on which the maximum in-plane shearing stress must tend to produce clockwise rotation of the element as viewed from the positive plane of σ_3. Point S_2^*, on the other hand, defines plane $BFGC$, 45°clockwise from the plane of σ_1, on which the maximum in-plane shearing stress must tend to produce counterclockwise rotation of the element as viewed from the positive plane of σ_3. The interested reader may want to construct 45° triangular elements containing each of these two planes and show the maximum in-plane shearing stress properly directed on them.

Note that for this problem, the plane of σ_3 coincides with the Z plane.

EXAMPLE 7.11

A triaxial state of stress is shown in figure (a) below where $\sigma_1 = 100$ MPa, $\sigma_2 = 75$ MPa and $\sigma_3 = 50$ MPa. Use Mohr's circle to determine the absolute maximum shearing stress and the plane(s) on which it acts as well as the normal stress on this plane(s). Show a 45° triangular stress element(s) containing this plane(s).

SOLUTION

Three two-dimensional Mohr's circles (one in the plane of σ_1 and σ_2, a second in the plane of σ_2 and σ_3, and the third in the plane of σ_1 and σ_3) are constructed as shown in figure (b) below. Each of the three circles is drawn as though the third principal stress did not exist. Of course,

τ_{ABS}, is given by the radius of the largest circle and its magnitude is given by Equation 7.58, where $\sigma_1 = 100$ MPa and $\sigma_3 = 50$ MPa. Thus,

$$\tau_{ABS} = \frac{100 - 50}{2} = 25 \text{ MPa} \qquad \textbf{ANS.}$$

Point S_1 represents a plane located 45° counterclockwise from the plane of σ_1 (point P_1) and 45° clockwise from the plane of σ_3 (point P_3). In other words, this plane (*ABCD* in figure (a) below) must bisect the 90° angle between the principal planes of σ_1 and σ_3. On this plane, τ_{ABS} must be so directed that it tends to produce clockwise rotation of the stress element when viewing it as we face the positive plane of σ_2. A triangular stress element containing this plane and τ_{ABS} is shown in figure (c) below. A similar analysis of point S_2 leads us to conclude that the plane it represents also bisects the 90° angle between the principal planes of σ_1 and σ_3 (*EFGH* in figure (a) below) and is perpendicular to the plane defined by point S_1. A triangular stress element containing this plane and τ_{ABS}, properly directed, is shown in figure (d) below. The normal stress on these two planes is $\sigma_{AVE} = (\sigma_1 + \sigma_3)/2$. Therefore,

$$\sigma_{AVE} = \frac{100 + 50}{2} = 75 \text{ MPa} \qquad \textbf{ANS.}$$

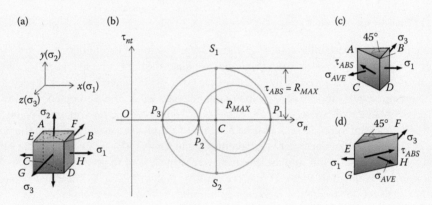

PROBLEMS

7.67 Refer to Example 7.9 and determine the direction cosines for σ_2.

7.68 Refer to Example 7.9 and determine the direction cosines for σ_3.

7.69 The state of triaxial stress at a point in a body is given as follows: $\sigma_x = 20$ ksi, $\sigma_y = 40$ ksi, $\sigma_z = -20$ ksi, $\tau_{xy} = -40$ ksi, $\tau_{yz} = 20$ ksi, and $\tau_{xz} = -60$ ksi. Determine the three principal stresses and the direction cosines for the maximum principal stress.

7.70 Let the state of stress at a point in a body be as follows: $\sigma_x = 70$, $\sigma_y = 10$, $\sigma_z = -20$, $\tau_{xy} = -40$, $\tau_{yz} = 10$, $\tau_{zx} = -20$ (units in MPa.) Determine the three principal stresses and the direction cosines for the intermediate principal stress.

7.71 The state of triaxial stress at a point in a body is given as follows: $\sigma_x = -25$ ksi, $\sigma_y = -10$ ksi, $\sigma_z = 20$ ksi, $\tau_{xy} = -40$ ksi, $\tau_{yz} = 15$ ksi, and $\tau_{xz} = -10$ ksi. Find the three principal stresses and the direction cosines for the minimum principal stress.

7.72 Let the state of stress at a point in a body be as follows: $\sigma_x = 60$ MPa, $\sigma_y = 15$ MPa, $\sigma_z = 20$ MPa, $\tau_{xy} = -30$ MPa, $\tau_{yz} = -10$ ksi, and $\tau_{xz} = 10$ ksi. Determine the three principal stresses and the direction cosines for the maximum principal stress.

7.73 The state of triaxial stress at a point in a body is given as follows: $\sigma_x = 30$ ksi, $\sigma_y = -10$ ksi, $\sigma_z = 20$ ksi, $\tau_{xy} = -20$ ksi, $\tau_{yz} = -15$ ksi, and $\tau_{xz} = 30$ ksi. Determine the three principal stresses and the direction cosines for the intermediate principal stress.

7.74 The state of triaxial stress at a point in a body is given as follows: $\sigma_x = 25$ MPa, $\sigma_y = 15$ MPa, $\sigma_z = -20$ MPa, $\tau_{xy} = -30$ MPa, $\tau_{yz} = -15$ MPa, and $\tau_{xz} = 10$ kMPa. Find the three principal stresses and the direction cosines for the minimum principal stress.

7.75 The principal stresses at a point in a body were found to be $\sigma_1 = 20$ ksi, $\sigma_2 = 10$ ksi, and $\sigma_3 = -10$ ksi. Use Mohr's circle to find the absolute maximum shearing stress and its plane(s) as well as the normal stress on this plane(s). Show a 45° triangular stress element(s) containing this plane(s).

7.76 Repeat Problem 7.75 if the principal stresses at a point in a body were found to be $\sigma_1 = 50$ MPa, $\sigma_2 = -10$ MPa and $\sigma_3 = -40$ MPa.

7.77 A two-dimensional state of stress is shown in the sketch. Use Mohr's circles to determine the absolute maximum shearing stress and its plane(s) as well as the normal stress on this plane(s). Show 45° triangular stress element(s) containing this plane(s). Given: $\sigma_x = 30$ ksi and $\sigma_y = 10$ ksi.

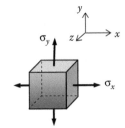

(Problems 7.77, 7.78, and 7.79)

7.78 In reference to the figure shown in Problem 7.77, a two-dimensional state of stress is shown in the sketch. Use Mohr's circles to determine the absolute maximum shearing stress and its plane(s) as well as the normal stress on this plane(s). Show 45° triangular stress element(s) containing this plane(s). Given $\sigma_x = 100$ MPa and $\sigma_y = -30$ MPa.

7.79 In reference to the figure shown in Problem 7.77, a two-dimensional state of stress is shown in the sketch. Use Mohr's circles to determine the absolute maximum shearing stress and its plane(s) as well as the normal stress on this plane(s). Show 45° triangular stress element(s) containing this plane(s). Given $\sigma_x = -10$ ksi and $\sigma_y = -15$ ksi.

7.80 A two-dimensional state of stress is shown in the sketch. Use Mohr's circles to determine the absolute maximum shearing stress and its plane(s) as well as the normal stress on this plane(s). Show 45° triangular stress element(s) containing this plane(s). Given $\sigma_x = -20$ MPa and $\sigma_z = -50$ MPa.

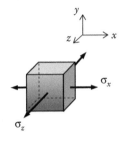

(Problems 7.80, 7.81, and 7.82)

7.81 In reference to the figure shown in Problem 7.80, a two-dimensional state of stress is shown in the sketch. Use Mohr's circles to determine the absolute maximum shearing stress and its plane(s) as well as the normal stress on this plane(s). Show 45° triangular stress element(s) containing this plane(s). Given $\sigma_x = -15$ ksi and $\sigma_z = 10$ ksi.

7.82 In reference to the figure shown in Problem 7.80, a two-dimensional state of stress is shown in the sketch. Use Mohr's circles to determine the absolute maximum shearing stress and its plane(s) as well as the normal stress on this plane(s). Show 45° triangular stress element(s) containing this plane(s). Given $\sigma_x = 80$ MPa and $\sigma_z = 40$ MPa.

7.83 A triaxial state of stress is shown in the sketch where $\sigma_1 = 35$ ksi, $\sigma_2 = 25$ ksi, and $\sigma_3 = 15$ ksi. Use Mohr's circle to determine the absolute maximum shearing stress and the plane(s) on which it acts as well as the normal stress on this plane(s). Show a 45° triangular stress element(s) containing this plane(s).

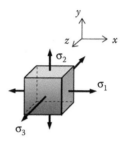

(Problems 7.83, 7.84, and 7.85)

7.84 In reference to the figure shown in Problem 7.83, a triaxial state of stress is shown in the sketch where $\sigma_1 = 90$ MPa, $\sigma_2 = 70$ MPa, and $\sigma_3 = -50$ MPa. Use Mohr's circle to determine the absolute maximum shearing stress and the plane(s) on which it acts as well as the normal stress on this plane(s). Show a 45° triangular stress element(s) containing this plane(s).

7.85 In reference to the figure shown in Problem 7.83, a triaxial state of stress is shown in the sketch where $\sigma_1 = 20$ ksi, $\sigma_2 = -10$ ksi, and $\sigma_3 = -30$ ksi. Use Mohr's circle to determine the absolute maximum shearing stress and the plane(s) on which it acts as well as the normal stress on this plane(s). Show a 45° triangular stress element(s) containing this plane(s).

7.86 The plane-stress condition at a point in a machine member is as shown in the sketch where $\sigma_x = 160$ MPa and $\sigma_y = 100$ MPa. The sense of τ_{xy} on X and Y planes is known to be negative but its magnitude is unknown. However, it is known that $\tau_{ABS} = 100$ MPa. Use Mohr's circle to determine the magnitude of the principal stresses and the magnitude of τ_{xy}.

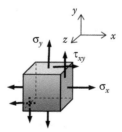

(Problems 7.86, 7.87, and 7.88)

7.87 In reference to the figure shown in Problem 7.86, the plane-stress condition at a point in a machine member is as shown in the sketch where $\sigma_x = -15$ ksi and $\tau_{xy} = 5$ ksi. The sense of σ_y is known to be negative but not its magnitude. However, it is known that $\tau_{ABS} = 15$ ksi. Use Mohr's circle to determine the magnitude of the principal stresses as well as the magnitude of σ_y.

7.88 In reference to the figure shown in Problem 7.86, the plane-stress condition at a point in a machine member is as shown in the sketch where $\sigma_y = 50$ MPa and $\tau_{xy} = -80$ MPa. The sense of σ_x is known to be positive but its magnitude is unknown. However, it is known that $\tau_{ABS} = 100$ MPa. Use Mohr's circle to determine the magnitude of the principal stresses and the magnitude of σ_x.

7.7 THIN-WALLED PRESSURE VESSELS

Numerous industrial applications require the use of containers for either storage or transmission of gases and liquids under high pressure. Examples include cylindrical or spherical tanks used for the storage of gaseous oxygen under high pressure and the piping used to deliver high-pressure liquids to hydraulic machines.

Two types of pressure vessels are generally identified as *thin walled* and *thick walled*. The distinction between the two types of vessels is based on the nature of the *circumferential* (also known as *hoop*) stress distribution over the thickness of the vessel. If the variation of this stress is such that it may be assumed approximately constant, the vessel is referred to as *thin walled*. If not, it is known as *thick walled*. Thin-walled vessels are discussed in this section and thick-walled vessels are discussed in Section 7.8. The distinction between thin-walled and thick-walled pressure vessels is also discussed in Section 7.8 after the development of appropriate equations.

The two most commonly used types of thin-walled pressure vessels are cylindrical and spherical. We will develop the stress equations for both these cases in the following paragraphs.

7.7.1 Cylindrical Vessels

Consider a thin-walled cylindrical vessel of thickness t as shown in Figure 7.24a. This vessel is subjected to internal fluid pressure p_1 and has covers at both ends. As a consequence of the internal pressure, the circumferential fibers of the vessel tend to expand because of tensile stress σ_c, tangent to the circumference, known as the *circumferential* or *hoop* stress. Moreover, because of the internal pressure, the longitudinal fibers tend to stretch because of the tensile stress σ_x known as the *longitudinal* stress. The stresses σ_c and σ_x are shown on a plane stress element properly oriented on the outside surface of the cylindrical vessel shown in Figure 7.24a, where one set of the element's planes is parallel to the axis of the vessel and the other set, perpendicular to this axis. Note that, because of geometric and loading symmetry, these two sets of planes are free of shearing stress and consequently, they are the principal planes and the stresses σ_c and σ_x are the principal stresses. If the stress element were taken on the inner surface instead of the outer surface of the cylinder, it would have to be treated as a three-dimensional stress element subjected to the same tensile stresses σ_c and σ_x in addition to a compressive radial stress $\sigma_r = p_1$, acting in the radial (r) direction. This compressive radial stress acts on a set of parallel planes, one of which is the inner concave surface of the cylinder. In the x-$y(r)$-$z(r)$ coordinate system shown, symbol (r) is used to signify a radial direction.

To develop an expression for the circumferential (hoop) stress σ_c, we construct a free-body diagram of a small portion of the cylindrical vessel with its fluid content as shown in Figure 7.24b. This free-body diagram was obtained by first slicing the cylinder into two halves using a cutting x-y plane, and then isolating a small segment from the rear half contained between two cutting planes at a distance dx apart and perpendicular to the axis of the cylinder. This free-body diagram is in equilibrium in the z direction, under the action of the internal fluid pressure p_1 and of the circumferential stresses σ_c, which are assumed to be uniformly distributed over the two areas $t \, dx$. The resultant force produced by the internal fluid pressure in the z direction is the product of p_1 and the

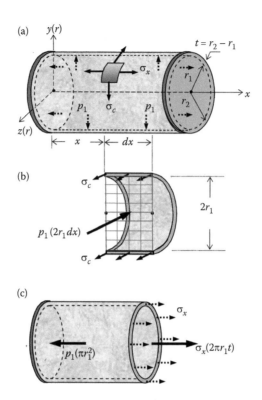

FIGURE 7.24 (a) A thin-walled cylindrical vessel of thickness t subjected to internal fluid pressure p_1. (b) The free-body diagram used in order to develop an expression for the circumferential (hoop) stress σ_c. (c) The free-body diagram used to determine the longitudinal stress σ_x.

projected area $2r_1\,dx$ (shown as a grid in teal color), namely, $p_1(2r_1\,dx)$. Moreover, the resultant force produced by the circumferential stress in the z direction is the product of σ_c and the two areas $2t\,dx$, that is, $\sigma_c\,(2t\,dx)$. Therefore, summing forces in the z direction we have

$$\sum F_z = 0: \quad \sigma_c(2t\,dx) - p_1(2r_1\,dx) = 0$$

Solving for σ_c, we obtain

$$\sigma_c = \frac{p_1 r_1}{t} \tag{7.59}$$

The longitudinal stress σ_x may be obtained using the free-body diagram shown in Figure 7.24c. This free-body diagram was obtained by cutting the cylindrical vessel into two parts using a cutting plane perpendicular to its axis and isolating the left part. It is in equilibrium in the x direction under the action of the internal pressure p_1 and the longitudinal stress σ_x, which are assumed uniformly distributed over the area $2\pi[(r_1 + r_2)/2]\,t$, where $(r_1 + r_2)/2$ is the mean radius of the vessel. However, a good approximation for this area is obtained by using the inner radius r_1 in the place of the mean radius $(r_1 + r_2)/2$. Thus, the resultant force produced by the longitudinal stress is $\sigma_x\,(2\pi r_1\,t)$. Besides, the resultant force produced by the internal pressure p_1 is the product of this pressure and the area πr_1^2, namely $p_1\,(\pi r_1^2)$. Therefore, summing forces in the x direction we have

$$\sum F_x = 0: \quad \sigma_x(2\pi r_1 t) - p_1(\pi r_1^2) = 0$$

Solving for σ_x yields

$$\sigma_x = \frac{p_1 r_1}{2t} \tag{7.60}$$

7.7.2 SPHERICAL VESSELS

A thin-walled spherical pressure vessel is shown in Figure 7.25a. This vessel has an inner radius r_1 and an outer radius r_2 such that the wall thickness is $t = r_2 - r_1$ and is subjected to internal fluid pressure p_1. To derive an expression for the tensile stress σ on the wall of the vessel, we need to construct a free-body diagram that exposes this stress. Such a free-body diagram, shown in Figure 7.25b, is obtained by using a horizontal cutting plane that slices the cylinder into two halves and discarding the upper one.

As indicated in Figure 7.25a, because of geometric and loading symmetry, the two sets of planes of a plane stress element taken at any point on the outside surface of the vessel are subject to a tensile stress σ which are equal in magnitude. Moreover, because of symmetry, these two sets of planes are free of shearing stress. Therefore, these two planes are principal planes and the stress σ is a principal stress.

The free-body diagram of the lower half of the spherical vessel, shown in Figure 7.25b, is in equilibrium in the vertical (i.e., $y(r)$) direction under the action of two forces. The first of these is the upward force $\sigma\,(2\pi r_1\,t\,)$ produced by the tensile stress σ acting over the thin annular area $2\pi r_1\,t$ and the second is the downward force $p_1(\pi r_1^2)$ produced by the internal fluid pressure p_1 acting over the projected area πr_1^2 (shown as a grid in teal color). Thus,

$$\sum F_y = 0: \quad \sigma(2\pi r_1 t) - p_1(\pi r_1^2) = 0$$

Solving for the tensile stress σ, we obtain

$$\sigma = \frac{p_1 r_1}{2t} \tag{7.61}$$

Note that the magnitude of the tensile stress σ in a spherical vessel is given by the same equation as that for the longitudinal stress σ_x in a cylindrical vessel.

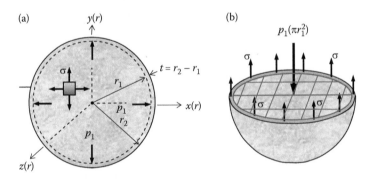

FIGURE 7.25 (a) A thin-walled spherical pressure vessel of thickness t subjected to internal fluid pressure p_1. (b) The free-body diagram of the lower half of the spherical vessel, used to determine the internal tensile stress σ acting over the thin annular area $2\pi r_1\,t$.

We should emphasize the fact that in Equations 7.59 through 7.61, the symbol r_1 is the inner radius of the vessel and that they are applicable only at points far removed from welded, riveted, or bolted joints where stress concentrations exist. Such complex stress systems, however, must be considered in the design of thin-walled pressure vessels. Moreover, the internal fluid pressure p_1 is the *gage pressure*, which is the difference between the internal pressure and the external atmospheric pressure.

EXAMPLE 7.12

A thin-walled cylinder with closed ends, for which $r_1 = 500$ mm and $r_2 = 520$ mm, is subjected to internal fluid pressure $p_1 = 2$ MPa. Determine (a) the absolute maximum shearing stress on the inner surface of the cylinder, (b) the absolute maximum shearing stress on the outer surface of the cylinder, and (c) the normal and shearing stresses on the wall of the cylinder on a plane inclined to its axis through 30°.

SOLUTION

a. A three-dimensional stress element on the inner surface of the cylinder is shown below. It is subjected to the following three principal stresses:

$$\sigma_1 = \sigma_c = \frac{p_1 r_1}{t} = \frac{2(0.50)}{(0.52 - 0.50)} = 50 \text{ MPa}$$

$$\sigma_2 = \sigma_x = \frac{p_1 r_1}{2t} = 25 \text{ MPa}$$

$$\sigma_3 = \sigma_r = p_1 = -2 \text{ MPa}$$

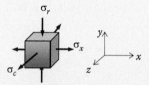

The absolute maximum shearing stress is given by Equation 7.58. Thus,

$$\tau_{ABS} = \frac{\sigma_1 - \sigma_3}{2} = \frac{50 - (-2)}{2} = 26 \text{ MPa} \qquad \textbf{ANS.}$$

b. A three-dimensional stress element on the outer surface of the cylinder is shown below. It is subjected to the following three principal stresses:

$$\sigma_1 = \sigma_c = \frac{p_1 r_1}{t} = \frac{2(0.50)}{(0.52 - 0.50)} = 50 \text{ MPa}$$

$$\sigma_2 = \sigma_x = \frac{p_1 r_1}{2t} = 25 \text{ MPa}$$

$$\sigma_3 = \sigma_r = 0$$

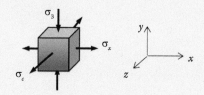

Using Equation 7.58 we obtain

$$\tau_{ABS} = \frac{\sigma_1 - \sigma_3}{2} = \frac{50 - 0}{2} = 25 \text{ MPa} \qquad \textbf{ANS.}$$

c. A two-dimensional stress element on the outside surface of the cylinder, similar to that in figure above, is shown in figure (a) below. The required plane, labeled B, is inclined to the σ_2 axis (the cylinder axis) or to the plane of σ_1, through the required $30°$ angle, shown clockwise in the diagram, although a counterclockwise angle would serve the same purpose.

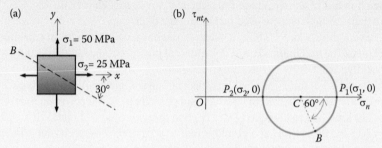

Mohr's circle for the plane stress element of figure (a) above is shown in figure (b) above. The coordinates of point B on the circle give the normal and shearing stresses on plane B. Thus,

$$\sigma_B = OC + R\cos((2)(30°)) = \frac{50 + 25}{2} + \left(\frac{50 - 25}{2}\right)\cos 60° = 43.75 \text{ MPa} \qquad \textbf{ANS.}$$

$$\tau_B = R \sin 60° = 10.83 \text{ MPa} \qquad \textbf{ANS.}$$

PROBLEMS

7.89 A gas storage thin-walled cylindrical tank with inner radius $r_1 = 2.5$ m and outer radius $r_2 = 2.6$ m is subjected to internal pressure $p_1 = 3$ MPa. Determine, on a section far removed from the two ends, (a) the maximum normal stress on the wall of the tank, (b) the absolute maximum shearing stress on the wall of the tank, and (c) the normal and shearing stresses on the wall of the tank on a plane inclined to the axis of the tank through a $45°$ angle.

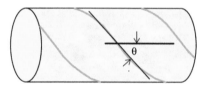

(Problems 7.89 and 7.90)

7.90 In reference to the figure shown in Problem 7.89, a thin-walled cylinder with closed ends is constructed, as shown in the sketch, by butt welding a thin plate along a helix that makes an angle $\theta = 30°$ with longitudinal lines parallel to the cylinder axis. If the inner radius is $r_1 = 50$ in., the internal pressure $p_1 = 300$ psi, and the normal and shearing stresses on the weld are not to exceed 15 and 10 ksi, respectively, determine the minimum permissible wall thickness.

7.91 A thin-walled cylinder with closed ends is constructed as shown in the sketch by butt welding a thin plate along a helix that makes an angle $\theta = 60°$ with longitudinal lines parallel to the cylinder axis. If $r_1 = 1.3$ m, $t = 25$ mm, and the normal and shearing stresses in the weld are not to exceed 150 and 100 MPa, respectively, determine the largest permissible internal pressure that may be applied to the cylinder.

7.92 A water storage cylindrical thin-walled stand tank is 20 m high, 2 m inside diameter, and has a wall thickness of 40 mm. The density of water is 1000 kg/m³. Determine the maximum normal and the absolute maximum shearing stress at the bottom of the tank when it is full of water. Ignore the effects of stress concentration.

7.93 A cylindrical tank is to be used for the storage of propane. The tank has a radius $r_1 = 8$ ft and a wall thickness $t = 7/8$ in. If the material has a yield strength in tension $\sigma_y = 36$ ksi, determine the maximum permissible internal pressure if the factor of safety with respect to yielding is 2.

7.94 A thin-walled cylindrical container has rigid plates fastened to its two ends as shown. The container is subjected to internal fluid pressure $p_1 = 400$ psi, to an axial tensile force $P = 100$ kips, and to a torque $Q = 0$, applied through rigid end plates. If $r_1 = 30.0$ in. and $r_2 = 30.5$ in., determine, at a point far removed from the two ends, the principal stresses and the absolute maximum shearing stress (a) on the outside surface and (b) on the inside surface.

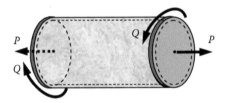

(Problems 7.94 and 7.95)

7.95 In reference to the figure shown in Problem 7.94, a thin-walled cylindrical container has rigid plates fastened to its two ends as shown. The container is subjected to internal fluid pressure $p_1 = 400$ psi, to an axial tensile force $P = 100$ kips, and to a torque $Q = 500$ kip·ft, applied simultaneously through the rigid end plates. If $r_1 = 30.0$ in. and $r_2 = 30.5$ in., determine, at a point far removed from the two ends, the principal stresses and the absolute maximum shearing stress (a) on the outside surface and (b) on the inside surface.

7.96 A toy balloon is inflated into a spherical shape with radius $r_1 = 150$ mm. If the thickness of the inflated balloon is $t = 0.15$ mm and the internal pressure is $p_1 = 0.05$ MPa, find the maximum tensile stress in the balloon skin.

7.97 A spherical container is made of an aluminum alloy which has an allowable tensile strength of 200 MPa. The container has radius $r_1 = 3.0$ m and wall thickness $t = 15$ mm. Find the maximum allowable internal pressure.

7.98 A spherical container is made of steel for which the ultimate strength is 50 ksi. The container has a radius $r_1 = 12$ ft and a wall thickness $t = \frac{3}{4}$ in. If the internal pressure is not to exceed 300 psi, determine the minimum factor of safety based on failure by rupture.

7.99 A spherical container with inner radius r_1 and wall thickness t is subjected to internal fluid pressure p_1. Show that the absolute maximum shearing stress is given by the expression $\tau_{ABS} = (p_1 r_1 + 2\, p_1 t)/4t$.

*7.8   THICK-WALLED CYLINDRICAL PRESSURE VESSELS

7.8.1   Stresses

As stated in the preceding section, when the circumferential stress on the wall of the cylinder cannot be properly assumed to be uniformly distributed across the wall thickness, the cylinder is treated as a thick-walled cylinder. Examples of thick-walled cylinders include gun barrels and the piping used to deliver high-pressure liquids to hydraulic machines.

Figure 7.26a shows a thick-walled cylindrical pipe that has open ends and is subjected to both internal pressure p_1 and external pressure p_2. The existence of these pressures gives rise to *circumferential (hoop)* stresses σ_c and *radial* stresses σ_r at any point in the wall thickness of the cylinder. Such a point is shown in Figure 7.26a as a differential element located at a distance r from the center of the cylindrical pipe. This differential element is bounded by two concentric cylindrical surfaces separated by a distance dr and by two longitudinal planes subtending the angle $d\theta$ between them. A plane stress element located at this point, with stresses σ_c and σ_r, is shown in Figure 7.26b. Owing to geometric and loading symmetries, no shearing stresses exist on the planes of σ_c and σ_r. It follows, therefore, that these two stresses are principal stresses. Note that both σ_c and σ_r vary with the radius r and that while the radial stress on the cylindrical surface located at a radius r is σ_r, it is $\sigma_r + d\sigma_r$ on the cylindrical surface located at $r + dr$. As in the case of thin-walled pressure vessels, the symbol (r) in the x-$y(r)$-$z(r)$ coordinate system shown is used to signify a radial direction. We note here, parenthetically, that radial stresses were not considered in the case of thin-walled pressure vessels because, while they exist, they are insignificantly small.

We now sum forces acting on the stress element of Figure 7.26b in the radial direction. Thus,

$$\sum F_r = 0: \quad (\sigma_r + d\sigma_r)(r + dr)d\theta\,dx - \sigma_r rd\theta\,dx - 2\sigma_c dr\,dx \sin\left(\frac{d\theta}{2}\right) = 0$$

Recalling that for very small angles, the function $\sin(d\theta/2)$ is approximately equal to the angle $(d\theta/2)$, and eliminating quantities of higher order, we obtain

$$\sigma_r dr + r\,d\sigma_r - \sigma_c dr = 0$$

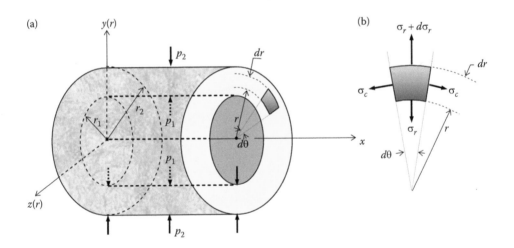

FIGURE 7.26   (a) Shows a thick-walled cylindrical pipe which has open ends and is subjected to both internal pressure p_1 and external pressure p_2. (b) Shows a free-body diagram used to determine the circumferential (hoop) stress σ_c and the radial stress σ_r at any point within the wall of the thick-walled cylinder.

Dividing by dr and rearranging the terms yields one relation between the two stresses σ_c and σ_r. Thus,

$$\sigma_c = \sigma_r + r\left(\frac{d\sigma_r}{dr}\right) \tag{7.62}$$

A second relation between the two unknown stresses σ_c and σ_r is obtained by considering the deformation of the cylindrical vessel in the longitudinal direction. Rigorous analyses have shown that longitudinal (axial) deformations are uniform for cylinders with open ends and may be assumed uniform for cylinders with closed ends at locations sufficiently removed from the two ends. Therefore, the assumption may be made that the longitudinal strain ε_x is constant. Applying the third of Equations 2.22 to the situation in hand, we obtain

$$\varepsilon_x = -\frac{\mu}{E}(\sigma_c + \sigma_r)$$

Since μ and E are constant material properties and the quantity ε_x is also assumed to be a constant, this expression may be written in the form

$$\sigma_c = -\sigma_r - \frac{E\varepsilon_x}{\mu} = -\sigma_r - 2K \tag{7.63}$$

The constant $2K$ in Equation 7.63 is equal to the quantity $(E\varepsilon_x)/\mu$, the exact value of which will be determined shortly from the boundary conditions. Thus, Equations 7.62 and 7.63 are two simultaneous equations in σ_c and σ_r and their solution is obtained in the following manner.

We first substitute the value of σ_c from Equation 7.63 into Equation 7.62 to obtain

$$2\sigma_r + r\left(\frac{d\sigma_r}{dr}\right) = -2K \tag{7.64}$$

Next, we multiply both sides of Equation 7.64 by r to obtain

$$2r\sigma_r + r^2\left(\frac{d\sigma_r}{dr}\right) = -2rK \tag{7.65}$$

The left-hand side of Equation 7.65 is the derivative with respect to r of the quantity $r^2\sigma_r$. Therefore,

$$\frac{d}{dr}(r^2\sigma_r) = -2rK \tag{7.66}$$

Integrating Equation 7.66 and solving for σ_r we obtain

$$\sigma_r = -K + \frac{C}{r^2} \tag{7.67}$$

where C is a constant of integration. Substituting from Equation 7.67 into Equation 7.63 yields

$$\sigma_c = -K - \frac{C}{r^2} \tag{7.68}$$

The two constants K and C in Equations 7.67 and 7.68 are found from the following two boundary conditions:

$$1.\ \text{For } r = r_1, \sigma_r = -p_1; \quad 2.\ \text{For } r = r_2, \sigma_r = -p_2 \tag{7.69}$$

We now substitute these conditions into Equation 7.67 to obtain two simultaneous equations in K and C. When solved, we obtain

$$K = \frac{p_2 r_2^2 - p_1 r_1^2}{r_2^2 - r_1^2}; \quad C = (p_2 - p_1)\left(\frac{r_1^2 r_2^2}{r_2^2 - r_1^2}\right) \tag{7.70}$$

If the values of K and C in Equations 7.70 are substituted into Equations 7.67 and 7.68, we obtain the equations for σ_r and σ_c in terms of pressures p_1 and p_2 and the radii r_1 and r_2 of the cylindrical vessel. Thus,

$$\sigma_r = \frac{p_1 r_1^2 - p_2 r_2^2}{r_2^2 - r_1^2} + (p_2 - p_1)\left(\frac{r_1^2 r_2^2 / r^2}{r_2^2 - r_1^2}\right) \tag{7.71}$$

$$\sigma_c = \frac{p_1 r_1^2 - p_2 r_2^2}{r_2^2 - r_1^2} - (p_2 - p_1)\left(\frac{r_1^2 r_2^2 / r^2}{r_2^2 - r_1^2}\right) \tag{7.72}$$

Note that the absolute maximum value of σ_r is the larger of the two pressures p_1 and p_2 and that the absolute maximum value of σ_c occurs at the inner surface of the cylinder, where r assumes its minimum value. Note also that, as in the case of thin-walled vessels discussed Section 7.6, the quantity p_1 represents the *gage* pressure.

The development of Equations 7.71 and 7.72 has assumed a cylinder with open ends, in which case the longitudinal stress σ_x does not exist. There are cases, however, where cylindrical vessels have closed ends and are subjected to internal and/or external pressures. In such cases, longitudinal stresses σ_x do exist, in addition to radial and circumferential stresses. In developing an expression for the longitudinal stress σ_x, we make the assumption that it is uniformly distributed on a transverse cross section sufficiently removed from the two ends.

To find the longitudinal stress σ_x, we consider the free-body diagram shown in Figure 7.27 which was obtained by cutting the cylindrical vessel into two parts using a transverse plane and retaining

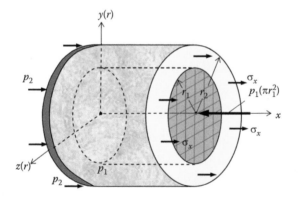

FIGURE 7.27   The free-body diagram of a part of a thick-walled cylinder used to determine the longitudinal stress σ_x in the cylinder.

only the part lying to the left of this cutting plane. Note that in an effort to simplify the diagram, only pressures p_1 and p_2 acting along the longitudinal direction (i.e., along the x axis) are shown.

To develop an expression for σ_x, we sum forces in the x direction. Thus,

$$\sum F_x = 0: \quad \pi(r_2^2 - r_1^2)\sigma_x + \pi r_2^2 p_2 - \pi r_1^2 p_1 = 0$$

Solving for σ_x we obtain

$$\sigma_x = \frac{p_1 r_1^2 - p_2 r_2^2}{r_2^2 - r_1^2} \tag{7.73}$$

7.8.2   DEFORMATIONS

The deformations of a cylindrical pressure vessel (both radial circumferential) with open ends due to pressures p_1 and p_2 become of major concern in certain industrial applications such as the shrink-fit operations. It is, therefore, desirable to relate these deformations to the geometry of the vessel and the pressures p_1 and p_2. Geometrically, the circumferential deformation δ_c is related to the radial deformation δ_r by the equation $\delta_c = 2\pi\delta_r$. Moreover, from the relation between deformation and strain, $\delta_c = 2\pi r \varepsilon_c$ in which ε_c is the circumferential strain. It follows, therefore, that

$$\delta_r = r\varepsilon_c \tag{7.74}$$

Using Equation 2.22, we write

$$\varepsilon_c = \frac{1}{E}(\sigma_c - \mu\sigma_r) \tag{7.75}$$

If we substitute Equation 7.75 into Equation 7.74, we obtain a general expression for the radial deformation δ_r in terms of σ_c and σ_r. Thus,

$$\delta_r = \frac{r}{E}(\sigma_c - \mu\sigma_r) \tag{7.76}$$

Substituting the values of σ_r and σ_c from Equations 7.71 and 7.72, respectively, into Equation 7.76 yields

$$\delta_r = \frac{(1-\mu)r}{E}\left[\frac{p_1 r_1^2 - p_2 r_2^2}{r_2^2 - r_1^2}\right] - \frac{(1+\mu)}{E}\left[\frac{(r_1^2 r_2^2 / r)(p_2 - p_1)}{r_2^2 - r_1^2}\right] \tag{7.77}$$

We now use the relation $\delta_c = 2\pi\delta_r$ along with Equation 7.77 to obtain

$$\delta_c = \frac{2\pi(1-\mu)r}{E}\left[\frac{p_1 r_1^2 - p_2 r_2^2}{r_2^2 - r_1^2}\right] - \frac{2\pi(1+\mu)}{E}\left[\frac{(r_1^2 r_2^2 / r)(p_2 - p_1)}{r_2^2 - r_1^2}\right] \tag{7.78}$$

7.8.3   SPECIAL CASES

We now consider two special cases of practical interest as follows:

7.8.3.1 Internal Pressure Only

There are many applications in which a cylindrical vessel is subjected only to internal pressure p_1. In such cases, the radial and circumferential stresses are obtained by setting $p_2 = 0$ in Equations 7.71 and 7.72, respectively. Thus,

$$\sigma_r = \frac{p_1 r_1^2}{r_2^2 - r_1^2}\left(1 - \frac{r_2^2}{r^2}\right) \tag{7.79}$$

$$\sigma_c = \frac{p_1 r_1^2}{r_2^2 - r_1^2}\left(1 + \frac{r_2^2}{r^2}\right) \tag{7.80}$$

Since the quantity r_2^2/r^2 is larger than unity, Equation 7.79 always yields a negative value for σ_r which, of course, indicates that σ_r is always a compressive stress with a maximum value of p_1 on the inner surface, where $r = r_1$. The circumferential stress σ_c, however, is always positive or tensile and, as in the case of σ_r, it assumes its maximum value of $[(r_2^2 + r_1^2)/(r_2^2 - r_1^2)]p_1$ at the inner surface where $r = r_1$.

If the cylindrical vessel has open ends, as, for example, in the case of a gun barrel, the walls are not subjected to longitudinal stress and $\sigma_x = 0$. If, on the other hand, the vessel has closed ends as, for example, in the case of an oxygen tank, the walls of the vessel are subjected to longitudinal stress σ_x given by Equation 7.73 by setting $p_2 = 0$ which yields

$$\sigma_x = \frac{p_1 r_1^2}{r_2^2 - r_1^2} \tag{7.81}$$

The radial and circumferential deformations experienced by a vessel with open ends, subjected to internal pressure only, are obtained by setting $p_2 = 0$ in Equations 7.77 and 7.78, respectively. Thus,

$$\delta_r = \frac{p_1 r_1^2}{E(r_2^2 - r_1^2)}\left[(1 - \mu)r + (1 + \mu)\frac{r_2^2}{r}\right] \tag{7.82}$$

$$\delta_c = \frac{2\pi p_1 r_1^2}{E(r_2^2 - r_1^2)}\left[(1 - \mu)r + (1 + \mu)\frac{r_2^2}{r}\right] \tag{7.83}$$

7.8.3.2 External Pressure Only

The stresses induced in a cylindrical vessel subjected only to external pressure p_2 (i.e., $p_1 = 0$) are obtained as special cases of the general expressions derived earlier. The radial and circumferential stresses are obtained from Equations 7.71 and 7.72, respectively, by setting $p_1 = 0$. Doing so yields

$$\sigma_r = -\frac{p_2 r_2^2}{r_2^2 - r_1^2}\left(1 - \frac{r_1^2}{r^2}\right) \tag{7.84}$$

$$\sigma_c = -\frac{p_2 r_2^2}{r_2^2 - r_1^2}\left(1 + \frac{r_1^2}{r^2}\right) \tag{7.85}$$

Note that σ_r and σ_c are both compressive stresses.

If the cylindrical vessel has open ends, the longitudinal stress $\sigma_x = 0$. If, however, the vessel has closed ends, the longitudinal stress is obtained from Equation 7.73 by setting $p_1 = 0$ to obtain

$$\sigma_x = -\frac{p_2 r_2^2}{r_2^2 - r_1^2} \tag{7.86}$$

The radial and circumferential deformations developed in the cylindrical vessel under the action of external pressure only are obtained from Equations 7.77 and 7.78, respectively, by setting $p_1 = 0$ to obtain

$$\delta_r = -\frac{p_2 r_2^2}{E(r_2^2 - r_1^2)} \left[(1 - \mu)r + (1 + \mu)\frac{r_1^2}{r} \right] \tag{7.87}$$

$$\delta_c = -\frac{2\pi p_2 r_2^2}{E(r_2^2 - r_1^2)} \left[(1 - \mu)r + (1 + \mu)\frac{r_1^2}{r} \right] \tag{7.88}$$

EXAMPLE 7.13

A cylindrical pipe has a 10-in. inner diameter and a 20-in. outer diameter. It is subjected only to internal pressure $p_1 = 5$ ksi. Construct the radial and circumferential stress variations across the wall thickness of the pipe.

SOLUTION

Since the pipe is subjected to internal pressure only, the radial and circumferential stresses are given by Equations 7.79 and 7.80, respectively. Thus,

$$\sigma_r = \frac{5(5^2)}{10^2 - 5^2}\left(1 - \frac{10^2}{r^2}\right) = 1.667\left(1 - \frac{100}{r^2}\right) \qquad \textbf{ANS.}$$

$$\sigma_c = \frac{5(5^2)}{10^2 - 5^2}\left(1 + \frac{10^2}{r^2}\right) = 1.667\left(1 + \frac{100}{r^2}\right) \qquad \textbf{ANS.}$$

A plot of these two equations is shown in the sketch. Note that while the radial stress σ_r is compressive everywhere, the circumferential stress σ_c is tensile throughout. Note also that the inner fibers of the pipe are heavily stressed when compared with the outer fibers and that the variations of both σ_r and σ_c are nonlinear.

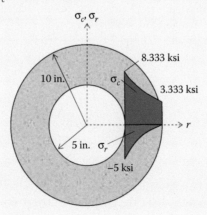

EXAMPLE 7.14

The cylindrical pipe described in Example 7.13 is subjected only to external pressure $p_2 = 5$ ksi. Construct the radial and circumferential stress variations across the wall thickness of the pipe.

SOLUTION

Since the pipe is subjected to external pressure only, the radial and circumferential stresses are given by Equations 7.84 and 7.85, respectively. Thus,

$$\sigma_r = -\frac{5(10^2)}{10^2 - 5^2}\left(1 - \frac{5^2}{r^2}\right) = -6.667\left(1 - \frac{25}{r^2}\right) \qquad \text{ANS.}$$

$$\sigma_c = -\frac{5(10^2)}{10^2 - 5^2}\left(1 + \frac{5^2}{r^2}\right) = -6.667\left(1 + \frac{25}{r^2}\right) \qquad \text{ANS.}$$

A plot of these two equations is shown in the sketch. Note that in this case both σ_r and σ_c are compressive stresses. The radial stress σ_r assumes a maximum value equal to p_2 on the outside surface, and the maximum circumferential stress occurs on the inner surface of the pipe.

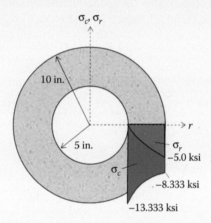

7.8.3.3 Shrink-Fitting Operations

Frequently, in order to increase their load-carrying capacities, composite cylinders are fabricated by shrinking an outer cylinder or jacket onto an inner cylinder as shown in Figure 7.28. Initially, the inside radius r_2 of the jacket is a little smaller (by a very small amount Δr) than the outside radius r_2 of the inner cylinder and the jacket is expanded by heating for easy placement on the cylinder. The inner radius of the cylinder is r_1 and the outer radius of the jacket is r_3. We need to develop an expression, in terms of Δr and the radii of the cylinder and the jacket, for the interface pressure p_i that is created between the jacket and the cylinder when the shrink-fitting operation is completed. Assume the jacket and cylinder are of the same material.

When the shrink-fitting process is completed, the interface pressure p_i expands the inside radius of the jacket by an amount δ_{rj} and shrinks the outside radius of the cylinder by an amount δ_{rc}. The jacket is subjected only to the internal pressure p_i and δ_{rj} is obtained from Equation 7.82. Thus,

$$\delta_{rj} = \frac{p_i r_2^2}{E(r_3^2 - r_2^2)}\left[(1 - \mu)r_2 + (1 + \mu)\frac{r_3^2}{r_2}\right] \qquad (7.89)$$

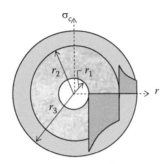

FIGURE 7.28 Diagram showing a composite cylinder fabricated by shrinking an outer cylinder or jacket, onto an inner cylinder. Also shown are the resulting tensile stress distribution in the jacket and the compressive stress distribution in the cylinder.

The cylinder is subjected only to the external pressure p_i. and δ_{rc} is given by Equation 7.87. Thus,

$$\delta_{rc} = -\frac{p_i r_2^2}{E(r_2^2 - r_1^2)}\left[(1 - \mu)r_2 + (1 + \mu)\frac{r_1^2}{r_2}\right] \tag{7.90}$$

The interface pressure p_i is found from the condition that the sum of the absolute magnitudes of δ_{rj} and δ_{rc} must be equal to the radial difference Δr. Therefore,

$$\delta_{rj} + \delta_{rc} = \Delta r \tag{7.91}$$

Substituting Equations 7.89 and 7.90 into Equation 7.91 and simplifying yields

$$\frac{2p_i r_2^3}{E}\left[\frac{r_3^2 - r_1^2}{(r_3^2 - r_2^2)(r_2^2 - r_1^2)}\right] = \Delta r \tag{7.92}$$

Solving Equation 7.92 for the interface pressure p_i, we obtain

$$p_i = \frac{(\Delta r)E}{2r_2^3}\left[\frac{(r_3^2 - r_2^2)(r_2^2 - r_1^2)}{r_3^2 - r_1^2}\right] \tag{7.93}$$

As stated earlier, the jacket is subjected to an internal pressure p_i only and, therefore, the circumferential stress distribution across its wall thickness is similar to that found for the pipe in Example 7.13. Moreover, the cylinder is subjected to an external pressure p_i only and the circumferential stress distribution across its wall thickness is similar to that found for the pipe in Example 7.14. These distributions are sketched qualitatively in Figure 7.28.

7.8.3.4 Internal Pressure on Thin-Walled Cylinders

The case of a cylindrical vessel subjected only to internal pressure p_1 was discussed earlier and Equations 7.79 through 7.81 were developed for σ_r, σ_c, and σ_x, respectively. We will now specialize these equations to the case of a cylindrical vessel with a relatively thin wall thickness.

Equation 7.80 that gives the circumferential stress σ_c at any point within the wall thickness of the cylindrical vessel is solved for the circumferential stress σ_{c1} on the inner surface ($r = r_1$) and for σ_{c2} on the outer surface ($r = r_2$). These solutions yield

$$\sigma_{c1} = \frac{p_1 r_1^2}{r_2^2 - r_1^2}\left(1 + \frac{r_2^2}{r_1^2}\right) \tag{7.94}$$

$$\sigma_{c2} = \frac{2p_1 r_1^2}{r_2^2 - r_1^2} \tag{7.95}$$

Dividing Equation 7.94 by Equation 7.95 we obtain the dimensionless ratio σ_{c1}/σ_{c2} as follows:

$$\frac{\sigma_{c1}}{\sigma_{c2}} = \frac{1}{2}\left[1 + \left(\frac{r_2}{r_1}\right)^2\right] \tag{7.96}$$

A graph of $(\sigma_{c1}/\sigma_{c2})$ as a function of (r_2/r_1) is shown in Figure 7.29. It is evident that as the ratio (r_2/r_1) increases (i.e., the wall thickness of the cylindrical vessel becomes larger), the ratio $(\sigma_{c1}/\sigma_{c2})$ increases. A range of values of (r_2/r_1) may, however, be chosen for which the ratio $(\sigma_{c1}/\sigma_{c2})$ is, for all practical purposes, sufficiently close to unity. For such cases, the circumferential stress distribution across the wall thickness may be assumed uniform without introducing appreciable error and the cylindrical vessel is said to be *thin walled*.

The range of values of (r_2/r_1), for which a cylindrical vessel may be assumed to be thin walled, (i.e., $\sigma_{c1}/\sigma_{c2} \approx 1$), depends on specific applications. If, for example, a 5% maximum variation in the circumferential stress is permissible across the wall thickness (i.e., $\sigma_{c1}/\sigma_{c2} = 1.05$), the ratio (r_2/r_1) may be no more than about 1.05. This means that the wall thickness (i.e., $t = r_2 - r_1$) cannot exceed a value of $0.05r_1$, or 5% of the inner radius. If on the other hand, a 10% maximum variation in circumferential stress is allowed, then (r_2/r_1) becomes about 1.09, and the wall thickness may be as large as 9% of the inner radius.

If the stress σ_{c1} is assumed to be equal to the stress σ_{c2}, Equation 7.94 yields the value of the uniform circumferential stress on the wall of a thin cylindrical vessel. Thus,

$$\sigma_c = \sigma_{c1} = \frac{p_1 r_1^2}{r_2^2 - r_1^2}\left(1 + \frac{r_2^2}{r_1^2}\right) = p_1\left(\frac{r_2^2 + r_1^2}{(r_2 - r_1)(r_2 + r_1)}\right) \tag{7.97}$$

Since the wall thickness is relatively small, r_2 may be assumed equal to r_1 and the quantities $r_2^2 + r_1^2$ and $r_2 + r_1$ in Equation 7.97 may be written, respectively, as $2r_1^2$ and $2r_1$. Besides, $r_2 - r_1 = t$, the wall thickness. Therefore, Equation 7.97 may be rewritten as follows:

$$\sigma_c = p_1\left(\frac{2r_1^2}{t(2r_1)}\right) = \frac{p_1 r_1}{t} \tag{7.98}$$

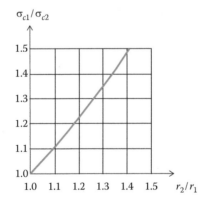

FIGURE 7.29 A graph of $(\sigma_{c1}/\sigma_{c2})$ as a function of (r_2/r_1) showing that the wall thickness (r_2/r_1) increases as the ratio $(\sigma_{c1}/\sigma_{c2})$ increases.

Equation 7.98 is identical with Equation 7.59 derived in Section 7.7 for thin-walled cylindrical vessels. Therefore, we may view the thin-walled cylindrical vessel as a special case of the thick-walled cylindrical vessel.

In general, the radial stress σ_r, because of its relatively small magnitude, is not as significant a stress quantity as the circumferential stress σ_c in the case of thin-walled cylindrical vessels. However, its value may be obtained from Equation 7.79. The longitudinal stress σ_x, when the thin-walled cylindrical vessel has closed ends, is obtained by specializing Equation 7.81. Thus,

$$\sigma_x = \frac{p_1 r_1^2}{r_2^2 - r_1^2} = \frac{p_1 r_1^2}{(r_2 - r_1)(r_2 + r_1)} = \frac{p_1 r_1^2}{t(2r_1)} = \frac{p_1 r_1}{2t} \tag{7.99}$$

Equation 7.99 is the same as Equation 7.60 derived earlier in Section 7.7 for a thin-walled cylindrical vessel.

PROBLEMS

7.100 A thick-walled cylinder with closed ends is subjected only to internal fluid pressure $p_1 = 8$ ksi. If the inner radius $r_1 = 5$ in. and the outer radius $r_2 = 10$ in. determine the maximum values of σ_c, σ, and σ_x at a section far removed from the ends of the cylinder.

7.101 Solve Problem 7.100 if the external pressure is increased from zero to 3 ksi.

7.102 A thick-walled cylinder with closed ends for which $r_1 = 250$ mm and $r_2 = 500$ mm is submerged in seawater to a depth of 20 m. Assume the density of seawater to be 1100 kg/m³ and determine the maximum values of σ_c, σ_r, and σ_x at a section sufficiently removed from the two ends of the cylinder. The gage pressure inside the cylinder is zero.

7.103 A thick-walled cylinder with $r_1 = 200$ mm is to be subjected only to the internal pressure $p_1 = 50$ MPa. If the allowable tensile stress in the cylinder is 100 MPa, determine the least permissible wall thickness of the cylinder.

7.104 A thick-walled cylindrical tank with $r_2 = 3$ ft has closed ends and is subjected only to the external pressure $p_2 = 20$ ksi. If the allowable compressive stress is 50 ksi, determine the least permissible wall thickness of the cylinder.

7.105 A composite thick-walled cylinder with open ends is fabricated by shrinking a jacket onto an inner cylinder. The following data are provided: $E = 30 \times 10^3$ ksi, $r_1 = 20$ in., $r_2 = 30$ in., $r_3 = 40$ in., and $\Delta r = 0.05$ in. Determine the maximum values of σ_c and σ_r.

7.106 The composite thick-walled cylinder of Problem 7.105 is subjected only to the internal pressure $p_1 = 30$ ksi. Determine the maximum value of σ_c.

7.107 A thick-walled cylindrical container with closed ends is subjected to internal pressure $p_1 = 60$ MPa ($p_2 = 0$) and to the axial tensile force $P = 1000$ kN ($Q = 0$) as shown. Let $r_1 = 250$ mm, $r_2 = 350$ mm, determine, at a point far removed from the two ends, the principal stresses and the absolute maximum shearing stress on the (a) outside surface and (b) inside surface.

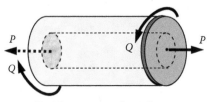

(Problems 7.107 and 7.108)

7.108 In addition to the loads given in Problem 7.107, the container is subjected to the torque $Q = 100 \text{ kN} \cdot \text{m}$ as shown. Assume the same geometry and determine, at a point far removed from the two ends, the principal stresses and the absolute maximum shearing stress on the (a) outside surface and (b) the inside surface.

7.109 A thick-walled cylinder with closed ends is subjected only to an internal pressure p_1. Let $r_1 = 36$ in., $r_2 = 46$ in., determine the maximum permissible value of p_1 if the allowable tensile stress is 22 ksi and the allowable shearing stress is 12 ksi.

*7.9   THEORIES OF FAILURE

For our purposes here, failure of a member is defined as one of two conditions:

1. Fracture of the member: This type of failure is characteristic of brittle material.
2. Initiation of inelastic (plastic) behavior in the member: This type of failure is characteristic of ductile material.

When a member is subjected to simple tension or compression, failure, as defined above, occurs when the applied axial stress reaches a limiting value σ_0. If the member is ductile, σ_0 represents the yield point or yield strength, σ_s, of the material; if the member is brittle, it represents the ultimate σ_u. However, when a member is subjected to a complex state of stress, the cause of failure, whether by *fracture* or by *yielding*, is unknown and attempts are made at predicting failure by one of the several theories that have been proposed over the years. It should be noted that a given material may fail in either a brittle or ductile manner, depending on such factors as temperature, load rate, and size. Therefore, in general, materials cannot be uniquely characterized as either *brittle* or *ductile*.

A failure theory is a *criterion* used in an effort to predict the failure of a member of a given material subjected to a complex stress condition, from knowledge of the properties of this material obtained from the simple tension, compression, or torsion tests. A few of the many theories of failure that have been proposed are discussed here. This discussion is divided into those theories applicable to brittle materials and those applicable to ductile materials.

7.9.1   BRITTLE MATERIAL

7.9.1.1   Maximum Principal Stress Theory

According to this theory, also known as *Coulomb's* theory, failure of a member subjected to plane stress occurs when the principal stress of the largest magnitude reaches the limiting value σ_0, which for brittle failure is equal to the ultimate strength, σ_u, as obtained from the simple tension test. Thus, on the basis of this theory, it is immaterial how complex or how simple the state of stress is. This member will fail by brittle action when the principal stress of largest magnitude reaches the critical value σ_u, which is assumed to be numerically the same for both tension and compression. Strictly speaking, this assumption is not valid because for most structural materials σ_u in tension is numerically less than that in compression. Even so, this theory constitutes a reasonable criterion to predict the failure of brittle members under complex stress conditions. Since in a given state of plane stress, the principal stress of the largest magnitude is either σ_1 or σ_3, where $\sigma_1 \geq \sigma_2 \geq \sigma_3$, algebraically, this theory may be represented mathematically as follows:

$$|\sigma_1| = \sigma_u \quad \text{or} \quad |\sigma_3| = \sigma_u \tag{7.100}$$

A graphical representation of Equations 7.100 for the case of *plane stress* is shown by the square area in Figure 7.30. *However, the gray area above the diagonal line AB is invalid because points*

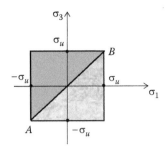

FIGURE 7.30 A graphical representation of Equation 7.100 in which only the *marble-colored area below the diagonal AB applies.*

within this area violate the restriction that, algebraically, $\sigma_1 \geq \sigma_2 \geq \sigma_3$. *Thus, only the marble-colored area below the diagonal AB applies.* This restriction does not impose a limitation on this theory because the marble-colored area below the diagonal AB accommodates all possible plane stress states in which σ_1 and σ_2 are both positive and $\sigma_3 = 0$; σ_1 is positive, $\sigma_2 = 0$ and σ_3 is negative; and, $\sigma_1 = 0$, σ_2 and σ_3 are both negative. Therefore, according to the maximum principal stress theory, a member will *not* fail by fracture as long as the plane stress state condition plots as a point within the marble colored area below the diagonal AB in Figure 7.30. On the other hand, failure by fracture occurs if the point lies outside this area.

7.9.1.2 Mohr's Theory

The maximum principal stress theory discussed above requires that the ultimate strength in tension be numerically equal to the ultimate strength in compression. This, of course, is not true for most structural materials, and, therefore, the maximum principal stress theory does not yield satisfactory answers. Mohr's theory, on the other hand, provides more satisfactory answers in predicting failure of materials for which the ultimate strength in tension is different from that in compression.

We will limit our discussion to the simplified Mohr's theory which is based solely on knowledge of the ultimate strength in tension, σ_{ut}, and the ultimate strength in compression, σ_{uc}. With this information, two Mohr's circles can be drawn, one with diameter σ_{ut} and center at A and the second with diameter σ_{uc} and center at B as shown in Figure 7.31. A plane stress state in which the two in-plane principal stresses are tensile (positive) and the out-of-plane principal stress is zero (i.e., σ_1 and σ_2 are positive and $\sigma_3 = 0$) will not lead to failure by fracture if the corresponding circle lies entirely within the confines of the circle with diameter σ_{ut} and center at A. In other words, no failure occurs if $\sigma_1 < \sigma_{ut}$ and $\sigma_2 < \sigma_{ut}$. Similarly, if the plane stress state is such that the two in-plane principal stresses are compressive (negative) and the out-of-plane principal stress is zero (i.e., $\sigma_1 = 0$, σ_2 and σ_3 are negative), then it will not lead to failure if the corresponding circle lies entirely within the confines of the circle with diameter σ_{uc} and center at B. In other words, no failure occurs if $|\sigma_2| < |\sigma_{uc}|$

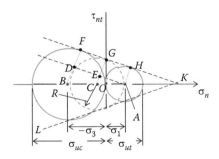

FIGURE 7.31 Graphical representation of Mohr's theory of failure.

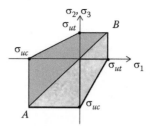

FIGURE 7.32 A second graphical representation of Mohr's theory of failure, representing Equation 7.103, in which only the area below the diagonal *AB* is valid.

and $|\sigma_3| < |\sigma_{uc}|$. Let us now plot the plane stress state with coordinates σ_1 (maximum stress) and σ_2 or σ_3 (minimum stress). Thus, according to Mohr's theory, *any plane stress state that lies within either of the two squares in the first and third quadrants of* Figure 7.32, *will not cause failure by fracture of the member on which it acts.*

Returning to Figure 7.31, we draw tangent lines *FK* and *LK* to the two circles of diameters σ_{ut} and σ_{uc}. If these two tangent lines are extended, they would intersect at point *K* on the σ_n axis. Let us now consider any plane stress state in which the two in-plane principal stresses are opposite in sign and the out-of-plane principal stress is zero (i.e., σ_1 is positive, $\sigma_2 = 0$ and σ_3 is negative). Such a stress condition may be analyzed by constructing a circle with center at *C* and radius $CG = (\sigma_1 - \sigma_3)/2$ as shown in Figure 7.31. According to Mohr's theory, failure by fracture occurs if σ_1 and σ_3 are sufficiently large for this circle to be tangent to both lines *FK* and *LK*. From Figure 7.31 we obtain

$$BD = \frac{\sigma_{uc}}{2} - \frac{\sigma_{ut}}{2}; \quad CE = R - \frac{\sigma_{ut}}{2} = \frac{\sigma_1 - \sigma_3}{2} - \frac{\sigma_{ut}}{2}$$

$$AB = \frac{\sigma_{uc}}{2} + \frac{\sigma_{ut}}{2}; \quad AC = \frac{\sigma_{ut}}{2} + OC = \frac{\sigma_{ut}}{2} - \left(\frac{\sigma_1 + \sigma_3}{2}\right) \tag{7.101}$$

Also, from similar triangles ACE and ABD we conclude that

$$\frac{BD}{AB} = \frac{CE}{AC} \tag{7.102}$$

Substituting Equations 7.101 into Equation 7.102 and simplifying, we obtain

$$\frac{\sigma_1}{\sigma_{ut}} = \frac{\sigma_3}{\sigma_{uc}} + 1 \tag{7.103}$$

When plotted on the $\sigma_1 - \sigma_3$ coordinate system of Figure 7.32, Equation 7.103 yields line $\sigma_{uc} - \sigma_{ut}$ in the fourth quadrant. Mathematically speaking, it is possible to obtain line $\sigma_{uc} - \sigma_{ut}$ in the second quadrant. *However, the gray area above the diagonal line AB is invalid because points within this area violate the requirement that, algebraically, $\sigma_1 \geq \sigma_2 \geq \sigma_3$. Thus, only the area below the diagonal AB applies.* This restriction does not impose a limitation on Mohr's theory because the marble-colored area below the diagonal *AB* accommodates all possible plane stress states in which σ_1 and σ_2 are both positive and $\sigma_3 = 0$; σ_1 is positive, $\sigma_2 = 0$ and σ_3 is negative; and, $\sigma_1 = 0$, σ_2 and σ_3 are both negative. Therefore, according to Mohr's theory, a member will *not* fail by fracture as long as the plane stress state condition plots as a point within the marble-colored area below the diagonal *AB* in Figure 7.32. On the other hand, failure by fracture occurs if the point lies outside this area.

7.9.2 DUCTILE MATERIALS

7.9.2.1 Maximum Shearing Stress Theory

This theory, also known as *Tresca's* criterion, states that a member subjected to any state of stress fails by *yielding* when the absolute maximum shearing stress, τ_{ABS}, reaches the critical value τ_o as obtained from the simple tension or compression test. This critical value τ_o is the maximum shearing stress in the tension or compression when failure by yielding takes place. For a uniaxial stress condition, the shearing stress is equal to one-half the axial normal stress, as may be concluded from Mohr's circle. Therefore, in the simple tension or compression test when yielding occurs $\tau_o = \sigma_y/2$.

If in a given plane-stress state in which the two in-plane principal stresses are tensile (positive) and the out-of-plane principal stress is zero (i.e., σ_1 and σ_2 are positive and $\sigma_3 = 0$), then it will not lead to failure by yielding if $\sigma_1 < \sigma_y$ and $\sigma_2 < \sigma_y$. Similarly, if the plane-stress state is such that the two in-plane principal stresses are compressive (negative) and the out-of-plane principal stress is zero (i.e., $\sigma_1 = 0$, σ_2 and σ_3 are negative), it will not lead to failure by yielding if $|\sigma_2| < |\sigma_y|$ and $|\sigma_3| < |\sigma_y|$. Thus, according to this theory, any plane-stress state that lies within either of the two squares in the first and third quadrants of Figure 7.33 will not cause failure by yielding of the member on which it acts. If, however, the plane-stress condition is such that the two in-plane principal stresses are opposite in sign and the out-of-plane principal stress is zero (i.e., σ_1 is positive, $\sigma_2 = 0$ and σ_3 is negative), the absolute maximum shearing stress, τ_{ABS}, for any combination of stresses is given by Equation 7.99, $\tau_{ABS} = (\sigma_1 - \sigma_3)/2$. Therefore, this theory may be stated in the form

$$\left| \sigma_1 - \sigma_3 \right| = \sigma_y \qquad (7.104)$$

When plotted on the $\sigma_1 - \sigma_3$ coordinate system of Figure 7.33, Equation 7.104 yields line $-\sigma_y - \sigma_y$ in the fourth quadrant. Mathematically speaking, it is possible to obtain line $-\sigma_y - \sigma_y$ in the second quadrant. *However, the gray area above the diagonal line AB is invalid because points within this area violate the requirement that, algebraically, $\sigma_1 \geq \sigma_2 \geq \sigma_3$. Thus, only the area below the diagonal AB applies.* This restriction does not impose a limitation on the maximum shearing stress theory because the marble-colored area below the diagonal AB accommodates all possible plane stress states in which σ_1 and σ_2 are both positive and $\sigma_3 = 0$; σ_1 is positive, $\sigma_2 = 0$ and σ_3 is negative; and $\sigma_1 = 0$, σ_2 and σ_3 are both negative.

7.9.2.2 The Energy of Distortion Theory

According to this theory, sometimes known as the *von Mises* theory, failure by *yielding* of a member subjected to combined loads takes place when the strain energy per unit volume used up in changing the shape of the member (energy of distortion) becomes equal to the energy of distortion per unit volume absorbed at failure by yielding in the uniaxial tension or compression test. Thus, this theory assumes that only a part of the total energy stored in a member subjected to combined stresses is responsible for failure by yielding. The remaining part, known as the *energy of volume change*, is responsible only for the change in volume of the member and does *not* contribute to failure by yielding.

As we will show in Chapter 8 (see Equation 8.55, in the general three-dimensional stress case, the energy of distortion per unit volume may be expressed in terms of the three principal stresses by

$$u_d = \frac{1 + \mu}{6E} [(\sigma_1 - \sigma_2)^2 + (\sigma_2 - \sigma_3)^2 + (\sigma_3 - \sigma_1)^2] \qquad (7.105)$$

where, as stated earlier, algebraically, $\sigma_1 \geq \sigma_2 \geq \sigma_3$, and μ is Poisson's ratio. Equation 7.105 provides a means of finding the energy of distortion per unit volume for any stress condition if the three

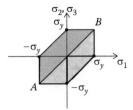

FIGURE 7.33   A diagram illustrating the *Maximum Shearing Stress Theory of Failure* for ductile material.

principal stresses are known. In the case of uniaxial tension or compression for which the principal stress is σ_1, failure by yielding occurs when this principal stress reaches the limiting value σ_y. Thus, the energy of distortion per unit volume at yielding in the uniaxial tension or compression is obtained from Equation 7.105 by setting $\sigma_1 = \sigma_y$ and $\sigma_2 = \sigma_3 = 0$ yielding

$$(u_d)_{uniaxial} = \left(\frac{1+\mu}{3E}\right)\sigma_y^2 \tag{7.106}$$

The quantity expressed in Equation 7.106 represents a unique property for a given material and is assumed here to be the same for both tension and compression.

The *energy of distortion* theory may now be formulated by stating that failure by yielding of a member subjected to a complex state of stress occurs when the energy of distortion per unit volume in the member, given by Equation 7.71, reaches the property stated in Equation 7.106. Therefore, setting these two quantities equal to each other and simplifying, we obtain a mathematical expression for the energy of distortion theory. Thus,

$$(\sigma_1 - \sigma_2)^2 + (\sigma_2 - \sigma_3)^2 + (\sigma_3 - \sigma_1)^2 = 2\sigma_y^2 \tag{7.107}$$

The relation expressed in Equation 7.107 may also be obtained on the basis of what has come to be known as the *octahedral shearing stress theory*. This theory states that failure of a member by yielding occurs when the octahedral shearing stress in the member reaches the octahedral shearing stress at yielding in the simple tension or compression test.

The *octahedral plane* is defined as one whose normal makes equal angles with the three principal directions. For a triaxial state of stress, it can be shown that the octahedral shearing stress, τ_{oct}, is given by

$$\tau_{oct} = \frac{1}{3}[(\sigma_1 - \sigma_2)^2 + (\sigma_2 - \sigma_3)^2 + (\sigma_3 - \sigma_1)^2]^{1/2} \tag{7.108}$$

In the uniaxial tension test for which the only nonzero principal stress is σ_1, failure by yielding occurs when this principal stress reaches σ_y. Thus, the octahedral shearing stress at failure by yielding in the simple tension test is obtained from Equation 7.108 by setting $\sigma_2 = \sigma_3 = 0$ and $\sigma_1 = \sigma_y$. This leads to $(\tau_{oct})_{uniaxial} = (\sqrt{2}/3)\sigma_y$. Replacing τ_{oct} in Equation 7.108 by this value and simplifying we obtain: $(\sigma_1 - \sigma_2)^2 + (\sigma_2 - \sigma_3)^2 + (\sigma_3 - \sigma_1)^2 = 2\sigma_y^2$. Obviously, this expression is identical to that given in Equation 7.107.

We now specialize Equation 7.107 to the two-dimensional case by setting any one of the three principal stresses equal to zero. If we set $\sigma_2 = 0$, we obtain

$$\sigma_1^2 - \sigma_1\sigma_3 + \sigma_3^2 \le \sigma_y^2 \tag{7.109}$$

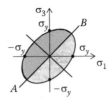

FIGURE 7.34 A two-dimensional representation of *The Energy of Distortion Theory.*

Note that the less than or equals sign is used instead of the equals sign of Equations 7.107 to indicate that a member subjected to a plane-stress state is safe as long as the quantity on the left side of Equation 7.109 is less than σ_y^2. Equation 7.109 represents an ellipse as shown in Figure 7.34. *However, as in previous cases, the gray area above the major axis AB is invalid because points within this area violate the requirement that, algebraically, $\sigma_1 \geq \sigma_2 \geq \sigma_3$. Thus, only the marble-colored area below this axis applies.*

EXAMPLE 7.15

A hollow shaft as shown below is subjected to a load P and a torque $Q = 20$ kip·ft. The material is cast iron for which $\sigma_u = \pm 10$ ksi. Use the maximum principal stress theory to find the maximum permissible compressive load P.

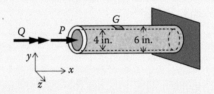

SOLUTION

At any point such as G on the outside surface of the shaft, we construct a stress element with two planes parallel and two planes perpendicular to the axis of the shaft as shown below. The stresses acting on this element are as follows:

$$\sigma_x = \frac{F}{A} = \frac{-P}{(\pi/4)(6^2 - 4^2)} = -0.0637\,P$$

$$\tau_{xy} = \frac{TR}{J} = \frac{20(12)(3)}{(\pi/32)(6^4 - 4^4)} = 7.0518 \text{ ksi}$$

The principal stresses are now found either from Equations 7.9 or from Mohr's circle solution. Thus,

$$\sigma_1 = -\frac{0.0637P}{2} + \sqrt{\left(\frac{0.0637P}{2}\right)^2 + 7.0518^2} \, ;$$

$$\sigma_2 = 0;$$

$$\sigma_3 = -\frac{0.0637P}{2} - \sqrt{\left(\frac{0.0637P}{2}\right)^2 + 7.0518^2}$$

Therefore, the principal stress of the largest magnitude is σ_3 and by the maximum principal stress theory, $|\sigma_3| \le \sigma_u$, we obtain

$$\frac{0.0637P}{2} + \sqrt{\left(\frac{0.0637P}{2}\right)^2 + 7.0518^2} = 10 \qquad (7.15.1)$$

The solution of Equation 7.15.1 for P yields

$$P = 78.920 \approx 78.9 \text{ kips} \qquad \textbf{ANS.}$$

EXAMPLE 7.16

A hollow shaft as shown below is subjected to a compressive load $P = 400$ kN and a torque Q. The material is cast aluminum for which $\sigma_{ut} = 100$ MPa and $\sigma_{uc} = 200$ MPa. Use Mohr's theory of failure to find the maximum permissible torque Q.

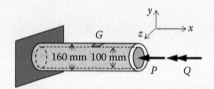

SOLUTION

At any point such as G on the outside surface of the shaft, we construct a stress element with two planes parallel and two planes perpendicular to the axis of the shaft as shown below. The stresses acting on this element are as follows:

$$\sigma_x = \frac{F}{A} = \frac{-P}{A} = \frac{-400(10^3)}{(\pi/4)(0.16^2 - 0.10^2)} = -32.647 \text{ MPa};$$

$$\tau_{xy} = \frac{TR}{J} = \frac{Q(10^3)(0.08)/10^6}{(\pi/32)(0.16^4 - 0.10^4)} = 1.467(10^{-3})Q \text{ MPa}$$

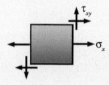

The principal stresses are now found either from Equations 7.16 and 7.17 or from Mohr's circle solution. Thus,

$$\sigma_1 = -\frac{32.647}{2} + \sqrt{\left(\frac{32.647}{2}\right)^2 + [1.467(10^{-3})Q]^2} \; ;$$

$$\sigma_2 = 0;$$

$$\sigma_3 = -\frac{32.647}{2} - \sqrt{\left(\frac{32.647}{2}\right)^2 + [1.467(10^{-3})Q]^2}$$

Applying Mohr's theory of failure as expressed by Equation 7.103 we obtain

$$-\frac{32.647}{200} + \frac{1}{100}\sqrt{\left(\frac{32.647}{2}\right)^2 + [1.467(10^{-3})Q]^2}$$

$$= -\frac{32.647}{400} - \frac{1}{200}\sqrt{\left(\frac{32.647}{2}\right)^2 + [1.467(10^{-3})Q]^2} + 1 \qquad (7.16.1)$$

Simplifying Equation 7.16.1 and solving for Q yields

$$Q = 47.9 \text{ kN} \cdot \text{m} \qquad \textbf{ANS.}$$

EXAMPLE 7.17

The plane-stress condition shown in figure (a) below exists at a critical point in an aluminum structural component. Determine if the structural component is safe on the basis of the maximum shearing stress theory and find the factor of safety knowing that the yield strength for aluminum in tension is (a) $\sigma_y = 80$ ksi and (b) $\sigma_y = 50$ ksi.

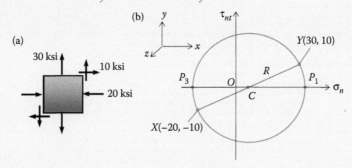

SOLUTION

Mohr's circle for the given state of stress is shown in figure (b) above, from which

$$OC = 5 \text{ ksi}; \quad R = 26.9 \text{ ksi}$$

Therefore,

$$\sigma_1 = 31.9 \text{ ksi}; \quad \sigma_2 = 0; \quad \sigma_3 = -21.9 \text{ ksi}$$

The maximum shearing stress theory expressed in Equation 7.104 is modified here to account for a factor of safety. Thus,

$$|\sigma_1 - \sigma_3| = \frac{\sigma_y}{F.S.} \Rightarrow F.S. = \frac{\sigma_y}{|\sigma_1 - \sigma_3|}$$

a. $F.S. = 80/53.8 = 1.487$. Therefore, the member *is* safe.   **ANS.**
b. $F.S. = 50/53.8 = 0.929$. Therefore, the member *is not* safe.   **ANS.**

EXAMPLE 7.18

Solve Example 7.17 on the basis of the energy of distortion theory.

SOLUTION

From Example 7.17 we have

$$\sigma_1 = 31.9 \text{ ksi}; \quad \sigma_2 = 0; \quad \sigma_3 = -21.9 \text{ ksi}$$

Equation 7.49, expressing the energy of distortion theory, is modified to introduce a factor of safety. Thus,

$$\sigma_1^2 - 2\sigma_1\sigma_3 + \sigma_3^2 = \left(\frac{\sigma_y}{F.S.}\right)^2 \Rightarrow F.S. = \frac{\sigma_y}{\sqrt{\sigma_1^2 - 2\sigma_1\sigma_3 + \sigma_3^2}}$$

a. $F.S. = \dfrac{80}{\sqrt{2195.8}} = 1.707$. Therefore, the member *is* safe.   **ANS.**

b. $F.S. = \dfrac{50}{\sqrt{2195.8}} = 1.067$. Therefore, the member *is* safe.   **ANS.**

Comparing these results with those obtained in Example 7.17, we conclude that the maximum shear stress theory of failure is more conservative that the energy of distortion theory.

PROBLEMS

7.110   The plane-stress condition shown is known to be at a critical point in a cast iron machine component. The fracture strength for cast iron is 20 ksi, assumed to be the same for both tension and compression. Use the maximum principal stress theory to determine if the machine component will fail by fracture.

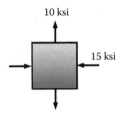

7.111 The plane-stress condition shown is known to be at a critical point in a cast iron machine component. The fracture strength for cast iron is 20 ksi, assumed to be the same for both tension and compression. Use the maximum principal stress theory to determine if the machine component will fail by fracture.

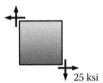

7.112 The plane-stress condition shown is known to be at a critical point in a cast iron machine component. The fracture strength for cast iron is 20 ksi, assumed to be the same for both tension and compression. Use the maximum principal stress theory to determine if the machine component will fail by fracture.

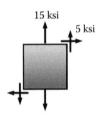

7.113 The plane-stress condition shown is known to be at a critical point in a cast iron machine component. The fracture strength for cast iron is 20 ksi, assumed to be the same for both tension and compression. Use the maximum principal stress theory to determine if the machine component will fail by fracture.

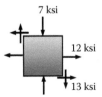

7.114 The plane-stress condition shown is known to be at a critical point in a cast iron machine component. The fracture strength for cast iron is 20 ksi, assumed to be the same for both tension and compression. Use the maximum principal stress theory to determine if the machine component will fail by fracture.

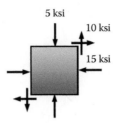

7.115 The plane-stress condition shown is known to be at a critical point in a cast aluminum machine component. The fracture strength for cast aluminum is 70 MPa in tension and 180 MPa in compression. Use Mohr's theory to determine if the machine component will fail by fracture.

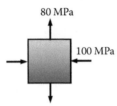

7.116 The plane-stress condition shown is known to be at a critical point in a cast aluminum machine component. The fracture strength for cast aluminum is 70 MPa in tension and 180 MPa in compression. Use Mohr's theory to determine if the machine component will fail by fracture.

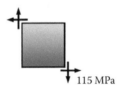

7.117 The plane-stress condition shown is known to be at a critical point in a cast aluminum machine component. The fracture strength for cast aluminum is 70 MPa in tension and 180 MPa in compression. Use Mohr's theory to determine if the machine component will fail by fracture.

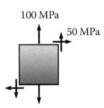

7.118 The plane-stress condition shown is known to be at a critical point in a cast aluminum machine component. The fracture strength for cast aluminum is 70 MPa in tension and 180 MPa in compression. Use Mohr's theory to determine if the machine component will fail by fracture.

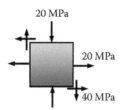

7.119 The plane-stress condition shown is known to be at a critical point in a cast aluminum machine component. The fracture strength for cast aluminum is 70 MPa in tension and 180 MPa in compression. Use Mohr's theory to determine if the machine component will fail by fracture.

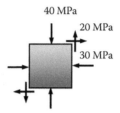

7.120 The hollow shaft ($d_o = 5$ in., $d_i = 4$ in.) shown in the sketch is made of cast aluminum ($\sigma_{ut} = \sigma_{uc} = \pm 8$ ksi) and is subjected to the combined action of a torque Q and a compressive force $P = 15$ kips. Use the maximum principal stress theory to find the maximum permissible torque Q.

(Problems 7.120 and 7.121)

7.121 Solve Problem 7.120 using Mohr's theory if $\sigma_{ut} = 8$ ksi and $\sigma_{uc} = 20$ ksi. All other conditions remain the same.

7.122 A solid circular shaft with diameter d = 100 mm made of cast iron ($\sigma_{ut} = \sigma_{uc} = 150$ MPa) is subjected to simultaneous action of torque $Q = 7$ kN·m and a compressive axial force P. Use the maximum principal stress theory to find the largest permissible force P so that no failure occurs.

7.123 Repeat Problem 7.122 using Mohr's theory if $\sigma_{ut} = 100$ MPa and $\sigma_{uc} = 200$ MPa. All other conditions remain the same.

7.124 The state of stress at a critical point in a steel structural component is given in the sketch. Use the maximum shearing stress theory to determine if the member will fail by yielding. If not, find the factor of safety. The yield strength for this steel is $\sigma_y = 30$ ksi.

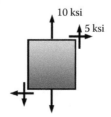

7.125 The state of stress at a critical point in a steel structural component is given in the sketch. Use the maximum shearing stress theory to determine if the member will fail by yielding. If not, find the factor of safety. The yield strength for this steel is $\sigma_y = 200$ MPa.

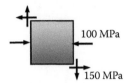

7.126 The state of stress at a critical point in a steel structural component is given in the sketch. Use the maximum shearing stress theory to determine if the member will fail by yielding. If not, find the factor of safety. The yield strength for this steel is $\sigma_y = 30$ ksi.

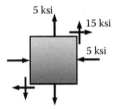

7.127 The state of stress at a critical point in a steel structural component is given in the sketch. Use the maximum shearing stress theory to determine if the member will fail by yielding. If not, find the factor of safety. The yield strength for this steel is $\sigma_y = 200$ MPa.

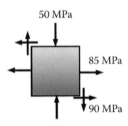

7.128 The state of stress at a critical point in a steel structural component is given in the sketch. Use the maximum shearing stress theory to determine if the member will fail by yielding. If not, find the factor of safety. The yield strength for this steel is $\sigma_y = 30$ ksi.

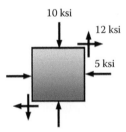

7.129 The state of stress at a critical point in a steel structural component is given in the sketch. Use the energy of distortion theory to determine if the member will fail by yielding. If not, find the factor of safety. The yield strength for this steel is σ_y = 30 ksi.

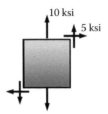

7.130 The state of stress at a critical point in a steel structural component is given in the sketch. Use the energy of distortion theory to determine if the member will fail by yielding. If not, find the factor of safety. The yield strength for this steel is σ_y = 200 MPa.

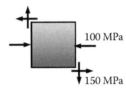

7.131 The state of stress at a critical point in a steel structural component is given in the sketch. Use the energy of distortion theory to determine if the member will fail by yielding. If not, find the factor of safety. The yield strength for this steel is σ_y = 30 ksi.

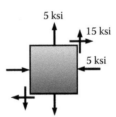

7.132 The state of stress at a critical point in a steel structural component is given in the sketch. Use the energy of distortion theory to determine if the member will fail by yielding. If not, find the factor of safety. The yield strength for this steel is σ_y = 200 MPa.

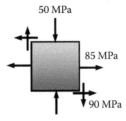

7.133 The state of stress at a critical point in a steel structural component is given in the sketch. Use the energy of distortion theory to determine if the member will fail by yielding. If not, find the factor of safety. The yield strength for this steel is σ_y = 30 ksi.

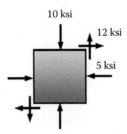

7.134 The ductile hollow shaft shown in the sketch is subjected to the combined action of torque Q and a tensile force $P = 300$ kN. The material has a yield strength in tension equal to 140 MPa. If $d_o = 180$ mm and $d_i = 120$ mm, use the maximum shearing stress theory to determine the largest permissible torque Q that may be applied.

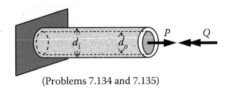

(Problems 7.134 and 7.135)

7.135 In reference to the figure shown in Problem 7.134, the ductile hollow shaft shown in the sketch is subjected to the combined action of torque $Q = 100$ kip·in. and a tensile force P. The material has a yield strength in tension equal to 35 ksi. If $d_o = 7$ in. and $d_i = 6$ in., use the maximum shearing stress theory to determine the largest permissible force P that may be applied.

7.136 A solid circular steel shaft is subjected to a bending couple $Q = 40$ kN·m and a tensile force $P = 100$ kN. The material has a yield strength in tension equal to 400 MPa. Find the least permissible diameter of the shaft by the maximum shearing stress theory.

7.137 A solid circular steel shaft with a diameter of 5 in. is subjected to a bending couple $Q = 50$ kip·ft and a tensile force P. The material has a yield strength in tension equal to 50 ksi. Find the largest permissible force P by the maximum shearing stress theory.

7.138 A solid 3 in. × 2 in rectangular shaft is subjected to a tensile force $P = 200$ kips and torque Q. The shaft is made of a ductile material for which the yield strength is 40 ksi. Use the maximum shearing stress theory to find the largest permissible value of Q.

7.139 A solid circular shaft is subjected to a torque $Q = 40$ kip · in and a bending moment $M = 80$ kip · in. Assume the material to be steel for which the yield strength in tension and compression is 30 ksi. Use the energy of distortion theory to find the least permissible diameter for the shaft.

REVIEW PROBLEMS

R7.1 Use a free-body diagram along with the equilibrium conditions to find the normal and shearing stresses on plane N defined by $\theta = 25°$. Check your answers using the stress transformation equations (Equations 7.6 and 7.9). Assume a standard coordinate system with x positive to the right, y positive upward, and z positive out of the page.

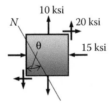

R7.2 Use appropriate equations to find the three principal stresses for the given plane stress system. Show the in-plane principal stresses on a stress element properly oriented relative to the X plane or x axis. Assume a standard coordinate system with x positive to the right, y positive upward, and z positive out of the page.

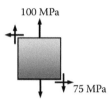

R7.3 A machine component may be idealized as a 3.5-in.-diameter circular shaft fixed at one end and attached to a rigid arm at the other. Let $P = 10$ kips, $a = 10$ in., $b = 8$ in., and $d = 12$ in., find the three principal stresses and the maximum in-plane shearing stress at point B on the shaft. Show the in-plane principal stresses and the maximum in-plane shearing stresses on stress elements properly oriented relative to the X plane or x axis.

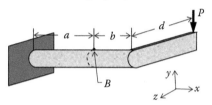

R7.4 Solve Problem R7.1 using Mohr's circle.
R7.5 Solve Problem R7.2 using Mohr's circle.
R7.6 A 6×12 ft sign is supported by a column with a hollow circular cross section the outside and inside diameters of which are 6 in. and 4 in. If the principal stress of the largest magnitude and the maximum in-plane shearing stress at point B are not to exceed 25 and 15 ksi, respectively, determine the largest magnitude of the uniform wind pressure p that the sign system can withstand.

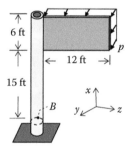

R7.7 Find the three principal stresses and their planes for the plane stress system obtained by superposing the stress elements shown in figures (a) and (b) below. Draw a rectangular stress element showing the two in-plane principal stresses properly oriented relative to the X plane.

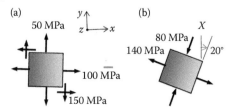

R7.8 An aluminum member of rectangular cross section is loaded as shown. If the allowable bending and shearing stresses at point B are, respectively, 15 and 10 ksi, determine the maximum permissible magnitude of P.

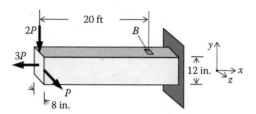

R7.9 A steel thin-walled cylindrical container has rigid plates fastened to its two ends as shown. The container is subjected to internal fluid pressure $p_1 = 450$ psi, to an axial tensile force $P = 300$ kips, and to a torque Q, applied through the rigid end plates. If $r_1 = 25.0$ in. and $r_2 = 25.5$ in., and the allowable tensile stress is 25 ksi, determine the maximum permissible torque Q. Ignore effects of stress concentration at the two ends.

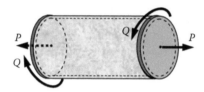

R7.10 A composite thick-walled cylinder is fabricated by shrinking a jacket onto an inner cylinder. The following data are provided: $E = 30 \times 10^3$ ksi, $r_1 = 30$ in., $r_2 = 40$ in., $r_3 = 50$ in., and $\Delta r = 0.05$ in. The cylinder is subjected to an internal pressure of 4.0 ksi. Determine the maximum value of σ_c at r_2 in the jacket.

R7.11 Let the state of stress at a point in a body be as follows: $\sigma_x = 100$ MPa, $\sigma_y = -80$ MPa, $\sigma_z = -50$ MPa, $\tau_{xy} = 40$ MPa, $\tau_{yz} = \tau_{xz} = -60$ MPa. Determine the three principal stresses and the direction cosines for the maximum principal stress.

R7.12 A triaxial state of stress is shown in the sketch where $\sigma_1 = 30$ ksi, $\sigma_2 = -20$ ksi, and $\sigma_3 = -40$ ksi. Use Mohr's circle to determine the absolute maximum shearing stress and the plane(s) on which it acts as well as the normal stress on this plane(s). Show 45° a triangular stress element(s) containing this plane(s).

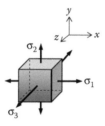

R7.13 The plane-stress condition at a point in a machine member is as shown in the sketch where $\sigma_x = 150$ MPa and $\sigma_y = -75$ MPa. The sense of τ_{xy} on the X and Y planes is known to be positive but its magnitude is unknown. However, it is known that the algebraically largest principal stress is 200 MPa. Use Mohr's circle to determine the magnitude of τ_{xy} and the other two principal stresses.

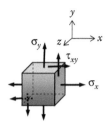

R7.14 The plane-stress condition shown is known to be at a critical point in a cast iron machine component. Use (a) the maximum principal stress theory ($\sigma_{ut} = \sigma_{uc} = 15$ ksi) and (b) Mohr's theory ($\sigma_{ut} = 15$ ksi; $\sigma_{uc} = 30$ ksi) to determine if the machine component will fail by fracture.

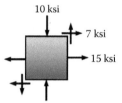

R7.15 A solid rectangular steel shaft is subjected to a torque Q and a compressive force $P = 120$ kN. The material has a yield strength in tension or compression is 400 MPa. Find the largest permissible value of Q by the maximum shearing stress theory. Let $b = 100$ mm. Ignore stress concentrations at the wall.

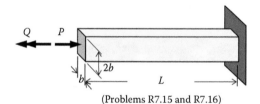

(Problems R7.15 and R7.16)

R7.16 In reference to the figure shown in Problem R7.15, a solid rectangular steel shaft with dimension $b \times 2b$ is subjected to a torque $Q = 50$ kip·in. and a compressive force $P = 150$ kips. The material has a yield strength in tension or compression of 70 ksi. Find the least permissible dimension b by the maximum shearing stress theory. Ignore stress concentrations at the wall.

8 Analysis of Strain

8.1 INTRODUCTION

We first introduced the concept of normal strain in Chapter 1 (see Equation 1.10) in connection with members subjected to axial loads and briefly introduced the concept of strain at a point (see Equation 1.25). In Chapter 2, we defined and discussed briefly the concept of shearing strain.

In this chapter, we expand our previous knowledge of strain. Thus, in Section 8.2, we review and expand our knowledge of normal and shearing strain and discuss their six components in the general, three-dimensional case. In Section 8.3, we deal with the subject of *plane strain*, consisting of two normal and one shearing strain components, because three of the six strain component reduce to zero. Also, in the case of plane strain, we develop the mathematical relations needed to determine the normal and shearing strains corresponding to axes inclined to the x and y axes, as well as the maximum in-plane normal and maximum in-plane shearing strains existing at the point in question. In Section 8.4, we introduce *Mohr's circle for strain*, which is an alternate semigraphical solution for the plane-strain problem. In Section 8.5, we analyze a three-dimensional stress system and show how Mohr's circle may be used to find the maximum normal and maximum shearing strains in the three-dimensional stress system. Finally, in Section 8.6, we discuss the maximum strain theory of failure to predict failure by fracture of brittle materials subjected to combined loads.

8.2 STRAIN AT A POINT: COMPONENTS OF STRAIN

When a deformable body is subjected to stresses, it experiences *deformations* and *distortions*. The term *deformations* refers to the geometric changes that take place in the dimensions (extensions or contractions) of the body, while the term *distortions* represents geometric changes in its shape.

Any line element in a stressed body is said to undergo deformation if its length increases or decreases. The term *average strain* of a line element is used to signify its unit deformation (i.e., deformation of the line element divided by its initial length). This unit deformation is known as *linear* or *normal* strain and denoted by the symbol ε_n. Therefore, if the initial length of a line element AB is L, as shown in Figure 8.1, and its deformation due to the applied loads P is δ (distance BC), an average value for the normal strain is given by

$$(\varepsilon_{AVE})_n = \frac{\delta}{L} \tag{8.1}$$

Consider the unloaded member shown in Figure 8.2a for which the cross-sectional geometry varies along its length in an arbitrary manner. In Figure 8.2b, this same member is subjected to a tensile axial load P. In order to define the strain at a point, we need to consider the deformation in the neighborhood of a point such as B. The length x, in Figure 8.2a, experiences a deformation δ and the length Δx a deformation $\Delta\delta$ when the load P is applied, as shown in Figure 8.2b. Thus, the strain ε_n at point B becomes

$$\varepsilon_n = \lim_{\Delta x \to 0} \frac{(\Delta x + \Delta\delta) - \Delta x}{\Delta x} = \lim_{\Delta x \to 0} \frac{\Delta\delta}{\Delta x} = \frac{d\delta}{dx} \tag{8.2}$$

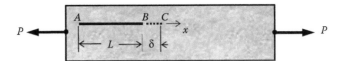

FIGURE 8.1 Definition of strain parameters.

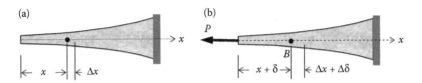

FIGURE 8.2 Definition of strain parameters for a member with varying cross-sectional properties.

At any point in a stressed body, such as point B in Figure 8.2b, the normal strain ε_n, can be completely defined by specifying its three rectangular components, ε_x, ε_y, and ε_z, along the three rectangular x, y, and z axes, respectively, as indicated in Figure 8.3. The component ε_x represents the normal strain of the body in the x direction, the component ε_y the normal strain in the y direction, and the component ε_z the normal strain in the z direction.

As was mentioned earlier, the term *distortion* signifies a change in the original shape of a stressed body. A more precise definition of distortion, however, is needed, and for this purpose, the change in angle between two initially perpendicular line elements is used. The change in the 90° angle between the two line elements n and t, as shown in Figure 8.4, represents a distortion and is given the name *shearing strain* and denoted by the symbol γ_{nt}.

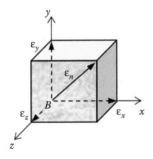

FIGURE 8.3 Three-dimensional strain element.

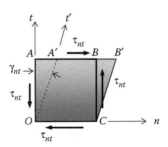

FIGURE 8.4 Strain element showing positive shear strain between lines n and t.

Note that two subscripts are assigned to the shearing strain γ, one for each of the two initially perpendicular line elements. In the three-dimensional case referred to an x–y–z coordinate system attached to a point such as B, as shown in Figure 8.3, three shearing strain components need to be considered: γ_{xy}, the shearing strain between the x and y axes; γ_{yz}, that between the y and z axes; and γ_{xz}, the shearing strain between the x and z axes.

The rectangular stress element $OABC$ shown in Figure 8.4 becomes the rhombus $OA'B'C$ under the action of the shearing stresses τ_{nt}. Thus, the shearing strain γ_{nt} represents the angle between the initial orientation of the t axis and its final orientation t', or the *change in the* $90°$ *angle that existed* between the n and t axes before the shearing stress was applied. For very small distortions, the angle γ_{nt} may be assumed equal to the tangent of this angle. Thus, referring to Figure 8.4, we have

$$\gamma_{nt} \approx \tan\gamma_{nt} = \frac{AA'}{OA} \tag{8.3}$$

8.2.1 Units and Sign Conventions

Note that, by definition as indicated by Equations 8.1 and 8.2, the normal strain ε_n is a dimensionless quantity. However, it is sometimes expressed in terms of units of length divided by units of length in order to give it a meaningful physical significance. Thus, for example, in the U.S. Customary system of measure, the unit in./in. is used. The sign convention for normal strain is such that *a positive normal strain represents an extension and a negative one, a contraction.*

As indicated in Equation 8.3, like the normal strain, ε, the shearing strain γ is a dimensionless quantity since it represents the ratio of two lengths. The sign convention for shearing strain is such that *a positive shearing strain represents a decrease in the* $90°$ *and a negative shearing strain represents an increase in this angle.* Thus, for example, the γ_{nt}, shown in Figure 8.4, is positive because it represents a decrease in the $90°$ angle between lines n and t. Note that this sign convention for shearing strain is consistent with the sign convention introduced earlier for shearing stress.

8.3 PLANE-STRAIN TRANSFORMATION EQUATIONS

The term *plane strain* is used to denote a condition in which all strains in a stressed body occur in a single plane. It should be emphasized, however, that when a member is subjected to stress, it experiences three-dimensional deformation, unless its deformation is prevented in one or more directions. Thus, in order to achieve a condition of plane strain, say in the x–y plane, the strain in the z direction must be prevented by physical restraint (i.e., by the application of force in the z direction). Under such conditions, $\varepsilon_z = \gamma_{yz} = \gamma_{xz} = 0$ and the only nonvanishing components of strain are ε_x, ε_y, and γ_{xy}.

Let us examine the strains produced by the plane-stress state acting on the stress element of Figure 8.5a. The assumption is made that strains in the z direction are zero and, therefore, we are dealing with a plane-strain system.

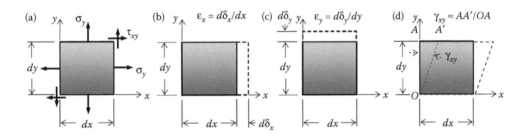

FIGURE 8.5   Illustration of individual normal and shear strains on a plane-strain element.

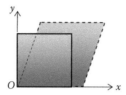

FIGURE 8.6   Illustration of the combined effect of normal and shear strains on a plane-strain element.

Under these conditions, the element experiences deformations in the x and y directions as well as a shearing strain in the x–y plane. The deformation in the x direction is depicted in Figure 8.5b and that in the y direction in Figure 8.5c. Note that, for simplicity in both of these figures, Poisson's effect has been excluded. The shearing strain in the x–y plane is shown in Figure 8.5d. Of course, all of these effects occur simultaneously, and the element assumes the shape of a parallelogram as shown schematically and, in an exaggerated manner, by the broken lines in Figure 8.6.

Let us assume that we know the strains ε_x, ε_y, and γ_{xy} in a given plane-strain state and that we wanted to determine, in terms of these quantities, the strains ε_n, ε_t, and γ_{nt}, where n and t are two orthogonal axes inclined to the x and y axes by the counterclockwise angle θ, as shown in Figure 8.7. The development of these desired relations is enhanced if we let the axes n and t be, respectively, the diagonals of two adjacent rectangular elements $OABC$ and $OCDG$ whose dimensions have been chosen such that the axes n and t are perpendicular to each other as shown in Figure 8.7.

If the stress state shown in Figure 8.5a is applied to the composite element $GABD$ of Figure 8.7, the element will deform and distort into the configuration shown by dashed parallelogram $G'A'B'D'$. The latter configuration is due to a deformation in the x direction equal to $\varepsilon_x(dx_1 + dx_2)$, a deformation in the y direction equal to $\varepsilon_y dy$ and a distortion due to the shearing strain γ_{xy}, which rotates line elements such as DB into position $D'B'$. Note that no generality is sacrificed by assuming that line element OC remains vertical while undergoing deformation to become line element OC'.

Let us now consider the deformation of line element OB, which, before deformation, coincided with the n axis. This deformation will be denoted by δ_n, which is obviously the composite effect of three different deformations; one in the x direction due to ε_x, a second in the y direction due to ε_y, and the third due to the distortion represented by γ_{xy}. These three deformations are depicted, separately, in Figure 8.8, where, since subscripting is not needed, the differential dx_1 has been replaced by dx. Figure 8.8a shows the change in length of line element OB due to the deformation $\varepsilon_x dx$. For very small deformations, line OB' will have essentially the same direction as line OB. Therefore, the angle $OB'C$ may be assumed equal to θ. Thus, for very small deformations, the

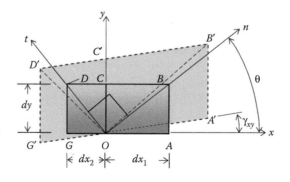

FIGURE 8.7   Strain element defining parameters for plane-strain transformation.

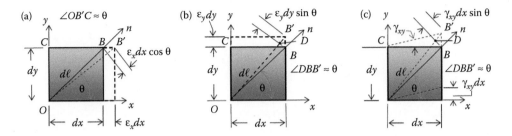

FIGURE 8.8   Illustration of plane normal and shear strains along a line.

increase in length of line element OB is $\varepsilon_x dx \cos\theta$ as shown in Figure 8.8a. Similarly, as shown in Figure 8.8b, the increase in length of line element OB due to the deformation $\varepsilon_y dy$ is $\varepsilon_y dy \sin\theta$. Finally, as shown in Figure 8.8c, the rotation of this line element due to the shearing strain γ_{xy} increases the length of line element OB by $\gamma_{xy} dx \sin\theta$. Adding together these three deformations, we obtain

$$\delta_n = \varepsilon_x dx \cos\theta + \varepsilon_y dy \sin\theta + \gamma_{xy} dx \sin\theta \tag{8.4}$$

If the initial length of line element OB is $d\ell$, then the strain ε_n of this is

$$\varepsilon_n = \frac{\delta_n}{d\ell} = \varepsilon_x\left(\frac{dx}{d\ell}\right)\cos\theta + \varepsilon_y\left(\frac{dy}{d\ell}\right)\sin\theta + \gamma_{xy}\left(\frac{dx}{d\ell}\right)\sin\theta \tag{8.5}$$

Since $dx/d\ell = \cos\theta$ and $dy/d\ell = \sin\theta$, we conclude that

$$\varepsilon_n = \varepsilon_x \cos^2\theta + \varepsilon_y \sin^2\theta + \gamma_{xy} \sin\theta\cos\theta \tag{8.6}$$

Using the trigonometric identities $\sin^2\theta = (1-\cos 2\theta)/2$, $\cos^2\theta = (1+\cos 2\theta)/2$ and $2\sin\theta\cos\theta = \sin 2\theta$, Equation 8.6 becomes

$$\varepsilon_n = \frac{1}{2}(\varepsilon_x + \varepsilon_y) + \frac{1}{2}(\varepsilon_x - \varepsilon_y)\cos 2\theta + \frac{1}{2}\gamma_{xy}\sin 2\theta \tag{8.7}$$

We will now develop the relation between γ_{nt} and the quantities ε_x, ε_y, and γ_{xy}. This development is made easier if we decompose the resultant distortion of Figure 8.7 into three separate distortions as shown in Figure 8.9. In Figure 8.9a, the deformation in the x direction, $\varepsilon_x(dx_1 + dx_2)$, is seen to change the 90° between lines OB and OD by the small angles BOB' and DOD'. Since both of these angles represent an increase in the 90° angle BOD, they are negative quantities according to our sign convention. Thus, if the initial length of line OB is $d\ell_1$ and that of line OD is $d\ell_2$, then assuming small deformations, the change in angle BOD ($\Delta \angle BOD$) due to the x deformation becomes

$$\Delta \angle BOD = -\frac{BG}{OB} - \frac{DE}{OD} = -\varepsilon_x\left(\frac{dx_1}{d\ell_1}\sin\theta + \frac{dx_2}{d\ell_2}\cos\theta\right) = -2\varepsilon_x \sin\theta\cos\theta = -\varepsilon_x \sin 2\theta \tag{8.8}$$

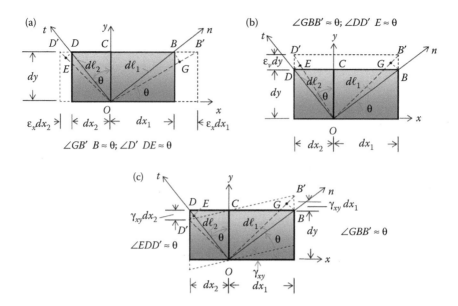

FIGURE 8.9 Decomposition of the resultant distortion into three separate distortions.

Similarly, Figure 8.9b shows that the deformation $\varepsilon_y dy$ changes the 90° angle between OB and OD by the small angles BOB' and DOD'. Since both of these angles represent decreases in the 90° between OB and OD, they are positive quantities according to our sign convention. Therefore, the change in angle BOD is

$$\Delta \angle BOD = \frac{BG}{OB} + \frac{DE}{OD} = \varepsilon_y \left(\frac{dy}{d\ell_1} \cos\theta + \frac{dy}{d\ell_2} \sin\theta \right) = 2\varepsilon_y \sin\theta\cos\theta = \varepsilon_y \sin 2\theta \qquad (8.9)$$

Finally, the change in the 90° angle between line OB and OD due to the shearing strain γ_{xy} is shown in Figure 8.9c. This change consists of the positive small angle BOB' and the negative small angle DOD'. Therefore, the change in angle BOD due to γ_{xy} becomes

$$\Delta \angle BOD = \frac{GB}{OB} - \frac{D'E}{OD} = \gamma_{xy} \left(\frac{dx_1}{d\ell_1} \cos\theta - \frac{dx_2}{d\ell_2} \sin\theta \right) = \gamma_{xy}(\cos^2\theta - \sin^2\theta) = \gamma_{xy} \cos 2\theta \qquad (8.10)$$

The resultant change in the 90° angle between line OB (n axis) and line OD (t axis) is the shearing strain γ_{nt} and is the algebraic sum of the three separate changes given by Equations 8.8, 8.9, and 8.10. Thus,

$$\gamma_{nt} = -\varepsilon_x \sin 2\theta + \varepsilon_y \sin 2\theta + \gamma_{xy} \cos 2\theta = -(\varepsilon_x - \varepsilon_y)\sin 2\theta + \gamma_{xy} \cos 2\theta \qquad (8.11)$$

If both sides of Equation 8.11 are divided by 2, we obtain

$$\frac{1}{2}\gamma_{nt} = -\frac{1}{2}(\varepsilon_x - \varepsilon_y)\sin 2\theta + \frac{1}{2}\gamma_{xy} \cos 2\theta \qquad \textbf{(8.12)}$$

The normal strain along the t axis may be obtained from Equation 8.7 by replacing the angle 2θ by the angle $(2\theta + \pi)$. Since $\sin(2\theta + \pi) = -\sin 2\theta$ and $\cos(2\theta + \pi) = -\cos 2\theta$, Equation 8.7 yields

$$\varepsilon_t = \frac{1}{2}(\varepsilon_x + \varepsilon_y) - \frac{1}{2}(\varepsilon_x - \varepsilon_y)\cos 2\theta - \frac{1}{2}\gamma_{xy}\sin 2\theta \tag{8.13}$$

A useful relation among normal strains is obtained if we add Equation 8.7 to Equation 8.13). This addition yields

$$\varepsilon_n + \varepsilon_t = \varepsilon_x + \varepsilon_y \tag{8.14}$$

We note that Equations 8.7, 8.12, 8.13, and 8.14 developed here for the plane-strain problem are mathematically similar, respectively, to Equations 7.6, 7.9, 7.10, and 7.116 developed in the preceding chapter for the plane-stress problem. Comparing these two sets of equations, we conclude that we may transform the plane-stress equations of Chapter 7 to the plane-strain equations of this chapter by making the following replacements:

$$\sigma_x \Rightarrow \varepsilon_x; \quad \sigma_y \Rightarrow \varepsilon_y; \quad \tau_{xy} \Rightarrow \frac{\gamma_{xy}}{2}; \quad \sigma_n \Rightarrow \varepsilon_n; \quad \sigma_t \Rightarrow \varepsilon_t; \quad \tau_{nt} \Rightarrow \frac{\gamma_{nt}}{2} \tag{8.15}$$

Therefore, all of the equations developed in Chapter 7 for *principal stresses, maximum in-plane shearing stresses*, as well as those for the *construction of Mohr's circle* may be transformed to corresponding strain equations as discussed in the following paragraphs.

8.3.1 PRINCIPAL STRAINS

The term *principal strains* is used to signify the maximum and minimum normal strains in a given plane-strain condition. The corresponding axes are called the *principal strain axes*. For an ideal material (homogeneous and isotropic), the principal strain axes coincide with the principal stress axes. As in the case of plane stress, we identify three principal strains, even in the two-dimensional case, and label them such that, algebraically, $\varepsilon_1 \geq \varepsilon_2 \geq \varepsilon_3$, where ε_1 is the largest principal stress, ε_3 is the least, and ε_2 is the intermediate principal strain. Of course, in the case of plane strain, one of the three principal strains is zero. Identification of the zero principal strain is not, in general, possible until all three principal strains are found. The angles defining the orientation of the principal axes with respect to the x and y axes are obtained from Equation 7.14 using the information in Equation 8.15. Thus,

$$2\theta_\varepsilon = \tan^{-1}\left[\frac{\gamma_{xy}}{\varepsilon_x - \varepsilon_y}\right] \tag{8.16}$$

The subscript ε is attached to the angle θ in Equation 8.16 to emphasize the fact that Equation 8.16 yields the orientation of the axes along which the normal strain ε is either a maximum or a minimum. Note that the angle is measured from the x axis. Also, using the information in Equation 8.15, we transform Equations 7.16 and 7.17 to obtain

$$\varepsilon_{MAX,MIN} = \frac{1}{2}(\varepsilon_x + \varepsilon_y) \pm R \tag{8.17}$$

The quantity R in Equation 8.17 is given by

$$R = \sqrt{\left(\frac{\varepsilon_x - \varepsilon_y}{2}\right)^2 + \left(\frac{\gamma_{xy}}{2}\right)^2} \qquad (8.18)$$

Note that Equation 8.18 was obtained by transforming Equation 7.15 using the information given in Equation 8.15. Note also that the shearing strain corresponding to the two in-plane principal axes is zero. This fact may be confirmed by retracing the steps followed in the case of principal stresses in Chapter 7 to show that the shearing stress is zero on principal planes. Thus, in order to show that the shearing strain corresponding to the two in-plane principal axes is zero, we substitute the values of sin 2θ and cos 2θ obtained from Equation 8.16 into Equation 8.12.

8.3.2 MAXIMUM IN-PLANE SHEARING STRAIN

The equation defining the orientation of the axes corresponding to the maximum in-plane shearing strain is obtained by transforming Equation 7.22. This leads to

$$2\theta_\gamma = \tan^{-1}\left[\frac{-(\varepsilon_x - \varepsilon_y)}{\gamma_{xy}}\right] \qquad (8.19)$$

The symbol θ_γ in Equation 8.19 is used to signify the angles defining the orientation of the axes corresponding to the maximum in-plane shearing strain. The equation providing the maximum in-plane shearing strain is obtained by transforming Equation 7.20. Thus,

$$|\lambda_{MAX}| = 2R = 2\sqrt{\left(\frac{\varepsilon_x - \varepsilon_y}{2}\right)^2 + \left(\frac{\gamma_{xy}}{2}\right)^2} \qquad (8.20)$$

EXAMPLE 8.1

The plane-strain state at a point in a body is defined as follows: $\varepsilon_x = -800 \times 10^{-6}$, $\varepsilon_y = -200 \times 10^{-6}$, and $\gamma_{xy} = -450 \times 10^{-6}$. Determine (a) the two in-plane principal strains and the orientation of the principal axes relative to the x axis and (b) the maximum in-plane shearing strain and the orientation of the corresponding axes.

SOLUTION

a. Using Equation 8.18, we find the quantity R. Thus,

$$R = \sqrt{\left(\frac{-800 - (-200)}{2}\right)^2 + \left(\frac{-450}{2}\right)^2} \times 10^{-6} = 375 \times 10^{-6}$$

Therefore, by Equation 8.17, we obtain

$$\varepsilon_{MAX,MIN} = \frac{1}{2}(-800 - 200) \times 10^{-6} \pm 375 \times 10^{-6} = -125 \times 10^{-6}, -875 \times 10^{-6}$$

Therefore, for this plane-strain condition, and using the requirement that, algebraically, $\varepsilon_1 \geq \varepsilon_2 \geq \varepsilon_3$, we have

$$\varepsilon_2 = -125 \times 10^{-6} \quad \varepsilon_3 = -875 \times 10^{-6} \quad \textbf{ANS.}$$

Since this is a plane-strain system, it follows that $\varepsilon_1 = 0$. Also, using Equation 8.16, we obtain

$$2\theta_\varepsilon = \tan^{-1}\left[\frac{-450}{-800 - (-200)}\right] = 36.870°; \quad \theta_\varepsilon = 18.435° \approx 18.4° \quad \textbf{ANS.}$$

Since this angle is positive, the orientation of the ε_3 axis is 18.4° counterclockwise from the x axis. We may confirm the fact that this angle defines the orientation of ε_3 and not that of ε_2 by substituting 36.870° for 2θ in Equation 8.7. This substitution would result in a value for $\varepsilon_n = -875 \times 10^{-6}$, which is the value of the principal strain labeled ε_3. Obviously, the orientation of ε_2 is 90° counterclockwise from the ε_3 axis. These geometric relations are shown in figure (a) below. Note that the direction of ε_1 is into the page and perpendicular to it.

b. The orientation of the axes corresponding to the maximum in-plane shearing strain, relative to the x axis, is given by Equation 8.19. Thus,

$$2\theta_\tau = \tan^{-1}\left[\frac{-(-800 - (-200))}{-450}\right] = -53.130°; \quad \theta_\tau = -26.565° \approx -26.6° \quad \textbf{ANS.}$$

Therefore, one of the two axes that experience the largest in-plane shearing strain is $-26.6°$ (clockwise) from the x axis. The second axis, of course, is perpendicular to the first. These two axes are shown in red in figure (a) below and given the labels f and g. Note that the sum of the angles θ_ε (18.4°) and θ_τ (26.6°) is 45.0° as it should be.

Substituting $2\theta_\tau = -53.130°$ into Equation 8.12, we have

$$\frac{1}{2}\gamma_{nt} = \frac{1}{2}(-800 - (-200)) \times 10^{-6} \sin(-53.130°)$$

$$= +(-450) \times 10^{-6} \cos(-53.130°) = -375.0 \times 10^{-6}$$

Thus, the maximum in-plane shearing strain becomes

$$\gamma_{MAX} = 2(-375.0 \times 10^{-6}) = -750.0 \times 10^{-6} \quad \textbf{ANS.}$$

Note that this same magnitude could be obtained from Equation 8.20 but not the sign.

Since the maximum in-plane shearing strain is negative, a square element at the intersection of lines f and g experiences an increase in the original 90° angle as shown, *in two equal halves*, by lines f' and g' in figure (b) below. Of course, the square element becomes a parallelogram as indicated, in an exaggerated form, by the dashed lines in figure (b) below. The sketch of figure (b) below is limited to the shearing strain γ_{MAX} and does *not* show any normal strains.

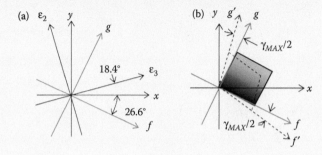

EXAMPLE 8.2

The state of plane strain at a point in a stressed body is defined as follows: $\varepsilon_x = -400 \times 10^{-6}$, $\varepsilon_y = 300 \times 10^{-6}$, and $\gamma_{xy} = 250 \times 10^{-6}$. Determine the strains for an element at the intersection of orthogonal n and t axes, where the n axis is $40°$ clockwise from the x axis.

SOLUTION

The normal strain along the n axis is given by Equation 8.7. Thus,

$$\varepsilon_n = \frac{1}{2}(-400 + 300) \times 10^{-6} + \frac{1}{2}(-400 - 300) \times 10^{-6} \cos 2(-40°)$$

$$+ \frac{1}{2}(250) \times 10^{-6} \sin 2(-40°)$$

$$= -233.878 \times 10^{-6} \approx -234.0 \times 10^{-6} \quad \textbf{ANS.}$$

The normal strain along the t axis is given by Equation 8.13. Thus,

$$\varepsilon_t = \frac{1}{2}(-400 + 300) \times 10^{-6} - \frac{1}{2}(-400 - 300) \times 10^{-6} \cos 2(-40°)$$

$$- \frac{1}{2}(250) \times 10^{-6} \sin 2(-40°)$$

$$= 133.878 \times 10^{-6} \approx 134.0 \times 10^{-6} \quad \textbf{ANS.}$$

The shearing strain experienced by the $90°$ angle between the n and t axes is found by using Equation 8.12, which after multiplication by 2 yields

$$\gamma_{nt} = -(-400 - 300) \times 10^{-6} \sin 2(-40°) + 250 \times 10^{-6} \cos 2(-40°)$$

$$= -645.953 \times 10^{-6} \approx -646.0 \times 10^{-6} \quad \textbf{ANS.}$$

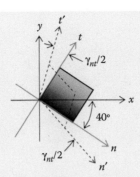

Since γ_{nt} is negative, it represents an increase in the $90°$ angle between the n and t axes. This shearing strain is shown as *two equal halves* and highly exaggerated in the sketch. Note that the sketch is limited to the shearing strain γ_{nt} and does *not* show any normal strains.

EXAMPLE 8.3

Refer to Example 8.2 and using Hooke's law in two dimensions developed in Chapter 2, find the normal stresses σ_n and σ_t on the N and T planes, respectively (i.e., along the n and t axes, respectively) and the shearing stress τ_{nt}. Let $E = 30 \times 10^6$ psi and $\mu = 0.3$. Show these stresses on a properly oriented stress element.

SOLUTION

From Example 8.2, we have $\varepsilon_n = -234.0 \times 10^{-6}$, $\varepsilon_t = 134.0 \times 10^{-6}$, and $\gamma_{nt} = -646.0 \times 10^{-6}$.

While Hooke's law in two dimensions is expressed in terms of x and y orthogonal axes, it can be used for any set of orthogonal axes such as n and t. Using Hooke's law in two dimensions expressed in Equations 2.23, we have

$$\sigma_n = \left(\frac{30 \times 10^6}{1 - 0.3^2} \right)(-234.0 + (0.3)(134.0)) \times 10^{-6} = -6389.011 \approx -6389 \text{ psi} \qquad \textbf{ANS.}$$

$$\sigma_t = \left(\frac{30 \times 10^6}{1 - 0.3^2} \right)(134.0 + (0.3)(-234.0)) \times 10^{-6} = 2103.297 \approx 2103 \text{ psi} \qquad \textbf{ANS.}$$

Using Equation 2.24, we find the value of the modulus of rigidity G. Thus,

$$G = \frac{30 \times 10^6}{2(1 + 0.3)} = 11.538 \times 10^6 \text{ psi}$$

Hooke's law in shear enables us to find the shearing stress τ_{nt}. Thus,

$$\tau_{nt} = (11.538 \times 10^6)(-646.0 \times 10^{-6}) = -7453.548 \approx -7454 \text{ psi} \qquad \textbf{ANS.}$$

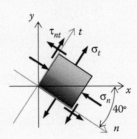

A properly oriented stress element is shown in the sketch above. *An alternate way to view Examples 8.2 and 8.3 is to say that the plane-stress state of Example 8.3 leads to the plane-strain state of Example 8.2.*

PROBLEMS

8.1 The state of plane strain at a point in a stressed body is defined by specifying the quantities ε_x, ε_y, and γ_{xy}. Refer to the companion sketch and determine ε_n, ε_t, and γ_{nt} for the plane-strain case:

$$\varepsilon_x = 300 \times 10^{-6}, \quad \varepsilon_y = -150 \times 10^{-6}, \quad \gamma_{xy} = 0, \quad \text{and} \quad \theta = 45°$$

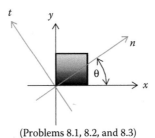

(Problems 8.1, 8.2, and 8.3)

8.2 In reference to the figure in Problem 8.1, the state of plane strain at a point in a stressed body is defined by specifying the quantities ε_x, ε_y, and γ_{xy}. Determine ε_n, ε_t, and γ_{nt} for the plane-strain case:

$$\varepsilon_x = 0, \quad \varepsilon_y = 0, \quad \gamma_{xy} = -400 \times 10^{-6}, \quad \text{and} \quad \theta = 30°$$

8.3 In reference to the figure in Problem 8.1, the state of plane strain at a point in a stressed body is defined by specifying the quantities ε_x, ε_y, and γ_{xy}. Determine ε_n, ε_t, and γ_{nt} for the plane-strain case:

$$\varepsilon_x = -300 \times 10^{-6}, \quad \varepsilon_y = 200 \times 10^{-6}, \quad \gamma_{xy} = 400 \times 10^{-6}, \quad \text{and} \quad \theta = -30°$$

8.4 Refer to Problem 8.1 and using Hooke's law in two dimensions of Chapter 2, determine the stresses σ_n, σ_t, and τ_{nt}. Let $E = 10.5 \times 10^3$ ksi and $\mu = 0.33$. Show these stresses on a properly oriented stress element.

8.5 Refer to Problem 8.3 and using Hooke's law in two dimensions of Chapter 2, determine the stresses σ_n, σ_t, and τ_{nt}. Let $E = 30 \times 10^3$ ksi and $\mu = 0.3$. Show these stresses on a properly oriented stress element.

8.6 The state of plane strain at a point in a stressed body is defined by specifying the quantities ε_x, ε_y, and γ_{xy}. Refer to the companion sketch and determine the principal strains and the orientation of the corresponding axes relative to the x axis for the following plane-strain case:

$$\varepsilon_x = 250 \times 10^{-6}, \quad \varepsilon_y = 150 \times 10^{-6}, \quad \text{and} \quad \gamma_{xy} = 300 \times 10^{-6}$$

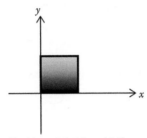

(Problems 8.6, 8.7, and 8.8)

8.7 The state of plane strain at a point in a stressed body is defined by specifying the quantities ε_x, ε_y, and γ_{xy}. Refer to the sketch in Problem 8.6 and determine the principal strains and the orientation of the corresponding axes relative to the x axis for the following plane-strain case:

$$\varepsilon_x = -200 \times 10^{-6}, \quad \varepsilon_y = 200 \times 10^{-6}, \quad \text{and} \quad \gamma_{xy} = 100 \times 10^{-6}$$

8.8 The state of plane strain at a point in a stressed body is defined by specifying the quantities ε_x, ε_y, and γ_{xy}. Refer to the sketch in Problem 8.6 and determine the principal strains and the orientation of the corresponding axes relative to the x axis for the following plane-strain case:

$$\varepsilon_x = 300 \times 10^{-6}, \quad \varepsilon_y = -150 \times 10^{-6}, \quad \text{and} \quad \gamma_{xy} = -400 \times 10^{-6}$$

8.9 Refer to Problem 8.7 and using Hooke's law in two dimensions of Chapter 2, determine the two in-plane principal stresses. Let $E = 29 \times 10^3$ ksi and $\mu = 0.31$. Show these stresses on a properly oriented stress element.

8.10 Refer to Problem 8.8 and using Hooke's law in two dimensions of Chapter 2, determine the two in-plane principal stresses. Let $E = 200$ GPa and $\mu = 0.33$. Show these stresses on a properly oriented stress element.

8.11 A thin rectangular plate lies in the x–y plane as shown. Before the plate is stressed, its dimensions were $b = 6$ in. and $h = 3$ in. After stressing, the dimension b increased to 6.0006 in. and the dimension h decreased to 2.9994 in. Assume that after stressing the corners of the plate are still perfectly square and that the plate experiences plane strain. Find the strains ε_n, ε_t, and γ_{nt}. What is the change in length of diagonal OA?

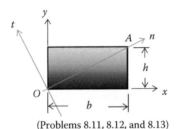

(Problems 8.11, 8.12, and 8.13)

8.12 In reference to the figure in Problem 8.11, a thin rectangular plate lies in the x–y plane as shown. Before the plate is stressed, its dimensions were $b = 700$ mm and $h = 300$ mm. Measurements of strains resulted in the following: $\varepsilon_x = -300 \times 10^{-6}$ and $\varepsilon_y = 200 \times 10^{-6}$. Assume that after stressing, the corners of the plate are still perfectly square and that the plate experiences plane strain. Find the strains ε_n, ε_t, and γ_{nt}. What is the change in length of diagonal OA?

8.13 In reference to the figure in Problem 8.11, a thin rectangular plate lies in the x–y plane as shown. Before the plate is stressed, its dimensions were $b = 20$ in. and $h = 8$ in. Measurements of strains resulted in the following: $\varepsilon_x = 300 \times 10^{-6}$, $\varepsilon_y = 150 \times 10^{-6}$, and $\gamma_{xy} = -200 \times 10^{-6}$. Assume that the plate experiences plane strain. Find the strains ε_n, ε_t, and γ_{nt} and the corresponding stresses σ_n, σ_t, and τ_{nt}. Show these stresses on a properly oriented stress element. Let $E = 30 \times 10^3$ ksi and $\mu = 0.3$.

8.14 The state of plane stress (*not* plane strain) at a point in a stressed body is shown in the sketch. Determine the strains ε_n, ε_t, ε_z, and γ_{nt} for of the following case:

$$\sigma_x = 150\,\text{MPa}, \quad \sigma_y = 200\,\text{MPa}, \quad \text{and} \quad \tau_{xy} = 100\,\text{MPa}$$

Let $E = 200$ GPa, $\mu = 0.3$, and $\theta = 30°$.

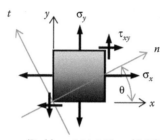

(Problems 8.14, 8.15, and 8.16)

8.15 In reference to the figure in Problem 8.14, the state of plane stress (*not* plane strain) at a point in a stressed body is shown in the sketch in Problem 8.14. Determine the strains ε_n, ε_t, ε_z, and γ_{nt} for the following case:

$$\sigma_x = -200\,\text{MPa}, \quad \sigma_y = 100\,\text{MPa}, \quad \text{and} \quad \tau_{xy} = -150\,\text{MPa}$$

Let $E = 200$ GPa, $\mu = 0.3$, and $\theta = -30°$.

8.16 In reference to the figure in Problem 8.14, the state of plane stress (*not* plane strain) at a point in a stressed body is shown in the sketch in Problem 8.14. Determine the strains ε_n, ε_t, ε_z, and γ_{nt} for of the following case:

$$\sigma_x = 15\,\text{ksi}, \quad \sigma_y = -20\,\text{ksi}, \quad \text{and} \quad \tau_{xy} = 10\,\text{ksi}$$

Let $E = 10 \times 10^3$ ksi, $\mu = 0.33$, and $\theta = 45°$.

8.4 MOHR'S CIRCLE FOR PLANE STRAIN

In Section 8.3, we discovered the many similarities between the plane-stress and plane-strain transformation equations that led us to conclude that we can create one set of equations from the other set by making use of the changes expressed in Equation 8.15. Because of these similarities, Mohr's circle construction, which was developed in Chapter 7 for the solution of the plane-stress problem, is equally applicable for the solution of the plane-strain problem. Only minor modifications as implied in Equation 8.15 become necessary.

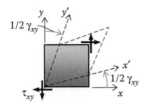

FIGURE 8.10   Sign convention for normal and shear strains for Mohr's circle.

The construction of the plane-strain Mohr's circle is similar to that of the plane-stress Mohr's circle. A coordinate system is established where the horizontal axis is labeled ε_n and the vertical axis $1/2\gamma_{nt}$. Point X is located by plotting the coordinates (ε_x, $-1/2\gamma_{xy}$) and point Y by plotting the coordinates (ε_y, $1/2\gamma_{xy}$). These two points are diametrically opposite on the circle and a line connecting them is a diameter of Mohr's circle. However, it is more meaningful and easier to interpret the plane-strain Mohr's circle if we construct it using a *special sign convention* for shearing strains similar to what was done in the case of the plane-stress Mohr's circle. This special sign convention requires that a clockwise rotation of a line element be plotted in the positive shearing strain region (above the ε_n axis) and a counterclockwise rotation be plotted in the negative region (below the ε_n axis). Such a special shearing strain sign convention is compatible with the special shearing stress sign convention as may be confirmed by examining Figure 8.10. In this figure, the stress element is subjected to a set of positive shearing stresses τ_{xy} resulting in the positive shearing strain γ_{xy}. Thus, a positive shearing strain, which represents a decrease in the 90° angel between the x and y axes, causes the x axis to rotate counterclockwise through $1/2\gamma_{xy}$ to position x' and, therefore, point X, in the plane-strain Mohr's circle, is a negative quantity. At the same time, a positive shearing strain causes the y axis to rotate clockwise through $1/2\gamma_{xy}$ to position y' and, therefore, point Y, in the plane-strain Mohr's circle, is a positive quantity. On the other hand, a negative shearing strain, which represents an increase in the 90° angle between the x and y axes, reverses the signs of points X and Y.

The circle shown in Figure 8.11 illustrates some of the concepts discussed above. This Mohr's circle was constructed for the case where the strains ε_x, ε_y, and γ_{xy} are all positive quantities. Interpretation and analysis of the plane-strain circle is performed in the same manner as was done for the case of plane stress.

8.4.1   PRINCIPAL STRAINS AND MAXIMUM IN-PLANE SHEARING STRAIN

In Figure 8.11, point P_1 represents the maximum normal strain, ε_1, and point P_2 the intermediate normal strain, ε_2, because for this particular case, ε_3 is zero. Note that ε_1 and ε_2 are the in-plane principal strains

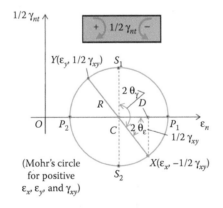

FIGURE 8.11   Mohr's circle for plane strain.

while ε_3 is the out-of-plane principal strain. Also, points S_1 and S_2 represent the maximum (positive) and minimum (negative) in-plane shearing strains. The magnitudes of these quantities are given by

$$\varepsilon_{MAX,MIN} = OC \pm R \tag{8.21}$$

$$1/2\gamma_{MAX,MIN} = \pm R \tag{8.22}$$

By examining the geometry of the circle in Figure 8.11, we conclude that

$$OC = \varepsilon_{AVE} = \frac{1}{2}(\varepsilon_x + \varepsilon_y) \tag{8.23}$$

$$R = \sqrt{\left(\frac{\varepsilon_x - \varepsilon_y}{2}\right)^2 + \left(\frac{\gamma_{xy}}{2}\right)^2} \tag{8.24}$$

The orientation of the maximum principal strain axis (represented by point P_1), measured from the x axis, is given by

$$2\theta_\varepsilon = \tan^{-1}\left[\frac{\gamma_{xy}}{2(\varepsilon_x - \varepsilon_y)}\right] \tag{8.25}$$

It is evident from Figure 8.11 that the orientation of the maximum shearing-strain axis, measured from the x axis, (represented by point S_1), is given by

$$2\theta_\gamma = 2\theta_\varepsilon + 90° \tag{8.26}$$

Equation 8.26 shows that the angle $2\theta_\gamma$ is 90° away from the angle $2\theta\varepsilon$ in a counterclockwise sense. In other words, in the element, the axis of maximum shearing strain, $1/2\gamma_{MAX}$, that experiences clockwise rotation is +45° from the principal axis of ε_1. Also, the orientation of the plane of $1/2\gamma_{MIN}$, corresponding to point S_2 (on the circle), is 90° from the principal axis of ε_1 in a clockwise sense. This, of course, means that in the actual element, the axis of $1/2\gamma_{MAX}$ that experiences counterclockwise rotation of the element is −45° from the axis of ε_1. Thus, the two axes of maximum and minimum shearing strains, $1/2\gamma_{MAX}$ and $1/2\gamma_{MIN}$, respectively, are perpendicular to each other. We should note here that an element contained between these two axis experiences not only the maximum shearing strain but also normal strains equal in magnitude to $OC = \varepsilon_{AVE} = 1/2(\varepsilon_x + \varepsilon_y)$.

8.4.2 INCLINED AXES

Mohr's circle can also be used to determine the strains relative to axes inclined at different angles to the x and y axes. Consider, for example, the circle constructed in Figure 8.12b based upon given positive values of ε_x, ε_y, and γ_{xy}, where the x–y coordinate system is indicated in Figure 8.12a. Let us assume that we needed to find the normal and shearing strains corresponding to the n–t coordinate system, which is inclined to the x–y system by the counterclockwise angle θ.

In order to find the strains corresponding to the n–t coordinate system, we locate point N on the circle, as shown in Figure 8.12b, by rotating from point X in a counterclockwise direction through the angle 2θ. The diametrically opposite point, T, represents the strains along the t axis.

The coordinates of point N provide the values of ε_n and $1/2\gamma_{nt}$. To find these coordinate, we need to find the angle 2α from the relation

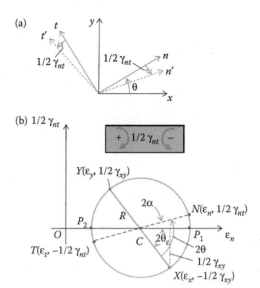

FIGURE 8.12 Strain transformation using Mohr's circle.

$$2\alpha = 2\theta - 2\theta_\varepsilon \tag{8.27}$$

From the geometry of the circle, Figure 8.11, the angle $2\theta\varepsilon$ is found from the relation

$$2\theta_\varepsilon = \tan^{-1}\left|\frac{1/2\gamma_{xy}}{1/2(\varepsilon_x - \varepsilon_y)}\right| = \tan^{-1}\left|\frac{\gamma_{xy}}{\varepsilon_x - \varepsilon_y}\right| \tag{8.28}$$

Note that only the magnitude of $2\theta_\varepsilon$ is needed for use in Equation 8.27. For the position of point N on the circle, both ε_n and $1/2\gamma_{nt}$ are positive. This, of course, means that ε_n represents an extension and that the shearing strain is such that the n axis rotates clockwise through $1/2\gamma_{nt}$ as shown in Figure 8.12a. For point T, on the other hand, ε_t is positive but $1/2\gamma_{nt}$ is negative. Therefore, ε_t represents an extension but the shearing strain is such that the t axis rotates counterclockwise through the angle $1/2\gamma_{nt}$ as shown in Figure 8.12a. Thus, we conclude that the shearing strain γ_{nt} is negative, according to the sign convention for shearing strains, because it represents an increase in the 90° angle between the n and t axes.

Having found the angle 2α, the magnitudes of ε_n, ε_t, and $1/2\gamma_{xy}$ are determined by computing the coordinates of points N and T. Thus,

$$\varepsilon_n = OC + R\cos 2\alpha; \quad \varepsilon_t = OC - R\cos 2\alpha; \quad \frac{1}{2}|\gamma_{nt}| = R\sin 2\alpha \tag{8.29}$$

The following examples illustrate some of the concepts discussed above.

EXAMPLE 8.4

The plane-strain condition at a point in a stressed body is defined as follows: $\varepsilon_x = 200 \times 10^{-6}$, $\varepsilon_y = -100 \times 10^{-6}$, and $\gamma_{xy} = -150 \times 10^{-6}$. Use Mohr's circle to determine: (a) The two in-plane principal strains. What is the third principal strain? (b) The maximum in-plane shearing strain. Show these strains on properly oriented elements.

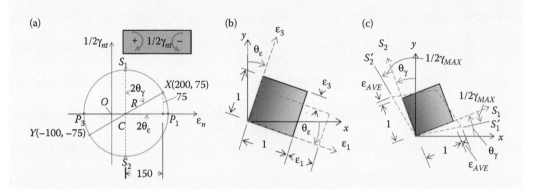

SOLUTION

Mohr's circle for the given plane-strain condition is constructed as shown in figure (a) above. Note that since γ_{xy} is negative, the 90° angle between the x and y axes increases. Therefore, the x axis rotates clockwise to its new position while the y axis rotates counterclockwise. Thus, point X on the circle is above the ε_n axis and point Y is below this axis. Note also that the common multiplier, 10^{-6}, has been ignored in the construction. The radius of the circle and the location of its center are computed, by analyzing its geometry, as follows:

$$OC = \frac{200 - 100}{2} = 50 \times 10^{-6}$$

$$R = \sqrt{150^2 + 75^2} = 167.705 \approx 167.7 \times 10^{-6}$$

a. The two in-plane principal strains are represented by points P_1 and P_3. Thus, using the circle geometry, we have

$$\varepsilon_1 = \varepsilon_{MAX} = OC + R = 217.7 \times 10^{-6}; \quad \varepsilon_3 = \varepsilon_{MIN} = OC - R = -117.7 \times 10^{-6} \text{ ANS.}$$

    Note that for this particular case, the out-of-plane principal strain is zero (i.e., $\varepsilon_2 = 0$).   **ANS.**

    The orientation of the ε_1 and ε_3 axes relative to the x axis is given by the clockwise angle $2\theta_\varepsilon$, which, from the circle, is

$$2\theta_\varepsilon = \tan^{-1}\left(\frac{75}{150}\right) = 26.565°; \quad \theta_\varepsilon = 13.283° \quad \textbf{ANS.}$$

    Figure (b) above shows a square element, of unit length on each side, located at the intersection of ε_1 and ε_3 axes. This element does *not* experience any shearing strain but undergoes the largest strain (ε_1) and the least strain (ε_3) as indicated by the dashed rectangular element shown in red.

b. The maximum in-plane shearing strain is represented by points S_1 and S_2.

    As observed from the circle, point S_1 is located at an angle $2\theta_\gamma$, counterclockwise from the X point, and S_2 is diametrically opposite to point S_1. The angle $2\theta_\gamma$ is found from the relation

$$2\theta_\gamma = 90° - 2\theta_\varepsilon = 90° - 26.565° = 63.435°; \quad \theta_\gamma = 31.717° \quad \textbf{ANS.}$$

The maximum in-plane shearing strain is equal in magnitude to $2R$ as may be seen from the circle. Thus,

$$|\gamma_{MAX}| = 2R = 2(167.7 \times 10^{-6}) = 335.4 \times 10^{-6} \quad \textbf{ANS.}$$

Note that both S_1 and S_2 have abscissas that are equal in magnitude to the dimension $OC = \varepsilon_{AVE}$ and, therefore, a square element, of unit length on each side, located at the intersection of axes S_1 and S_2, experiences not only the maximum shearing strain, γ_{MAX}, but also normal strains along both the S_1 and S_2 axes. These strains are shown, highly magnified, in figure (c) above, where the shearing strain is split equally between axes S_1 and S_2. Since point S_1 is positive (i.e., above the ε_n axis), it rotates clockwise to position S_1'. Also, since S_2 is negative (i.e., below the ε_n axis), it rotates counterclockwise to position S_2' as shown. Therefore, since the 90° angle between S_1 and S_2 increased, γ_{MAX} is a negative quantity. The final configuration of the square element is depicted, schematically, by the parallelogram shown in red dashed lines.

EXAMPLE 8.5

The state of plane strain at a point in a stressed body is defined as follows: $\varepsilon_x = -150 \times 10^{-6}$, $\varepsilon_y = 300 \times 10^{-6}$, and $\gamma_{xy} = 300 \times 10^{-6}$. Use Mohr's circle to determine the normal and shearing strains associated with a set of rectangular n and t axes located at 30° clockwise from the x and y axes as shown in figure (a) below. Show these strains on a properly oriented square element.

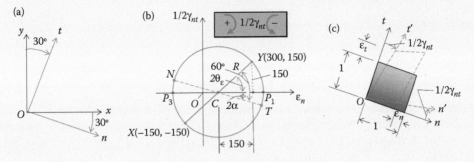

SOLUTION

Mohr's circle is constructed as shown in figure (b) above, ignoring the common factor 10^{-6}. Note that since γ_{xy} is positive, the x axis rotates counterclockwise (point X lies below the ε_n axis) while the y axis rotates clockwise (point Y lies above ε_n axis) as shown. Point N is located at $2 \times 30° = 60°$ clockwise from point X and point T, of course, is diametrically opposite to point N. The location of the center and the magnitude of the radius are

$$OC = \frac{-150 + 300}{2} = 75 \times 10^{-6}$$

$$R = \sqrt{105^2 + 150^2} = 212.123 \times 10^{-6}$$

The angle $2\theta_\varepsilon$ is found using the geometry of the circle. Thus,

$$2\theta_\varepsilon = \tan^{-1}\left(\frac{150}{150}\right) = 45.0°; \quad \theta_\varepsilon = 22.5°$$

Therefore, the angle 2α becomes

$$2\alpha = 60.0° - 45.0° = 15.0°; \quad \alpha = 7.5°$$

The coordinates of points N and T are:

$$\varepsilon_n = OC - R\cos15.0°$$
$$= -129.895 \approx -129.9 \times 10^{-6} \qquad \textbf{ANS.}$$

$$\varepsilon_t = OC + R\cos15.0°$$
$$= 279.895 \approx 279.9 \times 10^{-6} \qquad \textbf{ANS.}$$

$$\left|\frac{1}{2}\gamma_{nt}\right| = -R\sin15.0° = -54.901 \approx -54.9 \times 10^{-6}$$

$$\gamma_{nt} = -109.8 \times 10^{-6} \qquad \textbf{ANS.}$$

All of these strains are shown, magnified, in figure (c) above. The element at the intersection of the n and t axes becomes a parallelogram as indicated by the broken red lines.

EXAMPLE 8.6

A plane-stress state (*not* a plane-strain condition) is shown in figure (a) below. Use Mohr's circle to compute the two in-plane principal strains and their orientations relative to the x axis. Also find the third principal strain. Let $E = 30 \times 10^3$ ksi and $\mu = 0.3$. Show the two in-plane principal strains on a properly oriented element.

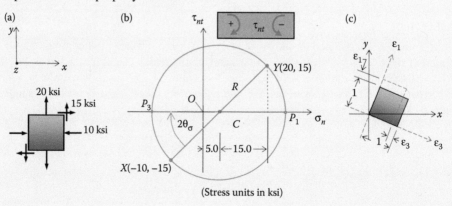

(Stress units in ksi)

SOLUTION

There are two possible ways to approach the solution of this problem. The first is to determine, using Hooke's law in two dimensions developed in Chapter 2, the strain ε_x, ε_y, and γ_{xy}, construct Mohr's circle for strain, and compute the principal strains. The second is to construct Mohr's circle for stress, determine the principal stresses, and, using Hooke's law in two dimensions, compute the principal strains. This is the approach that will be used for the solution of this example. Mohr's circle for stress is constructed as shown in figure (b) above. The location of the center and its radius, as well as the principal stresses, are computed as follows:

$$OC = \frac{-10 + 20}{2} = 5.0 \text{ ksi}; \quad R = \sqrt{15.0^2 + 15.0^2} = 21.213 \text{ ksi}$$

$$\sigma_1 = OC + R = 26.213 \text{ ksi}; \quad \sigma_3 = OC - R = -16.213 \text{ ksi}$$

Since a plane-stress state is assumed, $\sigma_2 = 0$. However, since this is *not* a plane-strain state, $\varepsilon_2 \neq 0$.

We now transform the above principal stresses into principal strains using Equations 2.22 after replacing ε_x by ε_1 and ε_y by ε_3. Thus,

$$\varepsilon_1 = \frac{1}{E}(\sigma_1 - \mu\sigma_{3)}) = \frac{1}{30 \times 10^3}(26.213 - 0.3(-16.213)) = 1036 \times 10^{-6} \qquad \textbf{ANS.}$$

$$\varepsilon_3 = \frac{1}{E}(\sigma_3 - \mu\sigma_1) = \frac{1}{30 \times 10^3}(-16.213 - 0.3(26.213)) = -803 \times 10^{-6} \qquad \textbf{ANS.}$$

$$\varepsilon_2 = -\frac{\mu}{E}(\sigma_1 + \sigma_3) = -\frac{0.3}{30 \times 10^3}(26.213 - 16.213) = -100 \times 10^{-6} \qquad \textbf{ANS.}$$

$$2\theta_\sigma = \tan^{-1}\left[\frac{15.0}{15.0}\right] = 45.0°; \quad \theta_\sigma = 22.5° \qquad \textbf{ANS.}$$

A square element at the intersection of the ε_1 and ε_3 axes does *not* experience shearing strains but becomes the red dashed rectangle shown, highly magnified, in figure (c) above.

8.4.3   DEVELOPMENT OF THE RELATION $G = E/2(1 + \mu)$

If we assume an ideal material (i.e., homogeneous and isotropic), the relation $G = E/2(1 + \mu)$, originally stated, without proof, as Equation 2.24 in Chapter 2, may now be easily developed. To this end, we use information obtained from Mohr's circles for stress and strain for the plane-stress state of pure shear.

Consider the plane-stress element of pure shear shown in Figure 8.13a and the associated stress Mohr's circle shown in Figure 8.13b. The resultant plane-strain condition is shown in Figure 8.14a and the corresponding strain Mohr's circle in Figure 8.14b.

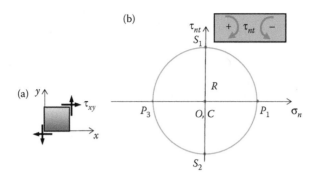

FIGURE 8.13 Plane-stress Mohr's circle for a pure shear stress element.

The principal stresses and principal strains are represented, respectively, by points P_1 and P_3 in Figures 8.13b and 8.14b. Thus,

$$\sigma_{1,3} = \pm R = \pm \tau_{xy} = \pm \frac{Tr}{J}; \quad \varepsilon_{1,3} = \pm R = \pm \frac{\gamma_{xy}}{2} = \pm \frac{\tau_{xy}}{2G} = \pm \frac{Tr}{2JG} \tag{8.30}$$

Using Hooke's law in two dimensions, expressed in Equations 2.23, and adapting the first of these equations to express it in terms of principal stresses and strains, we obtain

$$\sigma_1 = \frac{E}{1 - \mu^2}(\varepsilon_1 + \mu\varepsilon_3) \tag{8.31}$$

Substituting from Equation 8.30 for the values of σ_1, ε_1, and ε_3, Equation 8.31 becomes

$$\frac{Tr}{J} = \frac{E}{1 - \mu^2}\left[\frac{Tr}{2JG} + \mu\left(\frac{-Tr}{2JG}\right)\right] \tag{8.32}$$

Eliminating the common factor Tr/J, simplifying and solving for G, we obtain

$$G = \frac{E}{2(1 + \mu)} \tag{2.24}$$

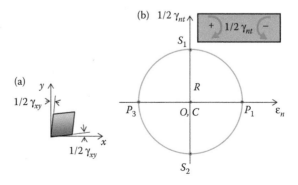

FIGURE 8.14 Plane-strain Mohr's circle for a pure shear stress element.

PROBLEMS

8.17 Solve Problem 8.1 using Mohr's circle.
8.18 Solve Problem 8.2 using Mohr's circle.
8.19 Solve Problem 8.3 using Mohr's circle.
8.20 Solve Problem 8.6 using Mohr's circle.
8.21 Solve Problem 8.7 using Mohr's circle.
8.22 Solve Problem 8.8 using Mohr's circle.
8.23 Solve Problem 8.11 using Mohr's circle.
8.24 Solve Problem 8.12 using Mohr's circle.
8.25 Solve Problem 8.13 using Mohr's circle.
8.26 Solve Problem 8.14 using Mohr's circle.
8.27 Solve Problem 8.15 using Mohr's circle.
8.28 Solve Problem 8.16 using Mohr's circle.
8.29 At a point on the outside surface of a member under the indicated loads, the resulting stresses are as shown on the companion stress element. For each of the following three cases, determine the two in-plane principal strains and the maximum in-plane shearing strain, stating their orientations relative to the x axis.

   The material for the tension member is such that $E = 30 \times 10^3$ ksi and $\mu = 0.3$.

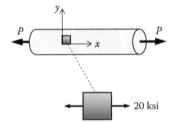

8.30 At a point on the outside surface of a member under the indicated loads, the resulting stresses are as shown on the companion stress element. For each of the following three cases, determine the two in-plane principal strains and the maximum in-plane shearing strain, stating their orientations relative to the x axis.

   The material for the shaft is such that $E = 70$ GPa and $\mu = 0.3$.

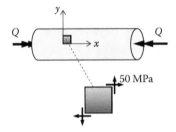

8.31 At a point on the outside surface of a member under the indicated loads, the resulting stresses are as shown on the companion stress element. For each of the following three cases, determine the two in-plane principal strains and the maximum in-plane shearing strain, stating their orientations relative to the x axis.

   The material for the member subjected to combined loads is such that $E = 10 \times 10^3$ ksi and $\mu = 0.33$.

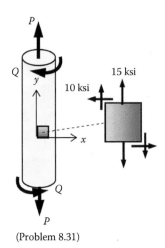

(Problem 8.31)

8.32 Refer to Problem 8.29 and find the strains ε_n, ε_t, and γ_{nt}, corresponding to an n–t coordinate system oriented at 30° clockwise from the x axis.

8.33 Refer to Problem 8.30 and find the strains ε_n, ε_t, and γ_{nt}, corresponding to an n–t coordinate system oriented at 30° clockwise from the x axis.

8.34 Refer to Problem 8.31 and find the strains ε_n, ε_t, and γ_{nt}, corresponding to an n–t coordinate system oriented at 30° clockwise from the x axis.

8.35 Let $L = 15$ ft, $p = 2$ kips/ft, $P = 20$ kips, $b = 3$ in., and $h = 6$ in. Select a point just to the right of B on the front surface of the beam and 2 in. above the neutral axis and determine the two in-plane principal strains and the maximum in-plane shearing strain, stating their orientations relative to the x axis. Let $E = 10 \times 10^3$ ksi and $\mu = 0.3$.

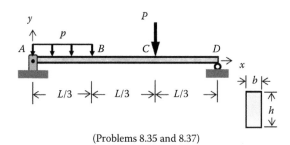

(Problems 8.35 and 8.37)

8.36 Repeat Problem 8.35 for a point just to the left of C and 2 in. below the neutral axis.

8.37 In reference to the figure in Problem 8.35, let $L = 6$ m, $p = 6$ kN/m, $P = 75$ kN, $b = 80$ mm, and $h = 150$ mm. Select a point midway between B and C on the front surface of the beam, 40 mm above the neutral axis, and determine the two in-plane principal strains and the maximum in-plane shearing strain, stating their orientations relative to the x axis. Let $E = 70$ GPa and $\mu = 0.33$.

8.38 Refer to Problem 8.35 and find the strains ε_n, ε_t, and γ_{nt}, corresponding to an n–t coordinate system oriented at 30° counterclockwise from the x axis.

8.39 Refer to Problem 8.36 and find the strains ε_n, ε_t, and γ_{nt}, corresponding to an n–t coordinate system oriented at 30° counterclockwise from the x axis.

8.40 Refer to Problem 8.37 and find the strains ε_n, ε_t, and γ_{nt}, corresponding to an n–t coordinate system oriented at 30° counterclockwise from the x axis.

*8.5   THREE-DIMENSIONAL HOOKE'S LAW

8.5.1   SUMMARY OF HOOKE'S LAWS IN ONE AND TWO DIMENSIONS

In Chapter 1, we introduced the relation that exists between stress and strain within the elastic limit for the material. This relation is known as one-dimensional or uniaxial Hooke's law. The equation representing this law is repeated here for convenience.

$$\sigma = E\varepsilon$$

(1.11, Repeated)

Another concept of interest introduced in Chapter 1 is the relation that exists between longitudinal and transverse strains. Thus, if the longitudinal strain is ε_x, then the transverse strains would be $\varepsilon_y = \varepsilon_z = -\mu\varepsilon_x$, where the constant μ is Poisson's ratio, defined by Equation 1.13, which is also repeated below for convenience.

$$\mu = \left|\frac{\varepsilon_T}{\varepsilon_L}\right|$$

(1.13, Repeated)

In Chapter 2, we developed Equations 2.22, which have come to be known as Hooke's law in two dimensions. Again, these equations are repeated here for convenience and ready reference. Thus,

$$
\left.
\begin{aligned}
\varepsilon_x &= \frac{\sigma_x}{E} - \mu\left(\frac{\sigma_y}{E}\right) = \frac{1}{E}(\sigma_x - \mu\sigma_y) \\[2mm]
\varepsilon_y &= \frac{\sigma_y}{E} - \mu\left(\frac{\sigma_x}{E}\right) = \frac{1}{E}(\sigma_y - \mu\sigma_x) \\[2mm]
\varepsilon_z &= -\frac{\mu}{E}(\sigma_x + \sigma_y); \quad \gamma_{xy} = \frac{\tau_{xy}}{G}
\end{aligned}
\right\}
$$

(2.22, Repeated)

Also, in Chapter 2, the first two of Equations 2.22 were solved simultaneously to obtain

$$
\left.
\begin{aligned}
\sigma_x &= \frac{E}{1-\mu^2}(\varepsilon_x + \mu\varepsilon_y) \\[2mm]
\sigma_y &= \frac{E}{1-\mu^2}(\varepsilon_y + \mu\varepsilon_x) \\[2mm]
\sigma_z &= 0
\end{aligned}
\right\}
$$

(2.23, Repeated)

The third relation in Equation 2.23 (repeated) was added to emphasize the fact that we are dealing with a plane-stress condition. Finally, Equation 2.24, which was given in Chapter 2 without proof, was developed recently in Section 8.4. This equation, relating the three material properties, is repeated here for completeness and convenience.

$$G = \frac{E}{2(1+\mu)}$$

(2.24)

8.5.2   HOOKE'S LAW IN THREE DIMENSIONS

We will now turn our attention to developing Hooke's law in three dimensions, referred to as the *generalized Hooke's law*. To this end, we refer to the three-dimensional stress element, whose

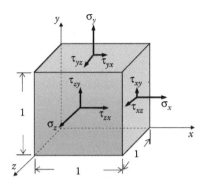

FIGURE 8.15 Three-dimensional stress element.

dimensions are unity on each side, as shown in Figure 8.15. Since, as has already been shown in previous chapters, $\tau_{xy} = \tau_{yx}$, $\tau_{yz} = \tau_{zy}$, and $\tau_{zx} = \tau_{xz}$, the stress element is subjected to six distinct stress components, namely, σ_x, σ_y, σ_z, τ_{xy}, τ_{yz}, and τ_{zx}. Each of the three normal stresses causes a normal strain, not only along its own axis but, because of Poisson's effect, also along the other two axes. On the other hand, each of the three shearing stresses causes a shearing strain only in the plain in which it acts and does not have any effect on any of the three normal strains. We will examine the effect of each of the three normal stresses separately and use *superposition* to obtain the resultant strain in each of the three coordinate directions. Of course, the use of superposition implies that the material is behaving linearly elastically.

The three-dimensional stress element of Figure 8.15 is decomposed into three two-dimensional stress elements as shown in Figure 8.16. Figure 8.16a represents a plane-stress state in the *x–y* plane, which results in the following strains:

$$\varepsilon_x = \frac{\sigma_x}{E}; \quad \varepsilon_y = \varepsilon_z = -\mu\frac{\sigma_x}{E}; \quad \gamma_{xy} = \frac{\tau_{xy}}{G} \tag{8.33}$$

The stress element shown in Figure 8.16b represents a plane-stress state in the *y–z* plane yielding the following strains:

$$\varepsilon_y = \frac{\sigma_y}{E}; \quad \varepsilon_x = \varepsilon_z = -\mu\frac{\sigma_y}{E}; \quad \gamma_{yz} = \frac{\tau_{yz}}{G} \tag{8.34}$$

Finally, the stress element of Figure 8.16c represents a plane-stress state in the *z–x* plane resulting in the following strains:

$$\varepsilon_z = \frac{\sigma_z}{E}; \quad \varepsilon_x = \varepsilon_y = -\mu\frac{\sigma_z}{E}; \quad \gamma_{zx} = \frac{\tau_{zx}}{G} \tag{8.35}$$

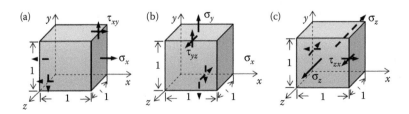

FIGURE 8.16 Decomposition of a three-dimensional stress element.

Adding up, algebraically, the normal strains in each of the three coordinate directions yields the following relations:

$$\varepsilon_x = \frac{\sigma_x}{E} - \mu\frac{\sigma_y}{E} - \mu\frac{\sigma_z}{E}$$

$$\varepsilon_y = \frac{\sigma_y}{E} - \mu\frac{\sigma_x}{E} - \mu\frac{\sigma_z}{E} \qquad (8.36)$$

$$\varepsilon_z = \frac{\sigma_z}{E} - \mu\frac{\sigma_x}{E} - \mu\frac{\sigma_y}{E}$$

Rewriting Equations 8.36 and adding the shearing strains given in Equations 8.33, 8.34, and 8.35 yields *Hooke's law in three dimensions*. Thus,

$$\varepsilon_x = \frac{1}{E}\left[\sigma_x - \mu(\sigma_y + \sigma_z)\right]; \quad \gamma_{xy} = \frac{\tau_{xy}}{G}$$

$$\varepsilon_y = \frac{1}{E}[\sigma_y - \mu(\sigma_x + \sigma_z)]; \quad \gamma_{yz} = \frac{\tau_{yz}}{G} \qquad (8.37)$$

$$\varepsilon_z = \frac{1}{E}[\sigma_z - \mu(\sigma_x + \sigma_y)]; \quad \gamma_{zx} = \frac{\tau_{zx}}{G}$$

A simultaneous solution of Equation 8.37 for the stresses σ_x, σ_y, and σ_z in terms of the strains ε_x, ε_y, and ε_z yields

$$\sigma_x = \frac{E}{(1+\mu)(1-2\mu)}[(1-\mu)\varepsilon_x + \mu(\varepsilon_y + \varepsilon_z)]$$

$$\sigma_y = \frac{E}{(1+\mu)(1-2\mu)}[(1-\mu)\varepsilon_y + \mu(\varepsilon_x + \varepsilon_z)] \qquad (8.38)$$

$$\sigma_z = \frac{E}{(1+\mu)(1-2\mu)}[(1-\mu)\varepsilon_z + \mu(\varepsilon_x + \varepsilon_y)]$$

Note that Equations 8.37 and 8.38 may be expressed in terms of principal stresses and principal strains by replacing σ_x, σ_y, and σ_z with σ_1, σ_2, and σ_3, respectively, and ε_x, ε_y, and ε_z with ε_1, ε_2, and ε_3, respectively. Note also that under such conditions, *all three shearing strains vanish*. Thus,

$$\varepsilon_1 = \frac{1}{E}[\sigma_1 - \mu(\sigma_2 + \sigma_3)]$$

$$\varepsilon_2 = \frac{1}{E}[\sigma_2 - \mu(\sigma_1 + \sigma_3)] \qquad (8.39)$$

$$\varepsilon_3 = \frac{1}{E}[\sigma_3 - \mu(\sigma_1 + \sigma_2)]$$

$$\sigma_1 = \frac{E}{(1+\mu)(1-2\mu)}[(1-\mu)\varepsilon_1 + \mu(\varepsilon_2 + \varepsilon_3)]$$

$$\sigma_2 = \frac{E}{(1+\mu)(1-2\mu)}[(1-\mu)\varepsilon_2 + \mu(\varepsilon_1 + \varepsilon_3)] \qquad (8.40)$$

$$\sigma_3 = \frac{E}{(1+\mu)(1-2\mu)}[(1-\mu)\varepsilon_3 + \mu(\varepsilon_1 + \varepsilon_2)]$$

8.5.3 Volume Change: Bulk Modulus of Elasticity

When an element is subjected to a three-dimensional stress system as shown in Figure 8.17a, it experiences deformations and distortions. If the element is of unit length on each side, then the deformations are represented by the normal strains ε_x, ε_y, and ε_z and the distortions by the shearing strains γ_{xy}, γ_{yz}, and γ_{zx} as may be found from Equations 8.37. Ignoring the distortions, the deformations may be represented as indicated in Figure 8.17b.

The initial volume of the cube is $V_1 = 1$ and its final volume is $V_2 = (1 + \varepsilon_x)(1 + \varepsilon_y)(1 + \varepsilon_z)$. Therefore, the change in volume per unit volume is $e = (1 + \varepsilon_x)(1 + \varepsilon_y)(1 + \varepsilon_z)-1$. Since the normal strains ε_x, ε_y, and ε_z are extremely small quantities, we may neglect their products in comparison to unity and reduce the quantity e to

$$e = \varepsilon_x + \varepsilon_y + \varepsilon_z \tag{8.41}$$

The quantity e in Equation 8.41 representing the change in volume per unit volume is known as the *dilation* of the material. If we assume that Hooke's law applies, the strains ε_x, ε_y, and ε_z may be replaced, using Equation 8.37, by the stresses σ_x, σ_y, and σ_z. After simplification, this replacement leads to

$$e = \frac{1 - 2\mu}{E}(\sigma_x + \sigma_y + \sigma_z) \tag{8.42}$$

A special practical case of Equation 8.42 results when all three stresses are equal in magnitude and compressive. An example of such a case occurs when a body is submersed deep in a liquid where it is subjected all around to the same compressive stress, or pressure p, known as *hydrostatic stress*. If we substitute the quantity p for each of the three stresses in Equation 8.42 and simplify we obtain

$$e = \frac{3(1 - 2\mu)}{E}p = \frac{p}{K} \tag{8.43}$$

where the quantity K is known as the *modulus of volume change* or the *bulk modulus of elasticity* and is equal to

$$K = \frac{E}{3(1 - 2\mu)} \tag{8.44}$$

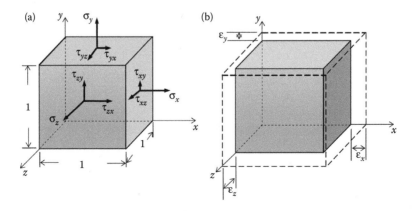

FIGURE 8.17 Three-dimensional stress element subjected to volume change and no distortion.

Under conditions of hydrostatic stress, and assuming an ideal material (homogeneous and isotropic), the normal hydrostatic strains, ε_h, are all negative and identical in magnitude in all three directions. Thus, Equation 8.41 yields

$$\varepsilon_h = -\frac{e}{3} \tag{8.45}$$

When a body is subjected to hydrostatic pressure, its volume decreases and the dilation is negative. Since the hydrostatic pressure is also negative, it follows from Equation 8.43 that the bulk modulus of elasticity K is positive. Since E is also positive, it follows from Equation 8.44 that the quantity $(1 - 2\mu)$ must be positive and, therefore, since Poisson's ratio μ is also a positive property of a material, we must come to the interesting theoretical conclusion that

$$0 < \mu < 0.5 \tag{8.46}$$

This theoretical conclusion is in agreement with experimental results that indicate that, for most materials of engineering interest, Poisson's ratio ranges between 0.25 and 0.35. Of special interest are two limiting theoretical values of μ. The first is $\mu = 0$, in which case, a positive deformation in one direction is not accompanied by negative deformations in the transverse directions. The second limiting value is $\mu = 0.5$, in which case, Equations 8.44 and 8.43 tell us, respectively, that the bulk modulus $K = \infty$ and the dilation $e = 0$. This, of course, means that the material is *rigid* and will not deform when subjected to compressive loads.

8.5.4 STRAIN ENERGY: ENERGY OF DISTORTION

When a deformable body is subjected to loads within the elastic range, elastic strain energy is stored in it. In the uniaxial case in which the only nonzero principal stress is σ_1, the strain energy per unit volume is given by Equation 1.20 to be $u = \sigma_1\varepsilon_1/2$. In the case of a triaxial stress condition represented by the principal stresses σ_1, σ_2, and σ_3, the stored strain energy per unit volume is, by superposition, the sum of the strain energies per unit volume due to each of the three principal stresses. Thus,

$$u = \frac{\sigma_1\varepsilon_1}{2} + \frac{\sigma_2\varepsilon_2}{2} + \frac{\sigma_3\varepsilon_3}{2} \tag{8.47}$$

Substituting from Equation 8.39 for the values of the three principal strains in terms of the principal stresses and simplifying, we obtain

$$u = \frac{1}{2E}[(\sigma_1^2 + \sigma_2^2 + \sigma_3^2) - 2\mu(\sigma_1\sigma_2 + \sigma_2\sigma_3 + \sigma_3\sigma_1)] \tag{8.48}$$

The total strain energy per unit volume stored in a body, given by Equation 8.48, consists of two component parts. One part represents the amount of energy per unit volume needed to distort the body, u_d, and is known as the *energy of distortion*. The second part represents the amount of energy per unit volume needed to change the volume of the body, u_v, and is known as the *energy of volume change*. Thus, the total energy per unit volume, u, may be expressed in terms of its two components as $u = u_d + u_v$. Solving for u_d, we obtain

$$u_d = u - u_v \tag{8.49}$$

Therefore, to determine the energy of distortion u_d, we need to find the energy of volume change u_v and subtract it from the total energy u given by Equation 8.48.

Let us consider the three-dimensional stress element shown in Figure 8.18a under the action of the three principal stresses σ_1, σ_2, and σ_3. To separate the total energy per unit volume into its two component parts, we need to resolve the state of stress of Figure 8.18a into the two states of stress shown in Figures 8.18b and 8.18c. The state of stress shown in Figure 8.18b represents a hydrostatic stress condition in which all three principal stresses are equal to σ_A. The stress σ_A needs to be adjusted in magnitude so that the stress state of Figure 8.18b produces the entire volume change in the unit element. It follows, then, that the stress state shown in Figure 8.18c would be responsible for all of the distortion of this element but none of the volume change. If the stress condition of Figure 8.18c is to produce *no volume change*, then this unit element does not experience a change in volume (i.e., $e = 0$) and by Equation 8.41, the algebraic sum of the three principal strains produced by the three principal stresses $\sigma_1 - \sigma_A$, $\sigma_2 - \sigma_A$, and $\sigma_3 - \sigma_A$ must be equal to zero. Thus, substituting from Equation 8.39 into Equation 8.41, we obtain

$$e = \frac{1}{E}\{(\sigma_1 - \sigma_A) - \mu[(\sigma_2 - \sigma_A) + (\sigma_3 - \sigma_A)]\} + \frac{1}{E}\{(\sigma_2 - \sigma_A) - \mu[(\sigma_1 - \sigma_A) + (\sigma_3 - \sigma_A)]\}$$

$$+ \frac{1}{E}\{(\sigma_3 - \sigma_A) - \mu[(\sigma_1 - \sigma_A) + (\sigma_2 - \sigma_A)]\} = 0 \tag{8.50}$$

Solving Equation 8.50 for σ_A, we obtain

$$\sigma_A = \frac{1}{3}(\sigma_1 + \sigma_2 + \sigma_3) \tag{8.51}$$

Equation 8.51 tells us that in order for the state of stress of Figure 8.18c to produce no volume change, the hydrostatic stress σ_A must be equal to the average of the three principal stresses. Under these conditions, the entire volume change in the unit element is produced by the state of stress shown in Figure 8.18b, and the strain energy of volume change per unit volume, u_v, may be obtained by adding the strain energies produced by the three principal stresses σ_A in a manner similar to Equation 8.47. Thus,

$$u_v = \frac{\sigma_A}{2}\left[\frac{1}{E}(\sigma_A - \mu(\sigma_A + \sigma_A))\right]$$

$$+ \frac{\sigma_A}{2}\left[\frac{1}{E}(\sigma_A - \mu(\sigma_A + \sigma_A))\right] + \frac{\sigma_A}{2}\left[\frac{1}{E}(\sigma_A - \mu(\sigma_A + \sigma_A))\right]$$

$$= \frac{3}{2}\left[\frac{1 - 2\mu}{E}\right]\sigma_A^2 \tag{8.52}$$

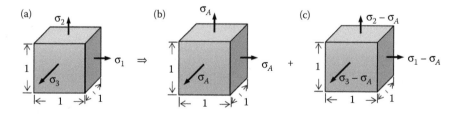

FIGURE 8.18 Three-dimensional stress element with the general stresses shown in (a) stresses separated into hydrostatic components (b), responsible for volume change, and the stress components responsible for distortion (c).

When we substitute the value of σ_A from Equation 8.51 into Equation 8.52 and simplify, we obtain

$$u_v = \frac{1 - 2\mu}{6E}(\sigma_1 + \sigma_2 + \sigma_3)^2 \qquad (8.53)$$

The energy of distortion, u_d, may now be obtained from Equation 8.49 by substituting for u its value from Equation 8.48. Thus,

$$u_d = \frac{1}{2E}[(\sigma_1^2 + \sigma_2^2 + \sigma_3^2) - 2\mu(\sigma_1\sigma_2 + \sigma_2\sigma_3 + \sigma_3\sigma_1)] - \frac{1 - 2\mu}{6E}(\sigma_1 + \sigma_2 + \sigma_3)^2 \qquad (8.54)$$

After rearranging terms and simplifying, Equation 8.54 reduces to

$$u_d = \frac{1 + \mu}{6E}[(\sigma_1 - \sigma_2)^2 + (\sigma_2 - \sigma_3)^2 + (\sigma_3 - \sigma_1)^2] \qquad (8.55)$$

EXAMPLE 8.7

A block of steel ($E = 30 \times 10^3$ ksi, $\mu = 0.3$) with initial dimensions $b = 4$ in., $w = 2$ in., and $h = 3$ in. as shown in the sketch is subjected to loads such that no shearing deformations are permitted. The dimensions b, w, and h, however, are free to change and measurements provided the following: $\Delta b = 0.005$ in., $\Delta w = -0.003$ in., and $\Delta h = 0.002$ in. Find (a) the principal stresses acting on the three faces of the block; (b) the change in volume of the block; and (c) the energy of distortion stored in the block.

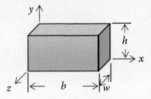

SOLUTION

a. The three principal strains are determined from the given deformations. Thus,

$$\varepsilon_1 = \varepsilon_x = \frac{0.005}{4} = 0.001250; \quad \varepsilon_2 = \varepsilon_y = \frac{0.002}{3} = 0.000667;$$

$$\varepsilon_3 = \varepsilon_z = -\frac{0.003}{2} = -0.001500$$

Substituting these values into Equation 8.40, we obtain

$$\sigma_1 = \frac{30 \times 10^3}{(1 + 0.3)(1 - 0.6)}[(1 - 0.3)(1250 \times 10^{-6}) + 0.3(667 - 1500) \times 10^{-6}]$$

$$= 36,063 \text{ psi} \approx 36.1 \text{ ksi} \qquad \textbf{ANS.}$$

$$\sigma_2 = \frac{30 \times 10^3}{(1 + 0.3)(1 - 0.6)}[(1 - 0.3)(667 \times 10^{-6}) + 0.3(1250 - 1500) \times 10^{-6}]$$

$$= 22,610 \text{ psi} \approx 22.6 \text{ ksi} \qquad \textbf{ANS.}$$

$$\sigma_3 = \frac{30 \times 10^3}{(1 + 0.3)(1 - 0.6)}[(1 - 0.3)(-15000 \times 10^{-6}) + 0.3(1250 + 667) \times 10^{-6}]$$

$$= -27,398 \text{ psi} \approx -27.4 \text{ ksi} \quad \textbf{ANS.}$$

b. The change in volume of the block is determined by finding the change of volume per unit volume from Equation 8.41 and multiplying it by the volume of the block. Thus,

$$\Delta V = eV = (0.00125 + 0.000667 - 0.001500)(4 \times 2 \times 3) = 0.010 \text{in.}^3 \quad \textbf{ANS.}$$

c. The energy of distortion is found from Equation 8.55. Thus,

$$u_d = \frac{1 + 0.3}{6(30 \times 10^3)}[(36.1 - 22.6)^2 + (22.6 + 27.4)^2 + (-27.4 - 36.1)^2]$$

$$= 48.5 \times 10^{-3} \text{ kip} \cdot \text{in./in.}^3 \quad \textbf{ANS.}$$

*8.6   MOHR'S CIRCLE FOR THREE-DIMENSIONAL STRAIN SYSTEMS

A three-dimensional strain conditions possesses three principal strains, ε_1, ε_2, and ε_3, labeled to satisfy the requirement that, algebraically, $\varepsilon_1 \geq \varepsilon_2 \geq \varepsilon_3$. The two-dimensional (plane) strain system may be considered a special case of the three-dimensional strain system in which one of the three principal strains is zero.

In analyzing the plane-strain case, we determined the maximum in-plane shearing strain, γ_{MAX}, given by Equation 8.20, or by the radius of Mohr's circle. However, this maximum in-plane shearing strain may not be the largest shearing strain at the point in a strained body under examination. In general, there is a shearing strain, known as the *absolute maximum shearing strain*, γ_{ABS}, which may be out of plane and larger in magnitude than the maximum in-plane shearing strain.

Consider, for example, a unit element subjected to the three principal stresses σ_1, σ_2, and σ_3 as shown in Figure 8.19a. Since no shearing stresses exist on principal planes, the resulting strains are the principal strains ε_1, ε_2, and ε_3 as shown in Figure 8.19b. We can identify three two-dimensional or plane-strain conditions in Figure 8.19b as follows:

A. *Plane-strain state in the plane of ε_1 and ε_2:* Since this is a plane-strain state in the plane of ε_1 and ε_2, we construct a two-dimensional strain element, obtained from Figure 8.19 when looking at

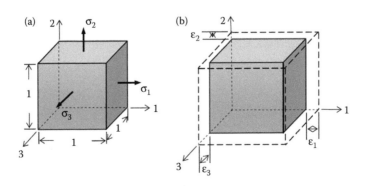

FIGURE 8.19   Three-dimensional stress element subjected to principal stresses.

the positive σ_3 plane, as shown in Figure 8.20a. The Mohr's circle corresponding to these two positive principal strains is shown in Figure 8.20b. The maximum in-plane shearing strain, of course, is equal in magnitude to twice the radius of the circle. Thus,

$$(\gamma_{MAX})_A = 2R = \varepsilon_1 - \varepsilon_2 \qquad (8.56)$$

where ε_1 and ε_2 are the values of the principal strains corresponding to points P_1 and P_2, respectively, and $(\gamma_{MAX})_A$ is the maximum in-plane shearing strain corresponding to point S_1 and point S_2. As deduced from the circle, point S_1 defines an axis 45° counterclockwise from the ε_1 (P_1) axis. Since S_1 is above the ε_n axis in the circle (positive), the S_1 axis rotates clockwise through $(1/2\gamma_{MAX})_A$ during distortion as shown in Figure 8.20c. Point S_2, on the other hand, defines an axis 45° clockwise from the ε_1 (P_1) axis. Since point S_2 is below the ε_n axis in Mohr's circle (negative), the S_2 axis rotates counterclockwise through $(1/2\gamma_{MAX})_A$ during distortion as shown in Figure 8.20c.

B. *Plane strain in the plane of ε_2 and ε_3:* Since this is a plane-strain state in the plane of ε_2 and ε_3, we construct a two-dimensional strain element, obtained from Figure 8.19 when looking at the positive σ_1 plane, as shown in Figure 8.21a. The Mohr's circle corresponding to these two positive principal strains is shown in Figure 8.21b. The maximum in-plane shearing strain, of course, is equal in magnitude to twice the radius of the circle. Thus,

$$(\lambda_{MAX})_B = 2R = \varepsilon_2 - \varepsilon_3 \qquad (8.57)$$

where ε_2 and ε_3 are the values of the principal strains corresponding to points P_2 and P_3, respectively, and $(\gamma_{MAX})_B$ is the maximum in-plane shearing strain corresponding to point S_1 and point S_2. As deduced from the circle, point S_1 defines an axis 45° counterclockwise from the ε_2 (P_2) axis. Since S_1 is above the ε_n axis in the circle (positive), the S_1 axis rotates clockwise through $(1/2\gamma_{MAX})_B$ during distortion as shown in Figure 8.21c. Point S_2, on the other hand, defines an axis 45° clockwise from the ε_2 (P_2) axis. Since point S_2 is below plane ε_n axis in Mohr's circle (negative), the S_2 axis rotates counterclockwise through $(1/2\gamma_{MAX})_B$ during distortion as shown in Figure 8.21c.

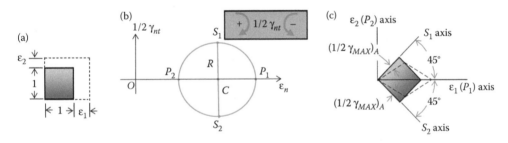

FIGURE 8.20 Two-dimensional strain element (ε_1, ε_2) and corresponding Mohr's circle.

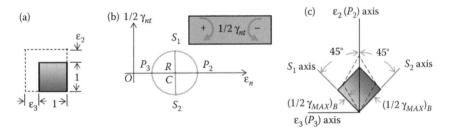

FIGURE 8.21 Two-dimensional strain element (ε_2, ε_3) and corresponding Mohr's circle.

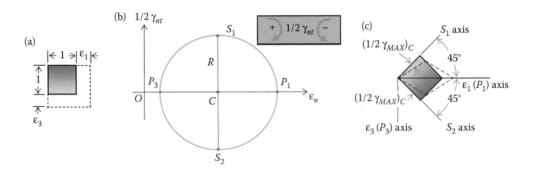

FIGURE 8.22 Two-dimensional strain element (ε_1, ε_3) and corresponding Mohr's circle.

C. *Plane strain in the plane of* ε_1 *and* ε_3: Since this is a plane-strain state in the plane of ε_1 and ε_3, we construct a two-dimensional strain element, obtained from Figure 8.19 when looking at the positive σ_2 plane, as shown in Figure 8.22a. The Mohr's circle corresponding to these two positive principal strains is shown in Figure 8.22b. The maximum in-plane shearing strain, of course, is equal in magnitude to twice the radius of the circle. Thus,

$$(\gamma_{MAX})_C = 2R = \varepsilon_1 - \varepsilon_3 \tag{8.58}$$

where ε_1 and ε_3 are the values of the principal strains corresponding to points P_1 and P_3, respectively, and $(\gamma_{MAX})_C$ is the maximum in-plane shearing strain corresponding to point S_1 and point S_2. As deduced from the circle, point S_1 defines an axis 45° counterclockwise from the ε_1 (P_1) axis. Since S_1 is above the ε_n axis in the circle (positive), the S_1 axis rotates clockwise through $(1/2\gamma_{MAX})_C$ during distortion as shown in Figure 8.22c. Point S_2, on the other hand, defines an axis 45° clockwise from the ε_1 (P_1) axis. Since point S_2 is below plane ε_n axis in Mohr's circle (negative), the S_2 axis rotates counterclockwise through $(1/2\gamma_{MAX})_C$ during distortion as shown in Figure 8.22c.

Comparing the magnitudes of the maximum in-plane shearing strains given by Equations 8.56, 8.57, and 8.58, we must conclude that the absolute maximum shearing strain, γ_{ABS}, is given by Equation 8.58 because it yields one half the difference between the largest and the smallest principal strains as long as these strains are labeled to satisfy the requirement that, algebraically, $\varepsilon_1 \geq \varepsilon_2 \geq \varepsilon_3$. In other words

$$\lambda_{ABS} = \varepsilon_1 - \varepsilon_3 \tag{8.59}$$

The three separate strain Mohr's circles shown in Figures 8.20, 8.21, and 8.22 may be combined into one diagram as shown in Figure 8.23. Such a diagram allows us to see at a glance the entire strain

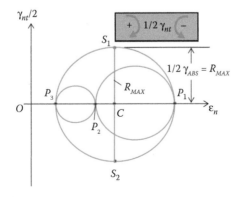

FIGURE 8.23 Combined Mohr's circle showing the entire strain system with maximum strain components.

system and to determine γ_{ABS} as twice the radius of the largest of the three circles (Equation 8.59). Let us consider the largest circle in Figure 8.23. Point S_1 represents one of the two axes experiencing the absolute maximum shearing strain. This axis is 45° counterclockwise from the ε_1 axis (point P_1) and 45° clockwise from the ε_3 axis (point P_3). In other words, the axis represented by point S_1 must bisect the 90° angle between the principal axes of ε_1 and ε_3, and, since S_1 is above the ε_n axis in the circle (positive), it must rotate clockwise during distortion. Point S_2 represents the second of the two axes experiencing the absolute maximum shearing strain. This axis is 45° clockwise from the ε_1 axis (point P_1) and 45° counterclockwise from the ε_3 axis (point P_3). In other words, the axis represented by point S_2 must bisect the 90° angle between the principal axes of ε_1 and ε_3, and, since S_2 is below the ε_n axis in the circle (negative), it must rotate counterclockwise during distortion.

EXAMPLE 8.8

The state of three-dimensional stress at a point in a body is defined by the following:

$$\sigma_x = -80 \text{ MPa}, \quad \sigma_y = 40 \text{ MPa}, \quad \sigma_z = -40 \text{ MPa}, \quad \tau_{xy} = -40 \text{ MPa},$$

$$\tau_{xz} = 120 \text{ MPa}, \quad \text{and} \quad \tau_{yz} = 80 \text{ MPa}$$

Determine (a) the principal strains and (b) the absolute maximum shearing strain. Show a sketch of the element before and after distortion using principal axes as references. Let $E = 75$ GPa and $\mu = 0.3$.

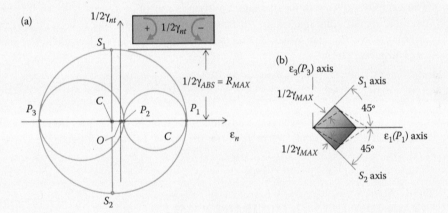

SOLUTION

a. The solution for this part of the problem is accomplished by first using Equation 7.53 to obtain the three stress invariants, which may then be substituted into Equation 7.52 to give the three principal stresses. Thus,

$$I_1 = \sigma_x + \sigma_y + \sigma_z = -80 \text{ MPa}$$

$$I_2 = \sigma_x\sigma_y + \sigma_y\sigma_z + \sigma_z\sigma_x - \tau_{xy}^2 - \tau_{yz}^2 - \tau_{zx}^2 = -24,000 \text{ MPa}^2$$

$$I_3 = \sigma_x\sigma_y\sigma_z + 2\tau_{xy}\tau_{yz}\tau_{zx} - \sigma_x\tau_{yz}^2 - \sigma_y\tau_{zx}^2 - \sigma_z\tau_{xy}^2 = 320,000 \text{ MPa}^3$$

Therefore, Equation 7.52 becomes

$$\sigma^3 + 80\sigma^2 - 24{,}000\sigma - 320{,}000 = 0 \qquad (8.8.1)$$

The solution of Equation 8.8.1 may be obtained by trial and error or by other methods for solving polynomials. In this particular case, Newton's method was used to obtain

$$\sigma_1 = 127.6\ \text{MPa}; \quad \sigma_2 = -12.9\ \text{MPa}; \quad \sigma_3 = -194.8\ \text{MPa}$$

Substituting these values into Equation 8.39, we obtain

$$\varepsilon_1 = \frac{1}{75 \times 10^9}[127.6 - 0.3(-12.9 - 194.8)] \times 10^6 = 2\,532 \times 10^{-6} \qquad \textbf{ANS.}$$

$$\varepsilon_2 = \frac{1}{75 \times 10^9}[-12.9 - 0.3(127.6 - 194.8)] \times 10^6 = 96.8 \times 10^{-6} \qquad \textbf{ANS.}$$

$$\varepsilon_3 = \frac{1}{75 \times 10^9}\left[-194.8 - 0.3(127.6 - 12.9)\right] \times 10^6 = -3\,056 \times 10^{-6} \qquad \textbf{ANS.}$$

b. The strain Mohr's circle for this strain system is shown in figure (a) above It is evident that the absolute maximum shearing strain is given by twice the radius of the largest circle, $2R_{MAX}$, in which P_1 represents the principal strain ε_1 and P_3, the principal strain ε_3. Thus,

$$\left|\gamma_{MAX}\right| = 2R_{MAX} = \varepsilon_1 - \varepsilon_3 = (2\,532 + 3\,056) \times 10^{-6} = 5588 \times 10^{-6} \qquad \textbf{ANS.}$$

Note that the same answer could have been obtained directly from Equation 8.59. The sketch of the element, before and after distortion, is shown in figure (b) above.

*8.7   STRAIN MEASUREMENTS: STRAIN ROSETTES

In general, strains serve as a means of calculating the stresses that act at a point in a given body. Thus, if the strains are known, or can be measured, the stresses can be obtained from Hooke's law. Unfortunately, there is no practical method available to measure shearing strains. However, in the case of *plane stress*, we are able to measure normal strains in a number of directions from which the shearing strain can be calculated. These normal strains may be measured by means of electric-resistance *strain gages*.

An electric-resistance strain gage consists of a *continuous* fine wire arranged into a grid and sandwiched between two small sheets of paper as shown, schematically, in Figure 8.24. This sandwich is cemented to the outside surface of the member for which the strain is needed, such that the wires line up with the direction in which the strain is to be measured. As the member is deformed, the wire in the strain gage changes in length giving rise to a change in its electrical resistance. This change in resistance is a measure of the strain to which the member is subjected and may be measured by an electric resistance measuring device such as a Wheatstone bridge.

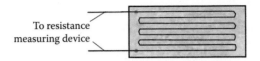

FIGURE 8.24   Electric resistance strain gage.

To be able to compute the shearing strain at a point, from measurements of normal strains, we need to make normal strain measurement in at least three different directions at this point. An arrangement of strain gages, measuring strains in three different directions at any point such as P, is shown in Figure 8.25 and is known as a *strain rosette*. The normal strains ε_a, ε_b, and ε_c, measured along axes a, b, and c, respectively, are substituted, separately, into Equation 8.7, to obtain a system of three simultaneous equations containing the three unknown quantities, ε_x, ε_y, and γ_{xy}. These three quantities may then be used to determine the corresponding stresses by means of Hooke's law for plane stress.

The two most commonly used strain rosettes are the *rectangular*, or 45°-*strain rosette*, and the *equiangular*, or 60°-*strain rosette*. The rectangular strain rosette is developed below and the equiangular rosette is dealt with in Example 8.9. The rectangular strain rosette consists of three strain gages placed so that their axes are at 45° with respect to each other. The direction of one of the three gages is taken as a reference and made to coincide with the x axis so that $\theta_a = 0°$. The other two directions are labeled as $\theta_b = 45°$ and $\theta_c = 90°$, both measured from the reference x axis. Applying Equation 8.7, separately, to each of the three axes of the strain rosette yields the following three simultaneous equations:

$$\varepsilon_0 = \varepsilon_x$$

$$\varepsilon_{45} = \frac{1}{2}\varepsilon_x + \frac{1}{2}\varepsilon_y - \frac{1}{2}\gamma_{xy}$$

$$\varepsilon_{90} = \varepsilon_y$$

When these simple simultaneous equations are solved for ε_x, ε_y, and γ_{xy} in terms of the measured quantities ε_0, ε_{45}, and ε_{90}, we obtain

$$\left. \begin{array}{l} \varepsilon_x = \varepsilon_0 \\ \varepsilon_y = \varepsilon_{90} \\ \gamma_{xy} = \varepsilon_0 + \varepsilon_{90} - 2\varepsilon_{45} \end{array} \right\} \qquad \textbf{(8.60)}$$

Having the strain components ε_x, ε_y, and γ_{xy}, we can construct Mohr's circle for strain and determine the principal strains and the maximum in-plane shearing strain. Of course, the principal stresses and the maximum in-plane shearing stress may be found using the two-dimensional Hooke's law. It should be pointed out that graphical constructions exist for obtaining the strain Mohr's circle

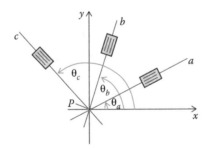

FIGURE 8.25   Strain gage arrangement for measuring shear strain.

directly from the three measured normal strain. These graphical constructions will not be discussed in this book, but the interested reader is referred to appropriate references in Appendix B.

EXAMPLE 8.9

The equiangular (60°) rosette is one in which the three strain axes make 60° angles with respect to each other. The direction of one of the three gages is taken as a reference and made to coincide with the x axis so that $\theta_a = 0°$. The other two directions are labeled as $\theta_b = 60°$ and $\theta_c = 120°$, both measured from the reference x axis. Derive equations, similar to those given in Equation 8.60, giving the values of ε_x, ε_y, and γ_{xy} in terms of the three normal strain measurements.

SOLUTION

Equation 8.7 is applied, separately, to each of the three axes of the strain rosette to obtain

$$\varepsilon_0 = \varepsilon_x$$

$$\varepsilon_{60} = \frac{1}{4}\varepsilon_x + \frac{3}{4}\varepsilon_y + \frac{\sqrt{3}}{4}\gamma_{xy}$$

$$\varepsilon_{120} = \frac{1}{4}\varepsilon_x + \frac{3}{4}\varepsilon_y - \frac{\sqrt{3}}{4}\gamma_{xy}$$

When the above three simultaneous equations are solved for ε_x, ε_y, and γ_{xy}, we obtain

$$\left.\begin{array}{l} \varepsilon_x = \varepsilon_0 \\[2mm] \varepsilon_y = \dfrac{1}{3}(2\varepsilon_{60} + 2\varepsilon_{120} - \varepsilon_0) \\[2mm] \gamma_{xy} = \dfrac{2}{\sqrt{3}}(\varepsilon_{60} - \varepsilon_{120}) \end{array}\right\} \qquad \textbf{ANS.}$$

PROBLEMS

8.41 A block of steel ($E = 30 \times 10^3$ ksi, $\mu = 0.33$) with initial dimensions $b = 5$ in., $w = 3$ in., and $h = 4$ in. as shown in the sketch is subjected to loads such that no shearing deformations are permitted. The dimensions b, w, and h, however, are free to change and measurements provided the following: $\Delta b = -0.004$ in., $\Delta w = 0.005$ in., and $\Delta h = -0.002$ in. Find (a) the principal stresses acting on the three faces of the block; (b) the change in volume of the block; and (c) the energy of distortion stored in the block.

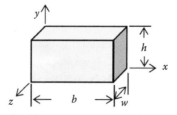

(Problems 8.41, 8.42, 8.43, and 8.44)

8.42 Solve Problem 8.41 if the block is aluminum ($E = 75$ GPa, $\mu = 0.28$) with initial dimensions $b = 240$ mm, $w = 80$ mm, and $h = 120$ mm. The dimension changes are $\Delta b = 0.10$ mm, $\Delta w = -0.15$ mm, and $\Delta h = 0.07$ mm.

8.43 In reference to the figure in Problem 8.41, a block of steel ($E = 30 \times 10^3$ ksi, $\mu = 0.33$) with initial dimensions $b = 4$ in., $w = 3$ in., and $h = 4$ in. as shown in the sketch is subjected to loads such that the three-dimensional stress system on the block is as follows: $\sigma_x = 15.0$ ksi, $\sigma_y = -10.0$ ksi, $\sigma_z = 5.0$ ksi, $\tau_{xy} = 3.0$ ksi, $\tau_{xz} = \tau_{yz} = 0$. Find (a) the changes that occur in the three dimensions of the block; (b) the absolute maximum shearing stress in the block; (c) the bulk modulus of elasticity; and (d) the total energy stored.

8.44 Solve Problem 8.43 for a magnesium block ($E = 45$ GPa, $\mu = 0.30$) with initial dimensions $b = 150$ mm, $w = 70$ mm, and $h = 100$ mm. The three-dimensional stress system on the block is as follows: $\sigma_x = -100$ MPa, $\sigma_y = 80$ MPa, $\sigma_z = -40$ MPa, $\tau_{xy} = 0$, $\tau_{xz} = \tau_{yz} = 50$ MPa.

8.45 An aluminum ($E = 10.3 \times 10^3$ ksi, $\mu = 0.30$) cubic element of dimension $b = 10$ in. on each side, is taken down to a depth of 15,000 ft below the surface of the ocean. Assume the density of ocean water to be 7.0 lb/ft^3 and determine (a) the change in the dimension b at this depth; (b) the total energy stored in the cube; (c) the energy of distortion stored in the cube; and (d) the bulk modulus of elasticity.

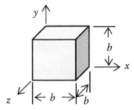

(Problems 8.45, 8.46, 8.47, and 8.48)

8.46 In reference to the figure in Problem 8.45, a steel ($E = 200$ GPa, $\mu = 0.28$) cubic element of dimension $b = 300$ mm on each side is taken to a depth h below the surface of the ocean. Measurements indicated that the dimension b decreased by 0.50 mm. Assume the mass of ocean water to 110 kg/m^3 and find (a) the depth h to which the block is taken; (b) the principal stresses in the block; (c) the energy of distortion in the block; and (d) the bulk modulus of elasticity.

8.47 In reference to the figure in Problem 8.45, an aluminum ($E = 10 \times 10^3$ ksi, $\mu = 0.30$) cubic element of dimension $b = 8$ in. on each side is subjected to triaxial loading. Laboratory measurements resulted in the following values: $\varepsilon_x = 250 \times 10^{-6}$, $\varepsilon_y = -200 \times 10^{-6}$, and $\varepsilon_z = 100 \times 10^{-6}$. No shearing strains were experience by the cube. Find (a) the principal stresses σ_1, σ_2, and σ_3; (b) the final volume of the cube; (c) the absolute maximum shearing stress; and (d) the bulk modulus of elasticity.

8.48 In reference to the figure in Problem 8.45, a steel ($E = 205$ GPa, $\mu = 0.33$) cubic element of dimension $b = 180$ mm on each side is subjected to triaxial loading. Laboratory measurements resulted in the following values: $\varepsilon_x = -200 \times 10^{-6}$, $\varepsilon_y = 300 \times 10^{-6}$, and $\varepsilon_z = -100 \times 10^{-6}$. No shearing strains were experience by the cube. Find (a) the principal stresses σ_1, σ_2, and σ_3; (b) the final volume of the cube; (c) the absolute maximum shearing stress; and (d) the energy of distortion stored in the cube.

8.49 The state of triaxial stress at a point in a body is given as follows: $\sigma_x = 20$ ksi, $\sigma_y = 40$ ksi, $\sigma_z = -20$ ksi, $\tau_{xy} = -40$ ksi, $\tau_{yz} = 20$ ksi, and $\tau_{xz} = -60$ ksi. Determine (a) the principal strains and (b) the absolute maximum shearing strain. Show a sketch of the element before and after maximum distortion using principal axes as references. Let $E = 10 \times 10^3$ ksi and $\mu = 0.28$.

8.50 Let the state of stress at a point in a body be as follows: $\sigma_x = 70$ MPa, $\sigma_y = 10$ MPa, $\sigma_z = -20$ MPa, $\tau_{xy} = -40$ MPa, and $\tau_{yz} = \tau_{xz} = 0$. Determine (a) the principal strains and (b) the absolute maximum shearing strain. Show a sketch of the element before and after maximum distortion using principal axes as references. Let $E = 200$ GPa and $\mu = 0.28$.

8.51 The state of triaxial stress at a point in a body is given as follows: $\sigma_x = -25$ ksi, $\sigma_y = -10$ ksi, $\sigma_z = 20$ ksi, $\tau_{xy} = -40$ ksi, $\tau_{yz} = 15$ ksi, and $\tau_{xz} = -10$ ksi. Find (a) the change in volume per unit volume and (b) the energy of distortion per unit volume stored in the body. Let $E = 30 \times 10^3$ ksi and $\mu = 0.33$.

8.52 Let the state of stress at a point in a body be as follows: $\sigma_x = 60$ MPa, $\sigma_y = 15$ MPa, $\sigma_z = 20$ MPa, $\tau_{xy} = -30$ MPa, $\tau_{yz} = -10$ ksi, and $\tau_{xz} = 10$ ksi. Find (a) the change in volume per unit volume and (b) the energy of distortion per unit volume stored in the body. Let $E = 70$ GPa and $\mu = 0.30$.

8.53 The state of triaxial stress at a point in a body is given as follows: $\sigma_x = 30$ ksi, $\sigma_y = -10$ ksi, $\sigma_z = 20$ ksi, $\tau_{xy} = -20$ ksi, $\tau_{yz} = 15$ ksi, and $\tau_{xz} = 30$ ksi. Determine (a) the principal strains and (b) the absolute maximum shearing strain. Show a sketch of the element before and after maximum distortion using principal axes as references. Let $E = 30 \times 10^3$ ksi and $\mu = 0.28$.

8.54 The state of triaxial stress at a point in a body is given as follows: $\sigma_x = 25$ MPa, $\sigma_y = 15$ MPa, $\sigma_z = -20$ MPa, $\tau_{xy} = -30$ MPa, $\tau_{yz} = -15$ MPa, and $\tau_{xz} = 10$ kMPa. Find (a) the absolute maximum shearing strain and (b) the bulk modulus of elasticity. Let $E = 70$ GPa and $\mu = 0.30$.

8.55 A rectangular rosette, as shown in the sketch, is attached at point P to the surface of a machine component under plane stress and the following readings were obtained: $\varepsilon_0 = 800 \times 10^{-6}$, $\varepsilon_{45} = 200 \times 10^{-6}$, and $\varepsilon_{90} = -400 \times 10^{-6}$. Determine (a) the principal strains and their directions relative to the x axis and (b) the maximum in-plane shearing strain. Show a sketch of the element before and after distortion using principal axes as references. Let $E = 30 \times 10^3$ ksi and $\mu = 0.30$.

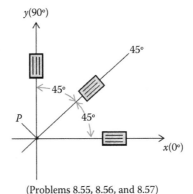

(Problems 8.55, 8.56, and 8.57)

8.56 In reference to the figure in Problem 8.55, a rectangular rosette, as shown in the sketch, is attached to the surface of a structural component subjected to plane stress and the following readings were obtained: $\varepsilon_0 = -200 \times 10^{-6}$, $\varepsilon_{45} = 300 \times 10^{-6}$, and $\varepsilon_{90} = -500 \times 10^{-6}$. Determine (a) the principal stresses and their directions relative to the x axis and (b) the maximum in-plane shearing stress. Let $E = 70$ GPa and $\mu = 0.3$.

8.57 In reference to the figure in Problem 8.55, a rectangular rosette, as shown in the sketch, is attached at point P to the surface of a machine component under plane stress and the following readings were obtained: $\varepsilon_0 = 0$, $\varepsilon_{45} = 500 \times 10^{-6}$, and $\varepsilon_{90} = 800 \times 10^{-6}$. Determine (a) the principal strains and their directions relative to the x axis and (b) the maximum in-plane shearing strain. Show a sketch of the element before and after distortion in reference to principal axes. Let $E = 30 \times 10^3$ ksi and $\mu = 0.3$.

8.58 An equiangular rosette (see Example 8.9), as shown in the sketch, is attached at point P to the surface of a structural component subjected to plane stress and the following readings were obtained: $\varepsilon_0 = 600 \times 10^{-6}$, $\varepsilon_{60} = 500 \times 10^{-6}$, and $\varepsilon_{120} = -200 \times 10^{-6}$. Determine (a) the principal stresses and their directions relative to the x axis and (b) the maximum in-plane shearing stress. Let $E = 10 \times 10^3$ ksi and $\mu = 0.3$.

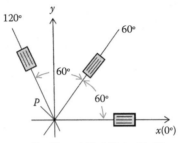

(Problems 8.58, 8.59, and 8.60)

8.59 In reference to the figure in Problem 8.58, an equiangular rosette (see Example 8.9), as shown in the sketch, is attached at point P to the surface of a machine component subjected to plane stress and the following readings were obtained: $\varepsilon_0 = -300 \times 10^{-6}$, $\varepsilon_{60} = -100 \times 10^{-6}$, and $\varepsilon_{120} = 200 \times 10^{-6}$. Determine (a) the principal strains and their directions relative to the x axis and (b) the maximum in-plane shearing strain. Show a sketch of the element before and after distortion using principal axes as references. Let $E = 75$ GPa and $\mu = 0.33$.

8.60 In reference to the figure in Problem 8.58, an equiangular rosette (see Example 8.9), as shown in the sketch, is attached at point P to the surface of a structural component subjected to plane stress and the following readings were obtained: $\varepsilon_0 = 0$, $\varepsilon_{60} = 200 \times 10^{-6}$, and $\varepsilon_{120} = 500 \times 10^{-6}$. Determine (a) the principal stresses and their directions relative to the x axis and (b) the maximum in-plane shearing stress. Let $E = 10 \times 10^3$ ksi and $\mu = 0.3$.

8.61 The rosette shown in the sketch is used to obtain normal strain readings at point P on the outside surface of a machine component under plane stress, along the α-, β-, and $\gamma(90°)$ directions. If $\alpha = 30°$ and $\beta = 60°$, develop expressions giving the strains ε_x, ε_y, and γ_{xy} in terms of ε_{30}, ε_{60}, and ε_{90}.

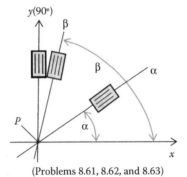

(Problems 8.61, 8.62, and 8.63)

8.62 In reference to the figure in Problem 8.61, the rosette shown in the sketch is used to obtain normal strain readings at point P on the outside surface of a structural component under plane stress, along the α-, β-, and γ(90°) directions. If $\alpha = -20°$ and $\beta = 50°$, develop expressions giving the strains ε_x, ε_y, and γ_{xy} in terms of ε_{-20}, ε_{50}, and ε_{90}.

8.63 In reference to the figure in Problem 8.61, the rosette shown in the sketch is used to obtain normal strain readings at point P on the outside surface of a machine component subjected to plane stress, along the α-, β-, and γ(90°) directions. If $\alpha = 0°$ and $\beta = 120°$, develop expressions giving the strains ε_x, ε_y, and γ_{xy} in terms of ε_0, ε_{120}, and ε_{90}.

8.64 A strain gage is attached to the outside surface of a steel strut ($E = 30 \times 10^3$ ksi, $\mu = 0.3$) as shown. If $P = 20$ kips and the cross-sectional area of the strut is 2.0 in.², determine the reading of the strain gage if (a) $\theta = 30°$ and (b) $\theta = 45°$.

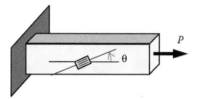

(Problems 8.64 and 8.65)

8.65 In reference to the figure in Problem 8.64, a strain gage is attached to the outside surface of an aluminum strut ($E = 75$ GPa, $\mu = 0.3$, $A = 3.6 \times 10^{-3}$ m²) as shown. If $\theta = 35°$, determine the magnitude of the force P corresponding to a strain reading of 150×10^{-6}.

8.66 A strain gage is attached to the outside surface of a 3-in.-diameter aluminum shaft ($E = 10 \times 10^3$ ksi, $\mu = 0.3$) as shown. Let $Q = 80$ kip · in., find the reading of the strain gage for (a) $\theta = 25°$ and (b) $\theta = 40°$.

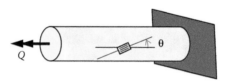

(Problems 8.66 and 8.67)

8.67 In reference to the figure in Problem 8.66, a strain gage is attached to the outside surface of a 100-mm-diameter steel shaft ($E = 200$ GPa, $\mu = 0.33$)as shown. If $\theta = 30°$, determine the torque Q corresponding to a strain measurement of -150×10^{-6}.

8.68 A strain gage is attached to the outside surface of a steel beam as shown. Let $P = 20$ kips, $\alpha = 40°$, $b = 2$ in., $h = 4$ in., $d = 24$ in., and $a = 1$ in., determine the reading of the strain gage for (a) $\theta = 20°$ and (b) $\theta = 40°$. Let $E = 30 \times 10^3$ ksi and $\mu = 0.3$.

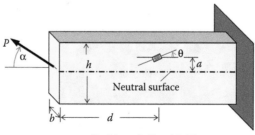

(Problems 8.68 and 8.69)

8.69 In reference to the figure in Problem 8.68, a strain gage is attached to the outside surface of a steel beam as shown and gave a reading of -500×10^{-6}. Let $b = 70$ mm,

$h = 120$ mm, $d = 1$ m, and $a = 25$ mm, determine the magnitude of the force P for $\alpha = 20°$ and $\theta = 30°$. Let $E = 200$ GPa and $\mu = 0.3$.

REVIEW PROBLEMS

R8.1 A thin rectangular plate lies in the x–y plane as shown. Before the plate is stressed, its dimensions were $b = 10$ in. and $h = 18$ in. Measurements of strains after stressing resulted in the following: $\varepsilon_x = 150 \times 10^{-6}$, $\varepsilon_y = -200 \times 10^{-6}$, and $\gamma_{xy} = 140 \times 10^{-6}$. Assume that the plate experiences plane stress. Find the strains ε_n, ε_t, and γ_{nt} and the corresponding stresses σ_n, σ_t, and τ_{nt}. Use the transformation equations. Let $E = 29 \times 10^3$ ksi and $\mu = 0.33$.

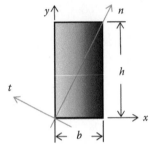

(Problems R8.1 and R8.2)

R8.2 In reference to the figure in Problem R8.1, a thin rectangular plate lies in the x–y plane as shown. Before the plate is stressed, its dimensions were $b = 150$ mm and $h = 250$ mm. Measurements of strains after stressing resulted in the following: $\varepsilon_x = -180 \times 10^{-6}$, $\varepsilon_y = 150 \times 10^{-6}$, and $\gamma_{xy} = -200 \times 10^{-6}$. Assume that the plate experiences plane strain and find the two in-plane principal strains and their orientation relative to the given axes. Also find ε_n and ε_t. Use the transformation equations. Let $E = 200$ GPa and $\mu = 0.3$.

R8.3 Solve Problem R8.1 using Mohr's circle.

R8.4 Solve Problem R8.2 using Mohr's circle.

R8.5 Let $L = 6$ m, $p = 20$ kN/m, $P = 500$ kN, $b = 80$ mm, and $h = 180$ mm. Select a point just to the left of B on the front surface of the beam and 50 mm above the neutral axis and determine the two in-plane principal strains stating their orientation with respect to the x axis. Also find the absolute maximum shearing strain. Show a sketch of the element before and after distortion using principal axes as reference. Let $E = 195$ GPa and $\mu = 0.33$.

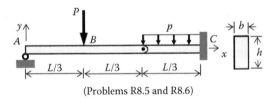

(Problems R8.5 and R8.6)

R8.6 Solve Problem R8.5 for a point just to the left of point C on the outside surface of the beam 35 mm below the neutral axis.

R8.7 A magnesium ($E = 6.5 \times 10^3$ ksi, $\mu = 0.33$) cubic element of dimension $b = 12$ in. on each side is subjected to triaxial loading. Laboratory measurements resulted in the following strain values: $\varepsilon_x = -300 \times 10^{-6}$, $\varepsilon_y = 250 \times 10^{-6}$, and $\varepsilon_z = 150 \times 10^{-6}$. No shearing strains were experienced by the cube. Find (a) the principal stresses σ_1, σ_2, and σ_3;

(b) the final volume of the cube; (c) the absolute maximum shearing stress; and (d) the bulk modulus of elasticity.

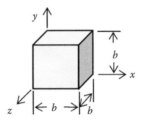

R8.8 The state of stress at a point in a body made of steel ($E = 30 \times 10^3$ ksi, $\mu = 0.28$) is described by the following information: $\sigma_x = -20$ ksi, $\sigma_y = -10$ ksi, $\tau_{xy} = 15$ ksi. If $\theta = -20°$, determine the quantities: ε_n, ε_t, ε_z, and γ_{nt}.

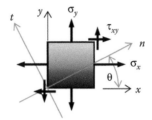

R8.9 At a point on the outside surface of a member subjected to combined loading, the stress system is as given in the companion stress element. Knowing that $E = 70$ GPa and $\mu = 0.3$, determine the two in-plane principal strains stating their orientation with respect to the x axis. Also find the absolute maximum shearing strain. Show a sketch of the element before and after distortion using the principal axes as reference.

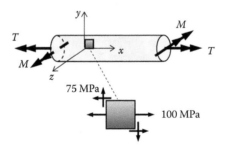

R8.10 An equiangular rosette (see Example 8.9), as shown in the sketch, is attached at point P to the surface of a structural component and the following readings were obtained: $\varepsilon_0 = -550 \times 10^{-6}$, $\varepsilon_{60} = 600 \times 10^{-6}$, and $\varepsilon_{120} = -300 \times 10^{-6}$. Determine (a) the principal stresses and their directions relative to the x axis and (b) the maximum in-plane shearing stress. Let $E = 15 \times 10^3$ ksi and $\mu = 0.3$.

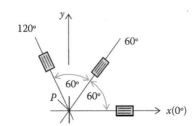

R8.11 A strain gage is attached to the outside surface of a 120- mm-diameter steel shaft ($E = 200$ GPa, $\mu = 0.3$) as shown. Let $Q = 25$ kN $\cdot$ m and $P = 800$ kN, find the reading of the strain gage for (a) $\theta = 30°$ and (b) $\theta = -50°$.

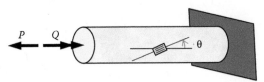

R8.12 The state of triaxial stress at a point in a body is given as follows: $\sigma_x = -35$ ksi, $\sigma_y = 20$ ksi, $\sigma_z = -20$ ksi, $\tau_{xy} = 15$ ksi, $\tau_{yz} = 20$ ksi, and $\tau_{xz} = -30$ ksi. Determine (a) the principal strains and (b) the absolute maximum shearing strain. Show a sketch of the element before and after maximum distortion using principal axes as references. Let $E = 30 \times 10^3$ ksi and $\mu = 0.28$.

R8.13 A rectangular rosette, as shown in the sketch, is attached at point P to the surface of a machine component and the following readings were obtained: $\varepsilon_0 = 500 \times 10^{-6}$, $\varepsilon_{45} = -150 \times 10^{-6}$, and $\varepsilon_{90} = 300 \times 10^{-6}$. Determine (a) the in-plane principal strains and their directions relative to the x axis and (b) the maximum in-plane shearing strain. Show a sketch of the element before and after maximum distortion using principal axes as references.

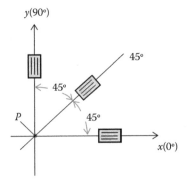

9 Columns

9.1 INTRODUCTION

A major concern of structural engineers and architects is the stability of a given structure. The structure may be designed properly from the standpoint of strength and deformation, but may possess certain characteristics that may render it unstable. This instability, if present, may lead to a complete, catastrophic failure of the entire structure, resulting in excessive property loss as well as possible loss of life.

A long, slender member that is designed to resist compressive axial loads is known as a *column*. Such a member is subject to instability under certain conditions and when used in a structure or a machine, must be properly designed if failure is to be avoided. Our main objective in this chapter is to study the behavior of columns and to learn how to design them properly.

Thus, in Section 9.2, we will develop the fundamental concepts of the stability of equilibrium. We will find that equilibrium of a given system may be classified as *stable, neutral,* or *unstable* depending upon its potential energy. If its potential energy is a minimum, the system is *stable,* implying that if its equilibrium is disturbed, the system will have a tendency to return to its original equilibrium position. If, on the other hand, the potential energy is constant, the system is in *neutral* equilibrium, implying that if its equilibrium is disturbed, the system will still be in equilibrium in the new position. Finally, if the potential energy is a maximum, the system will be *unstable,* in which case, if its equilibrium position is disturbed, the system will have a tendency to move further away from its original equilibrium position. In Section 9.3, we will discuss the ideal column theory, known as *Euler's column theory,* named after Leonard Euler, a Swiss mathematician who, over 200 years ago, developed the still-famous equation for the *critical buckling load* of long pin-ended columns. This basic fundamental equation is then used in Section 9.4 to introduce the concept of the *effective length* and derive the equations applicable to other end conditions. Section 9.5 is devoted to the analysis of more realistic column cases dealing with eccentrically loaded and initially curved columns. In Section 9.6, we will deal with some *empirical* column formulas that have been developed to fit experimental data and introduce the problem of designing centrically loaded columns. Finally, in Section 9.7, we will introduce the problem of designing eccentrically loaded columns.

9.2 STABILITY OF EQUILIBRIUM

In previous chapters, we considered equilibrium as a *state* of a system (structure or machine) enabling us to determine certain unknown reactive forces and did not concern ourselves with the *stability* of the system. Long and slender structural members subjected to compressive axial loads are susceptible to instability under certain conditions, which will be discussed later in this chapter. When such members are placed in a vertical position, they are known as *columns*.

9.2.1 THEORETICAL BACKGROUND

In order to better understand the behavior of columns and be able to design them properly to resist the action of compressive axial loads without failure, we need to know that equilibrium of a system can be *stable, unstable,* or *neutral.* Fortunately, we have mathematical criteria that enable us to determine whether a given *conservative* system is stable, unstable, or neutral. Figure 9.1 represents

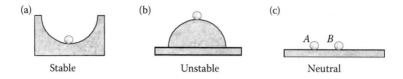

FIGURE 9.1   Types of equilibrium.

an aid in visualizing these three types of equilibrium. Figure 9.1a shows a small sphere at rest at the lowest point of a smooth concave cylindrical container. If displaced from this equilibrium position and released, the small sphere will naturally return to it. This type of equilibrium is said to be *stable* because even if forced from it and released, the sphere will always return to it. Note that when the sphere is disturbed from its *stable* equilibrium position to any other position inside the smooth container, its potential energy increases. This statement obviously means that when the small sphere is in the stable equilibrium position, its potential energy is a *minimum*. Figure 9.1b shows the small sphere carefully placed at the highest point of a smooth convex cylindrical body or dome. The sphere is said to be in a state of *unstable* equilibrium, because if disturbed from this equilibrium position and released, it will not return to it but will naturally continue its movement away from it. Note that when the sphere is moved from its *unstable* equilibrium position to any other position on the surface of the smooth dome, its potential energy decreases. This, of course, means that the potential energy of the sphere in its *unstable* equilibrium position is a maximum. Finally, Figure 9.1c represents a state known as *neutral* equilibrium, because, if displaced from the equilibrium position at *A* to any other position such as *B*, the small sphere will remain in that position with no tendency to return to or move farther away from the original equilibrium position. Note that, under such conditions, the potential energy of the sphere remains constant regardless of the position to which it is displaced.

The above concepts may be expressed mathematically for a conservative system with one degree of freedom (the only type of system dealt with in this book) represented by the variable θ, where the potential energy function is $U = f(\theta)$. The equilibrium positions of such a system are obtained from the condition that $dU/d\theta$. Whether these equilibrium positions are stable, unstable, or neutral may be ascertained by examining higher derivatives of the potential energy function. For a stable equilibrium position, the function U must have a *minimum* value and $d^2U/d\theta^2$ must be *positive*. For an unstable equilibrium position, the function U must have a *maximum* value and $d^2U/d\theta^2$ must be *negative*. For a neutral equilibrium position, the function U must have a *constant* value and *all* of its derivatives must be *zero*. These conclusions are depicted in Figure 9.2 and are summarized mathematically as follows:

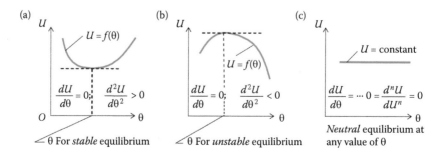

FIGURE 9.2   Potential energy functions for stable (a), unstable (b), and neutral (c) equilibrium.

Stable equilibrium:

$$\frac{dU}{d\theta} = 0; \quad \frac{d^2U}{d\theta^2} > 0 \tag{9.1}$$

Unstable equilibrium:

$$\frac{dU}{d\theta} = 0; \quad \frac{d^2U}{d\theta^2} < 0 \tag{9.2}$$

Neutral equilibrium:

$$\frac{dU}{d\theta} = \frac{d^2U}{d\theta^2} \cdots = \frac{d^nU}{dU^n} = 0 \tag{9.3}$$

9.2.2 Column Models

As stated earlier, a column is a slender structural member subjected to a compressive axial load and, generally, used in a vertical position. If the compressive axial load is sufficiently large, it will cause instability and failure (collapse) of the column. This compressive axial load is known as the *critical* or *failure load* for the column and is given the symbol P_{CR}. An example of a column of length L, fixed at end A and free at end B, is shown in Figure 9.3. If the compressive axial load reaches the critical value P_{CR}, the column will bend as shown by the curved line AB' and, in theory, it will continue to bend under the action of P_{CR} until complete failure occurs. Columns such as this may be analyzed and studied using a *mathematical model* along with the theoretical background presented above. This type of analysis provides insights into the behavior of real columns and makes it possible to predict the critical load for a given system.

A theoretical model for the column shown in Figure 9.3 is analyzed in Examples 9.1 and 9.2. The fixed support at A is represented by a coil spring, whose spring constant K represents the rotational resistance of the fixed support. In Example 9.1, the problem is solved using potential energy considerations and in Example 9.2, this same problem is solved using the conditions of equilibrium.

FIGURE 9.3 Column model.

EXAMPLE 9.1

The system shown in the companion below sketch represents a mathematical model for the column shown in the figure below. The vertical weightless rod AB, of length L, is supported at A by a frictionless hinge and a torsional spring of constant K. End B of the vertical rod is free and a weight W is placed on top of the rod as shown. Determine the critical weight, W_{CR}, in terms of K and L.

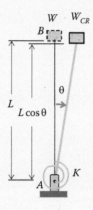

SOLUTION

In order to determine the equilibrium positions of the rod, we need to find the potential energy function U of the system for a given rotation θ. This potential energy function consists of two parts, potential energy of position U_g, measured from a reference at the top of the initial position of the rod, and the elastic potential energy U_e stored in the torsional spring. Thus,

$$U = U_g + U_e = -WL(1 - \cos \theta) + \frac{1}{2} K\theta^2 \qquad (9.1.1)$$

$$\frac{dU}{d\theta} = -WL \sin \theta + K\theta = 0 \qquad (9.1.2)$$

$$\frac{d^2U}{d\theta^2} = -WL \cos \theta + K \qquad (9.1.3)$$

One viable solution of Equation 9.1.2 is $\theta = 0°$. This equilibrium position will now be examined from the standpoint of stability.

Stable equilibrium: The position defined by $\theta = 0°$ is stable if the second derivative is larger than zero. Thus,

$$-WL \cos \theta + K > 0 \Rightarrow -WL(1) + K > 0 \Rightarrow \frac{WL}{K} < 1 \quad \text{or} \quad W < \frac{K}{L} \qquad (9.1.4)$$

Equation 9.1.4 indicates that the system is stable as long as $W < K/L$.

Unstable equilibrium: The position defined by $\theta = 0°$ is unstable if the second derivative is less than zero. Thus,

$$-WL \cos \theta + K < 0 \Rightarrow -WL(1) + K < 0 \Rightarrow \frac{WL}{K} > 1 \quad \text{or} \quad W > \frac{K}{L} \qquad (9.1.5)$$

Equation 9.1.5 indicates that the system is unstable if $W < K/L$.

Neutral equilibrium: The position defined by $\theta = 0°$ is neutral if the second derivative is equal to zero. Thus,

$$-WL \cos \theta + K = 0 \Rightarrow -WL(1) + K = 0 \Rightarrow \frac{WL}{K} = 1 \quad \text{or} \quad W = \frac{K}{L} \qquad (9.1.6)$$

Equation 9.1.6 gives the magnitude of the critical weight W. Note that it lies between the magnitudes needed for a stable equilibrium position (Equation 9.1.4) and that for an unstable position (Equation 9.1.5). Thus,

$$W_{CR} = \frac{K}{L} \qquad \textbf{ANS.}$$

The three types of equilibrium positions at $\theta = 0°$ may be further explained by examining Equation 9.1.2, which is rewritten in the following form:

$$\frac{WL}{K} = \frac{\theta}{\sin \theta} \qquad (9.1.7)$$

A plot of Equation 9.1.7 is shown in figure below for positive values of θ between $0°$ and $180°$. In the limit, when θ approaches zero, the ratio ($\theta/\sin \theta$) and hence the quantity WL/K becomes unity as shown. At $WL/K = 1$, the system is in a state of neutral equilibrium, which defines the magnitude of the critical load to be $W_{CR} = K/L$. For any value of WL/K above unity, the system is in a state of unstable equilibrium and, for any value of WL/K below unity, the system is in a state of stable equilibrium.

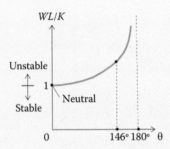

Note that $\theta = 0°$ may not be the only equilibrium position that this system possesses. Other equilibrium positions of the system, while *not* of interest for the problem at hand, depend upon the relative magnitudes of W, L, and K. For example, if we let $W = 250$ lb, $L = 1$ft, and $K = 55$ lb ft/rad and solve Equation 9.1.2, we obtain $\theta = 146°$. This position of equilibrium is shown in figure above. To determine if this position is stable, unstable, or neutral, we need to examine the sign of the second derivative of the potential energy function given by Equation 9.1.3. If we substitute the given numerical values along with $\theta = 146°$, we find that the second derivative is positive indicating that $\theta = 146°$ is a stable equilibrium position.

EXAMPLE 9.2

Solve Example 9.1 using the conditions of equilibrium.

SOLUTION

The free-body diagram of the column model of Example 9.1 is shown in the below sketch. There are two agents producing moments about the frictionless hinge at point A. The first is the applied weight W producing a moment, $(M_A)_A = W(L \sin \theta)$. The second is the coil spring developing a resisting moment, $(M_R)_A = K\theta$. Intuitively, we conclude that: (1) If the resisting moment is larger than the applied moment, the system for $\theta = 0°$ is in a state of *stable* equilibrium. (2) If, however, the resisting moment is less than the applied moment, the system for $\theta = 0°$ is in a state of *unstable* equilibrium. (3) If the resisting moment for $\theta = 0°$ is exactly equal to the applied moment, the system is in a state of *neutral* equilibrium. These three conditions are summarized below:

$$\text{Stable: } K\theta > WL \sin \theta \Rightarrow W < \frac{K\theta}{L \sin \theta}; \quad \theta \to 0°, \theta \approx \sin \theta; \quad W < \frac{K}{L}$$

$$\text{Unstable: } K\theta < WL \sin \theta \Rightarrow W > \frac{K\theta}{L \sin \theta}; \quad \theta \to 0°, \theta \approx \sin \theta, \quad W > \frac{K}{L}$$

$$\text{Neutral: } K\theta = WL \sin \theta \Rightarrow W = \frac{K\theta}{L \sin \theta}; \quad \theta \to 0°, \theta \approx \sin \theta, \quad W = \frac{K}{L}$$

This last value of W represents the critical weight. Below this value, the system is stable, and above it, the system is unstable. Therefore, to obtain the critical load, all we need do is apply the equilibrium equation, $\sum M_A = 0$, and solve for W. Thus, referring to the free-body diagram in the sketch above, we have

$$\sum M_A = 0 : K\theta - W_{CR}(L \sin \theta) = 0; \quad \theta \to 0°, \quad \theta \approx \sin \theta, \quad W_{CR} = \frac{K}{L} \qquad \textbf{ANS.}$$

Note that this answer is identical to that obtained in Example 9.1. In general, the equilibrium method is easier than the potential energy method because, unlike the latter, it is limited to a single equilibrium configuration (i.e., $\theta = 0°$). On the other hand, the potential energy method allows finding all equilibrium configurations (*if they are needed*) and determining if they are stable, unstable, or neutral.

PROBLEMS

9.1 A column model is shown in the sketch. Weightless rod *AB* of length *L* is supported at *A* by a frictionless hinge and at *C* by two identical springs, each with a spring constant *k* as shown. At their other ends, the springs are constrained to move in vertical frictionless guides. When in the vertical position, a weight *W* is placed on top of the rod. Use the potential energy method to find the critical weight W_{CR}.

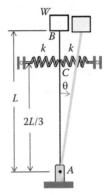

(Problems 9.1, 9.2, 9.3, and 9.4)

9.2 Solve Problem 9.1 by the equilibrium method.

9.3 The system described in Problem 9.1 has the following physical characteristics: $W = 150$ N, $L = 0.6$ m, and $k = 500$ N/m. Determine the positions of equilibrium for this system for values of θ between 0° and 180°. Examine these positions of equilibrium for stability and state if they are stable, unstable, or neutral.

9.4 The system described in Problem 9.1 has the following physical characteristics: $W = 200$ lb, $k = 400$ lb/ft, and $L = 1.5$ ft. Plot the quantity $(9W/8kL)$ versus θ and show all of the equilibrium positions between −180° and +180°. Determine if the positions are stable neutral or unstable.

9.5 A column model is shown in the sketch. Weightless rods *AB* and *BC*, each of length $L/2$, are hinged together at *B* and supported there with a coil spring of spring constant *K* as shown. Rod *AB* is supported at *A* by a frictionless hinge. Rod *BC* is supported at *C* by a frictionless hinge attached to a weight *W* that can slide freely in a frictionless vertical guide. Use the potential energy method to find the critical weight W_{CR}.

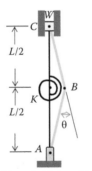

(Problems 9.5, 9.6, and 9.7)

9.6 Solve Problem 9.5 by the equilibrium method.

9.7 The system described in Problem 9.5 has the following physical characteristics: $W = 50$ lb, $L = 2$ ft, and $K = 75$ lb ft/rad. Determine the positions of equilibrium for this system for values of θ between $0°$ and $\pm180°$. Examine these positions of equilibrium for stability and state if they are stable, unstable, or neutral.

9.8 A column model is shown in the sketch. Weightless rods AB and BC, each of length $L/2$, are hinged together at B and supported there with a linear spring of spring constant k. The other end of the spring is attached to a roller that slides in a smooth vertical guide. Rod AB is supported at A by a frictionless hinge. Rod BC is supported at C by a frictionless hinge attached to a weight W that can slide freely in a frictionless vertical guide. Use the potential energy method to find the critical weight W_{CR}.

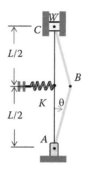

(Problems 9.8, 9.9, and 9.10)

9.9 Solve Problem 9.8 by the equilibrium method.

9.10 The system described in Problem 9.8 has the following physical characteristics: $W = 300$ N, $L = 2$ m, $k = 2$ kN/m. Determine the positions of equilibrium for this system for values of θ between $0°$ and $\pm180°$. Examine these positions of equilibrium for stability and state if they are stable, unstable, or neutral.

9.11 A column model consists of rigid rod AB of length L attached to a linear spring of spring constant k as shown. A weight W is placed on top of the rod. Write the potential energy function U in terms of θ, which need not be considered small. When $\theta = 0$, the spring is not stretched. Set to zero, the derivative of U with respect to θ and determine if the system is in equilibrium when $\theta = 0$. Is it possible for the spring to hold the system in equilibrium when $\theta = 90°$?

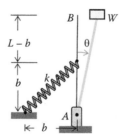

(Problems 9.11 and 9.12)

9.12 The system described in Problem 9.11 has the following physical characteristics: $W = 100$ lb and $L = 1$ ft. What value of k would make the position $\theta = 30°$ stable?

9.13 A structural system is modeled as the frame sown, consisting of two rigid angle segments connected by coil springs of spring constant K. Two identical forces P are applied as shown. Determine the critical load expressing it in terms of K and L. Use the potential-energy method.

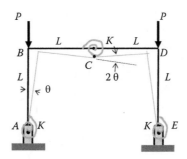

(Problems 9.13 and 9.14)

9.14 Refer to the system described in Problem 9.13. Let $P_{CR} = 6$ kips and $L = 20$ ft, determine the smallest permissible value of K consistent with stable equilibrium when $\theta = 0$.

9.3 EULER'S IDEAL-COLUMN THEORY

More than 200 years ago, Leonard Euler (1707–1783), a Swiss mathematician, laid the foundation for the study of column behavior. His name appears frequently in the literature of mathematics, science, and engineering, primarily because he holds the all-time record for mathematical productivity. Euler wrote over 80 volumes of mathematics, many of enduring interest and usefulness,

By definition, a *column* is a long and slender structural member designed to carry compressive loads. In this section, we will develop an equation that makes it possible to find the critical buckling load for an *ideal* column. Such a column, known as an *Euler* column, is shown in Figure 9.4 and described by the following restrictive assumptions:

1. The axial compressive load represented by P is applied at the end of the *weightless* column without eccentricity. This, of course, implies that the line of action of the applied loads coincides with the longitudinal axis of the column that passes through the centroid of all cross sections of the member.
2. The hinges (pins) at A and B are frictionless. This means that there is no resistance to rotation of the column at its two supports.
3. End B of the column where the compressive load P is applied is provided with a frictionless slider that moves within smooth guides. Thus, end B is free to move toward fixed end A as the load increases and the column is compressed.
4. The column is perfectly straight before the loads are applied and has the same cross-sectional area throughout its length (i.e., the column is *prismatic*).

FIGURE 9.4   Euler's ideal column.

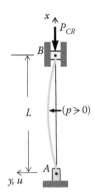

FIGURE 9.5   Euler's ideal column shown with a small disturbing force.

5. The column is fabricated of an ideal material that is homogeneous and isotropic and since it is prismatic, its flexural rigidity EI is constant.
6. The two hinges are constructed to allow the column to bend about any axis of the column's cross section.

It should be stated at the outset that such a column does not exist in practice. However, if care is exercised in the fabrication process, we may be able to approach the conditions stated in the above assumptions. We make these assumptions in order to simplify the mathematical solution of the problem.

If it were possible to construct and load an ideal column, it would not bend (buckle) but would continue to shorten axially under the action of an increasing compressive load P. To assure that the ideal column will buckle, a very small lateral disturbing force p may be introduced, as shown in Figure 9.5, and allowed to approach zero in the limit. It is assumed that this small lateral load is applied parallel to the strong axis of the column's cross section (the u axis) because the column will buckle about the weak axis (the v axis). If the compressive load P is less than the *critical load*, and the small lateral force p is applied, the column will bend as shown in Figure 9.5, but upon its removal. the column returns to a straight configuration indicating a *stable* equilibrium. If the compressive load is equal to the critical load P_{CR} when the small lateral force p is applied and removed, the column will remain in the slightly bent configuration. If we think of the small lateral disturbing force p as taking on a series of different values, the column will move from one bent *small* configuration to another, each corresponding to one of the p values. In other words, the column is in a state of indifferent or *neutral* equilibrium. The word *bifurcation* has been used to describe the fact that under the critical load, P_{CR}, the ideal column is in a state of *neutral* equilibrium either in the straight or in the slightly bent configuration. The Euler critical load, P_{CR}, represents the transition from a *stable* to an *unstable* equilibrium state for the ideal column. For compressive loads slightly less than P_{CR}, the straight configuration is stable and for compressive loads slightly larger than P_{CR}, the column is unstable and assumes a bent configuration.

The free-body diagram of segment AC at the lower end of the bent column is shown in Figure 9.6 at the moment when the compressive load has reached the critical value P_{CR}. Equilibrium tells us that at C, the internal axial force is $F = -P_{CR}$ and that the internal moment is $M = -P_{CR}\,u$ as shown. Note that in the case of columns, the internal bending moment is a function of the small lateral deflection of the column u. This contrasts with the internal moment in beams studied in Chapters 3, 4, 5, and 6, which are functions of the longitudinal coordinate x.

We now make use of Equation 5.8 developed in Chapter 5 for beam deflections and write it in the following form, replacing the deflection v by the deflection u. Thus,

$$EI\,u'' = M = -P_{CR}u \qquad\qquad (9.4)$$

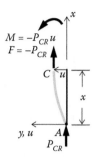

FIGURE 9.6   Free-body diagram of the lower end of Euler's ideal column.

Equation 9.4 may be rewritten by transposing the quantity $-P_{CR} u$ to the left-hand side and dividing by the constant flexural rigidity EI. Thus,

$$u'' + \left(\frac{P_{CR}}{EI}\right)u = 0 \tag{9.5}$$

For convenience, we set $P_{CR}/EI = k^2$ and rewrite Equation 9.5 in the form

$$u'' + k^2 u = 0 \tag{9.6}$$

Equation 9.6 is an ordinary, second-order, linear, and homogeneous differential equation with constant coefficients. Its general solution may be stated in the form

$$u = A \sin kx + B \cos kx \tag{9.7}$$

The values of the two constants A and B in Equation 9.7 are found from the two boundary conditions at the ends of the column. Thus,

$$1. \quad x = 0, \quad u = 0 \Rightarrow 0 = 0 + B(1) \Rightarrow B = 0 \tag{9.8}$$

$$2. \quad x = L, \quad u = 0 \Rightarrow 0 = A \sin kL + 0 \Rightarrow A \sin kL = 0 \tag{9.9}$$

In Equation 9.9, the constant A cannot be zero because, if this were the case, it would mean that the column is not deflected. As a matter of fact, the constant A is undefined and may have any value without violating Equation 9.9. Thus, the quantity $\sin kL$ must be zero to satisfy Equation 9.9. This implies that

$$kL = n\pi \ (n = 0, \pm 1, \pm 2,...) \tag{9.10}$$

Therefore,

$$k^2 L^2 = n^2 \pi^2, \quad \Rightarrow k^2 = \frac{n^2 \pi^2}{L^2} = \frac{P_{CR}}{EI} \tag{9.11}$$

Solving Equation 9.11 for P_{CR}, we obtain

$$P_{CR} = \frac{n^2 \pi^2 EI}{L^2} \tag{9.12}$$

The first nonzero value for n (i.e., $n = 1$) yields the smallest critical load and results in the equation known as *Euler's equation* for a column buckling load. Thus,

$$P_{CR} = \frac{\pi^2 EI}{L^2} \tag{9.13}$$

Equation 9.13 reveals that the Euler critical load for a column is directly proportional to the moment of inertia I. Since we are interested in the smallest compressive load that causes buckling of the column, we must use the minimum moment of inertia, which of course is I_v, the lesser of the two principal centroidal moments of inertia of its cross section (see Appendix C.2). Therefore, Euler's equation may be written in the form

$$P_{CR} = \frac{\pi^2 EI_v}{L^2} \tag{9.14}$$

Returning now to Equation 9.7 above and noting that $B = 0$, we can write the solution to the governing differential equation (Equation 9.6) as follows:

$$u = A \sin kx \tag{9.15}$$

Therefore, $n = 1$ yields the first mode of buckling of the column, which is depicted in Figure 9.5 and consist of a half sine wave. If $n = 2$ is chosen instead of $n = 1$, Equation 9.12 provides the second mode of buckling shown in Figure 9.7. Note that this mode, which consists of a full sine wave, can never occur unless external lateral forces p are applied as indicated in Figure 9.7. In such a case, the buckling load would be $4 P_{CR}$, where P_{CR} is Euler's buckling load.

As stated above, the ideal column buckles in the form of a half sine wave, but its amplitude A is indeterminate. This is due to the fact that no additional information is available outside of the two boundary conditions already used and the limitation imposed by the linear theory, which is based upon the approximation for the curvature used in developing the equation for the elastic curve. If the curvature were not approximated, it would be possible to determine the amplitude of the deflected shape, but this determination is of less interest than Equation 9.14 for Euler's critical load.

The normal stress that exists in the column when buckling occurs is referred to as the *critical stress* and is found by dividing the critical load by the cross-sectional area A of the column. Thus,

$$\sigma_{CR} = \frac{P_{CR}}{A} = \frac{\pi^2 E(I_v/A)}{L^2} = \frac{\pi^2 E}{(L/r_v)^2} \tag{9.16}$$

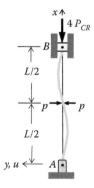

FIGURE 9.7   Second mode of buckling deformation of a column model.

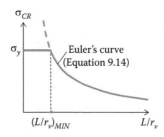

FIGURE 9.8 Theoretical design curve of an ideal Euler's column.

Note that in obtaining the last form of Equation 9.16, we made use of the fact that the radius of gyration $r_v = \sqrt{(I_v/A)}$ (see Appendix C.2). The quantity L/r_v is known as the *slenderness ratio* for a given column and, as will be seen later in this chapter, plays a significant role in the analysis and design of such a member.

A graph of Equation 9.16 is given in Figure 9.8 and labeled as *Euler's curve.* It is evident from Equation 9.16 that the only material property defining the *elastic behavior (buckling)* of an ideal column is the modulus of elasticity E of the material of which the column is fabricated. However, since Equation 9.16 is applicable only in the elastic range, the use of this equation is limited to values of the critical stress below the yield point or yield stress σ_y of the material (*assuming that σ_y represents the limit of elastic behavior*) and that part of Euler's curve above σ_y (broken line) does not apply. Also, as shown in Figure 9.8, σ_y defines a value, $(L/r_v)_{MIN}$, as a lower limit for the slenderness ratio below which Euler's equation does not apply. Equation 9.16 indicates that the critical stress is a direct function of the modulus of elasticity of the material. It follows, therefore, that each material of which columns are fabricated, requires its own Euler's curve similar to that shown in Figure 9.8.

EXAMPLE 9.3

An axially loaded, pin-ended column made of an aluminum alloy ($E = 75$ GPa, $\sigma_y = 200$ MPa) is to be 5 m in length and is to have a solid rectangular cross section as shown in the sketch. If the column is to carry a compressive load of 150 kN, determine the safe dimension b if a factor of safety of 2 is used. Check to see if the column is within the elastic range.

$$2b$$

$$b$$

SOLUTION

The minimum moment of inertia (see Appendix C.2) is $I_v = (1/12)(2b)b^3 = b^4/6$. Since the factor of safety is 2, the critical load becomes $P_{CR} = 2\,(150) = 300$ kN. Applying Euler's equation (Equation 9.14), we have

$$300 \times 10^3 = \frac{\pi^2(75 \times 10^9)(b^4/6)}{5^2}$$

Solving for the dimension b, we obtain

$$b = 0.0883 \text{ m} = 88.3 \text{ mm} \qquad \textbf{ANS.}$$

$$\sigma_{CR} = \frac{300 \times 10^3}{(0.0883)(2 \times 0.0883)} = 19.238 \text{ MPa} <<< \sigma_y$$

Since σ_{CR} is much less than the yield point of the material, the column is well within its elastic range.   **ANS.**

EXAMPLE 9.4

Refer to Example 9.3 and using the given material properties, construct a graph similar to that shown in Figure 9.8.

    a. Locate the particular point on this graph that represents the column of Example 9.3.
    b. Determine the minimum value of the slenderness ratio and the shortest length for this column below which Euler's equation no longer applies.

SOLUTION

    a. Equation 9.16 when specialized for the aluminum alloy of Example 9.3 becomes

$$\sigma_{CR} = \frac{\pi^2 E}{(L/r_v)^2} = \frac{\pi^2 (75 \times 10^9)}{(L/r_v)^2} = \frac{740\,220 \times 10^6}{(L/r_v)^2}$$

    A tabulation is prepared and the points plotted as shown. Of course, if needed, additional points may be computed and plotted. Note that the point (50, 296.1) falls outside of the range where Euler's equation applies.

    The specific point corresponding to the column of Example 9.3 may now be located on this graph by computing its slenderness ratio. Thus,

$$r_v = \sqrt{\frac{I_v}{A}} = \sqrt{\frac{(b^4/6)}{2b^2}} = \sqrt{\frac{b^2}{12}} = \sqrt{\frac{(0.0883)^2}{12}} = 0.02549 \text{ m}$$

$$\frac{L}{r_v} = \frac{5}{0.02549} = 196.155 \approx 196.2$$

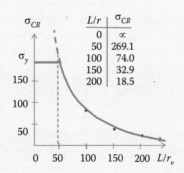

L/r	σ_{CR}
0	∞
50	269.1
100	74.0
150	32.9
200	18.5

The point on Euler's curve corresponding to the column of Example 9.3 is plotted with the coordinates (196.2, 19.2).

b. The minimum value of L/r_v is obtained from Equation 9.16 after replacing σ_{CR} with σ_y. Thus,

$$\left(\frac{L}{r_v}\right)_{MIN} = \sqrt{\frac{\pi^2 E}{\sigma_y}} = \sqrt{\frac{\pi^2(75 \times 10^9)}{200 \times 10^6}} = 60.837 \approx 60.8 \quad \text{(see point in graph above)} \qquad \textbf{ANS.}$$

The shortest column may be obtained from the known values of $(L/r_v)_{MIN}$ and r_v. Thus,

$$L_{MIN} = 60.837(0.02549) = 1.551 \text{ m} \approx 1.6 \text{ m} \qquad \textbf{ANS.}$$

EXAMPLE 9.5

An axially loaded, wide-flanged steel ($E = 29 \times 10^3$ ksi and $\sigma_y = 40$ ksi) column is to be pin-ended and have a length of 20 ft. Select the *lightest* wide flange section from Appendix E that will support a compressive load of 500 kips without buckling if a factor of safety of 2.5 is to be used.

SOLUTION

With the given factor of safety, the critical buckling load becomes $P_{CR} = 2.5(500) = 1250$ kips. Solving Euler's equation for I_v, we obtain

$$I_v = \frac{P_{CR}L^2}{\pi^2 E} = \frac{1250(20 \times 12)^2}{\pi^2(29 \times 10^3)} = 251.556 \text{ in.}^4 \approx 251.6 \text{ in.}^4$$

A check of Appendix E indicates that the *lightest* wide flange section with a minimum moment of inertia of at least 251.6 in.^{4} is a W18 $\times$ 119, which has an $I_v = 253$ in.^{4}

The required wide flange section is

$$\text{W18} \times 119 \qquad \textbf{ANS.}$$

This wide flange section has a cross-sectional area of 35.1 in.^{2} The critical stress is $\sigma_{CR} = 1250/35.1 \approx 35.6$ ksi, which is less than σ_y. Therefore, Euler's equation applies.

EXAMPLE 9.6

A pin-ended aluminum ($E = 75$ GPa, $\sigma_y = 150$ MPa) 3 m long, has the Z cross section shown in figure below. The thickness $t = 20$ mm is constant throughout. Determine the critical load for this column and justify the use of Euler's equation. What is the shortest length of this column for which Euler's equation applies?

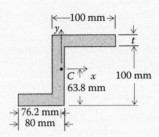

SOLUTION

The centroid of the cross is located as shown in figure above. The principal centroidal u and v axes are found and located as shown in figure below. The corresponding principal moments of inertia are computed and found to be $I_u = 14.8963 \times 10^{-6} \text{ m}^4$ and $I_v = 2.0165 \times 10^{-6} \text{ m}^4$ (see Appendix C.2). Thus, using Equation 9.14, we have

$$P_{CR} = \frac{\pi^2 (75 \times 10^9)(2.0165 \times 10^{-6})}{3^2}$$

$$P_{CR} = 165,850 \text{ N} \approx 165.9 \text{ kN} \qquad \textbf{ANS.}$$

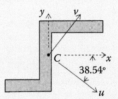

To justify the use of Euler's equation, we need to show that the critical stress is below the yield strength for the material. Thus,

$$\sigma_{CR} = \frac{P_{CR}}{A} = \frac{165.9 \times 10^3}{0.10(0.02) + 2(0.08)(0.02)} = 31.731 \times 10^6$$

$$\sigma_{CR} \approx 31.7 \text{ MPa} <<< \sigma_y$$

Since the critical stress is well below the yield strength for the give aluminum, the use of Euler's equation is permissible.

From Equation 9.16, after replacing σ_{CR} by σ_y, we obtain

$$\left(\frac{L}{r_v}\right)_{MIN} = \sqrt{\frac{\pi^2 E}{\sigma_y}} = \sqrt{\frac{\pi^2 (75 \times 10^9)}{200 \times 10^6}} = 60.837$$

$$L_{MIN} = 60.837 r_v = 60.837 \sqrt{\frac{I_v}{A}} = 60.837 \sqrt{\frac{2.0165 \times 10^{-6}}{0.0052}} = 1.198 \text{ m} = 1.2 \text{ m} \qquad \textbf{ANS.}$$

PROBLEMS

(Except where indicated, assume that Euler's equation applies.)

9.15 A column is to be fabricated of an aluminum alloy ($E = 10.5 \times 10^3$ ksi) and is to have a solid square cross section 2 in. on each side. If the length of the column is 4 ft, find the Euler critical load. A second column of the same material and length is to have a solid circular cross section and is to carry the same critical Euler load. Find the needed diameter for the circular column.

9.16 Repeat Problem 9.15 if the material is steel ($E = 200$ GPa) and the square cross section is to be 35 mm on each side and the length of the columns is 2 m.

9.17 A column is to be fabricated of steel ($E = 205$ GPa) and is to have a hollow circular cross section with outside diameter $d = 100$ mm and a wall thickness $t = 15$ mm as shown in figure (a) below. If the length of the column is 2 m, determine its Euler critical load. A second column of the same material and length is to have a solid rectangular cross section with outside dimensions a and $2a$ as shown in figure (b) below. If the rectangular column is to carry the same critical load as the circular column, determine the dimension a.

(a)

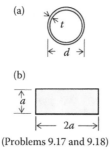

(b)

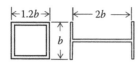

(Problems 9.17 and 9.18)

9.18 Repeat Problem 9.17 if the material is aluminum ($E = 10 \times 10^3$ ksi), $d = 5$ in., $t = 0.75$ in., and the length of the columns is 5 ft.

9.19 Two different cross sections, as shown, are being considered for use as an Euler column of a given length and a given volume. Determine on the basis of Euler's load-carrying capacity which of the two sections (the box or the I shape) is more effective. The constant thickness for both sections is $0.1b$.

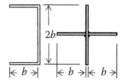

9.20 Two different cross sections, as shown, are being considered for use as an Euler column of a given length and a given volume. Determine on the basis of Euler's load-carrying capacity, which of the two sections (the channel or cruciform) is more effective. The constant thickness for both sections is $0.1b$.

9.21 A machine compression component 2 ft long has a solid square cross section 1.5 in. on each side and is made of an expensive material for which $E = 15 \times 10^3$ ksi. A proposal is made to reduce the cost by using a hollow square section of the same outside dimensions. If a 10% reduction in the Euler critical load is permissible, find the inside dimension of the hollow square and determine how much reduction in material cost would be achieved by using the hollow square section? Assume that the fabrication costs are the same for both.

9.22 Repeat Problem 9.21 if the machine component has a solid circular cross section 1.5 in. in diameter. All other conditions remain the same.

9.23 An aluminum ($E = 70$ GPa) Euler column, 10 m long, is fabricated by welding four plates to form the cross section shown. Let $a = 100$ mm and the thickness $t = 40$ mm, constant throughout. Determine the maximum allowable load if a factor of safety of 2 is to be used.

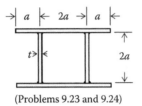

(Problems 9.23 and 9.24)

9.24 In reference to the figure in Problem 9.23, a steel ($E = 30 \times 10^3$ ksi) Euler column, 20 ft long, is fabricated by welding four plates to form the cross section shown. Let $a = 2$ in. and the thickness $t = 1$ in., constant throughout. Determine the factor of safety if a compressive axial load of 150 kips.

9.25 Two 127×13.4 steel ($E = 200$ GPa) channels are welded together to form the box section shown. If a compressive axial load of 150 kN is to be applied, determine the factor of safety. Let $L = 7$ m.

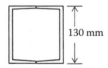

9.26 Two 127×13.4 steel ($E = 200$ GPa) channels are welded together to form the I section shown. If a factor of safety of 2 is needed, determine the maximum permissible Euler load. Let $L = 5$ m.

9.27 Two $5 \times 3 \times 1/2$ in. steel angle sections are to be welded together to fabricate a column. Assuming fabrication costs are identical, which of the three arrangements figure (a, b, or c) below is the one you would select for the application and why? Find the ratios of the Euler load-carrying capacities of the one you selected to the two not selected. The centroid location for this angle as well as its overall dimensions are shown in figure (a) below. The following properties are given for each angle: $A = 3.75$ in.^{2}, $I_x = 9.45$ in.^{4}, and $I_y = 2.58$ in.^{4}

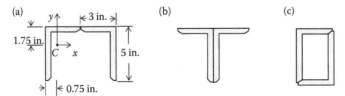

(Problems 9.27, 9.28, and 9.29)

9.28 A column is to be fabricated by welding two $5 \times 3 \times 1/2$ in. steel angle sections together as shown in figure (b) of Problem 9.27. Let $E = 29 \times 10^3$ ksi and $L = 15$ ft, determine the largest permissible Euler load if a factor of safety of 2.5 is required.

9.29 A column is to be fabricated by welding two $5 \times 3 \times 1/2$ in. steel angle sections together as shown in figure (c) of Problem 9.27. Find the factor of safety if the applied compressive axial load is 150 kips. Let $E = 29 \times 10^3$ ksi and $L = 12$ ft and assume Euler's equation applies.

9.30 Three identical pieces of magnesium are to be fastened together to fabricate a column. Assuming fabrication costs are identical, which of the three arrangements figure (a, b, or c) below is the one you would select for the application and why? Find the ratios of the Euler load-carrying capacities of the one you selected to the two not selected. Overall dimensions for each piece are shown in figure (b) below. Let $h = 10$ in.

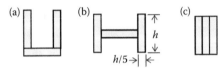

(Problems 9.30, 9.31, and 9.32)

9.31 A column is to be fabricated by fastening three identical pieces of magnesium together as shown in figure (a) of Problem 9.30. Let $E = 45$ GPa and $L = 4$ m, determine the largest permissible Euler load if a factor of safety of 2 is needed. Note that you need to convert I_y from Problem 9.30 to the metric system.

9.32 A column is to be fabricated by fastening three identical pieces of magnesium together as shown in figure (b) of Problem 9.30. Find the factor of safety if the applied load compressive axial load is 3,000 kN. Let $E = 45$ GPa and $L = 3$ m. Note that you need to convert I_y from Problem 9.30 to the metric system.

9.33 If a factor of safety of 2 is needed, determine the maximum permissible force P. Truss member BC is fabricated of an aluminum alloy ($E = 10 \times 10^3$ ksi) pipe, 3 in. outside diameter and 2.6 in. inside diameter.

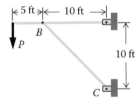

9.34 If a factor of safety of 2.4 is needed, determine the maximum permissible force P. Truss member BC is fabricated of a W356 × 38.7 steel ($E = 200$ GPa) section. Assume that the W356 × 38.7 section is oriented to buckle in the plane of the page.

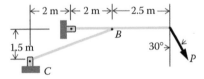

9.35 In reference to the figure in Problem 9.34, consider the frame shown in figure (a) below. Member *BD* has a box cross section fabricated by welding two $5 \times 3 \times 1/2$ in. steel angle sections ($E = 30 \times 10^3$ ksi) shown in figure (b) below. Determine the largest permissible force *P* that may be applied if a factor of safety of 2 is required. Assume that the section is oriented to force members *BD* to buckle in the plane of the page. The centroid for one of the angles is located in figure (b) below and the following properties are given for each angle: $A = 3.75$ in.^{2}, $I_x = 9.45$ in.^{4}, and $I_y = 2.58$ in.^{4}

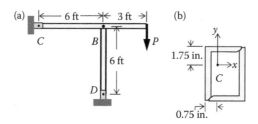

9.36 Determine Euler's critical load for the cross section indicated. Refer to Appendix C.2 for help in determining principal axes and principal moments of inertia. Let $a = 30$ mm, $b = 80$ mm, $c = 40$ mm, and $h = 120$ mm. The modulus of elasticity and the length for each column are indicated in the figures.

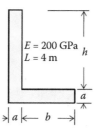

9.37 Determine Euler's critical load for the cross section indicated. Refer to Appendix C.2 for help in determining principal axes and principal moments of inertia. Let $a = 30$ mm, $b = 80$ mm, $c = 40$ mm, and $h = 120$ mm. The modulus of elasticity and the length for each column are indicated in the figures.

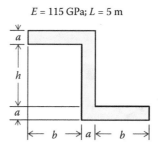

9.38 Determine Euler's critical load for the cross section indicated. Refer to Appendix C.2 for help in determining principal axes and principal moments of inertia. Let $a = 30$ mm, $b = 80$ mm, $c = 40$ mm, and $h = 120$ mm. The modulus of elasticity and the length for each column are indicated in the figures.

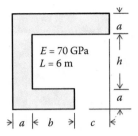

9.4 EFFECT OF END CONDITIONS

Columns with end supports different from the *pin-ended* Euler's column are used in many structural applications. Such columns are illustrated in Figure 9.9. The column in Figure 9.9a is fixed at one end and free at the other (*fixed-free* column), the one in Figure 9.9b is fixed at one end and pinned at the other (*fixed-pinned* column), and the column in Figure 9.9c is fixed at both ends (*fixed-fixed* column). We may determine the critical loads of such columns by writing the differential equations of their elastic curves and using the applicable boundary conditions, in a manner similar to that used for the Euler column. We will follow this process for the column of Figure 9.9a, the solution for which will help in developing the concept of the *effective length* of a column.

9.4.1 CRITICAL LOAD FOR COLUMN FIXED AT ONE END AND FREE AT THE OTHER

Such a column is shown in Figure 9.10a at the moment the compressive axial load reaches the critical value P_{CR}. The free-body diagram of segment AC of length x at the lower end of the buckled column is shown in Figure 9.10b. Equilibrium tells us that at C, the internal force is $F = P_{CR}$ and that the internal moment is $M = P_{CR} (\delta - u)$, where δ is the maximum deflection at the top of the column.

We are now able to write the differential equation for the bent shape of the column. Thus,

$$EI\,u'' = M = P_{CR}(\delta - u) \tag{9.17}$$

As in the case of Euler's equation, we introduce the notation $k = P_{CR}/EI$ and rewrite Equation 9.17) in the form

$$u'' + k^2 u = k^2 \delta \tag{9.18}$$

Equation 9.18 is a second-order, nonhomogeneous differential equation with constant coefficients. Its general solution is the sum of the homogeneous solution (see development of Euler's

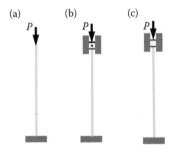

FIGURE 9.9 Column boundary conditions: (a) fixed-free, (b) fixed-pinned, and (c) fixed-fixed.

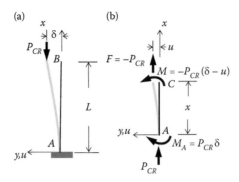

FIGURE 9.10   Buckled fixed-free column model and corresponding free-body diagram of its lower end.

equation), plus a particular solution based upon the term $k^2\delta$. It can be shown that the particular solution in this case is δ. Therefore, the general solution for Equation 9.18 becomes

$$u = A \sin kx + B \cos kx + \delta \tag{9.19}$$

$$u' = Ak \cos kx - Bk \sin kx \tag{9.20}$$

The three constants in Equations 9.19 and 9.20 are found from three boundary conditions as follows:

1. $x = 0, u = 0$; Equation 9.19 $\Rightarrow 0 = 0 + B(1) + \delta$ and

$$B = -\delta \tag{9.21}$$

2. $x = 0, u' = 0$; Equation 9.20 $\Rightarrow 0 = Ak(1) - 0$
   Since $k \neq 0$, it follows that

$$A = 0 \tag{9.22}$$

3. $x = L, u = \delta$; Equaiton 9.19 $\Rightarrow \delta = -\delta \cos kL + \delta$
   Simplifying, the third condition yields

$$\delta \cos kL = 0 \tag{9.23}$$

In Equation 9.23, the constant δ cannot be zero because, if this were the case, it would mean that the column is not deflected. As a matter of fact, the constant δ is undefined and may have any value without violating Equation 9.23. Thus, the quantity $\cos kL$ must be zero to satisfy Equation 9.23. This implies that

$$kL = \frac{n\pi}{2} \quad (n = 1, \pm3 \pm 5, \ldots) \tag{9.24}$$

Therefore,

$$k^2 L^2 = \frac{n^2 \pi^2}{4}, \quad \text{and} \quad k^2 = \frac{n^2 \pi^2}{4L^2} = \frac{P_{CR}}{EI} \tag{9.25}$$

Solving Equation 9.25 for P_{CR}, we obtain

$$P_{CR} = \frac{n^2 \pi^2 EI}{4L^2} \tag{9.26}$$

Obviously, the smallest critical load for this column is obtained by setting $n = 1$. The higher modes of this buckled column, obtained by using higher values on n, are not of interest at present. Therefore, the smallest critical load for a column fixed at one end and free at the other is

$$P_{CR} = \frac{\pi^2 EI_v}{4L^2} \tag{9.27}$$

9.4.2 EFFECTIVE LENGTH

The concept of the *effective length* of a column is very useful in finding the critical loads of columns with end conditions different from those of the pin-ended Euler's column. To illustrate, the fixed-free column is compared in Figure 9.11 to the pin-ended Euler's column, known as the *fundamental case*. While the shape of the buckled pin-ended Euler's column is a half sine wave, Figure 9.11a, that of the fixed-free column is a quarter sine wave, Figure 9.11b. In order to obtain a half sine wave for the fixed-free column, we need to add the mirror image of its buckled shape as shown in Figure 9.11b. This leads to the conclusion that while the pin-ended column has an effective length equal to its actual length ($L_e = L$), the fixed-free column has an effective length equal to twice its actual length ($L_e = 2L$). These conclusions are indicated in Figure 9.11. Another way to reach the same conclusions is to locate the *points of inflection* (i.e., points of zero moment) on the deflected shape and use the distance between them to define the effective length. Thus, in the case of the pin-ended column, these points are at the two ends of the column and the effective length is the full length of the column. In the case of the fixed-free column, one such point is at the free end and the other at the fictitious end after the mirror image of the deflected shape is added, resulting in a distance between the points of inflection, which is twice the actual length of the column.

Using the concept of the effective length of a column, we can modify Equation 9.14 slightly, and use this modified form to quickly determine the critical bucking loads of columns with end supports different from the fundamental Euler column. In view of the fact that the effective length of a column represents the length of an equivalent Euler column, we are able to rewrite Equation 9.14

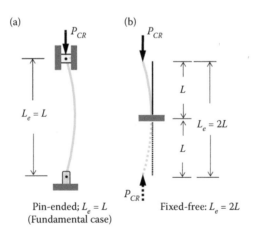

FIGURE 9.11 Concept of column effective length.

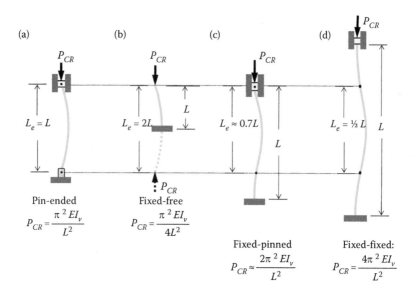

FIGURE 9.12   Critical buckling loads of columns with different end conditions.

in a more general form that may be used to find the critical load for any column as long as we can determine its effective length. Thus,

$$P_{CR} = \frac{\pi^2 EI_v}{L_e^2} \tag{9.28}$$

Applying Equation 9.28 to the case of the fixed-free column, which has an effective length $Le = 2L$, we find its critical load to be $P_{CR} = \pi^2 EI_v/4L^2$, which is identical to the critical load given by Equation 9.27 found above by a lengthy derivation.

Let us now consider the fixed-pinned and the fixed-fixed columns to determine their critical loads. All we need to do is compare their deflected shapes with that of the fundamental pin-ended column. This is done in Figure 9.12 where Figure 9.12a is the fundamental pin-ended column and Figure 9.12b is the fixed-free column, both of which have already been discussed. The fixed-pinned column is shown in Figure 9.12c, along with its deflected shape. A theoretical analysis of this column indicates that the first mode of failure possesses one point of inflection at a distance of $0.699L \approx 0.7L$ from the pinned end as shown. It follows, therefore, that the effective length of this column is $L_e = 0.7L$. Substituting this value into Equation 9.28 yields $P_{CR} \approx 2\pi^2 EI_v/L^2$. Similarly, the case of the fixed-fixed column is shown in Figure 9.12d along with its deflected shape. A theoretical analysis of this column reveals that the first mode of failure has two points of inflection symmetrically located at the center of the column at a distance of $1/2\,L$ as shown. Therefore, the effective length of this column is $L_e = 1/2\,L$ and Equation 9.28 yields $P_{CR} = 4\pi^2 EI_v/L^2$. For convenience, all of the above information is summarized in Figure 9.12.

EXAMPLE 9.7

Member *AB* is imbedded in a rigid wall at *B* and hinged at the center of rigid member *CD* and a horizontal load of 10 kips is applied at *D* as shown. Member *AB* is a hollow aluminum ($E = 10 \times 10^3$ ksi) pipe 3.5 in. outside diameter. If a factor of safety of 1.5 is needed, find the largest permissible inside diameter of the pipe.

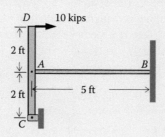

SOLUTION

The free-body diagram of rigid member CD (not shown) leads to the conclusion that the compressive force acting on member AB at the hinge at A is 20 kips. Obviously, column AB must be designed for $P_{CR} = 1.5 \times 20 = 30$ kips.

Member AB is a fixed-pinned column and according to Figure 9.12, its effective length is $L_e = 0.7 \times 5 = 3.5$ ft.

Solving Equation 9.28 for I_v yields

$$I_v = \frac{P_{CR}L_e^2}{\pi^2 E} = \frac{30(3.5 \times 12)^2}{\pi^2(10 \times 10^3)} = 0.53619 \text{ in.}^4$$

For a hollow circular cross section, $I_v = \pi/64(d_o^4 - d_i^4)$, where d_o and d_i are the outside and inside diameters, respectively. Thus,

$$\frac{\pi}{64}(3.5^2 - d_i^2) = 0.53619$$

Solving for d_i we obtain

$$d_i = 1.072 \text{ in.} \approx 1.1 \text{ in.} \qquad \textbf{ANS.}$$

PROBLEMS

9.39 A steel ($E = 30 \times 10^3$ ksi) machine component is idealized as shown where the support at A is a frictionless slider and that at B is a frictionless roller. The following information is known: $P = 5$ kips, $L = 18$ in., and member AB has a square cross section. Find the smallest size of the square if the required factor of safety relative to buckling is (a) 2 and (b) 3.

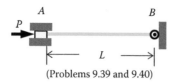

(Problems 9.39 and 9.40)

9.40 In reference to the figure in Problem 9.39, an aluminum ($E = 75$ GPa) machine component is idealized as shown where the support at A is a frictionless slider and that at B is a frictionless roller. The following information is known: $P = 30$ kN and member AB has a solid circular cross section with a diameter of 60 mm. If the needed factor of safety relative to buckling is 2.5, determine the largest permissible length L.

9.41 Member AB is subjected to the compressive force P as shown in figure (a) below. Its cross section is a hollow rectangle as shown in figure (b) below. Let $P = 10$ kips, $L = 2$ ft, and $E = 15 \times 10^3$ ksi. If a factor of safety of 2 relative to buckling is required, find the minimum permissible value of a.

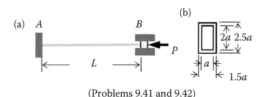

(Problems 9.41 and 9.42)

9.42 In reference to the figure in Problem 9.41, member AB is subjected to the compressive force P as shown in figure (a) above. Its cross section is a hollow rectangle as shown in figure (b) above. Let $L = 0.75$ m, $a = 50$ mm, and $E = 50$ GPa. If a factor of safety of 2.5 relative to buckling is required, find the maximum permissible compressive load P.

9.43 Steel ($E = 30 \times 10^3$ ksi) member ABC is hinged at A, fixed at C, and supported at B by rollers. The member has a hollow circular cross section with an outside diameter of 5 in. and an inside diameter of 4 in. If $L = 10$ ft and the needed factor of safety relative to buckling is 3, find the maximum permissible compressive load P.

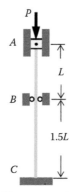

(Problems 9.43 and 9.44)

9.44 In reference to the figure in Problem 9.43, member ABC is hinged at A, fixed at C, and supported at B by rollers. The member has a steel ($E = 200$ GPa) W section placed so that its weak axis is perpendicular to the plane of the page. If $P = 300$ kN, $L = 3$ m, and the needed factor of safety relative to buckling is 2, select the lightest suitable W section from Appendix E.

9.45 Member ABC ($E = 30 \times 10^3$ ksi) is fixed at A against rotation to a frictionless slider, pinned at C, and supported at B by frictionless rollers as shown. The member is made of a standard 3-in. pipe (see Appendix E). If $L = 12$ ft and the required factor of safety relative to buckling is 2.5, find the maximum permissible compressive load P.

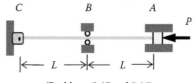

(Problems 9.45 and 9.46)

9.46 In reference to the figure in Problem 9.45, member *ABC* (*E* = 200 GPa) is fixed at *A* against rotation to a frictionless slider, pinned at *C*, and supported at *B* by frictionless rollers as shown. The member is made of a standard 102 mm pipe (see Appendix E). If *P* = 100 kN and the required factor of safety relative to buckling is 2, find the maximum permissible length *L*.

9.47 Two aluminum (*E* = 75 GPa) plates, 30 mm thick, are welded together to form the *T* section shown. The length of the column is 4 m and may be supported only at its two ends. If a factor of safety of 2.5 is needed, determine the critical load for each of the two following end conditions: (a) pinned-pinned and (b) fixed-free.

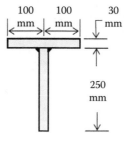

(Problems 9.47 and 9.48)

9.48 In reference to the figure in Problem 9.47, two steel (*E* = 205 GPa) plates are welded together to form the T section shown. The length of the column is 5 m and may be supported only at its two ends. If a factor of safety of 2 is needed, determine the critical load for each of the two following end conditions: (a) fixed-pinned and (b) fixed-fixed.

9.49 An aluminum (*E* = 10.5 × 10³ ksi) angle section is used for a column. The column is fixed at one end and free at the other and the support system allows the column to buckle in any direction. The column is to carry a compressive load of 30 kips. If a factor of safety of 2.3 is needed, determine the maximum permissible length of the column.

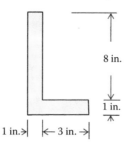

9.50 A steel (*E* = 200 GPa) Z section is used for a column. The column is fixed at one end and pinned at the other and the support system allows the column to buckle in any direction. The column is to have a length of 4 m. If a factor of safety of 2 is required, determine the maximum permissible compressive load that may be applied. Let *a* = 30 mm, *b* = 80 mm, and *h* = 120 mm.

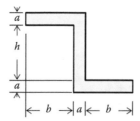

9.51 A magnesium ($E = 6.5 \times 10^3$ ksi) section is used for a column of length L. The column is fixed at both ends and the support system allows the column to buckle in any direction. The column is to carry a compressive load of 500 kips. If a factor of safety of 2 is needed, determine the largest permissible length L. Let $a = 2$ in.

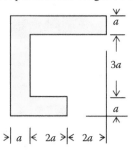

9.5   SECANT FORMULA

The "perfect" column discussed earlier does not exist. Its solution, however, provides some insights into the behavior of *real* columns. In order to obtain a closer approximation to the actual behavior of real columns, we will investigate two cases of interest. We will first develop the governing equation for an eccentrically loaded pin-ended column and then that for a pin-ended column that is initially bent in a sinusoidal shape.

9.5.1   Eccentrically Loaded Pin-Ended Column

Figure 9.13a shows a pin-ended column AB of length L subjected to an eccentric compressive force at B. We will make the assumption that despite the eccentricity e, the frictionless slider at B moves freely in the vertical direction. The lip at A is an integral part of the column and allows us to apply the compressive load eccentrically.

   The free-body diagram of the column when the critical load is reached is shown in Figure 9.13b. The maximum deflection of the column, which occurs at its center, is δ. The free-body diagram of a segment of length x at the lower end of the column is shown in Figure 9.13c. Equilibrium of this segment leads us to conclude that, at any section a distance x from the hinge at A, the axial force $F = -P$, and the bending moment $M = -P(e + u)$, where u is the deflection of the column at that point. Therefore, the differential equation for the deflected shape of the column becomes

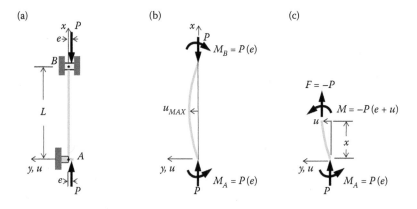

FIGURE 9.13   Eccentrically loaded pin-ended column model.

$$EI\,u'' = M = -P(e + u) \tag{9.29}$$

Letting $k^2 = P/EI_y$, Equation 9.29 may be rewritten in the following form:

$$u'' + k^2 u = -k^2 e \tag{9.30}$$

Equation 9.30 is a second-order, nonhomogeneous differential equation with constant coefficients. Its general solution is similar to the solution of Equation 9.18 and its general solution becomes

$$u = A \sin kx + B \cos kx - e \tag{9.31}$$

There are two boundary conditions enabling us to find the constants A and B as follows:

1. $x = 0$, $u = 0 \Rightarrow 0 = 0 + B(1) - e$ and $B = e$

$$B = e \tag{9.32}$$

2. $x = L$, $u = 0 \Rightarrow 0 = A \sin kL + e \cos kL - e$ and

$$A \sin kL = e(1 - \cos kL) \tag{9.33}$$

Substituting the trigonometric identities $\sin kL = 2 \sin kL/2 \cos kL/2$ and $\cos kL = 1 - 2 \sin^2 kL/2$ into Equation 9.33 and simplifying, we obtain

$$A = e \tan \frac{kL}{2} \tag{9.34}$$

Substituting into Equation 9.31 the values of the constants A and B found above, we obtain the deflection function for this column. Thus,

$$u = e\left[\tan \frac{kL}{2} \sin kx + \cos kx - 1 \right] \tag{9.35}$$

where $k = \sqrt{P/EI_y}$. We conclude, therefore, that unlike the ideal Euler's column, the eccentrically loaded column begins to exhibit deflections the moment a compressive force P is applied. This deflection increases as the load is increased.

9.5.2 Maximum Deflection

In view of the geometric and loading symmetry, the maximum deflection of the column occurs at $x = L/2$. Substituting this value for x in Equation 9.35 yields

$$u_{MAX} = e\left[\tan \frac{kL}{2} \sin \frac{kL}{2} + \cos \frac{kL}{2} - 1 \right] = e\left[\frac{\sin^2 kL/2 + \cos^2 kL/2}{\cos kL/2} - 1 \right] \tag{9.36}$$

Since $\sin^2 kL/2 + \cos^2 kL/2 = 1$ and $k = \sqrt{P/EI_y}$, Equation 9.36 becomes

$$u_{MAX} = e\left[\sec \sqrt{\frac{PL^2}{4EI_y}} - 1 \right] = e\left[\sec \sqrt{\frac{\pi^2 P}{(4\pi^2 EI_y/L^2)}} - 1 \right] = e\left[\sec \frac{\pi}{2} \sqrt{\frac{P}{(\pi^2 EI_y/L^2)}} - 1 \right] \tag{9.37}$$

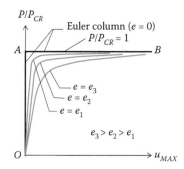

FIGURE 9.14 P/P_{cr} versus maximum centerline deflection for an eccentrically loaded pin-ended column.

The quantity $\pi^2 EI_v/L^2$ is equal to P_{CR}, the Euler critical buckling load. Equation 9.37 may, therefore, be written as

$$u_{MAX} = e\left[\sec\frac{\pi}{2}\sqrt{\frac{P}{P_{CR}}} - 1\right] \qquad (9.38)$$

Examination of Equation 9.38 reveals that the center deflection of the column becomes infinite when the quantity $(\pi/2)\sqrt{P/P_{CR}} = \pi/2$. This, of course, implies that $P/P_{CR} = 1$, and that $P = P_{CR}$. In other words, theoretically speaking, the eccentrically loaded column does not buckle until its maximum center deflection is infinite at which time the applied compressive load reaches the critical load for an ideal Euler column of identical length. Practically, however, this cannot take place because the material of the column would be in the plastic range of behavior much before this condition is reached. Thus, in the case of eccentrically loaded columns, we are interested, not in finding a critical load, but rather in finding the magnitude of the compressive load that can be applied without exceeding a given allowable center deflection or, as discussed later, in finding the magnitude of this load without exceeding a given allowable stress.

A schematic sketch of Equation 9.38 is shown in Figure 9.14 where the maximum center deflection is plotted on the horizontal axis and the dimensionless quantity P/P_{CR} on the vertical axis. As discussed earlier, the ideal Euler column remains straight until the critical buckling load is reached. This behavior is depicted by lines OA and AB in Figure 9.14 and indicates that as soon as P/P_{CR} reaches unity (point A), the deflection of the ideal Euler column becomes indeterminate and could reach infinity. Three qualitative curves are shown for three different values of the eccentricity e. Note that regardless of the value of the eccentricity, the curve approaches the ideal Euler column curve asymptotically as the center deflection becomes large.

9.5.3 Secant Formula

At any position along its length, the eccentrically loaded column is subjected to a combination of a constant compressive load P and a bending moment that increases as the center deflection increases. It follows therefore, that the maximum stress in such a column occurs at the position where the moment is maximum. This position, of course, is the center of the column where the maximum deflection occurs. Thus, using Equations 1.2 and 3.24, we obtain

$$\sigma_{MAX} = \frac{P}{A} + \frac{M_{MAX}\,c}{I_v} \qquad (9.39)$$

In Equation 9.39, A is the cross-sectional area of the column and I_v is the moment of inertia about its weak axis. The maximum moment in the column is $M_{MAX} = P(e + u_{MAX})$. Since $I_v = Ar_v^2$, Equation 9.39 becomes

$$\sigma_{MAX} = \frac{P}{A} + \frac{P(e + u_{MAX})c}{Ar_v^2} = \frac{P}{A}\left[1 + \frac{c}{r_v^2}(e + u_{MAX})\right] \tag{9.40}$$

Substituting from Equation 9.38 for u_{MAX} and simplifying, Equation 9.40 becomes

$$\sigma_{MAX} = \frac{P}{A}\left[1 + \frac{ec}{r_v^2}\sec\frac{\pi}{2}\sqrt{\frac{P}{P_{CR}}}\right] \tag{9.41}$$

Another and perhaps a more convenient form of Equation 9.41 is obtained by replacing P_{CR} using Equation 9.14, $P_{CR} = \pi^2 EI_v/L^2 = \pi^2 EAr_v^2/L^2$. After simplification, this leads to

$$\sigma_{MAX} = \frac{P}{A}\left[1 + \frac{ec}{r_v^2}\sec\left(\frac{1}{2}\frac{L}{r_v}\sqrt{\frac{P}{EA}}\right)\right] \tag{9.42}$$

The quantity ec/r_v^2 in Equation 9.42 is known as the *eccentricity ratio*, which is a dimensionless quantity measuring the eccentricity of loading relative to the cross-sectional properties of the column. The quantity L/r_v is the slenderness ratio of the column discussed earlier.

Equation 9.42 is known as the *secant formula* and applies only to pin-ended eccentrically loaded columns. However, it may be used for eccentrically loaded columns that are fixed at one end and free at the other by using the effective length of such a column ($L_e = 2L$) in place of the length L in Equation 9.42. However, whether the column is eccentrically loaded pin-ended or eccentrically loaded fixed-free, we need to insure that the maximum stress does not exceed the yield strength of the material.

Equation 9.42 is a transcendental equation. Therefore, in the solution of a given problem where all of the column geometric and material properties are known, it becomes necessary to solve for the quantity P/A by trial and error. Usually, it is necessary to find P/A in order not to exceed a given value of σ_{MAX}, which, generally speaking, is the yield strength for the material.

A qualitative graph of Equation 9.42, for a given value of E, is shown in Figure 9.15, where the slenderness ratio L/r_v is plotted on the horizontal axis and the quantity P/A on the vertical axis for three qualitative values of the eccentricity ratio ec/r_v^2. For comparison purposes, the curve for an Euler column of the same material is also shown. We note from this graph that, for a given L/r_v, the larger the eccentricity of loading, the less load the column can carry. This is especially true for

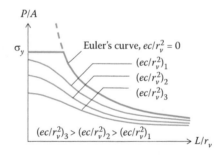

FIGURE 9.15 P/A versus slenderness ratio L/r_v for an eccentrically loaded pin-ended column.

small values of L/r_v, namely, for short and intermediate columns. For long columns, however, the eccentricity of the load becomes less pronounced as all curves approach the Euler curve.

9.5.4 Initially Bent Pin-Ended Columns

An initially curved column is shown in Figure 9.16a. For convenience, the bent shape is assumed sinusoidal of the form

$$u_o = A_o \sin \frac{\pi x}{L} \tag{9.43}$$

where u_o is the initial deflection at any point a distance x from the origin and A_o, the initial amplitude or midpoint deflection of the column as shown. The free-body diagram of a segment of length x at the lower end of the column is shown in Figure 9.16b. Equilibrium of this segment leads us to conclude that, at any section a distance x from the hinge at A, the axial force $F = -P$, and the bending moment $M = -P(u_o + u)$, where u is the deflection of the column due to the applied load at that point. Therefore, the differential equation for the deflected shape of the column becomes

$$EIu'' = M = -P(u_o + u) \tag{9.44}$$

Proceeding as in the case of the eccentrically loaded column, we obtain the following differential equation:

$$u'' + k^2 u = -k^2 A_o \sin \frac{\pi x}{L} \tag{9.45}$$

Equation 9.45 is a second-order, nonhomogeneous differential equation with constant coefficients. Its general solution may be written in the form

$$u = A \sin kx + B \cos kx - \left(\frac{k^2 L^2 A_o}{k^2 L^2 - \pi^2} \right) \sin \frac{\pi x}{L} \tag{9.46}$$

The constants A and B in Equation 9.46 are found from the two boundary conditions at the supports of the column. Thus,

1. $x = 0, u = 0 \Rightarrow B = 0$
2. $x = L, u = 0 \Rightarrow A \sin kL = 0$

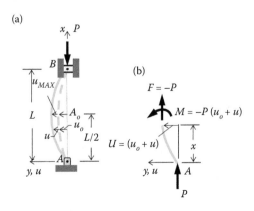

FIGURE 9.16 Initially bent pin-ended column model.

Substituting these values into Equation 9.46, we obtain

$$u = -\left(\frac{k^2 L^2 A_o}{k^2 L^2 - \pi^2}\right)\sin\frac{\pi x}{L} \tag{9.47}$$

Since the total displacement of the column is $U = u_o + u$ (see Figure 9.16), we obtain

$$U = A_o \sin\frac{\pi x}{L} - \left(\frac{k^2 L^2 A_o}{k^2 L^2 - \pi^2}\right)\sin\frac{\pi x}{L} = \left(1 - \frac{k^2 L^2}{k^2 L^2 - \pi^2}\right)A_o \sin\frac{\pi x}{L} \tag{9.48}$$

Since $k^2 = P/EI_v$, the quantity $k^2 L^2$ may be expressed in the following form:

$$k^2 L^2 = \frac{PL^2}{EI_v} = \frac{\pi^2 P}{\pi^2 (EI_v/L^2)} = \frac{\pi^2 P}{P_{CR}} \tag{9.49}$$

Substituting Equation 9.49 into Equation 9.48 and simplifying, we obtain

$$U = \left(\frac{A_o}{1 - P/P_{CR}}\right)\sin\frac{\pi x}{L} \tag{9.50}$$

The maximum total deflection of the column, U_{MAX}, occurs at midpoint ($x = L/2$) and becomes

$$U_{MAX} = \frac{A_o}{1 - P/P_{CR}} \tag{9.51}$$

Equation 9.51 shows that U_{MAX} becomes infinite when $P/P_{CR} = 1$ or $P = P_{CR}$. As in the case of an eccentrically loaded column, theoretically speaking, the initially curved column does not buckle until its maximum center deflection is infinite at which time the applied compressive load reaches the critical load for an ideal Euler column of identical length. Practically, however, this cannot take place because the material of the column would be in the plastic range of behavior much before this condition is reached. A qualitative plot of Equation 9.51 is shown in Figure 9.17. Note the similarities between this graph and that for the eccentrically loaded column shown in Figure 9.14.

Again, as in the case of eccentrically loaded columns, we are interested, not in finding a critical load, but rather in finding the magnitude of the compressive load that can be applied without exceeding a given allowable center deflection or, in finding the magnitude of this load without exceeding a given allowable stress.

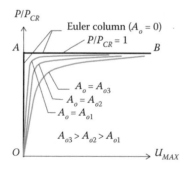

FIGURE 9.17 P/P_{cr} versus maximum centerline deflection for an initially bent pin-ended column.

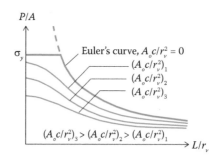

FIGURE 9.18 P/A versus slenderness ratio L/r_y for an initially bent pin-ended column.

The maximum stress in an initially curved column is given by Equation 9.39. Thus,

$$\sigma_{MAX} = \frac{P}{A} + \frac{M_{MAX}c}{I_v} = \frac{P}{A} + \frac{P(U_{MAX})c}{Ar_v^2} \tag{9.52}$$

Substituting the value of U_{MAX} from Equation 9.51 into Equation 9.52 and simplifying, we obtain

$$\sigma_{MAX} = \frac{P}{A}\left[1 + \frac{A_o c}{r_v^2}\left(\frac{1}{1 - P/P_{CR}}\right)\right] \tag{9.53}$$

Finally, substituting $P_{CR} = \pi^2 EA/(L/r_v)^2$ into Equation 9.53 leads to

$$\sigma_{MAX} = \frac{P}{A}\left[1 + \frac{A_o c}{r_v^2}\left(\frac{\pi^2 E}{\pi^2 E - P/A(L/r_v)^2}\right)\right] \tag{9.54}$$

The quantity $A_o c/r_v^2$ may be looked upon as a form of eccentricity ratio similar to ec/r_v^2. A qualitative graph of Equation 9.54 is given in Figure 9.18. Note the similarities between this graph and that for the eccentrically loaded column shown in Figure 9.15.

As stated earlier, ideal columns do not exist. Practical columns contain all sorts of imperfections that include initial eccentricity of loading despite attempts to the contrary, initial curvature despite all the care to fabricate a straight column, less than ideal hinges for the supports, and less than ideal (homogeneous and isotropic) material. Therefore, in the case of imperfect practical conditions, a column will begin to deflect as soon as the load is applied, resulting in the combined effect of a bending moment and a compressive force. To deal with such imperfections in practice, therefore, a *centrally* loaded column may be assumed to be *eccentrically loaded*. The extent of the assumed eccentricity would depend on the conditions that exist in a given situation but most often used values of the eccentricity ratios are less than one.

Let us now return to the secant formula expressed in Equation 9.42. A quantitative graph of this equation is plotted in Figure 9.19 (similar to the qualitative graph of Figure 9.15) for a specific steel whose modulus of elasticity $E = 29 \times 10^3$ ksi and whose yield strength $\sigma_y = 36$ ksi. Curves for six different values of the eccentricity ratio, in addition to Euler's curve ($ec/r_v^2 = 0$), are shown. Such graphs make it possible to find, without trial and error, the ratio P/A for given eccentricity and slenderness ratios. Therefore, to use a graph such as shown in Figure 9.19 in finding P/A for a specific set of conditions, we determine ec/r_v^2 and L/r_v corresponding to these conditions. We then locate a

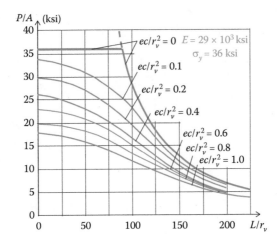

FIGURE 9.19 P/A versus slenderness ratio L/r_y for an eccentrically loaded pin-ended column with $E = 29 \times 10^3$ ksi and $\sigma_y = 36$ ksi.

point along the L/r_y axis corresponding to our specific value of L/r_y and draw a vertical line through it until it intersects the curve corresponding to our specific value of ec/r_v^2. The needed value of P/A is then read on the vertical axis. Note that interpolation between curves may become necessary. However, we should keep in mind, that such solutions are valid only for P/A values less than the yield strength of the material, which in this case is 36 ksi.

EXAMPLE 9.8

The hollow square cross section with outside dimensions of 50 mm × 50 mm shown is that for a pin-ended column 2 m long. A compressive load of 10 kN is applied eccentrically at point P a distance of 4 mm from the centroid C of the section which has a constant thickness $t = 5$ mm all around. Determine (a) the maximum deflection at the free end and (b) the maximum stress in the column. Assume $E = 100$ GPa and $\sigma_y = 20$ MPa.

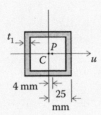

SOLUTION

The following properties are needed to determine the required quantities:

$$A = 0.050^2 - 0.040^2 = 0.0009 \text{ m}^2; \quad I_v = (1/12)(0.050^4 - 0.040^4) = 3.07 \times 10^{-7} \text{ m}^4$$

$$e = 0.004 \text{ m}; \quad P = 10 \text{ kN}; \quad L_e = 4 \text{ m}; \quad P_{CR} = \frac{\pi^2(100 \times 10^9)3.07 \times 10^{-7}}{4^2} = 18.937 \text{ kN}$$

$$r_v = \sqrt{\frac{3.07 \times 10^{-7}}{9.0 \times 10^{-4}}} = 0.01847 \text{ m}; \quad \frac{ec}{r_v^2} = \frac{0.004(0.025)}{0.01847^2} = 0.29313$$

a. The maximum deflection along the u axis at the free end of the column is given by Equation 9.38. Thus,

$$u_{MAX} = 0.004\left[\sec\frac{\pi}{2}\sqrt{\frac{10}{18.937}} - 1\right] = 0.00561 \text{ m} \approx 5.6 \text{ mm} \qquad \textbf{ANS.}$$

b. The maximum stress in the column is given by Equation 9.42. Thus,

$$\sigma_{MAX} = \frac{10}{0.0009}\left[1 + 0.29313\sec\left(\frac{4}{2(0.01847)}\right)\sqrt{\frac{10(10^3)}{100\times10^9(9.0\times10^{-4})}}\right]$$

$$\approx 19.0 \text{ MPa} \qquad \textbf{ANS.}$$

Since $\sigma_{MAX} < \sigma_y$, the above computations are valid.

EXAMPLE 9.9

A W12 × 50 steel ($E = 29 \times 10^3$ ksi, $\sigma_y = 36$ ksi) section is used for a pin-ended column 15 ft long. Structural requirements result in an eccentrically applied compressive load at point P a distance of 2 in. from the centroid of the section, as shown in the sketch. Assume that the column is braced so that buckling can occur only about the strong axis. (a) Use the curves in Figure 9.19 to determine the maximum allowable load-carrying capacity of the column assuming a factor of safety of 2. (b) Find the maximum deflection of this column due to the allowable load.

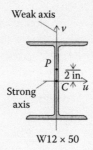

W12 × 50

SOLUTION

a. From Appendix E, we obtain the following properties for the W12 × 50:
$A = 14.7$ in.^{2}, $d = 12.19$ in., and about the strong axis, $I_u = 394$ in.^{4}, $r_u = 5.18$ in.
In order to use Figure 9.16, we need the following two quantities:

$$\frac{L}{r_u} = \frac{15\times12}{5.18} = 34.749 \approx 35; \quad \frac{ec}{r_u^2} = \frac{2(12.19/2)}{5.18^2} = 0.454 \approx 0.45$$

Note that we used the subscript u instead of v because, in this case, the column is forced to buckle about its strong axis. Since no curve is provided for $ec/r_u^2 = 0.45$, we interpolate between the curves corresponding to eccentricity ratios of 0.4 and

0.6 in order to get a value for P/A. This, of course, leads to an approximate but acceptable solution for purposes of this example. The approximate value obtained is $P/A = 24$ ksi. Therefore,

$$P = 24A = 24(14.7) = 352.8 \text{ kips}$$

Since a factor of safety of 2 is required, the maximum allowable load is

$$P_{ALL} = \frac{352.8}{2} = 176.4 \approx 176 \text{ kips} \qquad \textbf{ANS.}$$

b. Equation 9.38 enables us to find the maximum deflection at the center of the column. To use this equation, we need Euler's critical load for this column. Using 394 in.⁴ for I_y in Equation 9.14, we find $P_{CR} = 3480.6$ kips. Thus, Equation 9.38 yields

$$u_{MAX} = 2\left[\sec\frac{\pi}{2}\sqrt{\frac{176.4}{3480.6}} - 1\right] = 2\left[\sec 0.354\right]$$

$$= 2\left[1.066 - 1\right] = 0.132 \text{ in.} \qquad \textbf{ANS.}$$

PROBLEMS

9.52 A pin-ended steel ($E = 30 \times 10^3$ ksi, $\sigma_y = 40$ ksi) column of length L has a solid rectangular cross section as shown. A compressive force P is applied at the ends a distance e from the geometric center of the cross section. Let $P = 100$ kips, $L = 12$ ft, $e = 1$ in., $b = 4$ in., and $h = 6$ in., determine (a) the maximum deflection and (b) the maximum stress.

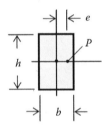

(Problems 9.52 and 9.53)

9.53 In reference to the figure in Problem 9.52, a pin-ended aluminum ($E = 200$ GPa, $\sigma_y = 400$ MPa) column of length L has a solid rectangular cross section as shown. A compressive force P is applied at the ends a distance e from the geometric center of the cross section. Let $L = 2$ m, $e = 30$ mm, $b = 100$ mm, and $h = 150$ mm. If the maximum deflection is limited to 12 mm, determine (a) the maximum permissible force P and (b) the maximum stress in the column.

9.54 A fixed-free steel ($E = 30 \times 10^3$ ksi, $\sigma_y = 40$ ksi) column of length L has a solid square cross section as shown. A compressive force P is applied at the free end a distance e from the geometric center of the cross section. Let $a = 1.5$ in., $P = 25$ kips, and $L = 24$ in. If the maximum deflection is 0.05 in., find (a) the eccentricity e and (b) the maximum stress in the column.

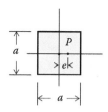

(Problems 9.54 and 9.55)

9.55 In reference to the figure in Problem 9.54, a pin-ended brass (E = 100 GPa, σ_y = 45 MPa) column of length L has a solid square cross section as shown. A compressive force P is applied at the free end a distance e from the geometric center of the cross section. Let a = 40 mm, e = 12 mm, and L = 0.75 m. If the maximum stress is found to be 76.5 MPa, find (a) the compressive force P and (b) the maximum deflection of the column.

9.56 A fixed-free brass (E = 15 × 10³ ksi, σ_y = 40 ksi) column of length L has a solid circular cross section of diameter D = 3.0 in. as shown. It is subjected to a compressive axial load P = 15 kips applied at a distance e = 0.3 in. from the centroid of the section. If the maximum stress in the column is limited to 20 ksi, find the largest allowable length L and the corresponding maximum deflection.

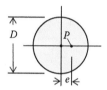

(Problems 9.56 and 9.57)

9.57 In reference to the figure in Problem 9.56, a pin-ended brass (E = 120 GPa, σ_y = 70 MPa) column of length L = 3 m has a solid circular cross section of diameter D as shown. It is subjected to a compressive axial load P = 60 kN applied at a distance e = 8 mm from the centroid of the section. If the maximum stress in the column is limited to 8 MPa, find the smallest allowable diameter D and the corresponding maximum deflection.

9.58 A W12 × 170 steel (E = 30 × 10³ ksi, σ_y = 40 ksi) wide flange section is used for a fixed-free column of length L and a compressive load P is applied eccentrically at a distance e from the centroid of the section as shown. Let P = 120 kips, e = 3.5 in., and L = 15 ft, determine (a) the maximum deflection of the column and (b) the maximum stress in the column.

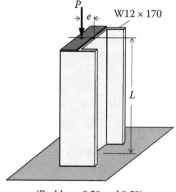

(Problems 9.58 and 9.59)

9.59 In reference to the figure in Problem 9.58, a W305 × 52.1 steel ($E = 205$ GPa, $\sigma_y = 150$ MPa) wide flange section is used for a fixed-free column of length L and a compressive load P is applied eccentrically at a distance e from the centroid of the section as shown. Let $e = 50$ mm and $L = 5$ m. If the maximum deflection of the column is to be limited to 60 mm, determine (a) the maximum permissible force P and (b) the maximum stress in the column.

9.60 A cast aluminum ($E = 10 \times 10^3$ ksi, $\sigma_y = 18$ ksi) machine component of length $L = 18$ in. has an I cross section as shown. During service, the component is subjected to an eccentric compressive load $P = 10$ kips applied at a distance $e = 3/4$ in. from the centroid of the I section. Assume the member to be a pin-ended column and find (a) the maximum deflection and (b) the maximum stress.

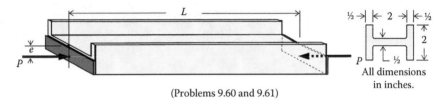

(Problems 9.60 and 9.61)

All dimensions in inches.

9.61 In reference to the figure in Problem 9.60, a cast aluminum ($E = 10 \times 10^3$ ksi, $\sigma_y = 18$ ksi) machine component of length $L = 18$ in. has an I cross section as shown. During service, the component is subjected to an eccentric compressive load P applied at a distance $e = 1/2$ in. from the centroid of the I section. Assume the member to be a pin-ended column and if the stress is limited to 8 ksi, (a) find the 0 maximum allowable load P. (b) What is the maximum deflection?

9.62 A standard 8-in. steel pipe (see Appendix E) of length $L = 15$ ft is used for a pin-ended column. It is subjected to an eccentrically applied load $P = 400$ kips as shown. Determine the maximum deflection and maximum stress if $e = 1.5$ in. Let $E = 30 \times 10^3$ ksi.

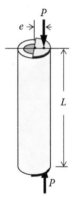

(Problems 9.62 and 9.63)

9.63 In reference to the figure in Problem 9.62, an extra strong 6-in. pipe (see Appendix E) of length $L = 12$ ft is used for a pin-ended column. It is subjected to an eccentrically applied load P as shown. If the maximum deflection and maximum stress are limited to 0.4 in. and 20 ksi, respectively, determine the largest allowable load P for an eccentricity $e = 1.0$ in. Let $E = 29 \times 10^3$ ksi and $\sigma_{MAX} = 25$ ksi.

9.64 A hollow square aluminum ($E = 75$ GPa) tube is used as a pin-ended column of length L. It is subjected to a load $P = 203.5$ kN applied eccentrically at a distance $e = 10$ mm from the centroid of the section as shown. The tube is 120 mm × 120 mm on the outside and its thickness $t = 20$ mm is constant all around. If the maximum stress is limited to 45 MPa, determine the maximum permissible length of the column.

(Problems 9.64 and 9.65)

9.65 A hollow square aluminum ($E = 10 \times 10^3$ ksi) tube is used as a pin-ended column of length $L = 10$ ft. It is subjected to a load $P = 10$ kips placed at a distance e from the centroid of the section as shown. The tube is 6 in. × 6 in. on the outside and its thickness $t = 1$ in. is constant all around. If the maximum deflection is 0.5 in., determine the corresponding eccentricity e of the load. What is the corresponding maximum stress?

9.66 A hollow rectangular steel ($E = 30 \times 10^3$ ksi) tube is used as a pin-ended column of length $L = 12$ ft. It is subjected to a load P applied eccentrically at a distance $e = 2$ in. from the centroid of the section as shown. Let $b = 8$ in., $h = 5$ in., and the thickness $t = 1$ in. be constant all around. If $\sigma_y = 36$ ksi and the factor of safety is 2, determine the maximum permissible load P. *Note that the factor of safety should be applied to the load and not to the stress.*

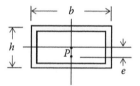

(Problems 9.66 and 9.67)

9.67 In reference to the figure in Problem 9.66, a hollow rectangular brass ($E = 105$ GPa) tube is used as a fixed-free column of length L. It is subjected to a load $P = 50$ kN applied eccentrically at a distance $e = 80$ mm from the centroid of the section as shown. Let $b = 300$ mm, $h = 180$ mm, and the thickness $t = 40$ mm be constant all around. If $\sigma_{ALL} = 30$ MPa, determine the maximum permissible length L.

9.68 Two S10 × 35 sections are welded together to form the lipped box section shown. It is used as a pin-ended column of length $L = 15$ ft and subjected to a load $P = 100$ kips eccentrically applied as indicated. Find (a) the factor of safety based on stress and (b) the maximum deflection of the column. Let $E = 30 \times 10^3$ ksi and $\sigma_y = 36$ ksi.

Two welded S sections
(Problems 9.68 and 9.69)

9.69 In reference to the figure in Problem 9.68, two S305 × 74.4 sections are welded together to form the lipped box section shown. It is used as a fixed-free column of length L and subjected to a load $P = 50$ kN eccentrically applied as indicated. If the maximum stress in the column is not to exceed 17.6 MPa, determine the largest permissible length L. Let $E = 200$ GPa and $\sigma_y = 100$ MPa.

9.70 Two C10 × 20 sections are welded together to form the box section shown. It is used as a pin-ended column of length $L = 10$ ft and subjected to a load P eccentrically applied as indicated. If $E = 29 \times 10^3$ ksi and $\sigma_y = 100$ ksi and the maximum stress is not to exceed 50 ksi, determine the allowable load P.

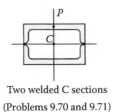

Two welded C sections

(Problems 9.70 and 9.71)

9.71 In reference to the figure in Problem 9.70, two C254 × 44.6 sections are welded together to form the box section shown. It is used as a fixed-free column of length L and subjected to a load $P = 134$ kN eccentrically applied as indicated. If $E = 200$ GPa, $\sigma_y = 250$ MPa, and the maximum deflection is not to exceed 15 mm, determine the largest permissible length L.

9.6 DESIGN OF CENTRICALLY LOADED COLUMNS

9.6.1 INELASTIC COLUMN BUCKLING

In the preceding discussions, we focused on the *elastic buckling* of columns for both ideal Euler centrally loaded columns and real eccentrically loaded columns. Inherent in these discussions is the assumption that the maximum stress in the column is well within the elastic range for the material so that the behavior of the column is purely elastic. If the elastic limit, however, is exceeded, the failure of a column is caused by *inelastic buckling*.

Consider the information contained in Figure 9.20, which, among other things, shows Euler's curve ABC as defined by Equation 9.16. This curve is only valid for centrally loaded columns. As

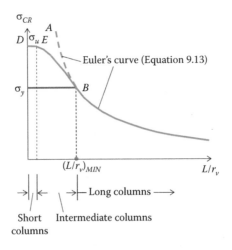

FIGURE 9.20 Column design curve illustrating inelastic behavior for short and intermediate columns and the elastic buckling of long columns.

stated earlier, segment AB does not apply because the stress in the column would be above the limit of elastic behavior. We can determine the value of the slenderness ratio, $(L/r_v)_{MIN}$, below which Euler's equation is not valid by setting σ_{CR} in Equation 9.16 equal to σ_y. This operation leads to

$$\left(\frac{L}{r_v}\right)_{MIN} = \sqrt{\frac{\pi^2 E}{\sigma_y}} \qquad (9.55)$$

The value of $(L/r_v)_{MIN}$ obtained from Equation 9.55 for a given material defines the lower limit of applicability of Euler's equation to that particular material. Those columns with L/r_v larger than $(L/r_v)_{MIN}$ are referred to as *long columns* (see segment BC in Figure 9.20). A long column fails by elastic buckling, which implies that if the Euler critical load is removed after failure, the column returns to its original unbent shape. On the other hand, those columns with very small values of L/r_v are not subject to instability and, therefore, do not buckle. They are referred as *short columns*, also known as *struts* (see segment DE in Figure 9.20). These members do not fail by buckling as do long columns, but by *inelastic action*, and the ultimate compressive strength, σ_u, of the material represents the highest value that the critical stress in those members can attain. Finally, between long and short columns is a range known as *intermediate columns* (see segment EB in Figure 9.20), which, because of their length, are subject to instability and fail by inelastic buckling since the stress is above the limit for elastic behavior. In other words, the stress in intermediate columns exceeds the elastic-limit stress before buckling takes place. Several theories have been proposed to describe and predict the inelastic buckling of such columns. Among them are the *tangent-modulus* theory, the *reduced-modulus* theory, and the *Shanley* theory. However, the methods that we use today for the design of intermediate columns are based, for the most part, on curves and equations obtained experimentally.

9.6.2 EMPIRICAL EQUATIONS

As stated above, with the exception of very long members, the design of columns is primarily accomplished using equations that are based upon experimental observations, known as *empirical equations*. As may have been surmised from our preceding discussion, the buckling of columns is a very complex problem and because of this, a very large number of empirical column equations have been proposed by many investigators since the time of Euler. We will introduce below three sets of such equations relating to three different materials: *structural steel*, *aluminum*, and *structural wood*. The use of such equations requires that we keep certain considerations in mind as follows:

1. As implied above, different materials require different empirical equations.
2. In general, empirical equations relate the allowable stress, σ_{ALL}, to the slenderness ratio, L_e/r_v. Sometimes, the empirical equation is stated in terms of the critical stress. In such a case, we need to introduce a factor of safety in order to express it in terms of the allowable stress. The needed factor of safety is either prescribed by codes or selected by the designer on the basis of judgment and experience.
3. Care must be exercised in evaluating the column-end conditions in order to determine the correct effective length, L_e, to use in a given situation.
4. Practically all empirical equations are valid for restricted values of the slenderness ratio L_e/r_v.
5. In general, the design of columns requires a trial-and-error method as illustrated in the examples that follow. This is so because the design equations are valid only within given ranges of the slenderness ratio, which is not known at the outset and must be assumed in order to select an appropriate, but tentative, design equation. In general, the first assumed slenderness ratio is not satisfactory and the process is repeated until a satisfactory value is found.

9.6.3 STRUCTURAL STEEL

The American Institute of Steel Construction (AISC), the organization primarily responsible for structural steel specifications in this country, employs an empirical equation for intermediate and short columns and Euler's equation for long columns. A factor of safety that varies with the slenderness ratio is used for short and intermediate columns and a fixed factor of safety for long columns where Euler's equation applies.

Equation 9.55, when modified to account for the effective length of the column, becomes

$$\left(\frac{L_e}{r_v}\right)_{MIN} = \sqrt{\frac{\pi^2 E}{\sigma_y}} \tag{9.56}$$

where the quantity $(L_e/r_v)_{MIN}$ represents the least value of the effective slenderness ratio for which Euler's equation applies. To account for a number of unknown factors, including residual stresses in rolled steel sections, AISC specifications require that σ_y in Equation 9.56 be replaced by $1/2\sigma_y$. In other words, Euler's equation is no longer valid when the stress in the column exceeds $1/2\sigma_y$. Under these conditions, AISC uses the symbol C_c to denote the separation between long columns and those that are intermediate or short. Thus, making these changes in Equation 9.56, we obtain

$$C_c = \sqrt{\frac{2\pi^2 E}{\sigma_y}} \tag{9.57}$$

For values of L_e/r_v larger than C_c, Euler's equation is valid, and for those less than C_c, Euler's equation does not apply and an empirical equation of parabolic form is used. For both ranges of the slenderness ratio, a factor of safety is used as indicated below.

9.6.3.1 Short and Intermediate Columns

This range of structural steel columns is characterized by an effective slenderness ratio less than C_c and the following parabolic empirical equation is used:

$$\sigma_{ALL} = \frac{\sigma_y}{F.S.}\left[1 - \frac{1}{2}\left(\frac{L_e/r_v}{C_c}\right)^2\right] \dots \dots \frac{L_e}{r_v} < C_c \tag{9.58}$$

The factor of safety applied in this range of effective slenderness ratios is variable, given by

$$F.S. = \frac{5}{3} + \frac{3}{8}\left(\frac{L_e/r_v}{C_c}\right) - \frac{1}{8}\left(\frac{L_e/r_v}{C_c}\right)^3 \tag{9.59}$$

9.6.3.2 Long Columns

Note that when $L_e/r_v = C_c$, Equation 9.59 yields $F.S. = 23/12$. AISC assumes this value for the factor of safety to apply unchanged for long columns where Euler's equation is valid. Thus, for long columns, AISC uses the following modified Euler's equation:

$$\sigma_{ALL} = \frac{12\pi^2 E}{23(L_e/r_v)^2} \dots \dots \frac{L_e}{r_v} \geq C_c \tag{9.60}$$

A plot of Equations 9.58 and 9.60 is shown in Figure 9.21 with factors of safety provided by Equation 9.59 for a specific structural steel for which $E = 29 \times 10^3$ ksi and $\sigma_y = 50$ ksi.

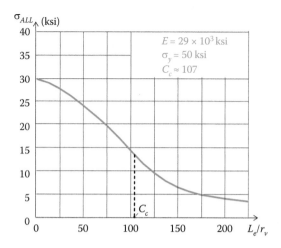

FIGURE 9.21　Allowable stress versus slenderness ratio L_e/r_y for steel columns with $E = 29 \times 10^3$ ksi and $\sigma_y = 36$ ksi.

It shows a smooth transition from long columns ($L_e/r_y \geq C_c \approx 107$) to short and intermediate columns ($L_e/r_y < C_c \approx 107$). Of course, different structural steels with different values of E and σ_y have design curves different from that shown in Figure 9.21.

EXAMPLE 9.10

An S20 $\times$ 75 steel section ($E = 30 \times 10^3$, $\sigma_y = 36$ ksi) is used for a fixed-free column 5 ft long. Determine the largest compressive centrally applied load that the column may carry. The column is not supported along its length and, therefore, can buckle in any direction.

SOLUTION

The structural properties for the given S20 $\times$ 75 needed for the solution are obtained from Appendix E. These are $A = 22.0$ in.^{2} and $r_y = 1.16$ in. Also, since the column is fixed-free, the effective length is $L_e = 2L = 10$ ft.

Therefore,

$$\text{Equation 9.57} \Rightarrow C_c = \sqrt{\frac{2\pi^2(30 \times 10^3)}{36}} = 128.3; \quad \frac{L_e}{r_y} = \frac{10(12)}{1.16} = 103.4$$

Since $L_e/r_y < C_c$, the column is in the intermediate/short range and Equations 9.59 and 9.58 apply. Thus,

$$F.S. = \frac{5}{3} + \frac{3}{8}\left(\frac{103.4}{128.3}\right) - \frac{1}{8}\left(\frac{103.4}{128.3}\right)^3 = 1.903; \quad \sigma_{ALL} = \frac{36}{1.903}\left[1 - \frac{1}{2}\left(\frac{103.4}{128.3}\right)^2\right] = 12.8 \text{ ksi}$$

The largest permissible load, P_{ALL}, becomes

$$P_{ALL} = \sigma_{ALL}A = 12.8(22.0) = 281.6 \approx 282 \text{ kips} \qquad \textbf{ANS.}$$

EXAMPLE 9.11

A structural steel ($E = 30 \times 10^3$ ksi, $\sigma_y = 40$ ksi) push rod of rectangular cross section is to serve as a machine component of length $L = 20$ in. that carries a centrally applied compressive load $P = 4$ kips. The rod may be assumed to be a pin-ended column and is to be designed using the AISC specifications. Determine the minimum required dimension b.

SOLUTION

Since the column is pin-ended, its effective length is $L_e = L = 20$ in. We now determine the least radius of gyration and corresponding slenderness ratio in terms of the unknown dimension b. Thus,

$$r_v = \sqrt{\frac{I_v}{A}} = \sqrt{\frac{(b^4/6)}{2b^2}} = \frac{b}{\sqrt{12}}; \quad \frac{L_e}{r_v} = \frac{20}{(b/\sqrt{12})} = \frac{20\sqrt{12}}{b}$$

Using Equation 9.57, we determine the value of C_c. Thus,

$$C_c = \sqrt{\frac{2\pi^2(30 \times 10^3)}{40}} = 121.673$$

The remaining steps consist of assuming values for the unknown dimension b and checking to see if the assumption is satisfactory.

First trial:

$$b = 0.5 \text{ in.}; \quad \frac{L_e}{r_v} = \frac{20\sqrt{12}}{0.5} = 138.564$$

Since $L_e/r_v > C_c$, the allowable stress is given by Equation 9.60. Thus,

$$\sigma_{ALL} = \frac{12\pi^2(30 \times 10^3)}{23(138.564)^2} = 8.046; \quad P_{ALL} = \sigma_{ALL}(A) = 8.046(0.5)(1.0) = 4.023 \text{ kips}$$

Since the computed allowable load is less than the applied load of 5 kips, the assumption of $b = 0.5$ in. is *not satisfactory* and another trial assumption for b larger than 0.5 in. is needed.

Second trial:

$$b = 0.75 \text{ in.}; \quad \frac{L_e}{r_v} = \frac{20\sqrt{12}}{0.75} = 92.376$$

Since $L_e/r_v < C_c$, the allowable stress is given by Equation 9.58 with a factor of safety given by Equation 9.59.

$$F.S. = \frac{5}{3} + \frac{3}{8}\left(\frac{92.376}{121.673}\right) - \frac{1}{8}\left(\frac{92.376}{121.673}\right)^3 = 1.897$$

$$\sigma_{ALL} = \frac{40}{1.897}\left[1-\frac{1}{2}\left(\frac{92.376}{121.673}\right)^2\right] = 15.009; \quad P_{ALL} = \sigma_{ALL}(A) = 15.009(0.75)(1.5) = 11.257 \text{ kips}$$

The value obtained for P_{ALL} is much larger than the required 5-kip capacity and the value $b = 0.75$ in. may represent an over-design. Therefore, several additional trials are needed before a satisfactory solution is obtained. However, all computations are performed as shown above. This was done and the value finally obtained is

$$b = 0.528 \text{ in.} \quad \textbf{ANS.}$$

EXAMPLE 9.12

A W14 × 82 steel ($E = 29 \times 10^3$, $\sigma_y = 50$ ksi) section is used for a fixed-fixed column of length L. The column is not supported along its length and, therefore, can buckle in any direction. If the column is to support a compressive centrally applied load $P = 650$ kips, determine the maximum permissible length L.

SOLUTION

The needed structural properties for the W14 × 82 section are obtained from Appendix E and are $A = 24.1$ in.^{2} and $r_v = 2.48$ in. Since the column is fixed-fixed, its effective length $L_e = 0.5L$.
Using Equation 9.57, we compute the value of C_c. Thus,

$$C_c = \sqrt{\frac{2\pi^2(29 \times 10^3)}{50}} = 106.999 \approx 107.0$$

Let us now assume that our column is in the long range ($L_e/r_v > C_c$) where Euler's equation applies. Thus,

$$\sigma_{ALL} = \frac{12\pi^2(29 \times 10^3)}{23(L_e/r_v)^2} = \frac{650}{24.1}; \quad \frac{L_e}{r_v} = 74.409 \approx 74.4$$

Since the computed value of L_e/r_v is less than C_c, we conclude that our assumption is wrong and our column must be in the intermediate/short range. Therefore, Equation 9.58 with a factor of safety given by Equation 9.59 must be used. However, to use these equations, we need to assume a certain length for the column and repeat the process by trial and error until a satisfactory value for the length is found.

First trial:

$$L = 15 \text{ ft}; \quad L_e = 7.5 \text{ ft}; \quad \frac{L_e}{r_v} = \frac{7.5 \times 12}{2.48} = 36.290; \quad F.S. = \frac{5}{3} + \frac{3}{8}\left(\frac{36.290}{107.0}\right) - \frac{1}{8}\left(\frac{36.290}{107.0}\right)^3 = 1.789$$

$$\sigma_{ALL} = \frac{50}{1.789}\left[1-\frac{1}{2}\left(\frac{36.290}{107.0}\right)^2\right] = 26.349 \text{ ksi}; \quad P_{ALL} = \sigma_{ALL}(A) = 26.349(24.1) = 635.0 \text{ kips}$$

The value found for P_{ALL} is below the required value of 650 kips. Therefore, a shorter column is needed.

Second trial:

$$L = 12 \text{ ft}; \quad L_e = 6 \text{ ft}; \quad \frac{L_e}{r_v} = \frac{6 \times 12}{2.48} = 29.032; \quad F.S. = \frac{5}{3} + \frac{3}{8}\left(\frac{29.032}{107.0}\right) - \frac{1}{8}\left(\frac{29.032}{107.0}\right)^3 = 1.766$$

$$\sigma_{ALL} = \frac{50}{1.766}\left[1 - \frac{1}{2}\left(\frac{29.032}{107.0}\right)^2\right] = 27.270 \text{ ksi}; \quad P_{ALL} = \sigma_{ALL}(A) = 27.270(24.1) = 657.2 \text{ kips}$$

On the basis of the above two trials, we conclude that the correct column length is somewhere between 12 and 15 ft. Therefore, several additional trials are needed before a satisfactory solution is obtained. However, all computations are performed as shown above. A few additional trials were made and the value finally obtained is

$$L = 12.9 \text{ ft}; \quad P_{ALL} = 650.7 \text{ kips} \quad \textbf{ANS.}$$

Aluminum alloys: The Aluminum Association provides the specifications needed for the design of aluminum-alloy columns. Unlike the case of structural steel where only two ranges of the slenderness ratio are specified, aluminum alloys require specifications for three distinct ranges of slenderness ratio: short, intermediate, and long as indicated in the qualitative graph of Figure 9.22. The quantities K_1, K_2, K_3, R_1, and R_2 in the equations of Figure 9.22 are constants that differ from one aluminum alloy to another. Note that for short columns the allowable stress is represented by a horizontal straight line. For intermediate columns, the allowable stress is given by a sloping straight line and for long columns, Euler's equation is used with an appropriate factor of safety.

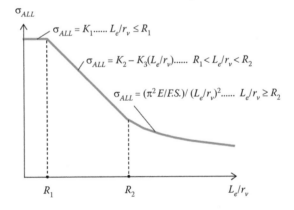

FIGURE 9.22 Qualitative graph of allowable stress versus slenderness ratio L_e/r_y for aluminum alloys.

The specific equations for two commonly used structural aluminum alloys provided by the Aluminum Association are stated below expressed in terms of the symbols used in this book:

2014–T6 Aluminum Alloy

$$\sigma_{ALL} = 28 \text{ ksi} \dots\dots \frac{L_e}{r_v} \leq 12 \tag{9.61}$$

$$\sigma_{ALL} = \left(30.7 - 0.23\left(\frac{L_e}{r_v}\right)\right) \text{ ksi} \dots\dots 12 < \frac{L_e}{r_v} < 55 \tag{9.62}$$

$$\sigma_{ALL} = \frac{54,000}{(L_e/r_v)^2} \text{ ksi} \dots\dots \frac{L_e}{r_v} \geq 55 \tag{9.63}$$

Note that the quantity $\pi^2 E/F.S.$ in Euler's equation is replaced by 54,000 ksi for this aluminum alloy.

6061–T6 Aluminum Alloy

$$\sigma_{ALL} = 19 \text{ ksi} \dots\dots \frac{L_e}{r_v} \leq 9.5 \tag{9.64}$$

$$\sigma_{ALL} = \left(20.2 - 0.126\left(\frac{L_e}{r_v}\right)\right) \text{ ksi} \dots\dots 9.5 < \frac{L_e}{r_v} < 66 \tag{9.65}$$

$$\sigma_{ALL} = \frac{51,000}{(L_e/r_v)^2} \text{ ksi} \dots\dots \frac{L_e}{r_v} \geq 66 \tag{9.66}$$

Note that the quantity $\pi^2 E/F.S.$ in Euler's equation is replaced by 51,000 ksi for this aluminum alloy.

EXAMPLE 9.13

The hollow square cross section shown is that for a 2014–T6 aluminum alloy fixed-pinned column of length $L = 10$ ft subjected to a centrically applied load $P = 120$ kips. Determine the minimum permissible dimension d if it is available in increments of 0.1 in.

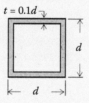

SOLUTION

In view of the fact that the column is made of 2014–T6 aluminum alloy, Equations 9.61 through 9.63 apply. In order to select the correct equation, we must know the slenderness ratio L_e/r_v. To find this ratio, we need to assume values for the dimension d, checking the validity of these assumptions until a satisfactory value is found. Thus,

$$L_e = 0.7L = 7 \text{ ft}; \quad I_v \frac{1}{12}\left[d^4 - (0.8d)^4\right] = 0.0492d^4 \text{ in.}^4; \quad A = d^2 - (0.8d)^2 = 0.36d^2 \text{ in.}^2$$

$$r_v = \sqrt{\frac{I_v}{A}} = \sqrt{\frac{0.0492d^4}{0.36d^2}} = 0.36968d \text{ in.}; \quad \frac{L_e}{r_v} = \frac{7 \times 12}{0.36968d} = \frac{227.224}{d}$$

First trial:

$$d = 5.0 \text{ in.}; \quad \frac{L_e}{r_v} = \frac{227.224}{5.0} = 45.4$$

Therefore, Equation 9.62 is used to find the allowable stress. Thus,

$$\sigma_{ALL} = 30.7 - 0.23(45.4) = 20.258 \text{ ksi}; \quad P_{ALL} = \sigma_{ALL}(A) = 20.258 \times 0.36(5.0)^2 = 182.3 \text{ kips}$$

Since $P_{ALL} \ggg 120$ kips, we need a smaller value for the dimension d.

Second trial:

$$d = 4.0 \text{ in.}; \quad \frac{L_e}{r_v} = \frac{227.224}{4.0} = 56.806$$

Equation 9.63, therefore, needs to be used to find the allowable stress. Thus,

$$\sigma_{ALL} = \frac{54,000}{(56.806)^2} = 16.734 \text{ ksi}; \quad P_{ALL} = \sigma_{ALL}(A) = 16.734 \times 0.36(4.0)^2 = 96.4 \text{ kips}$$

Since the needed load capacity is 120 kips, it follows that the correct value for the dimension d is somewhere between 4.0 and 5.0 in. Therefore, several additional trials are needed before a satisfactory solution is obtained. However, all computations are performed as shown above. A few additional trials were made and the value finally obtained, to the nearest 0.1 in., is

$$d = 4.3 \text{ in.}; \quad P_{ALL} = 123.4 \text{ kips} \quad \textbf{ANS.}$$

Structural woods: The design of structural wooden columns is based on empirical equations provided by the American Forest and Paper Association (AFPA). These equations make it possible to compute the allowable stress for sections in terms of the ratio L_e/d, where d is the dimension of the section along the principal axis of failure.

A qualitative graph for the allowable stress in wooden columns versus the ratio L_e/d is given in Figure 9.23. The quantity σ_c represents the allowable stress for a short block in compression parallel to the grain. Short columns are those for which $L_e/d < 11$. Note that at $L_e/d = 11$, the graph in Figure 9.23 possesses a small discontinuity. Long columns, where Euler's equation applies, are those for which $L_e/d \geq R_c$, where R_c is the value of L_e/d when the allowable stress is equal to $2/3\sigma_c$. In other words, R_c represents the smallest value of L_e/d for which Euler's equation, with a factor of safety, is

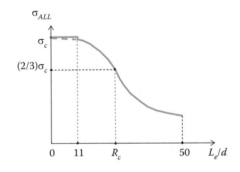

FIGURE 9.23 Allowable stress versus slenderness ratio L_e/d for structural woods.

valid. The factor of safety that AFPA uses with Euler's equation is 2.74. Therefore, we can find an expression for R_c by setting Euler's equation with this factor of safety equal to $2/3\sigma_c$ and solving for L_e/d. Thus, since for a rectangular cross section with a least dimension d, $r_v = d/\sqrt{12}$, we have

$$\frac{\pi^2 E}{2.74(L_e/r_v)^2} = \frac{\pi^2 E}{2.74(12)(L_e/d)^2} = \frac{0.30E}{(L_e/d)^2} = \frac{2}{3}\sigma_c$$

from which

$$R_c = \frac{L_e}{d} = 0.671\sqrt{\frac{E}{\sigma_c}} \tag{9.67}$$

Finally, the graph in Figure 9.23 depicts a range between $L_e/d = 11$ and $L_e/d = R_c$, which represents the intermediate range of wooden columns and is described by AFPA by a fourth-degree parabolic function.

The three ranges of wooden column behavior are defined mathematically by the following three equations:

$$\sigma_{ALL} = \sigma_c \cdots \frac{L_e}{d} \leq 11 \tag{9.68}$$

$$\sigma_{ALL} = \sigma_c\left[1 - \frac{1}{3}\left(\frac{L_e/d}{R_c}\right)^4\right] \cdots 11 < \frac{L_e}{d} \leq R_c \tag{9.69}$$

$$\sigma_{ALL} = \frac{0.30E}{(L_e/d)^2} \cdots R_c \leq \frac{L_e}{d} \leq 50 \tag{9.70}$$

EXAMPLE 9.14

The rectangular cross section shown is that for a fixed-fixed structural wood column of length L, which is to carry a centrically applied load of 25 kips. If $\sigma_c = 1.5$ ksi and $E = 2.0 \times 10^3$ ksi, determine the maximum permissible length L if the dressed dimensions are (a) $b = 3.5$ in. and $h = 5.5$ in. and (b) $b = 3.5$ in. and $h = 3.5$ in.

SOLUTION

Using Equation 9.67, we compute the value of R_c. Thus,

$$R_c = 0.671\sqrt{\frac{2.0 \times 10^3}{1.5}} = 24.5$$

a. Now that we know the value of $L_e/d = 24.5$ separating long from intermediate columns, we begin the process by making an assumption relative to slenderness ratio as follows:

First trial:

Assume $\frac{L_e}{d} > 24.5$; Equation 9.70 $\Rightarrow \sigma_{ALL} = \frac{25}{(3.5)(5.5)} = \frac{0.30(2.0 \times 10^3)}{(L_e/d)^2}$; $\frac{L_e}{d} = 21.5 < R_c$

Since the computed value is less than R_c, the assumption made is not valid. Therefore, the column must have an L_e/d ratio less than $R_c = 24.5$.

Second trial:

Assume $\dfrac{L_e}{d} < 24.5$; Equation 9.69 $\Rightarrow \sigma_{ALL} = \dfrac{25}{(3.5)(5.5)} = 1.5\left[1 - \dfrac{1}{3}\left(\dfrac{L_e/d}{24.5}\right)^4\right]$; $\dfrac{L_e}{d} = 19.5 < R_c$

Since the computed value of L_e/d is less than R_c, the second assumption is correct and L_e/d is in fact equal to 19.5 and $L_e = 19.5\ d = 19.5\ (3.5) = 68.25$ in. Therefore, since the column is fixed-fixed, we have

$$L = 2L_e = 2(68.25) = 136.5\ \text{in.} = 11.375 \approx 11.4\ \text{ft} \qquad \textbf{ANS.}$$

b. Since the sectional dimension along the axis of failure in part (b) is the same as that in part (a), namely, $d = 3.5$ in., we conclude that the same answer found there is still valid here. Therefore,

$$L \approx 11.4\ \text{ft} \qquad \textbf{ANS.}$$

PROBLEMS
Structural steel

9.72 An S24 × 100 steel ($E = 29 \times 10^3$ ksi, $\sigma_y = 40$ ksi) section is used for a pin-ended column 18 ft long. The column is so supported that it can buckle in any direction. Find the maximum centrically applied load that the column can carry.

9.73 Select the most suitable S steel ($E = 30 \times 10^3$ ksi, $\sigma_y = 36$ ksi) section from among the 15-in. nominal height group given in Appendix E to carry a centrically applied load of 100 kips. The column is pin-ended with a length of 14 ft. The column is so supported that it can buckle in any direction.

9.74 An S24 × 100 steel ($E = 29 \times 10^3$ ksi, $\sigma_y = 50$ ksi) section is used for a pin-ended column to carry a centrally applied load of 200 kips. The column is so supported that it can buckle in any direction. Find the maximum allowable length of this column.

9.75 A W457 × 177 steel ($E = 205$ GPa, $\sigma_y = 270$ MPa) section is used for a fixed-free column 6 m long. The column is so supported that it can buckle in any direction. Find the maximum centrically applied load that the column can carry.

9.76 Select the most suitable W steel ($E = 200$ GPa, $\sigma_y = 240$ MPa) section from among the 610-mm nominal height group given in Appendix E to carry a centrically applied load of 1,000 kN. The column is fixed-free with a length of 7 m. The column is so supported that it can buckle in any direction.

9.77 A W406 × 84.8 steel ($E = 205$ GPa, $\sigma_y = 340$ MPa) section is used for a fixed-free column to carry a centrically applied load of 1200 kN. The column is so supported that it can buckle in any direction. Find the maximum allowable length L.

9.78 Select the most suitable steel ($E = 30 \times 10^3$ ksi, $\sigma_y = 36$ ksi) W section from among the 18-in. nominal height group given in Appendix E to carry a centrically applied load of 450 kips. The column is fixed-fixed with a length of 28 ft. The column is so supported that it can buckle in any direction.

9.79 Select the most suitable steel ($E = 200$ GPa, $\sigma_y = 240$ MPa) W section from among the 305-mm nominal height group given in Appendix E to carry a centrically applied load of 700 kN. The column is fixed-fixed with a length of 4 m. The column is so supported that it can buckle in any direction.

9.80   Select the most suitable steel ($E = 29 \times 10^3$ ksi, $\sigma_y = 42$ ksi) S section from among the 20-in. nominal height group given in Appendix E to carry a centrically applied load of 500 kips. The column is fixed-fixed with a length of 18 ft. The column is so supported that it can buckle in any direction.

9.81   Select the most suitable steel ($E = 200$ GPa, $\sigma_y = 240$ MPa) S section from among the 508-mm nominal height group given in Appendix E to carry a centrically applied load of 800 kN. The column is fixed-fixed with a length of 8 m. The column is so supported that it can buckle in any direction.

9.82   A fixed-free steel ($E = 30 \times 10^3$ ksi, $\sigma_y = 36$ ksi) column, 6 ft long, has a rectangular cross section as shown. Let $b = 3$ in. and $h = 5$ in., determine the maximum permissible centrically applied compressive load. The column may buckle in any direction.

(Problems 9.82 and 9.83)

9.83   In reference to the figure in Problem 9.82, a fixed-free steel ($E = 200$ GPa, $\sigma_y = 300$ MPa) column, 3 m long, has a rectangular cross section as shown. If it is to support a centrically applied compressive force of 5000 kN and $h = 3/2b$, determine the least acceptable value of the dimension b to the nearest millimeter. The column may buckle in any direction.

9.84   A pin-ended steel ($E = 205$ GPa, $\sigma_y = 260$ MPa) column, 2.5 m long, has a hollow square cross section with dimension b on each side and constant thickness $t = 0.04b$ as shown. If the column is to carry a centrically applied compressive force of 500 kN, find the least acceptable value of the dimension b.

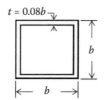

(Problems 9.84 and 9.85)

9.85   In reference to the figure in Problem 9.84, a fixed-fixed steel ($E = 30 \times 10^3$ ksi, $\sigma_y = 42$ ksi) column, 12 ft long, has a hollow square cross section with dimension b on each side and constant thickness $t = 0.04b$ as shown. Let $b = 6$ in., determine the maximum permissible centrically applied compressive load.

9.86   An 8-in., extra-strong pipe (see Appendix E) is used for a fixed-free steel ($E = 30 \times 10^3$ ksi, $\sigma_y = 50$ ksi) column, 15 ft long. Determine the maximum permissible centrically applied compressive load.

9.87   A fixed-free steel ($E = 200$ GPa, $\sigma_y = 300$ MPa) column, 4 m long, has a hollow circular cross section with outside diameter d_o and inside diameter $d_i = 0.9d_o$. If the column is to carry a centrically applied compressive force of 1000 kN, find the least acceptable value of the dimension d_o to the nearest millimeter.

9.88 Two C12 × 30 steel ($E = 29 \times 10^3$ ksi, $\sigma_y = 36$ ksi) sections are welded together to form the box section shown. This box section is used for a pin-ended column of length L, which is to carry a centrically applied compressive force of 150 kips. Determine the maximum permissible length L assuming that the column may buckle in any direction.

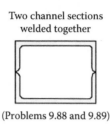

Two channel sections
welded together

(Problems 9.88 and 9.89)

9.89 In reference to the figure in Problem 9.88, select the most suitable steel ($E = 30 \times 10^3$ ksi, $\sigma_y = 42$ ksi) channel section from among the 10 in. nominal height group given in Appendix E, two of which are welded together to form the box section shown in the sketch to be used for fixed-fixed column with a length of 20 ft. The column is subjected to a centrically applied compressive force of 250 kips and may buckle in any direction.

9.90 An L203 × 203 × 25.4 steel ($E = 205$ GPa, $\sigma_y = 340$ MPa) angle section is used for a pin-ended column 4.5 m long. Determine the maximum permissible centrically applied compressive load that the column may carry if it can buckle in any direction.

9.91 An L6 × 6 × 1 steel ($E = 29 \times 10^3$ ksi, $\sigma_y = 36$ ksi) angle section is used for a fixed-pinned column to carry a centric load of 50 kips. Find the maximum permissible length of this column to the nearest 0.1 ft assuming it can buckle in any direction.

9.92 The cross section shown is that for a steel ($E = 200$ GPa, $\sigma_y = 300$ MPa) fixed-pinned column of length $L = 6$ m. Let $a = 30$ mm, $b = 80$ mm, and $h = 120$ mm, determine the largest permissible centrically applied compressive load that the column may carry. Assume that the column may buckle in any direction.

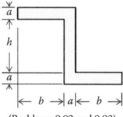

(Problems 9.92 and 9.93)

9.93 In reference to the figure in Problem 9.92, the cross section shown is that for a steel ($E = 75$ GPa, $\sigma_y = 150$ MPa) pin-ended column of length L designed to carry a centrically applied load of 120 kN. Let $a = 30$ mm, $b = 80$ mm, and $h = 120$ mm, determine the largest permissible length L the column may have. Assume that the column may buckle in any direction.

9.94 The cross section shown is that for a steel ($E = 205$ GPa, $\sigma_y = 350$ MPa) fixed-fixed column of length $L = 10$ m. Let $a = 30$ mm, $b = 80$ mm, $c = 40$ mm, and $h = 120$ mm and determine the largest permissible centrically applied compressive load that the column may carry. Assume that the column may buckle in any direction.

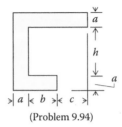

(Problem 9.94)

Aluminum alloys

9.95 A 2014–T6 aluminum alloy pin-ended column has a solid circular cross section 5 in. in diameter and 6 ft long. Find the maximum permissible centrically applied compressive load that this column may carry.

9.96 A 6061–T6 aluminum alloy fixed-pinned column has a solid circular cross section 6 in. in diameter. If the column is to carry a centrically applied compressive load of 400 kips, determine the largest permissible length of the column to the nearest 0.1 ft.

9.97 The hollow square cross section shown is that for a 2014–T6 aluminum alloy pin-ended column 8 ft long. If the column is to carry a centrically applied compressive force of 30 kips, determine the least permissible dimension b to the nearest 0.1 in.

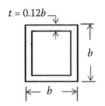

(Problems 9.97 and 9.98)

9.98 In reference to the figure in Problem 9.97, the hollow square cross section shown is that for a 6061–T6 aluminum alloy fixed-free column designed to carry a compressive centrically applied force of 50 kips. If $b = 4.0$ in., determine the largest permissible length of this column to the nearest 0.1 ft.

9.99 Three identical 2014–T6 aluminum plates are joined together securely to form the I section shown, which is used for a fixed-fixed column 20 ft long. If $b = 12.0$ in. and $t = 1.0$ in., determine the maximum permissible centrically applied compressive load that may be applied to the column. Assume the column may buckle in any direction.

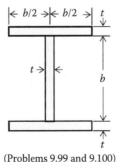

(Problems 9.99 and 9.100)

9.100 In reference to the figure in Problem 9.99, three identical 6061–T6 aluminum plates are joined together securely to form the I section shown, which is used for a fixed-free column 10 ft long. If the column is to be subjected to a centrically applied compressive force of 30 kips and $t = 0.1b$, determine the least permissible value of b to within 0.1 in. Assume the column may buckle in any direction.

9.101 Three 2014–T6 aluminum plates are joined together securely to form the U section shown, which is used for a fixed-pinned column 15 ft long. If $b = 8.0$ in. and $t = 0.8$ in., determine the maximum permissible centrically applied compressive load that may be applied to the column. Assume the column may buckle in any direction.

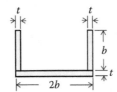

(Problems 9.101 and 9.102)

9.102 In reference to the figure in Problem 9.101, three 6061–T6 aluminum plates are joined together securely to form the U section shown, which is used for a fixed-fixed column 24 ft long. If the column is to be subjected to a centrically applied compressive force of 50 kips and $t = 0.1b$, determine the least permissible value of b to within 0.1 in. Assume the column may buckle in any direction.

9.103 A 6061–T6 aluminum alloy fixed-pinned column has the cross section shown where $b = 12$ in. and $t = 1.0$ in. If the column is to carry a centrically applied compressive load of 250 kips, determine the largest permissible length of the column. Assume the column may buckle in any direction.

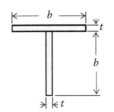

(Problems 9.103 and 9.104)

9.104 In reference to the figure in Problem 9.103, a 2014–T6 aluminum alloy pin-ended column has the cross section shown where $b = 10$ in. and $t = 1$ in. If the column has a length $L = 8$ ft, determine the maximum permissible centrically applied compressive force that may be applied to the column. Assume the column may buckle in any direction.

9.105 A 2014–T6 aluminum alloy fixed-pinned column has the cross section shown where $a = 2$ in. If the column has a length $L = 15$ ft, determine the maximum permissible centrically applied compressive force that may be applied to the column. Assume the column may buckle in any direction.

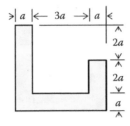

9.106 A 6061–T6 aluminum alloy fixed-fixed column has the cross section shown where $a = 1.5$ in. If the column is to carry a centrically applied compressive load of 100 kips, determine the largest permissible length of the column to the nearest 0.1 ft. Assume the column may buckle in any direction.

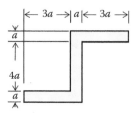

Structural woods

9.107 The rectangular cross section shown is that for a structural wooden, pin-ended column 15 ft long. The specific wood used has the following properties: $E = 2 \times 10^3$ ksi and $\sigma_c = 1.8$ ksi. If the column is to support a centrically applied compressive load of 15 kips, find the least permissible dimension b to the nearest 0.1 in. The column may buckle in any direction.

(Problems 9.107 and 9.108)

9.108 In reference to the figure in Problem 9.107, the rectangular cross section shown, with $b = 4$ in. is that for a structural wooden, fixed-free column. The specific wood used has the following properties: $E = 1.6 \times 10^3$ and $\sigma_c = 2$ ksi. If the column is to support a centrically applied compressive load of 50 kips, find the largest permissible length, to the nearest 0.1 ft, the column may have. The column may buckle in any direction.

9.109 Four identical pieces of structural pine, each with dressed dimensions of $1\frac{1}{2} \times 5\frac{1}{2}$ in., are glued and nailed securely together to form the box section shown and used for a pin-ended column 10 ft long. The wood used has the following properties: $E = 1.8 \times 10^3$ ksi and $\sigma_c = 1.4$ ksi. Find the maximum permissible centrically applied load that the column may carry.

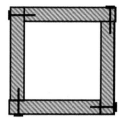

(Problems 9.109 and 9.110)

9.110 In reference to the figure in Problem 9.109, four identical pieces of structural pine, each with dressed dimensions of $2 \ 1/2 \times 5 \ 1/2$ in., are glued and nailed securely together to form the box section shown and used for fixed-free column. The wood used has the following properties: $E = 1.7 \times 10^3$ ksi and $\sigma_c = 1.5$ ksi. If the column is to support a centrically applied compressive load of 35 kips, find the largest permissible length of the column.

9.111 Two identical pieces of structural Douglas fir, each with dressed dimensions of 38×140 mm, are glued and nailed securely to two other identical pieces of structural Douglas fir, each with dressed dimensions of 38×184 mm to form the box section shown and used for a fixed-pinned column 4 m long. The Douglas fir used has the following properties: $E = 11$ GPa and $\sigma_c = 8$ MPa. Find the maximum permissible centrically applied load that the column may carry.

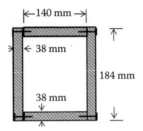

(Problems 9.111 and 9.112)

9.112 In reference to the figure in Problem 9.111, two identical pieces of structural Douglas fir, each with dressed dimensions of 38×140 mm, are glued and nailed securely to two other identical pieces of structural Douglas fir, each with dressed dimensions of 38×184 mm to form the box section shown and used for a fixed-fixed column. The Douglas fir used has the following properties: $E = 11$ GPa and $\sigma_c = 8$ MPa. If the column is to carry a centrically applied compressive load of 230 kN, find the maximum allowable length of the column to the nearest 0.001 m.

9.113 Two identical pieces of structural pine, each with dressed dimensions of $1\ 1/2 \times 5\ 1/2$ in. are glued and nailed securely to one piece of structural pine with dressed dimensions $1\ 1/2 \times 9\ 1/4$ in. to form the U section shown and used for a pin-ended column 12 ft long. The pine used has the following properties: $E = 1.6 \times 10^3$ ksi and $\sigma_c = 1.3$ ksi. Determine the largest allowable centrically applied compressive load that this column can carry.

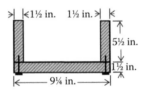

9.114 Two identical pieces of structural Douglas fir, each with dressed dimensions of 64×286 mm, are glued and nailed securely together to form the inverted T section shown and used for a fixed-pinned column. The Douglas fir used has the following properties: $E = 12$ GPa and $\sigma_c = 10$ MPa. If the column is to support a compressive centrically applied load of 150 kN, determine the largest permissible length of the column to the nearest 0.001 m. Assume the column is supported forcing it to buckle about the strong axis.

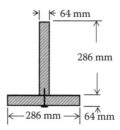

9.115 Three identical pieces of structural pine, each with dressed dimensions of 1 1/2 × 5 1/2 in., are glued and nailed securely together to form the I section shown and used for a fixed-fixed column 12 ft long. The pine used has the following properties: $E = 1.7 \times 10^3$ ksi and $\sigma_c = 1.4$ ksi. Find the maximum permissible centrically applied compressive load that the column can support.

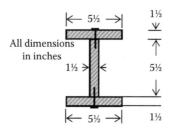

9.7   DESIGN OF ECCENTRICALLY LOADED COLUMNS

The secant equation developed earlier forms the basis for a rational design of eccentrically loaded columns and was extensively used in the past. However, because of complexities involved in the use of this equation, two other methods have been introduced and are currently widely used in the design of such columns. These two methods, known as the *allowable-stress method* and *interaction method*, are discussed in the following sections.

9.7.1   ALLOWABLE-STRESS METHOD

At any position along its length, the eccentrically loaded column is subjected to a combination of a constant compressive load P and a bending moment $M = Pe$, where e is the eccentricity of the compressive load. The maximum stress, σ_{MAX}, in such a column, therefore, is the sum of two stresses: the *axial compressive stress*, $\sigma_P = P/A$, and the *bending compressive stress*, $\sigma_M = Mc/I = M/S$, where A is the cross-sectional area of the column and S its section modulus with respect to the bending axis. According to the allowable-stress method, this maximum stress cannot exceed the allowable stress, σ_{ALL}, obtained for a centrally loaded column of the same material given by the empirical equations discussed in Section 9.6. In other words,

$$\frac{P}{A} + \frac{M}{S} = \sigma_{MAX} \leq \sigma_{ALL} \tag{9.71}$$

The use of Equation 9.71 in the design of eccentrically loaded columns implies that the stresses are well within the elastic limit of the material. It also requires that the same allowable stress be used for both the *axial compressive* stress and the *bending compressive* stress. However, because of the fact that the allowable axial compressive stress is generally less (in order to allow for the possibility of buckling) than the allowable bending compressive stress, this latter requirement leads to very conservative designs especially for short and intermediate columns.

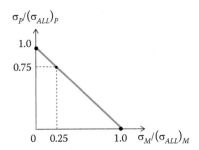

FIGURE 9.24   Axial load and bending moment interaction curve.

9.7.2   INTERACTION METHOD

This second method eliminates the weakness of *over-design* inherent in the allowable-stress method by assigning different allowable stress values for the two components of stress in the eccentrically loaded column. This is accomplished in two steps: First, we divide each term in Equation 9.71 by σ_{ALL} to obtain

$$\frac{\sigma_P}{\sigma_{ALL}} + \frac{\sigma_M}{\sigma_{ALL}} \leq 1 \tag{9.72}$$

As stated above, $\sigma_P = P/A$ and $\sigma_M = Mc/I = M/S$. Second, we replace σ_{ALL} in the first term by the allowable stress, $(\sigma_{ALL})_P$, for axial stress due to the compressive load P and in the second term by the allowable stress, $(\sigma_{ALL})_M$, for bending stress due to the moment M. Thus,

$$\frac{\sigma_P}{(\sigma_{ALL})_P} + \frac{\sigma_M}{(\sigma_{ALL})_M} \leq 1 \tag{9.73}$$

Equation 9.73 is known as an *interaction equation* because, in any given situation, it adjusts the relative importance of the axial force versus that of the bending moment. A plot of Equation 9.73 is shown in Figure 9.24 and reveals this interaction between the contribution of the axial force and that of the bending moment. Thus, for example, if the axial force is zero, only the bending moment exists and the design is based solely on bending action. If, however, the bending moment is zero, only the axial force exists and the design is based solely on compressive action. On the other hand, if both the axial force and bending moment exist simultaneously, then, as seen from Figure 9.24, Equation 9.73 makes the needed adjustments between these two factors. Thus, for example, if $\sigma_P/(\sigma_{ALL})_P = 0.75$, then the contribution of the bending moment is given by a value of $\sigma_M/(\sigma_{ALL})_M = 0.25$. This example is shown in Figure 9.24 and illustrates the interaction that exists between the compressive axial force and the bending moment in the buckling of an eccentrically loaded column.

As in the allowable stress method, the interaction method may be used for any material for which the allowable stresses may be found from empirical equations or are specified by the applicable codes.

EXAMPLE 9.15

Using the *allowable-stress* method, select a suitable wide flange steel ($E = 30 \times 10^3$ ksi, $\sigma_y = 42$ ksi) section from among the 12-in. nominal depth sections given in Appendix E. The selected section is to serve as a fixed-pinned column of length $L = 20$ ft and support an eccentrically applied compressive force $P = 75$ kips with an eccentricity $e = 4$ in. as shown.

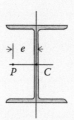

SOLUTION

Since the column is fixed-pinned, its effective length is $L_e = 0.7(20) = 14$ ft. Also, using Equation 9.57, we find $C_c = 118.7$. In order to proceed with the solution, we need to assume a given W12 section and perform the needed computations to ascertain if the assumed section is acceptable. If not, the process is repeated until an acceptable W12 section is found. Thus,

First trial:
W12 × 50: $A = 14.7$ in.^{2}, $S_y = 13.9$ in.^{3}, $r_y = 1.96$ in. Therefore,

$$\frac{L_e}{r_y} = \frac{14(12)}{1.96} = 85.7 < C_c$$

Equation 9.59 yields the applicable factor of safety and Equation 9.58 the needed allowable stress. Thus,

$$F.S. = \frac{5}{3} + \frac{3}{8}\left(\frac{85.7}{118.7}\right) - \frac{1}{8}\left(\frac{85.7}{118.7}\right)^3 = 1.890; \quad \sigma_{ALL} = \frac{42}{1.890}\left[1 - \left(\frac{85.7}{118.7}\right)^2\right] = 16.430 \text{ ksi}$$

By Equation 9.71, we have

$$\frac{75}{14.7} + \frac{75(4)}{13.9} = 26.7 > \sigma_{ALL}$$

Therefore, the W12 × 50 section is not acceptable and a stronger section is needed.

Second trial: The W12 × 53 section was also tried but computations similar to those in the first trial led to the conclusion that this section is unacceptable. Note: Properties of this W section are not available in Appendix E but can be found in the *Manual of Steel Construction, 8th Edition* (American Institute of Steel Construction, Chicago, IL, 1980).

Third trial:
W12 × 58: $A = 17.0$ in.^{2}, $S_y = 21.4$ in.^{3}, $r_y = 2.51$ in. Therefore,

$$\frac{L_e}{r_y} = \frac{14(12)}{2.51} = 66.9 < C_c$$

As in the first trial, we use Equation 9.59 to find the applicable factor of safety and Equation 9.58 to find the needed allowable stress:

$$F.S. = \frac{5}{3} + \frac{3}{8}\left(\frac{66.9}{118.7}\right) - \frac{1}{8}\left(\frac{66.9}{118.7}\right)^3 = 1.856; \quad \sigma_{ALL} = \frac{42}{1.856}\left[1 - \frac{1}{2}\left(\frac{66.9}{118.7}\right)^2\right] = 19.035 \text{ ksi}$$

By Equation 9.71, we have

$$\frac{75}{17.0} + \frac{75(4)}{21.4} = 18.430 < \sigma_{ALL}$$

Therefore, the W12 × 58 section is acceptable for the given application.   **ANS.**

EXAMPLE 9.16

Solve Example 9.15 using the *interaction* method. Let $(\sigma_{ALL})_M = 24$ ksi.

SOLUTION

As in Example 9.15, the effective length is $L_e = 0.7(20) = 14$ ft. Also, using Equation 9.57, we find $C_c = 118.7$. As in the case of Example 9.15, we need to assume a W12 section and perform the needed computations to ascertain if the assumed section is acceptable. If not, the process is repeated until an acceptable W12 section is found. Thus,

First trial:
W12 × 50: $A = 14.7$ in.^{2}, $S_y = 13.9$ in.^{3}, $r_y = 1.96$ in. Therefore,

$$\frac{L_e}{r_y} = \frac{14(12)}{1.96} = 85.7 < C_c$$

We now use Equation 9.59 to find the applicable factor of safety and Equation 9.58 to find the needed allowable stress for axial compression due to P:

$$F.S. = \frac{5}{3} + \frac{3}{8}\left(\frac{85.7}{118.7}\right) - \frac{1}{8}\left(\frac{85.7}{118.7}\right)^3 = 1.890; \quad (\sigma_{ALL})_P = \frac{42}{1.890}\left[1 - \frac{1}{2}\left(\frac{85.7}{118.7}\right)^2\right] = 16.430 \text{ ksi}$$

By Equation 9.73, we have

$$\frac{(75/14.7)}{16.430} + \frac{75(4)/13.9}{24} = 1.210 > 1$$

Therefore, the W12 × 50 section is *not* acceptable and a stronger section is needed.

Second trial: The W12 × 58 section was also tried, but computations similar to those in the first trial led to the conclusion that this section would be overly conservative.

Third trial:
W12 × 53: $A = 15.6$ in.^{2}, $S_y = 19.2$ in.^{3}, $r_y = 2.48$ in. (See note in Example 9.15 for this W section). Therefore,

$$\frac{L_e}{r_y} = \frac{14(12)}{2.51} = 66.9 < C_c$$

We now use Equation 9.59 to find the applicable factor of safety and Equation 9.58 to find the needed allowable stress for axial compression due to P:

$$F.S. = \frac{5}{3} + \frac{3}{8}\left(\frac{67.7}{118.7}\right) - \frac{1}{8}\left(\frac{67.7}{118.7}\right)^3 = 1.857; \quad (\sigma_{ALL})_P = \frac{42}{1.857}\left[1 - \frac{1}{2}\left(\frac{67.7}{118.7}\right)^2\right] = 18.939 \text{ ksi}$$

By Equation 9.73, we have

$$\frac{(75/15.6)}{18.939} + \frac{75(4)/19.2}{24} = 0.905 < 1$$

Therefore, the W12 × 53 section is the one to use for this application.   **ANS.**

Note that this wide flange obtained by the *interaction* method is smaller than the one required by the *allowable-stress* method of Example 9.15.

PROBLEMS

Structural steel

9.116   An S24 × 100 steel ($E = 29 \times 10^3$ ksi, $\sigma_y = 40$ ksi) section as shown, is used for a pin-ended column 15 ft long. Use the *allowable-stress* method to find the maximum eccentrically applied ($e = 2.0$ in.) load P that the column can carry.

(Problems 9.116 and 9.117)

9.117   In reference to the figure in Problem 9.116, use the *interaction* method to select the most suitable S steel ($E = 30 \times 10^3$ ksi, $\sigma_y = 36$ ksi) section from among the 12-in. nominal height group given in Appendix E to carry an eccentrically applied ($e = 1.5$ in.) load $P = 40$ kips as shown. The column is fixed-pinned with a length of 20 ft and the allowable bending stress is 22 ksi.

9.118   Use the *allowable-stress* method to select the most suitable W steel ($E = 200$ GPa, $\sigma_y = 240$ MPa) section from among the 610-mm nominal height group given in Appendix E to carry an eccentrically applied ($e = 40$ mm) load of 1000 kN as shown. The column is fixed-pinned with a length of 7 m. The column is so supported that it is forced to buckle about the strong axis.

(Problems 9.118 and 9.119)

9.119 In reference to the figure in Problem 9.118, a W406 × 84.8 steel ($E = 205$ GPa, $\sigma_y = 340$ MPa) section is used for a pin-ended column to carry an eccentrically applied ($e = 30$ mm) load of 1200 kN as shown. The column is so supported that it is forced to buckle about the strong axis. Use the *interaction* method to find the maximum allowable length L, to the nearest 0.1 m, if the bending allowable stress is 150 MPa.

9.120 Use the *interaction* method to select the most suitable steel ($E = 30 \times 10^3$ ksi, $\sigma_y = 36$ ksi) W section from among the 18-in. nominal height group given in Appendix E to carry an eccentrically applied ($e = 2.0$ in.) load of 50 kips as shown. The column is fixed-pinned with a length of 16 ft and the allowable stress in bending is 24 ksi.

(Problems 9.120 and 9.121)

9.121 Solve Problem 9.120 by the *allowable-stress* method.

9.122 An S508 × 142.8 steel ($E = 200$ GPa, $\sigma_y = 240$ MPa) section is used for a fixed-fixed column with a length of 8 m to carry an eccentrically applied ($e = 50$ mm) load P as shown and forced to buckle about the strong axis. Use the *allowable-stress* method to find the maximum permissible load P.

(Problems 9.122 and 9.123)

9.123 Solve Problem 9.122 by the *interaction* method if the bending allowable stress is 150 MPa.

9.124 Two S15 × 50 steel ($E = 30 \times 10^3$ ksi, $\sigma_y = 37$ ksi) sections are welded together to form the lipped box section shown. It is used as a pin-ended column of length L and subjected to a load $P = 100$ kips eccentrically applied as indicated. Use the *allowable-stress* method to find the maximum permissible length L to the nearest 0.1 ft.

Two welded S sections
(Problems 9.124 and 9.125)

9.125 Solve Problem 9.124 using the *interaction* method if the bending allowable stress is 22 ksi.

9.126 Two C10 × 20 steel ($E = 29 \times 10^3$ ksi, $\sigma_y = 36$ ksi) sections are welded together to form the box section shown. It is used as a fixed-fixed column of length $L = 15$ ft and subjected to a load P eccentrically applied as indicated. Use the *allowable-stress* method to find the maximum permissible load P.

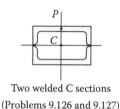

Two welded C sections
(Problems 9.126 and 9.127)

9.127 In reference to the figure in Problem 9.126, two C254 × 44.6 steel ($E = 200$ GPa, $\sigma_y = 250$ MPa) sections are welded together to form the box section shown. It is used as a fixed-free column of length L and subjected to a load $P = 120$ kN eccentrically applied as indicated. Use the *interaction* method to find the largest allowable length L (to nearest 0.1 m) if the bending allowable stress is 130 MPa.

9.128 A pin-ended steel ($E = 205$ GPa, $\sigma_y = 300$ MPa) column, 7.0 m long, has a hollow square cross section with dimension b on each side and constant thickness $t = 0.08b$ as shown. If the column is to carry an eccentrically applied compressive force of 20 kN as shown, find the least acceptable value of the dimension b to the nearest 0.001 m. Use the *allowable-stress* method.

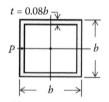

(Problems 9.128 and 9.129)

9.129 In reference to the figure in Problem 9.128, a fixed-fixed steel ($E = 30 \times 10^3$ ksi, $\sigma_y = 42$ ksi) column, 12 ft long, has a hollow square cross section with dimension b on each side and constant thickness $t = 0.08b$ as shown. Let $b = 6$ in. and using the *interaction* method with an allowable bending stress of 25 ksi, determine the maximum permissible eccentrically applied compressive load P.

9.130 A fixed-free steel ($E = 200$ GPa, $\sigma_y = 300$ MPa) column, 4 m long, has a hollow circular cross section with outside diameter d_o and inside diameter $d_i = 0.9d_o$. If the column is to carry an eccentrically applied compressive force $P = 150$ kN as shown, find the least acceptable value of the dimension d_o to the nearest 0.001 m. Use the *allowable-stress* method.

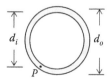

9.131 The steel ($E = 30 \times 10^3$ ksi, $\sigma_y = 42$ ksi) Z section shown is that for a fixed-fixed column 24 ft long. Design considerations require that the compressive load P be placed on the strong axis (i.e., the u axis) at a distance e from the centroid of section as shown. Use the *interactive* method with a bending allowable stress of 25 ksi to find the maximum permissible load P assuming that $e = 2.0$. Let $a = 1.0$ in., $b = 4.0$ in., and $h = 6.0$ in.

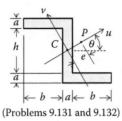

(Problems 9.131 and 9.132)

9.132 Solve Problem 9.131 if $e = 4.0$ in. instead of 2.0 in. All other conditions are as stated in Problem 9.136.

Aluminum alloys

9.133 The hollow square cross section shown is that for a 2014–T6 aluminum alloy pin-ended column 8 ft long. If the column is to carry an eccentrically applied compressive force $P = 12$ kips, use the *allowable-stress* method to determine the least permissible dimension b to the nearest 0.1 in.

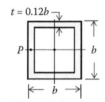

(Problems 9.133 and 9.134)

9.134 In reference to the figure in Problem 9.133, the hollow square cross section shown is that for a 6061–T6 aluminum alloy fixed-free column designed to carry a compressive eccentrically applied force $P = 15$ kips as shown. If $b = 5.0$ in., use the *interactive* method, with an allowable bending stress of 18 ksi, to determine the largest permissible length of this column to the nearest 0.1 in.

9.135 Three 2014–T6 aluminum plates are joined together securely to form the U section shown, which is used for a fixed-fixed column 20 ft long. If $b = 12.0$ in. and $t = 1.5$ in., use the *allowable-stress* method to find the maximum permissible eccentric compressive load P that may be applied to the column as shown.

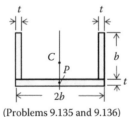

(Problems 9.135 and 9.136)

9.136 In reference to the figure in Problem 9.135, three 6061–T6 aluminum plates are joined together securely to form the U section shown, which is used for a fixed-free column 12 ft long. If the column is to be subjected to an eccentrically applied compressive force $P = 30$ kips as shown and $t = 0.1b$, use the *interactive* method to determine the least permissible value of b to the nearest 0.1 in. The bending allowable stress is 20 ksi.

9.137 Three identical 2014–T6 aluminum plates are joined together securely to form the I section shown, which is used for a fixed-pinned column 15 ft long. If $b = 8.0$ in. and $t = 0.8$ in., use the *allowable-stress* method to determine the maximum permissible eccentrically ($e = 2.0$ in.) applied load compressive load P that may be applied to the column as shown. The column is so supported that it is forced to buckle about the strong axis.

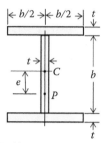

(Problems 9.137 and 9.138)

9.138 In reference to the figure in Problem 9.137, three 6061–T6 aluminum plates are joined together securely to form the I section shown, which is used for a fixed-fixed column 24 ft long. If the column is to be subjected to an eccentrically ($e = 0.2b$) applied compressive force $P = 25$ kips as shown and $t = 0.1b$, use the *interactive* method to determine the least permissible value of b to within 0.1 in. The column is so supported that it is forced to buckle about the strong axis. The bending allowable stress is 22 ksi.

9.139 A 6061–T6 aluminum alloy fixed-pinned column has the cross section shown where $b = 12$ in. and $t = 1.5$ in. A compressive load $P = 25$ kips is applied as shown. Use the *allowable-stress* method to determine the largest permissible length of the column to the nearest 0.1 ft. The column is so supported that it is forced to buckle about the strong axis.

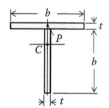

(Problems 9.139 and 9.140)

9.140 In reference to the figure in Problem 9.140, a 2014–T6 aluminum alloy pin-ended column has the cross section shown where $b = 10$ in. and $t = 1$ in. If the column has a length $L = 8$ ft, use the *interactive* method to determine the maximum permissible eccentrically applied compressive force P that may be applied to the column as shown. The column is so supported that it is forced to buckle about the strong axis. The bending allowable stress is 24 ksi.

9.141 The 6061–T6 aluminum section shown is that for a fixed-fixed column 20 ft long. Design considerations require that the compressive load P be placed on the strong axis (i.e., the u axis) at a distance e from the centroid of the section as shown. Use the *allowable-stress* method to find the maximum permissible load P assuming that $e = 2.5$. Let $a = 2.0$ in.

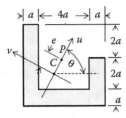

Structural woods

9.142 The rectangular cross section shown is that for a structural wooden, pin-ended column 15 ft long. The specific wood used has the following properties: $E = 2 \times 10^3$ ksi and $\sigma_c = 1.8$ ksi. If the column is to support an eccentrically ($e = 0.3b$) applied compressive load $P = 15$ kips as shown, use the *allowable-stress* method to find the least permissible dimension b to the nearest 0.1 in. The column is supported so as to buckle about the strong axis. Let $e = 0.3b$.

(Problems 9.142 and 9.143)

9.143 In reference to the figure in Problem 9.142, the rectangular cross section shown, with $b = 150$ mm, is that for a structural wooden, fixed-free column. The specific wood used has the following properties: $E = 12$ GPa and $\sigma_c = 10$ MPa. If the column is to support an eccentrically ($e = 0.08$ m) applied compressive load $P = 80$ kN, use the *interaction* method to find the largest permissible length of the column to the nearest millimeter. The column is forced to buckle about the strong axis. The bending allowable stress is 20 MPa. Let $e = 0.08$ m. The bending allowable stress is 20 MPa.

9.144 Four identical pieces of structural pine, each with dressed dimensions of 1 1/2 × 5 1/2 in., are glued and nailed securely together to form the box section shown and used for pin-ended column 10 ft long. The structural pine used has the following properties: $E = 1.8 \times 10^3$ ksi and $\sigma_c = 1.4$ ksi. Use the *interaction* method to find the maximum permissible eccentric load P applied as shown. The bending allowable stress is 2 ksi.

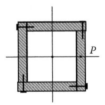

(Problems 9.144 and 9.145)

9.145 In reference to the figure in Problem 9.144, four identical pieces of structural pine, each with dressed dimensions of 2 1/2 × 5 1/2 in., are glued and nailed securely together to form the box section shown and used for fixed-free column. The structural pine used has the following properties: $E = 1.7 \times 10^3$ ksi and $\sigma_c = 1.5$ ksi. If the column is to support an eccentrically applied compressive load $P = 35$ kips, use the *allowable-stress* method to find the largest permissible length of the column to the nearest inch.

9.146 Two identical pieces of structural Douglas fir, each with dressed dimensions of 28 × 100 mm, are glued and nailed securely to two other identical pieces of Douglas fir, each with dressed dimensions of 28 × 184 mm to form the box section shown and used for a fixed-pinned column 4 m long. The Douglas fir used has the following properties: $E = 11$ GPa and $\sigma_c = 8$ MPa. Use the *allowable-stress* method to find the maximum permissible eccentric load applied as shown.

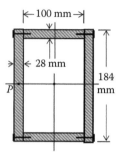

(Problems 9.146 and 9.147)

9.147 In reference to the figure in Problem 9.146, two identical pieces of structural Douglas fir, each with dressed dimensions of 28 × 100 mm, are glued and nailed securely to two other identical pieces of wood, each with dressed dimensions of 28 × 184 mm to form the box section shown and used for a fixed-fixed column. The Douglas fir used has the following properties: $E = 11$ GPa and $\sigma_c = 8$ MPa. If the column is to carry an eccentrically applied compressive load $P = 40$ kN, use the *interaction* method to find the maximum allowable length of the column to the nearest millimeter. The bending allowable stress is 10 MPa.

9.148 Two identical pieces of pine, each with dressed dimensions of 1 1/2 × 5 1/2 in. are glued and nailed securely to one piece of pine with dressed dimensions 1 1/2 × 9 1/4 in. to form the U section shown and used for a pin-ended column 12 ft long. The pine used has the following properties: $E = 1.6 \times 10^3$ ksi and $\sigma_c = 1.3$ ksi. Use the *allowable-stress* method to determine the largest permissible eccentric load P applied as shown.

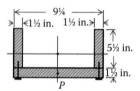

9.149 Two identical pieces of Douglas fir, each with dressed dimensions of 64 × 286 mm, are glued and nailed securely together to form the inverted T section shown and used for a fixed-pinned column. The Douglas fir used has the following properties: $E = 12$ GPa and $\sigma_c = 10$ MPa. If the column is to support a compressive eccentric load $P = 150$ kN applied as shown, use the *interaction* method to determine the largest permissible length of the column to the nearest millimeter. The bending allowable stress is 10 MPa.

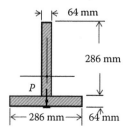

REVIEW PROBLEMS

R9.1 A column model is shown in the sketch. Weightless rods AB and BC are hinged together at B and supported there with two linear springs of spring constant k. The other ends of the springs are attached to rollers that slide in smooth vertical guides. Rod AB of length L is supported at A by a frictionless hinge. Rod BC of length L is supported at C by a frictionless hinge attached to a weight W that can slide freely in a frictionless vertical guide. (a) Use the potential energy method to find the critical weight W_{CR}. (b) Let $W = 100$ lb, $L = 2$ ft, and $k = 250$ lb/ft. Determine the positions of equilibrium for this system for values of θ between $0°$ and $90°$. Examine these positions of equilibrium for stability and state if they are stable, unstable, or neutral.

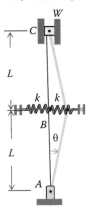

R9.2 Consider the frame shown in figure (a) below. Member BD has a channel cross section fabricated by welding two $6 \times 4 \times 3/4$-in. steel angle sections as shown in figure (b) below. Determine the largest permissible force P that may be applied if a factor of safety of 2.5 is required. Assume Euler's equation applies and that the channel cross section is free to buckle about its weak axis. The centroid for one of the angles is located in figure (b) below and the following properties are given for each angle: $A = 6.94$ in.², $I_x = 24.5$ in.⁴, and $I_y = 8.68$ in.⁴. Let $E = 30 \times 10^3$ ksi and $\sigma_y = 20$ ksi.

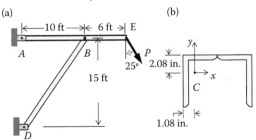

R9.3 Determine Euler's critical load for the cross section indicated. Refer to Appendix B.2 for help in determining principal axes and principal moments of inertia. Let $a = 25$ mm, $b = 115$ mm, and $h = 75$ mm. The modulus of elasticity and length of the column are, respectively, $G = 70$ GPa and $L = 6$ m.

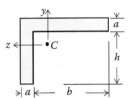

R9.4 Member BC in the frame of figure (a) below, made of steel ($E = 30 \times 10^3$ ksi), has the cross section shown in figure (b) below. Let $a = 1.0$ in., $b = 6.0$ in., and $c = 4.0$ in., determine the largest permissible force P that may be applied if a factor of safety of 2 is required. Assume Euler's equation applies. Assume also that the joints at B and C are designed to allow member BC to buckle in any direction. Refer to Appendix B.2 for help in determining principal axes and principal moments of inertia.

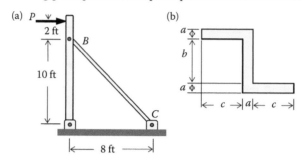

R9.5 Member ABC ($E = 200$ GPa, $\sigma_y = 150$ MPa) is fixed at A against rotation to a frictionless slider, attached at C to a frictionless hinge, and supported at B by frictionless rollers as shown. The member has W356 × 101.2 wide flange cross section (see Appendix E). If $L = 8$ m and the required factor of safety relative to buckling is 2.5, find the maximum permissible compressive load P. Assume the section is oriented such that buckling takes place in the plane of the page.

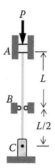

R9.6 Member BD of the frame of figure (a) below has the cross section shown in figure (b) below. Let $P = 100$ kips and $b = 2.0$ in., determine the largest permissible length L if a factor of safety of 2.0 is required. Assume that Euler's equation applies. Assume also that member BD is so supported that it can buckle about the weak axis. Refer to Appendix B.2 for help in determining principal axes and principal moments of inertia. Let $E = 10 \times 10^3$ ksi and $\sigma_y = 15$ ksi.

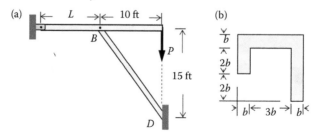

R9.7 Two C305 × 44.2 sections are welded together, back to back, to form the I section shown. It is used as a fixed-pinned column of length $L = 6$ m and subjected to a load

P eccentrically applied as indicated. If $E = 200$ GPa, $\sigma_y = 250$ MPa, $e = 50$ mm, and the maximum deflection is not to exceed 20 mm, determine the largest permissible load *P*.

Two welded C sections

R9.8 Two $8 \times 6 \times 1$ unequal-leg angles are welded together to form the T section shown and is used for a pin-ended column to support a load $P = 30$ kips as shown. The centroid for one of the angles is also located in the sketch and the following properties are given for each angle: $A = 13.0$ in.^{2}, $I_x = 80.8$ in.^{4}, and $I_y = 38.8$ in.^{4}. If the maximum stress is limited to 15 ksi, and the column is so supported to buckle about the strong axis, determine the maximum permissible length of the column. What is the maximum deflection? Let $E = 30 \times 10^3$ ksi.

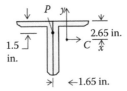

R9.9 Select the most suitable steel ($E = 30 \times 10^3$ ksi, $\sigma_y = 36$ ksi) W section from among the 14-in. nominal height group given in Appendix E to carry a centrally applied load of $P = 100$ kips. The column is fixed-fixed with a length of 18 ft. The column is so supported that it can buckle in any direction.

R9.10 Two $S457 \times 104.1$ steel sections ($E = 205$ GPa, $\sigma_y = 300$ MPa) are welded together to form the lipped box section shown. It is used as a fixed-free column of length *L* and subjected to a centrally applied load $P = 1000$ kN. Determine the maximum permissible length *L*.

R9.11 A 2014–T6 aluminum alloy fixed-fixed column has the cross section shown. If the column has a length $L = 15$ ft and is to support a centrally applied compressive force $P = 25$ kips, determine the least permissible dimension *a*. Assume the column may buckle in any direction.

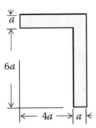

R9.12 Three identical pieces of Douglas fir, each with dressed dimensions of 64×286 mm, are glued and nailed securely together to form the I section shown and used for a fixed-pinned column. The Douglas fir used has the following properties: $E = 12$ GPa and

$\sigma_c = 10$ MPa. If the column is to support a centrically applied compressive load of 200 kN, determine the largest permissible length of the column.

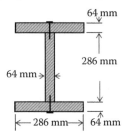

R9.13 Use the *allowable stress* method to select the most suitable steel ($E = 30 \times 10^3$ ksi, $\sigma_y = 36$ ksi) channel section from among the 10-in. nominal height group given in Appendix E to form, by welding back to back, the I section shown. This section is used to support an eccentrically applied ($e = 1.0$ in.) load $P = 120$ kips as shown. The column is fixed-pinned with a length of 20 ft and is supported to force buckling about the strong axis.

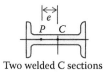

Two welded C sections

R9.14 The 6061–T6 aluminum ($E = 70$ GPa, $\sigma_y = 175$ MPa) Z section shown is that for a fixed-pinned column of length L. Design considerations require that the compressive load $P = 200$ kN be placed on the strong axis (i.e., the u axis) at a distance $e = 40$ mm. from the centroid of section as shown. Use the *interaction* method with a bending allowable stress of 120 MPa to find, to the nearest centimeter, the maximum permissible length L. Let $a = 30$ mm.

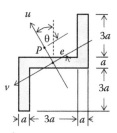

R9.15 Two identical pieces of pine, each with dressed dimensions of $11/2 \times 91/2$ in. are glued and nailed securely together to form the angle section shown and used for a pin-ended column 12 ft long. Design considerations require that the compressive load P be placed on the strong axis (i.e., the u axis) at a distance $e = 2.5$ in. from the centroid of section as shown. The pine used has the following properties: $E = 1.6 \times 10^3$ ksi and $\sigma_c = 1.3$ ksi. Use the *allowable-stress* method to determine the largest permissible eccentric load P applied as shown.

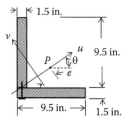

10 Excel Spreadsheet Applications

10.1 INTRODUCTION

The use of computer applications to solve analysis and design problems has become an integral part of twenty-first century engineering practice as powerful computer systems have become widely available to solve complicated mathematical and numerical problems. Spreadsheet applications are a type of computer program that makes it easy to manipulate numerical data and solve numerical engineering problems. Microsoft™ Excel® is one of the most widely used computer applications in this category because of its ease of use, flexibility, computational power, programmability, and availability.

While spreadsheet applications can typically be easily mastered to solve simple problems, creating applications to solve most engineering problems requires more than some basic knowledge of these programs. Engineering problem-solving typically requires finding the solutions to complicated mathematical equations, the creation of charts and graphs, and, sometimes, the implementation of iterative solution techniques. Developing spreadsheet applications at this level of problem-solving ability requires a more extensive knowledge of the capabilities of the software being used.

This chapter introduces some of the tools and techniques that can be used to solve engineering–mechanics problems, and demonstrates their use through a selection of examples that the authors hope would be useful to the readers. In the following paragraphs, a distinction is made between the person developing the spreadsheet application (the developer) and the person using the application to solve specific problems (the user).

10.2 SPREADSHEET APPLICATIONS CONCEPTS AND TECHNIQUES

The simplest way to describe a spreadsheet application is that it is a computer program specifically designed to manipulate tabulated data. However, what turns these kinds of applications into powerful tools is their ability to integrate a variety of information representation techniques to solve problems. The authors of this text assume that the reader has a basic understanding of spreadsheet application development. At a minimum, this includes understanding the difference between absolute and relative cell referencing, knowledge of what a range is, the ability to reference cells across different worksheets, the ability to create a formula, a chart, or a drawing object, and knowledge of the most commonly used mathematical functions. Readers who are not comfortable with any of the aforementioned topics and skills can refer to any of the many introductory Excel textbooks and online tutorials that are widely available. Some titles are listed in the reference section (Appendix B) of this textbook.

One of the most crucial steps in building a spreadsheet application is planning, which includes defining the scope of the application, identifying required input parameters, formulating a solution strategy, and outlining expected results and their presentation formats. Defining a scope is essential, because, without a clearly established scope, the project is open-ended, and would be difficult to complete. The fact of the matter is that when a scope is clearly defined, it becomes a lot easier to identify required input, define desired output, and formulate the solution steps. Knowing the types of input and output is necessary so that input can be properly solicited, and output is adequately presented.

The remaining part of this section is devoted to the discussion of a number of general topics that will help the spreadsheet-application developer to produce a robust and user-friendly application. In particular, the following topics will be addressed: using styles, protecting formula cells, data validation, using controls, automatic sorting, using predefined Excel functions, and defining new functions.

10.2.1 Using Styles

A common mistake spreadsheet-application developers make is not visually differentiating between input cells and other types of cells. The inability to tell the difference between input and an intermediate calculation or output leads to a great deal of confusion for the user and the possibility of typing in cells that contain formulae, essentially destroying the application. The newer versions of Excel provide a practical solution to this problem: the use of predefined cell styles. Figure 10.1 shows a snapshot of the various styles that are available for use in Excel 2010. The program also allows the user to define customized styles. A consistent use of predefined styles leads to the development of more user-friendly and robust spreadsheet applications.

10.2.2 Protecting Formula Cells

Another way of improving the user-friendliness and robustness of an application is to make sure that cells containing formulae are protected from accidental deletion or modifications. This can be done by clicking on *Protect Sheet* in the *Review* tab of the ribbon as shown in Figure 10.2. Figure 10.3 shows some of the options that can be used while protecting a worksheet, including the use of a password and making exceptions to allow specific actions to be performed by the user even when the sheet is protected.

Worksheet protection is designed to work with locked cells to allow the developer to exclude those cells that must be modified by the user of the application. Cells are locked or unlocked through the "Format Cells" dialog, as shown in Figure 10.4. Locked-cell contents can also be hidden by checking the "Hidden" box on the "Protection" tab of the "Format Cells" dialog.

10.2.3 Data Validation

Data validation is a way of ensuring that users enter only data relevant to the problem at hand. As such, data validation is a way of making sure that the application is of utmost user-friendliness and robustness. For example, consider the case of an application that requires computing the real square root of a number that can be entered by the user in cell I5. Without data validation, the user is

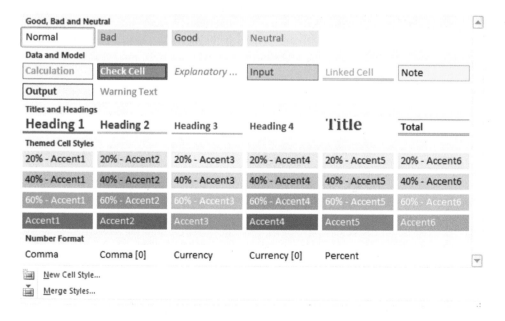

FIGURE 10.1 Microsoft Excel™ default cell styles.

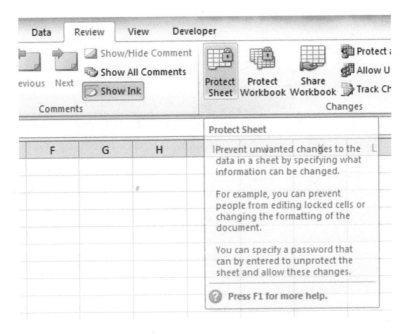

FIGURE 10.2   Activating worksheet protection in Microsoft Excel™.

allowed to enter any number in this cell; entering a negative number will produce the output shown in Figure 10.5.

A well-developed application should not produce such output; the developer can prevent the user from entering a negative number by limiting the input value to a decimal that is greater than or equal to 0, as shown in Figure 10.6. Should an attempt to enter a negative value be made, the program will not accept it and will display a message such as the one shown in Figure 10.7.

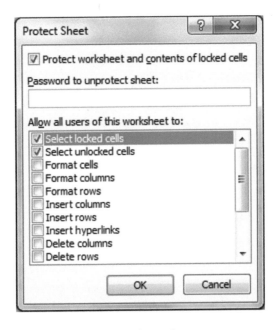

FIGURE 10.3   Microsoft Excel™ Worksheet protection options.

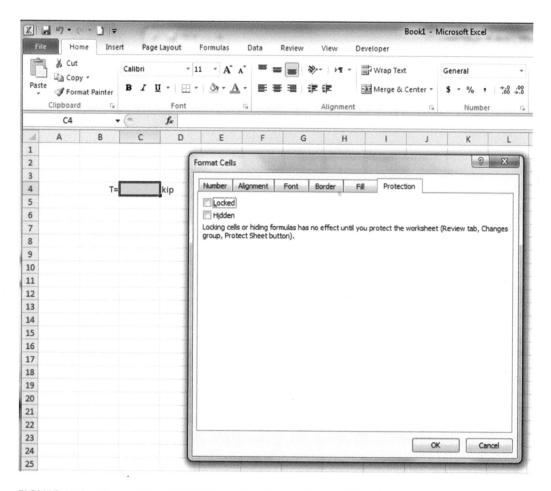

FIGURE 10.4   Microsoft Excel™ Cell Protection tab in the Format Cells window.

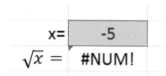

FIGURE 10.5   Sample Microsoft Excel™ display of invalid results.

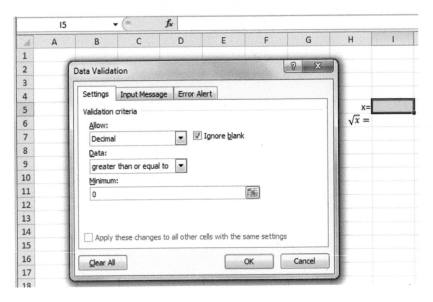

FIGURE 10.6   Microsoft Excel™ Data Validation window.

Data validation can be accessed by selecting the "Data" tab on the "Ribbon" and clicking on the data "Validation" icon while the cell containing the entry to be validated is selected. Data validation can be performed in many different ways, including limiting entries to whole or decimal numbers that meet specific requirements, forcing the user to select from a predefined list of values, limiting entries to a specific type, such as a date or a time, limiting the length of a text entry, or forcing the value to satisfy a developer-defined requirement. These types of data validation means are shown in Figure 10.8, and a sample implementation of the list validation method is shown in Figure 10.9, with its corresponding user view shown in Figure 10.10.

10.2.4   USING EXCEL FUNCTIONS

Excel comprises a healthy collection of functions that are provided in order to simplify application development, and enhance the computational power of the program. Some of these functions perform arithmetic evaluations like calculating the square root (SQRT), the natural logarithm (LN), or the cosine (COS) of a number while others perform more complex operations like finding the rank of a number in a list (RANK) or computing the sum of the products of the elements of two or more vector arrays (SUMPRODUCT). This section of the textbook describes the built-in functions, in alphabetical order, that are used in the presented examples.

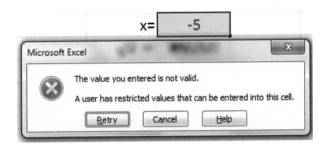

FIGURE 10.7   Sample Microsoft Excel™ data validation message.

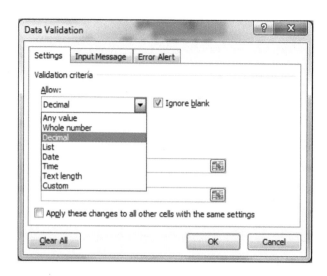

FIGURE 10.8   Sample Microsoft Excel™ data validation settings.

AND: This logical function returns "True" (1) if all of the passed arguments individually evaluate to other than 0 and "False" (0) if any one of the passed arguments evaluates to 0.

Function syntax: AND(Expression1, Expression2, Expression3, …)

Returns True if Expression1 evaluates to True, Expression2 evaluates to True, Expression3 evaluates to True, … Returns False otherwise.

ATAN: This function returns the inverse tangent of a number.

Function syntax: ATAN(Value)

The argument "Value" must be a real number, and the function returns an angle, in radians, between $-\pi/2$ and $\pi/2$.

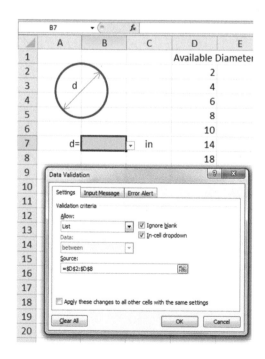

FIGURE 10.9   Setting list validation parameters in Microsoft Excel™.

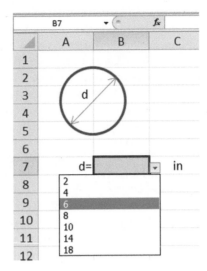

FIGURE 10.10   Sample input using the List Validation method.

CHOOSE: This function returns the specified entry in a given list. For example, if the range of cells B2 through F2 has the entries shown in Figure 10.11 and cell C4 has the number 3 in it, then entering the expression =*CHOOSE(C4,B2,C2,D2,E2,F2)* in cell C5 will display *Copper* in that cell. If the number 5 is entered in cell C4, then cell C5 will have *Wood* displayed in it.

Function syntax: CHOOSE(Index, Entry1, Entry2, Entry3, ...)
The index is an integer between 1 and the number of entries specified. The function returns the entry associated with the number specified by *Index*.

COS: This function returns the cosine of an angle.

Function syntax: COS(Value)
The argument "Value" must be a real number and is taken as being in radians.

IF: The "IF" function makes it possible to assign cell contents based on the applicability of a given condition. For example, if the square root of a number that is stored in cell A1 is to be computed in cell B1, all one has to do is enter the expression: =*SQRT(A1)* in cell B1. If the number in A1 is positive, then its square root will be computed and displayed in B1. However, if the number in A1 is negative, cell B1 will display the following: "NUM!", indicating that the

C5		f_x	=CHOOSE(C4,B2,C2,D2,E2,F2)		
	A	B	C	D	E	F
1	**No:**	1	2	3	4	5
2	**Material:**	Aluminum	Concrete	Copper	Steel	Wood
3						
4		Material Selected:	3			
5		Material Used:	Copper			

FIGURE 10.11   Sample implementation of the Microsoft Excel™ CHOOSE function.

argument that was passed to the SQRT function was not valid. To avoid this error, a developer may enter in cell B1 the following expression: =IF(A1 >=0, SQRT(A1), "No square root").

Function syntax: IF (Condition, Expression1, Expression2)

The function will evaluate "Condition." If Condition is True, the function returns Expression1; otherwise the function returns Expression2.

IFERROR: This function allows the developer to provide an alternative way of evaluating an expression that may result in an error. For example, say the cell A5 is to be used to display the square root of a number that is stored in A4. This can be accomplished by entering the following expression in A5: =SQRT(A4). However, if the number in A4 is negative, the result in A5 will be displayed as: "#NUM!" indicating that an unacceptable numeric argument was passed to the function. Moreover, if the cell in A4 contained a non-numeric entry, the result in A5 will be displayed as: "#VALUE!" indicating that the wrong type of argument was passed to the function. These conditions can be identified and dealt with adequately by entering the following expression in A5:

=IFERROR(SQRT(A4), IFERROR(SQRT(-A4)&"i", "Must enter a numeric value"))

This expression will produce the following results: if A4 contains 9, A5 will display "3"; if A4 contains -9, A5 will display "3i"; and if A4 contains a non-numeric value, then A5 will display "Must enter a numeric value".

Function syntax: IFERROR(Expression1, Expression2)

The function will evaluate Expression1. If no error is produced then, the function will return Expression1, otherwise the function will evaluate and return Expression2. If an error is produced during the evaluation of Expression2, the function will fail, and an error message is displayed in the cell that contains it.

RADIANS: This function transforms an angle from degrees to radians.

Function syntax: RADIANS(Value)

The function will return the equivalent in radians of the argument "Value," which is in degrees.

RANK.EQ: This function finds the ranking order of a number in a given list of numbers. For example, if a list that is stored in the range A1:E1 contains: 10, 4, 7, 8, 9, then entering the expression: =RANK.EQ(A1,$A$1:$E$1) in A2, and pressing the "Enter" key, will display 1 in A2, indicating that the number in A1, which is 10, is ranked first, that is, the highest in the list. Similarly, entering the expression =RANK.EQ(C1,$A$1:$E$1) in C2 and pressing the Enter key will display 4 in C2, indicating that the number 7 is ranked fourth in the list, and so on. This function is very valuable when automatic sorting of numbers is required. The "RANK.EQ" function is an improved version of the older function "RANK," which may not be supported by Excel in future versions of the program.

Function syntax: RANK.EQ(Value, Range, Order)

The function will return the rank of the provided value within the given range. If the "Order" is 0 or omitted, then the values are ordered in a descending range (the largest value in the range being ranked 1), and if the Order is 1, then the values are ordered in an ascending order (the smallest value in the range being ranked 1).

SIN: This function returns the sine of an angle.

Function syntax: SIN(Value)

The argument "Value" must be a real number and is taken as being in radians

TRANSPOSE: This function converts a vertical range of cells into a horizontal range, and vice versa. Consider the numbers shown in the range B2:F2 in Figure 10.12. To show this list vertically in the range I3:I7, select the destination vertical range, making sure the same number of cells—as in the original horizontal range—are selected, enter the

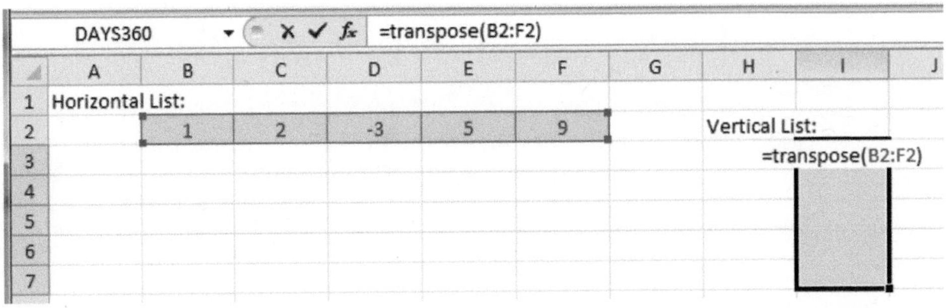

FIGURE 10.12   Sample implementation of the Microsoft Excel™ TRANSPOSE function.

expression: *=TRANSPOSE(B2:F2)* and then press the key combination: Ctrl–Shift–Enter. This key combination tells Excel that the destination range is to be treated as an array, and the formula in any of the cells in the range I3:I7 is displayed as "{=TRANSPOSE(B2:F2)}." The addition of the braces (the "curly" brackets) confirms the range is being treated as an array, and the resulting array is shown in Figure 10.13.

Function syntax: TRANSPOSE(Range)

The user must select the same number of cells in the destination range as there are in the original range, and then press the key combination Ctrl–Shift–Enter for this function to execute properly.

VLOOKUP: This function allows the developer to look up values in a table where the entries for any specifically desired item are listed in a vertical column. For example, if a stress analysis problem is to be limited to the consideration of a number of rectangular sections with specific dimensions, the developer can enter the information relating to the dimensions and cross-sectional properties of these sections in a table, and then use the "VLOOKUP" function to look them up and use them in a given situation. Assume the data is entered as shown in the Figure 10.14.

With the data table shown in Figure 10.13 defined, the expression *=VLOOKUP(3,$A2:$F7,4)* will return the value 7.25, which corresponds to the height, h (entered in column 4 of the table), associated with the 2 × 8 section, which is the height of the cross section (the 2 × 8), corresponding to the number 3 in the first column of the table. Similarly, the expression *=VLOOKUP(4,$A2:$F7,2)* will display 4 × 4, which is the item in column 2 of the table corresponding to the number 4 in the first column of the table.

Function syntax: VLOOKUP(Lookup Value, Range, Column Index Number, Range Lookup)

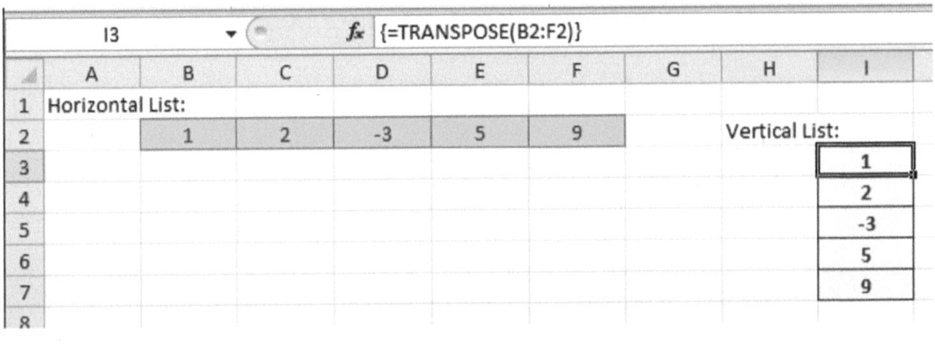

FIGURE 10.13   Results of the implementation of the Microsoft Excel™ TRANSPOSE function.

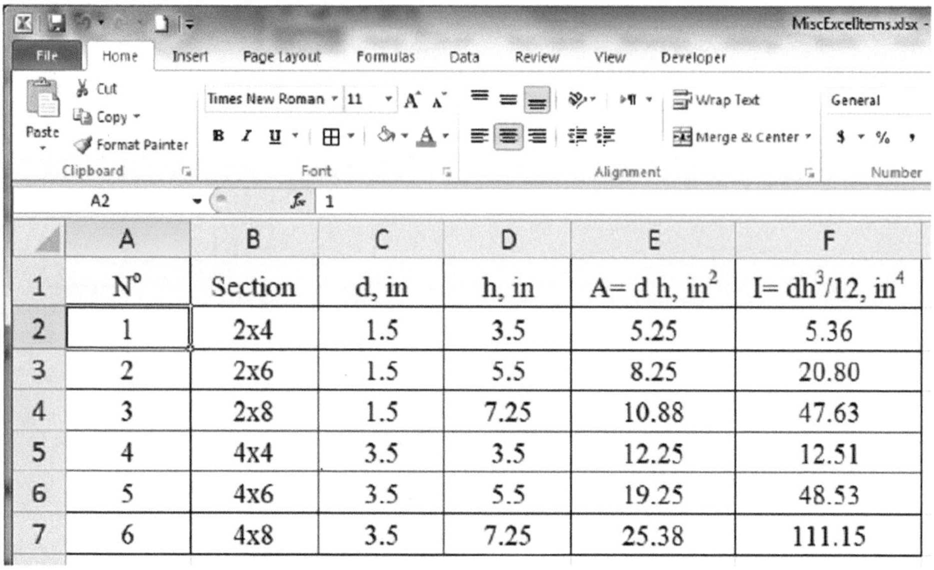

	A	B	C	D	E	F
1	N°	Section	d, in	h, in	A= d h, in^2	I= dh^3/12, in^4
2	1	2x4	1.5	3.5	5.25	5.36
3	2	2x6	1.5	5.5	8.25	20.80
4	3	2x8	1.5	7.25	10.88	47.63
5	4	4x4	3.5	3.5	12.25	12.51
6	5	4x6	3.5	5.5	19.25	48.53
7	6	4x8	3.5	7.25	25.38	111.15

FIGURE 10.14 Sample data setup for Microsoft Excel™ VLOOKUP function.

The function will return the value in the range that corresponds to the "Lookup Value" in the first column, and the column index number. If the "Range Lookup" is 0, the function returns an exact match only. If the Range Lookup is 1 or omitted, the function returns the exact or closest match. If the Range Lookup is 1 or omitted, the values in column 1 of the table must be ordered.

10.2.5 Automatic Real-Time Sorting

If done manually, sorting is a command that can easily be performed in Excel in a number of different ways, the easiest of which is to select the range of the data to be sorted, click the "Sort" icon in the "Data" tab on the "Ribbon," and select the desired sorting method. In contrast, automating the operation is not as straightforward as one would think. Automatic sorting may be necessary in some applications, when the user is allowed to enter values in any order, while the solution technique requires these same values to be in a specifically sorted order. One way to achieve automatic real-time sorting is by creating a function in Visual Basic, in which a sorting algorithm is implemented. However, this approach is limited to people who have an adequate knowledge of Visual Basic programming. The method presented in this textbook does not require any programming experience and can easily be applied to order any given list of numbers. The procedure utilizes three built-in functions: RANQ.EQ, IFERROR, and VLOOKUP. Say that a list of numbers, such

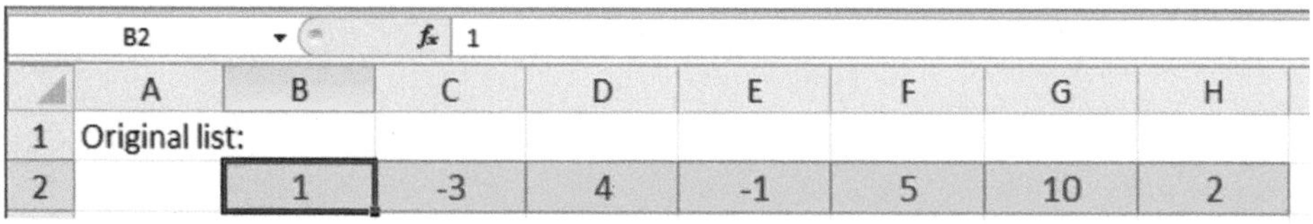

	A	B	C	D	E	F	G	H
1	Original list:							
2		1	-3	4	-1	5	10	2

FIGURE 10.15 Data for automatic real time sorting example.

| K2 | | | f_x {=TRANSPOSE(B2:H2)} | | | | | | | | | | | |

	A	B	C	D	E	F	G	H	I	J	K	L	M	N	O
										Actual Rank	Original List	Ascending Rank	Descending Ordered List	Descending Rank	Ascending Ordered List
1	Original list:														
2		1	-3	4	-1	5	10	2		5	1	1	10	7	-3
3										7	-3	2	5	6	-1
4										3	4	3	4	5	1
5										6	-1	4	2	4	2
6										2	5	5	1	3	4
7										1	10	6	-1	2	5
8										4	2	7	-3	1	10

FIGURE 10.16 Simple automatic sorting procedure output.

as the one shown in Figure 10.15, needs to be sorted. Notice how the numbers are not in any specific order and how the list includes positive and negative numbers, as well as duplicates.

The ordering process involves finding the rank of each entry in the list, and then using the "VLOOKUP" function to find the original entry associated with each rank value. In reference to Figure 10.15, note how the original list, which is horizontal in this case, is first transferred into a vertical column of cells, using the "TRANSPOSE" function, and how the rank associated with each value in the list is created just to the left of its corresponding value.

To determine the rank associated with each value, enter the formula $=RANK.EQ(K2,\$K\$2:\$K\$8)$ in cell J2, and then copy and paste it in the range J3:J8. The rank values must then be listed again in a new column in an ascending or descending order. If the rank values are listed in an ascending order, such as shown in column L in Figure 10.16, a descending ordered list will be produced. The descending ordered list can be obtained by entering the formula $=VLOOKUP(L2,\$J\$2:\$K\$8,2,0)$ in cell M2, and then copying and pasting it into the range M2:M8. In contrast, listing the rank values in a descending order (column N in Figure 10.16), will produce an ascending ordered list. To create the ascending ordered list, enter the formula $=VLOOKUP(N2,\$J\$2:\$K\$8,2,0)$ in cell O2, and then copy and paste it into the range O2:O8. The resulting descending and ascending ordered lists are shown in columns M and O, respectively, in Figure 10.16.

This simple procedure works perfectly if no duplicate numbers exist in the original list. However, and as illustrated in Figure 10.17, if duplicate numbers are part of the original list, the procedure will fail. This is because the largest rank will be smaller than the number of entries in the list causing the VLOOKUP function to result in an error.

Since it is reasonable to expect that some of the lists that may be dealt with in a mechanics-of-materials problem may have duplicate entries, it is necessary to find a way to make the procedure

| F2 | | | f_x -3 | | | | | | | | | | | |

	A	B	C	D	E	F	G	H	I	J	K	L	M	N	O
										Actual Rank	Original List	Ascending Rank	Descending Ordered List	Descending Rank	Ascending Ordered List
1	Original list:														
2		1	-3	4	-1	-3	10	2		4	1	1	10	7	#N/A
3										6	-3	2	4	6	-3
4										2	4	3	2	5	-1
5										5	-1	4	1	4	1
6										6	-3	5	-1	3	2
7										1	10	6	-3	2	4
8										3	2	7	#N/A	1	10

FIGURE 10.17 Output demonstrating the limitations of the simple automatic sorting procedure.

K2   *fx* {=TRANSPOSE(B2:H2)}

	A	B	C	D	E	F	G	H	I	J	K	L	M	N	O	P	Q	R	S	T	U
										Actual Rank	Original List	Ordered List		Descending Rank	0	1	2	3	4	5	6
1	Original list:																				
2		1	-3	4	-1	-3	10	2		4	1	-3		7		-3	-1	1	2	4	10
3										6	-3	-3		6	-3	-1	1	2	4	10	
4										2	4	-1		5	-1	1	2	4	10		
5										5	-1	1		4	1	2	4	10			
6										6	-3	2		3	2	4	10				
7										1	10	4		2	4	10					
8										3	2	10		1	10						

FIGURE 10.18 Output demonstrating the effectiveness of the enhanced automatic sorting procedure.

work for any random set of numbers. The procedure described herein can do just that, as shown in Figure 10.18.

The logic behind this procedure is easily seen once one realizes that the existence of duplicated entries reduces the value of the highest-ranked entry. For instance, and in reference to the example shown in Figure 10.17, the highest actual rank is six, while the number of entries is seven. This is because two of the numbers are the same, and accordingly ranked at the sixth position. As a consequence, looking up the seventh value in the list of entries using the VLOOKUP function produces an error since there is no "7" in the range of lookup values. By realizing that the seventh value is actually ranked sixth, it becomes possible to work around the problem. Suppose it is possible to detect an imminent failure of the VLOOKUP function whenever a given rank value is used, if the function can be executed at a lower rank value instead, it will eventually find an appropriate entry. To implement this, a matrix is created, having a number of rows equal to the number of entries in the list, and a number of columns equal to the expected number of duplicate entries. Ideally, the number of columns used is taken equal to the number of entries in the list to be ordered, to make sure that the case when all the numbers in the list are the same, is covered.

The implementation is shown here in Figure 10.19. The "IFERROR" function is used to detect a failure of the VLOOKUP function, and return an empty string when the error occurs. If no error occurs, then the function will return the entry value associated with the given rank value.

The formula $=IFERROR(VLOOKUP(\$N2-O\$1,\$J\$2:\$K\$8,2,0),"")$ is entered in Cell O2 and then copied and pasted in the range O2:U8. Say that this formula is entered in the cell in Column O on the row associated with a rank value n. Then the cell in question will contain the entry value, if it exists, associated with the rank n, and subsequent cells on the same row will contain the entry values associated with the rank values $n-1$, $n-2$, $n-3$, and so on, if they exist. Once this is done, it becomes clear that taking the smallest looked-up entry value in each row will produce an ascending ordered list if the rank values are entered in a descending order. Similarly, if the rank values are entered in an ascending order, a descending ordered list is produced.

10.2.6 Defining New Functions

A procedure is a collection of programming code instructions to perform a specific task. A function can be defined as a procedure that returns a value. In Excel, new functions are created using the Visual Basic for Applications (VBA) programming language. The VBA programming platform is

DAYS360   X ✓ fx =IFERROR(VLOOKUP($N2-O$1,$J$2:$K$8,2,0),"")

	A	B	C	D	E	F	G	H	I	J	K	L	M	N	O	P	Q	R	S	T	U
										Actual Rank	Original List	Ordered List		Descending Rank	0	1	2	3	4	5	6
1	Original list:																				
2		1	-3	4	-1	-3	10	2		4	1	-3		=IFERROR(VLOOKUP($N2-O$1,$J$2:$K$8,2,0),"")				1	2	4	10
3										6	-3	-3		6	VLOOKUP(lookup_value, table_array, col_index_num, [range_lookup])				4	10	
4										2	4	-1		5	-1	1	2	4	10		
5										5	-1	1		4	1	2	4	10			
6										6	-3	2		3	2	4	10				
7										1	10	4		2	4	10					
8										3	2	10		1	10						

FIGURE 10.19 Implementation of the enhanced automatic sorting procedure.

accessed by clicking the Visual Basic icon under the "Developer" tab on the Ribbon or by pressing the Alt-F11 key combination.

The intention of the authors is not to teach programming in this textbook. As a matter of fact, no Visual Basic programming at all was involved in the development of the sample applications being presented herein. Nonetheless, it must be noted that, in many cases, the use of programming can greatly enhance the usefulness and practicality of a spreadsheet application. For instance, let us assume that the computation of the maximum bending stress in a beam requires that the moment of largest magnitude from three different locations along the beam is known. Say that these moments are labeled $M1$, $M2$, and $M3$. Since any of the moments can be positive or negative, using the built-in Excel function $MAX(M1, M2, M3)$ will not work, because it will return the largest positive moment or the negative moment of least magnitude if no positive moments are available. Moreover, calling the function with the absolute values of the moments, as in $MAX\ ABS(M1), ABS(M2), ABS(M3))$, will always return a positive number, and cause the user to lose track of the fact that the moment may be negative. Since knowing the sign of the moment is important in determining which side of the beam is under tension and which is under compression, it is clear that this approach will not work. The information can be obtained without programming through a series of manipulations of the provided numbers or nested IF statements, but the simplest way to obtain the information is by creating a Visual Basic function, such as the following:

```
Function MaxMagnitude(M1 As Double, M2 As Double, M3 As Double) As Double
    Application.Volatile
    MaxMagnitude = M1
    If Abs(MaxMagnitude) < Abs(M2) Then MaxMagnitude = M2
    If Abs(MaxMagnitude) < Abs(M3) Then MaxMagnitude = M3
End Function
```

With this code in place, entering the expression $= MaxMagnitude(5, -10, 3)$ in a cell will display -10 in that cell. The statement *Application.Volatile* instructs Excel to recalculate the function every time a recalculation occurs in the spreadsheet. The use of functions is highly recommended for situations where in-cell formulae become long and cumbersome in such a way that they become difficult to follow and validate. It must be noted that Excel applications that have functions and procedures defined in them must be saved as macro-enabled workbooks, for the functions and procedures to work. These workbooks have the extension ".XLSM" to make sure the user knows they have executable code in them as they inherently have computer security risks associated with them.

10.2.7 Conditional Formatting

Conditional formatting is a visualization tool that can be used to enhance the output presentation in a spreadsheet application. As the name implies, conditional formatting consists of creating sets of rules to format cells based on conditions pertaining to their contents or the contents of other cells. Figure 10.20 shows a summary of the most commonly used rules. These rules are accessible by clicking on the "Conditional Formatting" icon on the "Home" tab of the Ribbon. For example, in an application that deals with the analysis of stress in a structural element, conditional formatting can be used to highlight cells containing stresses that exceed a certain value. In other applications, conditional formatting can be used to highlight input cells that contain unacceptable or otherwise out-of-range values.

10.2.8 Using Controls

Controls can be used to add interactivity to an Excel application. Quite often, adding controls can be done without any programming requirements on the part of the developer. Figure 10.21 shows a

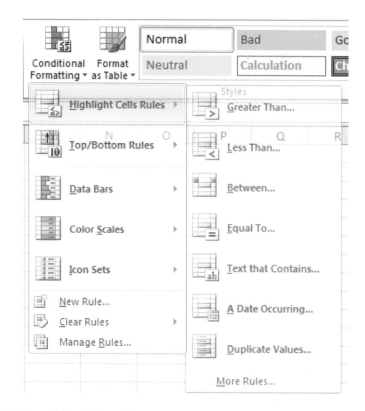

FIGURE 10.20   Microsoft Excel™ sample conditional formatting options.

selection of available controls, which are accessible through the Developer tab on the Ribbon. Excel provides two types of controls. Form-Controls are easy to use, and many of them do not require any major effort, beyond the initial setup, or Visual Basic programming to function. ActiveX-Controls are more sophisticated, but they do require some programming, to be of any use. By examining both types of controls, it becomes clear that Form-Controls are quite appropriate for developing most spreadsheet applications in mechanics.

Some of the Form-Controls that the reader might find useful are: the Combo-Box, the Check-Box, the Spin-Button, the Option-Button, and the Scroll-Bar. A Combo-Box allows the developer to limit an input parameter to a specific range of values. A Check-Box can be used to set the

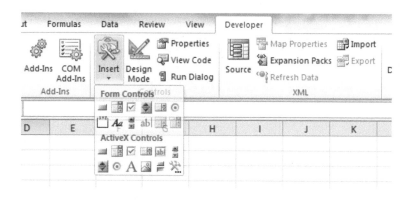

FIGURE 20.21   Microsoft Excel™ controls.

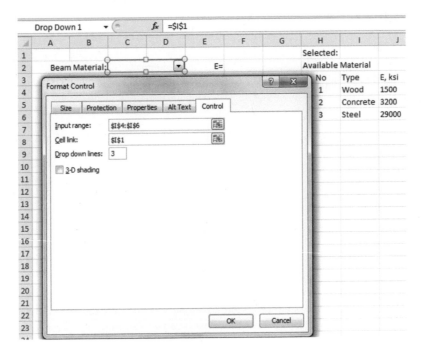

FIGURE 10.22   Implementation of a Combo-Box control.

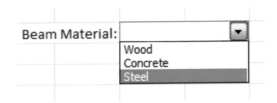

FIGURE 10.23   Sample Combo-Box control.

value of a cell to TRUE (1) or FALSE (0); a Spin-Button varies the value in a cell, between a developer-specified minimum value and a maximum value, using a developer-specified increment; an "Option" button must be part of a group of similar buttons to be of any use, and forces the user to make one selection out of many options; and the scroll bar allows the user to vary the value in a cell between a developer-specified minimum value and a maximum value, while providing a visual representation of the position of the selected value with respect to the minimum and maximum possible values. Figures 10.22 and 10.23 show an implementation of the Comb-Box control. In this implementation, the developer is limiting the user to select one type of material out of three that are made available. Once the user makes a selection, the associated reference number is automatically entered in cell I1 (1 for Wood, 2 for Concrete, and 3 for Steel). The number can then be used to determine the modulus of elasticity E of the material, by entering the expression: =VLOOKUP(I1,H4:J6,3) in cell F2. For the selection shown in Figure 10.23, cell F2 will display 29000.

10.3   EXAMPLE 1: DRAWING SHEAR AND MOMENT DIAGRAMS

Drawing shear and moment diagrams in a spreadsheet application can be frustrating if one important fact is not recognized: the shear or moment function being plotted can have different functional

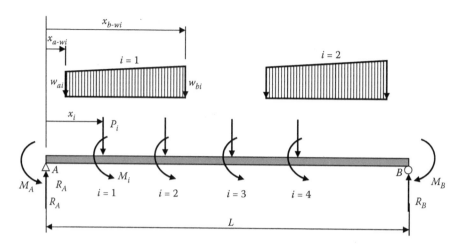

FIGURE 10.24 Example of a simply supported beam segment for shear and moment diagrams.

values at some locations along the member under consideration. This, of course, applies to all the points of application of concentrated forces and moments. To illustrate the procedure, it is proposed to create a spreadsheet application to draw the shear and moment diagrams for a simply supported beam segment subjected to the general set of loads shown in Figure 10.24. The loads consist of concentrated forces and moments at four possible locations along the beam, two end moments, and two distributed trapezoidal loads. This loading configuration will allow the user to plot the shear and moment diagrams of a beam in most situations possible in design. The choice of up to four concentrated loads and moments was made because it is typical to treat the loading from five or more concentrated forces that are applied within one span, as a distributed load. The ability to include end moments is being provided so the tool can be used to plot shear and moment diagrams for individual members that are part of a larger structure. In these cases, the end moments would be obtained from an analysis of the whole structure.

A screenshot of the completed application is shown in Figure 10.25. The figure shows the loading, shear, and moment diagrams of a simply supported beam that is 16 m long, and subjected to a concentrated force of 75 kN applied at 2 m from the left support and a trapezoidal load that varies from 50 kN/m, at a distance of 4 m measured from the left support, to 20 kN/m at a distance of 12 m measured from the left support. Figure 10.25 actually shows all the user gets to see out of the application. All necessary computations are performed in cells that are eventually hidden away. Input is limited to the units being used, the span length, and the locations and magnitudes of distributed loads and concentrated forces and moments. Program output includes a loading diagram, the shear force diagram, the moment diagram, and the magnitudes of distributed loads, shear force and bending moment at a specific point of interest along the beam. A slider control is made available to allow the user to select the point of interest anywhere between 0 and the span length.

The spreadsheet is set up in such a way that only cells that can accept input are accessible. This is done by protecting the worksheet to make sure the user does not inadvertently alter a cell that contains a formula. Output consists of providing the individual reactions associated with each load, the total reactions, a graphical representation of input loads, the shear diagram, the moment diagram, and the values of distributed load, shear, and moment at a given distance x.

The approach that was used to develop this spreadsheet application is actually very straightforward and follows the steps below:

1. User entries are first validated to make sure all loads are applied within the span. This is necessary to ensure the adequacy and accuracy of obtained results and input values that are found to be out of range are displayed in red by using conditional formatting.

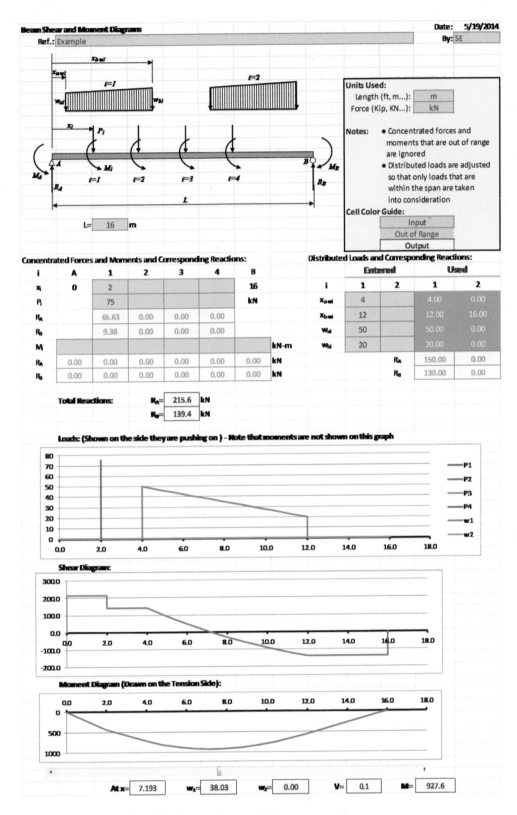

FIGURE 10.25   Screenshot of the shear and moment diagrams example.

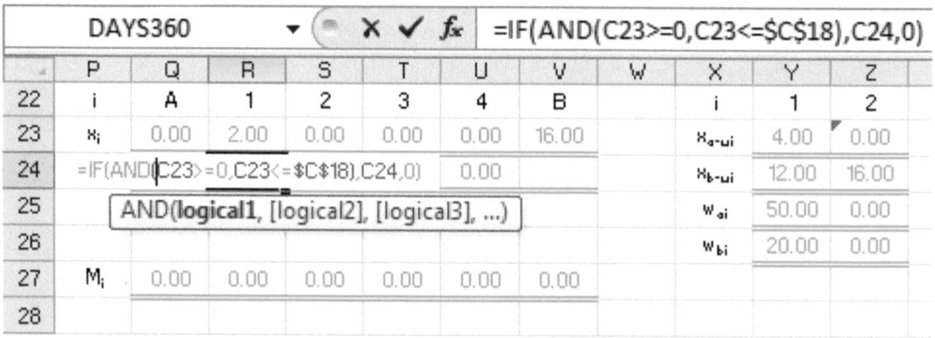

DAYS360				▼	⊙	✗ ✓ f_x	=IF(AND(C23>=0,C23<=$C$18),C24,0)			
	P	Q	R	S	T	U	V	W	X	Y	Z
22	i	A	1	2	3	4	B		i	1	2
23	x_i	0.00	2.00	0.00	0.00	0.00	16.00		x_{a-ui}	4.00	0.00
24	=IF(AND(C23>=0,C23<=$C$18),C24,0)					0.00			x_{b-ui}	12.00	16.00
25	AND(**logical1**, [logical2], [logical3], ...)								w_{ai}	50.00	0.00
26									w_{bi}	20.00	0.00
27	M_i	0.00	0.00	0.00	0.00	0.00	0.00				
28											

FIGURE 10.26 Data validation for the shear and moment diagrams example.

The validation is performed by copying acceptable input values into another area of the spreadsheet and using the validated values for all subsequent computations. For example, if the user specifies that a concentrated load is to be applied to a 16 m long beam at a distance of 18 m measured from the left support, this load will be ignored by the program as it falls outside the boundaries of the beam. The formula used to accomplish the validation of the first concentrated load is illustrated in Figure 10.26, and the same procedure is used for the other four possible load locations. For the distributed load, the program only takes the portion of the load that is applied to the beam if the start of the load or the end of the load falls outside of the span of the beam. This is done by interpolating between the provided loads for $x = 0$ and $x = L$ as applicable, where L is the beam span length. The provided graphical representation of the loads helps the user verify that the intended loads are being used to produce the desired shear and moment diagrams.

2. The coordinates of all the points of interest along the beam are collected in one column. These include the start of the beam at coordinate 0, the beam span length, and all the validated distances identifying the locations of distributed and concentrated loads and moments. While doing this, the developer must keep in mind that shear and moment diagrams may have different functional values at the same location along the beam. This will happen at points of application of concentrated forces and moments, including support locations. The most obvious way to deal with this issue is to include the location in the list a number of times. For example, the internal shear force just to the left of the point of application of a concentrated force is different from the one just to the right of the force. Consequently, the point must be listed at least twice. By doing so, a value of the shear just to the left of the force is associated with one of the locations, and the value of the shear just to the right of the force is associated with the second location. However, it can be challenging to get this to work properly as the developer must use lengthy formulae with complicated IF function calls to make sure that the forces are computed correctly and only at the required locations. One easy fix for this problem is to physically define locations just to the left and just to the right of every concentrated load location by slightly modifying the original coordinates. For instance, if a concentrated load is defined to be acting at a distance of 2 m measured from the left support, points of interest will be taken as 2.0000 m, (0.9999)(2) = 1.9998 m, and (1.0001)(2) = 2.0002 m. The applied changes are large enough to identify if a point is to the left, to the right, or exactly at the point of application of the load, yet too small to make a difference on the shear and moment diagrams. The approach is illustrated in Figure 10.27. The reader is also invited to note how the same approach is used at the points just to the right of the left support location and just to the left of the right support location. Also, the range of

			BG41		f_x	=0.9999*BG40
	BD	BE	BF	BG	BH	BI	BJ

	BD	BE	BF	BG	BH	BI	BJ
37	Ref	No	Rank	org. x	**Ordrd x**	Desc. Rank	0
38	A	1	41	0.0000	**0.0000**	43	
39		2	40	0.0001	**0.0000**	42	
40	x1	3	37	2.0000	**0.0000**	41	0
41		4	38	1.9998	**0.0001**	40	1E-04
42		5	36	2.0002	**1.6000**	39	1.6
43	x2	6	31	4.0000	**1.9998**	38	2
44		7	34	3.9996	**2.0000**	37	2
45		8	30	4.0004	**2.0002**	36	2
46	x3	9	19	8.0000	**3.2000**	35	3.2
47		10	22	7.9992	**3.9996**	34	4
48		11	18	8.0008	**4.0000**	33	
49	x4	12	8	12.0000	**4.0000**	32	
50		13	10	11.9988	**4.0000**	31	4
51		14	7	12.0012	**4.0004**	30	4
52	B	15	1	16.0000	**4.8000**	29	4.8
53		16	3	15.9984	**4.8000**	28	4.8
54	x_{a-u1}	17	31	4.0000	**5.6000**	27	5.6
55		18	31	4.0000	**6.4000**	26	6.4
56		19	29	4.8000	**6.4000**	25	6.4
57		20	27	5.6000	**7.1933**	24	7.193
58		21	26	6.4000	**7.2000**	23	7.2
59		22	23	7.2000	**7.9992**	22	7.999
60		23	21	8.0000	**8.0000**	21	8
61		24	17	8.8000	**8.0000**	20	
62		25	15	9.6000	**8.0000**	19	8
63		26	14	10.4000	**8.0008**	18	8.001
64		27	12	11.2000	**8.8000**	17	8.8
65	x_{b-u1}	28	8	12.0000	**9.6000**	16	
66		29	10	11.9988	**9.6000**	15	9.6
67	x_{a-u2}	30	41	0.0000	**10.4000**	14	10.4
68		31	41	0.0000	**11.2000**	13	11.2
69		32	39	1.6000	**11.2000**	12	11.2
70		33	35	3.2000	**11.9988**	11	
71		34	28	4.8000	**11.9988**	10	12
72		35	25	6.4000	**12.0000**	9	
73		36	19	8.0000	**12.0000**	8	12
74		37	15	9.6000	**12.0012**	7	12
75		38	13	11.2000	**12.8000**	6	12.8
76		39	6	12.8000	**14.4000**	5	14.4
77		40	5	14.4000	**15.9984**	4	
78		41	1	16.0000	**15.9984**	3	16
79	x_{b-u2}	42	3	15.9984	**16.0000**	2	
80	x	43	24	7.193333	**16.0000**	1	16

FIGURE 10.27 Identification of locations along the beam for correctly drawing shear and moment diagrams.

application of each one of the two distributed loads is divided into ten segments, and the point of interest that is specified by the user is also included in the list to make sure the values of distributed load, shear force, and bending moment associated with it are also calculated.

3. The points of interest collected in step two are then set up for automatic sorting, using the procedure outlined in Section 10.2 of this textbook. The sorting can also be automated, using a VBA function, but the procedure outlined in this text does not require any knowledge of programming. The sorted values are shown in Figure 10.27 in Column BH of

the worksheet. Many of the numbers are repeated but this will not have any effect on the results.

4. With all the points of interest identified and sorted, the loading, shear, and moment diagrams can now easily be obtained by using simple arithmetic manipulations.

a. First, a table of load locations and magnitudes, such as the one shown in Figure 10.28, is produced. Concentrated loads are entered in front of the coordinates where they are applied, and distributed loads are evaluated at each length increment within the range in which they are applied, using a simple interpolation. The developer must make sure the concentrated loads are only entered once in their corresponding column, as allowing otherwise will complicate the creation of the values to use to draw the shear diagram. This can easily be achieved by entering the following expression: =IF($O38 = P$33,IF(SUM(P37:P$38) = 0,P$35,""),"") in the cell corresponding to the first concentrated load column and the first x coordinate row and then copying the formula to cells corresponding to the other three concentrated-load columns and all remaining x coordinate rows.

b. Once all of the values of the loads are entered in their proper locations, the values of internal shear force can easily be obtained. For the shear due to the concentrated loads, the first entry is 0, the second entry is equal to the reaction at the left support, and each subsequent entry at any given point is equal to the value of shear at the previous point

	O	P	Q	R	S	T	U	V	W	X	Y	Z	AA
33	x1	2.00	4.00	8.00	12.00	4.00	0.00			x1		2.00	4.00
34	x2					12.00	16.00			x2			
35	Pi or wa	75.00	0.00	0.00	0.00	50.00	0.00			Pi or wa		75.00	0.00
36	wb					20.00	0.00			wb			
37	x	P1	P2	P3	P4	w1	w2			x	V_{Total}	V_{P1}	V_{P2}
38	0.0000					0	0			0.0	0.0	0.0	0.0
39	0.0000						0			0.0	215.6	65.6	0.0
40	0.0000						0			0.0	215.6	65.6	0.0
41	0.0001						0			0.0	215.6	65.6	0.0
42	1.6000						0			1.6	215.6	65.6	0.0
43	1.9998						0			2.0	215.6	65.6	0.0
44	2.0000	75					0			2.0	140.6	-9.4	0.0
45	2.0002						0			2.0	140.6	-9.4	0.0
46	3.2000						0			3.2	140.6	-9.4	0.0
47	3.9996						0			4.0	140.6	-9.4	0.0
48	4.0000		0			0	0			4.0	140.6	-9.4	0.0
49	4.0000		0			50	0			4.0	140.6	-9.4	0.0
50	4.0000		0			50	0			4.0	140.6	-9.4	0.0
51	4.0004					50	0			4.0	140.6	-9.4	0.0
52	4.8000					47	0			4.8	101.8	-9.4	0.0
53	4.8000					47	0			4.8	101.8	-9.4	0.0
54	5.6000					44	0			5.6	65.4	-9.4	0.0
55	6.4000					41	0			6.4	31.4	-9.4	0.0
56	6.4000					41	0			6.4	31.4	-9.4	0.0
57	7.1933					38.03	0			7.2	0.1	-9.4	0.0
58	7.2000					38	0			7.2	-0.2	-9.4	0.0
59	7.9992					35	0			8.0	-29.3	-9.4	0.0
60	8.0000			0		35	0			8.0	-29.4	-9.4	0.0
61	8.0000			0		35	0			8.0	-29.4	-9.4	0.0
62	8.0000			0		35	0			8.0	-29.4	-9.4	0.0
63	8.0008					35	0			8.0	-29.4	-9.4	0.0
64	8.8000					32	0			8.8	-56.2	-9.4	0.0

FIGURE 10.28 Associating locations along the beam segment with magnitudes of applied loads.

minus any applied concentrated force at the current point. For the shear due to the distributed loads, the first entry is 0, the second entry is equal to the reaction at the left support, and the principle of the graphical method is used to determine the value of the shear at subsequent points; i.e., the change in shear between two consecutive points is equal to the negative of the area under the loading diagram between those points. The areas, in this case, would be evaluated numerically. The shear force due to an applied moment is constant along the length of the beam and equal to the left support reaction. It is important to determine the values of shear for each individual load separately, as it makes it easier to check the results. The principle of superposition can then be used to obtain the shear values for the combined applied loads. Figure 10.29 shows a summary of the computed internal shears forces.

c. Once all of the values of internal shear forces are determined, the values of internal moment can be obtained using the principle of the graphical method, that is, the change in moment between two consecutive points is equal to the area under the shear diagram between those points. The areas, in this case, would again be evaluated numerically. The internal moments due to the applied external moments can be determined easier using equilibrium. It is important to determine the values of bending moment for each individual load separately as it makes it easier to check the results. The principle of

X	Y	Z	AA	AB	AC	AD	AE	AF	AG	AH	AI	AJ	AK
x1		2.00	4.00	8.00	12.00	4.00	0.00	0	2.00	4.00	8.00	12.00	16
x2						12.00	16.00						
Pi or wa		75.00	0.00	0.00	0.00	50.00	0.00						
wb						20.00	0.00						
X	V_{Total}	V_{P1}	V_{P2}	V_{P3}	V_{P4}	V_{u1}	V_{u2}	V_{MA}	V_{M1}	V_{M2}	V_{M3}	V_{M4}	V_{MB}
0.0	0.0	0.0	0.0	0.0	0.0	0.0	0.0	0.0	0.0	0.0	0.0	0.0	0.0
0.0	215.6	65.6	0.0	0.0	0.0	150.0	0.0	0.0	0.0	0.0	0.0	0.0	0.0
0.0	215.6	65.6	0.0	0.0	0.0	150.0	0.0	0.0	0.0	0.0	0.0	0.0	0.0
0.0	215.6	65.6	0.0	0.0	0.0	150.0	0.0	0.0	0.0	0.0	0.0	0.0	0.0
1.6	215.6	65.6	0.0	0.0	0.0	150.0	0.0	0.0	0.0	0.0	0.0	0.0	0.0
2.0	215.6	65.6	0.0	0.0	0.0	150.0	0.0	0.0	0.0	0.0	0.0	0.0	0.0
2.0	140.6	-9.4	0.0	0.0	0.0	150.0	0.0	0.0	0.0	0.0	0.0	0.0	0.0
2.0	140.6	-9.4	0.0	0.0	0.0	150.0	0.0	0.0	0.0	0.0	0.0	0.0	0.0
3.2	140.6	-9.4	0.0	0.0	0.0	150.0	0.0	0.0	0.0	0.0	0.0	0.0	0.0
4.0	140.6	-9.4	0.0	0.0	0.0	150.0	0.0	0.0	0.0	0.0	0.0	0.0	0.0
4.0	140.6	-9.4	0.0	0.0	0.0	150.0	0.0	0.0	0.0	0.0	0.0	0.0	0.0
4.0	140.6	-9.4	0.0	0.0	0.0	150.0	0.0	0.0	0.0	0.0	0.0	0.0	0.0
4.0	140.6	-9.4	0.0	0.0	0.0	150.0	0.0	0.0	0.0	0.0	0.0	0.0	0.0
4.0	140.6	-9.4	0.0	0.0	0.0	150.0	0.0	0.0	0.0	0.0	0.0	0.0	0.0
4.8	101.8	-9.4	0.0	0.0	0.0	111.2	0.0	0.0	0.0	0.0	0.0	0.0	0.0
4.8	101.8	-9.4	0.0	0.0	0.0	111.2	0.0	0.0	0.0	0.0	0.0	0.0	0.0
5.6	65.4	-9.4	0.0	0.0	0.0	74.8	0.0	0.0	0.0	0.0	0.0	0.0	0.0
6.4	31.4	-9.4	0.0	0.0	0.0	40.8	0.0	0.0	0.0	0.0	0.0	0.0	0.0
6.4	31.4	-9.4	0.0	0.0	0.0	40.8	0.0	0.0	0.0	0.0	0.0	0.0	0.0
7.2	0.1	-9.4	0.0	0.0	0.0	9.5	0.0	0.0	0.0	0.0	0.0	0.0	0.0
7.2	-0.2	-9.4	0.0	0.0	0.0	9.2	0.0	0.0	0.0	0.0	0.0	0.0	0.0
8.0	-29.3	-9.4	0.0	0.0	0.0	-20.0	0.0	0.0	0.0	0.0	0.0	0.0	0.0
8.0	-29.4	-9.4	0.0	0.0	0.0	-20.0	0.0	0.0	0.0	0.0	0.0	0.0	0.0
8.0	-29.4	-9.4	0.0	0.0	0.0	-20.0	0.0	0.0	0.0	0.0	0.0	0.0	0.0
8.0	-29.4	-9.4	0.0	0.0	0.0	-20.0	0.0	0.0	0.0	0.0	0.0	0.0	0.0
8.0	-29.4	-9.4	0.0	0.0	0.0	-20.0	0.0	0.0	0.0	0.0	0.0	0.0	0.0
8.8	-56.2	-9.4	0.0	0.0	0.0	-46.8	0.0	0.0	0.0	0.0	0.0	0.0	0.0

FIGURE 10.29   Computation of shear values along the beam.

	AN	AO	AP	AQ	AR	AS	AT	AU	AV	AW	AX	AY	AZ	BA
33	x1		2.00	4.00	8.00	12.00	4.00	0.00	0	2.00	4.00	8.00	12.00	16
34	x2						12.00	16.00						
35	Pi or w a		75.00	0.00	0.00	0.00	50.00	0.00	0.00	0.00	0.00	0.00	0.00	0.00
36	w b						20.00	0.00						
37	x	M_{Total}	M_{P1}	M_{P2}	M_{P3}	M_{P4}	M_{u1}	M_{u2}	M_{MA}	M_{M1}	M_{M2}	M_{M3}	M_{M4}	M_{MB}
38	0.0	0	0.000	0.000	0.000	0.000	0.000	0.000	0.000	0.000	0.000	0.000	0.000	0.000
39	0.0	0.000	0.000	0.000	0.000	0.000	0.000	0.000	0.000	0.000	0.000	0.000	0.000	0.000
40	0.0	0.000	0.000	0.000	0.000	0.000	0.000	0.000	0.000	0.000	0.000	0.000	0.000	0.000
41	0.0	0.022	0.007	0.000	0.000	0.000	0.015	0.000	0.000	0.000	0.000	0.000	0.000	0.000
42	1.6	345.000	105.000	0.000	0.000	0.000	240.000	0.000	0.000	0.000	0.000	0.000	0.000	0.000
43	2.0	431.207	131.237	0.000	0.000	0.000	299.970	0.000	0.000	0.000	0.000	0.000	0.000	0.000
44	2.0	431.250	131.250	0.000	0.000	0.000	300.000	0.000	0.000	0.000	0.000	0.000	0.000	0.000
45	2.0	431.278	131.248	0.000	0.000	0.000	300.030	0.000	0.000	0.000	0.000	0.000	0.000	0.000
46	3.2	600.000	120.000	0.000	0.000	0.000	480.000	0.000	0.000	0.000	0.000	0.000	0.000	0.000
47	4.0	712.444	112.504	0.000	0.000	0.000	599.940	0.000	0.000	0.000	0.000	0.000	0.000	0.000
48	4.0	712.500	112.500	0.000	0.000	0.000	600.000	0.000	0.000	0.000	0.000	0.000	0.000	0.000
49	4.0	712.500	112.500	0.000	0.000	0.000	600.000	0.000	0.000	0.000	0.000	0.000	0.000	0.000
50	4.0	712.500	112.500	0.000	0.000	0.000	600.000	0.000	0.000	0.000	0.000	0.000	0.000	0.000
51	4.0	712.556	112.496	0.000	0.000	0.000	600.060	0.000	0.000	0.000	0.000	0.000	0.000	0.000
52	4.8	809.480	105.000	0.000	0.000	0.000	704.480	0.000	0.000	0.000	0.000	0.000	0.000	0.000
53	4.8	809.480	105.000	0.000	0.000	0.000	704.480	0.000	0.000	0.000	0.000	0.000	0.000	0.000
54	5.6	876.380	97.500	0.000	0.000	0.000	778.880	0.000	0.000	0.000	0.000	0.000	0.000	0.000
55	6.4	915.120	90.000	0.000	0.000	0.000	825.120	0.000	0.000	0.000	0.000	0.000	0.000	0.000
56	6.4	915.120	90.000	0.000	0.000	0.000	825.120	0.000	0.000	0.000	0.000	0.000	0.000	0.000
57	7.2	927.616	82.563	0.000	0.000	0.000	845.054	0.000	0.000	0.000	0.000	0.000	0.000	0.000
58	7.2	927.616	82.500	0.000	0.000	0.000	845.116	0.000	0.000	0.000	0.000	0.000	0.000	0.000
59	8.0	915.819	75.008	0.000	0.000	0.000	840.811	0.000	0.000	0.000	0.000	0.000	0.000	0.000
60	8.0	915.795	75.000	0.000	0.000	0.000	840.795	0.000	0.000	0.000	0.000	0.000	0.000	0.000
61	8.0	915.795	75.000	0.000	0.000	0.000	840.795	0.000	0.000	0.000	0.000	0.000	0.000	0.000
62	8.0	915.795	75.000	0.000	0.000	0.000	840.795	0.000	0.000	0.000	0.000	0.000	0.000	0.000
63	8.0	915.772	74.993	0.000	0.000	0.000	840.779	0.000	0.000	0.000	0.000	0.000	0.000	0.000
64	8.8	881.575	67.500	0.000	0.000	0.000	814.075	0.000	0.000	0.000	0.000	0.000	0.000	0.000

FIGURE 10.30 Computation of moment values along the beam.

superposition can then be used to obtain the moment values for the combined applied loads. Figure 10.30 shows a summary of the computed internal moments.

5. With the values of loads, internal shear, and internal moment evaluated and listed next to their corresponding locations, drawing the loading, shear, and moment diagrams, is a simple operation of inserting a scatter chart with straight lines for each function. The developer needs to make sure the "Show data in hidden rows and columns" is checked in the "Select Data Source" dialog for each one of the charts. This ensures the charts are displayed correctly even if the columns where all the computations are performed are hidden from view.

6. If desired, the developer can add a slider to allow the user to look up loading, shear, and moment values at a specific point along the beam. Alternatively, the developer can designate the cell containing the x-value as an input cell, and allow the user to enter the value directly. Protecting the worksheet is highly recommended to make sure cells containing formulae are not accidentally modified by the user.

10.4 EXAMPLE 2: DRAWING MOHR'S CIRCLE

As described in Chapter 7 for plane stress, the state of stress at any given point on a cross section can be represented by the normal stresses σ_x and σ_y, and the shear stress τ_{xy}. These stresses are shown in their positive directions on a stress element in Figure 10.31.

The state of stress at the same point can also be represented on a rotated element by the normal stresses σ'_x and σ'_y, and the shear stress $\tau_{x'y'}$. This element would be rotated by an angle θ counterclockwise from the original element as shown in Figure 10.32.

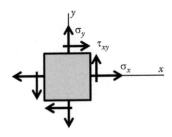

FIGURE 10.31 Plane stress element showing positive stress directions.

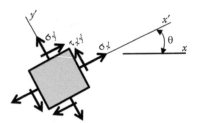

FIGURE 10.32 Rotated plane stress element.

The stresses on the inclined plane can be found geometrically by plotting Mohr's circle. This circle is centered at the point P: $((\sigma_x + \sigma_y)/2, 0)$ and has a radius:

$$R = \sqrt{\left(\frac{\sigma_x - \sigma_y}{2}\right) + \tau_{xy}^2} \qquad (10.1)$$

An Excel implementation of the Mohr's circle approach is shown in Figure 10.33. This geometric approach allows for the evaluation of the state of stress at any angle as well as the evaluation of the principal stresses and the maximum in-plane shear stress at the point under consideration.

The principal stresses σ_1 and σ_2 are given by the following equations:

$$\sigma_1 = \left(\frac{\sigma_x + \sigma_y}{2}\right) + R \qquad (10.2)$$

$$\sigma_2 = \left(\frac{\sigma_x + \sigma_y}{2}\right) - R \qquad (10.3)$$

The angle to the principal plane containing σ_1 and measured from the x axis with counterclockwise being positive is given by

$$\theta_p = \frac{1}{2} \cdot \arctan\left(\frac{-\tau_{xy}}{\sigma_x - (\sigma_x + \sigma_y/2)}\right) \qquad (10.4)$$

The maximum in-plane shear stress is given by

$$t_{\text{in-plane-max}} = R \qquad (10.5)$$

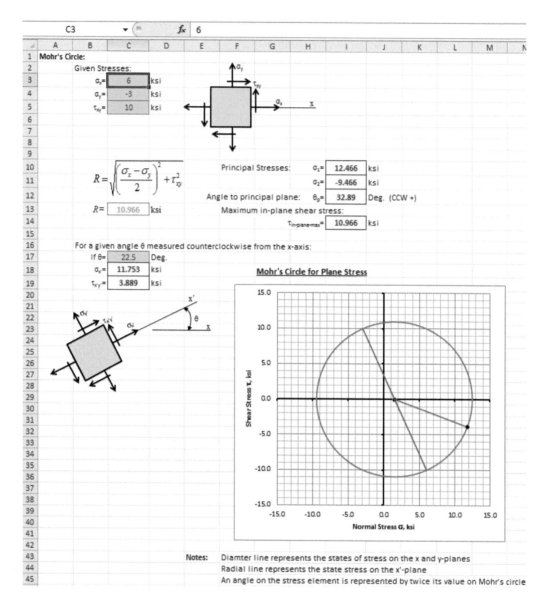

FIGURE 10.33   Microsoft Excel™ implementation of Mohr's circle stress analysis.

Note that in Excel, arctangent is obtained using the "ATAN" function and returns the result in radians. The formulae that were used to compute the principal stresses, the angle to the principal plane containing σ_1, and the maximum in-plane shear stress are shown in Figure 10.34.

10.4.1 PLOTTING MOHR'S CIRCLE

If a circle is plotted on an x–y coordinate system, each value on the x axis is associated with two values on the y axis, representing points on the perimeter of the circle. These points are symmetrically positioned with one another, with respect to the x axis; that is, they share the same x coordinate and have opposite y coordinates. It follows that the easiest way to draw a circle using a scatter chart is to define two functions: one for the top half of the circle and one for the bottom half, and plot them on the same graph. For Mohr's circle, the x axis represents normal stresses and the y axis represents

	H	I	J
	I10	f_x =(C3+C4)/2+C13	
9			
10		$\sigma_1=$ =(C3+C4)/2+C13	ksi
11		$\sigma_2=$ =(C3+C4)/2-C13	ksi
12	Angle to principal plane:	$\theta_p=$ =-DEGREES(ATAN(-C5/(C3-(C3+C4)/2)))/2	Deg. (CCW +)
13			
14		$\tau_{in\text{-}plane\text{-}max}=$ =C13	ksi

FIGURE 10.34   Calculation of principal stresses and angle to the principal plane.

shear stresses. Knowing the center point coordinates and the radius, R, of the circle, any point on the perimeter of the circle can be identified by the coordinates (σ', τ'), which can be related by the angle θ a line of length R is rotated from the x axis and with respect to the center of the circle at the point $P:((\sigma_x + \sigma_y)/2, 0)$. Two different values of τ', τ_1' and τ_2', are used to differentiate the points on the top half of the circle and points on the bottom half. The functions σ', τ_1', and τ_2' are defined as follows:

$$\sigma'(\theta) = \left(\frac{\sigma_x + \sigma_y}{2} \right) + R\cos(\theta) \tag{10.6}$$

$$\tau_1'(\theta) = R\sin(\theta) \tag{10.7}$$

$$\tau_2'(\theta) = -R\sin(\theta) \tag{10.8}$$

Consequently, if θ is varied from $0°$ to $180°$, and the corresponding values of σ', τ_1', and τ_2' are computed, these values can be used to draw Mohr's circle. The implementation of this approach is shown in Figure 10.35. The cells O12 to O48 of the worksheet shown in this figure have the numbers 0 to 180 using an increment of 5. Cell P12 has the expression =($C$3 + $C$4)/2 + $C$13*COS(RADIANS(O12)); cell Q12 has the expression =$C$13*SIN(RADIANS(O12)), and cell R12 has the expression = -$C$13*SIN(RADIANS(O12)). The contents of cells P12, Q12, and R12 are then copied all the way down to row 48 to produce the whole table, which is shown for this example in Figure 10.36. Mohr's circle can now easily be plotted by creating a scatter chart with smooth lines and no markers using σ' for the x axis and τ_1' and τ_2' as functions for the y axis to create the top- and bottom-half circles, respectively.

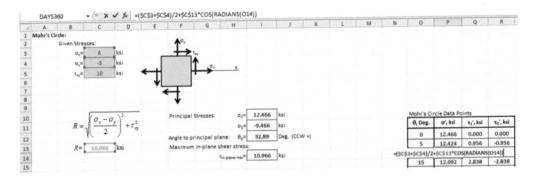

FIGURE 10.35   Implementation of approach for drawing Mohr's circle.

Mohr's Circle Data Points			
θ, Deg.	σ', ksi	τ_1', ksi	τ_2', ksi
0	12.466	0.000	0.000
5	12.424	0.956	-0.956
10	12.299	1.904	-1.904
15	12.092	2.838	-2.838
20	11.805	3.751	-3.751
25	11.438	4.634	-4.634
130	-5.549	8.400	-8.400
135	-6.254	7.754	-7.754
140	-6.900	7.049	-7.049
145	-7.483	6.290	-6.290
150	-7.997	5.483	-5.483
155	-8.438	4.634	-4.634
160	-8.805	3.751	-3.751
165	-9.092	2.838	-2.838
170	-9.299	1.904	-1.904
175	-9.424	0.956	-0.956
180	-9.466	0.000	0.000

FIGURE 10.36   Data points for drawing Mohr's circle.

10.4.2   STATE OF STRESS ON THE x' PLANE

The state of stress on a given x' plane defined by an angle θ measured counterclockwise from the x axis can be obtained using Equations 10.6 and 10.7. The normal and shear stresses acting on the plane normal to the x' axis are given by

$$\sigma_{x'} = \left(\frac{\sigma_x + \sigma_y}{2}\right) + R\cos(2(\theta - \theta_p)) \tag{10.9}$$

$$\tau_{x'y'} = R\sin(2(\theta - \theta_p)) \tag{10.10}$$

The implementation of these equations is shown in Figure 10.37.

C17	▼	f_x	22.5	
	B		C	
15				
16	For a given angle θ measured counterclockwise from the x-axis:			
17		If θ=	22.5	Deg.
18		σ_x=	=($C$3+$C$4)/2+$C$13*COS(RADIANS(2*(C17-I12)))	ksi
19		τ_{xy}=	=-$C$13*SIN(RADIANS(2*(C17-I12)))	ksi

FIGURE 10.37   Assessment of the state of stress along the x' axis.

10.4.3 REPRESENTATION OF THE STATE OF STRESS ON THE X PLANE, THE Y PLANE, AND THE X′ PLANE ON MOHR'S CIRCLE

Mohr's circle is not complete if the diameter line representing the state of stress on the x plane and the y plane and the radial line representing the state of stress on the x' plane are not shown. These lines can easily be added to the circle by following the following steps:

1. Create the data points for the diameter and radial lines. These can be created under the table that was used to obtain the data for Mohr's circle itself and the implementation is shown in Figure 10.38.

   In the above $\sigma_{min} = min(\sigma_x, \sigma_y)$; $\sigma_{max} = max(\sigma_x, \sigma_y)$, $\sigma_{avg} = (\sigma_x + \sigma_y)/2$, cell C3 contains σ_x, cell C4 contains σ_y, cell C5 contains τ_{xy}, and cells C18 and C19 contain σ_x' and $\tau_{x'y'}$ respectively. The right-most column (Column Q) contains the values of the corresponding shear stresses τ_{xy} assumed to be positive if it turns the element clockwise. The resulting calculations are shown in Figure 10.39.

2. The data obtained in step 1 above can now be plotted on the Mohr's circle chart by right-mouse clicking on the chart, and selecting "Select Data" as shown in Figure 10.40.

P50		▼ (*m*	f_x	=MIN(C3,C4,(C3+C4)/2)
⊿	O	P		Q
49	For plotting σ_x			
50	σ_{min}	=MIN(C3,C4,(C3+C4)/2)		=IF(P50=C3,-C5,C5)
51	σ_{avg}	=C3+C4+(C3+C4)/2-P50-P52		0
52	σ_{max}	=MAX(C3,C4,(C3+C4)/2)		=IF(P52=C3,-C5,C5)
53	For plotting $\sigma_{x'}$			
54	σ_{avg}	=(C3+C4)/2		0
55	$\sigma_{x'}$	=C18		=-C19
56				

FIGURE 10.38 Formulation of the data points for the diameter and radial lines on Mohr's circle.

P50			▼ (*m*	f_x	=MIN(C3,C4,(
⊿	N	O	P	Q	R
46		170	-9.299	1.904	-1.904
47		175	-9.424	0.956	-0.956
48		180	-9.466	0.000	0.000
49		For plotting σ_x			
50		σ_{min}	-3.000	10	
51		σ_{avg}	1.500	0.000	
52		σ_{max}	6	-10	
53		For plotting $\sigma_{x'}$			
54		σ_{avg}	1.500	0	
55		$\sigma_{x'}$	11.753	-3.889	

FIGURE 10.39 Data points for the diameter and radial lines on Mohr's circle.

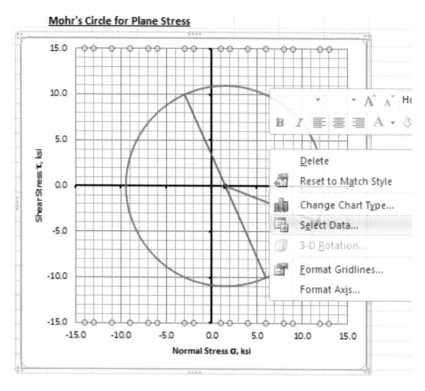

FIGURE 10.40   Adding diameter and radial lines data to Mohr's circle.

On the "Select Data Source" form that comes up, click the "**Add**" icon and fill out the information for the state of stress on the *x* and *y* planes, as shown in Figure 10.41, and then press the "**OK**" button.

While still on the "Select Data Source" form, click the "**Add**" icon again and fill out the information for the state of stress on the *x'* plane, as shown in Figure 10.42 and then press the "**OK**" button.

Once back on the "Select Data Source" form, click the "**OK**" button and the chart should look like the one shown in Figure 10.32, with the state of stress on all the planes of interest being adequately represented.

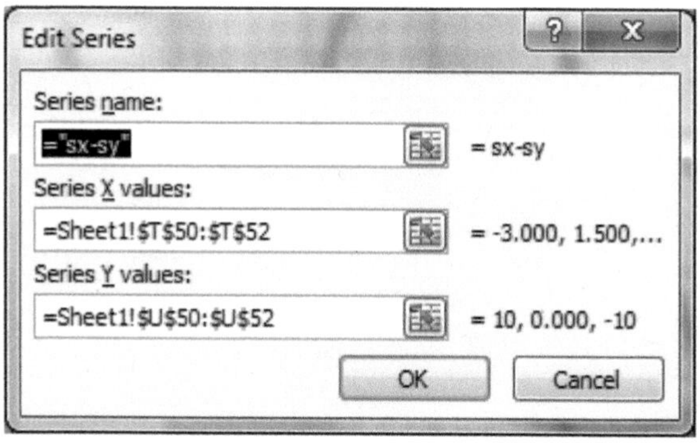

FIGURE 10.41   Entering data series for the Mohr's circle diameter line.

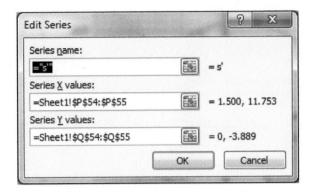

FIGURE 10.42 Entering data series for the Mohr's circle radial line.

10.5 EXAMPLE 3: PRINCIPAL STRESSES IN THREE-DIMENSIONAL STRESS ELEMENTS

The state of stress on a three-dimensional element is shown in Figure 10.43. The evaluation of principal stresses acting on this element requires elaborate mathematical manipulations to compute the eigenvalues of a 3×3 matrix. Furthermore, finding the directions of the principal stresses necessitates that the eigenvectors that are associated with the same matrix be obtained. The advantage of using a spreadsheet application such as Excel to solve these problems is that eigenvalues and eigenvectors can be obtained numerically.

The general formulation of the principal stress problem is given by

$$[[S] - \sigma_i[I]]\{X_i\} = \{0\} \tag{10.11}$$

In Equation 10.11, $[S]$ represents the state of stress at the point under consideration expressed in the x, y, and z directions, σ_i a principal stress, $[I]$ is the identity matrix, and $\{X_i\}$ is a vector representing the direction of the principal stress σ_i. The matrix $[S]$ is given by

$$[S] = \begin{bmatrix} \sigma_x & \tau_{xy} & \tau_{xz} \\ \tau_{yx} & \sigma_y & \tau_{yz} \\ \tau_{zx} & \tau_{zy} & \sigma_z \end{bmatrix} \tag{10.12}$$

Note that since $\tau_{xy} = \tau_{yx}$, $\tau_{xz} = \tau_{zx}$, and $\tau_{yz} = \tau_{zy}$, the matrix $[S]$ is symmetric. This makes it easier to find all of its eigenvalues and eigenvectors by using matrix deflation in conjunction with the power

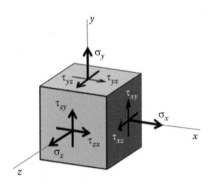

FIGURE 10.43 Three-dimensional stress element.

	M3		f_x	=H6					
	G	H	I	J	K	L	M	N	O
1			Date:	=TODAY()		Find the Largest ei			
2			By:	SE					
3							[S]=	=H6	=M4	=M5
4	a, MPa, psi, or ksi):	Mpa						=H9	=H7	=N5
5								=H10	=H11	=H8
6		$\sigma_x=$	-80	=IF($H$4<>"", $H$4,""					
7		$\sigma_y=$	40	=IF($H$4<>"", $H$4,""					

FIGURE 10.44   Implementation of the state of stress matrix [S].

method. The reader is referred to the numerical methods textbook by Al-Khafaji and Tooley (1986) for a treatment of this method as applied to the computation of principal stresses. The implementation involves the following steps:

1. In an area of the worksheet away from the print area, create the matrix [S] using the provided normal and tangential stresses, as shown in Figure 10.44.
2. Implement the power method to find the first principal stress (the largest eigenvalue). To do so, start by assuming the vector {1, 0, 0} in the cells Q11, R11, and S11, as an initial guess for the eigenvector associated with the first principal stress. The next guess is obtained by multiplying the matrix [S] by the assumed vector and then dividing the elements of the obtained vector by the element of largest magnitude, as shown in Figure 10.45. The element of largest magnitude is the first estimate of the principal stress (eigenvalue). This

	L12		f_x	{=TRANSPOSE(MMULT($M$3:$O$5,TRANSPOSE(Q11:S11)))}				
	K	L	M	N	O	P	Q	R	S
1	Find the Largest eigenvalue by solving: [[S] − σ_{p1} [I]]{X$_1$}={0}						Normalized Eigen Vector		
2						{X$_1$}=	-0.6116	-0.4402	0.6574
3		[S]=	-80	-40	80	σ_{p1} {X$_1$}^{T}{X$_1$}=	-72.858	-52.440	78.314
4			-40	40	120		-52.440	-37.744	56.367
5			80	120	-40		78.314	56.367	-84.178
6									
7									
8						σ_{p1}	Eigen Vector:		
9	Largest Eigen Value				Eigen Value:	-194.781	-0.930	-0.670	1.000
10	i	x_{1i}	x_{2i}	x_{3i}		max |x|	$x_{1i\text{-normalized}}$	$x_{2i\text{-normalized}}$	$x_{3i\text{-normalized}}$
11	0						1	0	0
12	1	-80.000	-40.000	80.000		80.000	-1.000	-0.500	1.000
13	2	180.000	140.000	-180.000		180.000	1.000	0.778	-1.000
14	3	-191.111	-128.889	213.333		213.333	-0.896	-0.604	1.000
15	4	175.833	131.667	-184.167		-184.167	-0.955	-0.715	1.000

	Q1013		f_x	=Q1011/SQRT(Q1011*Q1011+R1011*R1011+S1011*S1011)				
	K	L	M	N	O	P	Q	R	S
1006	995	181.212	130.429	-194.781		-194.781	-0.930	-0.670	1.000
1007	996	181.212	130.429	-194.781		-194.781	-0.930	-0.670	1.000
1008	997	181.212	130.429	-194.781		-194.781	-0.930	-0.670	1.000
1009	998	181.212	130.429	-194.781		-194.781	-0.930	-0.670	1.000
1010	999	181.212	130.429	-194.781		-194.781	-0.930	-0.670	1.000
1011	1000	181.212	130.429	-194.781		-194.781	-0.930	-0.670	1.000
1012									
1013						Normalized:	-0.612	-0.440	0.657

FIGURE 10.45   Implementation of the power method to obtain the first principal stress.

process needs to be repeated enough times until the estimates of the eigenvalue and eigen-vector stabilize. Depending on the situation, convergence may be achieved in as few as ten iterations or as many as 800. This implementation uses a constant number of 1000 iterations to ensure convergence does take place. The eigenvalue estimate obtained at the thousandth iteration is used as the principal stress and the corresponding eigenvector is normalized by dividing each element of the vector by the sum of the squares of all the elements. The normalized eigenvector represents the direction cosines associated with this principal stresses.

3. Create the first deflated matrix and use it to obtain the next largest eigenvalue and its cor-responding eigenvector. The deflated matrix $[Q]$ shown in Figure 10.46 is obtained using the following equation:

$$[Q_1] = [S] - \sigma_{p1}\{X_1\}^T\{X_1\} \tag{10.13}$$

In this equation, $[S]$ is the matrix representing the initial state of stress problem, σ_{p1} is the principal stress obtained in step 2 above, and $\{X_1\} = \{\cos \alpha, \cos \beta, \cos \gamma\}$ is the normal-ized eigenvector associated with σ_1, where α, β, and γ are the angles from the x, y, and z axes respectively to the line of action of σ_{p1}.

The power method is used again to find the next principal stress, which is now the largest eigenvalue of the deflated matrix . To do so, start by assuming the vector $\{0, 1, 0\}$ in the cells AB11, AC11, and AD11 as an initial guess for the eigenvector associated with the second principal stress. The next guess is obtained by multiplying the matrix by the assumed vector, and then dividing the elements of the obtained vector by the element of largest magnitude. Here again, the eigenvalue estimate obtained at the thousandth itera-tion is used as the principal stress, and the corresponding eigenvector is normalized by dividing each element of the vector by the sum of the squares of all the elements. The normalized eigenvector represents the direction cosines associated with this principal stresses.

	AB12		▼		f_x	=IF($AA12<>0,W12/$AA12,AB11)			
	V	W	X	Y	Z	AA	AB	AC	AD
1							Normalized Eigen Vector		
2		Deflated Matrix: $[Q_1] = [S] - \sigma_{p1}\{X_1\}^T\{X_1\}$				$\{X_2\}=$	0.0807	0.7919	0.6053
3		$[Q_1]=$	-7.142	12.440	1.686	$\sigma_{p2}\{X_2\}^T\{X_2\}=$	0.830	8.153	6.232
4			12.440	77.744	63.633		8.153	80.050	61.188
5			1.686	63.633	44.178		6.232	61.188	46.771
6									
7									
8						σ_{p2}	Eigen Vector:		
9	Second largest Eigen Value				Eigen Value:	127.651	0.102	1.000	0.764
10	i	X_{1i}	X_{2i}	X_{3i}		max $|x|$	$X_{1i\text{-normalized}}$	$X_{2i\text{-normalized}}$	$X_{3i\text{-normalized}}$
11	0						0	1	0
12	1	12.440	77.744	63.633		77.744	0.160	1.000	0.818
13	2	12.678	131.818	100.062		131.818	0.096	1.000	0.759
14	3	13.033	127.244	97.330		127.244	0.102	1.000	0.765

FIGURE 10.46 Deflated matrix and first iterations for evaluating the second principal stress and direction cosines.

4. The procedure described in step 3 can be repeated one more time to find the last principal stress and corresponding direction cosines. In this case, the new deflated matrix is obtained using the equation:

$$[Q_2] = [Q_1] - \sigma_{p2}\{X_2\}^T\{X_2\} \tag{10.14}$$

The power method is used with an initially assumed eigenvector of $\{0, 0, 1\}$ and the thousandth estimate of the eigenvalue is again taken as the last principal stress with the corresponding normalized eigenvector representing the direction cosines of this stress. Figure 10.47 shows the first few iterations of the method.

5. It is best to represent the principal stresses graphically once they are obtained. Unfortunately, Excel 2010 does not provide an option for producing 3D scatter plots. Consequently, the options available for representing the results graphically are to (1) create a VBA procedure that produces a 3D chart, (2) use an add-in (a program developed by a third party to perform specific tasks and designed to work within Excel), or (3) plot the 2D projections of the principal stresses on the x–y, y–z, and x–z planes. The last option is used in this implementation as it is the simplest and safest approach. Nine graphs are needed to represent the three principal stresses.

The first step is to identify the largest principal stress, σ_1, the middle principal stress, σ_2, and the smallest principal stress, σ_3. Note that in this example, the calculated principal stresses σ_{p1}, σ_{p2}, and σ_{p3} are stored in cells P1011, AA1011, and AL1011, respectively. If the maximum principal stress σ_1 is to be stored in cell Q1015, that cell would contain the formula: $=MAX(\$P\$1011,\$AA\$1011,\$AL\$1011)$ as shown in Figure 10.48.

The cell AM1015 containing σ_3 has the expression: $=MIN(\$P\$1011,\$AA\$1011,\$AL\$1011)$, and the cell AB1015 containing σ_2 has the expression: $=SUM(\$P\$1011,\$AA\$1011,\$AL\$1011)-Q1015-AM1015$.

	AL9			f_x	=AL1011				
	AG	AH	AI	AJ	AK	AL	AM	AN	AO
1							Normalized Eigen Vector		
2		Deflated Matrix: $[Q_2] = [S] - \sigma_{p2}\{X_2\}^T\{X_2\}$				$\{X_3\}=$	0.7870	-0.4232	0.4488
3		$[Q_2]=$	-7.972	4.287	-4.546				
4			4.287	-2.305	2.445				
5			-4.546	2.445	-2.592				
6									
7									
8						σ_{p3}	Eigen Vector:		
9	Smallest Eigen Value				Eigen Value:	-12.870	1.000	-0.538	0.570
10	i	x_{1i}	x_{2i}	x_{3i}		max |x|	$x_{1i\text{-normalized}}$	$x_{2i\text{-normalized}}$	$x_{3i\text{-normalized}}$
11	0						0	0	1
12	1	-4.546	2.445	-2.592		-4.546	1.000	-0.538	0.570
13	2	-12.870	6.921	-7.339		-12.870	1.000	-0.538	0.570
14	3	-12.870	6.921	-7.339		-12.870	1.000	-0.538	0.570
15	4	-12.870	6.921	-7.339		-12.870	1.000	-0.538	0.570

FIGURE 10.47 Deflated matrix and first iterations for evaluating the third principal stress and direction cosines.

	Q1015		▼	fx	=MAX($P$1011,$AA$1011,$AL$1011)			
	L	M	N	O	P	Q	R	S
1006	181.212	130.429	-194.781		-194.781	-0.930	-0.670	1.000
1007	181.212	130.429	-194.781		-194.781	-0.930	-0.670	1.000
1008	181.212	130.429	-194.781		-194.781	-0.930	-0.670	1.000
1009	181.212	130.429	-194.781		-194.781	-0.930	-0.670	1.000
1010	181.212	130.429	-194.781		-194.781	-0.930	-0.670	1.000
1011	181.212	130.429	-194.781	$\sigma_{p1}=$	-194.781	-0.930	-0.670	1.000
1012								
1013					Normalized:	-0.612	-0.440	0.657
1014								
1015					$\sigma_1=$	127.651	Ref=	2
1016						x	y	z
1017					Direction Cosines:	0.081	0.792	0.605

FIGURE 10.48   Identification of smallest, intermediate, and largest principal stresses.

To be able to associate each of the principal stresses with the corresponding direction cosines vector, a reference number is created for each one of them to identify which of the calculated principal stresses is used for each output one. For example, the principal stress σ_2 is actually the calculated principal stress σ_p3. Consequently, the reference number associated with σ_2 is 3, as can be seen in cell AD1015 in Figure 10.49, and the expression that was used to find it is shown in the formula bar in the same figure. With this reference number set, the direction cosines associated with s2 can be copied to cells AB1017 to AD1017 using the "CHOOSE" function.

Cell AB1017 has the expression: =CHOOSE(AD1015,$Q$1013,$AB$1013,$AM$1013); cell AC1017 has the expression: =CHOOSE(AD1015,$R$1013,$AC$1013,$AN$1013); and in cell AD1017 the expression =CHOOSE(AD1015,$S$1013,$AD$1013,$AO$1013) is used. The direction cosines are then copied in pairs of points to create y vs. x, y versus z, and x versus z scatter graphs representing the direction of each principal stress on each plane. Figure 10.50 shows an implementation of the copying process for the principal stress σ_3. The IF function is used to make sure that if the principal stress is 0 then no arrows should show on its graph.

A screenshot of the proposed spreadsheet solution for the three-dimensional principal stress problem is shown in Figure 10.51. The reader should note that during the development process special care is taken to make sure that the application can handle all possible input cases. For instance, if the user only specifies one normal stress, and lets all other stresses be zero, the application should recognize that and produce the correct direction for the principal stress. This is the reason why the initial estimates of the eigenvectors were chosen as was described before.

	AD1015			▼	fx	=IF(AND(AB1015=$P$1011,AB1015<>Q1015),1, IF(AB1015=$AA$1011,2,3))							
	W	X	Y	Z	AA	AB	AC	AD	AE	AF	AG	AH	AI	AJ
1009	13.002	127.651	97.573		127.651	0.102	1.000	0.764			998	-12.870	6.921	-7.339
1010	13.002	127.651	97.573		127.651	0.102	1.000	0.764			999	-12.870	6.921	-7.339
1011	13.002	127.651	97.573	$\sigma_{p2}=$	127.651	0.102	1.000	0.764			1000	-12.870	6.921	-7.339
1012														
1013					Normalized:	0.081	0.792	0.605						
1014														
1015					$\sigma_2=$	-12.870	Ref=	3						
1016						x	y	z						
1017					Direction Cosines:	0.787	-0.423	0.449						Dir
1018	x	y		z	y		z	x				x	y	
1019	0	0		0	0		0	0				0	0	
1020	0.787	-0.423		0.449	-0.423		0.449	0.787				-0.612	-0.440	0

FIGURE 10.49   Matching principal stresses with appropriate direction cosines.

	AH1020			f_x	=IF(AM1015<>0,AM1017,0)			
	AG	AH	AI	AJ	AK	AL	AM	AN	AO
1010	999	-12.870	6.921	-7.339		-12.870	1.000	-0.538	0.570
1011	1000	-12.870	6.921	-7.339	$\sigma_{p3}=$	-12.870	1.000	-0.538	0.570
1012									
1013						Normalized:	0.787	-0.423	0.449
1014									
1015						$\sigma_3=$	-194.781	Ref=	1
1016							x	y	z
1017						Direction Cosines:	-0.612	-0.440	0.657
1018		x	y		z	y		z	x
1019		0	0		0	0		0	0
1020		-0.612	-0.440		0.657	-0.440		0.657	-0.612

FIGURE 10.50 Implementation of the data copying process for creating scatter graph representations of the principal stress σ_3.

10.6   EXAMPLE 4: COMPUTATION OF COMBINED STRESSES

Consider a structural member subjected to a combination of forces and moments as shown in Figure 10.52. It is desirable to compute the stresses at a given point due to the combined action of the internal forces and moments shown. This section shows how Excel can be used to numerically evaluate such stresses for a circular or rectangular cross section and plot the stress distribution diagram.

Among other things, this Excel implementation example shows how controls can be used to enhance the user-friendliness of an application. It also shows how computationally heavy problems can easily be solved by carefully selecting how to represent data.

Figure 10.53 shows a screen shot of the completed application. The user specifies the units, internal loads, and geometry of the cross section and the program draws a scaled representation of the cross section, computes the minimum and maximum stresses, and draws a three-dimensional representation of the stress distribution on the cross section.

The user also has the option to specify the coordinates of a point on the cross section and the program calculates the stress at that point due to each of the applied internal loads, as well as the total stress. The program will also accurately show the location of the point on the graphical representation of the cross section. The technique used to draw the geometry of the cross section and the point of interest is very similar to the one used to draw Mohr's circle in Section 10.4, and, thus, will not be presented in this section.

"Option" button controls are used to allow the user to specify the type of cross section (Circular or Rectangular). As shown in Figure 10.54, the control can be linked to a cell on the worksheet where the number of the checked button in the group is entered. Since the user must make a selection between two options (Circular or Rectangular cross section), two controls must be inserted but both must be linked to the same cell. Say that the two controls are linked to cell $N$1 as shown in the figure. This cell will contain the number 1 if the first control is selected and 2 if the second control is selected. Controls are numbered in the order they are created.

The contents of the cell that is linked to the controls can now be used to determine which cross-section option was selected by the user. For example, the formula used to compute the area of the cross section will depend on whether a circular or a rectangular cross section was selected. In this case, the cross-sectional area is computed in cell H21 by inserting the formula =IF(N1=1,PI()*B11 *B11/4,E11*E12), as shown in Figure 10.55. In this formula, it is assumed that the diameter, d, of the circular cross section is stored in cell B11, and the width, b, and height, h, of the rectangular cross section are stored in E11 and E12, respectively. The same approach is used to calculate the

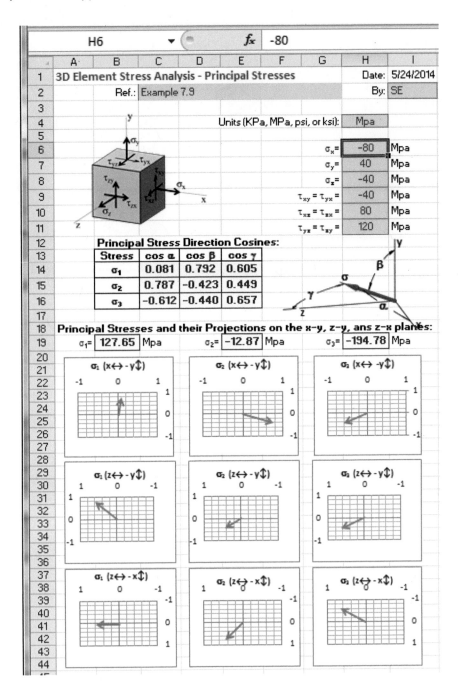

FIGURE 10.51 Screenshot of the proposed spreadsheet solution for the three-dimensional principal stress problem.

moments of inertia of the cross section with respect to the z and y axes. The moment of inertia, I_z, is calculated in cell H22 using the formula $=IF(N1=1,(PI()*(B11/2)^4)/4,(E11*E12^3)/12)$, and the moment of inertia, I_y, is calculated in cell H23 using the formula $=IF(N1=1,(PI()*(B11/2)^4)/4,(E12*E11^3)/12)$. Since the scope of this application only encompasses the axial force and the bending moments, no other properties of the cross section are needed. For other cases, the same approach can be used to evaluate any required cross-sectional parameters.

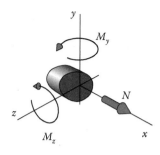

FIGURE 10.52   Section of a structural member subjected to a combination of forces and moments.

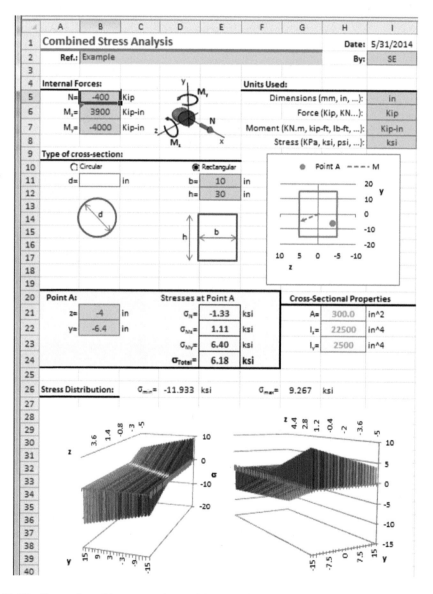

FIGURE 10.53   Screenshot of the proposed spreadsheet solution for analyzing stresses in members subjected to a combination of forces and moments.

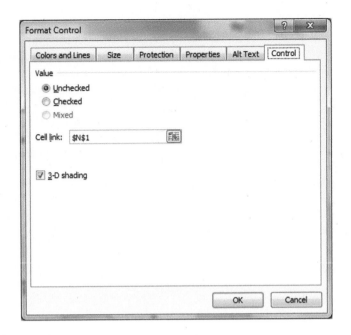

FIGURE 10.54 Implementation of a Microsoft Excel™ Option Button control.

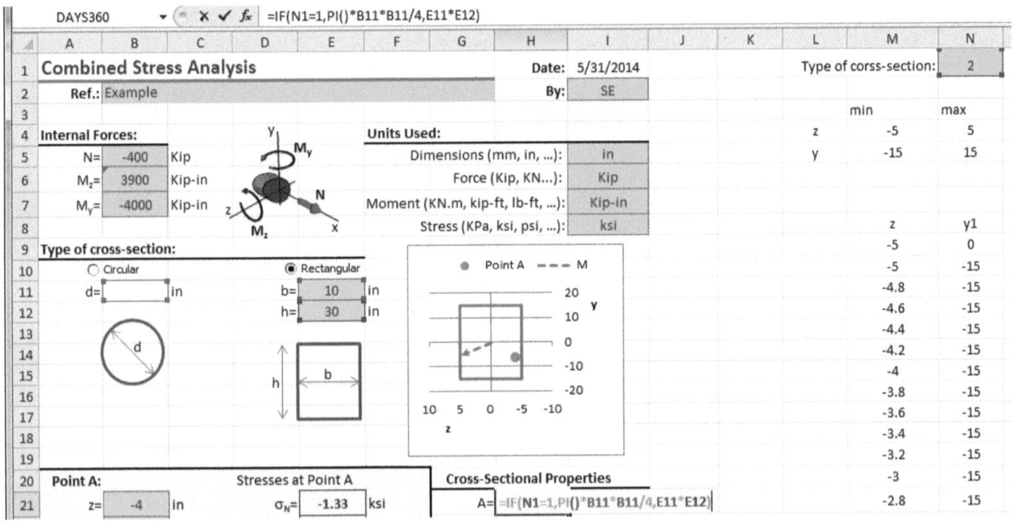

FIGURE 10.55 Screenshot of Combined Stress Analysis application showing the evaluation of cross-sectional properties.

The computation of the stresses at the point specified by the user is very straightforward. The stress σ_N, due to the normal force, is calculated in cell E21 using the formula =B5/H21. The stress σ_{Mz}, due to the bending moment with respect to the z axis is calculated in cell E22 using the formula $= -B6*B22/H22$ and the stress σ_{My}, due to the bending moment with respect to the y axis is calculated in cell E23 using the formula =B7*B21/H23. The total stress is simply the sum of the three. In all of these cases, a positive stress indicates the point in question is under tension.

Calculating the minimum and maximum stresses on the cross section is a little more challenging as it requires that bending moment stresses be computed at a number of points. To accomplish this, tables for computing the stresses due to each individual moment is created on one side of the work-sheet away from the main user-application interaction area. The first column of this table contains a listing of z-values varying from $-b/2$ to $+b/2$ ($-d/2$ to $d/2$ for the circular cross section) in a chosen increment. In this example, the range is divided into 50 segments. The header row of the table contains a listing of y-values varying between $-h/2$ and $+h/2$ ($-d/2$ to $+d/2$ for the circular cross section) again in a chosen increment. The range was divided into 20 segments for the implementation in this example. With the table set up this way, it becomes easy to calculate the stresses due to each of the moments.

The implementation of this technique for computing the stresses due to the moment Mz is shown in Figure 10.56. The stress at the first point (at $z = -b/2$ or $-d/2$, $y = -h/2$ or $-d/2$) is evaluated by inserting the formula $= -\$B\$6*R\$8/\$H\$22$ in cell R9 and then filling down and to the right to cover all the specified y and z coordinates.

The stresses due to the moment M_y are computed in a second table created just to the right of the first one, and a third table is created to compute the total combined stress at every point. These tables are shown in Figure 10.57 and 10.58, respectively. The reader is invited to note that only the first row of the σ_{Mz} table and the first column of the σ_{My} table are required to produce the σ_{Total} table. However, filling all the columns and rows of the first two tables simplifies the creation of the third table a little. The total stresses for a rectangular cross section are computed for all the points by simply adding all the stresses from all the individual internal loads.

	DAYS360		▼		X	✓	f_x	=-$B$6*R$8/$H$22					
	Q	R	S	T	U	V	W	X	Y	Z	AA	AB	AC	AD
7	z↓ y↘	σ_{Mz} calculated for fraction s of y=												
8		-15	-13.5	-12	-10.5	-9	-7.5	-6	-4.5	-3	-1.5	0	1.5	3
9	-5	=-$B$6*R$8/$H$22		2.08	1.82	1.56	1.3	1.04	0.78	0.52	0.26	0	-0.26	-0.52
10	-5	2.6	2.34	2.08	1.82	1.56	1.3	1.04	0.78	0.52	0.26	0	-0.26	-0.52
11	-4.8	2.6	2.34	2.08	1.82	1.56	1.3	1.04	0.78	0.52	0.26	0	-0.26	-0.52
12	-4.6	2.6	2.34	2.08	1.82	1.56	1.3	1.04	0.78	0.52	0.26	0	-0.26	-0.52
13	-4.4	2.6	2.34	2.08	1.82	1.56	1.3	1.04	0.78	0.52	0.26	0	-0.26	-0.52
14	-4.2	2.6	2.34	2.08	1.82	1.56	1.3	1.04	0.78	0.52	0.26	0	-0.26	-0.52
15	-4	2.6	2.34	2.08	1.82	1.56	1.3	1.04	0.78	0.52	0.26	0	-0.26	-0.52
16	-3.8	2.6	2.34	2.08	1.82	1.56	1.3	1.04	0.78	0.52	0.26	0	-0.26	-0.52

FIGURE 10.56   Computation of the stresses due to the moment M_z.

	DAYS360		▼		X	✓	f_x	=$B$7*$M9/$H$23		
	AH	AI	AJ	AK	AL	AM	AN	AO	AP	AQ
7						σ_{My} calculated for fraction s of y=				
8	9	10.5	12	13.5	15	-15	-13.5	-12	-10.5	-9
9	-1.56	-1.82	-2.08	-2.34	-2.6	=$B$7*$M9/$H$23	8	8	8	8
10	-1.56	-1.82	-2.08	-2.34	-2.6	8	8	8	8	8
11	-1.56	-1.82	-2.08	-2.34	-2.6	7.68	7.68	7.68	7.68	7.68
12	-1.56	-1.82	-2.08	-2.34	-2.6	7.36	7.36	7.36	7.36	7.36
13	-1.56	-1.82	-2.08	-2.34	-2.6	7.04	7.04	7.04	7.04	7.04
14	-1.56	-1.82	-2.08	-2.34	-2.6	6.72	6.72	6.72	6.72	6.72
15	-1.56	-1.82	-2.08	-2.34	-2.6	6.4	6.4	6.4	6.4	6.4
16	-1.56	-1.82	-2.08	-2.34	-2.6	6.08	6.08	6.08	6.08	6.08
17	-1.56	-1.82	-2.08	-2.34	-2.6	5.76	5.76	5.76	5.76	5.76

FIGURE 10.57   Computation of the stresses due to the moment M_y.

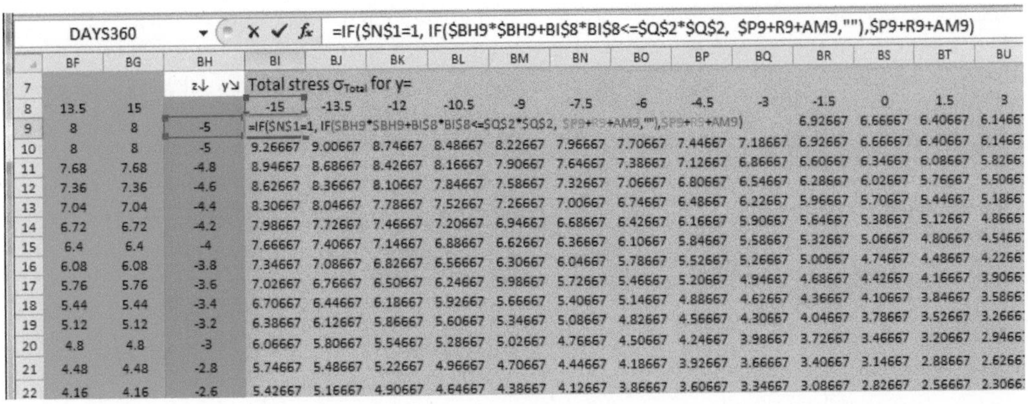

DAYS360 ▾ ✕ ✓ fx =IF($N$1=1, IF($BH9*$BH9+BI$8*BI$8<=$Q$2*$Q$2, $P9+R9+AM9,""),$P9+R9+AM9)

	BF	BG	BH	BI	BJ	BK	BL	BM	BN	BO	BP	BQ	BR	BS	BT	BU
7			z↓ y↘	Total stress σ_Total for y=												
8	13.5	15		-15	-13.5	-12	-10.5	-9	-7.5	-6	-4.5	-3	-1.5	0	1.5	3
9	8	8	-5	=IF($N$1=1, IF($BH9*$BH9+BI$8*BI$8<=$Q$2*$Q$2, $P9+R9+AM9,""),$P9+R9+AM9)									6.92667	6.66667	6.40667	6.1466
10	8	8	-5	9.26667	9.00667	8.74667	8.48667	8.22667	7.96667	7.70667	7.44667	7.18667	6.92667	6.66667	6.40667	6.1466
11	7.68	7.68	-4.8	8.94667	8.68667	8.42667	8.16667	7.90667	7.64667	7.38667	7.12667	6.86667	6.60667	6.34667	6.08667	5.8266
12	7.36	7.36	-4.6	8.62667	8.36667	8.10667	7.84667	7.58667	7.32667	7.06667	6.80667	6.54667	6.28667	6.02667	5.76667	5.5066
13	7.04	7.04	-4.4	8.30667	8.04667	7.78667	7.52667	7.26667	7.00667	6.74667	6.48667	6.22667	5.96667	5.70667	5.44667	5.1866
14	6.72	6.72	-4.2	7.98667	7.72667	7.46667	7.20667	6.94667	6.68667	6.42667	6.16667	5.90667	5.64667	5.38667	5.12667	4.8666
15	6.4	6.4	-4	7.66667	7.40667	7.14667	6.88667	6.62667	6.36667	6.10667	5.84667	5.58667	5.32667	5.06667	4.80667	4.5466
16	6.08	6.08	-3.8	7.34667	7.08667	6.82667	6.56667	6.30667	6.04667	5.78667	5.52667	5.26667	5.00667	4.74667	4.48667	4.2266
17	5.76	5.76	-3.6	7.02667	6.76667	6.50667	6.24667	5.98667	5.72667	5.46667	5.20667	4.94667	4.68667	4.42667	4.16667	3.9066
18	5.44	5.44	-3.4	6.70667	6.44667	6.18667	5.92667	5.66667	5.40667	5.14667	4.88667	4.62667	4.36667	4.10667	3.84667	3.5866
19	5.12	5.12	-3.2	6.38667	6.12667	5.86667	5.60667	5.34667	5.08667	4.82667	4.56667	4.30667	4.04667	3.78667	3.52667	3.2666
20	4.8	4.8	-3	6.06667	5.80667	5.54667	5.28667	5.02667	4.76667	4.50667	4.24667	3.98667	3.72667	3.46667	3.20667	2.9466
21	4.48	4.48	-2.8	5.74667	5.48667	5.22667	4.96667	4.70667	4.44667	4.18667	3.92667	3.66667	3.40667	3.14667	2.88667	2.6266
22	4.16	4.16	-2.6	5.42667	5.16667	4.90667	4.64667	4.38667	4.12667	3.86667	3.60667	3.34667	3.08667	2.82667	2.56667	2.3066

FIGURE 10.58 Computation of the total combined stresses.

The total stresses for a circular cross section are only computed for those points that fall within the cross section using the formula shown in Figure 10.58. Cells corresponding to points that fall outside the perimeter of the cross section are left blank. This will help ensure two things. First, an estimate of the minimum and maximum stresses on both types of cross-sections can be easily obtained with the *MIN* and *MAX* Excel functions using the whole range of possible y- and z-values as an argument. Second, the three-dimensional representation of the stress distribution on a circular cross section will capture the shape of the cross section if the stress functional-values on points that are outside of it are left blank. The stress distributions are created by selecting the data in the range BH8:CC61, inserting 3D-Column charts and formatting them as desired.

10.7 EXCEL SPREADSHEET APPLICATION PROJECTS

This section presents problem statements for the development of spreadsheet applications to solve specific problems in mechanics of materials. Although some hints may be provided, the readers are expected to use techniques and skills learned from the previous sections to develop the spreadsheet applications on their own.

10.7.1 PROJECT 10.1: CROSS-SECTIONAL PROPERTIES OF A GENERAL SHAPE

Develop a spreadsheet application to compute the coordinates of the centroid and the cross-sectional properties A, I_z, I_y, and I_{zy}, of a general shape defined by up to 20 connected points. The application should show a to-scale depiction of the cross section.

Hints:

1. Divide the cross section into a number of small elements, say 100 along the z axis and 100 along the y axis, and use numerical integration to compute the required properties. A larger number of segments will yield more accurate results but using too many segments will lead to excessive errors due to truncation and rounding-off.
2. Create separate worksheets, to compute the needed parameters for each element (area, y coordinate, z coordinate, …)
3. A screenshot from a sample implementation is shown in Figure 10.59.

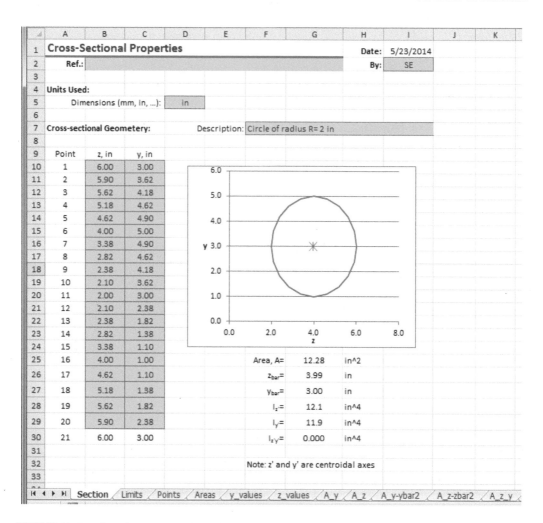

FIGURE 10.59   Sample screenshot from Project 10.1.

10.7.2   PROJECT 10.2: STRESSES IN STATICALLY INDETERMINATE SYSTEMS

Develop an Excel spreadsheet application to find the end reactions of a statically indeterminate member that can be subjected to axial (including the effect of temperature) or torsional loading. Consider the following loads: (1) a concentrated force placed anywhere along the member, (2) a concentrated torque placed anywhere along the member, (3) a uniformly distributed axial force along a part of the member, and (4) a uniformly distributed torque along a part of the member. The member can have up to three different segments, each with its own cross-sectional and material properties.

10.7.3   PROJECT 10.3: INELASTIC ANALYSIS

Develop an Excel spreadsheet application to find the maximum load that would produce a collapse mechanism on a fixed beam. Consider up to four concentrated forces of equal magnitude and a trapezoidal load applied anywhere within the span of the beam. The concentrated loads, when present, are equidistant from each other and the supports. The user should have the option to select which of the loads is to vary, to control collapse, while holding the other load constant. The application should draw the moment diagram for the case when collapse is imminent.

Answers to Even-Numbered Problems

(Note: Only partial answers are given for certain problems.)

CHAPTER 1

1.2   (a) $F = 15$ kN...$0 < x < 2$ m; (b) $F_1 = 6$ kN

1.4 $F_1 = -8$ kN

1.6 $F_2 = -6$ kN

1.8 $F = 2x^3$

1.10 $F_2 = 4x_2 + (x_2^4/4) - 8$

1.12 $F_B^L = 15$ kN; $F_B^R = 10$ kN

1.14 $F_B^L = -16$ kN; $F_B^R = -6$ kN

1.16 $\sigma_{AB} = 16.6$ ksi; $\sigma_{CD} = -20.9$ ksi

1.18 $P = 23.4$ kN

1.20 $P = 12.3$ kips

1.22 $\sigma_{AC} \approx 29.8$ ks

1.24   (b) $\sigma_{AB} \approx -37.7$ MPa

1.26 $P \approx 1.0$ kip

1.28 $P \approx 23.6$ kN

1.30 $\sigma_x \approx 4.1$ ksi

1.32 $\sigma_x \approx 31.8$ MPa

1.34 $\sigma_n = 100.0$ MPa; $\tau_{nt} = 57.7$ MPa; $\sigma_t = 33.3$ MPa; $\tau_{tn} = 57.7$ MPa

1.36 $\sigma_n = 2.3$ ksi; $\tau_{nt} = 0.6$ ksi $\sigma_t = 0.2$ ksi; $\tau_{tn} = 0.6$ ksi

1.38 $p \approx 6.7$ kN/m

1.40   (b) $\delta = -0.00750$ in.; $\Delta D \approx 0.00135$ in.

1.42   (b) $\mu = 1/3$

1.44   (b) $u_R \approx 24.8 \times 10^{-3}$ kip·in./in.^{3}

1.46   (c) $E \approx 114.3$ GPa

1.48   (c) $E = 56.7$ GPa

1.50   (a) $\sigma_U = 101$ ksi

1.52   (b) %ROA = 57.7

1.54   (a) $\sigma_U = 5050$ ps

1.56   (b) $\sigma_p = 195$ ksi

1.58 $P = 72.0$ kN

1.60 $\sigma_{ALL} \approx 33.3$ ksi

1.62 $P_{MAX} = 67.5$ kN

1.64 $P_{MAX} = 3.9$ kN

1.66 $P_{MAX} = 636.2$ kN

1.68 $\delta_{AB} = 0.10$ in.

1.70 $\delta_{EB} \approx -0.04$ in.

1.72 $P \approx 14.9$ kips

1.74 $\delta_{SPR} = 80.0$ mm

1.76 $\delta_D \approx 0.414$ in.

1.78 $A_2 = 1212.8$ mm^2

1.80 $a/b = 1.6$

1.82 (b) $\delta_B = \delta_{TUBE} \approx 38.8$ mm

1.84 $\delta \approx 0.048$ in.

1.86 $\delta \approx 1.6$ mm

1.88 (b) $\delta = 2.427$ in.; $\sigma = 194$ psi

1.90 (a) $P = 45$ N; (c) $\delta = 6.429 \times 10^{-6}$ m

1.92 $P = 84$ kN

1.94 $P \approx 24.7$ kips; $\delta_{PLATE} = 0.02$ in.

1.96 $P \approx 1189.9$ kN

1.98 $P = 139.3$ kips; $\delta_M = 0.029$ in.

1.100 $P \approx 6291.1$ kN

1.102 $F_S = 24$ kips; $e = 0.0013$ in.

1.104 $a/b = 1/4$; $\sigma_B \approx 3.7$ MPa

1.106 (a) $d = 0.582$ in.; (b) $a = 4.482$ in.

1.108 (a) $\phi \approx 29.8°$; (b) $a = 301$ mm

1.110 $P \approx 288.6$ kips; $\sigma_B \approx 2.8$ ksi

1.112 $P \approx 37.4$ kN

1.114 $P = 255.4$ kips

1.116 $A_X = 0.00019$ m^2; $A_Y = 0.00379$ m^2

1.118 $(\sigma_{TOTAL})_{BC} = 5.8$ MPa

1.120 $(\sigma_{TOTAL})_A = 61.1$ ksi

1.122 $P = 5.6$ kN

1.124 $P = 284.3$ kips

1.126 $P = 19.9$ kips

1.128 $P = 105.3$ kips

1.130 $P = 27.3$ kips

1.132 $r = 0.473$ in.

1.134 $\sigma_{MAX} = 21.9$ ksi

1.136 (a) $\sigma_{MAX} \approx 12.3$ ksi; (b) $\Delta_{MAX} = 0.022$ in.

1.138 (a) $\sigma_{MAX} = 31.3$ ksi; (c) $IF = 121$

1.140 (a) $\sigma_{MAX} \approx 2.9$ ksi; (b) $\Delta = 0.081$ in.

1.142 $v = 40.3$ ft/s

R1.2 $\sigma_n \approx 14.5$ ksi; $\tau_{nt} \approx 14.5$ ksi
$\sigma_t \approx 14.5$ ksi; $\tau_{tn} \approx 14.5$ ksi

R1.4 $P \approx 112.5$ kN

R1.6 $a = 4.444$ in.

R1.8 $\delta_A = 0.155$ in.

R1.10 (a) $\delta_A = 0.563$ in.; (c) $\delta_D = 0.744$ in.

R1.12 $\sigma_{ROD} = 316.7$ MPa; $\delta_{SPR} = 1.2$ mm

R1.14 $\sigma_A \approx -5.2$ MPa; $\sigma_S \approx 86.7$ MPa

R1.16 $P = 80.0$ kips

CHAPTER 2

2.2 $T_{BC} = 10$ kip$\cdot$ft; $T_{CD} = 20$ kip$\cdot$ft

2.4 $T_C^L = 1.5$ kN$\cdot$m; $T_C^R = -6.0$ kN$\cdot$m

2.6 $T_B^L = -30$ kip$\cdot$ft; $T_B^R = 30$ kip$\cdot$ft

2.8 (a) $T_{BC} = 50 - (100/3)(x - 1.5)^3$

2.10 $T_{AB} = 2x_1^3 - 40$; $b = 5.0$ in.

2.12 $L \approx 29.2$ m

2.14 (b) $u = 66.3$ lb·in./in.^{3}
2.16 (a) $G = 209.4$ GPa; (c) $\tau_U = 2904.6$ MPa
2.18 (b) $\theta_A = 33.534 \times 10^{-3}$ rad, ccw when viewed from A to C
2.20 (b) $\sigma_n = -75.8$ MPa; $\tau_{nt} = -90.3$ MPa
2.22 $T_{MAX} \approx 2.1$ kN·m
2.24 $D = 2.294$ in.
2.26 $\sigma_n = -7.4$ MPa; $\tau_{nt} = -4.3$ MPa
2.28 $Q_1 \approx 66.3$ N·m; $Q_3 \approx -1259.1$ N·m
2.30 $\tau_{MAX} = 530.5$ MPa
2.32 $\tau_{MAX} = 6.5$ ksi
2.34 $\theta_{E/A} = 6.124 \times 10^{-2}$ rad, ccw when viewed from E to A
2.36 $Q = 5.4$ kip·in.
2.38 $\theta_{A/C} = 27.108 \times 10^{-4}$ rad
2.40 $T_{MAX} \approx 1.0$ kN·m; $q \approx 16.8$ kN·m/m
2.42 $Q_{MAX} \approx 46.4$ N·m
2.44 $Q \approx 18.2$ kN·m
2.46 $F \approx 6.2$ kips
2.48 (a) $\tau_{MAX} = 115.0$ MPa; (b) $\theta_{MAX} = 9.2 \times 10^{-2}$ rad
2.50 (a) $\tau_S \approx 2.2$ ksi
2.52 $Q \approx 76.6$ kN·m
2.54 $Q \approx 219.2$ kN·m
2.56 $\tau_{MAX} = \tau_{CD} \approx 23.9$ kN·m
2.58 $\tau_{AB} \approx 39.2$ ksi; $\tau_{BC} \approx 15.7$ ksi
2.60 $\tau_{AB} = 0.542$ ksi; $\theta_B = 1.62 \times 10^{-3}$ rad
2.62 $\tau_{MAX} = 240.1$ MPa; $\theta = 2.29 \times 10^{-1}$ rad
2.64 $d_{MIN} \approx 29.0$ mm
2.66 $d_{MIN} = 0.893$ in.
2.68 $(\tau_{MAX})_{CD} = 10.1$ ksi; $\theta_{CD} = 1.217 \times 10^{-1}$ rad
2.70 $\tau \approx 58.9$ MPa
2.72 $\sigma \approx 28.3$ MPa; $\tau \approx 45.3$ MPa
2.74 $\sigma_n \approx 24.3$ ksi; $\tau_{nt} \approx 6.5$ ksi
2.76 $\sigma_n \approx 8.7$ ksi; $\tau_{nt} \approx -7.7$ ksi
2.78 $\sigma \approx -6.4$ ksi; $\tau \approx 20.5$ ksi
2.80 $\varepsilon_x = 636.6$ μ; $\gamma_{xy} = 1693.4$ μ
2.82 $P \approx 19.6$ kips; $Q \approx 9.4$ kip·in.
2.84 (a) $\varepsilon_n = 6900$ μ; (b) $\gamma_{nt} = 5200$ μ
2.86 $\sigma_n \approx 24.9$ ksi; $\sigma_t \approx 20.1$ ksi
2.88 (a) $P = 84.1$ kW; (b) $P = 43.4$ kW
2.90 $D = 87.5$ mm
2.92 $\omega \approx 314$ rpm
2.94 $W = 392.7$ N
2.96 $\tau_{MAX} \approx 69.7$ MPa; $\theta_{MAX} = 1.24 \times 10^{-2}$ rad
2.98 $\tau \approx 7.8$ ksi; $\theta = 4.70 \times 10^{-2}$ rad
2.100 $Q \approx 7.3$ kN·m
2.102 $\tau_{MAX} \approx 14.0$ MPa; $\theta = 2.533 \times 10^{-2}$ rad
2.104 $Q \approx 165\ 4$ kip·in.
2.106 $\tau_{MAX} \approx 7.1$ ksi; $\theta = 1.054 \times 10^{-3}$ rad
2.108 $Q = 16.2$ kip·in.
2.110 Circular shaft
2.112 (a) $\tau_{MAX} \approx 5.1$ ksi

2.114 $\tau_{MAX} \approx 11.8$ ksi
2.116 $a = 1.854$ in.
2.118 (a) $\tau_{MAX} \approx 29.7$ ksi; (b) $\theta = 0.266$ rad
2.120 (a) $\tau \approx 7.8$ MPa; (c) $\theta = 2.34 \times 10^{-3}$ rad
2.122 (a) $\tau_{MAX} = 1.519$ ksi; $\theta = 0.131$ rad
2.124 $a = 0.322$ in.
2.126 $b_{MIN} = 0.587$ in.
2.128 (a) $G = 6 \times 10^3$ ksi; (d) $u_R = 12.0 \times 10^{-3}$ kip·in./in.^{3}
2.130 (a) $d = 0.684$ in.; (b) $d = 0.238$ in.
2.132 See equation in problem statement
2.134 (a) $T \approx 1084.1$ kip·in.
2.136 $\theta = 6.96 \times 10^{-1}$ rad
2.138 (b) For $d = 0.02$ m: $\theta = 9.14 \times 10^{-2}$ rad
2.140 $R_i = 3.217$ in.
2.142 (a) $T_p \approx 93.5$ kN·m; $\theta_p = \infty$
R2.2 (b) $T_B^L = 74$ kN·m; (c) $T_B^R = 74$ kN·m
R2.4 $(Q_1)_{MAX} = 3.75$ kN·m
R2.6 $Q_1 = 196.4$ kip·in.; $Q_2 = 7.9$ kip·in.
R2.8 (b) $\sigma_n \approx 73.9$ MPa; $\tau_{nt} \approx 42.7$ MPa
R2.10 $d_{MIN} = 1.983$ in.
R2.12 $(d_i)_{MAX} \approx 5.7$ in.
R2.14 $P \approx 175.8$ kN; $Q \approx 4.9$ kN·m
R2.16 $Q_{MAX} \approx 14.2$ kip·in.

CHAPTER 3

3.2 (b) $M_C^R = 80$ kN·m; $V_C^R = -40$ kN
3.4 (a) $M_C^L = 62.5$ kN·m; $V_C^L = -20$ kN
3.6 (b) $M_C^R = -4pL^2/81$; $V_C^R = -(P + 2pL/9)$
3.8 (a) $M_C^L = -pL^2/8$; $V_C^R = pL/2$
3.10 (b) $M_C^L = 20$ kN·m; $V_C^L = 10$ kN
3.12 (a) $M_B^R = 0$; $V_B^R = 2.5$ kN
3.14 (b) $M_A = -24$ kN·m; $V_A = 12$ kN
3.16 $0 < \times < 8$ ft: $M = 21x - x^2$; $V = 21 - 2x$
3.18 $0 < \times < 3/4L$: $M = 1/4[P + 3pL/2]x - px^2/2$; $L < x < (3/2)L$: $V = p(3L/2 - x)$
3.20 $6 < \times < 12$ ft: $M = (3x - 18)$ kip·ft; $V = 3$ kips
3.22 $M_C^L = -100$ kip·ft; $M_C^R = -115$ kip·ft
3.24 $M_{MAX} \approx 32.3$ kN·m, at 3.464 m from A
3.26 $M_{MAX} = 100$ kN·m, under 75-kN load
3.28 $M_{MAX} = 104$ kip·ft, at 8 ft from A
3.30 $M_{MAX} \approx 68.1$ kip·ft, at 8.25 ft from A
3.32 $V_{MAX} = 18$ kips; $M_{MAX} = -63$ kip·ft, both at the fixed support
3.34 $V_{MAX} = -40$ kips; $M_{MAX} = -415$ kip·ft, both at the fixed support
3.36 $V_{MAX} = 50$ kN; $M_{MAX} = 100$ kN·m, both 2 m from left support
3.38 $V_{MAX} = 39$ kips; $M_{MAX} = -207$ kip·ft, both at the right support
3.40 $M_{MAX} = -10pa^2$, over entire overhang
3.42 $(\sigma_T)_{MAX} \approx 167.7$ MPa; $(\sigma_C)_{MAX} \approx 90.3$ MPa
3.44 $(\sigma_T)_{MAX} \approx 160.3$ MPa; $(\sigma_C)_{MAX} \approx 128.1$ MPa
3.46 $(\sigma_T)_{MAX} = (\sigma_C)_{MAX} \approx 208.3$ MPa
3.48 $(\sigma_T)_{MAX} \approx 13.1$ ksi; $(\sigma_C)_{MAX} \approx 7.8$ ksi

3.50 $(\sigma_T)_{MAX} \approx 12.0$ ksi; $(\sigma_C)_{MAX} \approx 17.0$ ksi

3.52 $(\sigma_T)_{MAX} \approx 21.9$ ksi; $(\sigma_C)_{MAX} \approx 17.5$ ksi

3.54 $(\sigma_T)_{MAX} \approx 19.0$ ksi; $(\sigma_C)_{MAX} \approx 37.9$ ksi

3.56 $(\sigma_T)_{MAX} \approx 42.4$ ksi; $(\sigma_C)_{MAX} \approx 25.4$ ksi

3.58 $(\sigma_T)_{MAX} = (\sigma_C)_{MAX} \approx 3.0$ ksi

3.60 $(\sigma_T)_{MAX} = (\sigma_C)_{MAX} \approx 277.8$ MPa

3.62 $(\sigma_T)_{MAX} \approx 54.2$ ksi; $(\sigma_C)_{MAX} \approx 32.5$ ksi

3.64 $(\sigma_T)_{MAX} \approx 20.4$ MPa; $(\sigma_C)_{MAX} \approx 28.8$ MPa

3.66 $(\sigma_T)_{MAX} \approx 13.6$ ksi; $(\sigma_C)_{MAX} \approx 19.2$ ksi

3.68 $(\sigma_T)_{MAX} \approx 3.8$ ksi; $(\sigma_C)_{MAX} \approx 6.3$ ksi

3.70 $p \approx 7.1$ kips/ft

3.72 $P \approx 18.8$ kips

3.74 $P \approx 77.0$ kips

3.76 $\tau_{MAX} = 0.0516(6.5^2 - y^2) -1.5 < y < 6.5$ in.; $\tau_{MAX} = \tau_{NA} = 2.180$ ksi

3.78 $\tau = 0.03265(5^2 - y^2)\ldots 3 < y < 5$ in.; $\tau_{MAX} = \tau_{NA} = 2.905$ ksi

3.80 (a) $\tau \approx 2.2$ ksi; (b) $\tau \approx 2.5$ ksi

3.82 (a) $\tau \approx 0.3$ MPa; (b) $\tau \approx 0.6$ MPa

3.84 (a) $\tau \approx 2.6$ MPa; (b) $\tau \approx 5.2$ MPa

3.86 (a) $\tau \approx 9.5$ MPa; (b) $\tau \approx 11.6$ MPa

3.88 (a) $\tau_O \approx 0.18$ MPa; (b) $\tau_E \approx 0.35$ MPa

3.90 $\tau_{MAX} = \tau_{NA} \approx 5.3$ ksi

3.92 $\tau_{MAX} = \tau_{NA} \approx 1.76$ MPa

3.94 $\tau_{MAX} = \tau_{NA} \approx 5.24$ MPa

3.96 $\tau_{MAX} = \tau_{NA} \approx 11.6$ MPa

3.98 $\tau_{MAX} = 1.40$ MPa, at junction between vertical and horizontal rectangles

3.100 $V = 2.726$ kN

3.102 $p \approx 1.33$ in.

3.104 $q = 0.428$ kips/in.

3.106 (a) $\tau_A \approx 6.3$ ksi; (b) $\tau_{NA} \approx 6.7$ ksi

3.108 (b) $\tau_{NA} \approx 5.2$ ksi

3.110 (a) $\tau_B \approx 1.8$ ksi; (b) $\tau_{NA} \approx 2.0$ ksi

3.112 (a) $\tau_A = 0$; $\tau_B \approx 1.1$ ksi

3.114 $\sigma_E \approx 184.7$ MPa; $\sigma_G \approx 362.8$ MPa

3.116 $\sigma_D \approx 64.7$ MPa; $\sigma_E \approx -54.6$ MPa

3.118 $P_{MAX} \approx 59.5$ kips

3.120 $\sigma_n \approx 45.1$ MPa; $\tau_{nt} \approx -96.8$ MPa

3.122 $\sigma_E \approx 289.7$ MPa; $\tau_E \approx 241.4$ MPa

3.124 $\sigma_B \approx 429.0$ psi; $\tau_B \approx 275.8$ psi

3.126 $\sigma_D \approx 150.9$ MPa; $\tau_D \approx 39.0$ MPa

3.128 $\sigma_B \approx -21.7$ MPa; $\sigma_D \approx -75.3$ MPa

3.130 $P \approx 40.8$ kips

3.132 (SI) S76 × 8.5

3.134 (SI) W305 × 44.6

3.136 (SI) C381 × 50.4

3.138 (SI) S305 × 60.7

3.140 (SI) L152 × 152 × 12.7

3.142 (US) W18 × 65

3.144 (SI) S203 × 34.2

3.146 (a) $\sigma \approx 262.5$ MPa; (b) $\sigma \approx 328.1$ MPa

3.148 $r \approx 16$ mm

3.150 $r \approx 25$ mm

3.152 $k \approx 1.3$

3.154 $r \approx 12.3$ mm

R3.2 $M = 20x - 2.5x^2 - 5x^3/36$

R3.4 $V_{MAX} = V_A = 18$ kips, at left support; $M_{MAX} = 54$ kip·ft, at 6 ft from A

R3.6 $(\sigma_T)_{MAX} \approx 26.1$ ksi; $(\sigma_C)_{MAX} \approx 15.7$ ksi

R3.8 $P \approx 17.9$ kips

R3.10 $p \approx 383$ mm

R3.12 $\sigma_H \approx 5.5$ ksi; $\tau_H \approx 0.5$ ksi

R3.14 $\sigma_n \approx -5.7$ ksi; $\tau_{nt} \approx 6.8$ ksi

R3.16 (US) W 24 × 84

R3.18 $\sigma_{MAX} \approx 8.9$ ksi

CHAPTER 4

4.2 $M_{MAX} = 70.7$ kN·m

4.4 $(\sigma_B)_{MAX} = 28.5$ MPa; $(\sigma_M)_{MAX} = 10.8$ MPa

4.6 $P_{MAX} = 45$ kN

4.8 $(\sigma_S)_{MAX} = 110.3$ MPa; $(\sigma_W)_{MAX} = 5.9$ MPa

4.10 $P = 316.5$ kN

4.12 $(\sigma_B)_{MAX} = 12.4$ MPa; $(\sigma_A)_{MAX} = 7.3$ MPa

4.14 $(\sigma_S)_{MAX} = 111.1$ MPa; $(\sigma_W)_{MAX} = 21.1$ MPa; $(\sigma_A)_{MAX} = 42.1$ MPa

4.16 $M_{MAX} = 117.9$ kN·m

4.18 $P_{MAX} = 6.4$ kN·m

4.20 $P_{MAX} = 83.3$ kips

4.22 $(\sigma_C)_{MAX} = 6.9$ MPa; $(\sigma_S)_{MAX} = 124.2$ MPa

4.24 $y_o = h/[\sigma_S/n\sigma_C) + 1]$; $y_o = 0.29$ m

4.26 $(\sigma_T)_{MAX} = -(\sigma_C)_{MAX} = 29.6$ ksi

4.28 $(\sigma_T)_{MAX} = -(\sigma_C)_{MAX} = 46.7$ ksi

4.30 $(\sigma_T)_{MAX} = 5.7$ ksi; $(\sigma_C)_{MAX} = 6.7$ ksi

4.32 $(\sigma_T)_{MAX} = 270.5$ MPa; $(\sigma_C)_{MAX} = 309.3$ MPa

4.34 $(\sigma_T)_{MAX} = -(\sigma_C)_{MAX} = 19.8$ ksi

4.36 $(\sigma_T)_{MAX} = 211.2$ MPa; $(\sigma_C)_{MAX} = 207.2$ MPa

4.38 $(\sigma_T)_{MAX} = -(\sigma_C)_{MAX} = 20.3$ ksi

4.40 $(\sigma_T)_{MAX} = 59.6$ ksi; $(\sigma_C)_{MAX} = 78.5$ ksi

4.42 $(\sigma_T)_{MAX} = 125.9$ MPa; $(\sigma_C)_{MAX} = 90.4$ MPa

4.44 $(\sigma_T)_{MAX} = 85.4$ ksi; $(\sigma_C)_{MAX} = 53.5$ ksi

4.46 $p_{MAX} = 6.1$ kN/m

4.48 $P_{MAX} = 1.2$ kips

4.50 $p_{MAX} = 10.6$ kN/m

4.52 $V_{DE} = 0.888\ V$

4.54 $V_{DE} = 1.10\ V$

4.56 (a) $V_{AB} = 0.08333\ V$; (b) $e = 0.14764a$

4.58 (a) $V_{AB} = 0.21548\ V$; (b) $e = 1.319$ in.

4.60 $T = 19.785$ kip·in.

4.62 $V_{BD} = 0.70711\ V$

4.64 $V_{AB} = 0.33333\ V$; $e = 0.57734a$

4.66 $\sigma = -878.9$ psi; $\tau = 214.8$ psi

4.68 $e = 2R$

4.70 (a) $V_{BD} = 0.26453\ V$; (b) $e = 0.74463a$

4.72 $\sigma = 1.5$ ksi; $\tau = 0.2$ ksi

4.74 $\tau = 1.733y - 0.085y^2$

4.76 $e_v = 0.222$ m; $e_u = 0.104$ m
4.78 $e_v = 0.024$ m; $e_u = 0.005$ m
4.80 $T = 4.59$ kip $\cdot$ in.
4.82 $(\sigma_T)_{MAX} = 36.9$ ksi; $(\sigma_C)_{MAX} = 17.4$ ksi
4.84 $P_{MAX} = 47.2$ kN
4.86 $(\sigma_T)_{MAX} = 18.8$ ksi; $(\sigma_C)_{MAX} = 8.4$ ksi
4.88 $(\sigma_T)_{MAX} = 11.3$ MPa; $(\sigma_C)_{MAX} = 5.7$ MPa
4.90 $f = 1.174$; $M_p = 810$ kip $\cdot$ in.
4.92 $P_y = 19.2$ kips; $P_p = 22.5$ kips
4.94 $p_y = 12.1$ kN/m; $p_p = 21.8$ kN/m
4.96 $f = 1.765$; $M_y = 4352$ kip $\cdot$ in.
4.98 $P_p = 213.4$ kips
4.100 $p_y = 6.4$ kN/m; $p_p = 9.5$ kN/m
4.102 $f = 2.157$; $M_p = 652.3$ kip $\cdot$ in.
4.104 $P_p = 23.3$ kips
4.106 $K = 2$
4.108 $P_p = 55$ kN
4.110 $K_1 = 1.5$; $K_2 = 1.0$
4.112 $K = 0.375$
4.114 $p_{p1}/p_{p2} = 5.491$
4.116 $b_{MIN} = 3.91$ in.
4.118 $(D_o)_{MIN} = 87.9$ mm; $(D_i)_{MIN} = 70.3$ mm
4.120 $w_{MIN} = 15.1$ mm; $d_{MIN} = 60.2$ mm
4.122 $w_{MIN} = 2.950$ in.; $d_{MIN} = 8.850$ in.
4.124 $w_{MIN} = 3.80$ in.; $d_{MIN} = 7.60$ in.
4.126 $P_{MAX} = 321.0$ kN
4.128 $w_{MIN} = 2.924$ in.; $d_{MIN} = 8.772$ in.
4.130 $P_{MAX} = 15.8$ kN
4.132 $(D_o)_{MIN} = 2.592$ in.; $(D_i)_{MIN} = 1.620$ in.
R4.2 $p_{MAX} = 4.5$ ksi
R4.4 $(\sigma_T)_{MAX} = -(\sigma_T)_{MAX} = 30.4$ ksi
R4.6 $(\sigma_T)_{MAX} = -(\sigma_T)_{MAX} = 30.4$ ksi
R4.8 $e = 0.340a$
R4.10 (a) $e = 76.1$ mm; (b) $\tau_{MAX} = 58.1$ MPa
R4.12 $f = 1.838$; $M_y = 6321.1$ kip $\cdot$ in.
R4.14 ABC; $p_p = 13.7$ kN/m
R4.16 $P = 1.9$ kN

CHAPTER 5

5.2 (a) $\rho_{MIN} = 7.250$ m; $M = 73.4$ N $\cdot$ m
5.4 $t = 0.4$ mm
5.6 (a) $\sigma_{MAX} = 50.4$ ksi; (b) $M = 0.6$ kip $\cdot$ in.
5.8 $\theta_{MAX} = PL^2/2EI$; $v_{MAX} = -PL^3/(3EI)$
5.10 $\theta_{MAX} = pL^3/(24EI)$; $v_{MAX} = -pL^4/(30EI)$
5.12 $\theta_{MAX} = -PL^2/(2EI)$; $v_{MAX} = -PL^3/(3EI)$
5.14 $\theta_A = -pL^3/(24EI)$; $v_{x = L/2} = -5pL^4/(384EI)$
5.16 $\theta_A = -7pL^3/(360EI)$; $\theta_B = -pL^3/(45EI)$; $v_{x = L/2} = 7pL^4/(480EI)$
5.18 $\theta_A = -QL/3EI$; $\theta_B = QL/6EI$; $v_{x = L/2} = -QL^2/(16EI)$
5.20 $\theta_A = 2QL/(3EI)$; $\theta_B = -5QL/(6EI)$; $v_{x = L/2} = 3QL^2/(16EI)$
5.22 $\theta_A = -pL^3/(90EI)$; $\theta_B = pL^3/(72E)$; $v_{x = L/2} = -89pL^4/(23040EI)$

5.24 $v_{MAX} = QL^2/(9\sqrt{3}EI)$

5.26 $v_{MAX} = 0.05795QL^2/EI$

5.28 $v_{MAX} = -0.00388pL^4/EI$

5.30   W12 × 19 U.S. section

5.32   W10 × 17 SI section

5.34 $p = 61.3$ kN/m

5.36 $p = 103.1$ kN/m

5.38 $v_{x = L} = pL^4/8EI$

5.40 $v_{x = 3a/2} = -9Qa^2/(16EI)$

5.42 $v_{MAX} = 2pL^4/(9\sqrt{3}EI)$

5.44 $v_{MAX} = \sqrt{3}Qa^2/(3EI)$

5.46 $C_1 = -QL/18$; $C_2 = 0$; $C_3 = -7QL/18$; $C_4 = QL^2/18$

5.48 $v_{MAX} = -0.01792PL^3/EI$

5.50 $v_{MAX} = -0.00327pL^4/EI$

5.52 $p = -p_o x^2/L$

5.54 $C_1 = 0$; $C_2 = -Q$; $C_3 = QL$; $C_4 = -QL^2/2$

5.56 $C_1 = C_2 = 0$; $C_3 = pL^3/60$; $C_4 = -pL^4/72$

5.58 $C_1 = Q/L$; $C_2 = 0$; $C_3 = QL/6$; $C_4 = 0$

5.60 $C_1 = -11Q/12L$; $C_2 = Q$; $C_3 = -31QL/90$; $C_4 = 0$

5.62 $C_1 = 4pL/15$; $C_2 = 0$; $C_3 = -236pL^3/945$; $C_4 = 0$

5.64 $v_A = -45PL^3/(128EI)$; $\theta_A = 25\ PL^2/(48EI)$

5.66 $v_A = 7QL^2/(15EI)$; $\theta_A = -23\ QL/(24EI)$

5.68 $v_B = -4PL^3/(81EI)$; $\theta_A = -PL^2/(9EI)$

5.70 $v_B = -1141PL^3/(41,742EI)$; $\theta_A = -1267PL^2/(20,736EI)$

5.72 $v_{MAX} = -0.07133QL^2/EI$

5.74   W14 × 34 U.S. section

5.76   W200 × 19.3 SI section

5.78 $v_A = -PL^3/(3EI)$; $\theta_A = PL^2/(2EI)$

5.80 $v_A = -pL^4/(30EI)$; $\theta_A = pL^3/(24EI)$

5.82 $v_A = -pL^4/8EI$; $\theta_A = pL^3/6EI$

5.84 $v_A = -169PL^3/384EI$; $\theta_A = 2PL^2/3EI$

5.86 $v_B = -19pL^4/(120EI)$; $\theta_A = 5pL^3/(24EI)$

5.88 $v_A = -79PL^3/(384EI)$; $\theta_A = PL^2/(3EI)$

5.90 $v_A = -73PL^3/(192EI)$; $\theta_A = 23PL^2/(48EI)$

5.92 $v_A = 131pL^4/(160EI)$; $\theta_A = -401pL^3/(192EI)$

5.94 $v_A = -14.1$ mm; $\theta_A = 0.0018$ rad

5.96 $v_A = -13.9$ mm; $\theta_A = 0.0054$ rad

5.98 $v_C = -QL^2/(16EI)$; $\theta_A = QL/6$

5.100 $v_C = -0.07407PL^3/EI$; $\theta_A = -32PL^2/(81EI)$

5.102 $x = L/3$; $v_{MAX} = QL^2/9\sqrt{3}EI$

5.104 $x = 0.45903L$; $v_{MAX} = 0.13763PL^3/EI$

5.106 $v_C = -PL^3/(64EI)$; $\theta_B = PL^2/(24EI)$

5.108 $v_C = -193PL^3/(6144EI)$; $\theta_B = 43PL^2/(512EI)$

5.110 $x = 0.91806$ m; $v_{MAX} = -0.71$ mm

5.112 $v_B = -0.7$ mm; $\theta_C = 0.00062$ rad

5.114 $v_A = -19QL^2/(54EI)$; $\theta_A = 5QL/(9EI)$

5.116 $v_A = -8PL^3/(243EI)$; $\theta_A = 19PL^2/(162EI)$

5.118 $x = 2L/\sqrt{27}$; $v_{MAX} = 4QL^2/(27\sqrt{27}EI)$

5.120 $x = 0.56872L$; $v_{MAX} = 0.00127PL^3/EI$

5.122 $\theta_A = -0.00075$ rad; $v_{x = L/2} = 0.24$ mm

5.124 $x = 1.127$ m; $v_{MAX} = -0.10$ mm

5.126 $A_y = 3Q/2L$; $B_y = -3Q/2L$; $M_A = Q/2$

5.128 $A_y = 17pL/8$; $B_y = -9pL/8$; $M_A = 5pL^2/8$

5.130 $A_y = 9Q/16L$; $C_y = -9Q/16L$; $M_C = Q/8$

5.132 $A_y = 163pL/216$; $C_y = 53pL/216$; $M_C = -377pL^2/216$

5.134 $B_y = 3Q/4L$; $C_y = -3Q/4L$; $M_C = Q/2$

5.136 $B_y = 17pL/8$; $C_y = 7pL/8$; $M_C = -pL^2/4$

5.138 $Q_A = pL^2/3$

5.140 (a) $\sigma_R = 101.5$ MPa; (c) $M_C = 4.8$ kN·m

5.142 $B_y = -9pL/8$; $M_A = 5pL^2/8$

5.144 $A_y = 27P/64$; $C_y = 101P/64$

5.146 $B_y = 3Q/4L$; $M_C = Q/2$

5.148 $C_y = pL/2$; $M_C = 0$

5.150 (a) $(\sigma_R) = 38.0$ ksi; (b) $v_b = -0.23$ in.

5.152 $P = 56.6$ kips

5.154 $P = 112.5$ kN

5.156 $B_y = (5PL^3 + 48EI\delta)/32L^3$; $M_C = (48EI - 11PL^3)/32L^2$

5.158 $p = 39.8$ kN/m

5.160 $A_y = 5P/16$; $C_y = 11P/16$; $M_C = -3PL/8$

5.162 $B_y = 11pL/8$; $C_y = -3pL/8$; $M_C = -pL^2/4$

5.164 (a) $\sigma_R = 38.0$ ksi; (b) $v_B = -0.23$ in.

5.166 $G_y = 3P/8$; $A_y = B_y = 13P/16$

5.168 $A_y = 39.4$ kips; $M_A = 244.2$ kip.ft; $C_y = 25.6$ kips;
$M_C = 211.7$ kip.ft; $B_y = 9.4$ kips on AB

5.170 $B_y = (22pL^4 - 3EI\delta)/8L^3$; $M_C = -(4pL^4 + 3EI\delta)/4L^2$

5.172 $A_y = -13P/8$; $B_y = 37P/8$; $M_A = -5PL/8$

5.174 $A_y = 48.75$ kips; $M_C = -168.75$ kip·ft; $v_B = -0.561$ in.

R5.2 $EIv' = -pL^2x - px^4/24L + 25pL^3/24$

R5.4 $p_{MAX} = 994$ kN/m

R5.6 $x = \sqrt{11}L/8$; $v_{MAX} = -11\sqrt{11}PL^3/6144EI$

R5.8 S24 × 106

R5.10 $v_A = 49PL^3/810EI$; $\theta_A = -37PL^2/144EI$

R5.12 $v_C = -283QL^2/1536EI$; $\theta_A = -115QL/768EI$

R5.14 $B_y = 37pa/8$; $M_C = -pa^2/2$; $V_C = 11pa/8$; W18 × 46

R5.16 $v_A = 49PL^3/810EI$; $\theta_A = -37PL^2/144EI$

R5.18 $B_y = 224P/121$; $A_y = -103P/121$; $M_A = 85PL/363$; $v_{MAX} = v_B = -14PL^3/363EI$

CHAPTER 6

6.2 $\theta_A = -3PL^2/EI$; $v_B = -3PL^3/2EI$

6.4 $\theta_A = 7PL^2/2EI$; $v_B = 2PL^3/3EI$

6.6 $\theta_A = -0.011$ rad; $v_A = -1.610$ in.

6.8 $\theta_A = -0.00565$ read; $v_A = -23.6$ mm

6.10 $\theta_A = (Q/6LEI)(L^2 - 3b^2)$; $v_B = (2Qab/6LEI)(b - a)$

6.12 $\theta_A = -23pL^3/1215$; $v_B = -56\,pL^4/10{,}935$

6.14 $\theta_A = -4PL^2/81EI$; $v_{x = L}/2 = -41PL^3/1296EI$

6.16 $\theta_A = 377QL/3888EI$; $v_B = 161QL^2/5832EI$

6.18 $\theta_B = -183pL^3/6144EI = -\theta_C$

6.20 $\theta_A = 0.02718$ rad; $v_C = 56.77$ mm

6.22 $\theta_C = 151Qa^2/48EI$; $v_D = 197Qa^2/24EI$

6.24 $\theta_A = QL/EI$; $v_A = QL^2/2EI$ ↑

6.26 $\theta_A = pL^3/24EI$; $v_A = pL^4/30EI$

6.28 $\theta_A = QL/6EI$; $v_B = QL^2/16EI$

6.30 $\theta_A = 5PL^2/81EI$; $v_B = 4PL^3/243EI$

6.32 $\theta_A = PL^2/12EI$; $v_C = 5PL^3/12EI$

6.34 $\theta_A = pa^3/12EI$; $v_B = pa^4/24EI$

6.36 $\theta_A = 0.41$ in.

6.38 $v = 0.495$ in.; $\theta = 22.2°$

6.40 $v = 0.937$ in.; $\theta = 15.0°$

6.42 $v = 1.341$ in.; $\theta = 18.921°$

6.44 $v = 1.519$ in.; $\theta = 18.921°$

6.46 $v = 1.628$ in.; $\theta = 10.2°$

6.48 $v = 0.287$ in.; $\theta = 38.1°$

6.50 $A_y = 299Q/216L$; $M_B = 263Q/216$

6.52 $V_B = -2465Q/2048L$; $M_B = 737Q/2048$

6.54 $F = 1537QL^2k/[1024(kL^3 + 3EI)]$

6.56 $F = 9ApL^4/(640AL^3 + 1920\,I\,l)$

6.58 $V_A = 699Q/512L$; $M_A = 375Q/3072$

6.60 $M_A = (6480EI\theta - 29pL^3)/1620L$

6.62 $v_{x = L/2} = 10.3$ mm $\downarrow$

6.64 $A_y = p_oL/24$; $V_B = 7p_oL/24$; $M_B = -p_oL^2/24$

6.66 $V_B = 17pL/8$; $M_B = -5pL^2/8$

6.68 $A_y = 9Q/16L$; $M_C = Q/8$

6.70 $V_C = 161Q/216L$; $M_C = -269Q/216$

6.72 $v_B = -0.561$ in.

6.74 $V_A = pL/2$; $M_A = -pL^2/12$

6.76 $V_A = P/2$; $M_A = -PL/8$

6.78 $V_A \approx 45.4$ kips

6.80 $V_A \approx 6.2$ kN; $M_A = 14.0$ kN·m ccw

6.82 $\Delta_{MAX} = 0.032$ in.; $IF = 114.04$

6.84 $\Delta_{MAX} = 2.976 \times 10^{-3}$ m; $W = 1.945$ kN

6.86 $\sigma_{MID} = 11.413$ MPa; $\Delta_{MID} = 0.641$ mm

6.88 $\Delta_{MAX} \approx 0.619$ in.; $\sigma_{MAX} \approx 8.048$ ksi

6.90 $\sigma_{MAX} = 31.5$ MPa; $\Delta_{MAX} = 109.1$ mm

6.92 $(\sigma_B)_{MAX} = 0.337$ MPa; $(\Delta_B)_{MAX} = 0.505$ mm

R6.2 $v_A = -4275PL^3/12EI$; $v'_D = 8531PL^2/24EI$

R6.4 $\theta_A = 187pL^3/192EI$; $v_C = 111pL^4/192EI$

R6.6 $v_B = -0.2369$ in.

R6.8 $F = 2049AL^3/[2048(AL^3 + 3I\,l)]$

R6.10 $v_p = 43pL^4/1230EI$

CHAPTER 7

7.2 $\sigma_n = 174.49$ MPa; $\tau_{nt} = 22.18$ MPa

7.4 $\sigma_n = 171.42$ MPa; $\tau_{nt} = 21.089$ MPa

7.6 $\sigma_n = -14.67$ ksi; $\tau_{nt} = -12.31$ ksi

7.8 $\sigma_n = -3.00$ ksi; $\tau_{nt} = 15.00$ ksi

7.10 $\sigma_n = 31.16$ ksi; $\tau_{nt} = -0.67$ ksi

7.12 $\sigma_1 = 144.34$ MPa; $\sigma_3 = -44.34$ MPa; $\sigma_2 = 0$ on Z plane

7.14 $\sigma_1 = 225.1$ MPa; $\sigma_3 = -75.1$ MPa; $\sigma_2 = 0$ on Z plane

7.16 $|\tau_{MAX}| = 15$ ksi; $\theta_\tau = 45.0°$; $\sigma_{MS} = -15$ ksi

7.18 $|\tau_{MAX}| = 11.314$ ksi; $\theta_\tau = 22.5°$; $\sigma_{MS} = -8.0$ ksi

7.20 $|\tau_{MAX}| = 5.59$ ksi; $\theta_\tau = -13.28°$; $\sigma_{MS} = 12.5$ ksi

7.22 $\sigma_1 = 5.639$ ksi; $\sigma_3 = -1.349$ ksi; $\sigma_2 = 0$ on Z plane

7.24 $\sigma_1 = 5.0$ MPa; $\sigma_3 = -306.8$ MPa; $\sigma_2 = 0$ on Y plane

7.26 $\sigma_n = -14.67$ ksi; $\tau_{nt} = 12.31$ ksi

7.28 $\sigma_n = 31.16$ ksi; $\tau_{nt} = -0.67$ ksi

7.30 $\sigma_1 = 3.31$ ksi; $\sigma_3 = -19.31$ ksi; $\sigma_2 = 0$ on Z plane; $\theta_\sigma = 22.5°$

7.32 $|\tau_{MAX}| = 11.31$ ksi; $\theta_\tau = 22.5°$; $\sigma_{MS} = -8.0$ ksi

7.34 $\sigma_1 = \sigma_2 = 0$ on Y and Z planes, respectively; $\sigma_3 = -82.662$ MPa on X plane

7.36 $\sigma_1 = 13.672$ MPa; $\sigma_3 = -42.648$ MPa; $|\tau_{MAX}| = 28.160$ MPa

7.38 $\sigma_x = -24.6$ MPa; $\sigma_y = 4.6$ MPa; $\tau_{xy} = -54.7$ MPa

7.40 $\sigma_x = 25.0$ ksi; $\sigma_y = 5.0$ ksi; $|\tau_{xy}| = 5.0$ ksi

7.42 $\sigma_1 = 33.47$ ksi; $\sigma_2 = 1.17$ ksi; $\sigma_D = 23.30$ ksi

7.44 $\sigma_D = 25.0$ ksi; $\sigma_E = 5.0$ ksi

7.46 $\sigma_x = \sigma_y = 15$ ksi; $|\tau_{xy}| = 5$ ksi

7.48 $\sigma_1 = 31.8$ ksi; $\sigma_2 = 6.4$ ksi; $\theta = 35.4°$

7.50 $\sigma_1 = 36.4$ ksi; $\sigma_2 = 8.6$ ksi; $\sigma_3 = 0$; $\theta \approx 4.5°$

7.52 $|\tau_{MAX}| = 12.7$ ksi; $\alpha = 9.7°$; $\sigma_{MS} = 19.1$ ksi

7.54 $|\tau_{MAX}| = 13.9$ ksi; $\alpha = 40.5°$; $\sigma_{MS} = 22.5$ ksi

7.56 $\sigma_1 = 2.6$ ksi; $\sigma_2 = 0.4$ ksi; $\theta = 31.7°$

7.58 $\sigma_1 = 15.6$ ksi; $\sigma_3 = -41.2$ ksi; $\theta = 31.7°$

7.60 Max. Comp. is $\sigma_3 = -2.15$ ksi; $|\tau_{MAX}| = 1.08$ ksi; $\sigma_{MS} = -1.08$ ksi

7.62 $\sigma_1 = 133.1$ MPa; $\sigma_3 = -23.3$ MPa; $\sigma_2 = 0$; $|\tau_{MAX}| = 78.2$ MPa; $\sigma_{MS} = 54.9$ MPa

7.64 $\sigma_1 = \sigma_2 = 0$; $\sigma_3 = -61.0$ ksi; $|\tau_{MAX}| = 30.5$ ksi; $\sigma_{MS} = -30.5$ ksi

7.66 $\sigma_1 = 47.1$ ksi; $\sigma_3 = -47.1$ ksi; $\sigma_2 = 0$; $|\tau_{MAX}| = 47.1$ ksi

7.68 $l_3 = -0.57935$; $m_3 = -0.63110$; $n_3 = 0.51582$

7.70 $\sigma_1 = 90.4$ MPa; $\sigma_3 = -37.2$ MPa; $l_2 = 0.23215$; $n_2 = 0.96573$

7.72 $\sigma_1 = 78.1$ MPa; $\sigma_2 = 18.0$ MPa; $l_1 = 0.86563$; $m_1 = -0.44723$; $n_1 = 0.22511$

7.74 $\sigma_2 = -8.9$ MPa; $\sigma_3 = -25.6$ MPa; $m_3 = 0.32279$; $n_3 = 0.94578$

7.76 $\tau_{ABS} = 45$ MPa

7.78 $\sigma_{AVE} = 35$ MPa; $\tau_{ABS} = 65$ MPa

7.80 $\sigma_{AVE} = -25$ MPa; $\tau_{ABS} = 25$ MPa

7.82 $\sigma_{AVE} = 40$ MPa; $\tau_{ABS} = 40$ MPa

7.84 $\sigma_{AVE} = 20$ MPa; $\tau_{ABS} = 70$ MPa

7.86 $\sigma_1 = 200$ MPa; $\sigma_2 = 60$ MPa; $\tau_{xy} = 63.2$ MPa

7.88 $\sigma_x = 128$ MPa; $\sigma_1 = 200$ MPa; $\sigma_3 = 0$

7.90 $t_{MIN} = 0.875$ in.

7.92 $\sigma_{MAX} = 4.905$ MPa; $\tau_{ABS} = 2.551$ MPa

7.94 (a) $\sigma_1 = 24.0$ ksi; $\sigma_2 = 13.052$ ksi; (b) $\sigma_2 = 13.052$ ksi; $\tau_{ABS} = 12.2$ ksi

7.96 $\sigma_{MAX} = 25.0$ MPa

7.98 $N = 1.736$

7.100 $(\sigma_c)_{MAX} = 13.333$ ksi (T) at $r = r_1$; $\sigma_x = 2.667$ ksi (T)

7.102 $(\sigma_c)_{MAX} = 0.576$ MPa (C); $(\sigma_r)_{MAX} = 0.216$ MPa (C)

7.104 $t_{MIN} = 1.658$ ft

7.106 $(\sigma_c)_{MAX} = 56,714$ psi (T) in jacket at $r = r_2$

7.108 (a) $\sigma_1 = 139.41$ MPa; $\sigma_2 = 53.41$ MPa; (b) $\sigma_2 = 63.48$ MPa; $\tau_{ABS} = 124.67$ MPa

7.110 Member does not fail

7.112 Member does not fail

7.114 Member fails

7.116 Member fails

7.118 Member does not fail

7.120 $Q \approx 8.3$ k·ft

7.122 $P = 1112$ kN

7.124 No failure; $F.S. = 2.121$
7.126 Member will fail
7.128 No failure; $F.S. = 1.224$
7.130 Member will fail
7.132 No failure; $F.S. = 1.045$
7.134 $Q = 63.6 \text{ kN} \cdot \text{m}$
7.136 $d = 100.6 \text{ mm}$
7.138 $Q \approx 31.6 \text{ kip} \cdot \text{in}$
R7.2 $\sigma_1 \approx 140.1 \text{ MPa}; \sigma_3 \approx -40.1 \text{ MPa}; \theta_\sigma \approx 28.2°$
R7.4 $\sigma_n = 4.785 \text{ ksi}; \tau_{nt} = 22.431 \text{ ksi}$
R7.6 $p_{MAX} \approx 26.6 \text{ lb/ft}^2$
R7.8 $P_{MAX} = 62.6 \text{ kips}$
R7.10 $(\sigma_c)_{r = r2} = 13.969 \text{ ksi}$
R7.12 $\tau_{ABS} = 35 \text{ ksi}; \sigma_{AVE} = -5 \text{ ksi}$
R7.14 (a) Member will fail; (b) Member will fail
R7.16 $b \approx 1.5 \text{ in.}$

CHAPTER 8

8.2 $\varepsilon_n \approx -173.2 \times 10^{-6}; \gamma_{nt} = -200 \times 10^{-6}$
8.4 $\sigma_n = \sigma_t \approx 1173.4 \text{ psi}; \tau_{nt} \approx -592.1 \text{ psi}$
8.6 $\varepsilon_1 \approx 358.1 \times 10^{-6}; \varepsilon_2 \approx 41.9 \times 10^{-6}$
8.8 $\varepsilon_1 \approx 376.0 \times 10^{-6}; \varepsilon_3 \approx -226.0 \times 10^{-6}$
8.10 $\sigma_t \approx 67.7 \text{ MPa}; \sigma_3 \approx -22.9 \text{ MPa}$
8.12 $\varepsilon_n \approx -222.4 \times 10^{-6}; \gamma_{nt} \approx -362.1 \times 10^{-6}$
8.14 $\varepsilon_n \approx 1094.2 \times 10^{-6}; \varepsilon_t \approx 131.3 \times 10^{-6}$
8.16 $\varepsilon_n \approx 1149 \times 10^{-6}; \varepsilon_t \approx -1484 \times 10^{-6}$
8.18 $\varepsilon_n \approx -173.2 \times 10^{-6}; \gamma_{nt} = -200 \times 10^{-6}$
8.20 $\varepsilon_1 \approx 358.1 \times 10^{-6}; \varepsilon_2 \approx 41.9 \times 10^{-6}$
8.22 $\varepsilon_1 \approx 376.0 \times 10^{-6}; \varepsilon_3 \approx -226.0 \times 10^{-6}$
8.24 $\varepsilon_n \approx -222.4 \times 10^{-6}; \varepsilon_t \approx 122.4 \times 10^{-6}$
8.24 $\varepsilon_n \approx -222.4 \times 10^{-6}; \gamma_{nt} = -362.1 \times 10^{-6}$
8.26 $\varepsilon_n \approx 1094.2 \times 10^{-6}; \varepsilon_t \approx 130.8 \times 10^{-6}$
8.28 $\varepsilon_n \approx 1148.5 \times 10^{-6}; \gamma_{nt} \approx -4655.0 \times 10^{-6}$
8.30 $\varepsilon_1 \approx 928.5 \times 10^{-6}; \varepsilon_3 \approx -928.5 \times 10^{-6}$
8.32 $\varepsilon_n \approx 450.0 \times 10^{-6}; \varepsilon_t \approx 16.7 \times 10^{-6}$
8.34 $\varepsilon_n \approx 709.6 \times 10^{-6}; \gamma_{nt} \approx 3260.4 \times 10^{-6}$
8.36 $\varepsilon_1 \approx 3334 \times 10^{-6}; \varepsilon_3 \approx -0.3 \times 10^{-6}$
8.38 $\varepsilon_n \approx -1640 \times 10^{-6}; \gamma_{nt} \approx 1953.8 \times 10^{-6}$
8.40 $\varepsilon_n \approx -1542.8 \times 10^{-6}; \varepsilon_t \approx -514.3 \times 10^{-6}$
8.42 (a) $\sigma_1 \approx 1.5 \text{ MPa}; \sigma_3 \approx -142.5 \text{ MPa}$; (b) $\Delta V = -2016 \text{ mm}^3$
8.44 (a) $\Delta h = 0.271 \text{ mm}$; (c) $K = 37.5 \text{ GPa}$; (d) $u \approx 392.5 \text{ kN} \cdot \text{m/m}^3$
8.46 (a) $h \approx 702.2 \text{ m}$; (b) $p \approx 0.758 \text{ MPa}$; (c) $u_d \approx 0$; (d) $K \approx 151.5 \text{ GPa}$
8.48 (a) $\sigma_1 \approx 46.2 \text{ MPa}; \sigma_3 \approx -30.8 \text{ MPa}$; (c) $u_d \approx 10.8 \text{ kN} \cdot \text{m/m}^3$
8.50 (a) $\varepsilon_1 \approx 492 \times 10^{-6}; \varepsilon_3 \approx -212 \times 10^{-6}$; (b) $|\gamma_{MAX}| = 704 \times 10^{-6}$
8.52 (a) $e = 542 \times 10^{-6}; u_d \approx 31.7 \text{ kN} \cdot \text{m/m}^3$
8.54 (a) $|\gamma_{MAX}| = 1487 \times 10^{-6}; K = 58.3 \text{ GPa}$
8.56 (a) $\sigma_1 = 0; \sigma_3 = -43.5 \text{ MPa}$; (b) $|\tau_{MAX}| = 8.5 \text{ MPa}$
8.58 (a) $\sigma_t \approx 8.2 \text{ ksi}$; (b) $\tau_{MAX} \approx 3.9 \text{ ksi}$
8.60 (a) $\sigma_1 = 5.52 \text{ ksi}; \sigma_2 = 1.08 \text{ ksi}$; (b) $|\tau_{MAX}| = 2.22 \text{ ksi}$
8.62 $\varepsilon_x = 0.868 \, \varepsilon_{-20} + 0.566 \, \varepsilon_{50} - 0.434 \, \varepsilon_{90}$

8.64   (a) $\varepsilon_{30} \approx 225 \times 10^{-6}$; (b) $\varepsilon_{45} \approx 117 \times 10^{-6}$
8.66   (a) $\varepsilon_{25} \approx -1503 \times 10^{-6}$; (b) $\varepsilon_{40} \approx -1932 \times 10^{-6}$
8.68   (a) $\varepsilon_{20} \approx -914 \times 10^{-6}$; (b) $\varepsilon_{40} \approx -648 \times 10^{-6}$
R8.2   (a) $\varepsilon_n \approx -25.8 \times 10^{-6}$; (b) $\varepsilon_t \approx -4.2 \times 10^{-6}$
R8.4   Same answers as in R8.2
R8.6 $\varepsilon_1 \approx 2130 \times 10^{-6}$; $\varepsilon_3 \approx -707 \times 10^{-6}$
R8.8 $\varepsilon_z \approx 280 \times 10^{-6}$; $\gamma_{nt} \approx 707 \times 10^{-6}$
R8.10 $\sigma_t = 1.117$ ksi; $|\tau_{MAX}| = 6.828$ ksi
R8.12 (a) $\varepsilon_1 \approx 1539 \times 10^{-6}$; (b) $|\gamma_{MAX}| \approx 3998 \times 10^{-6}$

CHAPTER 9

9.2 $W_{CR} = 8kL/9$
9.4 $\theta = 0$, stable; $\theta = \pm 67.98°$, unstable; $\theta = \pm 180°$, stable
9.6 $W_{CR} = 4K/L$
9.8 $W_{CR} = kL/4$
9.10 $\theta = 0$, stable; $\theta = \pm 180°$, unstable; $\theta = \pm 72.542°$, stable
9.12 $k = 1258.5$ lb/ft
9.14 $K_{MIN} = 40$ kip·ft/rad
9.16 $d = 39.95$ mm
9.18 $a = 3.439$ in.
9.20   Crusiform
9.22 $b = 0.844$ in.; Mat. Red. $\approx 32\%$
9.24   F.S. $= 1.645$
9.26 $P_{MAX} \approx 40.6$ kN
9.28 $P_{MAX} \approx 33.2$ kips
9.30   The U section
9.32   F.S. ≈ 2.328
9.34 $P_{MAX} \approx 30.1$ kN
9.36 $P_{CR} \approx 380.9$ kN
9.38 $P_{CR} \approx 695.7$ kN
9.40 $L = 1.253$ m
9.42 $P_{ALL} = 4705.1$ kN
9.44   (SI) W127 × 23.8
9.46 $L_{MAX} \approx 5.450$ m
9.48   (a) $P_{CR} \approx 1664.1$ kN; (b) $P_{CR} = 3328.2$ kN
9.50 $P_{MAX} \approx 1066.2$ kN
9.52   (a) $u_{MAX} \approx 0.348$ in.; (b) $\sigma_{MAX} \approx 12.6$ ksi
9.54   (a) $e \approx 0.047$ in.; (b) $\sigma_{MAX} \approx 15.4$ MPa
9.56 $L \approx 7.8$ ft; $u_{MAX} \approx 2.8$ in.
9.58 $u_{MAX} \approx 0.490$ in.; $\sigma_{MAX} \approx 8.1$ ksi
9.60 $u_{MAX} \approx 0.046$ in.; $\sigma_{MAX} \approx 14.9$ ksi
9.62 $u_{MAX} \approx 2.872$ in.; $\sigma_{MAX} \approx 151.2$ ksi
9.64 $L_{MAX} = 5.0$ m
9.66 $P_{MAX} = 123.1$ kips
9.68   (a) F.S. $= 2.515$; (b) $u_{MAX} = 0.254$ in.
9.70 $P_{ALL} = 212.5$ kips
9.72 $P_{ALL} \approx 25.7$ kips
9.74 $L_{MAX} = 15.7$ ft
9.76   (SI) W610 × 241.0
9.78   (US) W18 × 71

9.80   (US) S20 × 96
9.82 $P_{ALL} \approx 83.8$ kips
9.84 $P_{ALL} = 501$ kN
9.86 $P_{ALL} \approx 85.3$ kips
9.88 $L_e \approx 28.8$ ft
9.90 $P_{ALL} \approx 971.6$ kN
9.92 $P_{ALL} \approx 504.5$ kN
9.94 $P_{ALL} \approx 646.8$ kN
9.96 $L_{MAX} \approx 8.6$ ft
9.98 $L_{MAX} \approx 6.8$ ft
9.100 $b_{MIN} \approx 6.7$ in.
9.102 $P_{ALL} \approx 50.1$ kips
9.104 $P_{ALL} \approx 398.7$ kips
9.106 $L_{MAX} \approx 26.2$ ft
9.108 $L_{MAX} = 2.85$ ft
9.110 $L_{MAX} \approx 9.4$ ft
9.112 $L_{MAX} \approx 4.0$ m
9.114 $L_{MAX} \approx 2.710$ m
9.116 $P_{MAX} \approx 40.1$ kips
9.118 (SI) W610 × 81.8
9.120 (US) W18 × 40
9.122 $P_{MAX} \approx 1868$ kN
9.124 $L_{MAX} \approx 32.1$ ft
9.126 $P_{ALL} \approx 95.6$ kips
9.128 $b_{MIN} \approx 87$ mm
9.130 $d_o \approx 203$ mm
9.132 $P_{ALL} \approx 42.2$ kips
9.134 $L_{MAX} \approx 13.8$ ft
9.136 $b = 6.7$ in.
9.138 $b = 3.7$ in.
9.140 $P_{MAX} \approx 173.3$ kips
9.142 $b_{MIN} \approx 5.3$ in.
9.144 $P_{MAX} \approx 19.6$ kips
9.146 $P_{MAX} \approx 43.4$ kN
9.148 $P_{ALL} \approx 11.2$ kips
R9.2 $P_{MAX} \approx 10.2$ kips
R9.4 $P_{ALL} \approx 67.6$ kips
R9.6 $L = 8.297$ ft
R9.8 $L_{MAX} \approx 14.5$ ft; $u_{MAX} \approx 5.805$ in.
R9.10 $L = 13.484$ m
R9.12 $L = 4.493$ m
R9.14 $L \approx 3.03$ m

Appendix A: SI Units

A.1 FUNDAMENTAL UNITS IN MECHANICS

Quantity	Unit	Symbol
Mass	Kilogram	kg
Length	Meter	m
Time	Second	s

A.2 SELECTION OF DERIVED UNITS AND CONVERSION FACTORS

Quantity	Symbol	U.S. to SI
Length	m	1 ft = 0.3048 m
Area	m^2	$1\ ft^2 = 0.0929\ m^2$
Volume	m^3	$1\ ft^3 = 0.0283\ m^3$
Moment of inertia	m^4 or mm^4	$1\ in.^4 = 4.162 \times 10^{-7}\ m^4 = 4.162 \times 10^5\ mm^4$
Section modulus	m^3 or mm^3	$1\ in.^3 = 1.639 \times 10^{-5}\ m^3 = 1.639 \times 10^7\ mm^3$
Density	kg/m^3	$1\ lb/ft^3 = 16.03\ kg/m^3$
Force	N (Newton)	1 lb = 4.448 N
Moment or torque	$N \cdot m$	$1\ lb \cdot ft = 1.356\ N \cdot m$
Stress or pressure	N/m^2 (Pascal)	$1\ psi = 6.895 \times 10^3\ Pa = 6.895 \times 10^{-3}\ MPa$
Energy or work	$m \cdot N$ (Joule)	$1\ ft \cdot lb = 1.356\ m \cdot N$ (Joules)
Coefficient of thermal expansion	$m/m/°C$	$1\ in./in./°F = 1.8\ m/m/°C$

A.3 SI UNIT PREFIXES

Multiplication Factor	Prefix	Symbol	Pronunciation (U.S.)	Meaning (U.S.)
$1\ 000\ 000\ 000 = 10^9$	giga	G	jig'a (*a* as in *a*bout)	One billion times
$1\ 000\ 000 = 10^6$	mega	M	as in *mega*phone	One million times
$1\ 000 = 10^3$	kilo	k	as in *kilo*watt	One thousand times
$100 = 10^2$	hecto	h	heck'toe	One hundred times
$10 = 10^1$	deka	da	deck'a (*a* as in *a*bout)	Ten times
$0.1 = 10^{-1}$	deci	d	as in *deci*mal	One tenth of
$0.01 = 10^{-2}$	centi	c	as in *senti*ment	One hundredth of
$0.001 = 10^{-3}$	milli	m	as in *mili*tary	One thousandth of
$0.000\ 001 = 10^{-6}$	micro	μ	as in *micro*phone	One millionth of
$0.000\ 000\ 001 = 10^{-9}$	nano	n	nan'oh (*an* as in *an*t)	One billionth of

Appendix B: Selected References

Al-Khafaji, A. W. and Tooley, J. R. *Numerical Methods in Engineering Practice*. Holt, Rinehart, and Winston, Inc., New York, 1986.

American Institute of Steel Construction, Inc. *Manual of Steel Construction*, AISC, Chicago, IL, 1980.

Arges, K. P. and Palmer, A. E. *Mechanics of Materials*. McGraw-Hill Book Co., New York, NY, 1963.

ASME. *Journal of Applied Mechanics*. 1934, 1935, 1936, and 1951.

Au, T. *Elementary Structural Mechanics*. Prentice-Hall, Inc., Englewood Cliffs, NJ, 1963.

Bauld, R. *Mechanics of Materials*. Brooks/Cole Publishing Co., Engineering Division, Monterey, CA, 1982.

Beedle, L. S. *Plastic Design of Steel Frames*. John Wiley and Sons, Inc., New York, NY, 1958.

Beer, F. P. and Johnston, E. R. *Vector Mechanics for Engineers: Statics and Dynamics*. McGraw-Hill Book Co., New York, NY, 1977.

Beer, F. P., Johnston, E. R., Dewolf, J., and Mazurek, D. *Mechanics of Materials*. McGraw-Hill Book Co., New York, NY, 2014.

Boresi, A. P., Sidebottom, O., Seely, F. B., and Smith, J. O. *Advanced Mechanics of Materials*. John Wiley and Sons, Inc., New York, NY, 1978.

Burgreen, D. *Elementary Thermal Stress Analysis*. C. P. Press, Jamaica, NY, 1971.

Byars, E. F. and Snyder, R. D. *Engineering Mechanics of Deformable Bodies*. Internal Textbook Co., Scranton, PA, 1969.

Chajes, A. *Principles of Structural Stability Theory*. Prentice-Hall, Inc., Englewood Cliffs, NJ, 1974.

Chou, P. C. and Pegano, N.. *Elasticity*. Van Nostrand Reinhold Co., New York, NY, 1967.

Crandall, S. H., Dahl, N. C., and Lardner, T. J., eds. *An Introduction to the Mechanics of Solids*. McGraw-Hill Book Co., New York, NY, 1978.

Den Hartog, J. P. *Advanced Strength of Materials*. McGraw-Hill Book Co., New York, NY, 1952.

Den Hartog, J. P. *Strength of Materials*. Dover Publications, Inc., New York, NY, 1961.

Drucker, D. C. *Introduction to Mechanics of Deformable Solids*. McGraw-Hill Book Co., New York, NY, 1967.

Dugdale, D. S. and Ruiz, D. *Elasticity for Engineers*. McGraw-Hill Book Co. Ltd., London, UK, 1971.

Eisenstadt, M. M. *Introduction to Mechanical Properties of Materials*. Macmillan Publishing Co., New York, NY, 1971.

Ferguson, P. M. *Reinforced Concrete Fundamentals*. John Wiley and Sons, Inc., New York, NY, 1973.

Fong, T. J., ed. *Fatigue Mechanisms*. ASTM STP 675, ASTM, Philadelphia, 1979.

Ford, H. *Advanced Mechanics of Materials*. John Wiley and Sons, Inc., New York, NY, 1963.

Freudenthal, A. M. *Introduction to Mechanics of Solids*. John Wiley and Sons, Inc., New York, NY, 1966.

Frocht, M. M. Photoelastic studies in stress concentration, *Mechanical Engineering*, August, 485–489, 1936.

Frocht, M. M. *Photo-Elasticity*. John Wiley and Sons, Inc., New York, NY, 1941.

Frye, C. *Microsoft Excel 2010 Step by Step*. Microsoft Press, Redmond, WA, 2010.

Fuchs, H. O. *Metal Fatigue in Engineering*. John Wiley and Sons, Inc., New York, NY, 1980.

Gere, J. M. and Goodno, B. J. *Mechanics of Materials*. Cengage Learning, Samford, CT, 2012.

Ginsberg, J. H. and Jenin, J. *Statics and Dynamics*. John Wiley and Sons, Inc., New York, NY, 1977.

Gol'denblat, I. I. *Some Principles of the Mechanics of Deformable Media*. P. Noordhoff, Gronigen, Holland, 1962.

Hibbeler, R. C. *Engineering Mechanics: Statics*. Prentice-Hall, Upper Saddle River, NJ, 2012.

Hibbeler, R. C. *Mechanics of Materials*. Prentice-Hall, Upper Saddle River, NJ, 2014.

Higdon, A., Ohlsen, E. H., Stiles, W. B., Weese, J. A., and Riley, W. F. *Mechanics of Materials*. John Wiley and Sons, Inc., New York, NY, 1976.

Hoadley, A. *Essentials of Structural Design*. John Wiley and Sons, Inc., New York, NY, 1964.

Holman, J. P., *What Every Engineer Should Know about Excel*. CRC Press, Boca Raton, FL, 2006.

Ilyushin, A. A. and Lenskii, V. S. *Strength of Materials*. Pergamon Press, Inc., Elmsford, NY, 1967.

Jacobsen, L. S. Torsional stress concentrations in shafts of circular cross section and variable diameter, *Trans ASME*, 47, 619–641, 1925.

Jones, R. M. *Mechanics of Composite Materials*. Scripta Book Co., Washington, DC, 1975.

Levinson, J. *Mechanics of Materials*. Prentice-Hall, Inc., Englewood Cliffs, NJ, 1970.

Lin, T. H. *Theory of Inelastic Structures*. John Wiley and Sons, Inc., New York, NY, 1968.

Lubahn, J. D. and Felgar, R. P. *Plasticity and Creep of Metals*. John Wiley and Sons, Inc., New York, NY, 1961.

Luxmoore, A. R. and Owen, D. R. J., eds. *Numerical Methods in Fracture Mechanics.* University College Swansea, West Glamorgan, U.K., 1978.

McCormac, J. C. *Structural Analysis.* Intext Educational Publishers, New York, NY, 1975.

Meriam, J. T. *Statics—SI Version.* John Wiley and Sons, Inc. New York, NY, 1975.

Miller, F. E. and Doeringsfeld, H. A. *Mechanics of Materials.* International Textbook Co., Scranton, PA, 1966.

Mott, R. L. *Applied Strength of Materials.* Prentice-Hall, Inc., Englewood Cliffs, NJ, 1978.

Neville, A. M. *Properties of Concrete.* Pitman Publishing Pty Ltd., Carlton, Victoria, Australia, 1973.

Nichols, R. *Composite Construction Materials Handbook.* Prentice-Hall, Inc., Englewood Cliffs, NJ, 1976.

Ogibalov, P. M., Malinin, N. I., Netrebko, V. P., and Kishkin, B. P. *Structural Polymers, Testing Methods, Vols. I and II.* John Wiley and Sons, Inc., New York, NY, 1974.

Penton's controls & systems, *Machine Design,* ISSN:0024-9114, Reference Issues: 1975, 1976, and 1977.

Polakowski, N. H. and Ripling, E. *Strength and Structure of Engineering Materials.* Prentice-Hall, Inc., Englewood Cliffs, NJ, 1966.

Popov, E. P. *Mechanics of Materials (SI Version).* Prentice-Hall, Inc., Englewood Cliffs, NY, 1978.

Popov, E. P. *Engineering Mechanics of Solids.* Prentice-Hall, Inc., Englewood Cliffs, NY, 1990.

Radon, J. C., ed. *Fracture and Fatigue.* Pergamon Press, Inc., Elmsford, NY, 1980.

Roark, R. J. *Formulas for Stress and Strain.* McGraw-Hill Book Co., New York, NY, 1965.

Sandor, B. I. *Strength of Materials.* Prentice-Hall, Inc., Englewood Cliffs, NJ, 1978.

Seely, F. B. and Smith, J. O. *Resistance of Materials.* John Wiley and Sons, Inc., New York, NY, 1956.

Shames, I. H. *Mechanics of Deformable Solids.* Prentice-Hall, Inc., Englewood Cliffs, NY, 1964.

Shanley, F. R. *Strength of Materials.* McGraw-Hill Book Co., New York, NY, 1957.

Shigley, J. E. *Mechanical Engineering Design.* McGraw-Hill Book Co., New York, NY, 1972.

Sloane, A. *Mechanics of Materials.* Macmillan Publishing Co., New York, NY, 1952.

Smith, J. O. and Sidebottom, O. M. *Inelastic Behavior of Load-Carrying Members.* John Wiley and Sons, Inc., New York, NY, 1965.

Snyder, R. D. and Byars, E. F. *Engineering Mechanics: Statics and Strength of Materials.* McGraw-Hill Book Co., New York, NY, 1973.

Timoshenko, S. P. *Theory of Elastic Stability.* McGraw-Hill Book Co., New York, NY, 1936.

Timoshenko, S. P. *History of Strength of Materials.* McGraw-Hill Book Co., New York, NY, 1953.

Timoshenko, S. P. and Gere, J. M. *Mechanics of Materials.* Van Nostrand Reinhold Co., New York, NY, 1972.

Timoshenko, S. P. and Goodier, J. N. *Theory of Elasticity.* McGraw-Hill Book Co., New York, NY, 1970.

Ugural, A. C. and Fenster, S. K. *Advanced Strength and Applied Elasticity.* Elsevier Scientific Publishing Co., Inc., New York, NY, 1981.

Van Vlack, L. H. *Materials for Engineering.* Addison-Wesley Publishing Co., Inc., Reading, MA, 1982.

Volterra, E. and Gaines, J. H. *Advanced Strength of Materials.* Prentice-Hall, Inc., Englewood Cliffs, NJ, 1971.

Walkenbach, J. *Excel 2010 Bible.* Wiley Publishing, Inc., Indianapolis, IN, 2010.

White, R. N., Gergely, P., and Sexsmith, R. G. *Structural Engineering.* John Wiley and Sons, Inc., New York, NY, 1972.

Appendix C: Properties of Plane Areas

The evaluation of centroids, moments of inertia, and other cross-sectional properties are typically covered in introductory mechanics courses, such as statics. This section of the textbook offers a review of the subject matter and the cross-sectional properties of selected plane areas are provided in Appendix C.3.

C.1 CENTROID LOCATION

C.1.1 CENTROID OF AN AREA

In order to develop the methods used to locate the centroid of an area, we need to consider first the property of a *rigid* body known as the *center of gravity.*

A *rigid* body may be defined as a collection of particles whose distances remain fixed with respect to each other. Let the differential weight of each particle be dW, which represents the pull of gravity on the particle and, therefore, must be directed downward (i.e., along the negative z axis) toward the center of the Earth as shown in Figure C.1a. Of course, all of the forces dW are parallel to each other and the total weight W of the body is the resultant of all of the forces dW in the body (i.e., $W = \int dW$) and acts through point G, the *center of gravity* of the body. The coordinates of point G are $\bar{x}, \bar{y}$ (shown in Figure C.1a) and $\bar{z}$ (shown in Figure C.1b). Note that Figure C.1b is obtained from Figure C.1a, by rotating the coordinate system through 90° about the y axis.

To determine the coordinates $\bar{x}, \bar{y}$, and $\bar{z}$, we make use of the fact that the moment of the resultant force $W = \int dW$ about a given coordinate axis is equal to the sum of the moments of all the differential forces dW about the same axis. Thus, referring to Figure C.1a, we obtain $\bar{x}$ and $\bar{y}$ by writing

$$\bar{x} \int dW = \int x \, dW \Rightarrow \bar{x} = \frac{\int x \, dW}{\int dW} \tag{C.1}$$

$$\bar{y} \int dW = \int y \, dW \Rightarrow \bar{y} = \frac{\int y \, dW}{\int dW} \tag{C.2}$$

Referring now to Figure C.1b, we obtain $\bar{z}$ by writing

$$\bar{z} \int dW = \int z \, dW \Rightarrow \bar{z} = \frac{\int z \, dW}{\int dW} \tag{C.3}$$

Consider now a very thin plate, as shown in Figure C.2, which has a constant thickness and a constant weight density, γ, expressed in terms of *weight per unit area*. Therefore, a differential area dA of this plate has a differential weight $dW = \gamma \, dA$. If the product $\gamma \, dA$ is substituted for dW

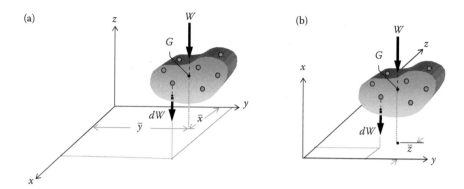

FIGURE C.1

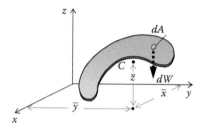

FIGURE C.2

in Equations C.1 and C.2, we find that the constant γ may be dropped and the resulting equations become

$$\bar{x} = \frac{\int x\,dA}{\int dA} \tag{C.4}$$

$$\bar{y} = \frac{\int y\,dA}{\int dA} \tag{C.5}$$

$$\bar{z} = \frac{\int z\,dA}{\int dA} \tag{C.6}$$

Equations C.4 through C.6 locate a point in space, inside or outside the area defined by the contour of the thin plate, known as the centroid C of this area. The centroid of an area is a purely geometric property and may be thought of as the point where we can assume the entire area to be concentrated if it were possible to shrink the area into a single point. In view of the fact that a plane area is a two-dimensional quantity, it is usually placed in a plane defined by a two-dimensional coordinate system say, the y–z system. In such a case, Equation C.4 yields $\bar{x} = 0$ and the centroid of the area is fully defined by finding the coordinates $\bar{y}$ and $\bar{z}$ using Equations C.5 and C.6. We will restate these two equations in a form that is more convenient for determining the centroid of an area. Thus,

$$\bar{y} = \frac{\int y\,dA}{\int dA} = \frac{\int y_e\,dA}{A} \tag{C.7}$$

$$\bar{z} = \frac{\int z\,dA}{\int dA} = \frac{\int z_e\,dA}{A} \tag{C.8}$$

The quantities y_e and z_e in Equations C.7 and C.8, respectively, signify the y and z coordinates, respectively, of the centroid of the element of area dA. Note that $\int dA$ was replaced by its equivalent value A.

Generally speaking, it is possible to avoid double integration, thus simplifying the process, by selecting the differential element of area as a thin rectangle parallel either to the y or to the z axis. Depending upon the function defining the given area, it is sometimes possible to determine both centroid coordinates using the same differential element of area. Some of these ideas are illustrated in Example C.1.

Because of the mathematical similarity between the moment of a force about the z axis given by $\int y\,dW$ and the integral $\int y\,dA$ in Equation C.7, this latter integral is referred to as the *first moment of area* about the z axis. Also, for a similar reason, the integral $\int z\,dA$ is known as the *first moment of area* about the y axis. In Appendix C.2, we will deal with integrals of the form $\int x^2\,dA$ and $\int y^2\,dA$, referred to as *second moments of areas* and known as *moments of inertia of areas*.

C.1.2 CENTROID OF A COMPOSITE AREA

A composite area is defined as one consisting of two or more simple areas. Consider, for example, the area shown in Figure C.3a. This area may be thought as a composite area consisting of the three rectangles A_1, A_2, and A_3 as shown in Figure C.3b, which also shows the corresponding centroids C_1, C_2, and C_3 of the three individual rectangles. We may determine the distances $\bar{y}$ and $\bar{z}$ locating the centroid of the composite area from the coordinates of the centroids C_1, C_2, and C_3 of the individual rectangles. For example, the principle of moments applied with respect to the z axis states that the moment of the total area $\Sigma A = A_1 + A_2 + A_3$ about the z axis must be equal to the sum of the moments produced by the three individual areas about the same axis. We may, therefore, write

$$\left(\sum A \right)\bar{y} = A_1 y_1 + A_2 y_2 + A_3 y_3 = \sum Ay$$

(a)   (b)

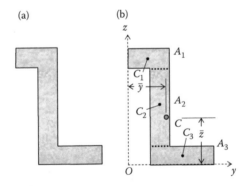

FIGURE C.3

Solving for $\bar{y}$, we obtain

$$\bar{y} = \frac{\sum Ay}{\sum A} \qquad \qquad \textbf{(C.9)}$$

Similarly, we may show that

$$\bar{z} = \frac{\sum Az}{\sum A} \qquad \qquad \textbf{(C.10)}$$

Note that the composite area in Figure C.3a could be subdivided in other ways different from that shown in Figure C.3b. However, the location of its centroid as obtained from Equations C.9 and C.10 remains the same. Also, while only three component areas were used in this illustration, it should be pointed out that the sums indicated in Equations C.9 and C.10 are valid for any finite number of component areas. Furthermore, these sums are algebraic and can accommodate not only additions but subtractions as well.

EXAMPLE C.1

Determine the coordinates $\bar{y}$ and $\bar{z}$ for the parabolic area shown in below Figure (a).

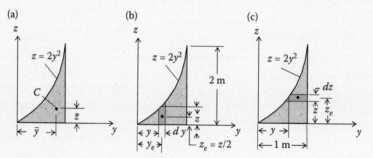

SOLUTION

To determine the dimension $\bar{y}$, we think of the parabolic area as a composite area consisting of an infinite number of differential vertical strips of size $z\,dy$ as shown in above Figure (b). Each differential element of area has the following properties: $dA = z\,dy = 2y^2\,dy$; $y_e = y$. Note that z was replaced by $2y^2$ in order to express everything in terms of the single variable y. Equation C.7 is most conveniently applied by evaluating two separate integrals. Thus,

$$\int y_e dA = \int_0^1 y(2y^2 dy) = \frac{1}{2}\,\text{m}^3; \quad A = \int dA = \int_0^1 2y^2 dy = \frac{2}{3}\,\text{m}^2$$

Applying Equation C.7, we obtain

$$\bar{y} = \frac{\displaystyle\int y dA}{A} = \frac{(1/2)}{(2/3)} = \frac{3}{4}\,\text{m} \qquad \textbf{ANS.}$$

Similarly, to find $\bar{z}$, we assume the entire area to consist of an infinite number of differential horizontal strips of area $dA = (1-y)dz = \{1-[(1/2)z]^{1/2}\}dz$ with $z_e = z$, as shown in Figure (c)
in previous page. Evaluating the upper quantity in Equation C.8, we obtain

$$\int z_e dA = \int z\left[1-\left(\frac{1}{2}z\right)^{1/2}\right]dz = \left[\frac{z^2}{2} - \frac{2}{5}\left(\frac{1}{2}\right)^{1/2} z^{5/2}\right]_0^2 = \frac{4}{10}\,m^3$$

Using the value $A = 2/3$ m^2 as obtained above and applying Equation C.8, we obtain

$$\bar{z} = \frac{\int z_e dA}{A} = \frac{(4/10)}{(2/3)} = \frac{6}{10}\,m \qquad \textbf{ANS.}$$

It is important to note that the coordinate $\bar{z}$ can also be obtained using the differential vertical strip of Figure (b) in previous page. In such a case, $dA = z\,dy = 2y^2\,dy$. However, as shown in
Figure (b) in previous page, $z_e = 1/2z$. Evaluating the upper integral in Equation C.8, we obtain

$$\int z_e dA = \int_0^1 y^2(2y^2 dy) = 2\int_0^1 y^4 dy = \left[\frac{2}{5}y^5\right]_0^1 = \frac{2}{5} = \frac{4}{10}\,m^3$$

This value is, of course, identical with the value obtained above using the differential horizontal strip of Figure (c) in previous page. Therefore, when applying Equation C.8 using the
area $A = 2/3$ m^2 found earlier, we obtain

$$\bar{z} = \frac{6}{10}\,m \qquad \textbf{ANS.}$$

This value of $\bar{z}$ is the same as obtained earlier on the basis of the differential horizontal
strip shown in Figure (c) in previous page. We should note here that the differential horizontal
strip shown in Figure (c) in previous page could also be used to determine the quantity $\bar{y}$.
It is left as an exercise for the student to show that this type of solution leads to the same value
for $\bar{y}$ obtained above.

EXAMPLE C.2

Find the coordinates $\bar{y}$ and $\bar{z}$ for the composite area shown in below Figure (a).

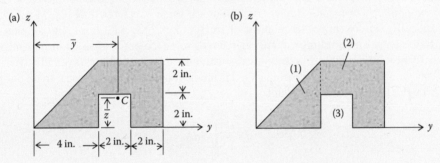

SOLUTION

The given composite area is subdivided into three simple geometric areas as shown in Figure (b) in previous page, where area (1) is a right-angle triangle, area (2) is a solid square, and area (3) is a hollow square. To find the coordinates $\bar{y}$ and $\bar{z}$, we make use of Equations C.9 and C.10. Thus,

$$\bar{y} = \frac{\sum Ay}{\sum A} = \frac{A_1 y_1 + A_2 y_2 - A_3 y_3}{A_1 + A_2 - A_3}$$

$$= \frac{((1/2)4 \times 4)((2/3)4) + (4 \times 4)((2/3)4 + 2) - ((2 \times 2)((2/3)4 + 1)}{(1/2)4 \times 4 + 4 \times 4 - 2 \times 2}$$

$$= 4.733 \text{ in.} \quad \textbf{ANS.}$$

$$\bar{z} = \frac{\sum Az}{\sum A} = \frac{A_1 z_1 + A_2 z_2 - A_3 z_3}{A_1 + A_2 - A_3}$$

$$= \frac{((1/2)4 \times 4)((2/3)4) + (4 \times 4)(2) - (2 \times 2)(1)}{(1/2)4 \times 4 + 4 \times 4 - 2 \times 2}$$

$$= 2.467 \text{ in.} \quad \textbf{ANS.}$$

C.2   MOMENT OF INERTIA OF AREA: PRINCIPAL AXES

C.2.1   DEFINITIONS

Consider the arbitrary area shown in Figure C.4. The *rectangular moments of inertia*, or the *second moments*, of this area with respect to the y and z axes, in the same plane as the area, are defined, respectively, by

$$I_y = \int z^2 dA \tag{C.11}$$

$$I_z = \int y^2 dA \tag{C.12}$$

In the future, we will use exclusively the simplified term *moment of inertia* instead of *rectangular moment of inertia* or *second moment of area* unless there is a compelling reason not to.

Occasionally, we determine the moment of inertia of an area with respect to an axis perpendicular to the area. The resulting quantity is referred to as the *polar moment of inertia*, which is traditionally given the symbol J to distinguish it from the moments of inertia I_y and I_z. Referring to the area shown in Figure C.4, we define its polar moment of inertia with respect to an out-of-plane x axis through point O by the expression

$$J_O = \int r^2 dA \tag{C.13}$$

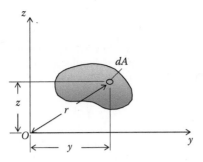

FIGURE C.4

Since $r^2 = z^2 + y^2$, the quantity J_O may be expressed in terms of I_y and I_z. Thus,

$$J_O = \int r^2 dA = \int (z^2 + y^2)dA = \int z^2 dA + \int y^2 dA$$
$$J_O = I_y + I_z \tag{C.14}$$

As Equation C.14 indicates, the polar moment of inertia of an area, with respect to an out-of-plane axis through any point, is equal to the sum of the moments of inertia with respect to any two orthogonal in-plane axes intersecting at the same point. Also, from the mathematical definition of the area moment of inertia, we conclude that it is always a positive quantity and that its unit is a length raised to the fourth power. Therefore, in the U.S. Customary system, units such as in.^{4} and ft^4 are common, and in the SI system, the most commonly used units are mm^4 and m^4.

Another quantity that is very useful in the study of mechanics is the *radius of gyration*. Let us imagine that an area A can be concentrated at a single point, as shown in Figure C.5. The perpendicular distances k_y and k_z of this point from the coordinate axes, known as the *radii of gyration* of the area A, are chosen to satisfy the following relationships:

$$I_y = k_y^2 A; \quad k_y = \sqrt{\frac{I_y}{A}} \tag{C.15}$$

$$I_z = k_z^2 A; \quad k_z = \sqrt{\frac{I_z}{A}} \tag{C.16}$$

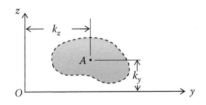

FIGURE C.5

The quantities I_y and I_z in Equations C.15 and C.16 are the moments of inertia of the area with respect to the y and z axes, respectively. A similar expression may be written for the polar radius of gyration of the area with respect to any axis perpendicular to the plane of the area. Thus, with respect to an x axis through point O in Figure C.5, for example,

$$k_O = \sqrt{\frac{J_O}{A}} \tag{C.17}$$

As may be evident from Equations C.15 through C.17, the radius of gyration of area has a unit of length. Therefore, in the U.S. Customary system, such units as the in. and the ft are common, and in the SI system, the units commonly used are the mm and the m.

C.2.2 MOMENT OF INERTIA BY INTEGRATION

The area moment of inertia for a given geometric shape may be determined by direct integration using the fundamental definition given in Equations C.11 and C.12. The procedure consists of selecting an element of area dA and expressing it in terms of the chosen coordinate variables. Since a differential element of area is a two-dimensional quantity, its mathematical expression may contain two variables, which would lead to a double integration process. However, a careful selection of the differential element of area would generally permit finding the moment of inertia using a single integration. Example C.3 illustrates some of these concepts.

EXAMPLE C.3

Use integration to find the moments of inertia I_y and I_z of the shaded area shown. Determine also the radius of gyration of the area with respect to the z axis through point O.

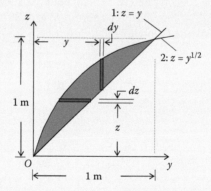

SOLUTION

To find the moment of inertia I_y, we establish a horizontal element of area $dA = (y_1 - y_2)$ $dz = (z - z^2)dz$ as shown. Thus,

$$I_y = \int z^2 dA = \int_0^1 (z^3 - z^4)dz = \left(\frac{z^4}{4} - \frac{z^5}{5} \right) \Bigg]_0^1$$

$$= 0.05 \text{ m}^4 \quad \textbf{ANS.}$$

To find the moment of inertia I_z, we construct a vertical element of area $dA = (z_2 - z_1)dy = (y^{1/2} - y)dy$ as shown. Thus,

$$I_z = \int y^2 dA = \int_0^1 (y^{5/2} - y^3)dy = \left(\frac{2}{7}y^{7/2} - \frac{1}{4}y^4\right]_0^1$$

$$= 0.036 \text{ m}^4 \quad \textbf{ANS.}$$

The radius of gyration with respect to the z axis through point O is given by Equation C.16. To find it, therefore, we need to find the area A. Using the horizontal element of area, we obtain

$$A = \int (y_1 - y_2)dz = \int_0^1 (z - z^2)dz = \left(\frac{1}{2}z^2 - \frac{1}{3}z^3\right]_0^1 = 0.1667 \text{ m}^2$$

Therefore,

$$k_z = \sqrt{\frac{I_z}{A}} = \sqrt{\frac{0.036}{0.1667}} = 0.465 \text{ m} \quad \textbf{ANS.}$$

C.2.3 PARALLEL-AXIS THEOREM

The parallel-axis theorem for the area moment of inertia is a very useful tool because it relates moments of inertia with respect to two parallel axes, one of which passes through the *centroid of the area*.

Let us consider the area shown in Figure C.6. The Y–Z coordinate system has its origin at the centroid C of this area. The y–z coordinate system has its origin at point O such that the y axis is at a distance b from the Y axis and parallel to it and the z axis is at a distance a from the Z axis and parallel to it. From the geometry of Figure C.6, we conclude that $y = Y + a$ and that $z = Z + b$. We now use the fundamental definition to write the moment of inertia of the area with respect to the y axis. Thus,

$$I_y = \int z^2 dA = \int (Z + b)^2 dA = \int Z^2 dA + b^2 \int dA$$

$$+ 2b \int Z dA \tag{C.18}$$

In the last form of Equation C.18, the first integral is I_y, the moment of inertia of the area with respect to its centroidal Y axis; the second integral is the area A and, therefore, the second term is the product

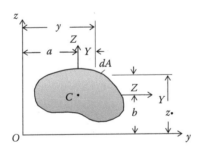

FIGURE C.6

Ab^2; by Equation C.7, the third integral is equal to the product $A\bar{Z}$, where $\bar{Z} = 0$ because it is measured from a centroidal Y axis and, therefore, the third term vanishes. Thus, Equation C.18 may be written as

$$I_y = I_Y + Ab^2 \tag{C.19}$$

Similarly,

$$I_z = I_Z + Aa^2 \tag{C.20}$$

Each of the two relations given in Equation C.19 or C.20 expresses the fact that *the moment of inertia of an area with respect to any axis is equal to the moment of inertia of the area with respect to a parallel centroidal axis plus the product of the area and the square of the distance between the two axes.* This important relationship is known as the parallel-axis theorem and may be expressed in the following general form:

$$I = I_C + Ad^2 \tag{C.21}$$

In Equation C.21, I represents the area moment of inertia with respect to any axis in the plane of the area A, I_C the moment of inertia with respect to an in-plane parallel centroidal axis, and d the distance between the two axes.

Using the same process that led to Equation C.21, we can develop a parallel-axis theorem for polar moments of inertia. Thus,

$$J = J_C + Ad^2 \tag{C.22}$$

In Equation C.22, J represents the polar moment of inertia with respect to any axis perpendicular to the area A, J_C the polar moment of inertia with respect to a parallel centroidal axis, and d the distance between the two axes.

Examination of Equations C.21 and C.22 reveals that the moments of inertia assume their least values with respect to centroidal axes. This conclusion is significant in discussing the theories underlying the bending of beams and the buckling of columns.

C.2.4 MOMENT OF INERTIA OF COMPOSITE AREAS

A composite area is one that may be broken down into two or more simple geometric shapes. The procedure used for finding the moment of inertia of such an area is based upon the fundamental definitions given in Equations C.11 and C.13. Consider, for example, Equation C.11, $I_y = \int z^2\, dA$. The integrand, $z^2 dA$, represents the differential quantity dI_y, the moment of inertia of the element of area dA with respect to the y axis. Therefore, $I_y = \int dI_y$. For a composite area consisting of a finite number of components, the integration is replaced by a summation and a general expression may be written as

$$I = \int dI \Rightarrow \sum_{i=1}^{n} I_i \tag{C.23}$$

In Equation C.23, I represents the moment of inertia of the composite area with respect to some axes and I_i the moment of inertia with respect to the same axis of the ith component of the composite area. In other words, Equation C.23 states that the moment of inertia of a composite area is equal to the algebraic sum of the moments of inertia of the component areas about the same axis. In finding the moment of inertia of a component area with respect to the desired axis, it becomes necessary to use the parallel-axis theorem as expressed in Equation C.21. Therefore, Equation C.23 may be rewritten in the following more useable form:

$$I = \sum_{i=1}^{n} (I_C + Ad^2)_i \qquad \text{(C.24)}$$

In applying Equation C.24, it becomes desirable to have access to the area moments of inertia for simple geometric shapes. A set of area properties, including moments of inertia for a selected group of simple geometric shapes is given in Appendix C.3.

EXAMPLE C.4

1. Determine the moments of inertia I_Y and I_Z for the Z section where Y and Z are centroidal axes.
2. Use the answers obtained in part (a) and the parallel-axis theorem to find the moments of inertia I_y and I_z where the y and z axes have their origin at point O as shown.

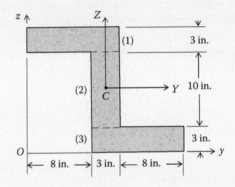

SOLUTION

1. We divide the Z section into three rectangular areas as shown, where rectangle (1) is the upper horizontal one, rectangle (2) is the vertical one and rectangle (3) is the lower horizontal rectangle. Note that rectangles (1) and (3) are identical from the standpoint of their contribution to the moment of inertia and will be so treated in the computations. By Equation C.24, we have

$$I_Y = \sum_{i=1}^{3} (I_C + Ad^2)_{iY} = 2(I_C + Ad^2)_{1Y} + (I_C + Ad^2)_{2Y}$$

$$= 2\left[\left(\frac{1}{12}\right)(11)(3)^3 + 33(6.5)^2\right] + \frac{1}{12}(3)(10)^3 = 3088 \text{ in.}^4 \qquad \textbf{ANS.}$$

$$I_Z = \sum_{i=1}^{3} (I_C + Ad^2)_{iZ} = 2(I_C + Ad^2)_{1Z} + (I_C + Ad^2)_{2Z}$$

$$= 2\left[\left(\frac{1}{12}\right)(3)(11)^3 + 33(4.0)^2\right] + \frac{1}{12}(10)(3)^3 = 1744.0 \text{ in.}^4 \qquad \textbf{ANS.}$$

Note that the centroidal moment of inertia for a single rectangle was obtained by consulting Appendix C.3.

2. The parallel-axis theorem is now applied to find the moments of inertia I_y and I_z. Thus, by Equation C.21,

$$I_y = (I_C + Ad^2)_y = 3088 + 96(8)^2 = 9232 \text{ in.}^4 \qquad \textbf{ANS.}$$

$$I_z = (I_C + Ad^2)_z = 1.744.0 + 96(9.5)^2 = 10408.0 \text{ in.}^4 \qquad \textbf{ANS.}$$

C.2.5 PRODUCT OF INERTIA

In the solution of many structural problems, it becomes necessary to find the so-called *principal axes of inertia* for a given cross-sectional area. Every cross-sectional area possesses two in-plane *principal axes of inertia*; with respect to one, the moment of inertia is a maximum, and with respect to the other, the moment of inertia is a minimum. These two moments of inertia are known as the *principal moments of inertia* of the area. As we will see shortly, the mathematical relationships that are involved in finding the principal axes and principal moments of inertia contain the integral $\int yz \, dA$, which carries the name *product of inertia* and is given the symbol I_{yz}. Thus, by definition

$$I_{yz} = \int yz \, dA \qquad \textbf{(C.25)}$$

Equation C.25 expresses a mathematical property of an area with respect to the specific y–z coordinate system as illustrated in Figure C.7. Examination of Equation C.25 together with Figure C.7 reveals that the magnitude and sign of the product of inertia for a given area depend upon the location of the perpendicular axes relative to the area. Thus, unlike the moments of inertia I_y and I_z, which can have only positive values, the product of inertia I_{yz} may have positive, zero, or negative values. As may be seen from the mathematical definition (see Equation C.25), the product of inertia, like the moments of inertia, has units of length raised to the fourth power. Therefore, the same units used for moments of inertia apply equally well for products of inertia in both the U.S. Customary or SI systems.

The mathematical definition for product of inertia given in Equation C.25 leads us to conclude that *if either y or z is an axis of symmetry for the area, the product of inertia I_{yz} vanishes.* This conclusion follows from the fact that for every element of area dA on one side of the axis of symmetry,

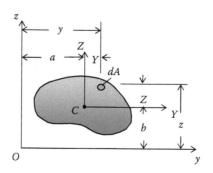

FIGURE C.7

there is an equal element of area on the other side, canceling its effect. As will be seen shortly, this property is very useful in finding the products of inertia of composite areas.

A parallel-axis theorem for products of inertia may be developed using Figure C.7. Thus,

$$I_{yz} = \int yzdA = \int (Y + a)(Z + b)dA = \int YZdA + b\int YdA + a\int ZdA + ab\int dA \qquad (C.26)$$

The first integral on the far right-hand side of Equation C.26 is I_{YZ}, the second and third integrals represent first moments of areas with respect to centroidal axes and, therefore, must each be zero, and the fourth integral is, of course the area A. Therefore, Equation C.26 reduces to

$$I_{yz} = (I_{YZ})_C + A(ab) \qquad (C.27)$$

The quantity $(I_{YZ})_C$ in Equation C.27 represents the product of inertia with respect to centroidal Y and Z axes. Equation C.27 expresses the parallel axis theorem for products of inertia. This theorem, when combined with the property of zero product of inertia corresponding to axes of symmetry, is a very useful tool to find the product of inertia of simple and composite areas.

EXAMPLE C.5

1. Find the product of inertia I_{yz} of the shaded area shown.
2. Use the results obtained in part (a) along with the parallel-axis theorem to find the product of inertia I_{YZ}.

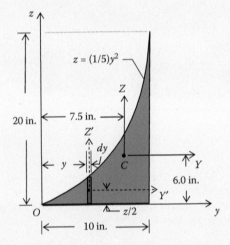

SOLUTION

1. While either a horizontal or a vertical differential element of area may be used, a vertical one was chosen of area $dA = z \, dy = (1/5)y^2 \, dy$. The product of inertia I_{yz} for the entire area may be found by summing the differential products of inertia dI_{yz} for the infinite number of vertical differential elements of areas dA in the entire shaded area. In other words $I_{yz} = \int dI_{yz}$, where dI_{yz} is found by the parallel-axis theorem. Thus, adapting Equation C.27 to the vertical differential area dA, we have

$$dI_{yz} = dI_{YZ'} + dA(ab) \qquad (C.5.1)$$

In Equation C.5.1, the quantity $dI_{Y'Z'} = 0$ because it is the product of inertia with respect to axes of symmetry. Also, the quantities a and b locating the centroid of the element of area dA from the y–z coordinate system are $a = y$ and $b = z/2 = (1/10)y^2$. Therefore, Equation C.5.1 becomes

$$dI_{yz} = 0 + \left(\frac{1}{5}\right)y^2 dy(y)\left(\frac{1}{10}y^2\right) = \frac{1}{50}y^5 dy$$

Integrating, we obtain

$$I_{yz} = \int dI_{yz} = \int_0^{10} \frac{1}{50}y^5 dy = \left(\frac{1}{300}y^6\right]_0^{10} = 3333.3 \text{ in.}^4 \qquad \textbf{ANS.}$$

2. Using the parallel-axis theorem for product of inertia given in Equation C.27, we conclude that

$$I_{YZ} = I_{yz} - A(ab) \qquad (C.5.2)$$

In Equation C.5.2, $a = 7.5$ in. and $b = 6.0$ in. as shown in the above sketch and

$$A = \int z\,dy = \int_0^{10} \frac{1}{5}y^2 dy = \frac{1}{15}\left(y^3\right]_0^{10} = 66.667 \text{ in.}^2$$

Therefore, Equation C.5.2 becomes

$$I_{YZ} = 3333.3 - 66.667(7.5)(6.0)$$

$$I_{YZ} = 333.3 \text{ in.}^4 \qquad \textbf{ANS.}$$

EXAMPLE C.6

Find the centroidal product of inertia I_{YZ} for the Z section of Example C.4, which, for convenience, is repeated in the sketch shown.

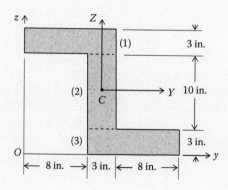

SOLUTION

As was done in Example C.4, we divide the Z section into three rectangular areas as shown, where rectangle (1) is the upper horizontal one, rectangle (2) is the vertical one and rectangle (3) is the lower horizontal rectangle. Note that rectangles (1) and (3) are identical from the standpoint of their contribution to the product of inertia and will be so treated in the computations. By Equation C.27, we have

$$I_{YZ} = 2(I_{YZ})_1 + (I_{YZ})_2 = 2\left[(I_{YZ})_C + A(a)(b)\right]_1 + (I_{YZ})_2$$
$$= 2\left[0 + 33(-4.0)(6.5)\right] + 0 \approx -1,716.0 \text{ in.}^4 \qquad \textbf{ANS.}$$

C.2.6   PRINCIPAL AXES AND PRINCIPAL MOMENTS OF INERTIA

As stated earlier, every cross-sectional area possesses two in-plane *principal axes of inertia*; with respect to one, the moment of inertia is a maximum, and with respect to the other, the moment of inertia is a minimum. These two moments of inertia are known as the *principal moments of inertia* of the area. Finding the principal axes and principal moments of inertia for a given area may be accomplished by first expressing its moment of inertia with respect to an arbitrary axis.

Consider the moment of inertia of the area shown in Figure C.8 with respect to the arbitrary n axis, which is inclined to the y axis through the positive counterclockwise angle θ. Note that the t axis, normal to the n axis, is inclined to the z axis through the same angle θ. Examination of the geometry in Figure C.8 leads us to conclude that the variables t and n may be expressed in terms of the variables y and z by the following relationships:

$$t = z\cos\theta - y\sin\theta; \quad n = y\cos\theta + z\sin\theta \qquad (C.28)$$

By definition (see Equations C.11 and C.12), the moment of inertia of the area about the n axis is

$$I_n = \int t^2 dA = \int (z\cos\theta - y\sin\theta)^2 dA$$
$$= \cos^2\theta \int z^2 dA + \sin^2\theta \int y^2 dA - 2\sin\theta\cos\theta \int yz dA \qquad (C.29)$$

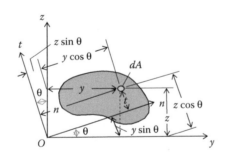

FIGURE C.8

Examination of the far right-hand side of Equation C.29 reveals that the first integral is I_y, the second is I_z, and the third is I_{yz}. Therefore, we may restate Equation C.29 as follows:

$$I_n = I_y \cos^2 \theta + I_z \sin^2 \theta - I_{yz}(2 \sin \theta \cos \theta) \qquad \text{(C.30)}$$

Using the trigonometric identities $\sin^2 \theta = (1 - \cos 2\theta)/2$, $\cos^2 \theta = (1 + \cos 2\theta)/2$ and $(2 \sin \theta \cos \theta) = \sin 2\theta$, we rewrite Equation C.30 in the form

$$I_n = \frac{1}{2}(I_y + I_z) + \frac{1}{2}(I_y - I_z)\cos 2\theta - I_{yz}\sin 2\theta \qquad \textbf{(C.31)}$$

The product of inertia I_{nt} is found in a manner similar to that used in developing Equation C.31. Thus,

$$
\begin{aligned}
I_{nt} &= \int nt\,dA = \int (z\cos\theta - y\sin\theta)(z\cos\theta + y\sin\theta)\,dA \\
&= \sin\theta\cos\theta\left[\int z^2\,dA - \int y^2\,dA\right] + (\cos^2\theta - \sin^2\theta)\int yz\,dA \\
&= (I_y - I_z)\sin\theta\cos\theta + I_{yz}(\cos^2\theta - \sin^2\theta)
\end{aligned}
\qquad \text{(C.32)}
$$

Using the three trigonometric identities stated above, we can restate Equation C.32 in the form

$$I_{nt} = \frac{1}{2}(I_y - I_z)\sin 2\theta + I_{yz}\cos 2\theta \qquad \textbf{(C.33)}$$

Retracing the steps leading to Equation C.31, we may obtain the moment of inertia I_t. However, we may obtain this moment of inertia more directly by replacing θ in Equation C.31 by the quantity $(\theta + 90°)$. Doing this leads us to conclude that

$$I_t = \frac{1}{2}(I_y + I_z) - \frac{1}{2}(I_y - I_z)\cos 2\theta + I_{yz}\sin 2\theta \qquad \textbf{(C.34)}$$

We note that Equations C.31, C.33, and C.34 are mathematically similar, respectively, to the plane stress transformation Equations 7.6, 7.9, and 7.10. This property enables us to analyze moments and products of inertia in a manner similar to that used in analyzing the plane stress problem discussed in Chapter 7. We will, therefore, proceed here in a manner similar to that used in Chapter 7.

If we differentiate I_n in Equation C.31 with respect to θ and set the result equal to zero, we obtain the values of θ for which I_n is either a maximum or a minimum. Thus,

$$2\theta = \tan^{-1}\left[\frac{-I_{yz}}{1/2(I_y - I_z)}\right] \qquad \textbf{(C.35)}$$

Equation C.35 defines two values of 2θ differing by $180°$ or two values of θ differing by $90°$. Corresponding to one of these two value of θ, the moment of inertia I_n is a maximum and, corresponding to the second, the moment of inertia I_n is a minimum. These two special values of moments of inertia, denoted here by I_u (maximum) and I_v (minimum), are known as principal moments of inertia and the corresponding u and v axes are the principal axes of inertia of the area.

Without losing generality, we may assume that the magnitude of I_y is larger than that of I_z. In such a case, Equation C.35 defines two angles for which the tangents are negative. One of these two angles occurs in the second quadrant and the second angle in the fourth quadrant as shown in Figure C.9. The value of R shown in Figure C.9 is found by the Pythagorean theorem. Thus,

$$R = \sqrt{(1/2(I_y - I_z))^2 + I_{yz}^2} \tag{C.36}$$

Substituting the functions of $\sin\theta$ and $\cos\theta$ given in Figure C.9 for the fourth quadrant into Equation C.31, we obtain

$$I_u = I_{max} = \frac{1}{2}(I_y + I_z) + R \tag{C.37}$$

Replacing R in Equation C.37 by its equivalent value from Equation C.36, we obtain

$$I_u = I_{max} = \frac{1}{2}(I_y + I_z) + \sqrt{(1/2(I_y - I_z))^2 + I_{yz}^2} \tag{C.38}$$

Substituting the functions of $\sin\theta$ and $\cos\theta$ given in Figure C.9 for the second quadrant into Equation C.31, we obtain

$$I_v = I_{min} = \frac{1}{2}(I_y + I_z) - R \tag{C.39}$$

$$I_v = I_{min} = \frac{1}{2}(I_y + I_z) - \sqrt{(1/2(I_y - I_z))^2 + I_{yz}^2} \tag{C.40}$$

Note that when we substitute the $\sin\theta$ and $\cos\theta$ functions given in Figure C.9 for either the second or fourth quadrant into Equation C.33, we conclude that the product of inertia is zero. This, of course, means that the product of inertia I_{nt} is zero with respect to the principal axes of inertia u and v. In other words, $I_{uv} = 0$. Recalling that the product of inertia is zero with respect to a set of orthogonal axes if at least one of these two axes is an axis of symmetry, we conclude that *an axis of symmetry is a principal axis of inertia*. However, this does not mean that principal axes of inertia are necessarily axes of symmetry.

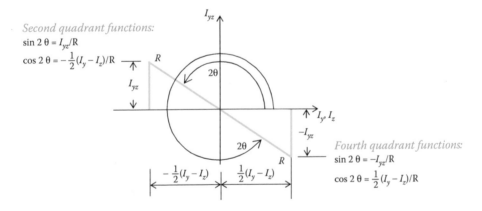

FIGURE C.9

EXAMPLE C.7

Use the results obtained in Examples C.4 and C.6 to find the principal centroidal axes and principal centroidal moments of inertia for the Z section shown. Note that neither of the two centroidal Y or Z axis is an axis of symmetry. It follows, therefore, that neither of these two axes is a principal axis of inertia.

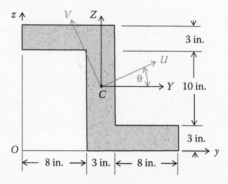

SOLUTION

For convenience, the needed results obtained in Examples C.4 and C.6 are summarized below:

$$I_Y = 3,088 \text{ in.}^4, \quad I_Z = 1,744 \text{ in.}^4, \quad \text{and} \quad I_{YZ} = -1,716 \text{ in.}^4$$

Using Equation C.35, we have

$$2\theta = \tan^{-1}\left[\frac{-I_{yz}}{1/2(I_y - I_z)}\right] = \tan^{-1}\left[\frac{1716}{1/2(3088 - 1744)}\right] = 68.614° \Rightarrow \theta \approx 34.3° \quad \textbf{ANS.}$$

Since θ is positive, it is a counterclockwise angle measured from the Y axis locating the U principal centroidal axis of inertia as shown in the sketch. Of course the V principal centroidal axis is perpendicular to the U principal centroidal axis as shown.

The principal moment of inertia I_U is obtained using Equation C.38 applied to the centroidal axes. Thus,

$$I_U = \frac{1}{2}(I_Y + I_Z) + \sqrt{(1/2(I_Y - I_Z))^2 + I_{YZ}^2}$$

$$= \frac{1}{2}(3088 + 1744) + \sqrt{\left(\frac{1}{2}(3088 - 1744)\right)^2 + (-1716)^2}$$

$$I_U \approx 4258.9 \text{ in.}^4 \quad \textbf{ANS.}$$

The principal moment of inertia I_V is found using Equation C.40 applied to the centroidal axes. Thus,

$$I_V = \frac{1}{2}(I_Y + I_Z) - \sqrt{(1/2(I_Y - I_Z))^2 + I_{YZ}^2}$$

$$= \frac{1}{2}(3088 + 1744) - \sqrt{(1/2(3088 - 1744))^2 + (-1716)^2}$$

$$I_V \approx 573.1 \text{ in.}^4 \quad \textbf{ANS.}$$

C.2.7 MOHR'S CIRCLE FOR MOMENTS AND PRODUCTS OF INERTIA

Equations C.31, C.33, and C.34 may be used to find I_n, I_t, and I_{nt} for any arbitrary n–t coordinate system defined by the angle θ relative to the y–z coordinate system. We may, however, find these quantities using a semigraphical method known as Mohr's circle, named after the German engineer Otto Mohr (1835–1918) who first introduced it. We will develop and use Mohr's circle in the following paragraphs.

Let us first move the first term on the right-hand side of Equation C.31 to the left-hand side and square the resulting equation, which we then add to the square of Equation C.33. These operations lead to

$$(I_n - 1/2(I_y + I_z))^2 + (I_{nt} + 0)^2 = (1/2(I_y - I_z))^2 + I_{yz}^2 \tag{C.41}$$

Examination of Equation C.41 shows that if we replace the term $(I_n - 1/2(I_y + I_z))^2$ by $(y-a)^2$ and the term $(I_{nt} + 0)^2$ by $(z-b)^2$ and recognize the right-hand side as R^2 (see Equation C.36), we obtain the equation $(y-a)^2 + (z-b)^2 = R^2$, which is the familiar equation of a circle in the y–z plane having a center at (a,b) and a radius R. Therefore, if we establish a coordinate system with the horizontal axis measuring I_n and a vertical axis measuring I_{nt}, Equation C.41 plots as a circle with radius R and a center at C on the I_n axis at a distance from the origin, point O. These two properties are given by

$$R = \sqrt{(1/2(I_y - I_z))^2 + I_{yz}^2} \qquad \textbf{(C.36 Repeated)}$$

$$OC = 1/2(I_y + I_z) \qquad \textbf{(C.42)}$$

Let us consider the area A shown in Figure C.10a, for which I_y, I_z, and I_{yz} are known or can be determined, where the y–z coordinate system is also shown. An arbitrary n–t coordinate system, defined by the angle θ with respect to the y–z coordinate system, is also shown in Figure C.10a. After establishing an I_n–I_{nt} coordinate system, as shown in Figure C.10b, we can construct Mohr's circle by locating the center C using Equation C.42 and a radius R given by Equation C.36. However, we can construct Mohr's circle more directly by establishing two *diametrically opposite points* on the circumference of the circle as follows:

When $\theta = 0$ in Figure C.10a, the n axis coincides with the y axis and by Equations C.31 and C.33, we conclude that $I_n = I_y$ and $I_{nt} = I_{yz}$. These two values (i.e., I_y and I_{yz}) locate point Y in Figure C.10b.

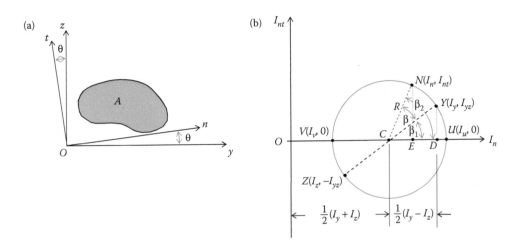

FIGURE C.10

When $\theta = 90°$ in Figure C.10a, the n axis coincides with the z axis and by Equations C.31 and C.33, we conclude that $I_n = I_z$ and $I_{nt} = -I_{yz}$. These two values (i.e., I_z and $-I_{yz}$) locate point Z in Figure C.10b. It can be shown (see Example C.8) that points Y and Z on the circumference of Mohr's circle are diametrically opposite and their angular separation is $180°$. Therefore, the straight line connecting these two points is a diameter of Mohr's circle and its intersection with the I_n axis locates point C, the center of Mohr's circle. Once the diameter is determined and the center located, Mohr's circle can be constructed. Usually, only a free-hand sketch of the circle is sufficient since the required values are obtained *not* by measurement but by a trigonometric analysis.

Example C.8 also shows that *any angle in the actual area is doubled when represented in Mohr's circle*. Thus, any axis such as n, which is located by the counterclockwise angle θ from the y axis in Figure C.10a, is represented by point N, located on the circumference of Mohr's circle by rotating from point Y through a counterclockwise angle of 2θ. We state this concept symbolically by the relationship

$$\beta = 2\theta \tag{C.43}$$

In Equation C.43, θ is the angle in the actual area and β is the corresponding angle in Mohr's circle.

Mohr's circle for moments and products of inertia allows us to accomplish two important objectives. The first is to locate the principal axes of inertia and determine the corresponding moments of inertia for a given cross-sectional area. These quantities are represented by point U (maximum value) and point V (minimum value) in Figure C.10b. The second objective is to find the moments and products of inertia for a given area with respect to inclined axes. Accomplishing these two objectives requires us to find certain geometric properties of Mohr's circle as follows:

$$OC = \frac{1}{2}(I_y + I_z) \tag{C.44}$$

$$R = \sqrt{(CD)^2 + (DY)^2} = \sqrt{\left(\frac{1}{2}(I_y - I_z)\right)^2 + I_{yz}^2} \tag{C.45}$$

$$I_u = I_{max} = OC + R \tag{C.46}$$

$$I_v = I_{min} = OC - R \tag{C.47}$$

$$\beta_1 = \tan^{-1}\left[\frac{I_{yz}}{1/2(I_y - I_z)}\right] \tag{C.48}$$

$$\beta_2 = \beta + \beta_1 = \tan^{-1}\left[\frac{I_{nt}}{I_n - OC}\right] \tag{C.49}$$

Note that all of the relations expressed in Equations C.44 through C.49 are obtainable directly from Mohr's circle and, therefore, there is no need to commit them to memory. Note also that they are the same in essence, if not in form, as those obtained earlier under Section C.2.6.

EXAMPLE C.8

Assume that I_y, I_z, and I_{yz} are known for the cross-sectional area A shown in Figure (a) below. The counterclockwise angle θ defines an n–t coordinate system relative to the y–z system. Construct Mohr's circle for this cross-sectional area and show that the point on the circle representing the n axis is located at a counterclockwise angle 2θ from the point representing the y axis. In other words, prove that any angle in the actual area is represented by twice this angle in Mohr's circle.

SOLUTION

As shown in Figure (b) below, we construct Mohr's circle for the given area, using the procedure discussed above. Note that point Y was located using the coordinate (I_y, I_{yz}) and point Z using the coordinates $(I_z - I_{yz})$. Of course, the center of the circle is at the intersection of I_n axis with the straight line connecting points Y and Z. We now locate point N, representing the n axis, at a counterclockwise angle β from point Y. Point T representing the t axis is diametrically opposite to point N. From the geometry of Figure (b) below, we conclude that

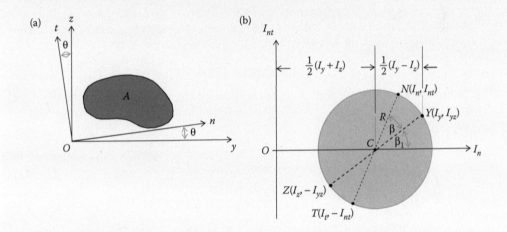

$$\sin(\beta_1 + \beta) = \frac{I_{nt}}{R} \tag{C.8.1}$$

If we substitute for I_{nt} its value from Equation C.33, we obtain

$$\sin(\beta_1 + \beta) = \frac{1/2(I_y - I_z)\sin 2\theta + I_{yz}\cos 2\theta}{R} \tag{C.8.2}$$

Also, from Figure (b) above

$$\sin \beta_1 = \frac{I_{yz}}{R}; \quad \cos \beta_1 = \frac{1/2(I_y - I_z)}{R} \tag{C.8.3}$$

From trigonometry, we know that

$$\sin(\beta_1 + \beta) = \cos \beta_1 \sin \beta + \sin \beta_1 \cos \beta \tag{C.8.4}$$

Substituting from Equations C.8.3 into Equation C.8.4, we obtain

$$\sin(\beta_1 + \beta) = \frac{1/2(I_y - I_z)\sin \beta + I_{yz}\cos \beta}{R} \tag{C.8.5}$$

Comparing Equations C.8.2 and C.8.5, we conclude that

$$\sin \beta = \sin 2\theta; \quad \text{and} \quad \cos \beta = \cos 2\theta \Rightarrow \beta = 2\theta \tag{C.8.6}$$

We conclude, therefore, that any angle measured in the actual area is doubled when represented in Mohr's circle. Thus, since points Y and Z in Mohr's circle represent, respectively, the y and z axes, which are separated by a 90° angle in the actual area, it follows that on the circumference of Mohr's circle, points Y and Z must be located so that the angle between them is $2 \times 90° = 180°$, that is, points Y and Z must be diametrically opposite. The same argument applies to points N and T on the circumference of Mohr's circle. Furthermore, the above analysis shows that the sign of β is the same as the sign of θ. We conclude, therefore, that the angle β in Mohr's circle has the same direction (clockwise or counterclockwise) as the angle θ in the actual area.   **ANS.**

EXAMPLE C.9

Use Mohr's circle to solve the problem stated in Example C.7.

SOLUTION

For convenience, the Z section of Example C.7 is repeated in Figure (a) in next page and its properties relative to the Y–Z coordinate system are repeated here:

$$I_Y = 3088 \text{ in.}^4, \quad I_Z = 1744 \text{ in.}^4, \quad \text{and} \quad I_{YZ} = -1716 \text{ in.}^4$$

Mohr's circle is constructed by first locating points $Y(3088; -1716)$ and $Z(1744; 1716)$ as shown in Figure (b) below, and connecting them with a straight line that intersects the I_n axis at point C, the center of the circle. The circle is then drawn, thus locating points U and V as shown.

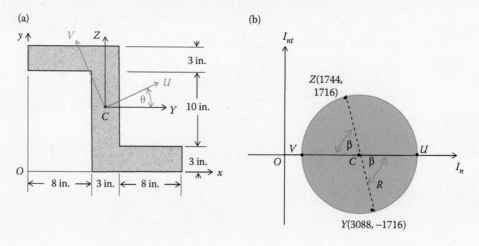

The following basic properties of the circle are now determined from the geometry. Thus,

$$OC = 1/2(I_Y + I_Z) \approx 2878.75 \text{ in.}^4$$

$$R = \sqrt{\left(\frac{1}{2}(I_y - I_z)\right)^2 + I_{yz}^2} \approx 1842.9 \text{ in.}$$

$$I_U = OC + R \approx 4258.9 \text{ in.}^4 \qquad \textbf{ANS.}$$

$$I_V = OC - R \approx 573.1 \text{ in.}^4 \qquad \textbf{ANS.}$$

$$\beta = \tan^{-1} \frac{I_{YZ}}{1/2(I_Y - I_Z)} \approx 68.614°$$

$$\theta = 1/2\,\beta \approx 34.3° \qquad \textbf{ANS.}$$

These answers are, of course, identical with those obtained in Example C.7.

C.3   PROPERTIES OF SELECTED PLANE AREAS

The properties of common cross-sections are provided in Table C.1.

TABLE C.1
Properties of Selected Plane Areas

Shape and Dimensions	Area	Centroid Location	Moment of Inertia	Radius of Gyration
Rectangle 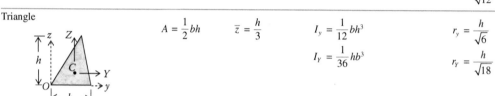	$A = bh$	$\bar{y} = \dfrac{b}{2}$ $\bar{z} = \dfrac{h}{2}$	$I_y = \dfrac{1}{3}bh^3$ $I_z = \dfrac{1}{3}hb^3$ $I_Y = \dfrac{1}{12}bh^3$ $I_Z = \dfrac{1}{12}hb^3$	$r_y = \dfrac{h}{\sqrt{3}}$ $r_z = \dfrac{b}{\sqrt{3}}$ $r_Y = \dfrac{h}{\sqrt{12}}$ $r_Z = \dfrac{b}{\sqrt{12}}$
Triangle 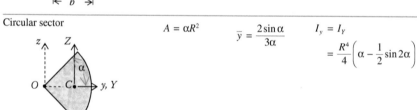	$A = \dfrac{1}{2}bh$	$\bar{z} = \dfrac{h}{3}$	$I_y = \dfrac{1}{12}bh^3$ $I_Y = \dfrac{1}{36}hb^3$	$r_y = \dfrac{h}{\sqrt{6}}$ $r_Y = \dfrac{h}{\sqrt{18}}$
Circular sector  $\alpha = \pi$ leads to full circle	$A = \alpha R^2$	$\bar{y} = \dfrac{2\sin\alpha}{3\alpha}$	$I_y = I_Y$ $= \dfrac{R^4}{4}\left(\alpha - \dfrac{1}{2}\sin 2\alpha\right)$	
	$A = \pi R^2$		$I_Z = I_Y = \dfrac{\pi R^4}{4}$	$r_Z = r_Y = \dfrac{R}{2}$
Ellipse 	$A = \pi ab$		$I_Y = \dfrac{\pi ab^3}{4}$ $I_Z = \dfrac{\pi ba^3}{4}$	$r_Y = \dfrac{b}{2}$ $r_Z = \dfrac{a}{2}$

(Continued)

TABLE C.1 (*Continued*)

Properties of Selected Plane Areas

Shape and Dimensions	Area	Centroid Location	Moment of Inertia	Radius of Gyration
*n*th-Degree parabolic quadrant 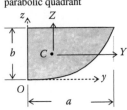	$A = \dfrac{nab}{n+1}$	$\bar{y} = \dfrac{(n+1)a}{2(n+2)}$ $\bar{z} = \dfrac{(n+1)b}{2n+1}$		
*n*th-Degree parabolic spandrel 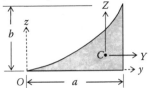	$A = \dfrac{ab}{n+1}$	$\bar{y} = \dfrac{(n+1)a}{n+2}$ $\bar{z} = \dfrac{(n+1)b}{2(2n+1)}$		

Appendix D: Typical Physical and Mechanical Properties of Selected Materials (U.S. Units and SI Units)

D.1 U.S. UNITS

Material	Specific Weight (lb/in.³)	Coefficient of Thermal Expansion (×10⁻⁶ in./in./°F)	Modulus of Elasticity (ksi)	Modulus of Rigidity (ksi)	Yield Strength Ten. (ksi)	Yield Strength Shear (ksi)	Ultimate Strength Ten. (ksi)	Ultimate Strength Comp. (ksi)	Ultimate Strength Shear (ksi)	Percent Elongation in 2 in.
Metals										
Carbon steels (e.g., ASTM A36)	0.284	6.5	29.0	11.5	36.0	20.5	58.0			23.0
High-strength low-alloy steels (e.g.,	0.284	6.5	29.0	11.5	46.0	26.0	67.0			21.0
ASTM A441, ASTM A572, G60)	0.284	6.5	29.0	11.5	60.0	34.0	75.0			18.0
Quenched and tempered steels (e.g., ASTM A514)	0.284	6.5	29.0	11.5	100.0	57.0	115.0			18.0
Weathering steels (e.g., ASTM A588)	0.284	6.5	29.0	11.5	50.0	28.5	70.0			21.0
Stainless steels (e.g., cold-worked Type 316)	0.290	9.9	27.0	10.5	150.0	85.5	200.0			25.0
Wrought aluminum (e.g., 6061-T6)	0.098	13.0	10.0	3.8	35.0	20.0	38.0		24.0	12.0
Cast aluminum (e.g., 195-T4)	0.102	12.7	10.0	3.8	16.0		32.0		26.0	8.5
Sand cast magnesium (e.g., AZ91B-F)	0.065	14.5	6.5	2.4	23.0		34.0			3.0
Titanium alloy (e.g., ASTM B265)	0.161	5.2	16.6		160.0		170.0			13.0
Copper alloy (e.g., ASTM B152)	0.322	9.8	17.0		40.0		45.0			15.0
Beryllium (e.g., hot-pressed industrial grade)	0.066	6.4	42.0		30.0		42.0			3.0
Lead (e.g., cast with 6% Sb)	0.393		3.5		2.8		6.6			22.0
Hard red brass alloy (e.g., 85% Cu,15% Zn, 0.07% Pb,0.06% Fe)	0.316	10.4	15.0		70.0		78.0			4.0
Nonmetals										
Commercial Woods										
Douglas fir	0.017		1.8					7.1	1.2	
Longleaf yellow pine	0.021		2.0					8.5	1.5	
White oak	0.024		1.6					7.0	1.9	
Concrete										
Low strength	0.083	5.6	3.1					3.0		
High strength	0.083	5.6	5.1					8.0		
Miscellaneous										
Glass fiber epoxy	0.066	12.8	3.0				15.0			4.0
Glass-reinforced polyester	0.058	34.0	1.5				19.3			2.0
Rigid foam polyurethane	0.022	35.0	0.1				2.8			5.5

D.2 SI UNITS

Material	Specific Weight (kN/m³)	Coefficient of Thermal Expansion (×10⁻⁶ m/m/°C)	Modulus of Elasticity (GPa)	Modulus of Rigidity (GPa)	Yield Strength Ten. (MPa)	Yield Strength Shear (MPa)	Ultimate Strength Ten. (MPa)	Ultimate Strength Comp. (MPa)	Ultimate Strength Shear (MPa)	Percent Elongation in 50 mm
Metals										
Carbon steels (e.g., ASTM A36)	77.1	11.7	200.0	79.3	248.2	141.3	400.0			23.0
High-strength	77.1	11.7	200.0	79.3	317.2	179.3	462.0			21.0
Low-alloy steels (e.g., ASTM A441, ASTM A572, G60)	77.1	11.7	200.0	79.3	413.8	234.4	517.1			18.0
Quenched and tempered steels (e.g., ASTM A514)	77.1	11.7	200.0	79.3	689.5	393.0	792.9			18.0
Weathering steels (e.g., ASTM A588)	77.1	11.7	200.0	79.3	344.8	196.5	482.7			21.0
Stainless steels (e.g., cold-worked Type 316)	78.7	17.8	186.2	72.4	1034.3	589.5	1379.0			25.0
Wrought aluminum (e.g., 6061-T6)	26.6	23.4	69.0	26.2	241.3	137.9	262.0		165.5	12.0
Cast aluminum (e.g., 195-T4)	27.7	22.9	69.0	26.2	110.3		220.6		179.3	8.5
Sand cast magnesium (e.g., AZ91B-F)	17.6	26.1	44.8	16.5	158.6		234.4			3.0
Titanium alloy (e.g., ASTM B265)	43.7	9.4	114.5		1103.2		1172.2			13.0
Copper alloy (e.g., ASTM B152)	87.4	17.6	117.2		275.8		310.3			15.0
Beryllium (e.g., hot-pressed industrial grade)	17.9	11.5	289.6		206.9		289.6			3.0
Lead (e.g., cast with 6% Sb)	106.7		24.1		19.3		45.5			22.0
Hard red brass alloy (e.g., 85% Cu,15% Zn, 0.07% Pb,0.06% Fe)	85.8	18.7	103.4		42.7		537.8			4.0
Nonmetals										
Commercial Woods										
Douglas fir	4.6		12.4					49.0	8.3	
Longleaf yellow pine	5.7		13.8					58.6	10.3	
White oak	6.5		11.0					48.3	13.1	
Concrete										
Low strength	22.5	10.0	21.4					20.7		
High strength	22.5	10.0	35.2					55.2		
Miscellaneous										
Glass fiber epoxy	17.9	23.0	20.7				103.4			4.0
Glass-reinforced polyester	15.7	61.2	10.3				133.1			2.0
Rigid foam polyurethane	6.0	63.0	0.7				19.3			5.5

Appendix E

E.1 DESIGN PROPERTIES FOR SELECTED WIDE-FLANGE (W SHAPES) STRUCTURAL STEEL SECTIONS (U.S. UNITS AND SI UNITS)

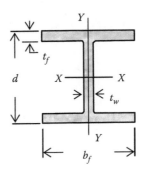

E.1.1 U.S. Units

Designation	Area A (in.²)	Depth d (in.)	Web Thickness t_w (in.)	Flange Width b_f (in.)	Flange Thickness t_f (in.)	X–X Axis I (in.⁴)	X–X Axis S (in.³)	X–X Axis r (in.)	Y–Y Axis I (in.⁴)	Y–Y Axis S (in.³)	Y–Y Axis r (in.)
W36 × 300	88.3	36.74	0.945	16.655	1.680	20300	1110	15.2	1300	156	3.83
× 280	82.4	36.52	0.885	16.595	1.570	18900	1030	15.1	1200	144	3.81
× 210	61.8	36.69	0.830	12.180	1.360	13200	719	14.6	411	67.5	2.58
× 194	57.0	36.49	0.765	12.115	1.260	12100	664	14.6	375	61.9	2.56
W33 × 241	70.9	34.18	0.830	15.860	1.400	14200	829	14.1	932	118	3.63
× 221	65.0	33.93	0.775	15.805	1.275	12800	757	14.1	840	106	3.59
× 152	44.7	33.49	0.635	11.565	1.055	8160	487	13.5	273	47.2	2.47
× 141	41.6	33.30	0.605	11.535	0.960	7450	448	13.4	246	42.7	2.43
W30 × 211	62.0	30.94	0.775	15.105	1.315	10300	663	12.9	757	100	3.49
× 191	56.1	30.68	0.710	15.040	1.185	9170	598	12.8	673	89.5	3.46
× 132	38.9	30.31	0.615	10.545	1.000	5770	380	12.2	196	37.2	2.25
× 124	36.5	30.17	0.585	10.515	0.930	5360	355	12.1	181	34.4	2.23
W27 × 178	52.3	27.81	0.725	14.085	1.190	6990	502	11.6	555	78.8	3.26
× 161	47.4	27.59	0.660	14.020	1.080	6280	455	11.5	497	70.9	3.24
× 114	33.5	27.29	0.570	10.070	0.930	4090	299	11.0	159	31.5	2.18
× 102	30.0	27.09	0.515	10.015	0.830	3620	267	11.0	139	27.8	2.15
W24 × 162	47.7	25.00	0.705	12.955	1.220	5170	414	10.4	443	68.4	3.05
× 146	43.0	24.74	0.650	12.900	1.090	4580	371	10.3	391	60.5	3.01
× 94	27.7	24.31	0.515	9.065	0.875	2700	222	9.87	109	24.0	1.98
× 84	24.7	24.10	0.470	9.020	0.770	2370	196	9.79	94.4	20.9	1.95
× 62	18.2	23.74	0.430	7.040	0.590	1550	131	9.23	34.5	9.80	1.38
× 55	16.2	23.57	0.395	7.005	0.505	1350	114	9.11	29.1	8.30	1.34

(Continued)

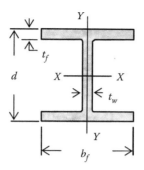

Designation	Area A (in.²)	Depth d (in.)	Web Thickness t_w (in.)	Flange		X–X Axis			Y–Y Axis		
				Width b_f (in.)	Thickness t_f (in.)	I (in.⁴)	S (in.³)	r (in.)	I (in.⁴)	S (in.³)	r (in.)
W21 × 147	43.2	22.06	0.720	12.510	1.150	3630	329	9.17	376	60.1	2.95
× 132	38.8	21.83	0.650	12.440	1.035	3220	295	9.12	333	53.5	2.93
× 93	27.3	21.62	0.580	8.420	0.930	2070	192	8.70	92.9	22.1	1.84
× 83	24.3	21.43	0.515	8.355	0.835	1830	171	8.67	81.4	19.5	1.83
× 57	16.7	21.06	0.405	6.555	0.650	1170	111	8.36	30.6	9.35	1.35
× 50	14.7	20.83	0.380	6.530	0.535	984	94.5	8.18	24.9	7.64	1.30
W18 × 119	35.1	18.97	0.655	11.265	1.060	2190	231	7.90	253	44.9	2.69
× 106	31.1	18.73	0.590	11.200	0.940	1910	204	7.84	220	39.4	2.66
× 71	20.8	18.47	0.495	7.635	0.810	1170	127	7.50	60.3	15.8	1.70
× 65	19.1	18.35	0.450	7.590	0.750	1070	117	7.49	54.8	14.4	1.69
× 46	13.5	18.06	0.360	6.060	0.605	712	78.8	7.25	22.5	7.43	1.29
× 40	11.8	17.90	0.315	6.015	0.525	612	68.4	7.21	19.1	6.35	1.27
W16 × 100	29.4	16.97	0.585	10.425	0.985	1490	175	7.10	186	35.7	2.51
× 89	26.2	16.75	0.525	10.365	0.875	1300	155	7.05	163	31.4	2.49
× 57	16.8	16.43	0.430	7.120	0.715	758	92.2	6.72	43.1	12.1	1.60
× 50	14.7	16.26	0.380	7.070	0.630	659	81.0	6.68	37.2	10.5	1.59
× 31	9.12	15.88	0.275	5.525	0.440	375	47.2	6.41	12.4	4.49	1.17
× 26	7.68	15.69	0.250	5.500	0.345	301	38.4	6.26	9.59	3.49	1.12
W14 × 730	215.0	22.42	3.070	17.890	4.910	14300	1280	8.17	4720	527	4.69
× 665	196.0	21.64	2.830	17.650	4.520	12400	1150	7.98	4170	472	4.62
× 426	125.0	18.67	1.875	16.695	3.035	6600	707	7.26	2360	283	4.34
× 398	117.0	18.29	1.770	16.590	2.845	6000	656	7.16	2170	262	4.31
× 132	38.8	14.66	0.645	14.725	1.030	1530	209	6.28	548	74.5	3.76
× 120	35.3	14.48	0.590	14.670	0.940	1380	190	6.24	495	67.5	3.74
× 82	24.1	14.31	0.510	10.130	0.855	882	123	6.05	148	29.3	2.48
× 74	21.8	14.17	0.450	10.070	0.785	796	112	6.04	134	26.6	2.48
× 68	20.0	14.04	0.415	10.035	0.720	723	103	6.01	121	24.2	2.46
× 61	17.9	13.89	0.375	9.995	0.645	640	92.2	5.98	107	21.5	2.45
× 53	15.6	13.92	0.370	8.060	0.660	541	77.8	5.89	57.7	14.3	1.92
× 48	14.1	13.79	0.340	8.030	0.595	485	70.3	5.85	51.4	12.8	1.91
× 38	11.2	14.10	0.310	6.770	0.515	385	54.6	5.87	26.7	7.88	1.55
× 34	10.0	13.98	0.285	6.745	0.455	340	48.6	5.83	23.3	6.91	1.53
× 26	7.69	13.91	0.255	5.025	0.420	245	35.3	5.65	8.91	3.54	1.08
× 22	6.49	13.74	0.230	5.000	0.335	199	29.0	5.54	7.00	2.80	1.04

(Continued)

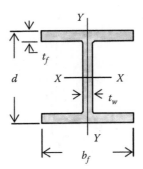

Designation	Area A (in.²)	Depth d (in.)	Web Thickness t_w (in.)	Flange Width b_f (in.)	Flange Thickness t_f (in.)	X–X Axis I (in.⁴)	X–X Axis S (in.³)	X–X Axis r (in.)	Y–Y Axis I (in.⁴)	Y–Y Axis S (in.³)	Y–Y Axis r (in.)
W12 × 190	55.8	14.38	1.06	12.67	1.735	1890	263	5.82	589	93	3.25
× 170	50	14.03	0.96	12.57	1.56	1650	235	5.74	517	82.3	3.22
× 58	17	12.19	0.36	10.01	0.64	475	78	5.28	107	21.4	2.51
× 50	14.7	12.19	0.37	8.080	0.64	394	64.7	5.18	56.3	13.9	1.96
× 45	13.2	12.06	0.335	8.045	0.575	350	58.1	5.15	50	12.4	1.94
× 35	10.3	12.5	0.3	6.56	0.52	285	45.6	5.25	24.5	7.47	1.54
× 30	8.79	12.34	0.26	6.52	0.44	238	38.6	5.21	20.3	6.24	1.52
× 22	6.48	12.31	0.26	4.030	0.425	156	25.4	4.91	4.66	2.31	0.847
× 19	5.57	12.16	0.235	4.005	0.35	130	21.3	4.82	3.76	1.88	0.822
W10 × 112	32.9	11.36	0.755	10.415	1.25	716	126	4.66	236	45.3	2.68
× 45	13.3	10.1	0.35	8.02	0.62	248	49.1	4.32	53.4	13.3	2.01
× 39	11.5	9.92	0.315	7.985	0.53	209	42.1	4.27	45	11.3	1.98
× 30	8.84	10.47	0.3	5.81	0.51	170	32.4	4.38	16.7	5.75	1.37
× 26	7.61	10.33	0.26	5.770	0.44	144	27.9	4.35	14.1	4.89	1.36
× 19	5.62	10.24	0.25	4.02	0.395	96.3	18.8	4.14	4.29	2.14	0.874
× 17	4.99	10.11	0.24	4.010	0.33	81.9	16.2	4.05	3.56	1.78	0.844
W8 × 67	19.7	9	0.57	8.28	0.935	272	60.4	3.72	88.6	21.4	2.12
× 58	17.1	8.75	0.51	8.220	0.81	228	52	3.65	75.1	18.3	2.1
× 28	8.25	8.06	0.285	6.535	0.465	98	24.3	3.45	21.7	6.63	1.62
× 21	6.16	8.28	0.25	5.270	0.4	75.3	18.2	3.49	9.77	3.71	1.26
× 18	5.26	8.14	0.23	5.25	0.33	61.9	15.2	3.43	7.97	3.04	1.23
× 15	4.44	8.11	0.245	4.015	0.315	48	11.8	3.29	3.41	1.7	0.876
× 13	3.84	7.99	0.23	4	0.255	39.6	9.91	3.21	2.73	1.37	0.843
W6 × 25	7.34	6.38	0.32	6.080	0.455	53.4	16.7	2.7	17.1	5.61	1.52
× 20	5.87	6.2	0.26	6.02	0.365	41.4	13.4	2.66	13.3	4.41	1.5
× 16	4.74	6.28	0.26	4.030	0.405	32.1	10.2	2.6	4.43	2.2	0.966
× 12	3.55	6.03	0.23	4	0.28	22.1	7.31	2.49	2.99	1.5	0.918
W5 × 19	5.54	5.15	0.27	5.030	0.43	26.2	10.2	2.17	9.13	3.63	1.28
× 16	4.68	5.01	0.24	5	0.36	21.3	8.51	2.13	7.51	3	1.27
W4 × 13	3.83	4.16	0.28	4.060	0.435	11.3	5.46	1.72	3.86	1.9	1

Source: Adapted from American Institute of Steel Construction, Inc. *Manual of Steel Construction*, AISC, Chicago, IL, 1980.

Note: The "W" in the designation column indicates a wide-flange section. The "W" is followed by the nominal height in inches and the weight in pounds per foot of length.

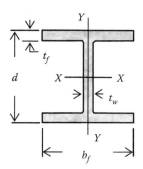

E.1.2 SI Units

Designation	Area A (10^{-3} m²)	Depth d ($\times 10^{-1}$ m)	Web Thickness t_w ($\times 10^{-3}$ m)	Flange Width b_f ($\times 10^{-1}$ m)	Flange Thickness t_f ($\times 10^{-2}$ m)	X–X Axis I ($\times 10^{-4}$ m⁴)	X–X Axis S ($\times 10^{-3}$ m³)	X–X Axis r ($\times 10^{-1}$ m)	Y–Y Axis I ($\times 10^{-5}$ m⁴)	Y–Y Axis S ($\times 10^{-4}$ m³)	Y–Y Axis r ($\times 10^{-2}$ m)
W914 × 446.3	56.97	9.332	24.003	4.230	4.267	84.495	18.190	3.861	54.11	25.56	9.73
× 416.5	53.16	9.276	22.479	4.215	3.988	78.667	16.879	3.835	49.95	23.60	9.68
× 312.4	39.87	9.319	21.082	3.094	3.454	54.942	11.782	3.708	17.11	11.06	6.55
× 288.6	36.77	9.268	19.431	3.077	3.200	50.363	10.881	3.708	15.61	10.14	6.50
W838 × 358.5	45.74	8.682	21.082	4.028	3.556	59.105	13.585	3.581	38.79	19.34	9.22
× 328.8	41.94	8.618	19.685	4.014	3.239	53.277	12.405	3.581	34.96	17.37	9.12
× 226.1	28.84	8.506	16.129	2.938	2.680	33.964	7.980	3.429	11.36	7.73	6.27
× 209.7	26.84	8.458	15.367	2.930	2.438	31.009	7.341	3.404	10.24	7.00	6.17
W762 × 313.9	40.00	7.859	19.685	3.837	3.340	42.872	10.865	3.277	31.51	16.39	8.86
× 284.1	36.19	7.793	18.034	3.820	3.010	38.168	9.799	3.251	28.01	14.67	8.79
× 196.4	25.10	7.699	15.621	2.678	2.540	24.016	6.227	3.099	8.16	6.10	5.72
× 184.5	23.55	7.663	14.859	2.671	2.362	22.310	5.817	3.073	7.53	5.64	5.66
W686 × 264.8	33.74	7.064	18.415	3.578	3.023	29.094	8.226	2.946	23.10	12.91	8.28
× 239.5	30.58	7.008	16.764	3.561	2.743	26.139	7.456	2.921	20.69	11.62	8.23
× 169.6	21.61	6.932	14.478	2.558	2.362	17.024	4.900	2.794	6.62	5.16	5.54
× 151.7	19.35	6.881	13.081	2.544	2.108	15.068	4.375	2.794	5.79	4.56	5.46
W610 × 241.0	30.77	6.350	17.907	3.291	3.099	21.519	6.784	2.642	18.44	11.21	7.75
× 217.2	27.74	6.284	16.510	3.277	2.769	19.063	6.080	2.616	16.27	9.91	7.65
× 139.8	17.87	6.175	13.081	2.303	2.223	11.238	3.638	2.507	4.54	3.93	5.03
× 125.0	15.94	6.121	11.938	2.291	1.956	9.865	3.212	2.487	3.93	3.42	4.95
× 92.2	11.74	6.030	10.922	1.788	1.499	6.452	2.147	2.344	1.44	1.61	3.51
× 81.8	10.45	5.987	10.033	1.779	1.283	5.619	1.868	2.314	1.21	1.36	3.40
W533 × 218.7	27.87	5.603	18.288	3.178	2.921	15.109	5.391	2.329	15.65	9.85	7.49
× 196.4	25.03	5.545	16.510	3.160	2.629	13.403	4.834	2.316	13.86	8.77	7.44
× 138.3	17.61	5.491	14.732	2.139	2.362	8.616	3.146	2.210	3.87	3.62	4.67
× 123.5	15.68	5.443	13.081	2.122	2.121	7.617	2.802	2.202	3.39	3.20	4.65
× 84.8	10.77	5.349	10.287	1.665	1.651	4.870	1.819	2.123	1.27	1.53	3.43
× 74.4	9.48	5.291	9.652	1.659	1.359	4.096	1.549	2.078	1.04	1.25	3.30
W457 × 177.0	22.65	4.818	16.637	2.861	26.924	91.154	37.854	2.007	105.31	73.58	6.83
× 157.7	20.06	4.757	14.986	2.845	23.876	79.500	33.429	1.991	91.57	64.56	6.76

(Continued)

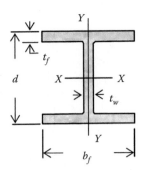

	Area	Depth	Web	Flange		X–X Axis			Y–Y Axis		
	A	d	Thickness	Width b_f	Thickness	I	S	r	I	S	r
Designation	$(10^{-3}$ m^2)	$(\times10^{-1}$ m)	t_w $(\times10^{-3}$ m)	$(\times10^{-1}$ m)	t_f $(\times10^{-2}$ m)	$(\times10^{-4}$ m^4)	$(\times10^{-3}$ m^3)	$(\times10^{-1}$ m)	$(\times10^{-5}$ m^4)	$(\times10^{-4}$ m^3)	$(\times10^{-2}$ m)
× 105.6	13.42	4.691	12.573	1.939	20.574	48.699	20.811	1.905	25.10	25.89	4.32
× 96.7	12.32	4.661	11.430	1.928	19.050	44.537	19.173	1.902	22.81	23.60	4.29
× 68.4	8.71	4.587	9.144	1.539	15.367	29.636	12.913	1.842	9.37	12.18	3.28
× 59.5	7.61	4.547	8.001	1.528	13.335	25.473	11.209	1.831	7.95	10.41	3.23
W406 × 148.8	18.97	4.310	14.859	2.648	25.019	62.018	28.677	1.803	77.42	58.50	6.38
× 132.4	16.90	4.255	13.335	2.633	22.225	54.110	25.400	1.791	67.85	51.46	6.32
× 84.8	10.84	4.173	10.922	1.808	18.161	31.550	15.109	1.707	17.94	19.83	4.06
× 74.4	9.48	4.130	9.652	1.796	16.002	27.430	13.273	1.697	15.48	17.21	4.04
× 46.1	5.88	4.034	6.985	1.403	11.176	15.609	7.735	1.628	5.16	7.36	2.97
× 38.7	4.95	3.985	6.350	1.397	8.763	12.529	6.293	1.590	3.99	5.72	2.84
W356 × 1085.9	138.71	5.695	77.978	4.544	124.714	595.209	209.754	2.075	1964.61	863.59	11.91
× 989.2	126.45	5.497	71.882	4.483	114.808	516.125	188.451	2.027	1735.68	773.47	11.73
× 633.7	80.65	4.742	47.625	4.241	77.089	274.712	115.856	1.844	982.30	463.75	11.02
× 592.1	75.48	4.646	44.958	4.214	72.263	249.738	107.499	1.819	903.22	429.34	10.95
× 196.4	25.03	3.724	16.383	3.740	26.162	63.683	34.249	1.595	228.09	122.08	9.55
× 178.5	22.77	3.678	14.986	3.726	23.876	57.440	31.135	1.585	206.03	110.61	9.50
× 122.0	15.55	3.635	12.954	2.573	21.717	36.711	20.156	1.537	61.60	48.01	6.30
× 110.1	14.06	3.599	11.430	2.558	19.939	33.132	18.353	1.534	55.77	43.59	6.30
× 101.2	12.90	3.566	10.541	2.549	18.288	30.093	16.879	1.527	50.36	39.66	6.25
× 90.7	11.55	3.528	9.525	2.539	16.383	26.639	15.109	1.519	44.54	35.23	6.22
× 78.8	10.06	3.536	9.398	2.047	16.764	22.518	12.749	1.496	24.02	23.43	4.88
× 71.4	9.10	3.503	8.636	2.040	15.113	20.187	11.520	1.486	21.39	20.98	4.85
× 56.5	7.23	3.581	7.874	1.720	13.081	16.025	8.947	1.491	11.11	12.91	3.94
× 50.6	6.45	3.551	7.239	1.713	11.557	14.152	7.964	1.481	9.70	11.32	3.89
× 38.7	4.96	3.533	6.477	1.276	10.668	10.198	5.785	1.435	3.71	5.80	2.74
× 32.7	4.19	3.490	5.842	1.270	8.509	8.283	4.752	1.407	2.91	4.59	2.64
W305 × 282.6	36	3.653	26.924	3.218	44.069	786.675	430.98	14.783	245.16	152.4	8.26
× 252.9	32.26	3.564	24.384	3.193	39.624	686.78	385.09	14.58	215.19	134.87	8.18
× 86.3	10.97	3.096	9.144	2.543	16.256	197.709	127.82	13.411	44.54	35.07	6.38
× 74.4	9.48	3.096	9.398	2.052	16.256	163.995	106.02	13.157	23.43	22.78	4.98
× 66.9	8.52	3.063	8.509	2.043	14.605	145.681	95.21	13.081	20.81	20.32	4.93
× 52.1	6.65	3.175	7.62	1.666	13.208	118.626	74.72	13.335	10.2	12.24	3.91
× 44.6	5.67	3.134	6.604	1.656	11.176	99.063	63.25	13.233	8.45	10.23	3.86

(Continued)

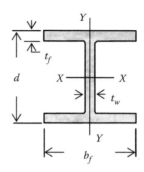

Designation	Area A (10⁻³ m²)	Depth d (×10⁻¹ m)	Web Thickness t_w (×10⁻³ m)	Flange		X–X Axis			Y–Y Axis		
				Width b_f (×10⁻¹ m)	Thickness t_f (×10⁻² m)	I (×10⁻⁴ m⁴)	S (×10⁻³ m³)	r (×10⁻¹ m)	I (×10⁻⁵ m⁴)	S (×10⁻⁴ m³)	r (×10⁻² m)
× 32.7	4.18	3.127	6.604	1.024	10.795	64.932	41.62	12.471	1.94	3.79	2.15
× 28.3	3.59	3.089	5.969	1.017	8.89	54.11	34.9	12.243	1.57	3.08	2.09
W254 × 166.6	21.23	2.885	19.177	2.645	31.75	298.021	206.48	11.836	98.23	74.23	6.81
× 66.9	8.58	2.565	8.890	2.037	15.748	103.225	80.46	10.973	22.23	21.79	5.11
× 58.0	7.42	2.52	8.001	2.028	13.462	86.992	68.99	10.846	18.73	18.52	5.03
× 44.6	5.7	2.659	7.62	1.476	12.954	70.759	53.09	11.125	6.95	9.42	3.48
× 38.7	4.91	2.624	6.604	1.466	11.176	59.937	45.72	11.049	5.87	8.01	3.45
× 28.3	3.63	2.601	6.35	1.021	10.033	40.083	30.81	10.516	1.79	3.51	2.22
× 25.3	3.22	2.568	6.096	1.019	8.382	34.089	26.55	10.287	1.48	2.92	2.14
W203 × 99.7	12.71	2.286	14.478	2.103	23.749	113.215	98.98	9.449	36.88	35.07	5.38
× 86.3	11.03	2.223	12.954	2.088	20.574	94.900	85.21	9.271	31.26	29.99	5.33
× 41.7	5.32	2.047	7.239	1.66	11.811	40.791	39.82	8.763	9.03	10.86	4.11
× 31.2	3.97	2.103	6.350	1.339	10.16	31.342	29.82	8.865	4.07	6.08	3.2
× 26.8	3.39	2.068	5.842	1.334	8.382	25.765	24.91	8.712	3.32	4.98	3.12
× 22.3	2.86	2.06	6.223	1.02	8.001	19.979	19.34	8.357	1.42	2.79	2.23
× 19.3	2.48	2.029	5.842	1.016	6.477	16.483	16.24	8.153	1.14	2.25	2.14
W152 × 37.2	4.74	1.621	8.128	1.544	11.557	22.227	27.37	6.858	7.12	9.19	3.86
× 29.8	3.79	1.575	6.604	1.529	9.271	17.232	21.96	6.756	5.54	7.23	3.81
× 23.8	3.06	1.595	6.604	1.024	10.287	13.361	16.71	6.604	1.84	3.61	2.45
× 17.9	2.29	1.532	5.842	1.016	7.112	9.199	11.98	6.325	1.24	2.46	2.33
W127 × 28.3	3.57	1.308	6.858	1.278	10.922	10.905	16.71	5.512	3.80	5.95	3.25
× 23.8	3.02	1.273	6.096	1.27	9.144	8.866	13.95	5.410	3.13	4.92	3.23
W102 × 19.3	2.47	1.057	7.112	1.031	8.763	4.703	8.95	4.369	1.61	3.11	2.54

Note: The "W" in the designation column indicates a wide-flange section. The "W" is followed by the nominal height in millimeters and the mass in kilograms per meter of length.

Numbers in this table represent direct conversions from U.S. Customary to SI units.

E.2   DESIGN PROPERTIES FOR SELECTED STRUCTURAL STEEL CHANNELS, C SHAPES (U.S. UNITS AND SI UNITS)

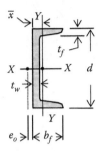

E.2.1   U.S. UNITS

Designation	Area A (in.²)	Depth d (in.)	Web Thickness t_w (in.)	Flange Width b_f (in.)	Flange Average Thickness t_f (in.)	$\bar{x}$ (in.)	Shear Center Location e_o	X–X Axis I (in.⁴)	X–X Axis S (in.³)	Y–Y Axis I (in.⁴)	Y–Y Axis S (in.³)
C15 × 50	14.7	15	0.716	3.716	0.650	0.798	0.583	404	53.8	11	3.78
× 40	11.8	15	0.52	3.52	0.65	0.777	0.767	349	46.5	9.23	3.37
× 33.9	9.96	15	0.4	3.400	0.650	0.787	0.896	315	42	8.13	3.11
C12 × 30	8.82	12	0.51	3.17	0.501	0.674	0.618	162	27	5.14	2.06
× 25	7.35	12	0.387	3.047	0.501	0.674	0.746	144	24.1	4.47	1.88
× 20.7	6.09	12	0.282	2.942	0.501	0.698	0.87	129	21.5	3.88	1.73
C10 × 30	8.82	10	0.673	3.033	0.436	0.649	0.369	103	20.7	3.94	1.65
× 25	7.35	10	0.526	2.886	0.436	0.617	0.494	91.2	18.2	3.36	1.48
× 20	5.88	10	0.379	2.739	0.436	0.606	0.637	78.9	15.8	2.81	1.32
× 15.3	4.49	10	0.24	2.6	0.436	0.634	0.796	67.4	13.5	2.28	1.16
C9 × 20	5.88	9	0.448	2.648	0.413	0.583	0.515	60.9	13.5	2.42	1.17
× 15	4.41	9	0.285	2.485	0.413	0.586	0.682	51	11.3	1.93	1.01
× 13.4	3.94	9	0.233	2.433	0.413	0.601	0.743	47.9	10.6	1.76	0.962
C8 × 18.75	5.51	8	0.487	2.527	0.39	0.565	0.431	44	11	1.98	1.01
× 13.75	4.04	8	0.303	2.343	0.390	0.553	0.604	36.1	9.03	1.53	0.854
× 11.5	3.38	8	0.22	2.26	0.39	0.571	0.697	32.6	8.14	1.32	0.781
C7 × 14.75	4.33	7	0.419	2.299	0.366	0.532	0.441	27.2	7.78	1.38	0.779
× 12.25	3.60	7	0.314	2.194	0.366	0.525	0.538	24.2	6.93	1.17	0.703
× 9.8	2.87	7	0.21	2.090	0.366	0.54	0.647	21.3	6.08	0.968	0.625
C6 × 13	3.83	6	0.437	2.157	0.343	0.514	0.38	17.4	5.8	1.05	0.642
× 10.5	3.09	6	0.314	2.034	0.343	0.499	0.486	15.2	5.06	0.866	0.564
× 8.2	2.40	6	0.2	1.92	0.343	0.511	0.599	13.1	4.38	0.693	0.492
C5 × 9	2.64	5	0.325	1.885	0.32	0.478	0.427	8.90	3.56	0.632	0.45
× 6.7	1.97	5	0.19	1.75	0.32	0.484	0.552	7.49	3	0.479	0.378
C4 × 7.25	2.13	4	0.321	1.721	0.296	0.459	0.386	4.59	2.29	0.433	0.343
× 5.4	1.59	4	0.184	1.584	0.296	0.457	0.502	3.85	1.93	0.319	0.283
C3 × 6	1.76	3	0.356	1.596	0.273	0.455	0.322	2.07	1.38	0.305	0.268
× 5	1.47	3	0.258	1.498	0.273	0.438	0.392	1.85	1.24	0.247	0.233
× 4.1	1.21	3	0.17	1.41	0.273	0.436	0.461	1.66	1.10	0.197	0.202

Source: Adapted from American Institute of Steel Construction, Inc. *Manual of Steel Construction*, AISC, Chicago, IL, 1980.

Note: The "C" in the designation column indicates an American Standard Channel. The "C" is followed by the nominal height in inches and the weight in pounds per foot of length.

The radius of gyration *r* may be found from the relation: $r = \sqrt{I/A}$.

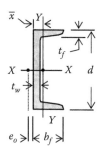

E.2.2 SI Units

Designation	Area A ($\times 10^{-4}$ m²)	Depth d ($\times 10^{-2}$ m)	Web Thickness t_w ($\times 10^{-3}$ m)	Flange Width b_f ($\times 10^{-2}$ m)	Flange Average Thickness t_f ($\times 10^{-3}$ m)	$\bar{x}$ ($\times 10^{-2}$ m)	Shear Center Location e_o ($\times 10^{-2}$ m)	X–X Axis I ($\times 10^{-7}$ m⁴)	X–X Axis S ($\times 10^{-5}$ m³)	Y–Y Axis I ($\times 10^{-8}$ m⁴)	Y–Y Axis S ($\times 10^{-6}$ m³)
C381 × 74.4	94.83	38.1	18.186	9.439	16.510	2.027	1.481	1681.6	88.16	457.9	61.9
× 59.5	76.13	38.1	13.208	8.941	16.51	1.974	1.948	1452.6	76.2	384.2	55.2
× 50.4	64.26	38.1	10.16	8.636	16.510	1.999	2.276	1311.1	68.83	338.4	51
C305 × 44.6	56.9	30.48	12.954	8.052	12.725	1.712	1.57	674.3	44.24	213.9	33.8
× 37.2	47.42	30.48	9.83	7.739	12.725	1.712	1.895	599.4	39.49	186.1	30.8
× 30.8	39.29	30.48	7.163	7.473	12.725	1.773	2.21	536.9	35.23	161.5	28.3
C254 × 44.6	56.9	25.4	17.094	7.704	11.074	1.648	0.937	428.7	33.92	164	27
× 37.2	47.42	25.4	13.36	7.33	11.074	1.567	1.255	379.6	29.82	139.9	24.3
× 29.8	37.94	25.4	9.627	6.957	11.074	1.539	1.618	328.4	25.89	117	21.6
× 22.8	28.97	25.4	6.096	6.604	11.074	1.61	2.022	280.5	22.12	94.9	19
C229 × 29.8	37.94	22.86	11.379	6.726	10.49	1.481	1.308	253.5	22.12	100.7	19.2
× 22.3	28.45	22.86	7.239	6.312	10.49	1.488	1.732	212.3	18.52	80.3	16.6
× 19.9	25.42	22.86	5.918	6.180	10.49	1.527	1.887	199.4	17.37	73.3	15.8
C203 × 27.9	35.55	20.32	12.37	6.419	9.906	1.435	1.095	183.1	18.03	82.4	16.6
× 20.5	26.06	20.32	7.696	5.951	9.906	1.405	1.534	150.3	14.8	63.7	14
× 17.1	21.81	20.32	5.588	5.74	9.906	1.45	1.77	135.7	13.34	54.9	12.8
C178 × 21.9	27.94	17.78	10.643	5.839	9.296	1.351	1.120	113.2	12.75	57.4	12.8
× 18.2	23.23	17.78	7.976	5.573	9.296	1.334	1.367	100.7	11.36	48.7	11.5
× 14.6	18.52	17.78	5.334	5.309	9.296	1.372	1.643	88.7	9.96	40.3	10.2
C152 × 19.3	24.71	15.24	11.1	5.479	8.712	1.306	0.965	72.4	9.5	43.7	10.5
× 15.6	19.94	15.24	7.976	5.166	8.712	1.267	1.234	63.3	8.29	36	9.2
× 12.2	15.48	15.24	5.08	4.877	8.712	1.298	1.521	54.5	7.18	28.8	8.1
C127 × 13.4	17.03	12.7	8.255	4.788	8.128	1.214	1.085	37.0	5.83	26.3	7.4
× 10.0	12.71	12.7	4.826	4.445	8.128	1.229	1.402	31.2	4.92	19.9	6.2
C102 × 10.8	13.74	10.16	8.153	4.371	7.518	1.166	0.98	19.1	3.75	18	5.6
× 8.0	10.26	10.16	4.674	4.023	7.518	1.161	1.275	16	3.16	13.3	4.6
C76 × 8.9	11.35	7.62	9.042	4.054	6.934	1.156	1.818	8.6	2.26	12.7	4.4
×7.4 × 6.1	9.48	7.62	6.553	3.805	6.934	1.113	0.996	7.7	2.03	10.3	3.8
	7.81	7.62	4.318	3.581	6.934	1.107	1.171	6.9	1.80	8.2	3.3

Note: The "C" in the designation column indicates a channel section. The "C" is followed by the nominal height in millimeters and the mass in kilograms per meter of length.

Numbers in this table represent direct conversions from U.S. Customary to SI units.

The radius of gyration r may be found from the relation: $r = \sqrt{I/A}$.

E.3 DESIGN PROPERTIES FOR SELECTED EQUAL-LEGS STRUCTURAL STEEL ANGLES (U.S. UNITS AND SI UNITS)

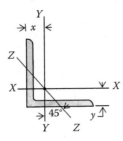

E.3.1 U.S. Units

| Designation | Unit Weight (lb/ft) | Area (in.²) | X–X and Y–Y Axes | | | | Z–Z Axis |
			I (in.⁴)	S (in.³)	r (in.)	x or y (in.)	r (in.)
L8 × 8 × 1–1/8	56.9	16.7	98	17.5	2.42	2.41	1.56
× 1	51	15	89	15.8	2.44	2.37	1.56
× 7/8	45	13.2	79.6	14	2.45	2.32	1.57
× 3/4	38.9	11.4	69.7	12.2	2.47	2.28	1.58
× 5/8	32.7	9.61	59.4	10.3	2.49	2.23	1.58
× 9/16	29.6	8.68	54.1	9.34	2.5	2.21	1.59
× 1/2	26.4	7.75	48.6	8.36	2.5	2.19	1.59
L6 × 6 × 1	37.4	11	35.5	8.57	1.8	1.86	1.17
× 7/8	33.1	9.73	31.9	7.63	1.81	1.82	1.17
× 3/4	28.7	8.44	28.2	6.66	1.83	1.78	1.17
× 5/8	24.2	7.11	24.2	5.66	1.84	1.73	1.18
× 9/16	21.9	6.43	22.1	5.14	1.85	1.71	1.18
× 1/2	19.6	5.75	19.9	4.61	1.86	1.68	1.18
× 7/16	17.2	5.06	17.7	4.08	1.87	1.66	1.19
× 3/8	14.9	4.36	15.4	3.53	1.88	1.64	1.19
× 5/16	12.4	3.65	13.0	2.97	1.89	1.62	1.20
L5 × 5 × 7/8	27.2	7.98	17.8	5.17	1.49	1.57	0.973
× 3/4	23.6	6.94	15.7	4.53	1.51	1.52	0.975
× 5/8	20	5.86	13.6	3.86	1.52	1.48	0.978
× 1/2	16.2	4.75	11.3	3.16	1.54	1.43	0.983
× 7/16	14.3	4.18	10	2.79	1.55	1.41	0.986
× 3/8	12.3	3.61	8.74	2.42	1.56	1.39	0.99
× 5/16	10.3	3.03	7.42	2.04	1.57	1.37	0.994
L4 × 4 × 3/4	18.5	5.44	7.67	2.81	1.19	1.27	0.778
× 5/8	15.7	4.61	6.66	2.4	1.2	1.23	0.779
× 1/2	12.8	3.75	5.56	1.97	1.22	1.18	0.782
× 7/16	11.3	3.31	4.97	1.75	1.23	1.16	0.785
× 3/8	9.8	2.86	4.36	1.52	1.23	1.14	0.788
× 5/16	8.2	2.4	3.71	1.29	1.24	1.12	0.791
× 1/4	6.6	1.94	3.04	1.05	1.25	1.09	0.795

Source: Adapted from American Institute of Steel Construction, Inc. *Manual of Steel Construction*, AISC, Chicago, IL, 1980.

Note: The "L" in the designation column indicates a structural steel angle. The "L" is followed by the length of the legs and their thickness in inches.

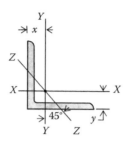

E.3.2 SI Units

| Designation | Unit Mass (kg/m) | Area ($\times10^{-3}$ m²) | X–X and Y–Y Axes | | | | Z–Z Axis |
			I ($\times10^{-6}$ m⁴)	S ($\times10^{-5}$ m³)	r ($\times10^{-2}$ m)	x or y ($\times10^{-2}$ m)	r ($\times10^{-2}$ m)
L203 × 203 × 28.6	84.6	10.774	40.791	28.677	6.147	6.121	3.962
× 25.4	75.9	9.677	37.044	25.891	6.198	6.02	3.962
× 22.2	66.9	8.516	33.132	22.942	6.223	5.893	3.988
× 19.1	57.9	7.355	29.011	19.992	6.274	5.791	4.013
× 15.9	48.6	6.2	24.724	16.879	6.325	5.664	4.013
× 14.3	44.0	5.600	22.518	15.305	6.35	5.613	4.039
× 12.7	39.3	5	20.229	13.7	6.35	5.563	4.039
L152 × 152 × 25.4	55.6	7.097	14.776	14.044	4.572	1.86	2.972
× 22.2	49.2	6.277	13.278	12.503	4.597	1.82	2.972
× 19.1	42.7	5.445	11.738	10.914	4.648	1.78	2.972
× 15.9	36	4.587	10.073	9.275	4.674	1.73	2.997
× 14.3	32.6	4.148	9.199	8.423	4.699	1.71	2.997
× 12.7	29.2	3.71	8.283	7.554	4.724	1.68	2.997
× 11.1	25.6	3.265	7.367	6.686	4.75	1.66	3.023
× 9.5	22.2	2.813	6.41	5.785	4.775	1.64	3.023
× 7.9	18.4	2.355	5.411	4.867	4.801	1.62	3.048
L127 × 127 × 22.2	40.5	5.148	7.409	8.472	3.785	1.57	2.471
× 19.1	35.1	4.477	6.535	7.423	3.835	1.52	2.477
× 15.9	29.8	3.781	5.661	6.325	3.861	1.48	2.484
× 12.7	24.1	3.065	4.703	5.178	3.912	1.43	2.497
× 11.1	21.3	2.617	4.162	4.572	3.937	1.41	2.504
× 9.5	18.3	2.329	3.638	3.966	3.962	1.39	2.515
× 7.9	15.3	1.955	3.088	3.343	3.988	1.37	2.525
L102 × 102 × 19.1	27.5	3.510	3.192	4.605	3.023	1.27	1.976
× 15.9	23.4	2.974	2.772	3.933	3.048	1.23	1.979
× 12.7	19.0	2.419	2.314	3.228	3.099	1.18	1.986
× 11.1	16.8	2.135	2.069	2.868	3.124	1.16	1.994
× 9.5	14.6	1.845	1.815	2.491	3.124	1.14	2.002
× 7.9	12.2	1.548	1.544	2.114	3.15	1.12	2.009
× 6.4	9.8	1.252	1.265	1.721	3.175	1.09	2.019

Note: The "L" in the designation column indicates a structural steel angle. The "L" is followed by the length of the legs and their thickness in millimeters.

Numbers in this table represent direct conversions from U.S. Customary to SI units.

E.4 DESIGN PROPERTIES FOR SELECTED AMERICAN STANDARD (S SHAPES), STRUCTURAL STEEL SECTIONS (U.S. UNITS AND SI UNITS)

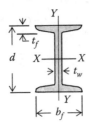

E.4.1 U.S. Units

Designation	Area A (in.²)	Depth d (in.)	Web Thickness t_w (in.)	Flange Width b_f (in.)	Flange Thickness t_f (in.)	X–X Axis I (in.⁴)	X–X Axis S (in.³)	X–X Axis r (in.)	Y–Y Axis I (in.⁴)	Y–Y Axis S (in.³)	Y–Y Axis r (in.)
S24 × 121	35.6	24.5	0.8	8.05	1.09	3160	258	9.43	83.3	20.7	1.53
× 106	31.2	24.5	0.62	7.87	1.09	2940	240	9.71	77.1	19.6	1.57
× 100	29.3	24	0.745	7.245	0.87	2390	199	9.02	47.7	13.2	1.27
× 90	26.5	24	0.625	7.125	0.87	2250	187	9.21	44.9	12.6	1.3
× 80	23.5	24	0.5	7	0.87	2100	175	9.47	42.2	12.1	1.34
S20 × 96	28.2	20.30	0.8	7.2	0.92	1670	165	7.71	50.2	13.9	1.33
× 86	25.3	20.3	0.66	7.06	0.92	1580	155	7.89	46.8	13.3	1.36
× 75	22	20	0.635	6.385	0.795	1280	128	7.62	29.8	9.32	1.16
× 66	19.4	20	0.505	6.255	0.795	1190	119	7.83	27.7	8.85	1.19
S18 × 70	20.6	18	0.711	6.251	0.691	926	103	6.71	24.1	7.72	1.08
× 54.7	16.1	18	0.461	6.001	0.691	804	89.4	7.07	20.8	6.94	1.14
S15 × 50	14.7	15	0.55	5.64	0.622	486	64.8	5.75	15.7	5.57	1.03
× 42.9	12.6	15	0.411	5.501	0.622	447	59.6	5.95	14.4	5.23	1.07
S12 × 50	14.7	12	0.687	5.477	0.659	305	50.8	4.55	15.7	5.74	1.03
× 40.8	12	12	0.462	5.252	0.659	272	45.4	4.77	13.6	5.16	1.06
× 35	10.3	12	0.428	5.078	0.544	229	38.2	4.72	9.87	3.89	0.98
× 31.8	9.35	12	0.35	5	0.544	218	36.4	4.83	9.36	3.74	1
S10 × 35	10.3	10	0.594	4.944	0.491	147	29.4	3.78	8.36	3.38	0.901
× 25.4	7.46	10	0.311	4.661	0.491	124	24.7	4.07	6.79	2.91	0.954
S8 × 23	6.77	8	0.441	4.171	0.426	64.9	16.2	3.1	4.31	2.07	0.798
× 18.4	5.41	8	0.271	4.001	0.426	57.6	14.4	3.26	3.73	1.86	0.831
S7 × 20	5.88	7	0.45	3.86	0.392	42.4	12.1	2.69	3.17	1.64	0.734
× 15.3	4.5	7	0.252	3.662	0.392	36.7	10.5	2.86	2.64	1.44	0.766
S6 × 17.25	5.07	6	0.465	3.565	0.359	26.3	8.77	2.28	2.31	1.3	0.675
× 12.5	3.67	6	0.232	3.332	0.359	22.1	7.37	2.45	1.82	1.09	0.705
S5 × 14.75	4.34	5	0.494	3.284	0.326	15.2	6.09	1.87	1.67	1.01	0.62
× 10	2.94	5	0.214	3.004	0.326	12.3	4.92	2.05	1.22	0.809	0.643
S4 × 9.5	2.79	4	0.326	2.796	0.293	6.79	3.39	1.56	0.903	0.646	0.569
× 7.7	2.26	4	0.193	2.663	0.293	6.08	3.04	1.64	0.764	0.574	0.581
S3 × 7.5	2.21	3	0.349	2.509	0.26	2.93	1.95	1.15	0.586	0.468	0.516
× 5.7	1.67	3	0.17	2.33	0.26	2.52	1.68	1.23	0.455	0.390	0.552

Source: Adapted from American Institute of Steel Construction, Inc. *Manual of Steel Construction*, AISC, Chicago, IL, 1980.

Note: The "S" in the designation column indicates an American standard section. The "S" is followed by the nominal height in inches and the weight in pounds per foot of length.

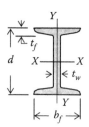

E.4.2   SI Units

Designation	Area A ($\times 10^{-3}$ m²)	Depth d ($\times 10^{-2}$ m)	Web Thickness t_w ($\times 10^{-3}$ m)	Flange Width b_f ($\times 10^{-2}$ m)	Flange Thickness t_f ($\times 10^{-3}$ m)	X–X Axis I ($\times 10^{-6}$ m⁴)	X–X Axis S ($\times 10^{-5}$ m³)	X–X Axis r ($\times 10^{-2}$ m)	Y–Y Axis I ($\times 10^{-7}$ m⁴)	Y–Y Axis S ($\times 10^{-6}$ m³)	Y–Y Axis r ($\times 10^{-2}$ m)
S610 × 180.0	22.97	62.23	20.32	20.45	27.69	1315	423	23.95	346.7	339.2	3.89
× 157.7	20.13	62.23	15.75	19.99	27.69	1224	393	24.66	320.9	321.2	3.99
× 148.8	18.9	60.96	18.92	18.4	22.1	995	326	22.91	198.5	216.3	3.23
× 133.9	17.1	60.96	15.88	18.1	22.1	937	306	23.39	186.9	206.5	3.3
× 119.0	15.16	60.96	12.7	17.78	22.1	874	287	24.05	175.6	198.3	3.4
S508 × 142.8	18.19	51.56	20.32	18.29	23.37	695	270	19.58	208.9	227.8	3.38
× 127.9	16.32	51.56	16.76	17.93	23.37	658	254	20.04	194.8	217.9	3.45
× 111.6	14.19	50.8	16.13	16.22	20.19	533	210	19.35	124	152.7	2.95
× 98.2	12.52	50.8	12.83	15.89	20.19	495	195	19.89	115.3	145	3.02
S457 × 104.1	13.29	45.72	18.06	15.88	17.55	385	169	17.04	100.3	126.5	2.74
× 81.4	10.39	45.72	11.71	15.24	17.55	335	146.5	17.96	86.6	113.7	2.9
S381 × 74.4	9.48	38.1	13.97	14.33	15.8	202	106.2	14.61	65.3	91.3	2.62
× 42.9	8.13	38.1	10.44	13.97	15.8	186	97.7	15.11	59.9	85.7	2.72
S305 × 74.4	9.48	30.48	17.45	13.91	16.74	127	83.2	11.56	65.3	94.1	2.62
× 60.7	7.74	30.48	11.73	13.34	16.74	113	74.4	12.12	56.6	84.6	2.69
× 52.1	6.65	30.48	10.87	12.9	13.82	95.3	62.6	11.99	41.1	63.7	2.49
× 47.3	6.03	30.48	8.89	12.7	13.82	91.2	59.6	12.27	39	61.3	2.54
S254 × 52.1	6.65	25.4	15.09	12.56	12.47	61.2	48.2	9.6	34.8	55.4	2.29
× 37.8	4.81	25.4	7.9	11.84	12.47	51.6	40.5	10.34	28.3	47.7	2.42
S203 × 34.2	4.37	20.32	10.44	10.59	10.82	27	26.5	7.87	17.9	33.9	2.03
× 27.4	3.49	20.32	6.88	10.16	10.82	24.0	23.6	8.28	15.5	30.5	2.11
S178 × 29.8	3.79	17.78	11.43	9.8	9.96	17.6	19.8	6.83	13.2	26.9	1.86
× 22.8	2.9	17.78	6.4	9.3	9.96	15.3	17.2	7.26	11	23.6	1.95
S152 × 25.7	3.27	15.24	11.81	9.06	9.12	10.9	14.4	5.79	9.6	21.3	1.71
× 18.6	2.37	15.24	5.89	8.46	9.12	9.2	12.1	6.22	7.6	17.9	1.79
S127 × 73.8	2.8	12.7	12.55	8.34	8.28	6.3	10	4.75	7.0	16.6	1.57
× 14.9	1.9	12.7	5.44	7.63	8.28	5.1	8.1	5.21	5.1	13.3	1.63
S102 × 14.1	1.80	10.16	8.28	7.1	7.44	2.8	5.6	3.96	3.8	10.6	1.45
× 11.5	1.46	10.16	4.9	6.76	7.44	2.5	5.0	4.17	3.2	9.4	1.48
S76 × 11.2	1.43	7.62	8.86	6.37	6.6	1.2	3.2	2.92	2.4	7.7	1.31
× 8.5	1.08	7.62	4.32	5.92	6.6	1.0	2.8	3.12	1.9	6.4	1.4

Note: The "S" in the designation column indicates an American standard section. The "S" is followed by the nominal height in millimeters and the mass in kilograms per meter of length.

Numbers in this table represent direct conversions from U.S. Customary to SI units.

E.5 DESIGN PROPERTIES FOR SELECTED STRUCTURAL STEEL PIPES (U.S. UNITS AND SI UNITS)

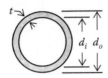

E.5.1 U.S. Units

Dimensions					Properties			
Nominal Diameter (in.)	Outside Diameter d_o (in.)	Inside Diameter d_i (in.)	Wall Thickness t (in.)	Unit Weight (lb/ft)	A (in.²)	I (in.⁴)	S (in.³)	r (in.)
Standard Weight								
1/2	0.84	0.622	0.109	0.85	0.25	0.017	0.041	0.261
1	1.315	1.049	0.133	1.68	0.494	0.087	0.133	0.421
1–1/2	1.9	1.61	0.145	2.72	0.799	0.31	0.326	0.623
2	2.375	2.067	0.154	3.65	1.07	0.666	0.561	0.787
2–1/2	2.875	2.469	0.203	5.79	1.7	1.53	1.06	0.947
3	3.500	3.068	0.216	7.58	2.23	3.02	1.72	1.16
3–1/2	4	3.548	0.226	9.11	2.68	4.79	2.39	1.34
4	4.500	4.026	0.237	10.79	3.17	7.23	3.21	1.51
5	5.563	5.047	0.258	14.62	4.3	15.2	5.45	1.88
6	6.625	6.065	0.28	18.97	5.58	28.1	8.50	2.25
8	8.625	7.981	0.322	28.55	8.4	72.5	16.8	2.94
10	10.750	10.020	0.365	40.48	11.9	161	29.9	3.67
12	12.75	12	0.375	49.56	14.6	279	43.8	4.38
Extra Strong								
1/2	0.84	0.546	0.147	1.09	0.32	0.02	0.048	0.25
1	1.315	0.957	0.179	2.17	0.639	0.106	0.161	0.407
1–1/2	1.9	1.5	0.2	3.63	1.07	0.391	0.412	0.605
2	2.375	1.939	0.218	5.02	1.48	0.868	0.731	0.766
2–1/2	2.875	2.323	0.276	7.66	2.25	1.92	1.34	0.924
3	3.500	2.900	0.300	10.25	3.02	3.89	2.23	1.14
4	4.5	3.826	0.337	14.98	4.41	9.61	4.27	1.48
5	5.563	4.813	0.375	20.78	6.11	20.7	7.43	1.84
6	6.625	5.761	0.432	28.57	8.4	40.5	12.2	2.19
8	8.625	7.625	0.5	43.39	12.8	106	24.5	2.88
10	10.75	9.75	0.5	54.74	16.1	212	39.4	3.63
12	12.750	11.75	0.5	65.42	19.2	362	56.7	4.33
Double-Extra Strong								
2	2.375	1.503	0.436	9.03	2.66	1.31	1.1	0.703
2–1/2	2.875	1.771	0.552	13.69	4.03	2.87	2.00	0.844
3	3.5	2.3	0.6	18.58	5.47	5.99	3.42	1.05
4	4.500	3.152	0.674	27.54	8.1	15.3	6.79	1.37
5	5.563	4.063	0.75	38.55	11.3	33.6	12.1	1.72
6	6.625	4.897	0.864	53.16	15.6	66.3	20.0	2.06
8	8.625	6.875	0.875	72.42	21.3	162	37.6	2.76

Source: Adapted from American Institute of Steel Construction, Inc. *Manual of Steel Construction*, AISC, Chicago, IL, 1980.

Note: The listed sections are available in conformance with ASTM specification A53 Grade B or A501. Other sections are made to these specifications.

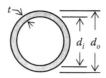

E.5.2   SI Units

Dimensions					Properties			
Nominal Diameter (mm)	Outside Diameter d_o ($\times 10^{-2}$ m)	Inside Diameter d_i ($\times 10^{-2}$ m)	Wall Thickness t ($\times 10^{-3}$ m)	Unit Mass (kg/m)	A ($\times 10^{-4}$ m^2)	I ($\times 10^{-8}$ m^4)	S ($\times 10^{-7}$ m^3)	r ($\times 10^{-3}$ m)
Standard Weight								
13	2.134	1.58	2.769	1.26	1.613	0.71	6.72	6.629
25	3.34	2.664	3.378	2.5	3.187	3.62	21.79	10.693
38	4.826	4.089	3.683	4.05	5.155	12.9	53.42	15.824
51	6.033	5.25	3.912	5.43	6.903	27.72	91.93	19.99
64	7.303	6.271	5.156	8.61	10.968	63.68	173.7	24.054
76	8.89	7.793	5.486	11.28	14.387	125.7	281.86	29.464
89	10.16	9.012	5.74	13.55	17.29	199.37	391.65	34.036
102	11.43	10.226	6.02	16.05	20.452	300.93	526.02	38.354
127	14.13	12.819	6.553	21.75	27.742	632.67	893.09	47.752
152	16.828	15.405	7.112	28.22	36	1169.96	1392.9	57.15
203	21.908	20.272	8.179	42.47	54.193	3017.67	2753.02	74.646
254	27.305	25.451	9.271	60.22	76.774	6701.3	4899.71	93.218
305	32.385	30.48	9.525	73.72	94.193	11612.2	7177.51	111.252
Extra Strong								
13	2.134	1.387	3.734	1.62	2.065	0.83	7.87	6.35
25	3.34	2.431	4.547	3.23	4.123	4.41	26.38	10.338
38	4.826	3.81	5.08	5.4	6.903	16.27	67.51	15.367
51	6.033	4.925	5.537	7.47	9.548	36.13	119.79	19.456
64	7.303	5.9	7.01	11.39	14.516	79.92	219.59	23.47
76	8.89	7.366	7.62	15.25	19.355	161.91	365.43	28.956
102	11.43	9.718	8.56	22.28	28.452	400	699.72	37.592
127	14.13	12.225	9.525	30.91	39.419	861.6	1217.55	46.736
152	16.828	14.633	10.973	42.5	54.193	1685.73	1999.21	55.626
203	21.908	19.368	12.7	64.55	82.58	4412.04	4014.82	73.152
254	27.305	24.765	12.7	81.43	103.871	8824.08	6456.48	92.202
305	32.385	29.845	12.7	97.32	123.871	15067.53	9291.43	109.982
Double-Extra Strong								
51	6.033	3.818	11.151	13.43	17.161	54.53	180.26	17.856
64	7.303	4.498	14.021	20.36	26	119.46	327.74	21.438
76	8.89	5.842	15.24	27.64	35.29	249.32	560.44	26.67
102	11.43	8.006	17.12	40.97	52.258	636.83	1112.68	34.798
127	14.13	10.32	19.05	57.35	72.903	1398.53	1982.83	43.688
152	16.828	12.438	21.946	79.08	100.645	2759.6	3277.4	52.324
203	21.908	17.463	22.225	107.73	137.419	6742.93	6161.51	70.104

Note: Numbers in this table represent direct conversions from U.S. Customary to SI units.

Appendix F: Design Properties for Selected Structural Wood Sections (U.S. Units and SI Units)

F.1 U.S. UNITS

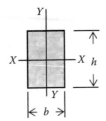

Nominal Dimensions $b \times h$ (in.)	Dressed Dimensions $b \times h$ (in.)	Unit Weight (lb/ft)	Area A (in.²)	X–X Axis		Y–Y Axis	
				I (in.⁴)	S (in.³)	I (in.⁴)	S (in.³)
8×12	7.5×11.5	21.0	86.25	950.5	165.3	404.3	107.8
8×10	7.5×9.5	17.3	71.25	535.9	112.8	334.0	89.1
8×8	7.5×7.5	13.7	56.25	263.7	70.3	263.7	70.3
6×12	5.5×11.5	15.4	63.25	697.1	121.2	159.4	58.0
6×10	5.5×9.5	12.7	52.25	393.0	82.7	131.7	47.9
6×8	5.5×7.5	10.0	41.25	193.4	51.6	104.0	37.8
6×6	5.5×5.5	7.4	30.25	76.3	27.7	76.3	27.7
4×12	3.5×11.25	9.6	39.38	415.28	73.83	40.20	22.97
4×10	3.5×9.25	7.9	32.38	230.84	49.91	33.05	18.89
4×8	3.5×7.25	6.2	25.38	111.15	30.66	25.90	14.82
4×6	3.5×5.5	4.7	19.25	48.53	17.65	19.65	11.23
4×4	3.5×3.5	3.0	12.25	12.51	7.15	12.51	7.15
3×12	2.5×11.25	6.8	28.13	296.63	52.73	14.65	11.72
3×10	2.5×9.25	5.6	23.13	164.89	35.65	12.04	9.64
3×8	2.5×7.25	4.4	18.13	79.39	21.90	9.44	7.55
3×6	2.5×5.5	3.3	13.75	34.66	12.60	7.16	5.73
3×4	2.5×3.5	2.1	8.75	8.93	5.10	4.56	3.65
2×12	1.5×11.25	4.1	16.88	177.98	31.64	3.16	4.22
2×10	1.5×9.25	3.4	13.88	98.93	21.39	2.60	3.47
2×8	1.5×7.25	2.6	10.88	47.63	13.14	2.04	2.72
2×6	1.5×5.5	2.0	8.25	20.80	7.56	1.55	2.06
2×4	1.5×3.5	1.3	5.25	5.36	3.06	0.98	1.31

Source: Adapted from National Forest Products Association (http://www.awc.org/CopyrightDisclaimer.php).
Note: All properties and weights are stated for dressed dimensions.

F.2   SI UNITS

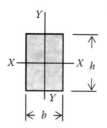

Nominal Dimensions $b \times h$ (mm)	Dressed Dimensions $b \times h$ (mm)	Unit Mass (kg/m)	Area A ($\times 10^{-3}$ m²)	X–X Axis		Y–Y Axis	
				I ($\times 10^{-6}$ m⁴)	S ($\times 10^{-5}$ m³)	I ($\times 10^{-7}$ m⁴)	S ($\times 10^{-5}$ m³)
203×305	190.5×292.1	31.24	55.645	395.627	270.877	1682.818	176.652
203×254	190.5×241.3	25.74	45.968	223.058	184.845	1390.208	146.008
203×203	190.5×190.5	20.38	36.290	109.760	115.201	1097.599	115.201
152×305	139.7×292.1	22.91	40.806	290.154	198.610	663.471	95.045
152×254	139.7×241.3	18.89	33.710	163.578	135.520	548.175	78.494
152×203	139.7×190.5	14.88	26.613	80.499	84.557	432.879	61.943
152×152	139.7×139.7	11.01	19.516	31.758	45.392	317.583	45.392
102×305	88.9×285.8	14.28	25.406	172.852	120.985	167.324	37.641
102×254	88.9×235.0	11.75	20.890	96.083	81.788	137.564	30.955
102×203	88.9×184.2	9.22	16.374	46.264	50.243	107.804	24.286
102×152	88.9×139.7	6.99	12.419	20.200	28.923	81.789	18.403
102×102	88.9×88.9	4.46	7.903	5.207	11.717	52.070	11.717
76×305	63.5×285.8	10.12	18.148	123.466	86.409	60.978	19.206
76×254	63.5×235.0	8.33	14.923	68.632	58.420	50.114	15.797
76×203	63.5×184.2	6.55	11.697	33.044	35.888	39.292	12.372
76×152	63.5×139.7	4.91	8.871	14.427	20.648	29.802	9.390
76×102	63.5×88.9	3.12	5.645	3.717	8.357	18.980	5.981
51×305	38.1×285.8	6.10	10.890	74.081	51.848	13.153	6.915
51×254	38.1×235.0	5.06	8.955	41.178	35.052	10.822	5.686
51×203	38.1×184.2	3.87	7.019	19.825	21.533	8.491	4.457
51×152	38.1×139.7	2.98	5.323	8.658	12.389	6.452	3.376
51×102	38.1×88.9	1.93	3.387	2.231	5.014	4.079	2.147

Source: Adapted from National Forest Products Association (http://www.awc.org/CopyrightDisclaimer.php).
Note: All properties and weights are stated for dressed dimensions.

Appendix G: Beam Slopes and Deflections for Selected Cases

Case	Beam	Slope and Deflection Equations
1		$v_a = (-Pa^2/6EI)(3a - x)$ $\theta = (-Pa^2)/(2EI)$; $v_b = (-Pa^2/6EI)(3x - a)$ $v_{MAX} = (-Pa^2/6EI)(3L - a)$
2		$\theta = (QL)/(EI)$ $v = (Qx^2)/(2EI)$ $v_{MAX} = (QL^2)/(2EI)$
3		$\theta = (-pL^3)/(6EI)$ $v = (-px^2/24EI)(x^2 + 6L^2 - 4Lx)$ $v_{MAX} = (-pL^4)/(8EI)$
4		$\theta = (-pL^3)/(24EI)$ $v = (-px^2/120LEI)(10L^3 - 10L^2 + 5Lx^2 - x^3)$ $v_{MAX} = (-pL^4)/(30EI)$
5		$\theta = (pL^3)/(60EI)$ $v = (-p/12EI)[(x^6/30L^2) - (L^3/5) + (L^4/6)]$ $v_{MAX} = (-pL^4)/(72EI)$
6		$\theta = (pa^3)/(6EI)$; $v_{MAX} = (-pa^3/24EI)(4L - a)$ $v_a = (-Px^2/24EI)(6a^2 - 4ax + x^2)$ $v_b = (-pa^3/24EI)(4x - a)$
7		$\theta_1 = [Pb(b^2 - L^2)]/(6EIL)$; $\theta_2 = [Pab(2L - b)]/(6EIL)$ $v_a = (Pbx/6EIL)(b^2 + x^2 - L^2)$ $v_b = -(Pb/6EIL)[(L/b)(x - a)^3 + (L^2 - b^2)x - x^3]$ $v_{MAX} = [(-Pb)(L^2 - b^2)^{3/2}]/[9(3)^{1/2}EI]$ at $x = [(L^2 - b^2)/3]^{1/2}$

(Continued)

Case	Beam	Slope and Deflection Equations

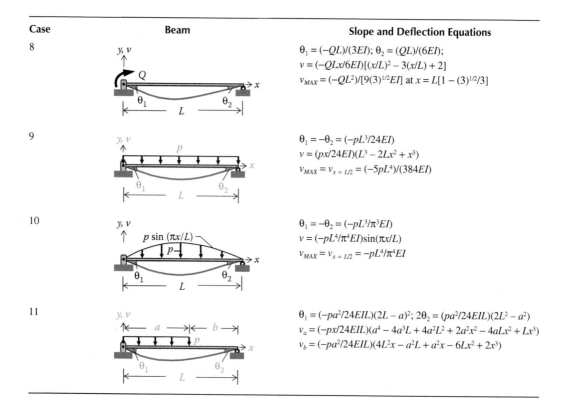

8

$\theta_1 = (-QL)/(3EI)$; $\theta_2 = (QL)/(6EI)$;

$v = (-QLx/6EI)[(x/L)^2 - 3(x/L) + 2]$

$v_{MAX} = (-QL^2)/[9(3)^{1/2}EI]$ at $x = L[1 - (3)^{1/2}/3]$

9

$\theta_1 = -\theta_2 = (-pL^3/24EI)$

$v = (px/24EI)(L^3 - 2Lx^2 + x^3)$

$v_{MAX} = v_{x = L/2} = (-5pL^4)/(384EI)$

10

$\theta_1 = -\theta_2 = (-pL^3/\pi^3EI)$

$v = (-pL^4/\pi^4EI)\sin(\pi x/L)$

$v_{MAX} = v_{x = L/2} = -pL^4/\pi^4EI$

11

$\theta_1 = (-pa^2/24EIL)(2L - a)^2$; $2\theta_2 = (pa^2/24EIL)(2L^2 - a^2)$

$v_a = (-px/24EIL)(a^4 - 4a^3L + 4a^2L^2 + 2a^2x^2 - 4aLx^2 + Lx^3)$

$v_b = (-pa^2/24EIL)(4L^2x - a^2L + a^2x - 6Lx^2 + 2x^3)$

Appendix H: Two-Dimensional Supports and Connections

Support or Connection	Reactive Force Components	Special Features

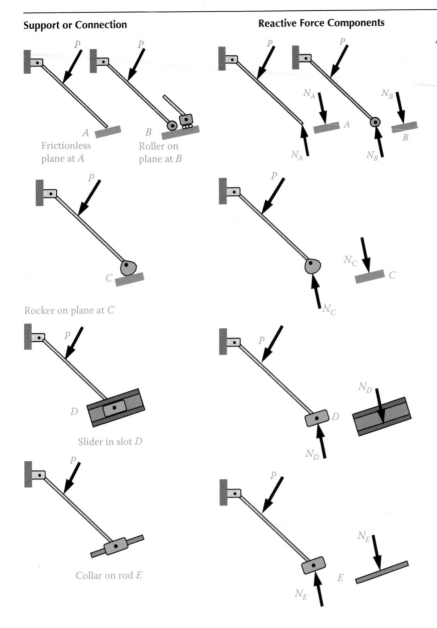

Frictionless plane at A

Roller on plane at B

Rocker on plane at C

Slider in slot D

Collar on rod E

All of the support systems shown on this page have the same common property of allowing translation along the plane or axis of support, but preventing it in a direction perpendicular to this plane or axis. Thus, for example, the roller support develops only one reaction component, N_B, and the slider in the slot, only one reaction component, N_D, both perpendicular to the direction of allowable translation.

(*Continued*)

Support or Connection	**Reactive Force Components**	**Special Features**
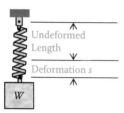 Spring with attached weight	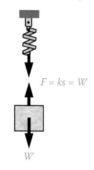 $F = ks = W$	The force, F, in a deformed spring is directed along the axis of the spring. The sense of this force is such that it is tension if the spring is stretched, and compression if the spring is shortened. Also, $F = ks$, where k is the spring constant equal to the force needed to deform the spring a unit distance and s is the deformation.
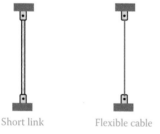 Short link Flexible cable	 F	The short link can resist tension or compression. The flexible cable, however, can resist only tension. In both cases, there is only one unknown quantity, namely, the the magnitude of the reactive force F because its direction is along the axis of the link or the cable.
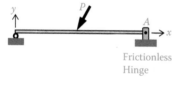 Frictionless Hinge	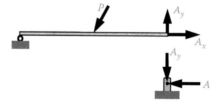	A frictionless hinge prevents any translation but permits rotation about the pin axis. The reaction at the hinge is usually expressed in terms of its x and y components. Thus, the reaction at hinge A consists of the two unknown quantities, A_x and A_y.
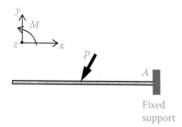 Fixed support	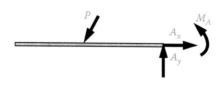	A fixed support prevents translation in any direction and rotation about a z axis at the support. Thus, the support reaction consists of two force components, A_x and A_y, and a moment component, M_A, as shown.

H.1   THREE-DIMENSIONAL SUPPORTS AND CONNECTIONS

Support or Connection	Reactive Force Components	Special Features

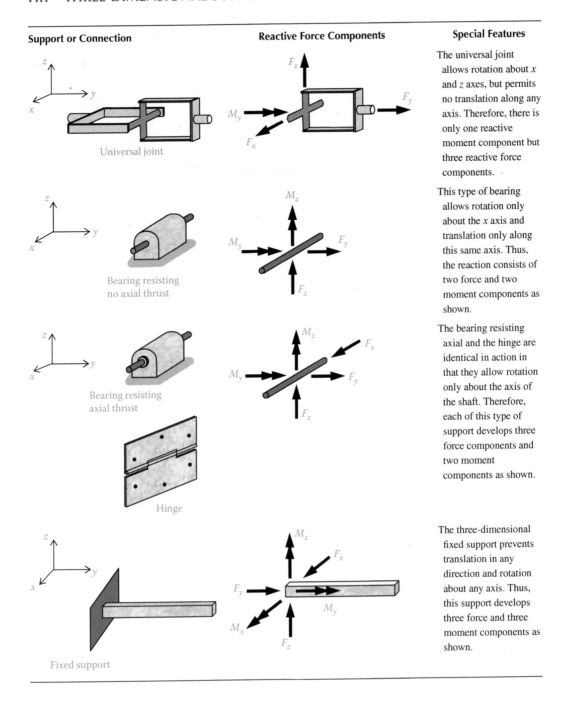

The universal joint allows rotation about x and z axes, but permits no translation along any axis. Therefore, there is only one reactive moment component but three reactive force components.

Universal joint

This type of bearing allows rotation only about the x axis and translation only along this same axis. Thus, the reaction consists of two force and two moment components as shown.

Bearing resisting no axial thrust

The bearing resisting axial and the hinge are identical in action in that they allow rotation only about the axis of the shaft. Therefore, each of this type of support develops three force components and two moment components as shown.

Bearing resisting axial thrust

Hinge

The three-dimensional fixed support prevents translation in any direction and rotation about any axis. Thus, this support develops three force and three moment components as shown.

Fixed support

Index

A

Absolute maximum shearing stress, 432, 506
Allowable stress, 29
    design, 199–202; *see also* Beams
AFPA (American Forest and Paper Association), 569
AISC (American Institute of Steel Construction), 563
American standard structural steel section properties, 689
    SI units, 690
    U.S. units, 689
Area–moment method, 279, 303; *see also* Deflections
            under symmetric loading
    beam subjected to concentrated couple, 306
    statically indeterminate beams, 333–341
    tangential deviations, 305
    theorems, 304
Average
    normal stress, 398
    strain, 475
Axial loads, 1; *see also* Impact loading; Internal axial
            force; Load-deformation relations; Statically
            indeterminate members; Strain; Stress
    answers to problems, 633–634
    review problems, 65–68

B

BC (Boundary conditions), 283
Beams, 149, 167, 279; *see also* Statically indeterminate
            beam; Symmetric bending
    allowable-stress design, 199
    assumption in beam analysis, 168
    deformation, 280
    examples, 152–153, 154–156, 162–165, 193–196, 200–202
    load, shear, and moment relationships, 160–166
    moment–curvature relationship, 279
    neutral axis, 167
    neutral surface, 167
    problems, 156–160, 165–166, 196–199
    shear and moment, 150–151, 153, 160
    sign convention, 151
    slopes, 695–696
    stress under load, 192–199
    symmetric bending, 280
Beams of mixed materials, 211; *see also* Bending loads
    analysis, 211–213
    application to reinforced concrete, 215
    basic assumption of beam theory, 211
    equivalent cross section, 212, 213
    example, 213–215, 216–217
    general principles, 211
    modular ratio, 216
    problems, 218–223
    shrinking or magnifying factor, 212
Bending; *see also* Beams; Symmetric bending
    equation, 170
    stress, 167
    types, 149

Bending loads, 211; *see also* Beams of mixed materials;
            Curved beams; Fatigue; Symmetric bending;
            Thin-walled open sections; Unsymmetric
            bending
    answers to problems, 638–639
    curved beams, 247
    elastoplastic behavior, 254
    example, 257–261
    fully plastic moment, 255
    internal normal stresses, 254
    moment redistribution, 257
    plastic hinge, 256–257
    problems, 261–264, 275–278
    pseudo mechanism, 256
    shape factor, 254–256
    simply-supported beam, 256
    statically indeterminate beam, 257
    stress–strain diagram, 254
    yield moment, 255
Breaking strength, *see* Fracture strength
Brittle materials, 26; *see also* Mechanical properties
Bulk modulus of elasticity, *see* Modulus of volume change

C

Cantilever parts, 307; *see also* Deflections under
            symmetric loading
    example, 307–313
    problems, 313–318
Castigliano's second theorem, 353; *see also* Deflection
    differential strain energy, 355
    example, 356–359
    free-body diagram, 355
    mathematical form of, 355
    problems, 359–361
    statically indeterminate beams, 373–380
    stored strain energy, 354
    strain energy, 356
Centrically loaded columns, 561; *see also* Column
    aluminum alloys, 567–568, 574–576
    column design curve, 561
    empirical equations, 562
    example, 564–567, 568–569, 570–571
    inelastic buckling, 561–562
    long columns, 563
    problems, 571–578
    short columns, 563
    stress versus slenderness ratio, 564, 567, 569
    structural steel, 563
    structural woods, 569–570, 576–578
Centroidal longitudinal principal plane, 149, 166; *see also*
            Beams
Centroid location, 651; *see also* Plane area properties
    center of gravity, 651
    centroid of area, 651–653
    properties of selected plane areas, 674–675
Centroid of composite area, 653–656; *see also* Plane area
            properties

Circular shafts, 78; *see also* Shafts; Torsional loads
  component parts, 81
  example, 84–90
  Hooke's law, 79
  material properties in shear, 82–83
  modulus of rigidity and rupture, 82
  problems, 90–95
  shaft distortion, 79
  shearing deformation, 79–81
  shearing strain, 78–79
  shearing stress, 79–81, 140
  shearing stress–strain diagram, 82
  sign convention, 79
  state of pure shear, 84
  stress determination, 84
  stress distribution, 80
  stress element, 83
  stress on inclined planes, 83–84
  torsional toughness, 83
Circumferential stress, 247, 441; *see also* Curved beams;
    Thin-walled pressure vessels
Coefficient of thermal expansion, 48; *see also* Statically
    indeterminate members
Column, 521; *see also* Centrically loaded columns;
    Eccentrically loaded columns; Euler's ideal-
    column theory; Secant formula
  answers to problems, 645–646
  buckled fixed-free column model, 542
  column boundary conditions, 541
  critical buckling loads of columns, 544
  critical load, 523, 541–543
  effective length, 543–544
  effect of end conditions, 541
  equilibrium types, 522
  example, 524–526, 544–545
  models, 523
  potential energy functions, 522
  problems, 527–529, 545–548, 589–592
  stability of equilibrium, 521–523
Compression test, *see* Tension test
Concentrated axial force, 1; *see also* Internal axial force
Concentrated torque, 69–70; *see also* Internal torque
Conditional formatting, 605; *see also* Excel spreadsheet
    applications
Continuity conditions, *see* Matching conditions (MC)
Conversion factors, 647; *see also* SI units
Coulomb's theory, *see* Maximum principal stress theory
Critical load, 523, 541
Critical stress, 533; *see also* Euler's ideal-column theory
Curved beams, 247; *see also* Bending loads
  bending moments, 247, 248
  circumferential strain, 248–250
  circumferential stresses, 247
  example, 250–252
  problems, 252–253
  radial stress, 247

D

Deflection, 343; *see also* Castigliano's second theorem;
    Impact loading; Singularity functions;
    Statically indeterminate beam; Unsymmetric
    bending loads
  answers to problems, 641–642

  review problems, 386–387
  for selected cases, 695–696
Deflections under symmetric loading, 279; *see also* Area-
    moment method; Double integration method;
    Statically indeterminate beam; Superposition
  answers to problems, 639–641
  cantilever parts, 307–318
  deflection methods, 279
  derivatives of deflection function, 295–296
  example, 296–298
  moment–curvature relationship, 280–281
  problems, 298–299, 338
Deformations, 20, 475; *see also* Strain at point; Stress
Derived units, 647; *see also* SI units
Distortions, 20, 475, 476; *see also* Stress; Strain at point
Double integration method, 279, 282; *see also* Deflections
    under symmetric loading
  beam curvature, 282
  deflected elastic curve, 283
  example, 283–288
  flexural rigidity, 282
  matching conditions, 283
  problems, 288–295
  second-order linear differential equation, 282
  statically indeterminate beams, 319–326
Ductility, 26–27; *see also* Mechanical properties

E

Eccentrically loaded columns, 578; *see also* Column
  allowable-stress method, 578
  axial load and bending moment interaction curve, 579
  example, 579–582
  interaction equation, 579
  interaction method, 578, 579
  maximum stress, 578
  problems, 582–588
Eccentricity ratio, 551
Effective length, 543
Elasticity modulus, 25; *see also* Mechanical properties
Elastic limit, 25; *see also* Mechanical properties
Elastoplastic stress–strain diagrams, 139; *see also*
    Torsional loads
Electric-resistance strain gage, 510, 511; *see also* Strain
    measurements
Empirical equations, 562
Endurance limit, 267; *see also* Fatigue
Endurance strength, 267; *see also* Fatigue
Energy absorption capacity, 27; *see also* Mechanical
    properties
Energy of distortion, 503–505; *see also* Failure theories;
    Three-dimensional Hooke's law
  theory, 460–462
Equal-leg design properties, 687
  SI units, 688
  U.S. units, 687
Euler, Leonard, 529; *see also* Euler's ideal-column theory
Euler's column theory, 521; *see also* Column
Euler's ideal-column theory, 529; *see also* Column
  critical stress, 533
  Euler's equation, 532
  Euler's ideal column, 529, 530, 533
  example, 533–536
  free-body diagram, 531

problems, 537–541
second mode of buckling deformation, 533
Excel functions, 597; *see also* Excel spreadsheet applications
 AND, 598
 ATAN, 598
 CHOOSE, 599
 IF, 599–600
 IFERROR, 600
 RADIANS, 600
 RANK. EQ, 600
 SIN, 600
 TRANSPOSE, 600–601
 VLOOKUP, 601–602
Excel spreadsheet applications, 593; *see also* Excel functions
 cell protection, 594, 596
 cell styles, 594
 combined stress computation, 626–631
 combo-box control, 607
 concepts and techniques, 593
 conditional formatting, 605, 606
 using controls, 605–607
 cross-sectional properties of shape, 631
 data validation, 594–598
 drawing Mohr's circle, 614–616
 drawing shear and moment diagrams, 607–614
 excel functions, 597–602, 604–605
 inelastic analysis, 632
 invalid result display, 596
 list validation, 598, 599
 planning, 593
 plotting Mohr's circle, 616–617
 projects, 631
 real-time sorting, 602–604
 state of stress on plane, 618–621
 stresses in 3D stress elements, 621–626, 627
 stresses in statically indeterminate systems, 632
 using styles, 594
 worksheet protection, 595

F

Factor of safety (FS), 29, 269
Failure load, *see* Critical load
Failure theories, 457; *see also* Stress analysis
 brittle material, 457
 ductile materials, 460
 energy of distortion theory, 460–462
 example, 462–465
 maximum principal stress theory, 457–458
 maximum shearing stress theory, 460, 461
 Mohr's theory, 458–459
 octahedral plane, 461
 octahedral shearing stress theory, 461
 problems, 465–471
Fatigue, 265; *see also* Bending loads
 bending moment, 266
 bending stress, 266
 complete stress reversal, 266
 crack propagation in fatigue failure, 265
 decomposition of stress variation, 269
 endurance limit, 267
 endurance strength, 267
 example, 270–273
 fatigue curves, 267, 268

fatigue specimen, 267
fractures, 265
Gerber parabola, 268–269
Goodman straight line, 269–270
problems, 273–275
R. R. Moore fatigue testing machine, 266
Soderberg straight line, 270
stress variation, 267
testing, 266
theories of fatigue failure, 268
Fictitious mechanism, *see* Pseudo mechanism
Flexural equation, *see* Bending—equation
Flexural rigidity, 282
Flexural stresses, *see* Bending—stress
Fracture strength, 26; *see also* Mechanical properties
Free-body diagram, 1; *see also* Axial loads
FS, *see* Factor of safety (FS)
Fully plastic moment, 255; *see also* Bending loads
Fully plastic torque, 140; *see also* Torsional loads
Function, 604; *see also* Excel spreadsheet applications

G

Gage pressure, 444; *see also* Thin-walled pressure vessels
Gerber parabola, 268–269; *see also* Fatigue
Goodman straight line, 269–270; *see also* Fatigue

H

Hooke's law, 79, 499; *see also* Circular shafts; Three-dimensional Hooke's law
Hoop stress, *see* Circumferential stress

I

Impact loading, 60, 113, 380; *see also* Axial loads; Deflection; Torsional loads
 analysis of, 61
 beam subjected to, 381
 elastic strain energy, 113
 example, 62–63, 115, 381–383
 impact factor, 62
 problems, 64–65, 116–120, 383–385
 shaft subjected to torque, 114
 shearing stress–strain diagram, 114
 strain energy, 61, 381
Internal axial force, 1, 2; *see also* Axial loads
 under concentrated axial loads, 2
 concentrated force, 1–3
 distributed forces, 4
 example, 5–7
 free-body diagram, 3
 problems, 7–8
 sign convention, 3–4
Internal normal stresses, 254; *see also* Bending loads
Internal torque, 69; *see also* Torsional loads
 concentrated torque, 69–70
 diagram, 73
 distributed torques, 71, 72
 example, 73–76
 problems, 76–78
 shaft subjected to concentrated torques, 71
 sign convention, 70–71
 variably distributed torque, 72

L

Left free body (LFB), 3
LFB, *see* Left free body (LFB)
Load-deformation relations, 34; *see also* Axial loads
  example, 36–39
  problems, 39–43
  strain at point, 34–36
Longitudinal strain, 26; *see also* Mechanical properties
Longitudinal stress, 441, 442; *see also* Thin-walled
    pressure vessels

M

Magnifying factor, 212
Matching conditions (MC), 283
Material properties, 677–678
Maximum principal stress theory, 457–458; *see also*
    Failure theories
Maximum shearing stress theory, 460, 461; *see also*
    Failure theories
MC, *see* Matching conditions (MC)
Mechanical properties, 21; *see also* Strain
  compression test, 21
  ductility, 26–27
  elasticity modulus, 25
  elastic limit, 25
  energy absorption capacity, 27
  fracture strength, 26
  MC-300PR, 23
  MTS extensometer, 23
  MTS universal testing machine, 22
  Poisson's ratio, 26
  proportional limit, 25
  Ramberg–Osgood equation, 25
  resilience modulus, 27, 28
  stress–strain diagrams, 24
  toughness modulus, 29
  ultimate strength, 26
  yield point, 25
  yield strength, 25
Membrane analogy, 123; *see also* Noncircular cross
    section shafts
Microsoft™ Excel®, 593; *see also* Excel spreadsheet
    applications
Modular ratio, 216
Modulus of elasticity, 25; *see also* Mechanical properties;
    Modulus of rigidity
Modulus of resilience, 27, 28; *see also* Mechanical
    properties
Modulus of rigidity, 82; *see also* Circular shafts
Modulus of rupture, 82
Modulus of toughness, 29; *see also* Mechanical properties
Modulus of volume change, 502; *see also* Three-
    dimensional Hooke's law
Mohr's circle, 389; *see also* Plane area properties;
    Strain analysis; Stress analysis
  combined, 508
  construction of, 410–412
  example, 416–419, 420–421, 491–495, 509–510
  45° triangular stress elements, 415
  inclined axes, 490–491
  for moments, 669–673
  for plane strain, 488–489, 496

for plane stress, 410, 411, 496
  problems, 421–427, 497–498
  relation $G = E/2(1 + \mu)$, 495
  sign conventions, 411, 489
  for strain, 475
  strains and in-plane shearing strain, 489–490
  strain transformation using, 491
  stress element, 413
  stresses and in-plane shearing stress, 412–419
  stress on inclined planes, 419–420
  for 3D strain systems, 506–509
  triaxial stress systems, 432–441
  2D strain element, 507, 508
Mohr's theory, 458; *see also* Failure theories
  graphical representation of, 458, 459
Moment–curvature relationship, 279, 280; *see also*
    Deflections under symmetric loading
  beam curvature, 280
  bending stress magnitude, 280
  elastic curve, 280
  example, 281
  strain magnitude, 280
Moment of inertia; *see also* Plane area properties
  of area, 656–658
  of composite areas, 660–662
  by integration, 658–659
Moment redistribution, 257; *see also* Bending loads
MTS Extensometer, 23; *see also* Mechanical properties
MTS Universal Testing Machine, 22; *see also* Mechanical
    properties

N

Necking, 26; *see also* Mechanical properties
Noncircular cross section shafts, 120; *see also* Shafts;
    Torsional loads
  analytical solutions, 120–123
  circular section with thin slit, 129
  distended thin membrane, 124, 131
  example, 123, 125–127, 130–131, 133–134
  experimental solutions, 123–127
  mathematical solutions, 122
  membrane analogy, 123
  narrow rectangles, 129–131
  problems, 134–139
  shear flow, 132
  special cases, 129
  torsion of thin rectangle, 127–129
  torsion of thin-walled tubes, 131–133
  warping, 121

O

Octahedral plane, 461
Octahedral shearing stress theory, 461; *see also*
    Failure theories

P

Parallel-axis theorem, 659–660; *see also* Plane area
    properties
Plane area properties, 651
  centroid location, 651–652
  centroid of composite area, 653–656

Mohr's circle for moments and products of inertia, 669–673

moment of inertia by integration, 658–659

moment of inertia of area, 656–658

moment of inertia of composite areas, 660–662

parallel-axis theorem, 659–660, 662–663

principal axes and principal moments of inertia, 665–668

product of inertia, 662–665

Plane-strain transformation equations, 477; *see also* Strain analysis

decomposition of distortion, 480

example, 482–486

in-plane shearing strain, 482

Mohr's circle for, 489

normal and shear strains, 477

parameters for, 478

plane strains, 479

principal strains, 481–482

problems, 486–488

shear strain effect on plane-strain element, 478

Plane-stress transformation equations, 392; *see also* Stress analysis

example, 394–396, 400–404

in-plane shearing stress, 398

normal stress, 398

plane stress system, 392

principal planes, 397

principal stresses, 396–398

problems, 404–409

rotated stress element, 394

shear stress system, 400

stress on inclined planes, 392–396

uniaxial stress system, 400

Plastic hinge behavior, 256–257; *see also* Bending loads

PNA (Plastic neutral axis), 255

Points of inflection, 543; *see also* Column

Poisson's ratio, 26; *see also* Mechanical properties

Power-transmission shaft design, 103–104; *see also* Torsional loads

Pressure vessels, 441; *see also* Thick-walled cylindrical pressure vessels; Thin-walled pressure vessels

Principal axes, 665–668; *see also* Beams; Plane area properties

of inertia, 149, 150

Principal moments of inertia, 665–668; *see also* Plane area properties

Product of inertia, 662, 669–673; *see also* Plane area properties

example, 663–665

parallel-axis theorem for, 662–663

principal axes of inertia, 662

product of inertia, 662

Progressive fracture, *see* Fatigue

Proportional limit, 25; *see also* Mechanical properties

Pseudo mechanism, 256; *see also* Bending loads

R

Radial stress, 247; *see also* Curved beams

Ramberg–Osgood equation, 25; *see also* Mechanical properties

Rectangular moments of inertia, *see* Moment of inertia—of area

Reinforced concrete beams, 211, 216

Resilience modulus, 27, 28; *see also* Mechanical properties

RFB (Right free body), 3

Rigid, 1; *see also* Axial loads

Rigidity modulus, 82; *see also* Circular shafts

Rupture, 82; *see also* Circular shafts

strength, *see* Fracture strength

S

Secant formula, 548; *see also* Column

eccentrically loaded pin-ended column, 548–549

eccentricity ratio, 551

example, 555–557

initial deflection, 552

initially bent pin-ended columns, 552–555

maximum deflection, 549–550

maximum moment, 551

maximum stress, 554

maximum total deflection, 553

problems, 557–561

Section modulus, 171

Shafts, 69; *see also* Circular shafts; Noncircular cross section shafts; Torsional loads

Shape factor, 255; *see also* Bending loads

Shear; *see also* Circular shafts; Noncircular cross section shafts; Thin-walled open sections

center, 234

flow, 132, 180

strain, 78–79

Shrinking factor, 212

Simply-supported beam, 256; *see also* Bending loads

Singularity functions, 343; *see also* Deflection

basic loading conditions and, 344

example, 345–350

free-body diagram, 345

operational rules, 344

problems, 350–353

statically indeterminate beams, 366–373

SI units, 647

Soap-film analogy, *see* Membrane analogy

Soderberg straight line, 270; *see also* Fatigue

Statically indeterminate beam, 257, 319; *see also* Beams; Bending loads; Deflections under symmetric loading

area–moment, 333–341

Castigliano's second theorem, 373–380

examples, 319–324, 327–331, 333–337, 366–370, 373–377

integrations, 319–326

problems, 324–326, 331–332, 337–341, 371–373, 377–380

singularity functions, 366–373

superposition, 327–332

Statically indeterminate members, 43; *see also* Axial loads

coefficient of thermal expansion, 48

example, 44–47, 49

problems, 49–56

redundants, 44

superposition principle, 44

temperature effects, 47–56

tensile stress creation, 48

thermal stresses, 48

Statically indeterminate shafts, 96; *see also* Torsional loads
    example, 96–99
    problems, 99–102
Statics, 1; *see also* Axial loads
Strain, 20, 510; *see also* Axial loads; Mechanical
        properties; Stress
    allowable stress, 29
    components, 475–477
    design considerations, 29
    example, 30
    normal, 20–21
    problems, 30–34
    rosette, 511
    sign convention, 21
    units, 21
Strain analysis, 475; *see also* Mohr's circle; Plane-strain
        transformation equations; Strain at point; Strain
        measurements; Three-dimensional Hooke's law
    answers to problems, 644–645
    review problems, 517–519
Strain at point; *see also* Strain analysis
    average strain, 475
    components of strain, 475–477
    deformations, 475
    distortions, 475, 476
    sign conventions, 477
    strain parameters, 476
    3D strain element, 476
Strain measurements, 510; *see also* Strain analysis
    electric-resistance strain gage, 510, 511
    example, 512
    problems, 512–517
    strain rosette, 511
Stress, 8; *see also* Axial loads; Strain
    average value, 11
    components, 390–392
    concentration, 56–60
    element, 10
    example, 14–16, 58–59
    on inclined planes, 11–13
    normal, 8–10
    at point, 389–390
    problems, 16–20, 59–60
    shearing, 9, 10–11
    sign convention, 13
    uniaxial stress condition, 11
    units, 13
Stress analysis, 389; *see also* Failure theories; Mohr's
        circle; Plane-stress transformation equations;
        Thick-walled cylindrical pressure vessels;
        Thin-walled pressure vessels; Three-
        dimensional stress systems
    answers to problems, 642–644
    normal stress, 390
    pressure vessels, 441
    review problems, 471–474
    shearing stress, 390
    sign convention, 392
    stress at point, 389–390
    stress components, 390–392
    triaxial stress system, 390
Stress concentration, 111, 202; *see also* Torsional loads
    example, 112–113, 204
    factors, 111, 203

    maximum shearing stress, 111
    problems, 204–207
Stress under combined loads, 104, 192–199; *see also*
        Beams; Torsional loads
    analysis of stress in shaft, 105
    example, 107–110
    general plane-stress condition, 106
    Hooke's law in two dimensions, 106–107
Structural steel
    channel properties, 685–686
    pipe properties, 691–692
Structural wood section properties, 693–694
Structure stability, 521; *see also* Column
Superposition, 44, 104, 106; *see also* Deflections under
        symmetric loading; Statically indeterminate
        members
    example, 300–301
    method, 299
    problems, 301–303
    statically indeterminate beams, 327–332
Supported beam, 256; *see also* Bending loads
Symmetric bending, 149; *see also* Beams; Bending loads
    answers to problems, 636–638
    bending equation, 170
    bending stress under symmetric loading, 166
    centroidal axes, 167
    centroidal longitudinal principal plane, 166
    centroidal principal axes of inertia, 150, 166
    compressive strain, 168
    elastic behavior, 168
    example, 171–173, 180–185, 204
    horizontal shearing stress, 179
    internal moment, 150, 169
    internal shear and moment, 150
    longitudinal surfaces normal to loads, 178–180
    maximum stress, 169, 171
    problems, 173–178, 186–192, 204–210
    section modulus, 171
    shear flow, 180
    shearing stress under symmetric loading, 178
    stress concentration, 202
    thin-walled open sections, 234–240
    vertical shearing stress, 180

T

Tension test, 21; *see also* Mechanical properties
Thermal stresses, 48; *see also* Statically indeterminate
        members
Thick-walled cylindrical pressure vessels, 447; *see also*
        Stress analysis
    cylinder fabrication by shrinking, 454
    deformations, 450
    example, 452–453
    external pressure, 451–453
    free-body diagram, 447, 449
    internal pressure, 451, 454–456
    problems, 456–457
    shrink-fitting operations, 453–454
    special cases, 450
    stresses, 447–450
Thin-walled open sections, 234; *see also* Bending loads
    beam with channel cross-subjected to load, 235
    example, 238–240, 241–242

internal shearing force, 236
problems, 242–247
shear center, 234, 236, 237, 241
shearing stress, 235–236
symmetric bending, 234–240
unsymmetric bending, 240–242
Thin-walled pressure vessels, 441; *see also* Stress analysis
circumferential stress, 441
cylindrical vessels, 441–443
example, 444–445
gage pressure, 444
longitudinal stress, 441, 442, 443
problems, 445–446
spherical vessels, 443–444
Three-dimensional Hooke's law, 499; *see also* Strain
analysis
energy of distortion, 503–505
energy of volume change, 503
example, 505–506
modulus of volume change, 502–503
3D stress element, 500, 502, 504
Three-dimensional strain conditions, 506–509; *see also*
Mohr's circle
Three-dimensional stress element, *see* Triaxial stress system
Three-dimensional stress systems, 427; *see also* Stress
analysis
absolute maximum shearing stress, 432
example, 430–432, 436–438
free-body diagram, 428
general, 428
maximum in-plane shearing stress, 432
Mohr's circle for, 432–441
principal stresses, 432
problems, 438–441
stress invariants, 430
surface traction, 429
3D stress condition, 435
three plane-stress systems, 432–434
Three-dimensional supports and connections, 699
Torques, 69; *see also* Torsional loads
Torsional loads, 69; *see also* Circular shafts; Internal
torque; Noncircular cross section shafts; Stress
under combined loads
answers to problems, 634–636
elastoplastic behavior, 139–142
elastoplastic stress–strain diagrams, 139
electric motor driving power tool, 70
example, 142–143
fully plastic torque, 140
impact loading, 113–120
power-transmission shaft design, 103–104
problems, 143–144, 145–148

statically indeterminate shafts, 96–102
stress concentration, 111–113
three-dimensional torque, 70
Torsional toughness, 83; *see also* Circular shafts
Toughness modulus, 29; *see also* Mechanical properties
Transverse strains, 26; *see also* Mechanical properties
Tresca's criterion, *see* Maximum shearing stress theory
Triaxial stress system, 390; *see also* Stress analysis
Two-dimensional supports and connections, 697–698

U

Ultimate strength, 26; *see also* Mechanical properties
Ultimate torque; *see also* Rupture
Units in mechanics, 647; *see also* SI units
Unsymmetric bending, 223; *see also* Bending loads
arbitrary centroidal axes, 223–225
in beam, 240
bending stresses under unsymmetric loading, 223
example, 226–228
principal centroidal axes, 225
problems, 229–233
shear center, 241
sign convention, 223
thin-walled open sections, 240–242
Unsymmetric bending loads, 361; *see also* Deflection
component deflections, 362
example, 363–364
problems, 364–366

V

VBA (Visual Basic for Applications), 604–605; *see also*
Excel spreadsheet applications
Volume change modulus, 502; *see also* Three-dimensional
Hooke's law
von Mises theory, *see* Energy of distortion—theory

W

Warping function, 121
Wide-flange structural steel section properties, 679
SI units, 682–684
U.S. units, 679–681

Y

Yield; *see also* Bending loads; Mechanical properties
moment, 255
point, 25
strength, 25
Young's modulus of elasticity, *see* Modulus of elasticity